Structure and Dynamics of Weakly Bound Molecular Complexes

NATO ASI Series

Advanced Science Institutes Series

A series presenting the results of activities sponsored by the NATO Science Committee, which aims at the dissemination of advanced scientific and technological knowledge, with a view to strengthening links between scientific communities.

The series is published by an international board of publishers in conjunction with the NATO Scientific Affairs Division

A	Life Sciences	Plenum Publishing Corporation
B	Physics	London and New York
C	Mathematical and Physical Sciences	D. Reidel Publishing Company Dordrecht, Boston, Lancaster and Tokyo
D	Behavioural and Social Sciences	Martinus Nijhoff Publishers
E	Applied Sciences	Dordrecht, Boston, Lancaster
F	Computer and Systems Sciences	Springer Verlag
G	Ecological Sciences	Berlin, Heidelberg, New York, London,
H	Cell Biology	Paris, and Tokyo

Series C: Mathematical and Physical Sciences Vol. 212

Structure and Dynamics of Weakly Bound Molecular Complexes

edited by

Alfons Weber

Molecular Spectroscopy Division,
National Bureau of Standards,
Gaithersburg, Maryland, U.S.A.

Springer-Science+Business Media, B.V.

Proceedings of the NATO Advanced Research Workshop on
Structure and Dynamics of Weakly Bound Molecular Complexes
Acquafredda di Maratea, Italy
September 21-26, 1986

Library of Congress Cataloging in Publication Data

NATO Advanced Research Workshop on Structure and Dynamics of Weakly Bound Molecular
Complexes (1986: Acquafredda di Maratea, Italy)
    Structure and dynamics of weakly bound molecular complexes / edited by Alfons Weber.
    p. cm.— (NATO ASI series. Series C, Mathematical and physical sciences; vol. 212)
    "Proceedings of the NATO Advanced Research Workshop on Structure and Dynamics of
Weakly Bound Molecular Complexes, Acquafredda di Maratea, Italy, September 21–26,
1986"—CIP t.p. verso.
"Published in cooperation with NATO Scientific Affairs Division."
Bibliography: p.
    Includes index.

    1. Molecular association—Congresses. 2. Complex compounds—Congresses.
I. Weber, Alfons, 1927–   . II. Title. III. Series: NATO ASI series. Series C,
Mathematical and physical sciences; no. 212.
QD461.N335 1986
541.2'24—dc 19

87–18793
CIP

ISBN 978-94-010-8261-7   ISBN 978-94-009-3969-1 (eBook)
DOI 10.1007/ 978-94-009-3969-1

TABLE OF CONTENTS

INTERNATIONAL ORGANIZING COMMITTEE

A. Weber (Chairman)
 Molecular Spectroscopy Division
 National Bureau of Standards
 Gaithersburg, MD 20899, USA

F. A. Gianturco
 Dipartimento di Chimica
 Universita degli Studi di Roma
 00100 Roma AD
 Italy

B. J. Howard
 Physical Chemistry Laboratory
 Oxford University
 Oxford, OX1 3QZ
 Great Britain

A. R. W. McKellar
 Herzberg Institute of Astrophysics
 National Research Council of Canada
 Ottawa, Ont. K1A 0R6
 Canada

A. van der Avoird
 Theoretical Chemistry Institute
 Faculty of Mathematical and
  Physical Sciences
 Katholieke Universiteit Nijmegen
 6525 ED Nijmegen
 The Netherlands

ACKNOWLEDGMENTS

The conduct of the NATO-Advanced Research Workshop on the "Structure and Dynamics of Weakly Bound Molecular Complexes" was made possible by the primary sponsorship and support of the NATO Scientific Affairs Division in Brussels, Belgium and special thanks are due to the late Dr. Mario DiLullo and to Dr. Craig Sinclair of this division. The Workshop was assisted by the generous support provided by the National Bureau of Standards, the Coblentz Society, the Laser Analytics Division of Spectra Physics, BOMEM, Inc., and the Standard Oil Research and Development Company.

The search for a suitable locale for the conduct of the Workshop was ably assisted by Dr. and Mrs. Tilo Kester of ITST v.z.w., in Overijse, Belgium. Their mediation in making the formal arrangements with the management of the Hotel Villa del Mare in Acquafredda di Maratea, Italy was of great help in getting the meeting off to a good start. Special thanks and commendations are also due to Mr. A. Guzzardi and Mrs. Susi Travisano and their staff of the Hotel Villa del Mare. Their solicitous attention to detail and the comfort of the participants made the Workshop an enjoyable event.

The planning of the scientific program was done in consultation with the members of the international organizing committee as well as friends and colleagues active in the field. To all of them as well as the reviewers of the manuscripts I express my gratitude for their thoughtful comments and suggestions.

Special thanks are due to Ms. R. A. Flynt for her excellent secretarial assistance throughout the preparative and follow-up stages of the Workshop and the preparation of this book.

Alfons Weber
National Bureau of Standards
Gaithersburg, MD 20899, USA

PREFACE

The study of weakly bound molecular complexes has in recent years brought this field of investigation to the forefront of physical and chemical research. The scope of the subject is wide and different terminology and nomenclature is current among the various subspecialties. Thus, the term "metal cluster" often connotes to the organic chemist a metal-organic compound, while the physicist will more likely think of groups of metal atoms held together by weak interatomic forces. Aggregates, clusters, complexes, van der Waals molecules, hydrogen-bonded molecules, etc. are terms currently in use, sometimes interchangeably while other times with well defined and mutually exclusive meanings. The subjects of this volume are the free, isolated van der Waals and hydrogen-bonded molecules. Owing to the present state of experimental knowledge these are mostly dimers, i.e., entities formed by two strongly bound molecules, an atom and a molecule, or two atoms held together by the weak hydrogen-bonding, or the still weaker van der Waals forces. Weakly bound complexes formed of more than two strongly bound sub-units, i.e., trimers, tetramers, etc., are now coming within reach of experimental observation and several papers in this book deal with them.

The study of van der Waals and hydrogen-bonded interactions has been pursued for several decades. Most of these investigations have, however, dealt with systems in the condensed phase in which bulk effects are commingled with and therefore mask the weak binary interactions. However, significant advances in both theoretical and experimental studies of isolated van der Waals and hydrogen-bonded molecules have taken place in recent years and the Pimentel (1) and Brinkman (2) reports recently issued by the U.S. National Academy of Sciences indicate that this field of research will retain a prominent position in the future. It was therefore deemed to be appropriate to convene a meeting of experts to discuss the status of the field and to assess the directions along which future advances would most likely occur.

The present book represents the proceedings of the NATO Advanced Research Workshop on the "Structure and Dynamics of Weakly Bound Molecular Complexes" held in Acquafredda di Maratea, Italy, September 21 through 26, 1986. The purpose of this Workshop was to bring together researchers from a variety of disciplines and with different backgrounds who are working on various aspects of the structures and dynamics of van der Waals and hydrogen-bonded complexes in the gaseous phase. Different experimental and theoretical approaches are used for the determination of the structures, potential functions, dynamics, and predissociation effects that characterize weakly bound complexes. Several key questions were prominent among the various problems addressed.

xv

The structures of over 100 molecules have now been determined by microwave and molecular beam electric resonance (MBER) spectroscopy (see, for example, the bibliography provided by S. Novick, pp 201-212). While some of these are predicted by accepted theory and phenomenological models, many complexes have quite unexpected structures. These findings therefore raise the question as to what are the underlying principles that determine how molecules stick together. This obviously has important implications for condensed phase work and the understanding of condensation phenomena. Here the interaction between the experimental spectroscopists and the quantum theoreticians is beneficial to answer the question "can structures be predicted from some fundamental principles or is each unique, requiring a separate spectroscopic study or ab initio calculation?"

The interaction between molecules is described by an intermolecular potential energy function - this interaction is most readily probed in weakly bound dimers. There are both theoretical and empirical approaches to this topic, each of which has its own problem areas. For example, how reliable are current ab initio calculations and how may they be improved to determine accurately that very small part of the total system energy that describes the intermolecular interactions? Further, how reliable are the current semi-empirical methods to map out the energy surfaces for small dimer complexes and, by extension, for larger complexes leading ultimately to clusters? On the other hand, can spectroscopy, perhaps in conjunction with other experimental data, be used to obtain unique anisotropic intermolecular potential surfaces?

Weakly bound complexes with internal energy less than their binding energy possess large amplitude internal degrees of freedom and execute tunneling motions between different equivalent conformations. These behaviors motivate the search for the best methods, analytical and group theoretical, to study the dynamics of these motions on the potential surface and to determine the most likely pathways out of the many possible for a conformation change to occur.

One very important dynamical process when complexes are excited by a photon of energy greater than that of the intermolecular bond is predissociation. This phenomenon manifests itself in broad spectral lines in a rotation-vibration band or in the break-off of the rotational branches at a particular J-value (rotational predissociation). Different rotation-vibration bands of a complex may show different line broadenings or branch break-off at different J-values. Among the questions that have been debated are the following: are the broad lines observed in some rotation-vibration bands a measure of vibrational predissociation or are they due to some internal energy redistribution effect?, as well as what are the underlying requirements for rapid predissociation? Much photodissociation work has been done with line tunable lasers with results that sometimes disagree with those obtained with high resolution continuously tunable lasers. Such discrepancies pose the question regarding the relative merits of these two techniques for studying the dissociation processes. Picosecond techniques for measuring the vibrational predissociation lifetimes of excited states have only recently been developed. It remains to be seen whether these lifetimes are consistent with those determined by high resolution, frequency domain spectroscopic methods.

This book is the result of the deliberations on these and other questions. The papers are contributions based on the invited talks. Also included are selected contributions based on poster papers presented at the Workshop. For convenience the papers are grouped in two sections, the first dealing with spectroscopy, and the determination of structures and potential functions while the second section deals with dynamical phenomena. This division is somewhat arbitrary since a fair number of papers in the first section deal with dynamical effects and a substantial spectroscopic content is in evidence in the second section.

The first four papers deal with the methods of microwave spectroscopy to deduce the structures of complexes held together by intermolecular forces or the conformations determined by intramolecular hydrogen-bonding. This section is followed by papers dealing with rotation-vibration spectra. The latter field has experienced remarkable progress during the last few years owing to the advances in experimental techniques by means of which it is now possible to fully resolve the rotation-vibration bands of, albeit simple, van der Waals and hydrogen-bonded molecules. Papers dealing with the high resolution visible and near ultraviolet spectra complete the spectroscopy section.

Information about the intermolecular potential is an important by-product of the spectroscopic studies, but other methods play very important roles as well. This is especially so for regions of the potential surface not accessible to spectroscopic study; here the beam scattering experiments provide a wealth of information. The interpretation of various models, the contributions of the different classical and quantum mechanical couplings that comprise the overall intermolecular potential, the relationships between bulk physical properties and the potential and how it can be determined by inversion techniques from macroscopic data, these and other questions are taken up in the last section of part I.

While dynamical effects are discernible in the spectra through linewidths and the break-off of the rotational branches, it is the interpretation of photodissociation experiments that has given cause for much discussion. The second part of the book deals with these and other aspects of the dynamics of predissociation, energy transfer, and highly excited vibrational states.

The program of the Workshop was designed to include all aspects of the study of free van der Waals and hydrogen-bonded molecular complexes, unperturbed by effects attributable to a surrounding medium, and the papers contained in this book reflect the spirit of this attempt. It is fair to point out however, that some of the topics taken up in this book are also presented in other publications, of which only a few of the most recent ones will be cited here [3-5]. Research in the field of weakly bound molecular complexes is moving along very rapidly and many new results have been reported in the literature since the conclusion of the Workshop. It is hoped that the present book will nevertheless be useful to researchers in the field for several years to come.

REFERENCES

1. Opportunities in Chemistry, (National Academy Press, Washington, D.C., 1985).

2. Physics Through the 1990's. Atomic, Molecular, and Optical Physics, (National Academy Press, Washington, D.C., 1986).

3. Van der Waals Systems, Topics in Current Chemistry, Vol. 93 (Springer-Verlag, Berlin-Heidelberg-New York, 1980).

4. Van der Waals Molecules, Faraday Disc. Chem. Soc., No. 73 (1982).

5. Hydrogen-Bonds, Topics in Current Chemistry, Vol. 120, P. Schuster, Editor (Springer-Verlag, Berlin-Heidelberg-New York, 1984).

A. Weber
Molecular Spectroscopy Division
National Bureau of Standards
Gaithersburg, MD 20899, USA

PART   I

SPECTROSCOPY, STRUCTURE, POTENTIALS

ROTATIONAL SPECTROSCOPY OF WEAKLY BOUND COMPLEXES:
CAPABILITIES AND LIMITATIONS

John S. Muenter
Department of Chemistry
University of Rochester
Rochester, New York 14627
U. S. A.

ABSTRACT. The vast majority of empirical information on the structure of weakly bound cluster molecules has come from spectroscopic studies of rotational energy levels. The emphasis of this talk will be on pure rotational transitions observed in the microwave and radio frequency portion of the spectrum. The principal experimental methods used to obtain these data are gas phase absorption, molecular beam electric resonance, and pulsed Fourier transform microwave studies of molecular beams. These techniques will be discussed in terms of their contrasting capabilities and shortcomings. Some of the basic assumptions and models commonly used to extract molecular properties from data will be considered. The usual risks of trying to see into the future will be accepted, and predictions will be made on what can be anticipated from these methods in succeeding years. Finally, the more general question of all experimental sources of rotational and structural information will briefly be considered in rotational, vibrational, and electronic spectroscopy of weakly bound complexes.

1. INTRODUCTION

As the first paper in both this conference and the structure section of the proceedings, the material presented is planned to serve as an introduction to succeeding papers. Thus this paper will be in the form of a review, rather than focusing on current research results. Since there is a natural tendency for papers to ignore the problems associated with their techniques or analyses, I will try to point out disadvantages and limitations while comparing different experiments. Perhaps a little controversy at the beginning of this workshop will lead to livelier discussion. The references given could not possibly be complete and only typical citations will be made, with apologies given now for those important papers not mentioned. For useful reviews see the first two references [1,2]. I will arbitrarily define weakly bound complexes as an assembly of two or more well defined chemical species held together by interactions weaker than a conventional

3

A. Weber (ed.), Structure and Dynamics of Weakly Bound Molecular Complexes, 3–21.
© 1987 by D. Reidel Publishing Company.

covalent bond. This definition encompasses atom-atom, atom-molecule,
and molecule-molecule bimolecular species, as well as a host of
possible larger complexes. Intramolecular hydrogen bonds will not be
discussed. Only electric dipole transitions will be considered in any
detail, and the majority of work discussed will relate to microwave and
radio frequency transitions.

From an historical prospective, the first work that falls within
these guidelines is the gas phase microwave absorption study of
carboxylic acid dimers by Costain and coworkers[3]. It was, however,
the ease of producing weakly bound species in molecular beam nozzle
expansions that lead to the very rapid growth of this field. First,
species with classic strong hydrogen bonds like hydrogen fluoride
dimer[4] and water dimer[5] were studied by molecular beam electric
resonance (MBER) spectroscopy. MBER studies of much more weakly bound
van der Waals complexes, such as Ar-HCl[6], quickly followed the
initial (HF)$_2$ work in Bill Klemperer's laboratory. These early studies
produced many surprising results. Some that come to mind are the large
amplitude tunneling motions in (HF)$_2$ and (H$_2$O)$_2$, the fact the HF-ClF is
not hydrogen bonded[7], and the relatively large dipole moment of
Ar-CO$_2$[8]. The next major experimental advance came from Bill Flygare
and coworkers with the development of pulsed molecular beam Fourier
transform microwave spectroscopy[9]. The combination of high
resolution and very high sensitivity in this technique has made it
extremely powerful and productive. While the majority of work has been
on molecular beams, many significant gas phase studies have been
carried out both with microwave absorption[10,11] and infrared
absorption[12,13].

2. EXPERIMENT

2.1. Molecular Beam Electric Resonance Spectroscopy

The principal features of MBER spectroscopy that distinguish it from
all other spectroscopic methods are the use of electrostatic state
selection and obtaining all spectroscopic information from molecular
beam intensity rather than from radiation intensity. These features
provide the very unusual characteristics of spectroscopic sensitivity
independent of both transition moment and transition frequency. In
ideal cases, transition moments of just a few millidebye are sufficient
for MBER study[14]. This has permitted the study of several van der
Waals complexes, such as Ar-CO$_2$, made up of an atom and a nonpolar
molecule[8,15,16]. The extremely broad frequency range available to
MBER studies of weakly bound clusters is demonstrated by measurements
having been made from 10^4 to 10^{14} Hz. For example, a transition in Ar-
CO$_2$ has been measured at 8.8 kHz[8] and a 120 THz transition has been
observed in (HF)$_2$[17]. There are several other very useful
consequences of detecting beam molecules rather than photons. Since
the basic signal source is from a mass spectrometer, the chemical
identity of the molecule absorbing the photons is seldom in doubt.
This is of obvious importance considering the indiscriminate formation

of all sorts of cluster molecules in a free jet expansion.

Because photons are not detected it is a simple matter to control the resolution of the experiment by tailoring the spectral distribution of the radiation source. Very broad bandwidths may be used, limited only by the requirement that the spectral density within the homogeneous linewidth is sufficient to induce the desired transition. 10 MHz linewidths are readily achieved for initial searches in a new molecular system. This means that one can search for new transitions at 1000 MHz/hour using a 10 second instrument response time with the same sensitivity that these lines can later be observed with 1 kHz linewidth. In this way, all of the advantages of very high resolution may be enjoyed without the usual difficulty of slow and laborious searching. This characteristic also makes it difficult to directly compare the sensitivity of MBER experiments to that of more conventional methods.

The inherent high resolution of MBER experiments permits observation of small hyperfine interactions. Precise Stark effect and dipole moment measurements are straightforward. These capabilities not only provide detailed information on the electronic structure of the molecule, but by comparing the vector and tensor properties of the complex with monomer properties, vital information on both the geometry and dynamics of the complex can be obtained.

One disadvantage of the MBER method is the relative complexity of the apparatus. The penalty for the advantages obtained from detecting molecules rather than photons is the crucial role played by the mass spectrometer molecular beam detector. The ionization efficiency and molecular ion fragmentation characteristics of the beam detector are the most important factors affecting instrument sensitivity. A "perfect" beam detector would increase sensitivity by several orders of magnitude.

The use of electrostatic state selection also has limitations. The selectivity of electrostatic quadrupole fields is quite good for second order Stark effect encountered in small linear molecules. However, the long bondlength associated with weak binding of a bimolecular complex means that complexes are normally slightly asymmetric prolate top molecules with rather small values for (B+C)/2. In addition, the majority of states in these molecules usually exhibit first order Stark effect in the focusing fields. The focusing of second order Stark effect states of complexes is hindered by higher order Stark effect aberrations, and the state selection process exhibits poor selectivity in the presence of states with first order Stark effects. This combination of conditions places a high premium on achieving the lowest possible rotational temperature in the beam expansion. As the temperature decreases, the signal to noise ratio is improved both by an increasing numerator and a decreasing denominator. The signal increases as a larger fraction of the complexes reside in a single rotational level and, at the same time, the noise decreases as higher K states with first order Stark effects are not present to be inadvertently focused into the detector.

An additional advantage of the focusing fields in the apparatus, in combination with the mass spectrometer beam detector, is the ability to

optimize beam conditions at the start of a new project. The many
source parameters, such as temperature, pressure, concentrations,
diluent gas, nozzle shape, etc., can first be adjusted while observing
the mass spectrum of the beam. In the absence of fragmentation, this
would be an ideal procedure. However, fragmentation of weakly bound
species in a mass spectrometer ion source will always be significant,
and many weakly bound molecules have been studied (knowingly or
unknowingly) using fragment ion signals. Since large clusters are
likely to be nonpolar, the focusing behavior of a beam indicates the
presence of higher polymers[18] and permits easy adjustment of
conditions before extensive spectroscopic searches are initiated.

MBER spectrometers have also proven to be very flexible machines.
A wide variety of different beam sources have been used. Zeeman
measurements are readily carried out[19] and have been done on at least
one van der Waals molecule[20]. It is possible to gain access to the
molecular beam at several different places during the MBER process and
this provides for a wide variety of different laser experiments on
stable[21] and weakly bound[17,22,23] species. It is obvious that many
more combinations of lasers and MBER spectrometers will be realized in
the future.

2.2. Pulsed Molecular Beam Fourier Transform Microwave Spectroscopy

The universal quest of all experimental spectroscopists is the
technique that provides simultaneously higher resolution and higher
sensitivity. To a very substantial extent, Bill Flygare satisfied this
goal with the development of pulsed beam FT microwave spectroscopy[9].
As molecular beam experiments go, the method is relatively simple. A
large vacuum chamber houses a commercial pulsed valve and a microwave
Fabry-Perot interferometer. On entering the interferometer, the pulse
of molecules is exposed to a pulse of microwave radiation to populate
excited states of transitions falling within the bandwidth of the
interferometer. After this excitation pulse has decayed, the free
induction decay of the molecules is observed. This time domain
spectrum is then transformed to the more familiar frequency domain.
High resolution is achieved through the collisionless directed flow of
the beam and the time of flight through the Fabry-Perot cavity.
Several factors combine to achieve high sensitivity. The pulsed beam
source provides a high density pulse of molecules with very low
translational and rotational temperatures. Source conditions can be
found such that a substantial fraction of the molecules in the beam
exist in the desired cluster. The noise in the spectrum is determined
by the sensitivity of the microwave receiver used to detect the
molecular emission. Low noise receivers have been highly developed for
radar and communication purposes and can readily be assembled from
commercial components. Signal to noise ratios are easily improved by
summing signals from a relatively large number of pulses. Finally, the
Hadamard advantage of simultaneously collecting data over an extended
spectral range is a trait of all FT methods.

Approximately ten of these instruments have been built and a large
number of weakly bound molecules have been studied[2]. The majority of

this work has been in the 4-18 GHz frequency region. Zeeman studies have been made on several cluster molecul s[24]. The excellent sensitivity of FT microwave spectroscopy as been used to study a variety isotopic species in natural abund nce to obtain additional information on structure and dynamics of omplexes[25]. Of particular interest are the isotopes of Kr and Xe th t have nuclear quadrupole moments since the isolated atom is spheri al and has an eQq identically equal to zero. Several measurements of induced field gradients at rare gas nuclei have been made in complexes[26]. The large number of systems studied in FT microwave experiments has permitted the observation of general trends and the development of models to explain general characteristics of hydrogen bonded molecules[27]. An extremely important recent development has been the observation of clusters larger than bimolecular by FT microwave spectroscopy[28].

The two most apparent difficulties with FT microwave experiments are the tedious sweep procedures and unusual lineshapes. Since the Fabry-Perot interferometer has a very high Q, the etalon spacing must be reset for each new spectral segment. The bandwidth of the Fabry-Perot and the sampling time of the analog to digital converter limit the width of each segment to a few MHz. It is also necessary to have appropriate microwave pulse height and width to achieve the desired π pulse to maximize emission in the transitions being sought. This, combined with the wide range of possible beam source parameters, can make initial searching a slow process. For these reasons, much of the early work was based on systems already studied by MBER methods.

The lineshape typically observed in an FT microwave experiment consists of closely spaced doublets. This was originally explained[29] as an inherent Doppler effect of the diverging molecular beam. It was thought that this undesirable feature could only be avoided by sacrificing a large amount of sensitivity. More recently, it appears that this phenomenon results from a nonuniform distribution of molecules across the beam caused by the formation of larger clusters in the higher density central portion of the beam[30]. At least in many cases, the doubling can be completely suppressed by using milder expansion conditions. If optimal source conditions are found, the peak intensity of the desired single transition can actually be greater than the doublet peak intensity[30].

The many similarities and differences of FT microwave and MBER experiments invite a more detailed comparison. A brief comparison would note that an MBER instrument is more complex but more flexible with regards to initial searches, while the ultimate sensitivity advantage of an FT microwave spectrometer is very desirable. This points out the complimentary nature of the two techniques; the two in combination could be more powerful and productive than the separate instruments. As yet no laboratory has both capabilities, but collaboration has begun between separate laboratories[31]. The frequency range of MBER is greater, for both low and high frequencies, because there is no need to detect the radiation. MBER exhibits narrower linewidth at low frequencies, while the microwave resolutions of the two methods are comparable. The difficulty of making Stark effect measurements in FT microwave resonant structures is a

significant limitation. Both methods can use a wide variety of
molecular beam sources and there are reasons to believe that MBER
experiments could improve their sensitivity by using pulsed sources to
achieve lower rotational temperatures. There are many obvious ways
that FT microwave instruments can be combined with lasers.

2.3. Gas Phase Microwave Absorption

While conventional gas phase microwave absorption produced the first
rotational spectroscopy of weakly bound molecules, it might appear that
the success of molecular beam techniques would totally displace gas
phase absorption. This very definitely has not been the case. It is
certainly easy to generate associated species in a nozzle expansion,
and the low rotational temperature obtained is a boon to ground state
studies, but there are also disadvantages in the beam configuration.
Even in the best of circumstances, a molecular beam is a tenuous medium
and the total number of molecules involved is small. It is not even
appropriate to consider something like a "long path length" in a beam
experiment[32]. Perhaps the greatest limitation of low temperature
beam expansions is the inability to study low lying excited states.
The beam expansions used to generate weakly bound molecules invariably
produce only the lowest energy configuration of the species in
question. For example, the very first MBER investigation of a hydrogen
bonded molecule, $(HF)_2$, studied HF-HF, DF-DF, and HF-DF, but could not
produce DF-HF[4]. All of these species have been observed in gas phase
samples using millimeter wave microwave absorption[33]. It can also be
useful to access high lying rotational levels not populated in an
expansion since centrifugal distortion properties are an important
source of intermolecular potential function information[34].
    Molecules with relatively strong hydrogen bonds are the most
logical candidates for study under equilibrium conditions. Obviously
one also wants to work at low temperatures. These two considerations
are somewhat mutually exclusive, however, since monomers such as H_2O,
HF, and HCN which exhibit strong hydrogen bonds have low vapor pressure
for the same reason. Therefore, it may be possible to study systems
with much weaker interactions if low enough temperatures can be
reached. This is clearly demonstrated by the gas phase infrared study
of Ar-HCl by Pine and Howard[35]. Millen, Legon, and coworkers have
studied a number of binary complexes containing H_2O, HF, and HCN[10]
and many low lying excited vibrational states were observed. Similar
results have been obtained for $HF-H_2CO$[36]. A significant number of
infrared studies have also been done on gas phase samples using
classical[12], laser difference frequency[35,37], color center
laser[38], and FTIR[39] methods.
    The resolution of gas phase studies are, of course, limited by
Doppler and pressure broadening and small hyperfine interactions can
not be resolved. Stark effect data may be available since most of the
microwave work has used Stark modulation, but the compination of
relatively high pressure, broad lines and complex spectra have
precluded dipole moment measurements in some gas phase work. Relative
intensity measurements provide estimates for low frequency vibrational

mode spacings, and absolute intensity measurements can produce accurate binding energies in ideal cases[40]. While the range of cluster molecules amenable to gas phase study is certainly more limited than those which can be produced in beam expansions, there still remains much to be done by this relatively simple technique.

2.4. Other Methods Yielding Rotational Energy Levels

While the major concern of this paper is pure rotational spectroscopy of complexes, vibrational and electronic spectroscopies are also sources of rotational information. Many elegant studies of electronic transitions in weakly bound systems have been made[41]. These studies have involved clusters larger than bimolecular, and are now being extended to the realm of much larger molecules[42]. The vibrational experiments already mentioned also contain extensive rotational information[12,35,37-40]. An obvious, but important, characteristic of electronic and vibrational transitions is the ability to study nonpolar complexes. The need of a permanent dipole moment for pure rotational transitions eliminates this spectroscopy from the study of symmetric complexes, such as dimers of nonpolar monomers, having a center of symmetry. Magnetic dipole transitions can be used in special cases and a molecular beam magnetic resonance study of $Ar-O_2$ has been made[43]. A small number of Raman experiments have also been conducted on beams of clusters[44]. Recently it has been possible to directly observe infrared laser absorption in pulsed molecular beams[45] with linewidths the order of 100 MHz. Bolometer studies of infrared beam excitation of clusters, with linewidths of only a few MHz, are also beginning to yield analyzed rotational structure[46].

A new method of pure rotational spectroscopy has grown out of the observation of direct infrared laser absorption of molecular beams. The microwave spectrum of Ne-OCS has been studied in microwave-infrared double resonance (MIRDR)[47]. In this experiment, the intensity of the IR absorption is monitored while modulated microwave radiation also irradiates the molecules. Because of the low rotational temperature, rotational transitions can cause the IR absorption signal to change by as much as 25%. This technique was applied to the ground vibrational state, where no vibrational state saturation was require by the laser. By using a laser with sufficient power to saturate the IR transition, the method can be extended to the excited vibrational state. An advantage of all double resonance methods is the ability to both simplify spectra and make assignments within extremely complex spectra.

3. ANALYSIS OF SPECTRA

3.1. Introduction

In many ways, rotational spectra of weakly bound complexes are very similar to spectra of conventional covalently bound molecules. The vast majority of rotational transitions in complexes have been analyzed with the conventional centrifugally distorted rigid rotor

Hamiltonian[48], after appropriate allowances have been made for
occasional occurrences of inversion tunneling or internal rotation.
Most spectra have been analyzed with the rigid rotor model augmented
with just a single centrifugal distortion of the length of the weak
bond, i.e. the centrifugal distortion is treated as that of a diatomic
molecule. Thus microwave transition frequencies yield the usual A, B,
and C rotational constants and one or more centrifugal distortion
constants. Since the inertial tensor of clusters are frequently very
prolate, the A rotational constant is often unknown or poorly
determined. That most weakly bound complexes behave as quasirigid
molecules indicates that intermolecular potential functions are, in
general, quite anisotropic. However, this certainly does not mean that
there are not many manifestations of large amplitude motions,
particularly.angular motions, in rotational spectra of complexes. The
challenge is to obtain structural and, where possible, potential
function information.

3.2. Structure Determination

The first assumption made in the structure determination process is
that the structure of the monomer units remain unchanged on the
formation of the weak bond. While there is evidence[49] that strong
hydrogen bond formation leads to slight lengthening of the proton donor
bond, or very small charge redistribution within a monomer[50], this
assumption is generally considered to be very good. This is fortunate,
since there would be little possibility of determining any complex
structure if the coordinates of every atom had to be found. Given
unchanged monomer units, it is just necessary to find the separation
and relative orientation of the constituents. This requires a range of
two to six structural parameters in going from the simple case of an
atom-linear molecule complex to the general bimolecular species. As
larger complexes are considered, each additional monomer unit can
require up to six additional parameters to specify the geometry. As
will be discussed below, isotopic data is much less useful in this type
of structure determination, as compared with conventional molecules,
and all possible sources of geometric information must be considered.
These general considerations explain why only structures for small,
relatively symmetric, complexes have been well determined.
      The simplest structure to consider is the atom-diatom with two
parameters, the separation, R, between the atom and the center of mass
of the diatom and the angle, θ, between the diatom axis and the axis
connecting the atom and diatom. Representing this type system, many
experimental and theoretical studies of rare gas-hydrogen halide, RG-
HX, molecules have been carried out[51]. Ar-HCl was also the first van
der Waals system studied[6] and this initial work immediately exhibited
manifestations of large amplitude bending in van der Waals molecules.
The separation R is well determined by (B+C)/2, which is directly
available from the rotational spectrum. However, the angle essentially
specifies the position of the proton (or deuteron), which has little
effect on moments of inertia or rotational constants. The unaltered
monomer assumption does provide angular information in the form of

projections of monomer properties onto the intertial axes of the complex. In the RG-HX case, the dipole moment of HX projects onto the complex axis as a vector giving $\langle\cos\theta\rangle$, while the halogen quadrupole interaction (eQq) projects as a second rank tensor giving $\langle\cos^2\theta\rangle$. Although all indications[51] specify a linear equilibrium geometry for these systems, the observed values for $\langle\cos\theta\rangle$ and $\langle\cos^2\theta\rangle$ are far from unity, different from each other, and show large isotopic variation. Another indication of large amplitude bending affects is the fact that R obtained from (B+C)/2 for Ar-HCl and Ar-DCl are different from one another. So while a significant amount of angular information is contained in the spectra, considerable care must be exercised in interpreting this data.

Values for $\langle\cos^2\theta\rangle$ from eQq projections are usually quite reliable. The field gradient at a nucleus depends on the $1/r^3$ electron distribution, weighting heavily the electron density close to the nucleus, and should be relatively insensitive to perturbations from weak interactions. Small corrections to the gradient arising from monomer moments or induced moments (Sternhiemer corrections) can be made. Halogen and nitrogen eQq values are large and accurately measurable. Unfortunately, deuterium eQq's are small and may be subject to experimental measuring errors. In cases where accurate deuterium eQq values have been obtained for RG-DX, the value of $\langle\cos^2\theta\rangle$ from both the halogen and deuterium eQq measurements agree to within 1%. Other hyperfine interactions can also be used, but there small size normally precludes quantitative results. One exception is the direct spin-spin interaction in HF which has been measured in several species. A further advantage of $\langle\cos^2\theta\rangle$ values is that they can be combined with rotational constant information since rotational constants project dominantly as P_2, the second Legendre polynomial. Values for $\langle\cos\theta\rangle$ are subject to much more uncertainty since electrostatic interactions can cause large changes in dipole moments. For example, the dipole moment of water dimer differs from the vector sum of the monomer moments by almost 0.5 D[5]. It is quite difficult to calculate the effects of these electrostatic interactions both because the necessary monomer multipole moments are seldom well enough known, and because the convergence properties of the multipole expansion are poor at the distances involved. More sophisticated models have been developed to deal with these problems[52].

The somewhat extreme angular effects in RG-HX molecules arise primarily from the very small moments of inertia for HX, which are essentially the reduced masses for bending vibrations in these complexes. Structural considerations for atom-linear molecule complexes having larger molecular constituents are more conventional, but still show the effects of large angular excursions. One direct measure of large amplitude motions is the inertial defect $\Delta = I_c - I_a - I_b$, since an atom-linear molecule must be a planar complex. For approximately T shaped rare gas-linear triatomic molecules, where A, B, and C are known, Δ is typically in the range of 1-5 amu·Å^2. Given a nonzero inertial defect, one can obviously calculate more than one pair of structural parameters from three data. One convenient set of expressions is $\sin\theta = (b/A)[(A-C)/(b-C)]$ and $R = \{k[(m+M)/(mM)](1/C-1/B)\}^{1/2}$, where b is the rotational constant of the linear monomer, in MHZ, m and M are the masses, in amu, of the rare gas and monomer molecule, and k=505379 MHz amu·A^2 for units conversion. These two expressions rely only on A and C of the complex. The first indicates the angle is determined primarily by A and, in the limit of θ=90, A=b as anticipated. R is dominated by C, but equally well could have been calculated using B or (B+C)/2. Depending on what combination of rotational constants are used, R can vary by 0.05 A or more and θ can change by several tenths of a degree. Clearly generalizations based on the geometry of clusters must be consistent to within this magnitude of uncertainty.

It is even more important to realize that the bond lengths and angles discussed above can be quite far from equilibrium values. Consider the rare gas-carbonyl sulfides molecules, where θ is 70.6, 73.5, and 74.4 degrees for Ne-OCS, Ar-OCS, and Kr-OCS respectively. It is tempting to conclude that while the angle is nearly the same for the Ar and Kr species, it is significantly different for Ne-OCS. However, a one dimensional model with plausible force constant scaling[47] shows that these angles are entirely consistent with a single equilibrium angle of 78 degrees for each complex. The model indicates a very reasonable bending amplitude of 15 degrees for Ne-OCS and successively smaller amplitudes for the Ar and Kr species. It can be very risky to consider changes in structures obtained from rotational constants and angular projections to also represent differences in equilibrium structures.

Structure determinations for bimolecular species more complicated than atom-linear molecules become progressively more difficult. Water dimer[5] and water-ethylene[52] provide two examples of more general bimolecular complexes which illustrate both the problems involved in structure determination and some of the ways these problems can be solved. Water dimer exhibits large vibrational amplitudes and electrostatic interactions as well as inversion tunneling. Water-ethylene has similar problems, with the tunneling motion through a barrier to internal rotation with a 1 kCal/mole height. (This is a very large barrier, given that it is comparable to the binding energy of the complex.) These structure determinations use rotational constants for more than one isotopic species, as well as projections of moments and hyperfine properties, to arrive at reasonably precise descriptions of the geometry of the complexes. Unfortunately, the rotational spectrum of a complex has not always led to a satisfactory structure determination, even in the case of the atom-linear molecule case. Ar-HCN exhibits enormous centrifugal distortion coefficients and several aspects of the rotational data suggest that the usual semi-rigid rotor approximations do not lead to a proper interpretation of the structure of the molecule[54]. The radio frequency spectrum of Ar-HCCH[15] gave a correct dipole moment and $\langle\cos^2\theta\rangle$ value for this complex, but the geometry calculated by assuming a small inertial defect is not correct. Recent infrared absorption studies [55] show an extremely large inertial defect, and the geometry of Ar-HCCH is not obvious at his time.

Fortunately, subtle aspects of structures do not always have to be identified to gain new insights into the behavior of weakly bound systems. General trends in the many structure determinations[1,2] have been discerned[27] and specific complexes exhibit readily recognized characteristics. For example, the fact that strongly hydrogen bonded dimers exhibit angular structures determined by the lone pair orientation of the proton acceptor[56] did not require extremely precise angle measurements in the early structure determinations. Another very pretty example is in the recent studies of complexes containing an HF or HCl molecule along with two or three argon atoms[28]. When a second Ar atom is added to Ar-HF, the qualitative structure is that of an argon dimer with the HF perpendicular to the dimer bond, proton toward the $(Ar)_2$. The third Ar atom forms a triangular $(Ar)_3$ and a C_{3v} complex with the proton toward the center of the triangle. The $Ar_n HCl$ complexes behave similarly. As these clusters grow, will the HX molecule always be on the outside? These observations do not require 0.01 Å or 1 degree resolution in structure determination.

The above paragraphs have primarily addressed questions of geometric structure. Also of considerable interest are questions of electronic structure, particularly how the formation of the week bond(s) might alter the electronic structure of the monomer units. In fact, there has been very little information that can point to any alteration of monomer electronic structure, other than that caused by electrostatic distortions of one monomer by the multipole moments of another monomer. A number of observations of hyperfine properties or dipole moments have been made to try to identify such effects. One clear cut case is the observation of nonspherical charge distributions in rare gas atoms. While nonzero eQq values have been measured for complexes containing ^{83}Kr and ^{131}Xe, the field gradients can be explained just by electrostatic effects[26]. A related eQq measurement has been done in the N_2-HF complex[50] where different eQq values were observed for the two nitrogen atoms. Analysis of this system suggests charge transfer of 0.03 e from one N atom to the other. However, several assumptions were made in this analysis and it is not known how large the uncertainties are in this number. The most significant monomer electronic change that has been measured is the HF bond length in HF complexes having strong hydrogen bonds[49]. The HF bond length is accessible through the large spin-spin hyperfine interaction and, while any one measurement could be questioned, systematic variations in several complexes are convincing evidence of bond lengthening[49].

The dipole moments of polar complexes composed of nonpolar monomer units might also indicate the existence of charge transfer on the formation of weak bonds[8,15,16]. While the moments observed range from 0.027 D for Ar-HCCH[15] to 0.10 D for Ar-NCCN[16], it is not possible to sort out any charge transfer. These moments can arise from at least three sources. The quadrupole moment of the molecule can induce a dipole in the Ar atom, the Ar can perturb the zero point bending vibrations of the linear molecule to give a slightly bent vibrationally averaged "linear" molecule within the complex, or there could be charge transfer. In the case of $Ar-CO_2$, where the CO_2

quadrupole moment has been well determined[57] the induced moment agrees very well with observation. This agreement could, of course, be fortuitous since higher electrostatic interactions have been ignored and the signs of the various sources of moments can certainly lead to cancellations. The size of the observed moments can quite plausibly be attributed to any of these explanations, and moments of this type cannot be used to invoke charge transfer.

2.3. Vibrational Energies, Potential Functions, and Dynamics

One of the principal reasons for studying weakly bound complexes is to obtain information about intermolecular forces and potential functions. In addition, vibrational properties of the complex, such as density of states and coupling between high and low frequency degrees of freedom, play a dominant role in intramolecular relaxation and unimolecular dissociation of complexes. A large portion of this conference is devoted to these latter subjects. Rotational spectroscopy of weakly bound species can supply much information on intermolecular interactions and intramolecular energy transfer. Many of the characteristics of clusters which make structure determination more difficult are exactly those features that supply potential function and dynamics information. Some of the behavior of $<\cos\theta>$ and $<\cos^2\theta>$ have already been mentioned. The bond length in Ar-HCl, determined from $(B+C)/2$, is quite different from the same R in Ar-DCl[6], implying angular dependent radial potential functions. The extreme centrifugal distortion of high J states, or "rotational saturation" of the geometry, in $(HF)_2$ and $(HCl)_2$ obviously contain potential function data[58]. While at present it is not realistic to invert spectroscopic data for a general bimolecular complex to obtain a potential function[51], it is relatively straightforward to evaluate vibrational averages and centrifugal distortions of molecular properties given an analytic potential function.

    In relating any empirical data to a potential function, it is always necessary to consider just what portions of the potential are sampled by the measurement in question. In general, rotational spectra sample the potential near the bottom, fairly close to the equilibrium minimum. However, this generality does not have to be satisfied for bending motions of monomer units with very small moments of inertia, such as hydrogen halides or water. In these cases, the zero point bending amplitude can approach ± 90 degrees and samples all of the space on that cut through the potential function. In contrast, stretching motions in weakly bound systems are quite similar to conventional covalent molecules when viewed in a reduced basis. For example, in a typical van der Waals complex the ratio of stretching amplitude to bond length is a few percent, the same as it is for a strongly bound molecule. While the stretching force constant and vibrational frequency are certainly much smaller than encountered in normal molecules, the ratio of rotational energies to vibrational energies are again quite similar for the two very different systems. The ratio of rotational constant to vibrational spacing, B/ω, is the order of 10^{-3} for both cases.

The small ratios of B/ω and amplitude to bond length mean that a bimolecular complex can be treated as a diatomic molecule for small perturbations such as centrifugal distortion of rotational energy levels. The expression for centrifugal distortion in an harmonic diatomic molecule, $D_J = 4B^3/\omega^2$, has been used to obtain estimates for the stretching frequency of many complexes. Where comparisons exist between the results of this simple model and detailed potential functions, the agreement is typically within a few wavenumbers. An extension to this kind of analysis is to fit B and D_J to a specific radial potential function such as a Lennard-Jones function. This then yields a dissociation energy for the complex. This result is very much less reliable because the rotational transition data, in contrast to the angular potential, samples the radial potential only near the minimum and are not very sensitive to the well depth. Thus determining a harmonic frequency, i.e. the curvature of the potential at R_e, from rotational data is much more reliable that determining a dissociation energy. This kind of analysis is also quite sensitive to the choice of the equilibrium bond length and the specific form used for the potential.

Complexes such as Ar-HCl are linear molecules, with a single quartic centrifugal distortion coefficient. In asymmetric top complexes, where transitions are observed for different J and K levels, there are five quartic constants[48] and these can provide additional vibrational information. In the first analysis of this type, both stretching and bending harmonic frequencies were obtained for Ar-CO_2[8]. This kind of analysis has also been done for Ar-O_3, Ar-SO_2[59], and Ar-ClCN[60]. The success of this type analysis depends, in part, on the simplicity of the molecules. In Ar-CO_2, the two low frequency vibrations are of different symmetry and are not easily mixed. The separation of high frequency from low frequency vibrations insures well determined low frequencies from centrifugal distortion constants. In Ar-O_3 and Ar-SO_2 there are three low frequency modes, two of the same symmetry. The off diagonal term in the force constant matrix mixing these two is small in Ar-O_3, but larger than one diagonal element in Ar-SO_2. This procedure obviously becomes rather limited as larger complexes are considered, even though much more centrifugal distortion data can be obtained from experiments sampling high J,K states. In these cases, the centrifugal distortion information has to be treated as just one more source of input to a general potential function determination.

Large amplitude motions in the form of inversion tunneling and internal rotation have been observed in several complexes. In the case of inversion, very precise inversion frequencies are measured, but it is difficult to obtain much quantitative information about the shape of the barrier. This is, of course, a classic spectroscopic problem. One notable exception, where a quantitative description of a barrier was obtained, was the detailed potential surface calculated from spectroscopic data for $(HF)_2$[61]. The heights of several barriers to internal rotation have been determined making the usual assumptions of barrier shape.

    The general problem of extracting an anisotropic potential from all
forms of data available for a pair of interacting molecules is an
extremely difficult one. However, in the specific case of rare gas-
hydrogen halide interactions, a large amount of work has been done to
produce quite detailed potential functions. Since it is not realistic
to invert observed data to recover the desired potential, it is
necessary to begin with some form of paramaterized function from which
observed properties can be calculated. If the assumed potential is to
be optimized, it is essential that all of the required calculations can
be done efficiently. To handle the basic effects of the anisotropy,
the angular and radial coordinates are separated. The facile angular
motion and relatively conventional radial motion suggests that this
separation be done in the spirit of the Born-Oppenheimer separation of
electronic and nuclear coordinates. This has been labeled BOARS, Born-
Oppenheimer angular-radial separation, and was developed by Holmgren,
Waldman, and Klemperer[62] and refined by Hutson and Howard[63]. Using
any potential surface, the angular wave equation is solved many times
with R as a parameter to obtain an effective radial potential function.
Spectroscopic observables can then be calculated efficiently enough
that paramaterized potential functions can be optimized by comparing
calculated and observed properties in a least squares
procedure[64,65,51].

    An important advantage to this approach is that it is not limited
to just a single source of data. For Ar-HCl[65] Hutson and Howard
included an analysis of the MBER data sufficiently precise that
centrifugal distortion of eQq was an important source of information on
the angular-radial coupling term in the potential and, at the same
time, included scattering, pressure broadening, and virial coefficient
data in the potential function optimization. In this way, data
sensitive to quite different portions of the potential were combined to
yield a single optimum surface. They also discussed the problems
associated with the paramaterization chosen and investigated several
functional forms for the intermolecular interaction. Since calculated
and observed properties were compared in a least squares sense, the
Jacobian derivative matrix, which describes the sensitivity of each
parameter of the potential to each data type, was calculated. Much
physical insight can be obtained by perusing this array of derivatives.
Perhaps equally important, a careful study of the derivative matrix
generates a healthy skepticism toward potential functions generated
from less complete procedures.

    A large number of RG-HX potentials have been calculated and several
generalizations may be made. One of the more interesting results is
the need to include a secondary minimum in Ne-HCl at 180 degrees,
corresponding to a linear HCl-Ne geometry. The Ne-HCl data[66] is
sensitive to this geometry because of the extreme angular motions in
this complex, where HCl is nearly a free rotor. Even though the
heavier rare gas atoms do not effectively sample this part of the
potential, this local minimum was included in the other RG-HCl
surfaces. Only in the case of Ar-HCl is the well depth really
accurately determined, owing to the inclusion of scattering and second
virial coefficient data in the optimization. Comparing predicted

second virial coefficients obtained from parameterized potentials of
the other RG-HX systems with experimental results would be very useful
in assessing the accuracy of the well depths of these potentials. The
RG-HF paper[51b] discusses the origin of the anisotropy of all of the
RG-HX interactions and shows that, for the HF interactions, induction
accounts for 70% of the anisotropy, but less that 40% of the anisotropy
was attributed to induction in RG-HCl complexes. This same analysis
found that induction and dispersion forces could account for the
attractive potential without need for postulating a weak or incipient
chemical bond.

The BOARS method and its modifications are not limited to the atom-
diatom system, but have also been used to construct a potential surface
for $(HF)_2$[61]. As previously mentioned, this work describes the height
and shape of the barrier to inversion. It also shows a nonlinear
hydrogen bond at equilibrium and an equilibrium separation of the
monomer units, R_e, more than a tenth of an angstrom shorter than that
determined from the rotational constants. This large difference
between R_0 and R_e points to the need for care in interpreting
structural data obtained from a minimum of rotational data. This $(HF)_2$
surface was constructed only from rotational data so the accuracy of
the dissociation was uncertain. This was a concern since the
calculated D_0 was less than 1100 cm^{-1}, quite small for a canonical
strong hydrogen bond. However, recent IR intensity measurements
confirm the low result, giving 1038(40) cm^{-1}[40b]. This also agrees
with the most recent ab initio calculations for $(HF)_2$[67].

The very detailed results which can be obtained for simple systems,
such as RG-HX, should not be taken as an indication that there is no
longer any controversy over the general origins and characteristics of
weak interactions. This notion can quickly be dispelled by reading a
pair of back to back comments published by the Chairman of this session
and by the concluding speaker of the structure section of this
meeting[68].

The first two portions of this section addressed internal energy
levels and potential functions for weak complexes. Any information of
this kind, including that obtained from rotational spectroscopy, is
essential to understanding the dynamics of these systems. Both more
detailed potential functions and a much better understanding of
coupling between low frequency and high frequency vibrational motion
will greatly contribute to our understanding of energy transfer within,
reactions between and within, and dissociation of clusters. Rotational
energy levels and rotational spectroscopy play a more direct role in
relaxation measurements when lifetime information is inferred from
spectroscopic linewidths. A large number of studies of excited states
of weak complexes have attributed experimental linewidths to relaxation
or dissociation processes[41,37,35,17,69,70]. Many of these studies,
particularly infrared predissociation experiments in the 10 micron
region, have observed signals much broader than rotational envelopes,
while others observe smaller broadening of individual rotational lines.
Intermediate cases, where linewidths are comparable to rotational
spacings, must exist. In all of these situations (with the possible
exception of the very broad lines) knowledge of rotational structure

and J,K assignments of the spectrum are needed for rational
interpretation of the data. Even the most qualitative models
describing dissociation mechanisms rely on knowing, or guessing, the
geometric structure of the dimer involved. Clearly, rotational
spectroscopy will continue to make many contributions to the
understanding of cluster dynamics.

3. PREDICTIONS

At the time I wrote the abstract for this paper it seemed like a good
idea to try to guess some of the things weak complexes were going to
divulge about themselves in the immediate future. Now, having read the
abstracts for this meeting, this seems to have been rather foolhardy.
Also one must consider all of the discussions and interactions which
will occur during the next five days, and then try to guess how this
communication will affect the many research programs represented here.
Clearly this is no easy task, which augurs well for the health and
vitality of our research interests.
    Some generalities are pretty clear. Many more complexes will be
studied and, while the dominance of bimolecular species will probably
continue, a growing number of larger clusters will be investigated.
Perhaps that Holy Grail of studying the transition from isolated
molecules to solvated molecules may begin to come into the realm of
feasibility, rather than existing only in the introductions to research
proposals. The last few years has seen a mushrooming of infrared
studies of complexes, both in the mid IR and far IR. These data will
be incorporated into ever improving potential functions and, in a more
qualitative way, lead to a better understanding of the interactions of
the vibrational degrees of freedom in complexes. It is not obvious,
however, that all of these efforts will yield an intuitive
understanding of the structures of all weakly bound molecules (say in
the range of 0.5 to 15 kCal/mol) at the same level that we understand
covalent molecules.
    A few specific experimental developments can safely be mentioned.
More laser experiments will be done in conjunction with molecular beam
microwave techniques and, as in the cases of the MIRDR method,
microwave experiments will be done in conjunction with molecular beam
laser methods. More and more pulsed beam experiments will be done and
I think pulsed beam MBER experiments can be more sensitive than cw
experiments (given fixed source chamber pumping speed). Pulsed beam
microwave FT spectrometers will become ever more automated to remove
some of the scanning problems, making this technique even more
productive. Given the current results coming from Gutowsky's
laboratory, bigger and bigger clusters will be studied. Various model
development, and more basic theory, will go along with the new
empirical data.
    So, a lot has been done, there remains much more to do, and this
conference will contribute greatly to these efforts!

REFERENCES

1. T. R. Dyke, Topics Current Chem., **120**, 86 (1984).
2. A. C. Legon, Ann. Rev. Phys. Chem., **34**, 275 (1983); A. C. Legon and D. J. Millen, Chem. Rev., **86**, 635 (1986).
3. C. C. Costain and G. P. Srivastava, J. Chem. Phys., **41**, 1620 (1964); E. M. Bellot and E. B. Wilson, Tetrahedron, **31**, 2896 (1975).
4. T. R. Dyke, B. J. Howard, and W. A. Klemperer, J. Chem. Phys., **56**, 2442 (1972); B. J. Howard, T. R. Dyke, and W. A. Klemperer, J. Chem. Phys., **81**, 5417 (1984).
5. T. R. Dyke and J. S. Muener, J. Chem. Phys., **60**, 2929 (1974); T. R. Dyke, K. M. Mack, and J. S. Muenter, J. Chem. Phys., **66**, 498 (1977); J. A. Odutola and T. R. Dyke, J. Chem. Phys., **72, 5062** (1980).
6. S. E. Novick, P. Davies, S. J. Harris, and W. A. Klemperer, J. Chem. Phys., **59**, 2273 (1973).
7. S. E. Novick, K. C. Janda, and W. A. Klemperer, J. Chem. Phys., **65**, 5115 (1976).
8. J. M. Steed, T. A. Dixon, and W. A. Klemperer, J. Chem. Phys., **70**, 4095 (1979).
9. T. J. Balle, E. J. Campbell, M. R. Keenan, and W. H. Flygare, J. Chem. Phys., **71**, 2723 (1979); ibid. **72**, 922 (1980); T. J. Balle and W. H. Flygare, Rev. Sci. Inst., **52**, 33 (1981).
10. A. C. Legon, D. J. Millen, and S. C. Rogers, Chem. Phys. Lett., **41**, 137 (1976); A. C. Legon, D. J. Millen, and P. J. Mjoberg, Chem. Phys. Lett., **47**, 589 (1977); J. W. Bevan, Z. Kisiel, A. C. Legon, D. J. Millen, and S. C. Rogers, Proc. Roy. Soc., **A372**, 441 (1980).
11. See references 33 and 36.
12. A. R. W. McKeller, Farad. Disc. Chem. Soc., **72**, 89 (1982); G. E. Ewing, Acct. Chem. Res., **8**, 185 (1975) and references therein.
13. See references 35, 37-40.
14. S. C. Wofsy, J. S. Muenter, and W. A. Klemperer, J. Chem. Phys., **53**, 4005 (1970).
15. R. L. DeLeon and J. S. Muenter, J. Chem. Phys., **72**, 6020 (1980).
16. W. L. Ebenstein and J. S. Muenter, J. Chem. Phys., **80**, 1417 (1984).
17. R. L. DeLeon and J. S. Muenter, J. Chem. Phys., **80**, 6092 (1984).
18. T. R. Dyke and J. S. Muenter, J. Chem. Phys., **57**, 5011 (1972).
19. F. deLeeuw and A. Dymanus, J. Mol. Spec., **48**, 427 (1973); B. Fabricant and J. S. Muenter, J. Chem. Phys., **66**, 5274 (1977).
20. S. L. Shostak and J. S. Muenter, to be published.
21. W. L. Ebenstein and J. S. Muenter, J. Chem. Phys., **80**, 1417 (1984).
22. G. T. Fraser, D. D. Nelson, A. Charo, and W. A. Klemperer, J. Chem. Phys., **82**, 2535 (1985).
23. M. D. Marshall, A. Charo, H. O. Leung, and W. A. Klemperer, J. Chem. Phys., **83**, 4924 (1985).
24. E. J. Campbell and W. G. Read, J. Chem. Phys., **78**, 6490 (1983)
25. B. L. Cousins, S. C. O'Brien, and J. M. Lisy, J. Phys. Chem., **88**, 5142 (1984).
26. M. R. Keenan, L. W. Buxton, E. J. Campbell, T. J. Balle, and W. H. Flygare, J. Chem. Phys., **73**, 3523 (1980); E. J. Campbell, L. W.

Buxton, and A. C. Legon, J. Chem. Phys., **78**, 3483 (1983), and references therein.

27. A. D. Buckingham and P. W. Fowler, J. Chem. Phys., **79**, 6426 (1983); A. C. Legon, NATO Adv. Workshop, Structure and Dynamics of Weakly Molecular Complexes, Paper M-2, 1987 (this volume).

28. H. S. Gutowsky, T. D. Klots, C. Huang, C. A. Schumuttanmaer, and T. Emmilsson, J. Chem. Phys., **83**, 4817 (1985); ibid. J. Am. Chem. Soc., **107**, 7174 (1985); see also papers RD2-RD4 41st Symp. on Mol. Spec. Ohio State Univ. (1986).

29. E. J. Campbell, L. W. Buxton, T. J. Balle, and W. H. Flygare, J. Chem. Phys. **74**, 829 (1981).

30. F. J. Lovas and R. D. Suenram, to be published.

31. D. D. Nelson, G. T. Fraser, F. J. Lovas, R. D. Suenram, and W. A. Klemperer, 41st Symp. Mol. Spec., Ohio State Univ., paper TF1, (1986).

32. Slit shaped beams can provide significantly longer path lengths, however one is still considering centimeters rather than kilometers possible in gas cells.

33. W. J. Lafferty, R. D. Suenram, and F. J. Lovas, to be published.

34. It should be mentioned that the rotational temperature in a molecular beam can be varied over a considerable range by varying source conditions, although this usually entails sacrifice in sensitivity.

35. B. J. Howard and A. S. Pine, Chem. Phys. Lett., **122**, 1 (1985).

36. F. J. Lovas, R. D. Suenram, S. Ross, and M. Klobukowski, to be published.

37. A. S. Pine and W. J. Lafferty, J. Chem. Phys., **78**, 2154 (1983); N. Ohashi and A. S. Pine, J. Chem. Phys., **81**, 73 (1984).

38. E. K. Kyro, M. Eliades, A. M. Gallegos, P. Shoja-Chagervand, and J. W. Bevan, J. Chem. Phys., **85**, 1283 (1986), and references therein.

39. B. A. Wofford, J. W. Bevan, W. B. Olson, and W. J. Lafferty, Chem. Phys. Lett., 124 , 579 (1986); B. A. Wofford, J. W. Bevan, and W. B. Olson, J. Chem. Phys., **85**, 105 (1986).

40. A. C. Legon, D. J. Millen, P. J. Mjoberg, and S. C. Rogers, Chem. Phys. Lett., **55**, 157 (1978); A. S. Pine and B. J. Howard, J. Chem. Phys., **84**, 590 (1986).

41. D. H. Levy, Adv. Chem. Phys., **47**, 323 (1981); Ann. Rev. Phys. Chem., **31**, 197 (1980).

42. D. H. Levy, NATO Adv. Workshop, Structure and Dynamics of Weakly Bound Molecular Complexes, Paper T-5, (1987) (this volume).

43. J. Mettes, B. Heymen, P. Verhoeve, J. Reuss, D. C. Laine, and G. Brooks, Chem. Phys., **92**, 9 (1985).

44. M. Maroncelli, G. A. Hopkins, J. W. Nibler, and T. R. Dyke, J. Chem. Phys., **83**, 2129 (1985).

45. G. D. Hayman, J. Hodge, B. J. Howard, J. S. Muenter, and T. R. Dyke, Chem. Phys. Lett., **118**, 12 (1985); C. M. Lovejoy, M. D. Schuder, and D. J. Nesbitt, Chem. Phys. Lett., **127**, 374 (1986).

46. K. W. Jucks, Z. S. Huang and R. E. Miller, 41st Symp. Mol. Spec., Ohio State Univ., Paper-TF5 (1986).

47. G. D. Hayman, J. Hodge, B. J. Howard, J. S. Muenter, and T. R. Dyke, J. Chem. Phys., submitted.

48. J. K. G. Watson, J. Chem. Phys., **48**, 4517 (1968).
49. A. C. Legon and D. J. Millen, Proc. Roy. Soc. **A404**, 89 (1986).
50. P. D. Soper, A. C. Legon, W. G. Read, and W. H. Flygare, J. Chem. Phys., **76**, 292 (1982).
51. J. M. Hutson and B. J. Howard, Mol. Phys., **45**, 769 (1982); ibid. **45**, 791 (1982); for experimental work see many references in these theoretical papers.
52. A. D. Buckingham and P. W. Fowler, Can. J. Chem., **63**, 2018 (1985).
53. K. I. Peterson and W. A. Klemperer, J. Chem. Phys., **81**, 3842 (1984); ibid. **85**, 725 (1986).
54. K. R. Leopold, G. T. Fraser, F. J. Lin, D. D. Nelson, and W. A. Klemperer, J. Chem. Phys., **81**, 4922 (1984).
55. D. G. Prichard, B. J. Howard, and J. S. Muenter, to be published.
56. R. Viswanathan and T. R. Dyke, J. Chem. Phys., **77**, 1166 (1982).
57. A. D. Buckingham, R. L. Disch, and D. A. Dunmur, J. Am. Chem. Soc., **90**, 3104 (1968).
58. A. S. Pine, W. J. Lafferty, and B. J. Howard, J. Chem. Phys., **81**, 2939 (1984).
59. J. S. Muenter, R. L. DeLeon, and A. Yokozeki, Farad. Disc., **73**, 63 (1982).
60. M. R. Keenan, D. B. Wozniak, and W. H. Flygare, J. Chem. Phys., **75**, 631 (1981).
61. A. E. Barton and B. J. Howard, Farad. Disc., **73**, 45 (1982).
62. S. L. Holmgren, M. Waldman, and W. A. Klemperer, J. Chem. Phys., **67**, 4414 (1977).
63. J. M. Hutson and B. J. Howard, Mol. Phys., **41**, 1123 (1980).
64. S. L. Holmgren, M. Waldman, and W. A. Klemperer, J. Chem. Phys., **69**, 1661 (1978).
65. J. M. Hutson and B. J. Howard, Mol. Phys., **43**, 493 (1981).
66. A. E. Barton, D. J. B. Howlett, and B. J. Howard, Mol. Phys., **41**, 619 (1980).
67. D. W. Micheal, C. E. Dykstra, and J. M. Lisy, J. Chem. Phys., **81**, 5998 (1984).
68. A. D. Buckingham and P. W. Fowler, J. Chem. Phys., **79**, 6426 (1983); F. A. Baiocchi, W. Reiher, and W. A. Klemperer, J. chem. Phys., **79**, 6428 (1983).
69. K. C. Janda, Chem. Rev., **86**, 507 (1986).
70. R. E. Miller, P. F. Vohralik, and R. O. Watts, J. Chem. phys., **80**, 5453 (1984).

PULSED-NOZZLE, FOURIER-TRANSFORM MICROWAVE SPECTROSCOPY OF HYDROGEN-BONDED DIMERS.

A.C. LEGON
Department of Chemistry,
University of Exeter,
Stocker Road, Exeter   EX4 4QD
United Kingdom

ABSTRACT. The technique of pulsed-nozzle, Fourier-transform microwave spectroscopy for observing the rotational spectra of hydrogen-bonded (and other weakly bound) dimers is discussed. The sensitivity of the technique for weakly bound species and its accompanying high resolution are illustrated by reference to examples. The molecular properties of dimers that can be determined from rotational spectra via the spectroscopic constants are outlined and, in particular, the determination of geometry from rotational constants is discussed by reference to $SO_2 \cdots HF$. The observed geometry for $SO_2 \cdots HF$ is then considered in the light of some simple rules for predicting the angular geometries of hydrogen-bonded dimers. Finally, some recent results for (SO_2, HCN) are presented and the observed angular geometry is compared with the predictions of the Buckingham-Fowler electrostatic model for weakly bound dimers.

1.   INTRODUCTION

To investigate the nature of weak interactions between pairs of closed shell molecules (including the hydrogen bonding interaction) it is desirable to obtain the properties of the dimer in isolation. Then the weak interaction is unperturbed by surrounding molecules of a solvent or lattice. Rotational spectroscopy is a key method for determining precisely a number of molecular properties and, moreover, has the advantage that it is conducted in the gas phase at low pressure. The molecules so investigated are therefore in effective isolation, as required.

The main problem in observing the rotational spectrum of a hydrogen-bonded dimer lies in achieving a detectable concentration of the species $B \cdots HA$ in the mixture of B and HA. When the hydrogen bond is relatively strong, as in $HCN \cdots HF$[1] and $H_2O \cdots HF$,[2] it is possible to use conventional Stark-modulation microwave spectroscopy with a cooled mixture of B and HA at thermal equilibrium in the absorption cell. For more weakly bound dimers, lower temperatures are required to give a favourable equilibrium constant for the formation of $B \cdots HA$ but usually

23

A. Weber (ed.), Structure and Dynamics of Weakly Bound Molecular Complexes, 23–42.

the complex condenses before a sufficient concentration is achieved.
Obviously, a method of effectively cooling the mixture but avoiding
condensation is then required.

Very low effective temperatures and high concentrations of B$\cdots$HA
species are produced when a mixture of B and HA diluted in, say, argon
is expanded supersonically and adiabatically through a nozzle into a
vacuum. A supersonically expanded gas was first used in an
investigation of the rotational spectrum of a weakly bound complex by
Klemperer and coworkers.[3] They detected rotational transitions of
HF$\cdots$HF in a supersonic molecular beam by electric resonance
spectroscopy (MBERS). Later, Flygare and his collaborators[4,5]
developed another type of microwave spectrometer employing a supersonic
molecular source. In the pulsed-nozzle, Fourier-transform microwave
spectrometer, a rotational transition of B$\cdots$HA is detected in emission
following the rotational polarization of a pulse of gas mixture through
its interaction with a synchronous microwave radiation pulse in a
spacious Fabry-Perot cavity. In this article the principles of the
pulsed-nozzle F-T method will be outlined and its sensitivity and
resolution illustrated.

The techniques employing a supersonically expanded gas pulse have
the well-known advantage of inherently high sensitivity with a
concomitant high resolution, even for very weakly bound dimers. A
disadvantage is that rotational spectra are restricted almost
invariably to the vibrational ground state as a result of very low
effective temperatures. The valuable information about the
intermolecular potential energy function that attends the observation
of vibrational satellites[6] associated with the low energy
intermolecular modes of the dimer is then lost. The observation of
these vibrational satellites is straightforward with a thermal
equilibrium source of dimers but unfortunately such sources are
appropriate only for relatively strongly bound dimers.

A distinct *chemical* advantage of using a supersonic expansion is
that dimers formed from almost any pair of molecules can be
investigated. The choice of B and HA is then dictated not by the
technique but by the investigator who can change the components
systematically to allow the variation in some particular property of
the hydrogen bond along the series to be investigated. For example,
in this article, some conclusions about the angular geometries of
hydrogen-bonded dimers that have been so obtained will be presented.

2.   THE PULSED-NOZZLE, FOURIER-TRANSFORM MICROWAVE SPECTROMETER

2.1. Principles of Operation

A schematic diagram of a pulsed-nozzle, Fourier-transform microwave
spectrometer is shown in Figure 1. The spectrometer is composed of
four main elements. The first is the pulsed-nozzle source of dimers
B$\cdots$HA. The second element consists of three sources of monochromatic
radiation having frequencies ν, ν-30 MHz and 30 MHz, respectively,
while the third element is the microwave Fabry-Perot cavity in which

the molecule-radiation interactions occur. The final element is the
signal detection/processing system. The following is an outline of
the principles of operation of the spectrometer. Details have been
given elsewhere.[7-10]

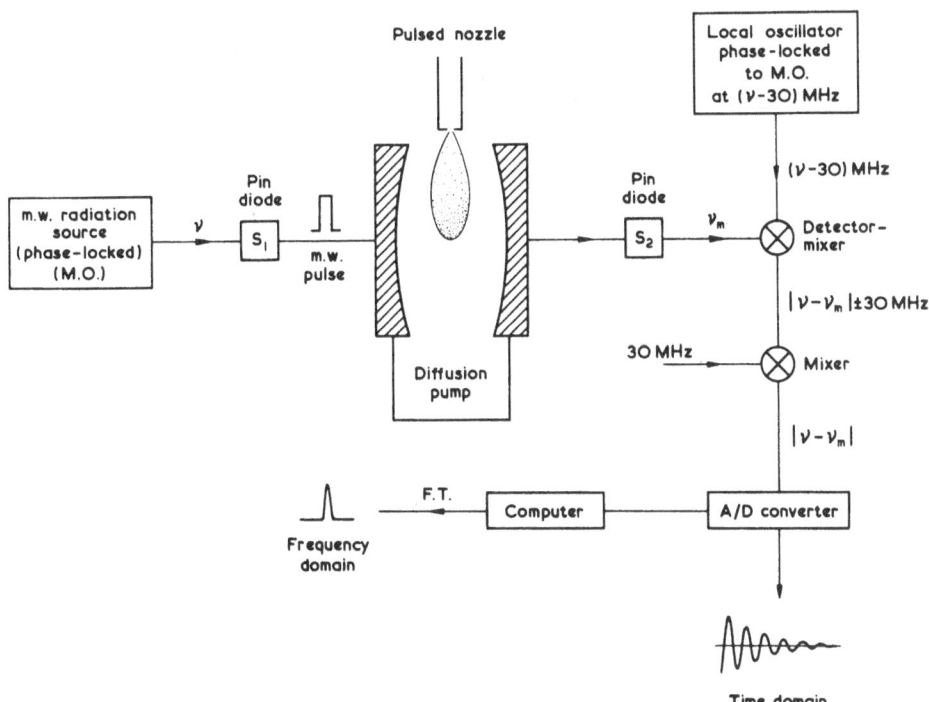

Figure 1. Schematic diagram of a pulsed-nozzle, Fourier-transform
microwave spectrometer.

2.1.1. *Sequence of gas and radiation pulses.* The successful operation
of the spectrometer depends on achieving the correct sequence and
timing of the gas and radiation pulses. First, a short pulse of the
gas mixture (about 2% each of B and HA diluted in argon and held at
approx. 2 atm. pressure) is propelled into the evacuated Fabry-Perot
cavity along a direction perpendicular to the cavity axis, as shown in
Figure 1. Typically, the gas pulse is produced by opening a solenoid
valve for about 1 ms and allowing the gas to be expanded through a
circular orifice (0.7 mm diameter) in a plate covering the outlet of
the valve.
      Secondly, after a delay, Δt, sufficient for the gas pulse to have
arrived between the Fabry-Perot mirrors, a pulse of monochromatic
microwave radiation (frequency ν) is allowed to enter the cavity by

opening the pin-diode switch S_1 for, typically, 1 μs. The high-Q
Fabry-Perot cavity is, of course, first critically tuned to accept
radiation of frequency ν. The microwave pulse length is chosen so
that the band of frequencies (≃ 1 MHz centred at ν) thereby generated
and carried into the cavity matches the cavity bandwidth ($\Delta\nu \simeq 1$ MHz).

2.1.2. *Molecule-radiation interactions.* If the species B···HA in the
gas pulse exhibits at least one rotational transition whose frequency
ν_m ralls within the cavity bandwidth, the radiation density resulting
from the microwave pulse is usually sufficient to induce a macroscopic
rotational polarization in the ensemble of molecules B···HA. The delay
Δt is obviously critical. If Δt is either too short or too long, the
microwave pulse will arrive in the cavity either before or after the
gas pulse.

The radiation that actually leads to the spectrum detected by the
spectrometer is the spontaneous coherent emission at frequency ν_m that
accompanies the subsequent decay of the rotational macroscopic
polarization with a characteristic time of T_2. The key to the
sensitivity of the technique lies in the inequality $T_2 \gg \tau_c$, where
τ_c is the characteristic time for dissipation of the polarizing pulse
within the Fabry-Perot cavity. Typical values are ∿ 100 μs for T_2
but only ∿ 0.2 μs for τ_c (for the instrument at the University of
Exeter). Consequently, the pin diode switch S_2, which isolates the
detector from the Fabry-Perot cavity, can be held closed until
e.g. 1 μs after the polarizing pulse has entered the cavity. During
that time, the background radiation decays through many half-lives
but the coherent spontaneous emission is only just beginning. Or
opening S_2, only the desired emission at ν_m reaches the detector and
t e intense polarizing radiation is sensibly absent.

2.1.3. *Detection of molecular emission.* The molecular emission or
frequency ν_m is detected by a superheterodyne method conducted in two
stages. First, ν_m is mixed in a balanced mixer/amplifier with the
radiation of frequency ν-30 MHz to give a mixed-down signal of
frequency $|\nu-\nu_m| \pm 30$ MHz. The output from the amplifier is
subsequently mixed in a double-balanced mixer with the radiation at
30 MHz to give a final signal of frequency $|\nu-\nu_m|$, that is the beat
frequency between the polarizing radiation and that emitted by the
molecules. The detected signal thus consists of a sinusoidally
oscillating voltage of frequency $|\nu-\nu_m|$ which is exponentially damped.
This is then digitized (typically at a rate of 0.5 μs per point for
512 points) and stored in a computer. The time-domain record so
obtained from the $J = 1 \leftarrow 0$ transition of $^{16}O^{12}C^{32}S$ is shown in Figure 2.
The presence of the additional, low frequency beat will be discussed
later. The emission from the $J = 1 \leftarrow 0$ transition of $^{16}O^{12}C^{32}S$ is very
strong and signal averaging is not essential to obtain a spectrum like
that shown in Figure 2. When averaging is necessary, however, it is
important to have identical phases in the repeated signals $|\nu-\nu_m|$.
Because the molecular emission signal ν_m stands in fixed phase
relationship to the polarizing radiation, the identical phases of
repeated signals $|\nu-\nu_m|$ is assured if the radiations of frequency

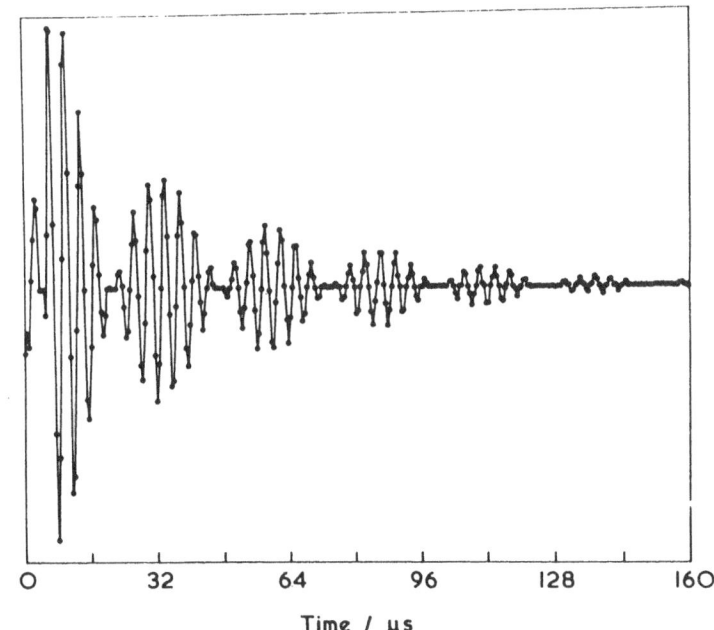

Figure 2. Emission signal from the $J = 1\leftarrow0$ transition of $^{16}O^{12}C^{32}S$. The polarizing radiation had a frequency of 12162.7100 MHz and the signal was digitized at a rate of 0.5 μs per point.

ν-30 MHz and 30 MHz used during the superheterodyne detection are also fixed in phase with respect to the polarizing radiation. The so-called phase-coherence of the ν and ν-30 MHz radiations is readily achieved by using a phase-locked master oscillator (frequency ν) and then phase-locking to this a second microwave source (the local oscillator in Figure 1) by means of a frequency stabilizer so that the difference between the oscillators is maintained at 30 MHz. A 30 MHz signal of correct phase is then automatically obtained by taking the beat of the two oscillators from a mixer.

The time-domain record of Figure 2 is related through Fourier transformation to the frequency-domain spectrum shown in Figure 3. The points in Figure 3 are offset from the polarizing frequency ν at a rate of 3.90625 kHz per point. The transition, which should be a singlet, appears as a doublet as a result of an instrumental effect discussed in detail elsewhere.[8,9] It is sufficient to note here that the true rotational transition frequency lies at the doublet centre. A method of calibrating the spectrometer to ensure an accuracy of 1 kHz in the measurement of frequencies is described in ref. 11.

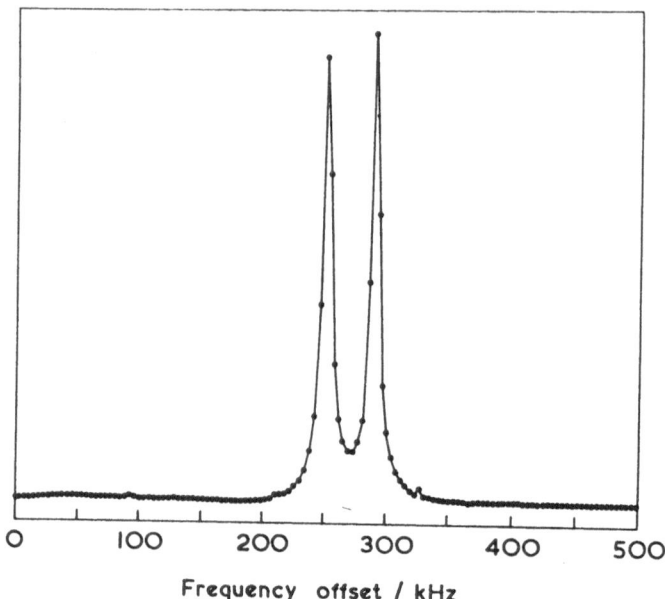

Frequency offset / kHz

Figure 3. Frequency-domain spectrum resulting from Fourier
transformation of the emission signal shown in Figure 2. Frequencies
are offset at a rate of 3.90625 kHz per point from ν = 12162.7100 MHz.
See text for discussion of the doubling effect.

2.2. Sensitivity and Resolution.

The high sensitivity of the pulsed-nozzle, F-T technique for even very
weakly hydrogen-bonded species may be gauged by a consideration of the
linear species OC$\cdots$HCN.[12] The weakness of the hydrogen bond in
OC$\cdots$HCN is illustrated by the value k_σ = 3.3 N m^{-1} of the
intermolecular stretching force constant[12] which should be compared,
for example, with k_σ = 18.2 N m^{-1} in HCN$\cdots$HF.[13] Nevertheless,
rotational transitions of the isotopic species $O^{13}C\cdots H^{12}C^{14}N$ can be
observed easily in their natural abundance of $\sim$ 1%, as shown for the
J = 4$\leftarrow$3 transition in Figure 4. The signal-to-noise ratio is high even
though only a few minutes was spent collecting the data. The three
hyperfine components that remain after allowance, as indicated, for the
instrumental doubling arise from ^{14}N- nuclear quadrupole coupling.
The high sensitivity of the technique arises for three reasons. First,
nozzle expansion leads to a gas pulse rich in dimer molecules in their
vibrational ground state which persist because they travel in
collisionless expansion in the molecule-radiation interaction region.
Secondly, detection in the time domain leads to a signal-to-noise
enhancement compared with frequency-domain detection.[9] The third

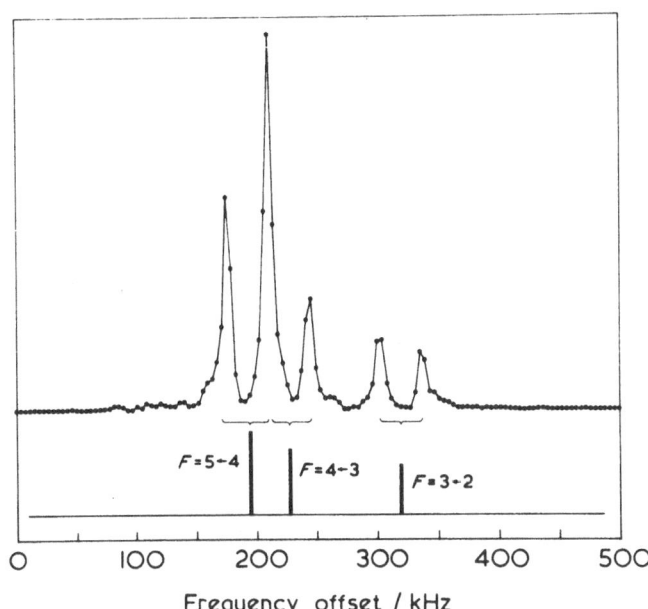

Frequency offset / kHz

Figure 4. $J = 4 \leftarrow 3$ transition of $O^{13}C \cdots H^{12}C^{14}N$ recorded in natural abundance showing ^{14}N-nuclear quadrupole hyperfine structure. Frequencies are offset from $\nu = 11707.3027$ MHz at a rate of 3.90625 kHz per point. Redrawn, with permission, from Ref. 12.

advantage accrues from the use of gas pulses rather than a continuous jet and has been discussed previously.[9]
    We note from Figure 4 that each individual component has a full-width at half-height of 10 kHz. Thus, a high resolution accompanies the high sensitivity for dimers and is characteristic of the technique. The various contributions to the line width have been discussed by Campbell *et al.*[8] A further illustration of the resolution is given in Figure 5 which shows the $J = 1 \leftarrow 0$ transition of $DC^{15}N \cdots HC^{15}N$ observed in a mixture of $DC^{15}N$ and $HC^{15}N$ containing 99% ^{15}N. This transition occurs at a relatively low frequency for microwave spectroscopy but the signal-to-noise ratio is still good and the three D-nuclear quadrupole components are well resolved.[14] The instrumental doubling effect, which is proportional to ν, is only just perceptible.

3.   PROPERTIES OF HYDROGEN-BONDED DIMERS FROM PULSED-NOZZLE, FOURIER-TRANSFORM MICROWAVE SPECTROSCOPY

Pulsed-nozzle, Fourier-transform microwave spectroscopy has now been applied to a number of hydrogen-bonded dimers $B \cdots HA$. By analysis of ground-state rotational spectra observed in this way it has been

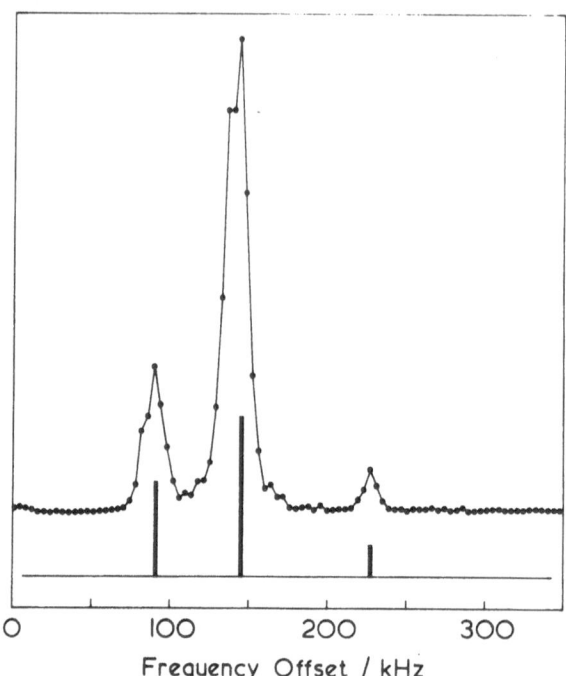

Figure 5. $J = 1 \leftarrow 0$ transition of $DC^{15}N \cdots HC^{15}N$ showing the three
D-nuclear quadrupole hyperfine components $F = 1 \leftarrow 1$, $2 \leftarrow 1$, $0 \leftarrow 1$ in order
of increasing offset from zero. Frequencies are offset from
$\nu = 3211.5150$ MHz at a rate of 3.90625 kHz per point. Redrawn, with
permission, from Ref. 14.

possible to determine a range of spectroscopic constants for the
dimers. As an example, we give in Table I such constants for three
isotopic species of the dimer formed by sulphur dioxide with hydrogen
fluoride.[15] In Table I, A_0, B_0 and C_0 are ground-state rotational
constants, Δ_J, Δ_{JK} and δ_J are centrifugal distortion constants, the
D_{gg} $(g = a, b, c)$ are nuclear spin-nuclear spin coupling constants
and the χ_{gg} are D-nuclear quadrupole coupling constants. We note from
Table I that $\Delta_0 = I_C^0 - I_b^0 - I_a^0$ has an essentially invariant value of
1.4 amu Å². Such a value is typical of a planar, weakly bound dimer
and provides immediately some information about the dimer geometry.
As an example of the hyperfine structure that leads to the D_{gg} and
χ_{gg} in Table I, we show in Figures 6(a) and (b) the $1_{10} \leftarrow 1_{01}$ transition
of the species $(^{32}SO_2, HF)$ and $(^{32}SO_2, DF)$, respectively.
    Given spectroscopic constants of the type shown in Table I, it has
been possible, with the aid of models of the dimer, to interpret them
in terms of various properties of the dimer. A summary of the types of
spectroscopic constant available from rotational spectroscopy and the
dimer properties to which they lead is given in Table II.[16]

TABLE I. Spectroscopic constants of (SO_2, HF)

Spectroscopic constant	$(^{32}SO_2, HF)$	$(^{32}SO_2, DF)$	$(^{31}SO_2, HF)$
A_0/MHz	16502.775(4)	16320.261(5)	16366.014(1)
B_0/MHz	2100.308(1)	2081.076(2)	2078.428(1)
C_0/MHz	1853.642(1)	1836.804(2)	1834.827(1)
Δ_J/kHz	18.31(2)	17.03(2)	17.12(16)
Δ_{JK}/kHz	-521.78(4)	-473.54(44)	-403.42(1)
δ_J/kHz	4.37(3)	4.05(3)	4.37(assumed)
D_{aa} or χ_{aa}/kHz	190(2)	227(3)	-
$(D_{bb}-D_{cc})$ or $(\chi_{bb}-\chi_{cc})/kHz$	-57(4)	68(13)	-
$\Delta_0/amu\ Å^2$	1.3958(2)	1.3290(4)	1.4026(2)

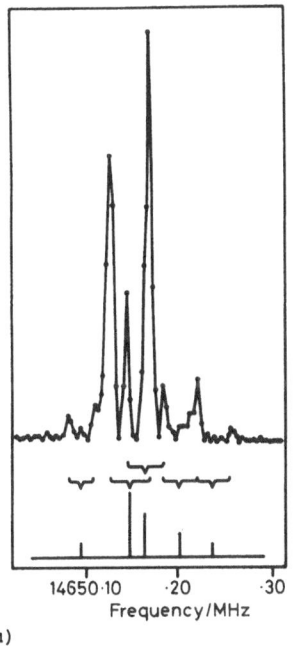

(a)

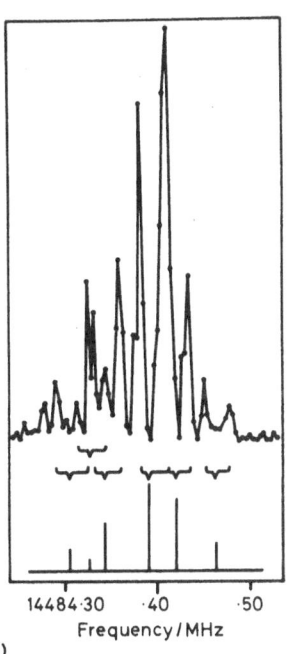

(b)

Figure 6. (a) $1_{10} \leftarrow 1_{01}$ transition of $(^{32}SO_2, HF)$. Each of the five H,^{19}F nuclear spin-nuclear spin hyperfine components is split into a doublet by an instrumental effect. (b) $1_{10} \leftarrow 1_{01}$ transition of $(^{32}SO_2, DF)$. The transition consists of six components arising from D-nuclear quadrupole coupling and D,^{19}F spin-spin coupling. Redrawn, with permission, from Ref. 15.

Furthermore, the investigation of selected series of dimers B···HA has allowed trends in properties to be deduced and some generalisations to be made. For example, the intermolecular stretching force constants

TABLE II

Spectroscopic constant/effect	*Dimer property*
Rotational constant, B_0.	{*Symmetry, geometry, nature of binding.*
Centrifugal distortion constants, D_J and Δ_J.	*Intermolecular force constant k_σ (strength of binding).*
Nuclear quadrupole and nuclear spin-nuclear spin coupling constants.	*Internal dynamics, electric charge distribution, HA bond lengthening.*
Stark effect.	*Electric dipole moment.*
Zeeman effect.	{*Magnetic properties, molecular electric quadrupole moment.*

k_σ have been determined via the centrifugal distortion constant D_J or Δ_J[17] for various related series B···HA, where B = Ar, N_2, CO, PH_3, H_2S, HCN, CH_3CN, H_2O and A is in turn F, Cl, CN or Br. The values[18] are collected together in Table III. It is clear from Table III that the strength of the hydrogen bond, as measured by k_σ, increases down each column (in which B varies but HA is held constant) but along a given row (B constant, A varies) the k_σ value decreases in the order F > Cl > CN ∿ Br. In fact, it has been possible[19] to assign to each B a *nucleophilicity* (or hydrogen-bond basicity) N and to each HA an *electrophilicity* (or hydrogen-bond acidity) E such that all of the k_σ values in Table III can be reproduced by a simple function of N and E. In a similar manner, the lengthening of the HF bond that accompanies formation of B···HF has been deduced[20] from the H,[19]F nuclear spin-nuclear spin coupling constants of B···HF and the D-nuclear quadrupole coupling constants of B···DF for the series of B given above. The lengthening is found to increase uniformly along this series.

Another generalisation[21] concerns the angular geometries of dimers B···HA and is discussed in more detail in the next section.

4.   ANGULAR GEOMETRIES OF HYDROGEN-BONDED DIMERS

By consideration of the angular geometries of a number of dimers (determined by the pulsed-nozzle, F-T or MBERS or Stark-modulation methods), some simple rules for predicting the angular geometry of a hydrogen-bond species B···HA have been proposed.[21] The rules are based on the electrostatic view that the electrophilic H atom in HA seeks out a nucleophilic (high electron density) region on B:

> *The gas-phase equilibrium geometry of a dimer B···HA can be obtained in terms of the nonbonding and π-bonding electron*

pairs on B as follows: (i) the axis of the HA molecule
coincides with the supposed axis of a nonbonding pair, as
conventionally envisaged, or, if B has no nonbonding pairs,
(ii) the axis of the HA molecule intersects the internuclear
axis of the atoms forming the π-bond and is perpendicular
to the plane of symmetry of the π-orbital.

TABLE III. Values of k_σ / N m^{-1} for series B$\cdots$HA.

B HA	HF	HCl	HCN	HBr
Ar	1.4	1.2	1.0	0.8
N_2	5.5	2.5	2.3	-
CO	8.5	3.9	3.3	3.0
PH_3	10.9	5.9	4.3	5.0
H_2S	12.0	6.8	4.7	5.9
HCN	18.2	9.1	8.1	7.3
CH_3CN	20.1	10.7	9.8	-
H_2O	24.9	12.5	11.1	-

Thus, in rule (i) the nonbonding electron pair, as conventionally
pictured, represents the nucleophilic region while in rule (ii) this
role is fulfilled by the π-bonding electron pair. When B carries both
nonbonding and π-bonding electron pairs, rule (i) is definitive.
    The rules can be illustrated by reference to a few selected
examples. Figure 7(a) shows the familiar π-electron density model
for ethylene alongside the observed geometry[22] for the ethylene-
hydrogen chloride dimer. The observed geometry is as predicted by
the model, with the HCl axis perpendicular to the plane of the
ethylene molecule. Cyclopropane is well known to behave in some ways
like an unsaturated hydrocarbon. Indeed, a model that exhibits
pseudo-π-electron density with bent C-C single bonds was proposed by
Coulson and Moffitt[23] and is shown in Figure 7(b) along with the
observed geometry of the dimer of cyclopropane with hydrogen
chloride.[24] We note that the observed geometry is as predicted by
the above rules.
    We now turn to consider the cases in which B exhibits nonbonding
electron pairs. When B is a linear molecule carrying only one
nonbonding electron pair, rule (i) predicts it to form a simple linear
dimer with, for example, HF and this is indeed observed. The
nonbonding pair model of HCN and the observed linear geometry[1] of
HCN$\cdots$HF are shown in Figure 8(a). A similar result obtains[25] for
OC$\cdots$HF, as shown in Figure 8(b). Likewise, the simple model for each
of NH_3 and PH_3 has a nonbonding pair with axial symmetry and in each
case the dimer with HCl is predicted to have C_{3v} symmetry, in agreement
with experiment[26,27] (see Figure 9).
    The situation is more interesting when the molecule B carries
more than one nonbonding pair. If an atom in B exhibits two
equivalent nonbonding pairs, there will be two equivalent conformers
of its hydrogen-bonded dimer with, say, HF. For example, a simple

34   A. C. LEGON

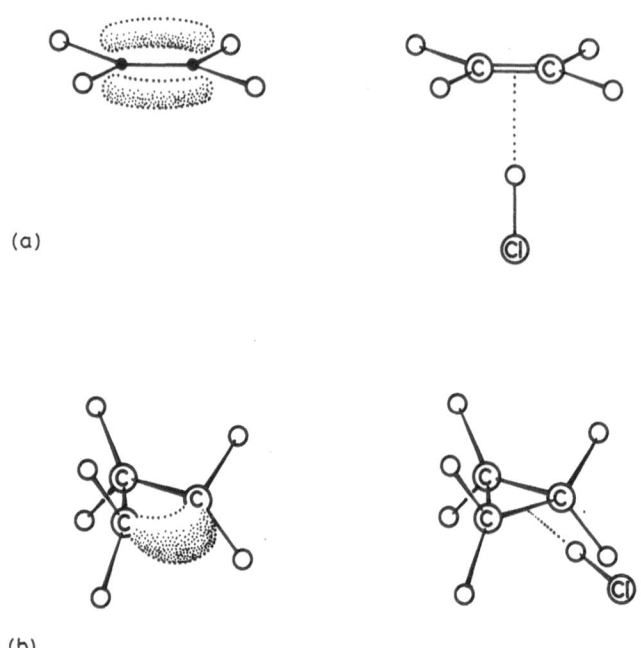

(a)

(b)

Figure 7.   (a)   The π-electron density model of ethylene and the
observed geometry of the ethylene-hydrogen chloride dimer.   (b)   A
pseudo-π-electron density model of cyclopropane and the observed
geometry of the cyclopropane-hydrogen chloride dimer.

model of H_2O has two equivalent nonbonding pairs disposed so that the
central O atom is surrounded tetrahedrally by four electron pairs,
as shown in Figure 10(a).   An analysis of the vibrational satellites
associated with the hydrogen-bond modes in the rotational spectrum of
$H_2O \cdots HF$ shows[2] that in the equilibrium conformation the configuration
at oxygen is pyramidal, as depicted in Figure 10(b), but the potential
energy barrier at the planar form is only 1.5 kJ mol^{-1} and in the
zero-point state the dimer is effectively planar.   On the other hand,
the analogous dimers $H_2S \cdots HX$, where X = F,[28] Cl,[29] CN[30] and Br[31] are
rigidly pyramidal in the zero-point state with no evidence of inversion
doubling and with the angle between the HX axis and the C_2 axis of H_2S
close to 90° [see Figure 10(b)].   These results are in accord with the
interpretation of the electronic structure of H_2S that has a
nonbonding electron pair axis making an angle of 90° with the H_2S
plane.   The simple model of formaldehyde also exhibits two equivalent
nonbonding pairs on the oxygen atom [Figure 11].   The observed angular

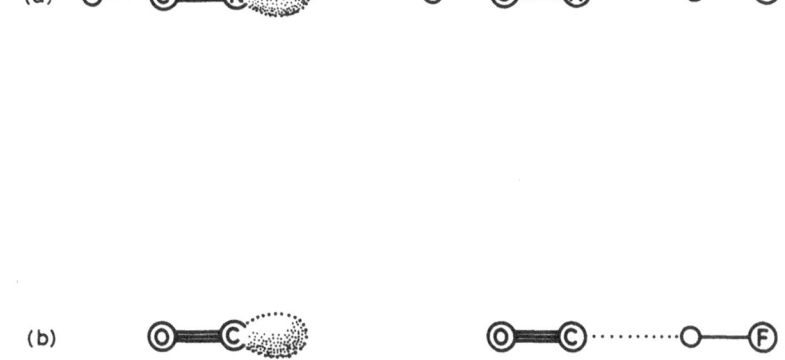

Figure 8. Nonbonding electron pair model for B and observed geometries of B···HF for (a) B=HCN and (b) B=CO.

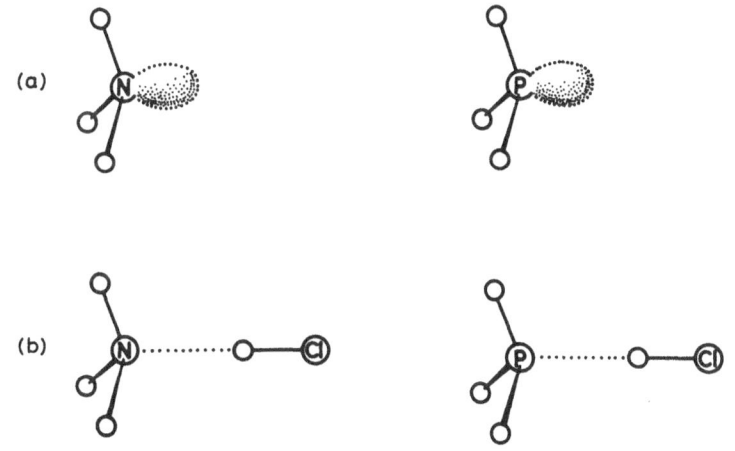

Figure 9. (a) Models of NH₃ and PH₃ showing nonbonding electron pairs. (b) Observed geometries of H₃N···HCl and H₃P···HCl.

geometry[32] of H₂CO···HF, also shown in Figure 11, is planar with an angle between the CO and HF directions consistent with the model and

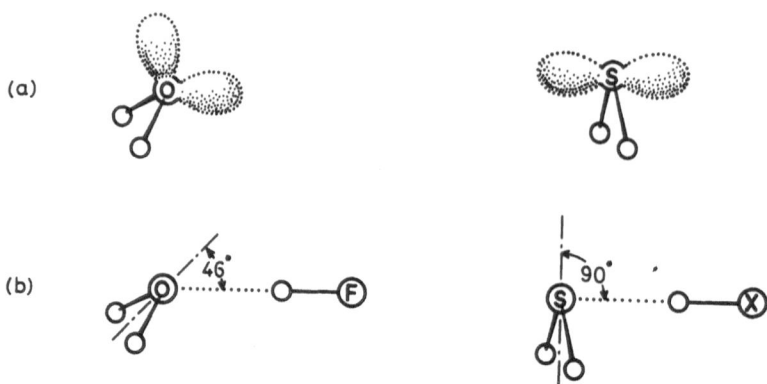

Figure 10. (a) Nonbonding electron pair models of H_2O and H_2S.
(b) Observed geometries of $H_2O\cdots HF$ and $H_2S\cdots HX$.

the above rules. In this case too, no inversion doubling is observed in
the zero-point state.

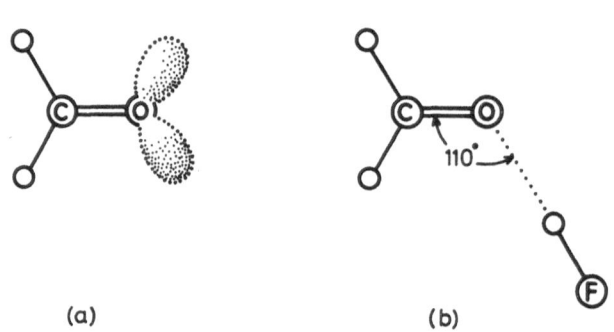

Figure 11. (a) Model of formaldehyde showing nonbonding electron
pairs. (b) The observed geometry of $H_2CO\cdots HF$.

The models of B illustrated in Figures 10 and 11 all show two
equivalent nonbonding pairs of electrons on B. The application of the
rules when B is sulphur dioxide is more complicated. The model
depicted in Figure 12 (a) assigns two inequivalent nonbonding pairs to
oxygen and these are presumably the most nucleophilic regions of SO_2.

The rules then predict that the hydrogen-bonded dimer formed with HF should have one of the two angular geometries shown in Figure 12(b), that is either with $\theta \approx 120°$ (*trans* form) or with the angle $\theta \approx 240°$ (*cis* form). The observed geometry[15] of (SO_2,HF), obtained by fitting the rotational constants A_0, B_0 and C_0 of Table I under the assumption of unchanged component geometries on dimer formation, is given in Table IV. It is clear that the observed isomer has the *cis* form, with the angle θ much closer to 240° than 120°. The geometry of the observed isomer of SO_2,HCl is also included in Table IV and is very similar[33] to that of (SO_2,HF). The failure of certain transitions in both (SO_2,HF) and (SO_2,HCl) to be fitted by the usual effective semirigid rotor analysis suggests that perhaps the energy difference between the *cis* and *trans* forms could be small in both cases.

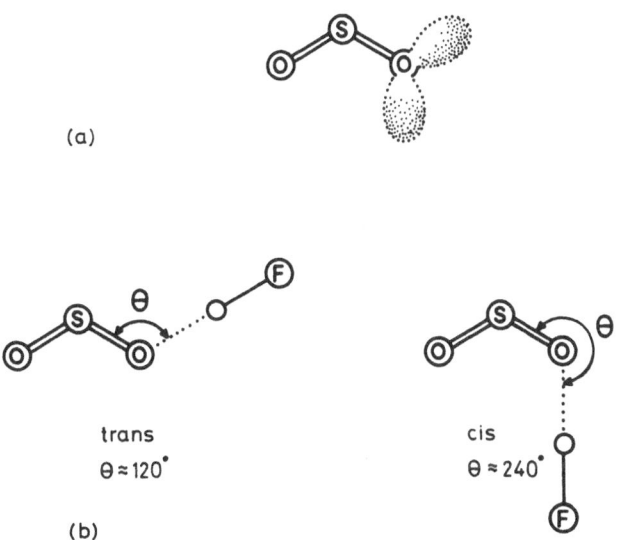

Figure 12. (a) Model of sulphur dioxide showing nonbonding electron pairs on oxygen. (b) Possible geometries of the hydrogen-bonded dimer $SO_2 \cdots HF$.

TABLE IV. Observed geometries of $SO_2 \cdots HA^a$

A	$r(O \cdots A)/Å$	θ/deg
F	2.818(6)	215.1(1)
Cl	3.379(7)	209.73(3)

a See Figure 12 for definition of angle θ.

5.   THE ANGULAR GEOMETRY OF (SO_2,HCN)

The above discussion and a recent, more extensive analysis[18]
demonstrate that the rules proposed for predicting the angular
geometries of hydrogen-bonded dimers B···HA are qualitatively
successful. Of course, the rules cannot predict in the case of a dimer
such as SO_2···HF which of the two forms, cis or trans (see Figure 12),
is the more stable. It does seem likely, however, that the energy
difference between the two forms is small and that as the hydrogen
bond in SO_2···HA becomes weaker the energy difference will decrease.
Since there is some slight evidence (see above) from the rotational
spectrum of SO_2···HF for the presence of another low-energy
conformation[15] (presumably the trans form), it is of interest to
investigate a more weakly bound species SO_2···HA with a view to
detecting both isomers. Table III makes it clear that in general
B···HCN is considerably more weakly bound than B···HF, if k_σ is taken
as a measure of hydrogen-bond strength.
    Recently, Buckingham and Fowler[34] have developed a quantitative
electrostatic model which predicts angular geometries of weakly bound
dimers in good agreement with experiment and which allows estimates of
binding energies. They find that for SO_2···HF the two lowest energy
forms have the cis and trans geometries, as shown in Figure 12, with
the trans form more stable but only by 3 kJ mol^{-1}. On the other hand,
for SO_2···HCN the cis form is found to be more stable than the trans
by as little as 0.4 kJ mol^{-1}.
    Clearly, it seems likely in view of the above considerations that
both cis and trans forms of SO_2···HCN would be observed in the
spectrum of a dilute mixture of SO_2 and HCN in argon when investigated
by pulsed-nozzle, Fourier-transform microwave spectroscopy. The
rotational spectra of several isotopic species of a dimer (SO_2,HCN)
have indeed been detected and analyzed by this technique to give the
spectroscopic constants shown in Table V[35].

TABLE V.   Spectroscopic constants of (SO_2,HCN)

Spectroscopic constant	($^{32}SO_2$,HC^{14}N)	($^{32}SC_2$,DC^{14}N)	($^{32}SO_2$,HC^{15}N)
A_0/MHz	8633.849(1)	8577(4)	8620.8173(8)
B_0/MHz	1848.890(2)	1743.127(4)	1820.716(1)
C_0/MHz	1615.863(2)	1536.345(4)	1594.795(1)
Δ_J/kHz	9.56(1)	8.94(6)	9.151(5)
Δ_{JK}/kHz	114.84(7)	100(3)	107.70(4)
δ_J/kHz	0.838(6)	0.75(4)	0.774(3)
δ_K/kHz	89(1)	89(assumed)	84.5(6)
P_b/amu Å^2	48.977	48.972	48.972

A detailed analysis of the rotational constants in Table V on
the basis of unperturbed monomer geometries reveals that the geometry
of the isomer of (SO_2,HCN) responsible for the observed spectrum has

neither the *cis*- nor the *trans*-planar form of the type shown in Figure 12(b). In fact, the weak interaction cannot be described by a hydrogen bond. The experimental geometry[35] of the dimer has C_s symmetry with the N atom of HCN nearest to the SO_2 subunit, as shown in Figure 13. The HCN axis is approximately perpendicular to the plane

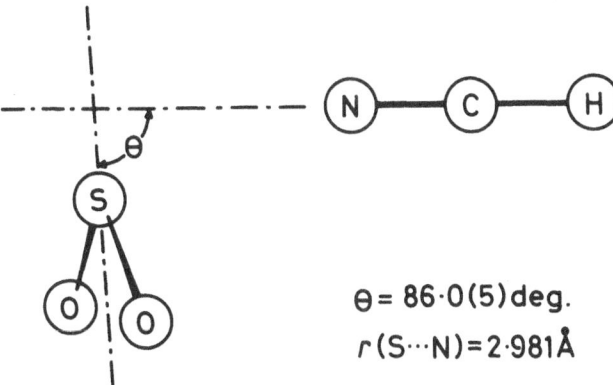

$\theta = 86 \cdot 0(5) \text{deg.}$

$r(S \cdots N) = 2 \cdot 981 \text{Å}$

Figure 13.   Observed geometry of (SO_2, HCN).

of the SO_2 molecule ($\theta = 86°$) and intersects the C_2 axis of the SO_2 subunit at a distance of 1.28 Å above the S nucleus. The quantity $\Delta_0 = I_c^0 - I_b^0 - I_a^0 = -19.115$ amu Å^2, compared with 1.4 amu Å^2 in $SO_2 \cdots HF$ (see Table I), demonstrates clearly that the nuclei of (SO_2, HCN) are not coplanar. In addition, the fact that $P_b = \frac{1}{2}(I_a^0 + I_c^0 - I_b^0) = \Sigma_i m_i b_i^2$ is almost identical to the corresponding quantity[36], $P_a = 49.05$ amu Å^2, in free SO_2 and is isotopically invariant establishes that the ac inertial plane is a symmetry plane, with only the oxygen atoms lying outside it, and that the SO_2 geometry is sensibly unchanged from the free molecule.

Qualitatively, the observed geometry of (SO_2, HCN) can be understood if the axial, nonbonding electron pair of HCN seeks out the region of minimum electron density of SO_2, which presumably lies in the region of the S atom. It is worth noting in this context that the component of the molecular electric quadrupole moment of SO_2 along the direction perpendicular to the SO_2 plane is large and positive[37] and also that presumably the S atom lies at the positive end of the electric dipole moment.

The Buckingham-Fowler electrostatic model has been applied to (SO_2, HCN).[34] In this model, each monomer molecule is represented by a set of point multipoles embedded in hard atomic spheres. The point multipoles (charges, dipoles, and quadrupoles) are calculated from an SCF wavefunction using a distributed multipole analysis and describe the charge distribution of the monomer. The short-range repulsion is represented by the hard spheres which are centred on the heavy atoms and assigned van der Waals radii. The equilibrium geometry of the

dimer is then that which minimises the electrostatic energy of
interaction between the two sets of multipoles, no two heavy atoms
being allowed to approach closer than the sum of their hard-sphere
radii. Although the *cis*-planar hydrogen-bonded form is predicted to
be lowest in energy when using van der Waals radii for the hard
spheres,[34] it has since been found that a minimum in the electrostatic
energy does occur for a geometry like that shown in Figure 13 and
indeed that if the N···S contact distance is allowed to fall below
3.0 Å this becomes the global minimum.[38] We show in Figure 14 a
comparison of the observed geometry[35] of (SO_2,HCN) with that predicted
by the Buckingham-Fowler model[38] when the N···S contact distance is
3 Å, both in projection on the *ac* inertial plane. The agreement is
excellent, the angle θ (see Figures 13 and 14 for definition) having
the values 86° (observed) and ≃ 80° (calculated).

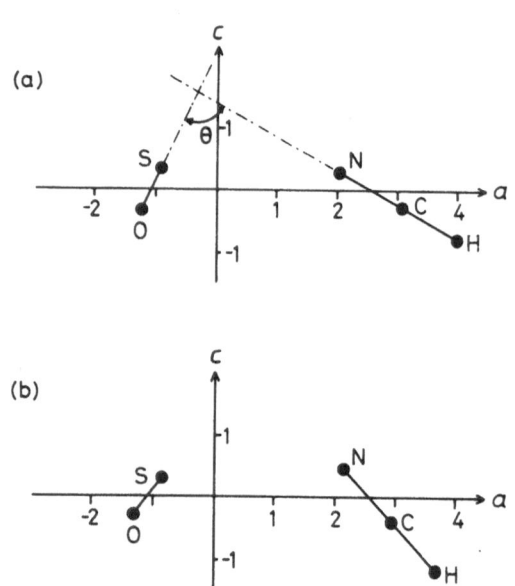

Figure 14.   Comparison of (a) observed geometry of (SO_2,HCN) and
(b) calculated geometry from Buckingham-Fowler model, both shown in
projection on the *ac* inertial plane.

ACKNOWLEDGEMENT

Research grants from S.E.R.C. are gratefully acknowledged. The author
thanks Dr. P.W. Fowler for permission to use his unpublished results.

6. REFERENCES

1. A.C. Legon, D.J. Millen, and S.C. Rogers, *Proc. R. Soc. London, Ser. A,* 370, 213, (1980).
2. Z. Kisiel, A.C. Legon, and D.J. Millen, *Proc. R. Soc. London, Ser A,* 381, 419, (1982).
3. T.R. Dyke, B.J. Howard, and W. Klemperer, *J. Chem. Phys.,* 56, 2442, (1972)
4. T.J. Balle, E.J. Campbell, M.R. Keenan, and W.H. Flygare, *J. Chem. Phys.,* 71, 2723, (1979).
5. T.J. Balle, E.J. Campbell, M.R. Keenan, and W.H. Flygare, *J. Chem. Phys.,* 72, 922, (1980).
6. For a recent discussion of the information available from vibrational satellites in rotational spectra of dimers see: A.C. Legon and D.J. Millen, *Chem. Revs.,* 86, 635, (1986).
7. E.J. Campbell, L.W. Buxton, T.J. Balle, and W.H. Flygare, *J. Chem. Phys.,* 74, 813, (1981).
8. E.J. Campbell, L.W. Buxton, T.J. Balle, M.R. Keenan, and W.H. Flygare, *J. Chem. Phys.,* 74, 829, (1981).
9. T.J. Balle and W.H. Flygare, *Rev. Sci. Instrum.,* 52, 33, (1981).
10. A.C. Legon, *Ann. Rev. Phys. Chem.,* 34, 275, (1983).
11. A.C. Legon and L.C. Willoughby, *Chem. Phys.,* 74, 127, (1983).
12. E.J. Goodwin and A.C. Legon, *Chem. Phys.,* 87, 81, (1984).
13. A.C. Legon, D.J. Millen, and L.C. Willoughby, *Proc. R. Soc. London, Ser. A,* 401, 327, (1985).
14. A.J. Fillery-Travis, A.C. Legon, L.C. Willoughby, and A.D. Buckingham, *Chem. Phys. Lett.,* 102, 126, (1983).
15. A.J. Fillery-Travis and A.C. Legon, *J. Chem. Phys.,* in press.
16. For a detailed discussion, see Ref. 10.
17. D.J. Millen, *Can J. Chem.,* 63, 1477, (1985).
18. A.C. Legon and D.J. Millen, *Accounts of Chemical Research,* to be published.
19. A.C. Legon and D.J. Millen, *J. Amer. Chem. Soc.,* submitted.
20. A.C. Legon and D.J. Millen, *Proc. R. Soc. London, Ser. A,* 404, 89, (1986).
21. A.C. Legon and D.J. Millen, *Faraday Discuss. Chem. Soc.,* 73, 71, (1982).
22. P.D. Aldrich, A.C. Legon, and W.H. Flygare, *J. Chem. Phys.,* 75, 2126, (1981).
23. C.A. Coulson and W.E. Moffitt, *Phil. Mag.,* 40, 1, (1949).
24. A.C. Legon, P.D. Aldrich, and W.H. Flygare, *J. Am. Chem. Soc.,* 102, 7584, (1980) and 104, 1486, (1982).
25. A.C. Legon, P.D. Soper, and W.H. Flygare, *J. Chem. Phys.,* 74, 4944, (1981).
26. E.J. Goodwin, N.W. Howard, and A.C. Legon, *Chem. Phys. Lett.,* in press.
27. A.C. Legon and L.C. Willoughby, *J. Chem. Soc., Chem. Commun.,* 997, (1982).
28. R. Viswanathan and T.R. Dyke, *J. Chem. Phys.,* 77, 1166, (1982).
29. E.J. Goodwin and A.C. Legon, *J. Chem. Soc., Faraday Trans. 2,* 80, 51, (1984).

30. E.J. Goodwin and A.C. Legon, *J. Chem. Soc., Faraday Trans. 2*, 80, 1669.
31. A.I. Jaman and A.C. Legon, *J. Mol. Structure*, in press.
32. F.A. Baiocchi and W. Klemperer, *J. Chem. Phys.*, 78, 3509, (1983).
33. A.J. Fillery-Travis and A.C. Legon, *Chem. Phys. Lett.*, 123, 4, (1986).
34. A.D. Buckingham and P.W. Fowler, *Can. J. Chem.*, 63, 2018, (1985).
35. E.J. Goodwin and A.C. Legon, *J. Chem. Phys.*, in press.
36. J.K.G. Watson, *J. Mol. Spectroscopy*, 48, 479, (1973).
37. A.W. Ellenbrock and A. Dymanus, *Chem. Phys. Lett.*, 42, 303, (1976).
38. P.W. Fowler, unpublished results.

STRUCTURE DETERMINATION OF WEAKLY BOUND COMPLEXES

T.R. Dyke
Department of Chemistry
University of Oregon
Eugene, Oregon 97403
USA

ABSTRACT. Radiofrequency and microwave spectra have been obtained for hydrogen bonded complexes such as $(H_2O)_2$, $NH_3 \cdot H_2O$ and $NH_3 \cdot H_2S$ with the molecular beam electric resonance method. The purpose of these experiments is to understand the structural factors influencing the formation of hydrogen bonds and to provide accurate data for models and ab initio calculations of hydrogen bonding. The information available from this work includes rotational and centrifugal distortion constants, electric dipole moments, and nuclear spin-spin and quadrupole coupling hyperfine constants. The geometries of these complexes are calculated principally from the rotational constants with some angular information from the hyperfine interaction results. Deuterium substitution data is particularly useful for increasing the number of structural parameters which can be determined. The large amplitude vibrational motions of the monomers in weakly bound complexes complicate this procedure in two ways. Firstly, interchange of indistinguishable nuclei by tunneling processes can lead to large splittings in the rotational spectra, as observed for $(H_2O)_2$, and to more subtle effects such as the absence of K-doubling and resulting linear Stark effects in $NH_3 \cdot H_2S$. Secondly, even if these effects are absent or have been accounted for, the spectroscopic constants used in a structure calculation reflect vibrational averages, typically for several isotopically distinct species.

1. INTRODUCTION

Precision structural results for hydrogen bonded and van der Waals complexes are useful for understanding intermolecular interactions and have important applications in modelling condensed phases, macromolecule and other complex systems. These results are further useful for comparison with ab initio calculations and for the construction of potential energy surfaces for these molecules. Rotational spectroscopy, particularly in molecular beams, has been a fruitful source of this information[1,2]. In addition, rotationally resolved electronic[3] and vibrational[4] spectra supply similar results.

The relatively low binding energy of the complexes studied gives

43

A. Weber (ed.), Structure and Dynamics of Weakly Bound Molecular Complexes, 43–56.
© 1987 by D. Reidel Publishing Company.

rise to large amplitude motions which complicate the interpretation of experimental results in terms of structure. To fully utilize the accuracy of the experimental data, it is necessary to calculate the energy levels of the complex from a potential energy surface which can be varied to fit the data. Except in the simplest cases of atom-diatom[5] and diatom-diatom[6] complexes, this has not been done because of the large computational effort required to calculate vibration-rotation energy levels to the necessary accuracy. Instead, simpler models, typically based on the rigid-rotor approximation are used to analyze the results in terms of effective structures. The effects of large amplitude vibrational motions will then cause the effective geometries obtained to deviate from the equilibrium geometries of the complexes. These effects will be discussed in this paper using examples from our molecular beam electric resonance studies of ammonia complexes with water and hydrogen sulfide.

A second result of large amplitude motion is that internal rotations of the monomers in the complex can interchange identical nuclei by tunneling processes. In the case of molecules such as water dimer, the resulting tunneling splittings give rise to a very complicated energy level pattern and rotational-tunneling spectrum. For molecules such as the ammonia-water and ammonia-hydrogen sulfide complexes, the a-type spectra appear to be very similar to that of a rigid rotor, but subtle effects such as the lack of K-doubling and first-order Stark effects can be seen for states in which the NH_3 is undergoing nearly unhindered internal rotation. Recent progress in understanding the effects of these tunneling processes on the rotational spectra will be discussed.

2.   EXPERIMENTAL

The results described below were obtained by the molecular beam electric resonance method. In these experiments, the complexes are formed in a seeded argon free jet expansion, and rotational temperatures of a few degrees Kelvin are expected under these conditions. For molecules with complex spectra, such as $(H_2O)_2$, the simplification of the spectrum achieved with such an extremely cold source is an important aid in assigning the transitions and most of the work discussed below was done under these conditions. However, it is worth noting that the information content of the spectrum can be increased by working with somewhat higher temperatures. As an example, the rotational temperature of $(H_2O)_2$ formed in a pure H_2O expansion is roughly 50 K and many more states are accessible using these source conditions.

After formation of the complexes, the molecular beam is rotationally state selected, allowed to interact with radiofrequency or microwave radiation, and then state analyzed and detected. The observed spectra have narrow linewidths, ~3 kHz for radiofrequency transitions and ~20 kHz for microwave transitions in our apparatus. The resulting information consists most importantly of accurate rotational constants for the ground vibrational state of the complex, typically for several isotopically distinct species. In addition, electric dipole moments

and nuclear hyperfine interactions can be calculated from these spectra. A sample radiofrequency spectrum is shown below in Figure 1.
This technique is discussed in detail in the paper by J.S. Muenter and
is further described elsewhere.[1]

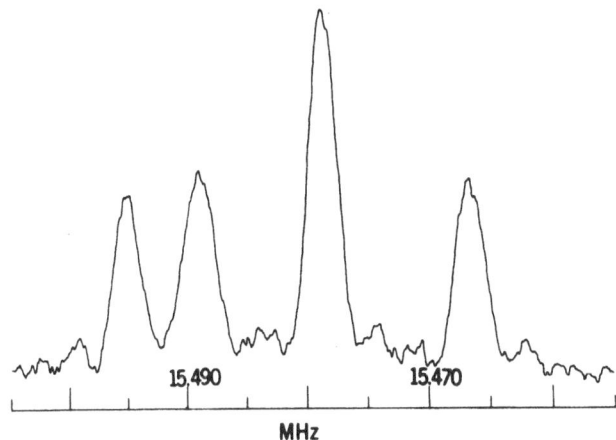

Figure 1. J = 1, M_J = 0 → 1 radiofrequency transition of
water dimer at a 600 volt/cm electric field strength. The
rovibrational E-symmetry is shown by the four line hyperfine
pattern indicative of a triplet nuclear spin state.

3. STRUCTURAL RESULTS AND THE EFFECTS OF LARGE AMPLITUDE MOTION

3.1. Vibrational Averaging Effects

The results of studies on the $NH_3 \cdot H_2O$[7] and $NH_3 \cdot H_2S$[8] molecules are given
in Table I. As mentioned above, information concerning geometry is

TABLE I. Structural data for $NH_3 \cdot H_2O$ and $NH_3 \cdot H_2S$

	$NH_3 \cdot H_2O$	$NH_3 \cdot H_2S$
R_{NX}/Å	2.983	3.634
θ_N/°	23.1	24.0
$\theta_d - \alpha$/°	4	1
χ_d/°	~0	~0
ω_s/cm^{-1}	168	104
k_s/mdyne/Å	0.14	0.072
μ_a/D	2.977	2.674

principally determined from rotational constant data for these molecules and deuterium substituted analogs. In addition, the nitrogen quadrupole coupling interaction has been used to determine the orientation of the ammonia unit in the complex. The experimental errors in the coordinates listed in Table I are small. However, since a rigid rotor model was used to calculate these results, their interpretation depends on the degree of zero-point vibrational motion for the isotopic species used in the calculation, and the derived coordinates will differ strongly from the equilibrium geometry in some instances.

To understand these effects, it is useful to express the dependence of the experimentally measured quantities on the structure of the complex:

$$(B+C)/2 \simeq B_o \{1 - [i_{yy}^d + i_{zz}^d \sin^2\theta_d \cos^2\chi_d + i_{xx}^d \cos^2\theta_d + i_{xz}^d \sin2\theta_d \cos\chi_d +$$

$$+ i_{xx}^N + i_{xx}^N \cos^2\theta_N + i_{zz}^N \sin^2\theta_N]/(2I_o) + \ldots\} \tag{1}$$

where B_o and I_o are the pseudo-diatomic rotational constant and moment-of-inertia for the complex, and the i's are the inertia tensor elements for the proton donor and for the ammonia monomer. Similarly, θ_d and θ_N specify the angle between the proton donor and ammonia monomer symmetry axes and the a-axis of the complex. χ_d gives the angle of rotation of the proton donor around its symmetry axis[7]. A similar expression is given in Eqn. (2) for the aa-component of the quadrupole coupling interaction:

$$(eqQ)_{aa} = (eqQ_N)^o <(3\cos^2\theta_N - 1)/2> \tag{2}$$

where eqQ_N is the free ammonia coupling constant.

To see the effect of vibrational averaging, the various terms in these equations, such as $\cos^2\theta$, can be expanded around the equilibrium geometry:

$$\cos^2\theta = \cos^2\theta_e - \sin2\theta_e \Delta\theta - \cos2\theta_e \Delta\theta^2 + \ldots \tag{3}$$

The expectation value of this function depends on θ_e and the vibrational motion:

$$\theta_e = 0°: \quad <\cos^2\theta> = \cos^2\theta_e - <\Delta\theta^2> + \ldots \tag{4}$$

$$\theta_e = 45°: \quad <\cos^2\theta> = \cos^2\theta_e - <\Delta\theta> + \ldots \tag{5}$$

Thus for θ_e near 0°, the vibrational average of $\cos^2\theta$ depends directly on the root-mean square angular deviation, while around 45° the average value of $\Delta\theta$ is the leading order vibrational coordinate. Inspection of Table I, shows that for the two ammonia complexes studied, Eqn. (3) gives an effective angle of 23 - 24° for the orientation of the ammonia relative to the a-axis of the complex. Consideration of Eqn. (4) then suggests that this should be interpreted as the root-mean-square angular displacement of the ammonia and that the *equilibrium* value of θ is much smaller. This is further supported by the reduction in the effective angle measured upon deuteration of the ammonia[7]. The proton donors, H_2O and H_2S for these complexes, are hydrogen bonded to the ammonia and their C_2 axes are slightly over 45° from the a-axis of the complex (in Table I, an angle α, which is one-half the monomer bond angle, has been subtracted from θ_d to show the deviation from a linear hydrogen bond). Eqn. (1) shows how this angle is expressed in $(B+C)/2$. In this case the effective angles determined from $(B+C)/2$ data for several different isotopic species will reflect the *average* angular displacement as suggested by Eqn. (5). In the harmonic oscillator approximation, $\langle\Delta\theta\rangle = 0$ and one expects the resulting effective angle to be quite close to the equilibrium angle. However, anharmonic effects will cause $\langle\Delta\theta\rangle$ to be non-zero and the anharmonic force constants for most complexes are unknown (not to mention the harmonic force constants!). A crude calculation for $(H_2O)_2$[9], based on a potential energy surface from *ab initio* theory, did verify that the average angular displacements are small for that complex, but this point must be treated cautiously.

To summarize this section, rotational spectroscopic data is of very high accuracy and the effective structural parameters extracted from such data can in principle be quite accurate. However, these structures are not equilibrium structures and may vary considerable from them as shown above. In order to make accurate comparisons with theoretical calculations, either the appropriate vibrational averaging must be done for the theoretical calculation, or the experimental results must be corrected for vibrational effects, which will at least require both harmonic and anharmonic force field information. In order to fully utilize the accuracy of the experimental data, directly fitting potential energy surfaces to the data will likely be the best route, although the energy level calculations involved are likely to be formidable.

3.2 Tunneling Effects

3.2.1 Dimers: The Water Dimer Rotational Tunneling Spectrum.

The weak binding of hydrogen bonded and van der Waals complexes leads to interesting tunneling effects in the rotational energy levels and spectra. Such effects can arise when a monomer substituent of the complex has identical nuclei which can interchange by an internal rotation or, in the case of dimers, when the monomers interchange their positions. As examples of the former case, ammonia internal rotation in $NH_3 \cdot H_2O$ and $NH_3 \cdot H_2S$ will be discussed later on. Tunneling splittings generated by the interchange of monomers in a dimer was first observed in the molecular beam electric resonance spectrum of $(HF)_2$[10]. Since the interchange of the HF monomers in this complex involves a reversal of the a-component of the electric dipole moment, the a-type transitions were found to be offset by a ±19.7472 GHz tunneling doubling from the frequencies expected for a rigid rotor. In understanding the tunneling behavior of $(HF)_2$, permutation inversion group theory[11] was found to be useful and a similar situation was found for $(H_2O)_2$[12] (also see the paper by J.T. Hougen). The $(H_2O)_2$ molecule represents a more complex example of tunneling behavior since both types of tunneling motion discussed above can occur in this complex, and our work on this dimer will be discussed below.

The first experimental work on the water dimer rotational spectrum was done with a pure H_2O expansion, resulting in a very complex micro-wave spectrum[13]. Although the spectrum was not completely assigned, series of rigid rotor transitions were observed and the structure of the molecule obtained from them[13]. In further work on partially deuterated species[9], it was found that most of the lines observed with the pure water expansion disappeared when a seeded argon beam source was used. Since the rotational temperatures in this expansion are expected to be only a few degrees Kelvin, most of the lines which disappeared are presumably from states with K>0 and are "frozen" out because of their large A-rotational constants. However, some transi-tions remained which are clearly perturbed by tunneling doublings and these lines, along with others for K=0 states showing no tunneling splittings, are given in Table II. The transition frequencies were

Table II. Water Dimer K=0 Frequencies

Transition* $J \rightarrow J'$	Frequency (MHz)
--- $(H_2O)_2$ ---	
$2^+ \rightarrow 1^-$	4863.16
$0^- \rightarrow 1^+$	7354.86
$1 \rightarrow 0$	12132.70
$1_+ \rightarrow 0_-$	12321.00
$3^+ \rightarrow 2^-$	17122.61
$2_+ \rightarrow 1_-$	24640.88
$4^+ \rightarrow 3^-$	29416.39
--- $(D_2O)_2$ ---	
$1^+ \rightarrow 0^-$	9692.86
$1 \rightarrow 0$	10864.58
$1 \rightarrow 0_+$	10864.54
$1^- \rightarrow 0^+$	12036.68
$1^- \rightarrow 2^+$	20557.68

*± refers to the lower and upper half of the tunneling doublets, respectively.

analyzed with a rotational energy level expression including tunneling doubling:

$$W = \tfrac{1}{2}(B+C)J(J+1) - D_J J^2(J+1)^2 \pm \tfrac{1}{2}[\nu + \delta J(J+1) + \gamma J^2(J+1)^2] \qquad (6)$$

and the molecular constants are shown in Table III.

Table III. Water Dimer Molecular Constants*

Type of Spectrum	(B+C)/2 MHz	ν MHz	δ MHz	γ MHz	μ Debye
		$---(H_2O)_2---$			
E_1	6160.59	0	0	0	2.644
E_2	6066.44	0	0	0	2.617
A/B	6073.98	19526.80	-24.34	0.10	2.617
		$---(D_2O)_2---$			
E	5432.21	0	0	0	2.610
E	5432.19	0	0	0	2.623
A/B	5432.47	1172.25	-0.34	-	2.620
A/B†	(5432)	(2150)	-	-	(2.627)

*D_J fixed at 0.04 MHz.
†Tentative. Based on radiofrequency spectra.

The tunneling behavior observed can be understood with regard to the group theoretical classification of the energy levels carried out previously[12]. The permutation-inversion group of water dimer is iso-morphic with the D_{4h} point group, and a correlation diagram for K=0 energy levels is shown in Figure 2. In drawing this diagram it was assumed that "feasible" permutations included the complete interchange of the two monomers as well as two-fold internal rotations of each monomer. Additionally, it was assumed that the vibrational-tunneling energies do not change appreciably for states labelled by different J quantum numbers. With this assumption, the correlation diagram shows that rotational transitions for the degenerate, E-symmetry levels will not be perturbed by tunneling splittings whereas transitions involving non-degenerate levels will show tunneling doublings for the a-type transitions shown.

These assumptions are verified by the the two types of transitions observed and by the observation of triplet nuclear spin states observed for transitions with E-type selection rules, as predicted by the group theoretical arguments. This is illustrated in Figure 1, which shows an E_1 radiofrequency transition with a four-line hyperfine structure char-acteristic of triplet spin states. The magnitude of the splittings is determined by the proton magnetic dipole-dipole interaction and a small contribution from the spin-rotation interaction. Since the coupling constants are known for H_2O^{14}, the dimer spin-spin constants can be

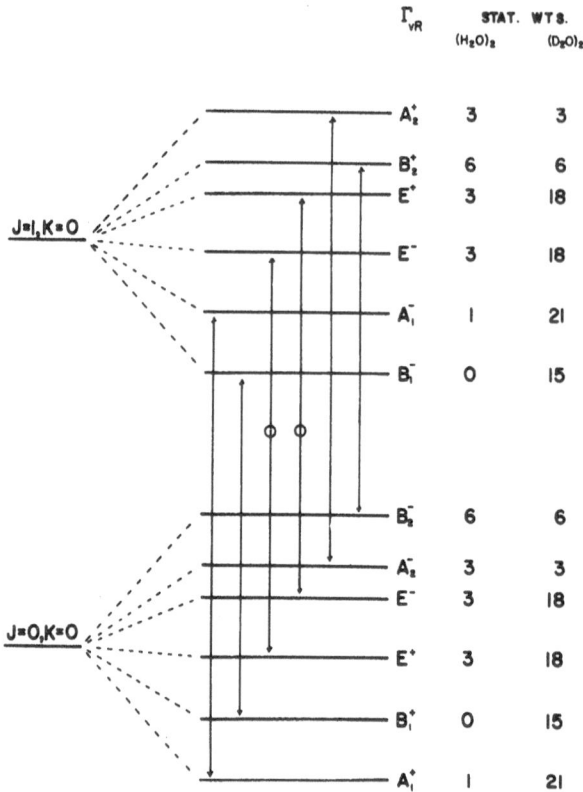

Figure 2. Correlation diagram for water dimer. The
tunneling sub-levels of each rotational state are
classified with a permutation-inversion group iso-
morphic with the D_{4h} point group.

related to the orientation of the monomers in the complex. For the
water dimer group, the total wavefunction must tranform like B_2. Using
symmetry adapted total wavefunctions[9,12], it is easy to show that $E(\pm)$
tunneling states correspond to a triplet spin state localized on monomer 1 and a singlet state on monomer 2, and the converse[15]. The
observed spin-spin interaction constant for the E_1 state in Figure 1
corresponds to one H_2O molecule with the vector connecting the hydrogens at an angle of 34° from the a-axis of the complex, compared to 37°
predicted for the proton donor from our structure. Similarly, the E_2
state constant gives an angle of 73° for this vector, in good agreement
for the proton acceptor ($\chi_a = 90°$) when the effects of vibrational
averaging are considered.

These results clearly show the complexities introduced to high
resolution spectra by tunneling effects. In the case of water dimer,
these effects are now reasonably well understood and the low tempera-

ture, $K = 0$, spectrum assigned. Group theoretical considerations have
been quite useful in this work; however, for further characterizing the
details of the tunneling motions and for determining the potential
energy surface, accurate methods for calculating the vibration-rotation
energy levels of dimers with polyatomic molecule constituents will have
to be developed. Additionally, although extensive ab initio calculations for this system have been reported[16,17], further work along these
lines would be useful in interpreting the observed tunneling splittings.

3.2.2 Internal Rotation Effects in Ammonia Complexes. Tunneling
motions of identical monomer units cause the dramatic effects discussed
in the previous section. For less symmetric bimolecular complexes,
internal rotations of its monomers can cause more subtle but nonethe-
less interesting results. As an example, the $J = 1 \rightarrow 2$ transition for
the $NH_3 \cdot H_2S$ complex discussed earlier shows a second transition which
is 3.96 MHz lower in frequency than the $K = 0$ line. When a non-zero
electric field is employed in the resonance region, linear Stark
effects are observed at all field strengths employed, and the transition can be assigned as $J = 1 \rightarrow 2$, $K = 1$. The absence of K-doubling
and presence of first-order Stark effects are puzzling, however, since
(B-C) is predicted to be about 36 MHz for the structure of Table I.
Thus quadratic Stark effects should have been observed and the zero
field frequency displaced by about 72 MHz to higher frequency from the
$K = 0$ transition. This is further borne out by radiofrequency transitions $(K = 1, \Delta J = 0, \Delta M_J = 1)$ observed from $J = 1$ through 9. All of
these show linear Stark effects at low electric fields and therefore
cannot have originated in states with appreciable K-doubling. This
data is summarized in Table IV in the form of electric dipole moments

TABLE IV. $NH_3 \cdot H_2S$ Rotational constant and
electric dipole moment results.

State	(B+C)/2 MHz	μ_a Debye
K=0, A	3233.068	2.6741
K=0, (E?)	3233.334	2.6739
J (K=1,E)		
1	3232.080*	2.676
4	-	2.660
5	-	2.648
6	-	2.642
7	-	2.625
8	-	2.616
9	-	2.598

* Calculated with $D_J = 14kHz$

calculated from the various transitions observed. Also included are a
third set of less intense microwave transitions observed and assigned
as K = 0 lines.

The origin of these effects can be understood by considering the
NH_3 constituent in the limit of free rotation. Since the ammonia
monomer is a symmetric top, K = 1 rotational levels will not exhibit K-
type doubling and will have first-order Stark effects. Thus, the
corresponding excited *internal rotor* states for ammonia complexes will
show this behavior in the low barrier limit. Similar effects have been
observed in ammonia complexes in which the excited states involve the
internal rotation of a second substituent with a larger moment-of-
inertia[18,19]. The result is less obvious in these cases, however,
since the internal rotor is itself an asymmetric top and generates most
of the asymmetry doubling observed for the complex.

The analysis of internal rotation problems for monomers has been
extensively treated, and our results can be modelled with a Hamiltonian
used by E.B. Wilson *et al*[20]:

$$H = BJ_b^2 + CJ_c^2 + A_f J_a^2 + (A_f + A_r)j_a^2 - 2A_f j_a J_a + V_3(1 - \cos3\alpha)/2 \qquad (7)$$

where A_r is the NH_3 monomer rotational constant defined with respect to
the a-axis of the complex, and A_f is an analogous constant for the
"framework" H_2S or H_2O part of the complex. j is the operator corresponding to the angular momentum of the ammonia internal rotor, while
the remaining angular momentum operators and rotational constants have
their usual meanings. It has also been assumed that the NH_3 symmetry
axis is aligned along the a-axis of the complex, and that the potential
function is adequately represented by a three-fold barrier, one-dimensional function. This last approximation is probably the most serious,
but is adequate for our purpose.

The torsion-rotation energy levels are calculated by diagonalization of the Hamiltonian of Eqn.(7) with an angular momentum basis set
labelled by J, K, M_J and m, where the latter quantum number designates
the projection of the NH_3 angular momentum along the a-axis of the
complex. A correlation diagram based on this is given in Figure 3.
The symmetry labels shown are with reference to the permutation group
isomorphic with the threefold rotation group. In the free rotor limit,
where m is a good quantum number and K nearly so, the lowest excited
internal rotor state of $NH_3 \cdot H_2S$ is the degenerate K = 1 level, in which
the angular momentum is generated by the NH_3 internal rotation (m = K
levels, E symmetry). This doubly degenerate state would not exhibit K
doubling and will have a first-order Stark effect, as does free ammonia. A second set of K = 1 levels with one unit of ammonia angular
momentum and, therefore, two units of "framework" angular momentum is
also present, at much higher energy because of the Coriolis-like term,
$-2A_f j_a J_a$, of Eqn. (7). As the threefold-barrier to internal rotation
is increased, states differing by 3 in the m quantum number are mixed
by the potential, and these two K = 1 levels come closer together as
$\langle j_a \rangle$ and the Coriolis term approach zero. Before these levels merge

completely, the states with K = ±1 are mixed by the asymmetry of the
framework to form the K-doublet of E symmetry shown in the diagram in
the high barrier limit. For a completely rigid rotor, this K-doublet
would further merge with the K-doublet of A symmetry correlating with
the m = 0, K = 1 free rotor states. Thus we have assigned the observed
K = 1 transitions to the lowest energy state of E-symmetry (for K = 1).
In addition, it is likely that the second series of K = 0 lines noted
in Table IV is from the lowest E-symmetry state with K = 0. In the
free rotor limit, this state would lie about 10 cm^{-1} above the lowest
E-state, but in the high barrier limit it is the lowest E-state. Further, in the free rotor limit, the intensity of this state would be
expected to be less than for the K = 0 A state, and the observed
intensity ratio is in fact roughly 1:3. However, the E-state might
also be completely "frozen" out of the expansion, and the observed
intensity ratio suggests the alternate possibility of singlet and
triplet H_2S tunneling states. Thus the exact nature of the K = 0 lines
is ambiguous, and it is likely that only rather difficult deuteration
and partial deuteration studies will resolve the ambiguity.

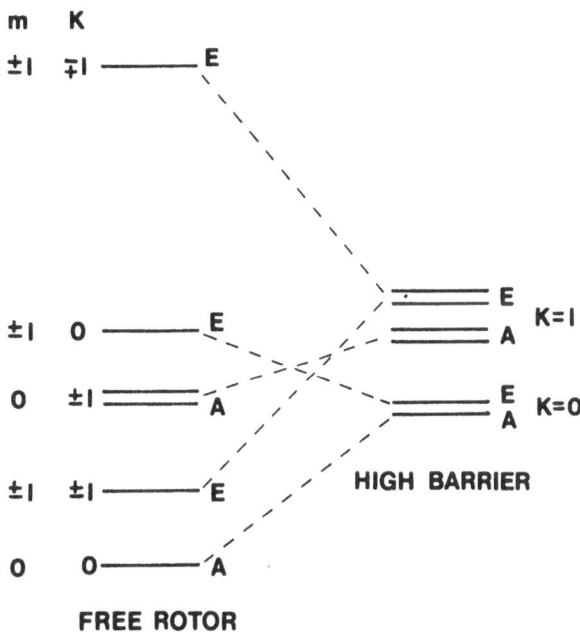

Figure 3. Correlation diagram for NH_3 complexes. Only
the few lowest free rotor states are presented. Free
rotor states differing by three in the m quantum number
will also be mixed with the states shown for a non-zero
threefold barrier.

The experimental results for K = 1 $NH_3 \cdot H_2S$ states show that this
complex is more to the free rotor side of the correlation diagram, and

only an upper limit to the barrier can be set. Since the K-doubling
increases like J(J+1), the radiofrequency Stark effect data in Table IV
for high J-states provides the lowest upper limit to the barrier height
parameter, V_3. An extended analysis of the Stark effects for this
case, using the eigenfunctions of the internal rotation Hamiltonian,
Eqn. (7), shows that the barrier height (V_3) for $NH_3 \cdot H_2S$ must be less
than ~850cm^{-1}. Any higher barriers would cause the Stark effect data
in Table IV to show larger changes in μ_a than observed, and eventually,
quadratic electric field dependence. This result is not too surprising
since it is comparable to the binding energy of the complex. By deu-
terating the complex, the framework (H_2S) A_f rotational constant will
be approximately halved, and the internal rotation quenched by barrier
heights as low as ~400 cm^{-1}. Such measurements are in progress.

    For a more strongly bound complex such as $NH_3 \cdot H_2O$, the threefold
barrier might be expected to be larger than for the H_2S complex, and a
similar study has been performed on the former molecule, including the
deuteration studies. The radiofrequency results are displayed as
dipole moments in Table V, but again only linear Stark effects were
observed. The dipole moments displayed in Table V do have a large J-

TABLE V. $NH_3 \cdot H_2O$ and $ND_3 \cdot D_2O$ dipole moments

State	$NH_3 \cdot H_2O$ μ_a (Debye)	$ND_3 \cdot D_2O$ μ_a (Debye)
J (K=1,E)		
1	2.971	-
3	2.911	2.922
4	2.871	2.854
5	2.823	2.801
6	2.768	2.713
7	2.703	-

dependence, which would be the first effect observed moving toward the
high barrier side of the correlation diagram. However, the decrease is
essentially the same for the deuterated molecule, and the change with J
is not rapid enough to be the result of the K-doubling beginning to
quench the internal rotation. Thus this decrease in μ_a with J is
almost certainly due to centrifugal distortion, particularly since a
decrease with J would be expected as the end-over-end rotation of the
complex forces the H_2O towards an anti-parallel orientation to the NH_3.
The somewhat smaller decrease with J observed for $NH_3 \cdot H_2S$ presumably
has a similar origin. As for $NH_3 \cdot H_2S$, an upper limit to the threefold
barrier height can be set, and from the $ND_3 \cdot D_2O$ result, $V_3 < 880$ cm^{-1}.
At first glance, a lower value for this limit might be anticipated,
given the $NH_3 \cdot H_2S$ result, but the calculation scales roughly like A_f
which is similar for D_2O and H_2S.
    It is obvious from these results that internal rotation effects

will be prevalent in weakly bound complexes, since a relatively high barrier is required to prevent this tunneling motion, at least for hydrides like ammonia. In the cases mentioned above, only upper limits to the barrier could be estimated from the a-type rotational spectrum. To obtain a precise determination of the internal rotation barrier and the finer details of the hindering potential, it is likely that vibration-rotation spectroscopy will have to be employed.

4. CONCLUSION

The work discussed above illustrates some of the effects and problems encountered in high resolution rotational spectra of weakly bound complexes caused by large amplitude motion. The tunneling effects are certainly the most dramatic in that they cause very large deviations from rigid rotor behavior and are therefore difficult to predict in advance of experiment, except in the sense that for hydride monomers, some tunneling effects will almost surely occur. In addition, group theoretical considerations are quite useful in outlining the possible behavior of these molecules and for assigning spectra. The larger problem of extracting the "structure" of non-rigid molecules from the spectral data has been primarily handled by rigid rotor models. With some caution concerning the vibrational averaging of experimentally determined quantities, these models are adequate for achieving a basic picture of the structure of weakly bound complexes. To fully utilize the accuracy of the spectral data, more difficult procedures involving directly fitting potential energy surfaces to the experimental results will likely be necessary.

5. ACKNOWLEDGEMENTS

I would like to thank the members of my research group who did much of the work discussed above, particularly P. Herbine, J.A. Odutola, G. Johnson, D. Prinslow, A. Hu and D. Kallimanis. I would also like to note that the first work on water dimer was done with J.S. Muenter in his lab at the University of Rochester. I have had many illuminating conversations with W. Klemperer and J.S. Muenter concerning this work. Much of the financial support for this research was from the Air Force Office of Scientific Research.

6. REFERENCES

1. T.R. Dyke, Topics in Current Chemistry, 120, 86 (1984).
2. A.C. Legon, Ann. Rev. Phys. Chem. 34, 275 (1983).
3. D.H. Levy, Ann. Rev. Phys. Chem. 31, 197 (1980).
4. G.D. Hayman, J. Hodge, B.J. Howard, J.S. Muenter and T.R. Dyke, Chem. Phys. Lett. 118, 12 (1985).
5. a)J.M. Hutson and B.J. Howard, Molec. Phys. 45, 769 (1982). b)S.L. Holmgren, M. Waldman, and W. Klemperer, J. Chem. Phys. 69, 1661 (1978); 67, 4419 (1977).

6. A.E. Barton and B.J. Howard, Faraday Disc. Chem. Soc. 73, 45 (1982).

7. P. Herbine and T.R. Dyke, J. Chem. Phys. 83, 3768 (1985).

8. P. Herbine, A. Hu, G. Johnson, and T.R. Dyke. To be published.

9. J.A. Odutola and T.R. Dyke, J. Chem. Phys. 72, 5062 (1980).

10. T.R. Dyke, B.J. Howard and W. Klemperer, J. Chem. Phys. 56, 2442 (1972). B.J. Howard, T.R. Dyke and W. Klemperer, J. Chem. Phys. 81, 5417 (1984).

11. H.C. Longuet-Higgins, Mol. Phys. 6, 445 (1963).

12. T.R. Dyke, J. Chem. Phys. 66, 492 (1977).

13. T.R. Dyke, K.M. Mack and J.S. Muenter, J. Chem. Phys. 66, 498 (1977).

14. T.R. Dyke and J.S. Muenter, unpublished data.

15. D.D. Nelson and W. Klemperer, private communication.

16. M.J. Frisch, J.E. Del Bene, J.S. Binkley and H.F. Schaeffer III, J. Chem. Phys. 84, 2279 (1986).

17. O. Matsuoka, E. Clementi and M. Yoshimine, J. Chem. Phys. 64, 1351 (1976).

18. G.T. Fraser, K.R. Leopold and W. Klemperer, J. Chem. Phys. 81, 2577 (1984).

19. G.T. Fraser, D.D. Nelson, Jr., G.J. Gerfen and W. Klemperer, J. Chem. Phys. 83, 5442 1986).

20. E.B. Wilson, Jr., C.C.Lin, and D.R. Lide, J. Chem. Phys. 23, 136 (1955).

MICROWAVE SPECTRA AND WEAK INTRAMOLECULAR HYDROGEN
BONDING IN 3-BUTENE-1-THIOL AND N-METHYLALLYLAMINE

K.-M. Marstokk and Harald Møllendal
Department of Chemistry
The University of Oslo
P. O. Box 1033 Blindern
N-0315 Oslo 3
Norway

ABSTRACT. The microwave spectra of 3-butene-1-thiol, $HSCH_2CH_2CH=CH_2$, and one deuterated species, $DSCH_2CH_2CH=CH_2$, reveal the existence of at least three conformations. The heavy-atom gauche form has an intramolecular hydrogen bond formed between the mercapto group hydrogen atom and the π-electrons of the double bond. Two other extended conformations were also identified. The hydrogen-bonded gauche conformation is 2.9(5) kJ mol^{-1} more stable than Extended I and 3.6(6) kJ mol^{-1} more stable than Extended II. The MW spectra of N-methylallylamine, $CH_3NHCH_2CH=CH_2$, and one deuterated species, $CH_3NDCH_2CH=CH_2$, demonstrate the existence of only one stable conformation. This rotamer is also stabilized by a very weak H bond formed between the methylamino group hydrogen atom and the π electrons of the double bond.

1. 3-BUTENE-1-THIOL

1.1. Introduction

The conformational properties of 3-buten-1-ol, $HOCH_2CH_2CH=CH_2$, have

57

A. Weber (ed.), Structure and Dynamics of Weakly Bound Molecular Complexes, 57–68.
© *1987 by D. Reidel Publishing Company.*

been studied by infrared[1,2] and microwave (MW)[3] spectroscopic methods
as well as by electron diff   on.[4] It was found both in solution[1,2]
and in the gas phase[3,4] that the most stable conformation of this
compound has an intramolecular hydrogen (H) bond formed between the
hydroxyl group hydrogen atom and the π electrons of the double bond.
It was estimated[3] that this rotamer is as least 3 kJ mol^{-1} more stable
than any other hypothetical form in the free state. 3-Butene-1-thiol,
HSCH$_2$CH$_2$CH=CH$_2$, was chosen for study in order to compare the ability
of the mercapto and the hydroxyl groups to form internal H bonds with
π electrons. Hydroxyl groups generally form stronger H bonds than
mercapto groups with the same acceptors. This is also found in this
study.

1.2. Experimental

3-Butene-1-thiol was synthesized largely as described in the
literature,[5,6] purified by gas chromatography and identified by IR,
PMR and ^{13}C NMR spectroscopy. Extensive spectral measurements were
made in the 26.5-38 GHz spectral region using an improved version of
the spectrometer described briefly in Ref. 7. The cell was cooled to
about -60 ^{0}C. The vapour pressure was approximately 1-2 Pa. The
deuterated species, DSCH$_2$CH$_2$CH=CH$_2$, was produced by direct exchange
with heavy water in the wave guide.

1.3. Results

The MW spectrum is quite dense at -60 ^{0}C. The strongest transitions of
the spectrum turned out to be a-type R-branch as well as b-type
Q-branch lines of the hydrogen-bonded gauche conformation shown in
Fig. 1. The strongest of these transitions have peak absorption
coefficients of roughly 4*10^{-7} cm^{-1} at this temperature.

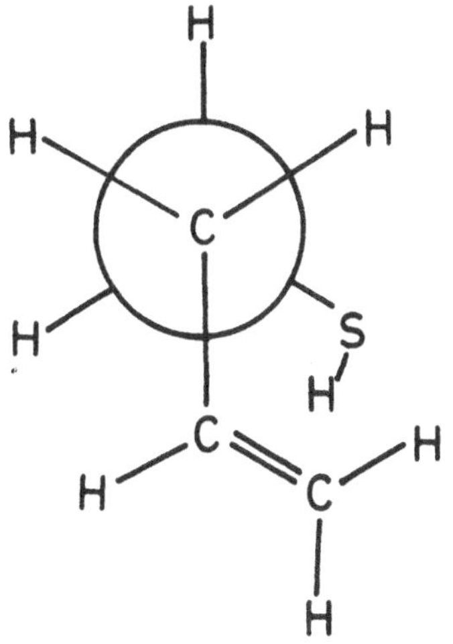

Fig. 1. The H-bonded <u>gauche</u> conformation viewed along the $-H_2C-CH_2-$ bond. The C-S and $-H_2C-CH=$ bonds are <u>gauche</u> to one another. The C-C-C-S dihedral angle is $65(3)^0$ from <u>syn</u>. The C-C-C=C dihedral angle is twisted $57(3)^0$ from <u>anti</u> in order to allow an intramolecular H bond to be formed between the mercapto group hydrogen atom and the double bond π electrons. The H-S-C-C dihedral angle is $50(5)^0$ from <u>syn</u>.

Bond-moment calculations[8] predict $\mu_a = 3.3*10^{-30}$ C m, $\mu_{\underline{b}} = 2.6*10^{-30}$ C m, and $\mu_{\underline{c}} = 1.7*10^{-30}$ C m, respectively. The $\underline{J}=8 \leftarrow 7$ and $\underline{J}=9 \leftarrow 8$ <u>a</u>-type <u>R</u>-branch lines were first found, and the <u>b</u>-type transitions were then readily assigned. No <u>c</u>-type lines were found although their positions can be very accurately predicted. The <u>c</u>-component of the dipole moment is thus presumed to be somewhat smaller then predicted above by the bond-moment method. Unfortunately, no dipole moment could be determined due to insufficient intensities of low $\underline{J}$ transitions. The resulting spectroscopic constants derived using transitions with a $\underline{J}$ value of maximum 30, is shown in Table I. This table also includes the spectroscopic constants of the deuterated species which was assigned in a straightforward manner. Several vibrationally excited states were also assigned and they are discussed in a forthcoming paper[9] which will give a more complete account of this work.

TABLE I. Spectroscopic constants of the ground vibrational state of the hydrogen-bonded gauche conformation of 3-butene-1-thiol.

Species	$HSCH_2CH_2CH=CH_2$	$DSCH_2CH_2CH=CH_2$
Number of transitions	117	78
Root-mean-square dev./MHz	0.065	0.078
A_0/MHz	6894.4387(34)	6703.6341(87)
B_0/MHz	2308.11700(84)	2302.6352(41)
C_0/MHz	1882.05016(78)	1868.0054(41)
Δ_J/kHz	2.2710(26)	2.118(30)
Δ_{JK}/kHz	-13.346(14)	-12.478(29)
Δ_K/kHz	33.22(13)	28.69(39)
δ_J/kHz	0.70608(68)	0.7076(11)
δ_K/kHz	3.885(25)	3.747(37)

Uncertainties represent one standard deviation.

Another prominent feature of the spectrum is the lumps of lines occurring every 2.8 GHz. They are modulated at low Stark voltages and were assigned as the a-type R-branch pile-ups of the two highly prolate extended conformations depicted in Fig. 2. Both these two conformations have the asymmetry parameter[10] κ -0.999 as well as similar dipole moment components along the a-axis calculated by the bond-moment method[8] to be approximately $4.4*10^{-30}$ C m for each of them. It is also possible for the molecule to take an extended conformation in which the S-H bond is anti to the $-H_2C-CH_2-$ bond. This hypothetical rotamer (not shown in Fig. 2) would also be very prolate and is predicted[8] to have a dipole moment component of about $2.6*10^{-30}$ C m along the a-axis. The reason why a successful identification of this third hypothetical conformation was not made is of course the

fact that μ_a is considerably smaller than for Extended I and II.
In addition, the anti arrangement of S-H bond is less favourable by
about 1.5 kJ mol^{-1} than the gauche arrangement, as shown in the case
of ethanethiol.[11] The ground-state pile-up of Extended I has B+C
2833.1 MHz for the parent species, and B+C 2795.5 MHz for the
deuterated species. The values found for Extended II were 2838.2 MHz
and 2793.4 MHz, respectively. The results for vibrationally excited
states are discussed in Ref. 9. No dipole moment could be determined
for any of the two extended forms due to low intensities.

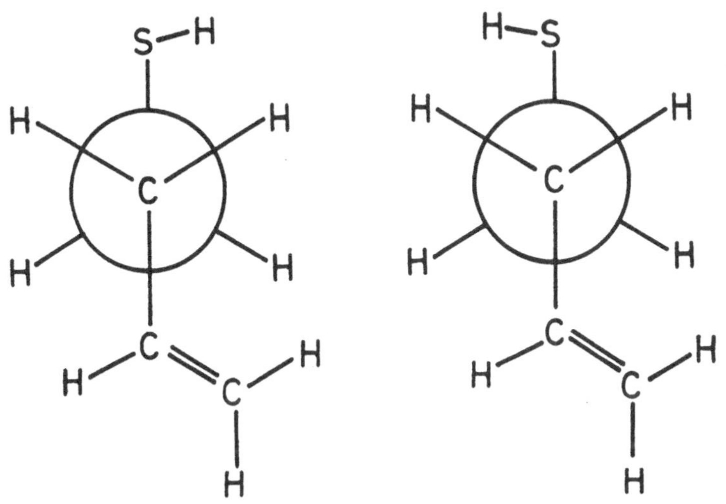

Fig. 2. Extended I and II differs from the H-bonded gauche of Fig. 1
in having an anti arrangement for the C-S and the -CH$_2$-CH= bonds. The
H-S-C-C dihedral angles are 60^0 from syn.

The energy differences between the three identified rotamers were
determined from relative intensity measurements. The H-bonded gauche
conformation was found to be 2.9(5) kJ mol^{-1} more stable than Extended
I and 3.6(6) kJ mol^{-1} more stable than Extended II. The dipole moments
calculated by the bond-moment method[8] were used to derive these energy

differences. The quoted uncertainties represent one standard deviation. The uncertainties arising from using calculated dipole moments are presumed to have been allowed properly for in the stated uncertainties.

The assignments reported above and in Ref. 9 include all strong lines of the spectrum as well as many weaker transitions. It is also likely that further conformations would possess sizable dipole moments. The fact that there are no relatively strong unassigned lines left, makes us conclude that the three rotamers assigned in this work are also the three most stable forms of 3-butene-1-thiol. It is conservatively estimated that the H-bonded gauche form is at least 3 kJ mol^{-1} more stable than any further hypothetical unassigned conformation.

The rotational constants (or the B+C combination determined for the two extended forms) do not suffice to determine a full geometrical structure for the three conformations. Assumptions have to be made in order to derive interesting structural parameters. Using selected parameters taken from recent accurate studies[9] fits of some parameters which depend strongly on the rotational constants, were made. The C-C-C=C dihedral angle was found to be 57(3)^0 from anti for the H-bonded gauche conformation and 62(3)^0 for both extended forms. The S-C-C-C was determined to be 65(3)^0 from syn, while this angle is almost exactely anti for both extended rotamers.[9] The H-S-C-C dihedral angle is 50(5)^0 from syn in the gauche rotamer. No fit was made for this angle for the two extended conformers.

1.4. Discussion

There is a considerable amount of evidence that a weak intramolecular H bond is indeed formed in the gauche conformation. The fact that this rotamer actually has the mercapto group hydrogen atom directed in the most favourable orientation for this kind of interaction, is one piece of evidence. Moreover, the distance between this hydrogen atom and the nearest carbon atom of the double bond is 260 pm which is about 30 pm shorter than the sum of the van der Waals distances of hydrogen and aromatic carbon.[12] Another evidence is that the _gauche_ conformation is considerably more stable than any other form of the molecule. This fact would be hard to explain if one had to exclude a stabilizing interaction between the mercapto group and the double bond. The H bond in the title compound is not as strong as that in the corresponding alcohol, and this has the following two consequences: The C-C-C=C dihedral angle is $75(3)^0$ in the H-bonded conformation of 3-buten-1-ol[3] in contrast to $57(3)^0$ in 3-butene-1-thiol. The angle difference in the two molecules of about 18^0 thus leads to a shorter distance between the proton and the π electrons in the alcohol and consequently a stronger interaction in $HOCH_2CH_2C=CH_2$ than in $HSCH_2CH_2CH=CH_2$. The stronger hydrogen bond in 3-buten-1-ol than in the _gauche_ conformer of 3-butene-1-thiol results in a much higher population of the _extended_ forms in the thiol than in the alcohol. In 3-buten-1-ol the population of _extended_ conformers was so low that they could not be detected by MW spectroscopy,[3] while large fractions of the gas consists of the two _extended_ forms for the 3-butene-1-thiol.

2. N-METHYLALLYLAMINE

2.1. Introduction

Allyl derivatives of the general form $X-CH_2-CH=CH_2$ normally takes _syn_ or _skew_ conformations as discussed in a recent paper.[13] In the _syn_ form, the X-C-C=C skeleton is planar and the C-X bond eclipses the double bond, whereas in the _skew_ form the C-X bond is rotated 120^0 out of this plane. If additional rotational isomerism around the C-X bond is possible, several _syn_ or _skew_ forms may exist. Allylamine, $H_2NCH_2CH=CH_2$, is one such example. Five rotameric forms, two _syn_ and three _skew_, are conceivable for this compound. The two _syn_ and two of the three possible _skew_ conformations have indeed been assigned by Botskor and coworkers.[14] There are small energy differences between

the four different rotamers of allylamine. In the related molecule
N-methylallylamine, $CH_3NHCH_2CH=CH_2$, the total number of possible <u>syn</u>
and <u>skew</u> rotamers are in fact no less than nine. Three of these are
<u>syn</u> conformations; three <u>skew</u> conformations arise when the CH_3NH-
moiety is rotated 120^0 out of plane in a clockwise manner, and the
final three <u>skew</u> conformations are formed when the said moiety is
rotated 120^0 in a counter-clockwise manner. It was found that the <u>skew</u>
rotamer shown in Fig. 3 is the most stable form of the molecule.
Additonal rotamers, which may or may not exist, are at least 3 kJ
mol^{-1} less stable than this form. The <u>skew</u> rotamer of Fig. 3 has the
methyl group <u>anti</u> to the CH_2-CH bond. A very weak H bond is presumably
formed between the methylamino group hydrogen atom and the π electrons
of the double bond

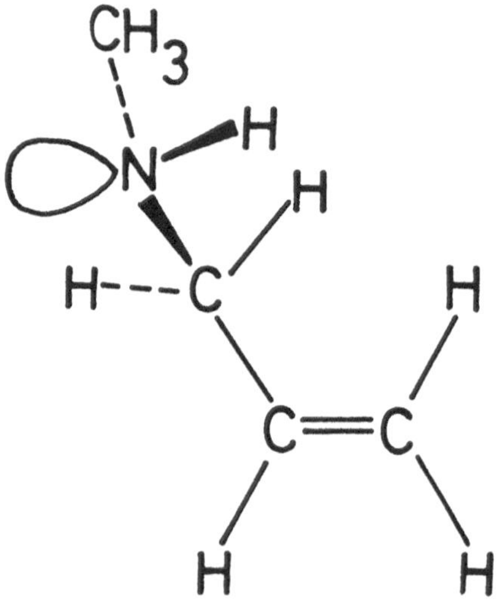

<u>Fig. 3</u>. The most stable rotameric form of N-methylallylamine. The
N-C-C=C dihedral angle is $123(3)^0$ from <u>syn</u> and the methyl group is
<u>anti</u> to the CH_2-CH bond. This conformation is at least 3 kJ mol^{-1} more
stable than any other rotameric form of the molecule.

2.2 Experimental

N-methylallylamine was purchased from Fluka A. G., Buchs, Switzerland. The compound was purified by gas chromatography before use. Studies were made in the 12.4-26.7 and 28.0-38.0 GHz spectral regions at dry-ice temperature and a pressure of about 1 Pa using the spectrometer described above.[7] The deuterated species was produced by direct exchange with heavy water in the wave guide.

2.3. Results

The microwave spectrum of N-methylallylamine is rich and of moderate intensity. The strongest lines which turned out to be high-J $\underline{b}$- and $\underline{c}$-type $\underline{Q}$-branch transitions of the conformation shown in Fig. 3, have peak absorption coefficients of roughly $4*10^{-7}$ cm^{-1} at dry-ice temperature. Over 200 ground-state transitions were assigned for this rotamer with a maximum J-value of 74. More details are given in Ref. 15. None of the observed transitions were split by quadrupole interactions of the ^{14}N nucleus. The ground-state spectroscopic constants of the parent and deuterated species are shown in Table II. Results for several excited states are published elsewhere.[15] No dipole moment could be determined due to insufficient intensities of low-J transitions. A total of about 600 ground and excited-state transitions were assigned for this conformer. Searches for the other possible rotamers among the relatively few and rather weak remaining unidentified transitions were negative. Other hypothetical rotamers are predicted to possess sizable dipole moments. Intensity considerations of unassigned lines lead us to conclude that the assigned conformer is at least 3 kJ mol^{-1} more stable than any further unidentified conformation. Only the N-C-C=C dihedral angle was fitted to the rotational constants with the remaining structural parameters taken from related highly accurate structures of related compounds. The N-C-C=C dihedral angle was found to be 123(3)^0.

66

K.-M. MARSTOKK AND H. MØLLENDAL

TABLE II. Spectroscopic constants of the
ground vibrational state of N-methylallylamine.

Species	$CH_3NHCH_2CH=CH_2$	$CH_3NDCH_2CH=CH_2$
Number of transitions	227	54
Root-mean-square dev./MHz	0.087	0.103
A_0/MHz	19998.241(12)	18687.038(52)
B_0/MHz	2235.7349(14)	2220.8662(46)
C_0/MHz	2203.1495(16)	2189.5416(47)
Δ_J/kHz	0.6441(16)	0.560(14)
Δ_{JK}/kHz	-23.329(33)	-19.774(69)
Δ_K/kHz	452.69(25)	429(11)
δ_J/kHz	-0.104892(76)	-0.093241(75)
δ_K/kHz	38.65(27)	12.58(37)
ϕ_J/Hz	0.00079(69)	-0.178(14)
ϕ_{JK}/Hz	0.029(27)	a
ϕ_{KJ}/Hz	1.18(27)	a
ϕ_K/Hz	-37.3(65)	a
φ_J/Hz	-0.000547(18)	-0.000377(15)
φ_{JK}/Hz	-0.462(53)	a
φ_K/Hz	0.0[b]	0.0[b]

Uncertainties represent one standard deviation. [a] Preset at
the value found for the parent species. [b] Preset at zero.

2.4. Discussion

This study shows that substitution of a hydrogen atom in the amino
group in allylamine with a methyl group has rather large
conformational consequences. Instead of four rotameric forms with
rather similar energies as found for allylamine,[14] the one rotamer
shown in Fig. 3 is the predominating form of N-methylallylamine. This
conformation is characterized by having ideal steric conditions in

that the large methylamino group is twisted out of the C-C=C plane and in having the methyl group in <u>anti</u> position to the $-CH_2-CH=$ bond. In addition, a weak H bond may be formed between the amino group hydrogen atom and the double bond π electrons, as the non-bonded distance between this atom and the nearest double-bond carbon atom is 263 pm which is about 30 pm shorter than the sum of the van der Waals radii of aromatic carbon and hydrogen.[12] Finally, in the conformation shown in Fig. 3 (the most stable rotamer) repulsion between the lone pair of the nitrogen nucleus and the π electrons of the double bond is minimal. N-Methylallylamine thus prefers this conformation because steric conditions are favourable, lone pair π electron repulsion is minimal, and because a weak hydrogen bond may stabilize it.

2.5 Acknowledgment

Mr. Marko Opresnik is thanked for synthesizing 3-butene-1-thiol. Norges Teknisk Naturvitenskapelige Forskningsråd is thanked for a travel grant to the NATO advanced research workshop at Maratea, Italy.

2.6. References

1. M. Oki and H. Iwamura <u>Bull. Chem. Soc. Jpn.</u> <u>32</u> (1959) 567.
2. W. Ditter and A. P. Luck <u>Ber. Bunsenges. Phys. Chem.</u> <u>75</u> (1971) 163.
3. K.-M. Marstokk and H. Møllendal <u>Acta Chem. Scand.</u> <u>A 35</u> (1981) 395.
4. M. Trætteberg and H. Østensen <u>Acta Chem. Scand.</u> <u>A 33</u> (1979) 491.
5. J.-M. Suzur, M.-P. Crozet and C. Dupuy <u>C. R. Acad. Sc. Paris, Series C,</u> <u>264</u> (1967) 610.
6. C. Walling and M. S. Pearson <u>J. Am. Chem. Soc.</u> <u>86</u> (1964) 2262.
7. K.-M. Marstokk and H. Møllendal <u>J. Mol. Struct.</u> <u>5</u> (1970) 205.
8. C. P. Smyth <u>Dielectric Behavior and Structure,</u> McGraw-Hill, New York 1955, p. 244.
9. K.-M. Marstokk and H. Møllendal <u>Acta Chem. Scand.,</u> in press.

10. W. Gordy and R. L. Cook Microwave Molecular Spectra, Wiley,
    New York 1984, p. 324.

11. F. Inagaki, I. Harada and T. Shimanouchi J. Mol. Spectrosc. 46
    (1973) 381.

12. L. Pauling The Nature of the Chemical Bond, 3rd Ed., Cornell
    University Press, New York 1960, p. 260.

13. Z. Smith, N. Carballo, E. B. Wilson, K.-M. Marstokk and H.
    Møllendal J. Am. Chem. Soc. 107 (1985) 1951.

14. a. G. Roussy, J. Demaison, I. Botskor and H. D. Rudolph J.
    Mol. Spectrosc. 38 (1971) 535; b. I. Botskor, H. D. Rudolph and
    G. Roussy Ibid. 52 (1974) 457; c. I. Botskor, H. D. Rudolph and
    G. Roussy Ibid. 53 (1974) 15; d. I. Botskor and H. D. Rudolph
    Ibid. 71 (1978) 430.

15. K.-M. Marstokk and H. Møllendal Acta Chem. Scand., submitted for
    publication.

HIGH RESOLUTION INFRARED SPECTROSCOPY OF VAN DER WAALS MOLECULES

Brian J. Howard
Physical Chemistry Laboratory
Oxford University
South Parks Road
Oxford OX1 3QZ
U.K.

ABSTRACT. The high resolution laser infrared spectroscopy of weakly bound complexes is shown to provide a valuable source of information on the structure and dynamics of such species. It also contains important information on the nature of the intermolecular potential energy surface. Some of the most detailed information is contained in the direct absorption spectroscopy of those dimers present in equilibrium with the monomers in low temperature, low pressure, long pathlength static cells. This is exemplified by the results on rare gas atoms bound to hydrogen chloride. The spectra contain a wealth of information on the structure and dynamics of such species. However these equilibrium spectra are limited to clusters containing monomers with open rotational structure in the infrared, otherwise the dimer spectrum can be completely obliterated by nearly continuous absorption of the monomer. Also the dimer spectra can be frequently very dense and difficult to analyse. It is shown that such problems can be overcome in part by direct absorption experiments in pulsed supersonic jets. The spectra for complexes of rare gases bound to each of nitrous oxide and carbonyl sulphides are considered. The derived structural parameters and their implication on the intermolecular potential energy surface are discussed. For these series of rare gas containing complexes (e.g. Rg–HCl, Rg–OCS and Rg–N_2O), it is shown that the long range attraction is increased on exciting the monomer vibration. At the same time the short range repulsion is also increased. As a result, the changes in the structures of the complexes on exciting the monomer vibration and the band origin shift on cluster formation are both determined by a delicate balance between changes in attractive and repulsive forces.

1. INTRODUCTION

The spectroscopy of van der Waals molecules provides a valuable source of information on the weak forces between molecules. Historically the first spectroscopic evidence for the formation of van der Waals complexes was to be found in extraneous features in the infrared spectrum of moderately high pressure gas mixtures of hydrogen halides [1-3].

A. Weber (ed.), Structure and Dynamics of Weakly Bound Molecular Complexes, 69–84.

For example, Rank et al. [2] found additional structure in the low temperature infrared spectrum of HCl in the presence of a high pressure of
Ar. Such features varied quadratically with pressure and were attributed to Ar-HCl. No rotational structure was observed owing to pressure
broadening and instrumental resolution. However subsequent high resolution studies [4] have confirmed the earlier assignment and have shown
the features to be unresolved P and R branches of the linear dimer.

Despite this early work, for a long time the only true progress in
the infrared spectroscopy of clusters was the elegant work of Welsh,
McKellar and coworkers who obtained rotationally resolved spectra of
many H_2 complexes [5,6]. During this period, considerable progress was
being made in the microwave and electronic spectroscopy of complexes,
primarily through the use of nozzle beam techniques [7,8].

The past few years have seen a resurgence of interest in the infrared region of the spectrum, especially as suitable tunable infrared
lasers have become available. These lasers in conjunction with low
temperature, low pressure static cell techniques [4] or molecular free
jets [9,10,11] have provided the sensitivity to obtain highly resolved
spectra of many complexes. Much of this progress will be reviewed
elsewhere in this publication and I shall concentrate on simple rare
gas-linear molecule systems I have investigated. In section 2 recent
progress in the absorption spectroscopy of rare gas HCl clusters will
be discussed. In section 3, the direct absorption spectroscopy of complexes in supersonic expansions will be presented. The results for
complexes of N_2O and OCS will be discussed. Throughout, the relative
merits of jet and static cell techniques will be compared. In section
4, the spectroscopic results will be reviewed. The rotational constants
and band origin shifts will be shown to provide valuable information on
the intermolecular potential energy surface and how it changes upon
vibrational excitation.

2. ABSORPTION SPECTROSCOPY IN STATIC CELLS

In a gas mixture at thermal equilibrium a small fraction of the molecules are present as van der Waals dimers. Unfortunately the concentration of such species is very low. Stogryn and Hirschfelder [12],
using model potentials, have shown that the concentration of dimers in
the vapour above a liquid near its boiling point rarely exceeds 0.1%.
At lower pressures, where the dimers can have resolved rotational structure, the percentage concentration is even less. However measurable
infrared absorption from the dimer can be obtained by using a sufficiently long absorption pathlength and a monomer with a sufficiently open
absorption spectrum; this reduces the risk of monomer absorptions ob-
literating large fractions of the dimer spectrum; under these conditions,
monomer lines are 100% absorbing and often appear very broad relative
to dimer lines.

2.1 Absorption spectrum of Ar-HCl

Figure 1 illustrates a portion of the absorption due to Ar-HCl in a

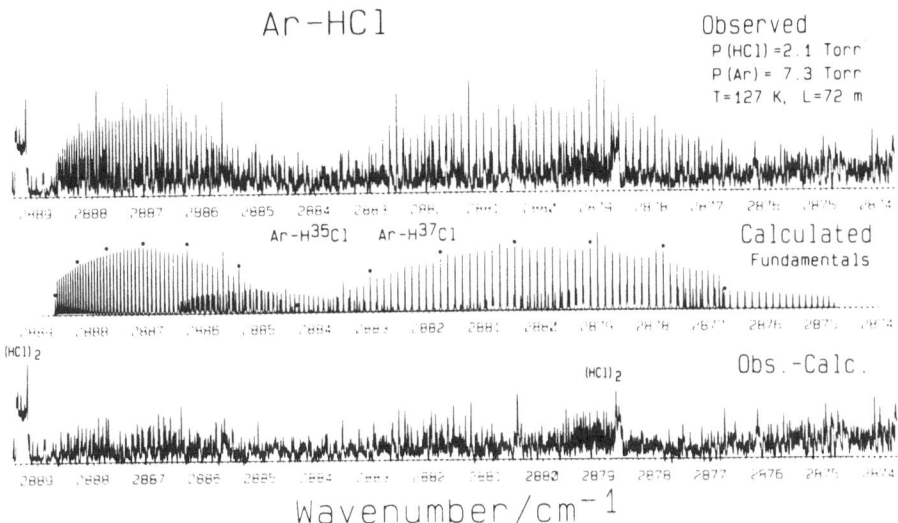

Figure 1.   Observed and calculated spectra of Ar-HCl in the fundamental region.   Dots on simulations are for J=0 mod 10.

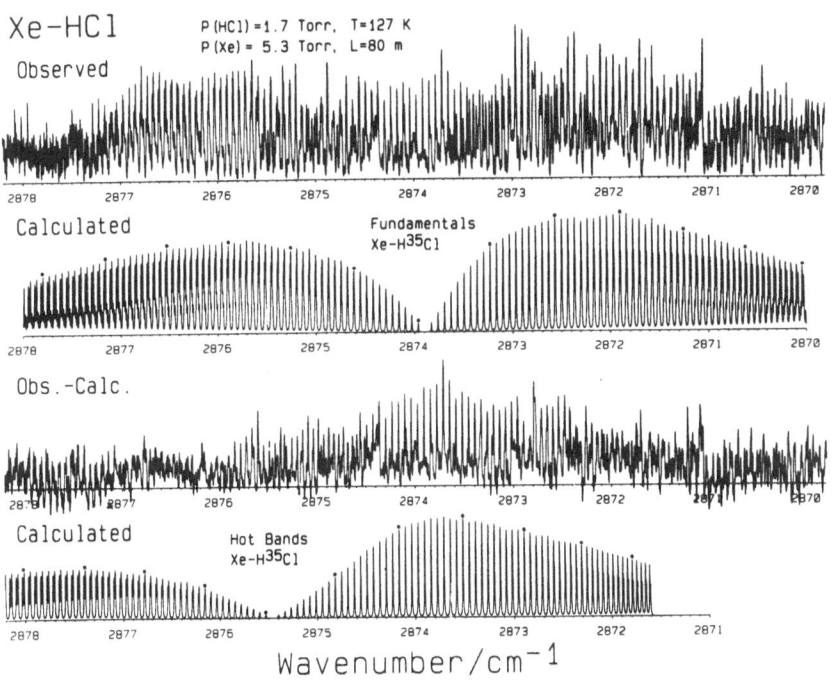

Figure 2.   Observed spectrum for Xe-HCl in the HCl band origin region together with calculated fundamental and hot band spectra.   Dots on simulation are for J=0 mod 10.

long-path White cell (L = 72m) cooled to 127K. It is a rich spectrum
with a wealth of information on the complex. The strongest feature is
a parallel band arising from the HCl stretch in this quasi-linear molecule. Just like other van der Waals molecules, Ar-HCl exhibits extreme
non-rigidity and, in order to fit the P and R branches to experimental
accuracy, a high order power series in $[J(J+1)]^n$ is required (n=0-5).
The bottom spectrum in Figure 1 illustrates the result of subtracting
the calculated fundamental spectrum. This residual spectrum contains
much structure due to excited states of the complex. Although as yet
not completely analysed, such hot bands can potentially provide much
added information on the nature of the complex in its excited states.

One noticeable feature of the original spectrum (and one reproduced
in the simulation) is the sudden cut-off at J~60. At this rotational
level the calculated rotational energy is about 180 cm^{-1}, much greater
than the calculated binding energy of 117 cm^{-1} [13]. The molecule is consequently exhibiting rotational predissociation with the rotational lifetime changing rapidly with J. The lower levels are metastable, bound by
a centrifugal barrier.

2.2 Absorption spectra of Xe-HCl

The absorption spectra of the other rare gas-HCl complexes (Rg = Ne, Kr,
Xe) have been similarly observed in long-path absorption cells. Figure
2 shows a portion of the Xe-HCl spectrum in the monomer band origin
region. The analysis is less obvious than in the case of Ar-HCl because
of more severe congestions and the presence of several Xe isotopes.

The complexity is best explained in terms of the simulation under-
neath. Xenon in natural abundance consists of four isotopes with abundances of greater than 10 percent and several others with non-negligible
abundances. As a result there are many isotopic forms of the dimer,
each with slightly different rotational constants. At high J, each
rotational line splits into its many isotopic components and the apparent
intensity of the P and R branches drops off more rapidly than expected.
However from a detailed analysis of the fit to the spectrum, in which
all the higher order centrifugal distortion constants are constrained to
be in their correct reduced mass ratios, the band origins of the different isotopes are determined to be slightly different. As a consequence
the isotopic separation of rotational lines is slightly more pronounced
in the R-branch than the P-branch.

After subtracting the calculated fundamental spectrum from that
experimentally observed, one is still left with strong features due to
hot bands. A simulation of the first hot band, obtained from a least
squares fit to line positions, is shown at the bottom of Figure 2. This
shows a more pronounced asymmetry in peak intensity between P and R
branches than the fundamental spectrum. Again the rotational lines
split into their isotopic components at high J. However here there is
an even greater isotopic shift of the band origin, so that while lines
quickly split into isotopic components in the R-branch, in the P-branch
the lines initially become better overlapped as J is increased. The
maximum overlap is at a value of J close to the position of maximum intensity for an individual isotope. As a consequence intense, narrow
P-branch lines are observed, with this branch having greater peak inten-

sity than that of the ground state even though it derives from an excited van der Waals vibrational state with its population depleted by a Boltzmann factor.

Why is the isotopic band origin shift for the hot band greater than that for the fundamental? The initial bathochromic shift of the HCl vibration on formation of the Xe-HCl complex can be explained in terms of a deeper potential well for Xe interacting with a vibrationally excited HCl molecule than with one in its ground vibrational state. Assuming the intermolecular potential to be identical for the different isotopes of Xe, the small isotopic dependence of the band origin for the fundamental arises from the small changes in the differences of the zero point energies in the two potential wells. This is almost certainly dominated by the change in the stretching vibrational contribution. The fact that the band origins of heavier isotopes are shifted to slightly lower frequency indicates that the effective van der Waals bond stretching force constant is slightly larger for an HCl molecule vibrationally excited. The hot bands arise from excited stretching states and since this excitation is approximately twice the stretching contribution to the zero point energy (assuming no severe mixing with bending motions). Thus the isotopic band origin shift is expected to be approximately three times as great in the first hot band as in the fundamental, as is observed.

2.3 Vibrational dependence of the intermolecular potential energy surfaces for Rg-HCl systems

From an analysis of the fundamental bands of the Rg-HCl spectra, it is possible to derive effective structural constants for the complexes. Table I shows the effective rare gas-hydrogen chloride centre of mass distance for both ground and excited vibrational states of HCl together with the observed band origin shift. In addition are shown the total well depth, ε'', from the optimised ground state potential surfaces [13].

TABLE I. Molecular constants for Ar-HCl, Kr-HCl and Xe-HCl (^{35}Cl isotope)

	Ar-HCl	Kr-HCl	Xe-HCl
$R''/\text{Å}$	4.0064	4.0824	4.2752
$R'/\text{Å}$	4.0252	4.0962	4.2756
$\Delta\nu/\text{cm}^{-1}$	-1.7688	-5.2314	-12.0801
$\varepsilon''/\text{cm}^{-1}$	180.5	213.9	263.6

In all cases there is a bathochromic shift of the HCl frequency on complex formation due to the excited vibrational state being more tightly bound than the ground state. However, while the well depths for the complexes scale well with the expected long range attractions and hence the rare gas polarisabilities, the band origin shifts which to first order reflect the changes in well depth on vibrational excitation show no such simple relationship; $\Delta\nu$ increases far more rapidly than expected by comparison with ε''. A part of the problem is that the repulsive forces as well as the long-range attraction increase between the two states. For an isolated HCl molecule the mean bond length increases by 0.019Å on vibrational excitation so that at fixed R the Rg-H distance on average decreases and the intermolecular repulsion must increase. At the same time the amplitude of the hydrogen vibrational motion increases so that when it is convoluted over the expected exponential Rg-H repulsion, a further increase in the effective repulsion can be expected.

This increase in repulsive forces is evident from the increase in the effective intermolecular distances on vibrational excitation. This is not as great as the expected 0.019 Å increase of the HCl bond-length because of a compensating increase in attraction. In fact it is the increase in attraction that is dominant in the band origin shift while the increase in repulsion is dominant in the change of bond-length.

A complete analysis of these effects requires an accurate determination of the intermolecular potential surface for the excited as well as the ground state of HCl. However some insight can be obtained from a simple model. It has been shown in simple rare gas-rare gas systems [14] that the long range C_6 coefficients are given quite accurately by the expression ε/R^6 where ε is the well depth and R_m is the interatomic d - tance at the potential minimum. If it is possible to separate off the bending motion (perhaps in some adiabatic sense) it might be expected that the changes in the long range C_6 attraction coefficient can be given by the change in bond length and the change in well depth via the following expressions

$$C_6'/C_6'' \simeq (\varepsilon'/\varepsilon'')(R_m'/R_m'')^6$$

where ' and " refer to respectively excited and ground states of HCl. Representing the change in well depth, $\varepsilon'-\varepsilon''$, by $\Delta\varepsilon$ and similarly for ΔR_m and ΔC_6

$$(C_6'/C_6'') \simeq (1 + \Delta\varepsilon/\varepsilon'')(1 + \Delta R_m/R_m'')^6$$

If the differences in the zero point energies (for van der Waals bending and stretching motion) for the ground and excited states can be neglected $\Delta\varepsilon$ is numerically the same as the band origin shift ($\Delta\varepsilon = -\Delta\nu$). Also assuming $\Delta R_m \simeq \Delta R_0$, one obtains

$$\Delta C_6/C_6'' \simeq (1 - \Delta\nu/\varepsilon'')(1 + \Delta R_0/R_m)^6 - 1$$

This analysis yields values of the percentage change in the long-range attraction ($\Delta C_6/C_6'$) for Ar, Kr and Xe of approximately 4.39%, 4.54% and 4.65% respectively. While this is obviously a very crude model it does indicate that there is substantial uniformity between the rare gas atoms in the changes in the long-range attraction on HCl vibrational excitation.

The precise band origin shifts require an accurate description of the changes in the intermolecular repulsion.

3. ABSORPTION SPECTROSCOPY IN SUPERSONIC JETS

As mentioned above, the spectra of van der Waals molecules in equilibrium cells contain a wealth of information on the complex, with all levels up to the dissociation limit being populated. However the great density of spectral lines, together with the presence of much stronger monomer features results in severe congestion in the spectrum together with difficulties in analysis.

Supersonic nozzle sources, which have been used to good effect in microwave/rotational and electronic spectroscopy [7,8,15], also provide a very suitable source of van der Waals molecules for infrared studies. In such jets there is sufficient conversion of the monomer to van der Waals dimer (conversion efficiency of about 10% being typical). Also the low translational and rotational temperatures result in simplified dimer spectra without severe contamination from many monomer transitions. The technique of laser excitation with bolometric detection is discussed elsewhere in this volume. I shall concentrate on recent work at Oxford on the absorption spectroscopy in pulsed free jets.

Figure 3, shows a schematic diagram of the equipment. Infrared radiation from a diode laser intersects the output of a pulsed nozzle. Any transient absorption of the infrared radiation is detected by a dual

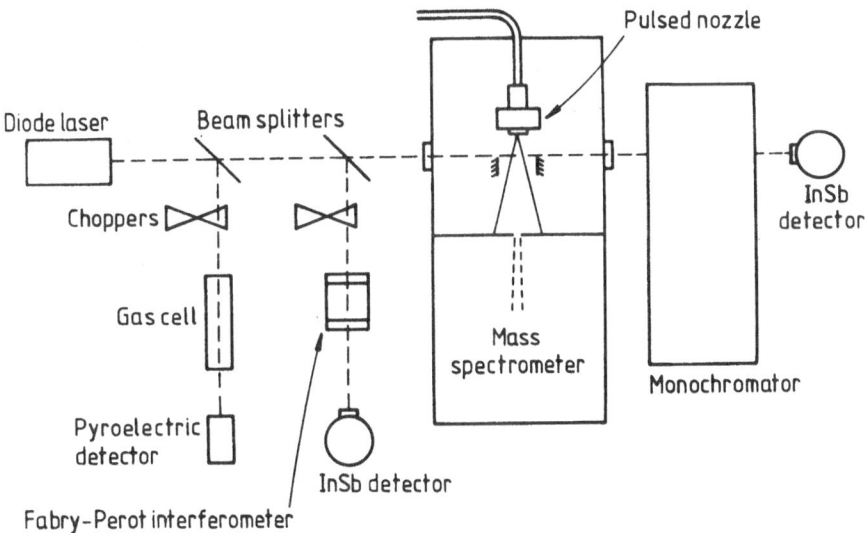

Figure 3. Schematic diagram for diode laser molecular beam spectrometer.

gate boxcar integrator. One gate is set to monitor the laser power just before the gas pulse while the second overlaps the pulse and the difference signal is averaged. Short duration, high density gas pulses are used so as to increase the instantaneous infrared absorption without overloading a pumping speed limited vacuum system. Also because of the dominant 1/f noise of a semi-conductor laser and detection system, the use of two adjacent short duration gates helps to discriminate against laser noise.

Figure 4 shows a typical absorption spectrum of Ar–OCS close to the ν_3 monomer vibration. The main feature observed is that due to a Q-branch ($^RQ_o(J)$) of a near T-shaped complex. In the same vicinity is the R(6) monomer line whose intensity is no greater than many dimer lines. This indicates the high conversion efficiency of monomer to dimer and the substantial cooling in the nozzle. As a result there is little complication of the spectrum from monomer lines and the dimer spectrum is generally well resolved. The price to be paid for these advantages

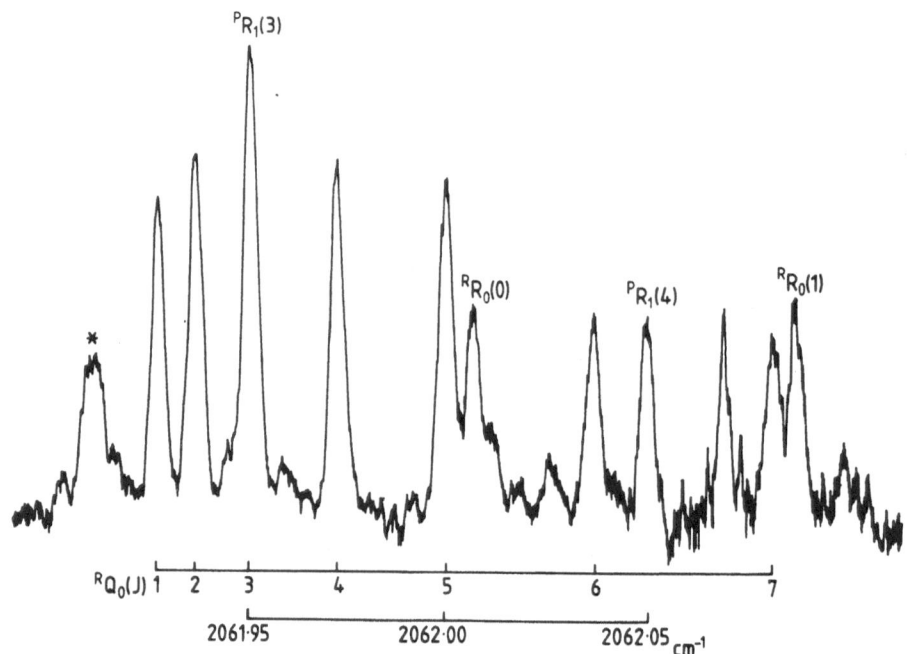

Figure 4. Ar–OCS ^RQ_o band in the OCS ν_3 region. * indicates the R(6) monomer ν_3 hot band with one quantum of ν_1 excited.

is that few excited vibrational states are sufficiently populated to be observed and the spectrum only consists of a few rotational lines so that the molecular constants are less precisely determined.

Similar spectra have also been obtained for a range of Rg–OCS complexes (Rg = Ne, Ar, Kr) and Rg–N$_2$O complexes (Rg = Ne, Ar, Kr, Xe). All these molecules are asymmetric tops with near T-shaped equilibrium

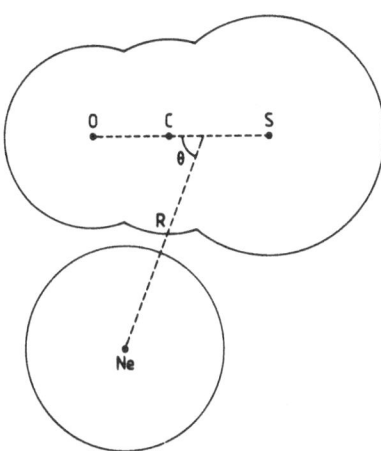

Figure 5. Ne-OCS coordinate system. The coordinates for Rg-OCS and Rg-N$_2$O are similarly defined. R connects the rare gas atom to the centre of mass of the linear triatomic molecule. θ is the angle subtended by the rare gas and external nitrogen atom in Rg-N$_2$O complexes.

geometries. The spectra have been fitted to provide ground and excited state spectroscopic constants. Assuming a structure similar to that in Figure 5, the effective structural parameters can be derived from the expressions

$$\sin\theta \ = \ (b/A)\,[\,(C-A)/(C-b)\,]^{\frac{1}{2}}$$

$$R \ = \ \{k\,[\,(m_{Rg} + m_{XYZ})/m_{Rg}m_{XYZ}\,]\,(1/C\,-\,1/b)\,\}^{\frac{1}{2}}$$

where A, B, C are the rotational constants of the complex, b is the XYZ monomer rotational constant in the vibrational state of interest, and k is a conversion factor (505379 MHz amu Å^2). The resulting structural constants for the OCS complexes are given in Table II. The corresponding results for the Rg-N$_2$O complexes are shown in Table III. It should be noted that both θ and 180°-θ will reproduce the observed rotational constants but only the acute angle as defined is compatible with isotopic substitution in the limited microwave data.

The first noticeable result is that within a given series, the complexes all have a very similar structure; the changes in intermolecular distances are completely explained in terms of the changes in the van der Waals radii of the rare gas atoms. In addition, for a given complex, there is very little change in the effective structure on vibrational excitation. This is probably in part due to the fact that the vibration being excited is largely orthogonal to the van der Waals bond. Although it is difficult to discern from the derived constants because of insufficient accuracy, it appears that there is greater increase in the van der Waals bond length in the Ne clusters than in those of the heavier

TABLE II.   Molecular constants for Ne–OCS, Ar–OCS
and Kr–OCS

	Ne–OCS	Ar–OCS	Kr–OCS
R''/Å	3.5379	3.7029	3.8070
θ''/deg	70.59	73.53	74.44
R'/Å	3.5410	3.7025	3.7810
θ'/deg	70.78	73.54	74.28
$\Delta\nu$/cm^{-1}	+0.1136	−0.4618	−0.8565

Table III.   Molecular constants for Ne–N$_2$O, Ar–N$_2$O,
Kr–N$_2$O and Xe–N$_2$O

	Ne–N$_2$O	Ar–N$_2$O	Kr–N$_2$O	Xe–N$_2$O
R''/Å	3.2448	3.4666	3.5926	3.7795
θ''/deg	82.31	82.92	83.02	82.96
R'/Å	3.2502	3.4661	3.5932	3.7809
θ'/deg	82.10	82.92	83.05	82.97
$\Delta\nu$/cm^{-1}	+0.3605	+0.1509	−0.1021	−0.4914

rare gases; a similar trend was also observed in the Rg–HCl dimers.   In
addition the band origins become progressively shifted to lower frequency
as the mass of the rare gas atom is increased.   These facts are discussed
in the following section.

4.   ANALYSIS OF SPECTROSCOPIC CONSTANTS

As noted in the previous section, there is very little change in the
effective structure of each of the Rg–N$_2$O or Rg–OCS complexes on vibrational excitation.   This may be in part due to the vibration excited
being almost orthogonal to the van der Waals bond.   However if one takes
the Rg–N$_2$O series, it will be noticed that there is an unusual band
origin shift, a hypsochromic shift being observed for the Ne and Ar
complexes but a bathochromic shift for the Kr and Xe species.   Can any
simple model explain the shifts in such a related series of molecules?
    Most models of long-range attraction [16,17] suggest that both the
dispersion and induction forces in these systems should scale as the
rare gas polarisability.   Thus the fractional change in the long-range
attraction coefficients should be the same for all complexes.
    Even though the vibration excited in N$_2$O or OCS is largely ortho-
gonal to the van der Waals bond, any repulsion between the rare gas and
atoms of the linear X–Y–Z molecule will be modulated by the vibration.
Since the amplitude of this motion increases on vibrational excitation,

any averaging over the supposed exponential atom-atom repulsions will
lead to an increase in the repulsive forces in the excited vibrational
state. As shown below, the importance of this effect is dependent upon
the repulsive strength parameter β.

4.1 Vibrational dependence of repulsion

Suppose the repulsion between two atoms is given by $A \exp(-\beta\rho)$ where ρ
is the instantaneous interatomic distance. If one atom (in the molecule)
now oscillates with some displacement x relative to some reference position R so that $\rho = R-x$, one can average over the vibration to give an
effective repulsive potential,

$$\langle A \exp-\beta\rho\rangle = \langle A \exp(-\beta(R-x))\rangle$$

$$= A \exp(-\beta R).\langle \exp+\beta x\rangle$$

$$\simeq A \exp(-\beta R)(1 + \beta\langle x\rangle + \beta^2\langle x^2\rangle/2)$$

where in the final line we have assumed small displacements x and expanded the exponential. The contribution from $\langle x\rangle$ reflects any change
in the average position of atoms due to monomer vibrational motion and
the term containing $\langle x^2\rangle$ reflects any change due to the amplitude of the
vibration within the dimer. Thus at a given intermolecular separation,
R, the ratio of the repulsive contributions to the potential in the two
vibrational states is given by

$$\frac{V_{rep}(v=1)}{V_{rep}(v=0)} = \frac{1 + \beta\langle x\rangle_1 + \frac{1}{2}\beta^2\langle x^2\rangle_1}{1 + \beta\langle x\rangle_o + \frac{1}{2}\beta^2\langle x^2\rangle_o}$$

where suffices 0 and 1 refer respectively to averaging in the ground and
excited states. Further expanding the quotient

$$V_{rep}(v=1)/V_{rep}(v=0) \simeq 1 + \beta[\langle x\rangle_1 - \langle x\rangle_o] + \beta^2[\langle x^2\rangle_1 - \langle x^2\rangle_o]/2$$

For a near harmonic motion $\langle x\rangle_{1,o}$ are approximately zero, if the reference configuration is taken as the equilibrium position of the nuclei.
Thus the change in repulsion depends almost completely upon the change
in the mean squared amplitude of the vibration and the repulsive strength
parameter β. The former term is essentially the same in all complexes
while β will depend upon the particular rare gas atom involved.

4.2 Model for band origin shift

The previous two ideas can be incorporated in a simple one-dimensional
model for the intermolecular interaction which in turn can explain the
shift in monomer frequency in the complex. A suitable form is the exponential $- R^{-6}$ Buckingham potential [17] which incorporates the correct
limiting long-range and short-range behaviour. Writing it in the form

$$V(R) = A \exp(-\beta R) - C_6/R^6$$

the well depth is given by

$$\varepsilon = (C_6/R_m^6)/(1 - 6/\beta R_m)$$

where R_m is the equilibrium intermolecular separation. If we now assume a small change ΔA and ΔC_6 in the repulsive and attractive coefficients, the change in the position of the radial minimum is given by, to first order, by

$$\Delta R_m = (\Delta A/A - \Delta C_6/C_6)/(\beta - 7/R_m)$$

and the change in the well depth is given by

$$\Delta\varepsilon = (C_6/R_m^6)(\Delta C_6/C_6 - 6\Delta A/\beta R_m A)$$

or

$$(\Delta\varepsilon/\varepsilon) = [\beta R_m/(\beta R_m - 6)](\Delta C_6/C_6 - 6\Delta A/\beta R_m A)$$

Thus the fractional change in the well depth is a fine balance between the changes in the attractive and repulsive contributions. However as explained above, for a given series of complexes (e.g. Rg-N$_2$O) it is to be expected that $\Delta C_6/C_6$ will be approximately independent of rare gas atom. Hence it is more appropriate to write the above expression in the form

$$(1 - 6/\beta R_m)(\Delta\varepsilon/\varepsilon) = \Delta C_6/C_6 - 6\Delta A/\beta R_m A$$

But from the previous section

$$\Delta A/A \simeq (\beta^2/2).\Delta x^2$$

In addition, if the zero point energy of a given complex is much less than the well depth, the observed band origin shift, which depends upon the difference in binding energies of ground and excited-state molecules, will be dominated by the change in well depths between the two states; in this approximation it is assumed that the change in zero point energy due to the van der Waals vibrations is negligible, which is probably not strictly valid for the Rg-HCl complexes. It is then reasonable to replace $\Delta\varepsilon$ by $-\Delta\nu$ (note the change of sign). Thus

$$(1 - 6/\beta R_m)(\Delta\nu/\varepsilon) = (3\beta/R_m).\Delta x^2 - \Delta C_6/C_6 \qquad (1)$$

The left hand side of the expression, which represents a scaled shift in frequency, should then be a linear function of $(3\beta/R_m)$ and such a graph should have an intercept of $\Delta C_6/C_6$.

Although potential surfaces have been determined for the $Rg-N_2O$ and $Rg-OCS$ systems [18], so that quite accurate values of ε are available, the values of β cannot be considered very reliable. Instead we present a model which appears to have good predictive powers.

4.3 Model for repulsive forces

The short-range repulsive forces between molecules have always been attributed to the overlap of the electron distributions of the interacting molecules [16]. It has also been shown [19] that the exponential repulsive parameter β depends quantitatively upon the tails of the wavefunctions of the interacting molecules. However from quantum defect theory [20] the asymptotic form of the electronic wavefunction depends upon the ionisation energy of the molecule and takes the form

$$\psi \propto \exp(-r/a_o n_{eff})$$

where a_o is the Bohr radius and n_{eff} is an effective principal quantum number such that $n^2_{eff} = R/I$, Rydberg constant/ionisation energy. From the overlap of the wavefunctions, the asymptotic form of the repulsion is $A \exp(-\beta R)$ where

$$\beta = (1/n_{1,eff} + 1/n_{2,eff})/a_o \qquad (2)$$

As a consequence, a knowledge of the ionisation energies of the interacting species provides a prescription for calculating β for each system. Table IV provides the information for the systems discussed in this section.

4.4 Quantitative relationship for the band origin shift as a function of rare gas atom

The expression for the scaled band origin shift shown in equation (1) is tested for the $Rg-N_2O$ series where the widest range of rare gas atoms is available. $\Delta\nu$ and R_m are taken directly from the spectroscopic data. ε for each system is taken from model potential surfaces [18] and is expected to be accurate to better than ± 10%. The repulsive parameter is estimated from expression (2) with values derived in Table IV. These results are illustrated in Figure 6. A very near linear relationship between $(1 - 6/\beta R_m)(\Delta\nu/\varepsilon)$ and (β/R_m) is observed. The intercept yields a value of $\Delta C_6/C_6$ of 2.3% and the slope indicates a value of $\Delta<x^2>$ of 0.007 Å^2.

In all the complexes of rare gases studied to date, the band origin always shifts to lower frequency as the rare gas series is descended. This may be in part due to the increase in the long-range attraction as the mass of the rare gas atom is increased. However attempts to model these shifts solely in terms of changes in attraction have met with only partial success [9]. Although the shifts always correlate well with the rare gas polarisability, α_{Rg}, and α_{Rg}/R^6, the expected form of

TABLE IV. Molecular parameters required for
calculating model repulsive forces

Molecule	Ionisation potential/eV	$1/n_{eff}$
N_2O	12.894	0.9735
OCS	11.17	0.9061
Ne	21.559	1.2588
Ar	15.755	1.0761
Kr	13.996	1.0142
Xe	12.127	0.9441

the long-range attraction, the relationship is non-linear and the non-zero intercept at zero polarisability is difficult to explain.

It is thus apparent that the repulsive forces are very important in all these systems. While the spectroscopic data are all compatible with a constant fractional change in the long-range attraction, the fine details of the frequency shifts depend upon the effects of repulsion, which can be very different. Neon is a very tightly bound, compact atom and repulsive interactions with it change rapidly with distance. As a consequence an oscillating atom is strongly repelled, so that on vibrational excitation there is a strong tendency for the intermolecular distance to increase. Xenon, on the other hand, is a much "softer" atom.

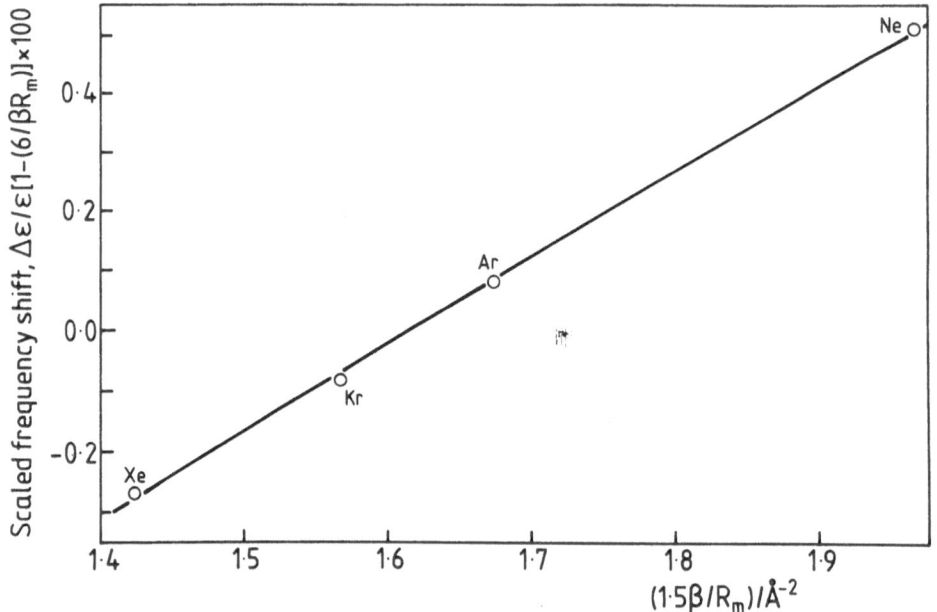

Figure 6. Correlation of the scaled band origin shift (see text) with the repulsive strength parameter in $Rg-N_2O$ complexes.

The (logarithm of the) repulsive forces change more slowly with distance and there is a much lower need for the intermolecular distance to increase. Returning to the expression for the band origin shift,

$$(\Delta\nu/\varepsilon) \; = \; [\beta R_m/(\beta R_m - 6)][(3\beta/R_m)\Delta x^2 - \Delta C_6/C_6]$$

several points can be made about attractive and repulsive contributions. First βR_m is almost independent of rare gas atom. For Rg-N_2O it is always close to 13.5 and the initial factor $[\beta R_m/(\beta R_m - 6)]$ takes a value close to 1.8. From the extrapolation of Figure 6, $\Delta C_6/C_6 \simeq 0.023$ so that without any change in repulsion $\Delta\nu/\varepsilon \simeq -4\%$. This would produce red-shifts of the order of 5 cm^{-1} for Ar and Kr. The fact that the observed shifts are very close to zero is a direct consequence of the repulsion. $(3\beta/R_m)$ varies from about 2.9 for Xe to about 3.9 for Ne, and with $\Delta x^2 \simeq 0.007$ Å^2, almost complete cancellation of the contribution from the $\Delta C_6/C_6$ term results.

Thus, in this limited range of complexes there are important changes in both the attractive and repulsive limbs of the intermolecular potential energy surface on exciting a vibration. There is generally just a minor change in the structure of the complex, with any increase in the intermolecular bond being greatest for neon complexes. The band origin shift also tends to be very small because of almost complete cancellation of the attractive and repulsive contributions, the final shift being a very sensitive balance between the two effects.

ACKNOWLEDGEMENT

This work has been supported by the SERC (UK) and the National Bureau of Standards (USA). I wish to thank Drs. A.S. Pine, G.D. Hayman, J. Hodge and Professors T.R. Dyke and J.S. Muenter for their contributions to this work.

5. REFERENCES

1. D.F. Smith, J. Mol. Spectrosc. 3, 473 (1959).
2. D.H. Rank, B.S. Rao and T.A. Wiggins, J. Chem. Phys. 37, 2511 (1962).
3. D.H. Rank, P. Sitaram, W.A. Glickman and T.A. Wiggins, J. Chem. Phys. 39, 2673 (1963).
4. B.J. Howard and A.S. Pine, Chem. Phys. Letts. 122, 1 (1985).
5. A.R.W. McKellar and H.L. Welsh, Can. J. Phys. 52, 1082 (1974).
6. A.R.W. McKellar, Far. Disc. Chem. Soc. 73, 89 (1982).
7. A.C. Legon, Ann. Rev. Phys. Chem. 34, 273 (1983).
8. D.H. Levy, Adv. Chem. Phys. 47, 323 (1981) ; Ann. Rev. Phys. Chem. 31, 197 (1980).
9. G.D. Hayman, J. Hodge, B.J. Howard, J.S. Muenter and T.R. Dyke, Chem. Phys. Letts. 118, 12 (1985).
10. R.E. Miller and R.O. Watts, Chem. Phys. Letts. 105, 409 (1984).
11. C.M. Lovejoy, M.D. Schuder and D.J. Nesbitt, J.Chem. Phys. (to be published, 1986).

12. D.E. Stogryn and J.O. Hirschfelder, J. Chem. Phys. 31, 1531 (1959).
13. J.M. Hutson and B.J. Howard, Mol. Phys. 45, 769 (1982).
14. G.C. Maitland and E.B. Smith, Chem. Phys. Letts. 22, 443 (1973).
15. S.E. Novick, P.B. Davies, S.J. Harris and W. Klemperer, J. Chem. Phys. 59, 2273 (1973).
16. H. Margenau and N.R. Kestner, Theory of Intermolecular Forces (Pergamin Press, Oxford, 1971).
17. G.C. Maitland, M. Rigby, E.B. Smith and W.A. Wakeham, Intermolecular Forces (Clarendon Press, Oxford, 1981).
18. A.M. Hough and B.J. Howard, Mol. Phys. (to be published).
19. P.R. Gellert, Part II Thesis (Oxford, 1984).
20. M.J. Seaton, Rept. Prog. Phys. 46, 167 (1983).

VIBRATION-ROTATION SPECTROSCOPY OF ArHCl BY FAR-INFRARED LASER AND MICROWAVE/FAR-INFRARED LASER DOUBLE RESONANCE SPECTROSCOPY

Ruth L. Robinson, Douglas Ray, Dz.-Hung Gwo, and
Richard J. Saykally
Department of Chemistry and Materials and Molecular
Research Division
University of California and Lawrence Berkeley Laboratories
Berkeley, CA 94720

ABSTRACT. Direct absorption by the low frequency van der Waals vibrations in ArHCl has been observed using the technique of intra-cavity far infrared laser spectroscopy and more recently with micro-wave far-infrared double resonance. Constants are reported for the Π bending state of $ArH^{37}Cl$. These constants favor, though not conclusively, a potential surface with a double minimum. Further work on other vibrational bands of ArHCl using double resonance methods will elucidate the nature of the surface directly.

INTRODUCTION

A great deal of effort has recently gone into the determination of anisotropic intermolecular potential energy surfaces. The status of this endeavor is revealed in recent work by Hutson and Howard,[1-3] and Hutson[4,5] on the prototypical binary van der Waals complex ArHCl. Data from a variety of sources, including virial coefficients, spectral line broadening, molecular beam scattering, and rotational spectroscopy, have been fit to empirical potential surfaces. Observables calculated from these anisotropic surfaces have been compared with available experimental results and the agreement has generally been impressive. In particular, two anisotropic surfaces, labelled M3 and M5, have produced equally good results. However, these two surfaces differ in a very fundamental regard; the M3 surface has only a single minimum at the linear ArHCl configuration while the M5 surface also has a secondary minimum at the more weakly bound linear ArClH geometry. The experimental observables currently available are not sensitive to even such gross differences in the anisotropy of the potential.

The nature of rovibrational eigenstates calculated from these two surfaces are quite different, however. These differences do not manifest themselves in properties which intrinsically require averaging over large angular regions, but they are quite apparent in spectroscopic observables, such as the molecular dipole moments, hyperfine constants, Coriolis coupling parameters, and in the arrangement of

85

A. Weber (ed.), Structure and Dynamics of Weakly Bound Molecular Complexes, 85–92.
© *1987 by D. Reidel Publishing Company.*

energy levels. The relative energies of Σ and Π vibrational states
are particularly sensitive to the details of the potential anisotropy,
since the nodal characters of their angular wavefunctions are so different.

Transitions among the rovibronic eigenstates of the van der Waals
potential are thus the most sensitive probe of its anisotropy. However, such spectra occur in the far infrared region of the spectrum
($10-350$ cm^{-1}), which is a very difficult region to work in. Consequently, very little has been achieved in the study of vibrations of
van der Waals bonds. Low resolution studies made at high pressures
and low temperatures, exemplified in the work of Boom and van der
Elsken,[6] constitute the essence of these studies. The observed
spectra exhibit hot bands, overtones, combination bands, and difference bands, all badly overlapped and without rotational resolution. The information content of such spectra is clearly limited by
the inability to extract any of its details.

We have developed a new far infrared laser experiment which em-
ploys an intracavity supersonic free jet to generate and cool van der
Waals complexes and which utilizes the Stark effect to tune rovibrational transitions into coincidence with the fixed laser frequency.
The design and operation of this experiment and its initial employment
for measurement of the lowest perpendicular bending vibration of
$ArH^{35}Cl$ have been described.[7] More recent experiments have been
directed at the assignment of the concommitant ^{37}Cl spectra, the
extension of both sets of data to higher rotational states, and the
measurement of the lowest stretching state and the lowest Σ bend. All
of these efforts have been impeded by difficulties in assignment of
the FIR laser Stark spectra. We have recently incorporated the capabilities for microwave-FIR double resonance measurements into this
spectrometer in order to take advantage of the wealth of microwave
spectroscopic data that exist for van der Waals complexes.[8] In the
rest of this paper we report the results of the analysis of the Π bend
of $ArH^{37}Cl$, describe the recent experimental modifications, and present some preliminary results obtained with the new double resonance
system.

EXPERIMENTAL

A schematic diagram of the apparatus is shown in Figure 1. A far-
infrared laser is pumped by a 3-meter cw carbon dioxide laser with
output powers of 50-120 watts on a single line selected with a
grating. The far infrared laser is divided into two regions by a
polypropylene beam splitter placed at Brewster's angle. The trans-
versely pumped gain region contains one of about sixty different small
molecules, e.g. methanol, at a pressure of 20-200 mTorr. The interaction region in the other half of the laser cavity is where the spectroscopy occurs. A supersonic free jet is formed by a 1/4" Pyrex tube
drawn out to a 50 micron diameter orifice. This nozzle is mounted
perpendicular to the laser mode on a positioner which allows the
nozzle to be moved in the plane perpendicular to the laser. Typically

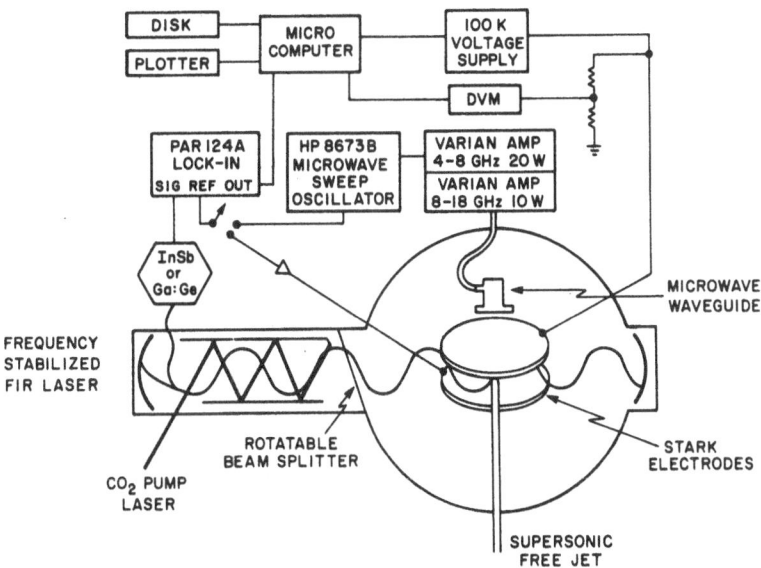

INTRACAVITY FAR-INFRARED LASER STARK SPECTROMETER

Figure 1. Schematic of the intracavity far-infrared laser/microwave
double resonance Stark spectrometer.

the nozzle is operated with a backing pressure of five atmospheres of
a premixed sample of 2% HCl (Matheson technical grade) in argon (House
grade), directed into the 24 inch diameter vacuum chamber which is
pumped by a 10 inch diffusion pump. The resulting chamber pressure is
about one millitorr and the Mach disk length is 6 cm.
    The far-infrared laser is discretely tunable in frequency, i.e.
by choosing from sixty different molecules for the laser and the different carbon dioxide pump laser lines, over 2,000 different laser
transitions are potentially available throughout the far-infrared
region, but these lines are not individually tunable. Tunability is
accomplished via the Stark effect on the molecule under investigation. To accomplish this, two aluminum electrodes, 20 inches diameter
and 1/2 inch thick are held at a 5 cm separation by precision Macor
spacers on either side of the area of interaction between the far
infrared laser and the free jet. Up to 100 kvolts has been applied to
one of these plates with a high voltage power supply (Glassman, Model
PG 200N2.5). When single resonance measurements are being made, modulation is accomplished by transforming the reference output of a lock-
in amplifier to an amplitude of 50-1200 volts peak-to-peak at a frequency of 70 kHz and applying it to the other electrode.
    A fraction of the far infrared laser power is reflected out of

the cavity with a 1/4 inch diameter mirror to an InSb detector operated at 4°K. The output of the detector is demodulated by a lock-in
amplifier and displayed as a function of electric field.

The limiting linewidths of single resonance spectra have different sources depending upon the particular experimental conditions. With the nozzle sufficiently removed from the far infrared
laser beam (~2 cm or more) we observe Doppler limited linewidths of
1.3 MHz HWHM. Under favorable conditions (high power laser, appropriate modulation depth) inverse Lamb dips are observed in our spectra[7] which exhibit a width of approximately .2 MHz HWHM.

When the nozzle is moved in closer to the FIR laser beam the
spectral linewidths and shapes change. The linewidth becomes broader
because of the electric field inhomogeneities induced by the presence
of the nozzle tip in the interaction region. In ArHCl, the quadrupole
structure becomes unresolvable. Peak positions change slightly due to
a combination of the perturbed electric field and alteration of the
effective cavity length of the FIR laser. The most dramatic effect is
that the intensity of a peak sometimes increases by two orders of magnitude. This has been determined to be partially an effect of putting
the laser on the sensitive part of its gain curve but the rest of the
effect must be a density effect since electric field effects have been
ruled out. When the laser is sampling the highest pressure region of
the jet, just at the nozzle, pressure broadening may be the dominant
source of spectral line broadening.

For the microwave/far infrared double resonance experiments a
microwave sweep oscillator (HP8673B) is amplified by a travelling wave
tube (Varian VZC6961K3 for frequencies of 4-8 GHz with 20 watts power
or a Varian VZM6990K1 for 8-18 GHz with 10 watts). The microwaves are
introduced from one of various sized horns placed as close to the
interaction region as possible without causing electric field breakdown. The method of modulation can be the Stark modulation, as in the
single resonance experiments, or microwave amplitude or frequency
modulation. Stark modulation is to be preferred at some microwave
frequencies where microwave pickup is a problem. Otherwise microwave
modulation is better because there is no single resonance background
signal in this case. Scans have been taken both at a fixed microwave
frequency while scanning the electric field and with frequency swept
at a fixed field.

FIR RESULTS

Using single resonance FIR laser spectroscopy, significant progress
has been recently made in characterizing the Π bending state of
ArHCl. The analysis for the chlorine 35 isotope has been previously
reported.[7] We have recently fit 68 lines of the chlorine 37 isotope. The preliminary constants are reported in Table I, along with
theoretical results of Hutson[5] from the M3 and M5 potential surfaces. The values of the ℓ-type doubling constant q_ℓ (-50.9 MHz for
ArH^{35}Cl and -41.7 MHz for ArH^{37}Cl) agree much more closely with that
obtained for the ArH^{35}Cl M5 potential (-52.2 MHz) than with the M3

value (+94.1 MHz). This appears to provide support for the argument
that the ArHCl potential contains a secondary minimum. However,
Hutson[5] has pointed out that the difference in the sign of q_θ for the
M3 and M5 potentials is not directly related to the existence or absence of a secondary minimum but is a result of the coincidence that
by using the M3 potential, the Σ stretching state is calculated to be
at higher energy than the Π bending state whereas in the M5 potential
these positions are reversed. The other molecular constants for the Π
bending state do not strongly favor one potential surface over the
other. Therefore the support for the M5 potential using the Π bending
state alone is rather weak. Work in progress on the Π bend of both
isotopes involves attempts to include higher J states using double
resonance techniques in order to obtain reliable values for the centrifugal distortion parameters, which provide useful information on
the radial part of the potential surface.

Table I.

Π Bend constants of $ArH^{37}Cl$ compared with $ArH^{35}Cl^7$ and the M3 and M5
calculations for $ArH^{35}Cl$. Numbers in parentheses indicate one standard error.

	$ArH^{37}Cl$	$ArH^{35}Cl$	M3	M5
ν_o (MHz)	1018504.5(1.0)	1018731.20(31)	998160.	1052000.
B (MHz)	1659.95(36)	1696.70(22)	1686.5	1735.5
q_θ (MHz)	-41.7(1.0)	-50.90(35)	+94.1	-52.5
μ (D)	.25908(38)	.26026(11)		
eqQ_{aa} (MHz)	1.67(11.)	-6.5(4.1)		
$eqQ_{bb}-eqQ_{cc}$ (MHz)	-61.3(5.7)	-74.4(2.2)		
$\langle P_1(\cos\theta)\rangle$	.2337(4)	.2348(1)	.216	.236
$\langle P_2(\cos\theta)\rangle$	-.02(16)	.096(61)	-.1261	-.0974

The low frequency stretch and low frequency Σ bend are more sensitive probes of the existence of a secondary minimum in the potential
than the Π bend because they do not have angular modes at the linear
configuration. Spectral lines have been observed in the regions of
the predictions[5] of the M5 potential for both these bands and various
tests of these lines have shown them to be due to ArHCl. Present work
on these bands includes double resonance experiments which will be
discussed in the next section.

DOUBLE RESONANCE RESULTS

Double resonance has been incorporated into the apparatus as a tool to aid in the assignment of quantum numbers in systems, such as ArHCl, where the ground vibrational state energy levels have been fully analyzed by microwave spectroscopy and thus we can probe known transitions with microwaves and use the observation of a double resonance signal to confirm the assignment of a vibrational transition.

A double resonance spectrum observed for the Π bend of ArH^{35}Cl is shown in Figure 2. The microwave transition observed is J,M = 11←00 in the ground vibrational state, and the FIR transition is Q(1), M = 1←1. This transition is shown in the energy level diagram in Figure 3 The electric field was fixed at 3.33 kV/cm, for this spectrum, and the microwave frequency was swept. The same FIR transition is shown in the single resonance spectrum in Figure 4, taken with the same conditions as Figure 2 except on a five times less sensitive scale. This spectrum was taken to maximize sensitivity at the expense of resolution, thus the center four quadrupole transitions, labelled b, c, d, and e are unresolved. The quadrupole peaks probed at 3.33 kV/cm in the double resonance spectrum are labelled and correspond to peaks b, c and d, respectively. The presence of Lamb dips (not shown) confirms that the pure FIR transitions are saturated, however we have not yet observed saturation in the microwave transition.

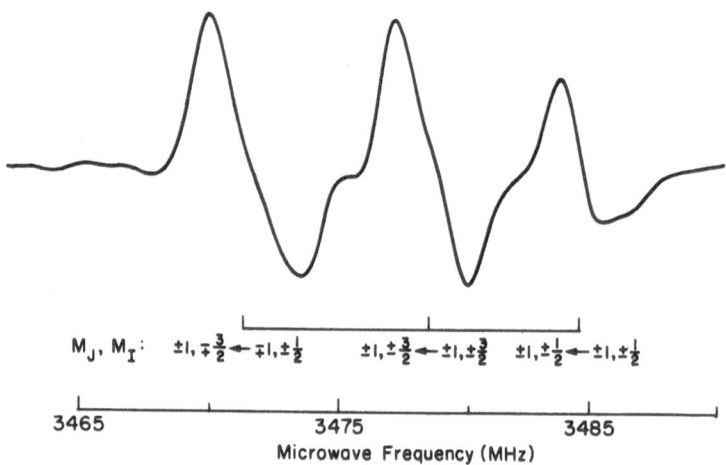

Figure 2. Double resonance spectrum. The far-infrared transition is Q(1) of the Π bend of ArHCl and the microwave transition is J,M = 11←00 in the ground state. The electric field was set at 3.33 kV/cm. Quadrupole structure for the ground state is unresolved. The labelled M_J, M_I transitions refer to quadrupole structure in the FIR transition.

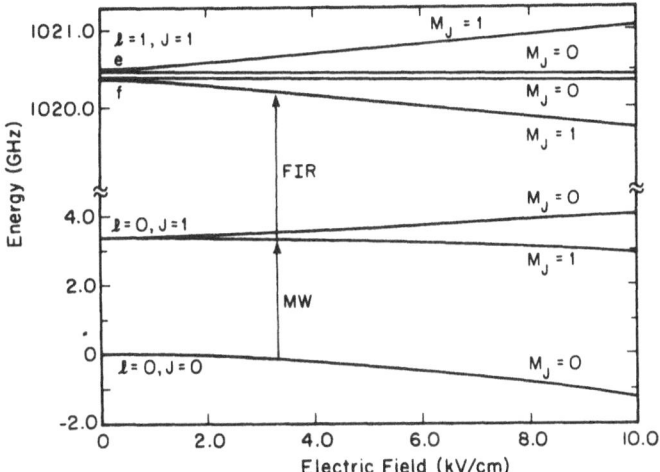

Figure 3. An energy level diagram of the π bend of ArHCl as a
function of electric field. The double resonance transition labelled
corresponds to the spectrum in Figure 2.

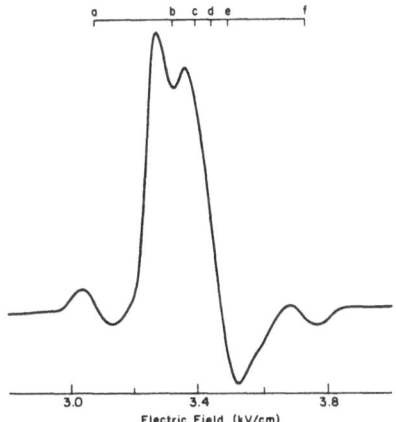

Figure 4. A single resonance spectrum of the same FIR transition as
shown in Figure 2 taken on a scale five times less sensitive. The
signal-to-noise ratio is approximately ten times greater for single
resonance. Quadrupole structure, though not completely resolved here,
is labelled as follows (M_J, M_I):

a $\pm 1, \mp -3/2 \leftarrow \pm 1, \mp 3/2$ d $\pm 1, \pm 1/2 \leftarrow \pm 1, \pm 1/2$

b $\pm 1, \mp -3/2 \leftarrow \pm 1, \mp 1/2$ e $\mp 1, \pm 1/2 \leftarrow \mp 1, \mp 3/2$

c $\pm 1, \pm 3/2 \leftarrow \pm 1, \pm 3/2$ f $\pm 1, \mp -1/2 \leftarrow \pm 1, \mp - 1/2$

Thus far the double resonance technique has been used to confirm assignments for $Q(1)$, $Q(2)$ and $R(0)$ lines in the Π bend of both $\mathrm{ArH}^{35}\mathrm{Cl}$ and $\mathrm{ArH}^{37}\mathrm{Cl}$. Present work on other bands involves scanning over all accessible ground state rotational transitions for each FIR transition observed in the predicted regions of the Σ stretch and Σ bend to determine the rotational assignments of these spectra.

CONCLUSION

The Π bending state has now been analyzed for both chlorine isotopes of ArHCl. The molecular constants favor a double well potential surface. The analysis of spectra tentatively ascribed to the Σ bend and Σ stretching states will determine the nature of the potential surface much more directly. The addition of double resonance techniques to our apparatus is expected to aid in the difficult analysis of these bands.

ACKNOWLEDGEMENTS

This work was supported by the Director, Office of Energy Research, Office of Basic Energy Sciences, Chemical Division of the U.S. Department of Energy under Contract No. DE-AC03-76SF00098.

REFERENCES

1. J.M. Hutson and B.J. Howard, Molec. Phys. **41**, 1123 (1980).
2. J.M. Hutson and B.J. Howard, Molec. Phys. **43**, 493 (1981).
3. J.M. Hutson and B.J. Howard, Molec. Phys. **45**, 769 (1982).
4. J.M. Hutson, J. Chem. Phys. **81**, 2357 (1984).
5. J.M. Hutson, J. Chem. Soc., Faraday Trans. 2, **84**, 1163 (1986).
6. E.W. Boom and J. van der Elsken, J. Chem. Phys. **73**, 15 (1980).
7. D. Ray, R.L. Robinson, D.-H. Gwo, and R.J. Saykally, J. Chem. Phys. **84**, 1171 (1986).
8. See, for example, the following review articles: A.C. Legon, Ann. Rev. Phys. Chem. **34**, 275 (1983); T.R.Dyke, Top. Curr. Chem. **120**, 85 (1984); W. Klemperer, J. Mol. Struct. **59**, 161 (1980).
9. S.E. Novick, P. Davies, S.J. Harris, and W. Klemperer, J. Chem. Phys. **59**, 2273 (1973) and S.E. Novick K.C. Janda, S.L. Holmgren, M. Waldman, and W.Klemperer, J. Chem. Phys. **65**, 1114 (1976).

VIBRATIONAL ANOMALIES AND DYNAMIC COUPLING
IN HYDROGEN-BONDED VAN DER WAALS MOLECULES

A. S. Pine
Molecular Spectroscopy Division
National Bureau of Standards
Gaithersburg, Maryland 20899
United States

ABSTRACT. High-resolution infrared spectra of the hydrogen halide
dimers and the rare gas-hydrogen halide complexes have been recorded
under thermal equilibrium conditions in a long path coolable White cell
using a tunable difference-frequency laser. Detailed and comprehensive
structural and dynamical information has been obtained from the fully
resolved rotational transitions between the ground and vibrationally
excited levels. Empirical potential energy surfaces have been
determined to explain dynamical phenomena such as rotational and
vibrational predissociation, vibrational shifts and large-amplitude
librational and tunneling motions. A number of these features, strongly
influenced by the large excursions of the light hydrogen atoms involved
in the van der Waals bond, are quite anomalous and require close
attention to dynamic coupling.

1. INTRODUCTION

In the past few years we have recorded the high-resolution infrared
absorption spectra of the hydrogenic stretching bands of the HF and DF
dimers [1,2], the HCl dimer [3] and the rare gas-HCl [4,5] and rare gas-
HF [6] complexes using a tunable difference-frequency laser system. The
molecules were studied under thermal equilibrium conditions, just above
the condensation temperatures of the hydrogen halides (T~210 K for HF
and T~130 K for HCl) in a long path (L~64 to 88 m) coolable White cell
at pressures of a few Torr to ensure that pressure broadening would be
less than the Doppler width of the molecules. Since the laser linewidth
(FWHM~1 MHz) was less than 1% of the Doppler width, the spectral
resolution was limited by the molecules with virtually no instrumental
distortion.
    The rotational levels and the low frequency (van der Waals)
vibrational modes of these complexes are highly excited at these
equilibrium temperatures, sampling most of the intermolecular potential
surfaces. Thus our thermal infrared spectra greatly supplement the
earlier molecular-beam microwave and radio frequency studies [7-11]
carried out for the ground state vibrations at low effective
temperatures (T<10 K). Those precision microwave measurements of the
frequencies of the lower rotational levels provided definitive
structures, useful information about the intermolecular forces and
anisotropies and the large-amplitude bending and tunneling motions. The
microwave results help to clarify and verify the more extensive infrared
spectra which yield the vibrational dependence of these quantities plus
a few other dynamic phenomena such as rotational and vibrational

93

A. Weber (ed.), Structure and Dynamics of Weakly Bound Molecular Complexes, 93–105.

predissociation and intermode coupling. Prior infrared studies [12-16]
of these complexes did not fully resolve the rotational fine structure
within the observed vibrational band contours, so the interpretations
were inconclusive. More recently, tunable infrared lasers have been
used to probe some of these complexes generated in adiabatically-cooled
molecular beams [17-20]. The supersonic jets greatly simplify the
spectrum and analysis of the complex by reducing the overlapping "hot"
bands and the contaminant spectra from the parent molecules or sibling
species and by permitting sub-Doppler resolution, with suitable jet,
probe and detector configurations, in a collision-free environment.
These molecular-beam infrared experiments nicely complement the thermal
equilibrium results by enhancing resolution and cleaning up the spectral
background at the cost of quenching the higher excitations which help to
characterize the potential surfaces.

The particular series of molecules discussed here provides a
systematic variation of the effect of electronic structure and atomic
mass on weak van der Waals interactions with binding energies ranging
from ~100 cm^{-1} to ~1000 cm^{-1}. Accurate binding energies have been
deduced from the observation of rotational predissociation in the rare
gas-hydrogen halide (Rg-HX) complexes [4-6]. For the HX-HX dimers the
binding energies are obtained from absolute infrared absorption
intensities [21], independent of a concentration measurement, assuming a
minimally-perturbed transition moment for the outer HX unit. The red
shifts of the high frequency vibrational modes of the complex relative
to the isolated monomers provide a direct measure of the increase of
binding energies upon vibrational excitation. The vibrational
dependence of the rotational constants indicates that the van der Waals
bond lengths generally contract upon excitation of the HX stretch for
all of the complexes except the Rg-HCl which exhibit a small expansion.
Vibrational predissociation has been observed for the "bound-H" stretch
of the HF dimer as an excess broadening of the transitions above the
expected Doppler and pressure widths. No excess broadening was resolved
for the "free-H" stretch of the HF dimer, or for any of the other
complexes, indicating lifetimes at least an order-of-magnitude longer
than the "bound-H". The tunneling frequency of the HX dimers between
equivalent conformations exhibits a marked decrease upon vibrational
excitation implying a substantial increase in the tunneling barrier.

These phenomena have not been explained quantitatively in terms of
the known electrostatic multipolar moments and polarizabilities of the
constituent molecules or atoms. Here we will discuss several dynamic
contributions which are enhanced due to the large-amplitude hydrogenic
motions in these H-bonded species. In particular we will examine the
anomalously large red shifts in the Rg-HX complexes, the vibrational
dependence of the tunneling in the HF dimer and the mode-specific
vibrational predissociation in the HF dimer.

2. RARE GAS-HYDROGEN HALIDE COMPLEXES

As an example of the type of data recorded with the difference-frequency
laser on a cooled sample, we show in Figure 1 the spectrum of Ar-HF in
the region just below the band center of the isolated HF monomer. The
experimental conditions are given on the figure. The spectrum is
actually dominated by the $^rP_0$ subbranch of the "free-H" stretching band
of the HF dimer which gives rise to the strong tunneling doublets of
alternating intensity [1]. Most of the other lines appear when Ar is
added. They are attributed to the ν_1 stretching fundamental of Ar-HF
and an accompanying hot band originating in the van der Waals stretch,

ν_3, as indicated by the simulated spectra below the observed trace. The relative strengths of the spectra of the two species reflect the shift in equilibrium concentration towards the HF dimer due to its much greater binding energy [21], countering the low HF to Ar partial pressures. However, one cannot arbitrarily raise the Ar pressure to enhance the Ar-HF species without increasing the pressure broadening (already comparable to the Doppler width here) which degrades the resolution. On the other hand, decreasing the HF pressure just reduces the signal-to-noise ratio at the available path lengths.

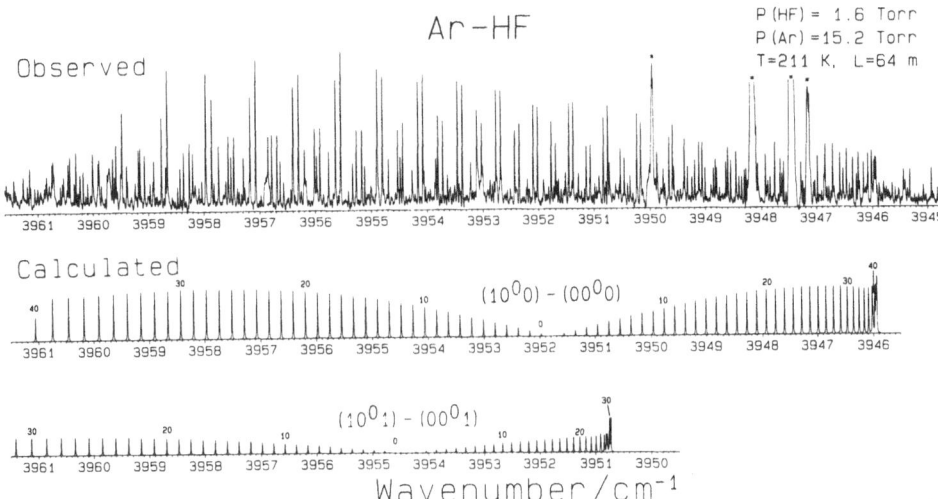

Figure 1. Observed and calculated spectra of Ar-HF for the (10^00)-(00^00) fundamental and (10^01)-(00^01) hot bands. The labels specify J" for the transitions. Simulated lineshapes are Voigt profiles with a Lorentzian half width of 100 MHz. The strong doublets of alternating intensity arise from the HF dimer $^rP_0$ subband of the "free-H" stretching vibration. The asterisks identify atmospheric water vapor lines.

The Ar-HF bands exhibit P-branch band heads indicative of a vibrational increase in the rotational constant B corresponding to a contraction of the van der Waals bond length when the HF stretch is excited. The ν_1 band center of Ar-HF is red shifted ~10 cm^{-1} from the band center of the isolated HF monomer (ν_0=3961.4229 cm^{-1}) implying an increased van der Waals attraction upon HF excitation or, equivalently, a decrease in the HF bonding upon complexation. The $\nu_1+\nu_3'-\nu_3$ hot band is blue shifted back towards the monomer as excitation of the van der Waals mode anharmonically weakens the intermolecular bond, thereby strengthening the intramolecular bond. Primed quantities refer to the v_1=1 level.

For the fundamental band of Ar-HF, the series can be followed out to J=40 where it abruptly disappears before the Boltzmann population would normally be depleted. This is a consequence of rotational predissociation as the molecular constituents can tunnel through the centrifugal barrier when the rotational energy exceeds the van der Waals

binding energy. For Ar-HF all the rotational levels for J>32 are
metastable, with tunneling rates that increase exponentially with J,
until J=40 where the lifetime broadening is comparable to the Doppler
and pressure broadening. The next higher J transitions are 5 to 10
times broader and are simply lost in the background. This rotational
predissociation cutoff is even more evident in Ar-HCl [4] where it
occurs at J=60 due to a smaller B value and a slightly higher
dissociation energy.

The spectral transitions in the ν_1 band are least-squares fit to a
power series in J(J+1) (with coefficients ν_v, B_v, $-D_v$, H_v, L_v and P_v)
for the v_1=0 and 1 levels [6]. Such high-order centrifugal distortion
constants are required in order to fit the data within experimental
precision. They signal a breakdown in convergence of the usual power
series expansion and are indicative of the extreme anharmonicity
encountered for rotational levels approaching dissociation. The J
assignments for ν_1 are made with regard to the known microwave B_0 and D_0
constants [10] which were included in the fit with 10^8 higher weight
than the infrared measurements.

Since the data in Figure 1 sample the van der Waals potential from
the zero-point level up to dissociation, it is possible to test or
refine existing potential models. Fitting to a physical potential model
might also decrease the number of parameters required to specify the
data and might have some predictive capabilities for the van der Waals
stretching modes, permitting the hot bands to be assigned in the absence
of microwave data for the lower level. For this we have used a model
radial Maitland-Smith vibrational potential (a modified Lennard-Jones
with 4 adjustable parameters) and the usual centrifugal potential (no
adjustable parameters) to fit the rotational energy levels in the v_1=0
and 1 states [4-6]. The rotational levels may be obtained directly from
the data by combination differences or, as done here, may be calculated
from the fitted rotational and distortion constants to smooth the data.
Here we are neglecting any coupling of radial and angular coordinates
and are averaging over the zero-point large-amplitude bending
(librational) motion.

The resulting upper and lower vibrational potentials for Ar-HF are
shown in Figure 2. The vibrational shift, ν_1-ν_0, from the monomer is
seen to be the difference in dissociation or binding energies, D_0-D_0',
in the two vibrational levels. Since the radial coordinate is the
center-of-mass spacing between the Ar and HF components, the vibrational
eigenvalues correspond to the van der Waals stretching mode, ν_3. This
enables us to predict the ν_3 ladders rather accurately and thereby
assign the ν_1+ν_3'-ν_3 hot bands as shown in Figure 1.

The results for the measurements and analysis of the rare gas-
hydrogen halide spectra are summarized in Table I. The vibrational
shifts, ν_1-ν_0, from the monomer are measured directly and are accurate
to better than 0.001 cm^{-1}. The zero-point dissociation energies, D_0,
and the van der Waals stretching frequencies, ν_3, both in the ν_1 ground
state are somewhat model dependent with estimated uncertainy of ~1% and
~3% respecively. Of course, their differences with the excited state
values, D_0' and ν_3', are just the red shifts for the fundamentals and
the subsequent blue shifts for the hot bands, which are known to much
higher precision. The $\Delta B=B_1-B_0$ values give a measure of the van der
Waals bond length change upon ν_1 excitation. Note that the Rg-HF
complexes contract and the Rg-HCl expand since $\Delta B_1/B_0$~-2$\Delta R_1/R_0$. The J_{rp}
are the observed (calculated for Xe-HF) rotational predissociation J
values.

Figure 2. Maitland-Smith
model effective radial
potentials for the van der
Waals coordinate in Ar-HF
for the ground and excited
H-F stretching vibrations.

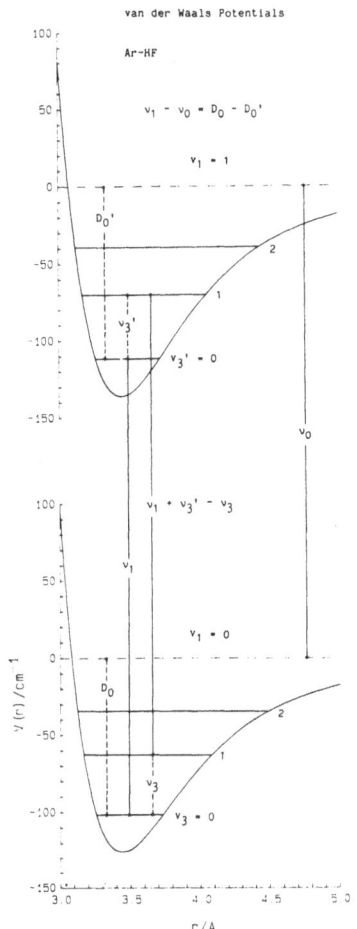

The red shifts for the Rg-HF complexes are considerably larger than for the corresponding Rg-HCl, exhibiting the same trend as the ΔR_1 values towards greater van der Waals binding as the HX vibration is excited. Both the red shifts and the binding energies, D_0, scale with the polarizability of the rare gas atom as expected. The surprisingly smaller binding, or zero-point dissociation, energies of the Rg-HF complexes compared to the corresponding Rg-HCl, arise principally from the larger zero-point energies for the van der Waals bending and stretching modes in the lighter HF complexes. The equilibrium van der Waals wells are actually deeper for Rg-HF than for Rg-HCl, in keeping with the greater Lewis acidity of HF.

Table I. Comparison of rare gas-hydrogen halide complexes.

	$\nu_1 - \nu_0$ (cm^{-1})	D_0 (cm^{-1})	ν_3 (cm^{-1})	$\Delta B_1/B_0$ %	J_{rp}
^{40}Ar-HF	-9.6538	101.7	39.1	0.346	40
^{84}Kr-HF	-17.5179	133.0	41.1	1.004	54
^{129}Xe-HF	-29.1852	180.9	43.4	1.753	68
^{40}Ar-HCl	-1.7690	114.7	31.3	-0.929	60
^{84}Kr-HCl	-5.2314	154.0	31.5	-0.673	
^{129}Xe-HCl	-12.0805	206.1	33.0	-0.123	
				$-\Delta R_1/R_0$	

Not shown or discussed here are the observations of the $\nu_1 + \nu_2$ stretch-bend combination bands in Ar-HCl [4,5] and Ar-HF [6] which provide considerable information on the frequency and amplitude of the librational motion, on the anisotropy of the potentials and on the Coriolis coupling of the angular-radial coordinates.

2.1. Red Shifts in Rg-HX Complexes

As seen in Table I, the red shifts, $\nu_1 - \nu_0$, for these atom-diatom complexes are anomalously large, amounting to 10 to 15% of the binding energies, D_0, for the Rg-HF complexes. In this section we would like to discuss the possible mechanisms responsible for this large increase in binding energy with vibrational excitation. We list several contributing factors in Table II which can be readily estimated for Ar-HF from known monomer properties to compare with measurement. The most obvious of these is the change in the dipole-induced dipole energy, ΔE_{did}, which can be estimated from Stark splitting measurements of the permanent dipole moments in v=0 [22] and v=1 [23] and the infrared fundamental band transition moment [24] of HF. From the known polarizability of argon [25] and the librationally-averaged $<P_2(\cos\theta)>$ from microwave hyperfine splittings [10] and with $R=R_0$, this contribution is only ~11% of the measured red shift.

The next higher electrostatic interaction term from the quadrupole-induced dipole energy, ΔE_{qid}, has not been measured directly. However, we can use the anharmonic change in the quadrupole moment of HF calculated by ab initio methods [26] and obtain another small contribution of ~13% to the red shift. Since the quadrupolar contribution is greater than the dipolar, it appears that the multipolar expansion is non-convergent and that much higher moments need to be considered. However, the relative magnitudes of these two terms depend on the rather arbitrary definition of the origin at the center-of-mass of the HF bond. The ordering would be reversed, for example, if the origin were at the midpoint of the HF bond consistent with the ionic character of the chemical bond. The vibrational changes in the higher-order moments of HF have not been measured either and their electronic structure calculation is less reliable, so we cannot estimate their contribution.

Table II. Vibrational shift contributions

$\nu_1 - \nu_0(HX)$	measured for Ar-HF	$[-9.7 \text{ cm}^{-1}]$
$\Delta E_{did} = -\alpha_{Rg} \Delta(\mu_{HX}^2) (3\cos^2\theta+1)/2R^6$		$[-1.1 \text{ cm}^{-1}]$
$\quad \Delta(\mu_{HX}^2) = <1\|\mu^2\|1> - <0\|\mu^2\|0>$		
$\quad \cong <1\|\mu\|1>^2 \div <0\|\mu\|0>^2 + <1\|\mu\|0>^2$		
$\Delta E_{qid} = -6\,\alpha_{Rg}\,\Delta(\mu_{HX}\theta_{HX})\,\cos^3\theta/R^7$		$[-1.3 \text{ cm}^{-1}]$
$\Delta E_{disp} = -C\,\alpha_{Rg}\,\Delta(\alpha_{HX})/R^6$		$[-2.2 \text{ cm}^{-1}]$
$\quad \alpha_{HX} = [(\alpha_{\|}+2\alpha_{\perp})+(\alpha_{\|}-\alpha_{\perp})(3\cos^2\theta-1)]/3$		
$\Delta E_{lib} \cong \nu_2' - \nu_2 \cong (\Delta B_{HX}/B_{HX})\nu_2/2$		$[-1.3 \text{ cm}^{-1}]$
$\Delta E_{kin} \cong (\nu_3^2/2\nu_0)(m_{Rg}m_X/m_H m_{Tot})$		$[+2.4 \text{ cm}^{-1}]$
$\Delta E_{wall}^C = \partial^2 V/\partial R^2\|_{Re} (\Delta R)^2$		
$\quad = -n(n+1)D_e(\Delta R/R_e)^2$		$\sim[-8.4 \text{ cm}^{-1}]$
$\Delta E_{wall}^Q = \Delta^+ - \Delta^-$		$\sim[-5.4 \text{ cm}^{-1}]$

Since a major portion of the binding energy is due to dispersion forces, we need to include the vibrational change in the polarizability of HF. This is related to the Raman cross section and has been calculated by ab initio methods [26], so that we obtain an estimate for ΔE_{disp} of ~23% of $\nu_1-\nu_0$. Thus the change in induction and dispersion energies from the known vibrational dependence of the HX monomer properties can account for less than half the observed increase in binding energy. In these multipolar and polarizability terms, the small vibrational changes in the average bending angle, θ, and the center-of-mass spacing, R_0, of the complex resulting from the increased binding energy should also be included for self-consistency. ΔR is obtained from the ΔB value and $\Delta\theta$ for Ar-HF can be inferred from a recent Stark measurement of the excited state dipole moment [20]. These effects contribute less than ~10% to the red shifts in Ar-HF and yield comparable blue shifts in the Rg-HCl complexes which expand when excited due to the anharmonicity of the HX bond.

Another small contribution to the change in D_0 comes from the change in zero-point energy of the doubly-degenerate librational mode, ν_2, which we denote ΔE_{lib}. Though we have measured ν_2' from the $\nu_1+\nu_2'$ combination band [6], there are as yet no far-infrared measurements of ν_2. For Ar-HCl there is virtually no difference between ν_2' [4] and ν_2 [27,28]. However, we can estimate ΔE_{lib} (~13% of $\nu_1-\nu_0$) from the known ΔB_{HF} value of the monomer as given in Table II, assuming no vibrational variation in the anisotropy of the potential.

All of these small contributions to the red shift are substantially countered by a blue shift arising from the kinetic coupling in a conventional F-G matrix harmonic force field calculation. Assuming a linear molecule, this ΔE_{kin} given in Table II yields a positive shift of ~25% of $\nu_1-\nu_0$ for Ar-HF; it should be somewhat smaller for the dynamically bent molecule. The blue shift arises from the added restoring force of the van der Waals "spring" against the large-amplitude hydrogenic stretching motion. Had the molecule been halogen-bonded, the m_X/m_H factor would be inverted and the effect reduced substantially.

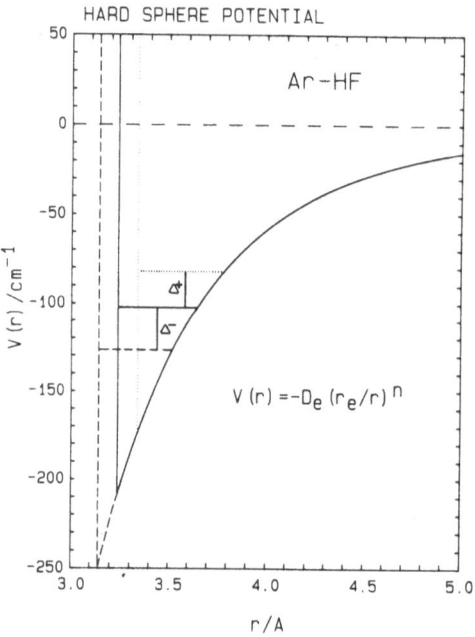

Figure 3. Hard-sphere radial potential illustrating coupling of the high and low frequency stretching modes via the repulsive wall.

Thus we are still left with most of the observed red shift to explain. We believe that the bulk of this effect can be ascribed to the repulsive wall motion against the very nonlinear van der Waals force. This is best illustrated by a hard sphere model such as shown in Figure 3. We have taken a potential of the form, $V(R) = -D_e/(R/R_e)^6$, where the solid line representing $v_1=0$ yields the observed dissociation energy, D_0, and rotational constant, B_0, for Ar-HF. The dashed and dotted positions of the wall correspond to the contraction and expansion

turning points of the hydrogen in the HF stretching motion, ignoring HF
bond anharmonicity. On the average, the contraction part of the cycle
yields increased bonding over the decreased bonding from the expansion.
Not shown are the effects of HF motion on the attractive portion of the
potential, but those are just the small terms mentioned above. This
dynamic repulsive wall effect must be properly averaged over the HF
vibration to obtain its contribution to red shift. Rough estimates of
the dynamic wall coupling may be obtained classically from the curvature
of the potential at the wall, denoted ΔE^C_{wall} in Table II, or quantum
mechanically by taking the difference of the zero-point levels at the
turning points, denoted ΔE^Q_{wall}. Both estimates yield the dominant
contributions to the red shifts, though they are both probably
overestimated by the artificial shape of the hard sphere potential.
However, the dynamic kinetic and wall coupling terms must be considered
when dealing with hydrogen-bonded systems.

3. HYDROGEN HALIDE DIMERS

The infrared spectra of the hydrogen halide dimers have been discussed
in detail elsewhere [1-3]. Here we will consider just two dynamic
phenomena--the vibrational dependence of the tunneling splittings and
the mode-specific vibrational predissociation. Neither of these effects
have been adequately explained. Both are strongly influenced by the
large-amplitude hydrogenic stretching motion and may possibly be
understood in terms of the dynamic coupling mechanisms discussed in the
previous section.

3.1. Dynamic Tunneling

The tunneling motion of the HF dimer was discovered in the original
microwave spectra of HF-HF and DF-DF [7]. There the tunneling frequency
for the K=0 subband of the ground vibrational state was found to be
~19.7 GHz for HF-HF and ~1.6 GHz for DF-DF, demonstrating a dramatic
isotopic variation for a doubling of the effective mass. The tunneling
is even more clearly evident in the infrared spectra of these species
[1,2] as a series of doublets with alternating intensities, as shown
also in Figure 1. The alternating intensities arise from the nuclear
spin statistical weights of the two HF components which are
indistinguishable under tunneling exchange.
    There are two high frequency HF stretching modes of the dimer,
denoted ν_1 for the outer or "free" H and ν_2 for the inner or "bound" H.
The ν_2 mode is further red shifted (~93.1 cm^{-1}) from the isolated HF
monomer than ν_1 (~30.5 cm^{-1}), consistent with the dynamic effects
previously discussed for hydrogen-bonded vibrations. The infrared
selection rules [27,28] establish that the observed tunneling subbands
are split by the difference of the tunneling frequencies in the ground
and ν_1 vibrations and the sum for ν_2. The measured tunneling splittings
in the excited ν_1 and ν_2 vibrations are reduced by a factor of ~3 from
that in the ground state for both the HF and DF dimers [1,2]. This
implies a substantially increased tunneling barrier for the excited
vibrations. There is also a significant K dependence of the splittings
due to centrifugal distortion or, alternatively, to a difference in the
effective A rotational constants in the two tunneling sublevels [2].
    The vibrational increase in the tunneling barrier and consequent
decrease in the splittings has been proposed to arise from the
difficulty in exchanging the localized vibrational excitation from one

monomer to the other during the tunneling [29]. Mills [30] has
formalized this mechanism as a breakdown in the Born-Oppenheimer
separation of the high and low frequency vibrations, though no
quantitative estimates were given. Phenomenologically, it is possible
to represent the tunneling effects by simple one dimensional quartic
potentials for the ν_1 and ν_2 levels as shown respectively by the upper
and lower solid curves in Figure 4. The ground state tunneling
potential is superimposed as the dashed curve shifted up by the monomer
frequency, ν_0. The wavenumbers are accurately scaled to reproduce the
observed tunneling splittings in the HF dimer, shown magnified at the
center of the figure, and are referenced at the zero-point level of ν_0
to represent the measured ν_1 and ν_2 red shifts. The tunneling

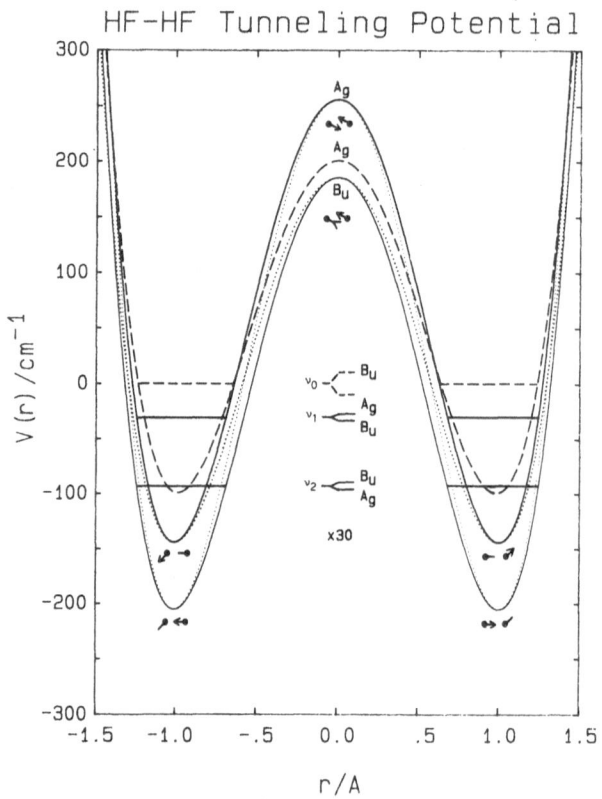

Figure 4. Phenomenological HF dimer tunneling potentials for the
excited "free" and "bound" H-F stretching vibrations (upper and lower
solid curves) scaled relative to the ground state potential (dashed
curve) offset by the HF monomer frequency.

coordinate is just the HF monomer bond length multiplied by the
difference in the angles of the two monomer axes relative to the center-
of-mass axis. The transition state at the r=0 barrier peak has C_{2h}
symmetry as determined from the observed selection rules [30,31] and
from chemical intuition. The A_g and B_u symmetries of the tunneling

levels are also unambiguously determined from the observed spin weights, rotational assignments, band types and selection rules. The peak-to-valley barrier heights for these model potentials are 300, 400 and 391 cm^{-1} for ν_0, ν_1, and ν_2 respectively [2]; the equilibrium angular displacement is taken to be 122° for all levels, ignoring ΔA effects.

At the transition state, we see from the empirical potentials in Figure 4 that the ν_2 potential (lower curve) is only ~14 cm^{-1} below the ν_0 (dashed curve); whereas ν_1 (upper) is ~56 cm^{-1} above ν_0. As discussed previously [2], these vibrational increases in the effective barriers are one to two orders-of-magnitude larger than those given by the anharmonic increase in dipole moment of HF or the transition moment contribution. However, the relative barrier peak positions are reasonably consistent with the kinetic contributions in a harmonic force field F-G matrix calculation if we take the symmetric and antisymmetric vibrational combinations as shown in Figure 4. The F-G matrix calculation yields a blue shift for the A_g symmetric mode and no shift at all for the B_u antisymmetric mode because the van der Waals "spring" is stretched and compressed by the symmetric mode but not the antisymmetric. The blue shift for the symmetric mode depends on the angular deviation from linearity where it can range up to $\sim(\nu_4{}^2/2\nu_0)(m_X/m_H)$ (which is ~100 cm^{-1} for FH-HF) where ν_4 is the van der Waals stretching frequency.

This ordering of the mode symmetries at the transition state appears inconsistent with the observed ν_1 and ν_2 tunneling symmetries unless the potentials correlate as the dotted curves in Figure 4. Unfortunately the dotted potentials would produce the wrong tunneling splittings--too small for ν_2 and too large for ν_1. This paradox can be resolved if we note that away from the transition state the molecule belongs to the C_s point group with both potentials of A' symmetry, giving rise to an avoided crossing resulting in the solid curves. This avoided crossing may also be responsible for a strong mixing of the mode character and, thereby, the exchange of localized excitation mentioned above.

It should be mentioned that the ordering of the mode symmetries at the transition state has not been definitely established by the above arguments. Only that the magnitude of the empirical blue shift at the barrier peaks can be accounted for by kinetic coupling and there is no inconsistency with observed selection rules and tunneling splittings. It is not clear at this time how other contributions, such as the dynamic wall coupling discussed in connection with the Rg-HX red shifts, would affect the transition state barrier. At the C_{2h} transition state, the A_g symmetric vibration is purely Raman active and the B_u antisymmetric vibration is purely infrared active. It is possible that the mode character at the transition state may be tested via combination bands of the high frequency HF stretches with the large-amplitude libration or incipient tunneling mode, ν_5'. These combination bands might be anomalously strong due to the large modulation of the transition moment orientation along the intermolecular axes, as for the stretch-bend combinations of the Rg-HX complexes [4,6].

3.2. Mode-specific Vibrational Predissociation

In our high-resolution spectra of the HF dimer [1,2], we observed that the linewidths of the transitions in the bound-H stretch, ν_2, were considerably broader than those for the free-H stretch, ν_1. After deconvolving the measured pressure broadening (of ~20 MHz/Torr), and

calculated Doppler widths (HW/e=114 MHz), we obtained a residual
Lorentzian half width of ~204±62 MHz for ν_2 of HF-HF which we attributed
to vibrational predissociation. For the ν_1 band there was no excess
broadening above the observed pressure broadening with an upper limit on
the uncertainty of ~10 MHz. Within this same uncertainty, no
vibrational predissociation broadening was resolved for either the ν_1 or
ν_2 bands of DF-DF [2] or HCl-HCl. [3]. Recently, using very-high-
resolution molecular beam techniques, Huang et al [19] have resolved a
finite vibrational predissociation half width of ~7±1 MHz for the $^rR_0^+(0)$
line of the ν_1 band of the HF-HF. Thus the ν_1 vibration lives about
30±10 times longer than ν_2 for the HF dimer.

This large difference in predissociation lifetimes for the two
vibrations was somewhat surprising. For a number of reasons, more
nearly equal lifetimes were expected for two modes so close in energy
(~63 cm^{-1} apart) compared to the large excess energy (~3000 cm^{-1}) above
dissociation. The usual statistical models of dissociation invoke a
rapid intramolecular vibrational equilibration, resulting in similar
lifetimes for all modes of similar excess energies [32]. Evidently the
density of vibrational levels is too low for the small HF dimer; and the
modes are too localized for the statistical models to apply. The
dipole-dipole van der Waals couplings, $\Delta\mu_1 \cdot \mu_2$ and $\mu_1 \cdot \Delta\mu_2$, are identical
for the two modes, and would also result in equal lifetimes. Again the
electrostatic coupling contributions are apparently of minor importance.
Energy gap theories [33] imply lifetimes controlled principally by the
excess energy, so would not adequately distinguish between the two
modes.

Following the initial observation of mode-specific vibrational
predissociation in the HF dimer [1,2], Halberstadt et al [34] performed
a close-coupling calculation on a simplified HF dimer model potential
and obtained a predissociation rate for the bound-H stretching mode in
reasonable agreement with experiment. However, they did not make
predictions for the free-H stretch which has now been measured [19].

We note that the dynamic kinetic and wall coupling mechanisms
discussed in the previous sections for hydrogen-bonded excitations may
provide a simple understanding of the differences in the predissociation
rates of the bound and free-H vibrations. These couplings are
proportional to the amplitude of the motion of the vibrating atoms
projected on the van der Waals coordinate. For such coupling, the ratio
of the predissociation rates for the ν_2 and ν_1 modes would be
~$(m_X/m_H)(\cos\theta_2/\cos\theta_1)$ where the angles are measured between the two HX
axes and the van der Waals axis. This ratio is ~30 for the HF dimer,
~15 for the DF dimer and much larger for the nearly orthogonal
configuration of the HCl dimer [3]. In fact, the substantial
centrifugal distortion of θ_1, limiting at 90°, observed for K rotation
about the a-axis of the HF dimer [2] should lead to a rapid decrease in
the ν_1 predissociation rate with K. The corresponding θ_2 for the ν_2
mode is close to zero for all accessible K levels and so would hardly
change. Clearly this dynamic coupling hypothesis for the mode-specific
vibrational predissociation is testable for the predicted isotopic and K
rotational variations, and we look forward to further ultra-high-
resolution molecular-beam studies of these interesting species.

REFERENCES

1. A.S. Pine and W.J. Lafferty, J. Chem. Phys. **78**, 2154 (1983).

2. A.S. Pine, W.J. Lafferty and B.J. Howard, J. Chem. Phys. **81**, 2939 (1984).
3. N. Ohashi and A.S. Pine, J. Chem. Phys. **81**, 73 (1984).
4. B.J. Howard and A.S. Pine, Chem. Phys. Lett. **122**, 1 (1985).
5. B.J. Howard and A.S. Pine, to be published.
6. G.T. Fraser and A.S. Pine, J. Chem. Phys. **85**, 2502 (1986).
7. T.R. Dyke, B.J. Howard and W. Klemperer, J. Chem. Phys. **56**, 2442 (1972).
8. S.E. Novick, P. Davies, S.J. Harris and W. Klemperer, J. Chem. Phys. **59**, 2273 (1973).
9. T.A. Dixon, C.H. Joyner, F.A. Baiocchi and W. Klemperer, J. Chem. Phys. **74**, 6539 (1981).
10. M.R. Keenan, L.W. Buxton, E.J. Campbell, A.C. Legon and W.H. Flygare, J. Chem. Phys. **74**, 2133 (1981).
11. F.A. Baiocchi, T.A. Dixon, C.H. Joyner and W. Klemperer, J. Chem. Phys. **75**, 2041 (1981).
12. D.F. Smith, J. Mol. Spectrosc. **3**, 473 (1959).
13. J.L. Himes and T.A. Wiggins, J. Mol. Spectrosc. **40**, 418 (1971).
14. D.H. Rank, P. Sitaram, W.A. Glickman and T.A. Wiggins, J. Chem. Phys. **39**, 2673 (1963).
15. D.H. Rank, B.S. Rao and T.A. Wiggins, J. Chem. Phys. **37**, 2511 (1962).
16. M. Larvor, J.-P. Houdeau and C. Haeusler, Can. J. Phys. **56**, 334 (1978).
17. C.M. Lovejoy, M.D. Schuder and D.J. Nesbitt, Chem. Phys. Lett. **127**, 374 (1986).
18. C.M. Lovejoy, M.D. Schuder and D.J. Nesbitt, J. Chem. Phys. to be published.
19. Z.S. Huang, K.W. Jucks and R.E. Miller, J. Chem. Phys. **85**, 3338 (1986).
20. Z.S. Huang, K.W. Jucks and R.E. Miller, J. Chem. Phys. to be published.
21. A.S. Pine and B.J. Howard, J. Chem. Phys. **84**, 590 (1986).
22. J.S. Muenter, J. Chem. Phys. **56**, 5409 (1972).
23. T.E. Gough, R.E. Miller and G. Scoles, Faraday Discuss. Chem. Soc. **71**, 77 (1981).
24. A.S. Pine, A. Fried and J.W. Elkins, J. Mol. Spectrosc. **109**, 30 (1985).
25. C.G. Gray and K.E. Gubbins, Theory of Molecular Fluids (Clarendon, Oxford, 1984).
26. S. Liu and C.E. Dykstra, J. Phys. Chem. to be published.
27. M.D. Marshall, A. Charo, H.O. Leung and W. Klemperer, J. Chem. Phys. **83**, 4924 (1986).
28. D. Ray, R.L. Robinson, D. Gwo and R.J. Saykally, J. Chem. Phys. **84**, 1171 (1986).
29. B.J. Howard, private communication.
30. I.M. Mills, J. Phys. Chem. **88**, 532 (1984).
31. J.T. Hougen and N. Ohashi, J. Mol. Spectrosc. **109**, 134 (1985).
32. P.J. Robinson and K.A. Holbrook, Unimolecular Reactions (Wiley, New York, 1972).
33. G.E. Ewing, J. Chem. Phys. **72**, 2096 (1980).
34. N. Halberstadt, Ph. Brechignac, J.A. Beswick and M. Shapiro, J. Chem. Phys. **84**, 170 (1986).

SLIT JET IR ABSORPTION SPECTROSCOPY OF MOLECULAR COMPLEXES

David J. Nesbitt*
Joint Institute for Laboratory Astrophysics, University of
Colorado and National Bureau of Standards and Department of
Chemistry and Biochemistry, University of Colorado, Boulder,
Colorado 80309 USA

ABSTRACT. The combination of high resolution ($\lesssim 10^{-3}$ cm^{-1}) cw tunable difference frequency generation, high sensitivity ($\lesssim 10^{-6}/\sqrt{Hz}$) direct absorption methods, and long path length (2.54 cm) pulsed slit expansions provide a powerfully general technique for studying weakly bound complexes in a cold molecular beam environment. Transient absorption of the narrow band laser provides a nonintrusive probe of the quantum state, velocity, temporal and spatial dependence of cluster formation in the pulsed molecular beam. High resolution fundamental, combination and hot band spectra of ArHF, HFN$_2$ and HFCO$_2$ complexes are presented. Detailed information on the molecular structure is determined for vibrationally excited states which sample the potential energy surface far from the ground state, near equilibrium geometry.

1. INTRODUCTION

Weak attractions between molecules and atoms control the rates and dynamics of a wide variety of chemical and physical phenomena.[1] Consequently, a major effort in the past several years has been directed toward a detailed characterization of the potential energy surfaces in which weakly bound molecular complexes are formed.[2] There has been an emphasis on simple molecular systems which can facilitate close comparison between theory and experiment; complexes of hydrogen halides with various inert gases, diatomic and small polyatomic molecules have therefore been of particular interest.[3-18] Pressure broadening,[19] crossed molecular beam[20] and transport studies[21] of the molecular constituents can address the behavior of "complexes" in highly vibrationally excited, i.e., unbound, scattering states. Microwave and radio frequency studies, using both molecular beam electric resonance detection[3-5,8] and Fourier transform coherent transient methods,[6,7] have greatly elucidated the _ground_ vibrational state of many weakly bound

---

*Staff Member, Quantum Physics Division, National Bureau of Standards.

A. Weber (ed.), Structure and Dynamics of Weakly Bound Molecular Complexes, 107–130.

complexes. Indeed, these high resolution rotational spectroscopic
methods have proven invaluable in characterizing the shape of the potential energy surface at near equilibrium geometries, and hence determining molecular structures. Infrared spectroscopy offers a direct
and complementary probe of the potential surface at molecular configurations _away_ from the global energy minimum. The application of IR
methods to the considerable range of bound, vibrationally _excited_
states of complexes has only very recently become feasible, and form
the focus of this paper.

In the past few years, there has been an explosive growth in the
field of vibrational spectroscopy of complexes due to the development
of several techniques with adequate sensitivity[17,18,22-32] and resolution to observe low concentrations of weakly bound species in either
cooled cell or supersonic beam environments. Pine and coworkers have
used the high spectral resolution of a tunable difference frequency
laser to investigate hydrogen halide dimers and van der Waals complexes in a cooled White cell, with 64-80 meter optical absorption
path lengths.[22,23] FTIR methods and mode hop scanning of a color
center laser have also been used to study several strongly bound complexes in long path length cells.[27] There have been several supersonic beam techniques developed for IR spectroscopy of complexes which
exploit the tremendous cooling in the expansion to promote efficient
synthesis and enhance populations in low quantum states. OPO and CO_2
laser vibrational predissociation spectroscopy has been used to study
complexes of hydrogen fluoride, water and several hydrocarbons.[1,28]
Bolometric methods pioneered by Scoles and coworkers that detect laser
induced warming or predissociative deflection of the species in the
molecular beam have shown excellent sensitivity for study of a variety
of cluster molecules.[29] Extensions of these methods by Miller and
coworkers look very promising for sub-Doppler Stark spectroscopy on
vibrationally excited states.[18] Direct diode laser absorption spectroscopy of rare gas-OCS complexes has been recently demonstrated.[30]
Far infrared laser-microwave double resonance[31] and intracavity laser
Stark spectroscopic methods[32] have been exploited to study low frequency, intermolecular vibrational modes in ArHCl.

Extension of direct absorption IR studies of weakly bound complexes into the molecular beam regime requires exceptional sensitivity
to compensate for the roughly 5 orders of magnitude decrease in path
length between a White cell and conventional pinhole expansion. Efforts in our laboratory have been directed toward combining near shot
noise limited detection of IR laser direct absorption and slit supersonic expansion methods to achieve a general probe for high resolution
($\Delta\nu \lesssim 0.001$ cm^{-1}) IR spectroscopy of weakly bound complexes.[24-26] Key
advantages of this approach are that 1) jet temperatures are low enough
to eliminate spectral congestion but can be controlled to populate very
low frequency vibrations in the complexes, 2) velocity narrowing in the
slit expansion permits sub-Doppler absorption linewidths in an unskimmed
molecular beam, and 3) the slit jet geometry provides density × absorption path lengths ~2 orders of magnitude longer than obtained in typical
pinhole expansions. This paper describes the application of these techniques to a series of small HF-containing complexes (ArHF, HFN_2 and

$HFCO_2$), in which fundamental, combination, and low frequency intermolecular hot bands are observed and serve to elucidate the shape of
the potential energy surface sampled in the vibrationally excited
state.

2. EXPERIMENTAL

Infrared spectra of molecular clusters, featuring sub-Doppler resolution at low (5-45 K) temperatures are obtained by high sensitivity,
direct absorption detection with a computer-controlled spectrometer
(Fig. 1) based on a tunable infrared laser and a pulsed slit supersonic expansion. Detection limits are $\leq 3 \times 10^9$ molecules/cm^3/ quantum
state for HF-containing complexes over a 2.5 cm absorption pathlength.

2.1. Tunable IR Laser Spectrometer

A difference frequency laser provides cw infrared light, tunable from
2.2-4.2 µm, with a 0.0015 cm^{-1} linewidth. The outputs of a single
frequency ring dye laser (0.25 W on Rhodamine 6G, horizontal polarization) and a single frequency Ar$^+$ laser (0.40 W, vertical polarization)
are combined with a polarization beamsplitter and focused with a 20 cm
focal length achromatic lens into a 90 degree phase matched, temperature tuned, 5 cm long $LiNbO_3$ crystal. This produces approximately 20
µW of IR at the difference between the Ar$^+$ and dye laser frequencies.

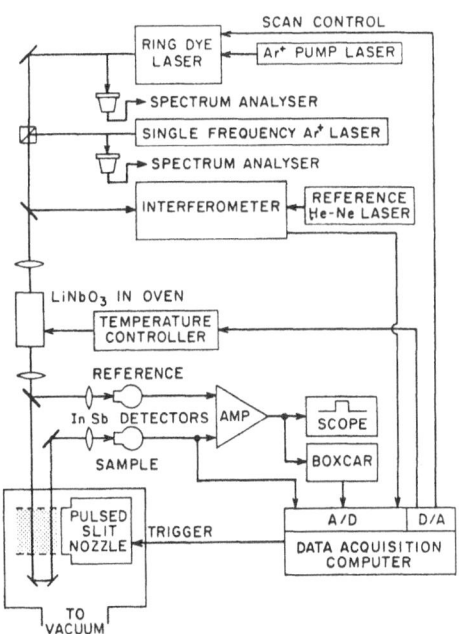

Fig. 1. Schematic of the
tunable IR laser spectrometer.

The linewidths of the passively stabilized dye and Ar^+ lasers are 46 ±
5 and 14 ± 5 MHz, respectively; the 50 ± 5 MHz infrared linewidth is
the uncorrelated sum of the visible linewidths.

A 50% beamsplitter divides the IR output into sample and reference beams that are focused with 2.5 cm focal length CaF_2 lenses onto
matched 0.04 mm^2 InSb photovoltaic detectors. IR-induced photocurrents
are amplified in matched, home-built amplifiers with 100 kΩ transimpe-
dance. The difference signal between sample and reference is amplified,
filtered with a 5 kHz center frequency bandpass, sampled by a boxcar
integrator and displayed on an oscilloscope. Imperfect subtraction
of fluctuations in infrared power leads to amplitude noise that corresponds to 0.01% absorption using a 500 μs boxcar gate and 5 kHz detection bandwidth. The observed absorption sensitivity is of the order
$10^{-6}/(Hz)^{1/2}$ at the detection frequency of 5 kHz, within a factor of
3 of the shot noise limit for 10 μW of infrared light at 2.6 μm.[26]

The visible frequencies are measured with a traveling Michelson
interferometer constructed after the design of Hall and Lee.[33] The
reference laser is a 633 nm He-Ne laser polarization-stabilized to
1 MHz as determined by measuring the beat frequency offset from an
iodine-stabilized He-Ne laser. Precision is increased 16-fold by
phase-locking to the sinusoidal interferometer signal and counting
the zero crossings of the phase-locked oscillator. The precision of
each measurement is ±0.001 cm^{-1}, which is further improved by signal
averaging to ≲0.0004 cm^{-1}. The absolute frequency accuracy of the
measurements is obtained from internal references absorbing in the IR;
for all results reported herein the infrared frequencies are measured
relative to HF monomer transitions.[34] The techniques used reproduce
the known frequency differences between reference lines separated by
80 cm^{-1} to within 0.0010 cm^{-1}; we take this as a conservative estimate of the absolute frequency uncertainty.

2.2. Pulsed Slit Valve Construction and Performance

Van der Waals and hydrogen bonded clusters are produced in a supersonic expansion from a pulsed slit valve of our own design (see Fig.
2). The nozzle has a 1.25 cm slit length and a variable slit width,
and produces a low-temperature expansion with a 2.5 cm double-pass
absorption length. The nozzle holder consists of a stainless steel
disk, 2.5 cm in diameter, into which a 350 μm, straight-walled slot
with a knife-edged lip is formed by electrodeless discharge machining.
Interchangeable nozzles are produced by cutting a 2.5 cm stainless
steel disk in half along a diameter, machining the cut edges flat,
and surface grinding a rectangular notch in each half. The two pieces
are bolted onto the nozzle holder in contact with each other; the two
notches form a sharp-edged slit with width uniform to 5 μm and with
mirror-smooth surfaces contacting the gas flow. For the experiments
reported in this paper, the slit width was 125 ± 12 μm; the thickness
is 375 μm and the length 1.25 cm.

A fluorocarbon elastomer seal cut from a length of O-ring material is held in contact with the knife-edged lip of the nozzle holder
by a leaf spring; the spring is compressed to maintain the closed

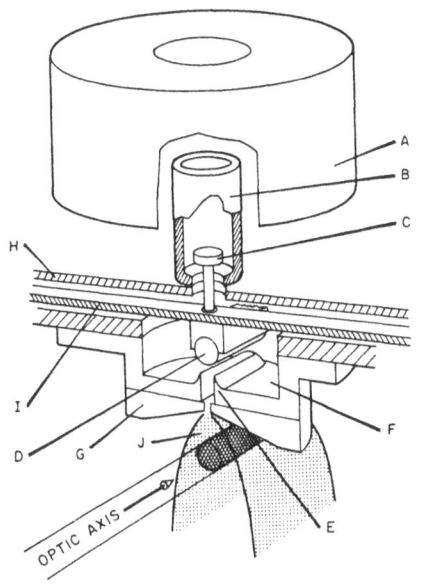

Fig. 2. Pulsed valve for slit supersonic expansion. The valve features variable slit width and produces gas pulses of 150-600 μs duration; linewidths are ≤50 MHz due to collimation of molecular velocities perpendicular to the axis of the nozzle. A current pulse through solenoid (A) accelerates plunger (B) which strikes rod (C) connected to an elastomer seal (D). The seal is lifted from the sharp-edged nozzle holder (E) permitting gas in chamber (F) to flow through interchangeable nozzles (G). Seal strikes stop (H) and is closed by leaf spring (I). Wedge-shaped jet (J) issuing from nozzle is probed by IR beam at variable distance from nozzle.

configuration. The valve is opened by passing a current pulse (30 A peak, 300 μs FWHM) through a solenoid. The resulting magnetic field gradient accelerates a ferromagnetic plunger against a rod connected to the seal. The elastomer seal is impulsively accelerated, travels approximately 250 μm before hitting a stop, and is then returned to the closed position by spring tension and recoil. The valve is acoustically quiet during operation, which is essential to reduce amplitude instability in the dual beam configuration.

The unskimmed wedge-shaped jet issuing from the nozzle is probed by two passes of the infrared sample beam at distances that can be varied from less than the 0.7 mm IR beam radius to several centimeters. Figure 3 depicts the time-dependent infrared absorption profile of a typical gas pulses. The absorption is due to the R(3) transition of the hydrogen-bonded species N_2HF, formed in a mixture of 2% N_2, 1% HF in Ar. The top trace is obtained 0.3 cm from the nozzle exit; the lower trace is obtained at 5.3 cm and shows clustering resulting in the concave structure on the top of the profile. The pulse fall time of 70 μs is consistent with the single-exponential decay of the pressure in the dead volume between the seal and the nozzle; the 10% to 90% rise time is 40 μs. The pulse width may be varied from 150 μs to over 1 ms, controlled by varying the distance between the seal and stop and by varying the current pulse. The pulse width is 450-600 μs for the experiments described in this paper.

Molecular velocity components parallel to the slit are quenched by collisions among molecules issuing from adjacent regions of the expansion. This collimation of velocities reduces the Doppler contribution to the linewidth measured along the slit axis.[35] The residual

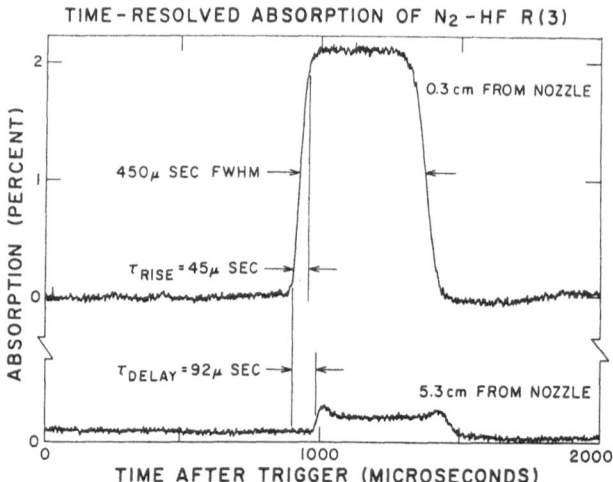

Fig. 3. Time-resolved IR absorption of typical gas pulse. The absorption is due to the R(3) transition of the ν_1 fundamental band of
the hydrogen-bonded species N_2HF; maximum absorbance is 2.1%. Pulse
rise and fall times are determined by dead volume in nozzle holder;
pulse width is variable (see text). Upper trace: 0.3 cm from nozzle
exit. Lower trace: 5.3 cm from nozzle exit. Travel time corresponds
to velocity of 6.1×10^4 cm s^{-1}. Dip in lower pulse is due to clustering during the travel; clustering is enhanced at higher backing
pressures and lower nozzle temperatures.

Doppler contribution is estimated from observed rotational temperatures to be $\lesssim 0.001$ cm^{-1} for our expansion conditions and geometry.
With the above apparatus and the probe beam less than 1 cm from the
nozzle, observed infrared linewidths for stable species (HF, methane,
etc.) are 0.0015 cm^{-1}, limited at present by the infrared laser linewidth. This represents a factor of 5 to 7 improvement over the direct
absorption Doppler widths from an unskimmed pinhole expansion, and results in a comparable increase in the peak absorption intensity. A
skimmed molecular beam also provides a small Doppler width, but only
at the expense of number density and pathlength; there is no corresponding enhancement in the peak absorption.

2.3. Clustering Behavior in the Expansion

The optimum conditions for formation of molecular complexes are determined by observing the time-dependent absorption of monomer or cluster
species as the composition or pressure is varied. This ability to
monitor concentrations of particular quantum states spectroscopically
permits a completely nonintrusive probe of the spatial and time-dependent clustering processes. In addition, due to the high frequency

resolution of the IR laser, one can monitor relative populations of various velocity groups resolved along the probe axis, and thereby obtain a full velocity map of the species in the beam. By way of example, Fig. 4 displays frequency scans across the HF R(0) absorption line in the slit as a function of backing pressure. The probe laser geometry is along the long axis of the linear expansion, and roughly 0.6 cm (x/d = 50) downstream from the slit; the HF/Ar ratio is 1:100. At low backing pressures the absorption lines are single, intense peaks, with sub-Doppler widths corresponding to perpendicular velocities of order 10^4 cm/s. As the backing pressure increases, there is a systematic reduction in the absorption at line center, which implies a selective removal of J=0 monomer species with low perpendicular velocities. At pressures above 500 Torr, this selective elimination of low velocity absorbers is sufficient to split the peak into fully resolved doublets. It is important to note that the peak absorption strength per unit pressure decreases by 500 fold over the same pressure range, indicating that the low velocity absorbers are not simply being redistributed into higher perpendicular velocity groups. Investigation of higher J states in the expansion verify that the absorption loss of R(0) is also not due to selective rotational excitation away from J=0.

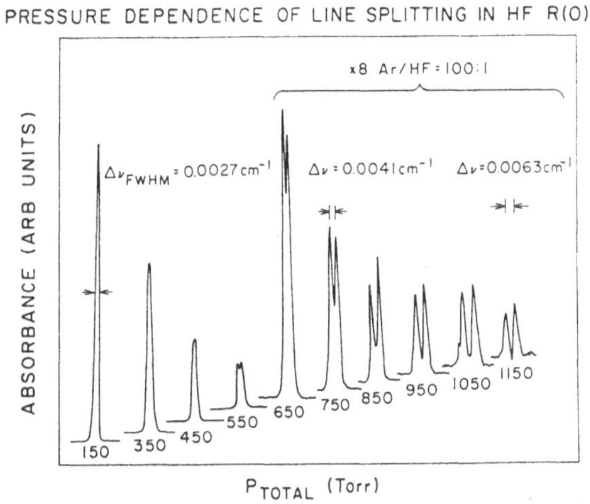

PRESSURE DEPENDENCE OF LINE SPLITTING IN HF R(0)

Fig. 4. Lineshape behavior on R(0) of HF monomer (1% HF/Ar) as a function of backing pressure. In the absence of clustering, the absorbance would grow linearly with pressure, instead a rapid drop in intensity is observed. The colder central region of the expansion clusters preferentially, which due to the Doppler effect greatly reduces absorption intensity at line center and produces the characteristic doubling of the lineshape. More than half of the HF is clustered at $\geq$300 Torr; at 1000 Torr more than 99% of the monomer is bound in complexes, but less than 1% in either ArHF or $(HF)_2$ (see Fig. 5).

   The severe loss of HF is by clustering in the expansion to form
ArHF, HF dimer, and higher molecular weight species. Due to extensive
fragmentation upon electron impact, accurate characterization of the
relative populations of various clusters in an expansion has been difficult to obtain. By way of contrast, absorption spectroscopy permits
a direct measure of the concentrations of HF monomer, $(HF)_2$ and ArHF
as a function of expansion conditions. The integrated absorption signals for HF monomer R(0) and ArHF R(7), normalized to backing pressure
are displayed in Fig. 5. In these plots, clustering behavior is evidenced by deviations from a horizontal line; nonlinear growth or removal
is indicated by positive or negative derivatives, respectively. The HF
monomer J=0 populations exhibit a monotonic depletion above 150 Torr of
the 1% HF/Ar mixture, i.e., the same pressure regime in which the ArHF
population grows. The ArHF signal strengths roll off at higher backing
pressures due either to sequential clustering to larger molecular weight
species or by loss of free HF precursors with which to form ArHF. The
absorption signal behavior for HF dimer, which is also observed in this
spectral region, is quite similar to ArHF. The signal strengths for
ArHF and HF dimer, corrected for partition functions, translate into
total number densities of less than 1% of the HF monomer number density
lost through clustering. This behavior indicates extensive sequential

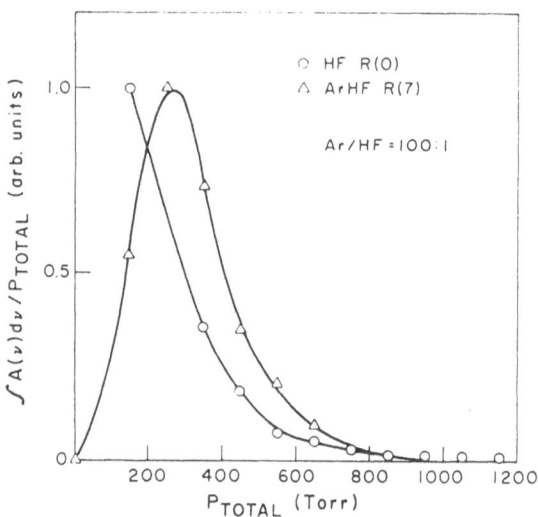

Fig. 5. Pressure dependence of integrated absorption signals for R(0)
of HF monomer and R(7) of ArHF. The ArHF data have been multiplied by
420. The behavior of $(HF)_2$ has also been measured and is quite similar. If one corrects for partition function effects, less than 1% of
the clustered HF monomer (see Fig. 4) can be accounted for in either
$(HF)_2$ or ArHF for the optimum backing pressure, indicating a predomi-
nance of higher oligomers in the beam.

clustering occurs which preclude substantial population in the small cluster, ArHF and $(HF)_2$ species even under optimized expansion conditions. Qualitatively similar behavior is noted for expansions of other hydrogen halides and small hydrocarbons such as acetylene and methane. These observations are consistent with recent scattering experiments on ammonia complexes by Buck and coworkers,[36] which indicate consistently low populations of dimer species in their beams.

3. RESULTS AND DISCUSSION

The direct absorption slit expansion methods described above have been used in our laboratories to investigate several complexes of hydrogen halides and small hydrocarbons. In this section results are presented for ArHF, HFN_2, and $HFCO_2$, which illustrate the richness of dynamical (and hence spectral) behavior that is exhibited in this series of three-, four-, and five-atom complexes.

3.1. ArHF ν_1 Fundamental and $\nu_1 + \nu_2$ Combination Band

The ArHF high frequency stretch fundamental spectrum $(10^0 0) \leftarrow (00^0 0)$ occurs near 3952 cm^{-1}, and exhibits a simple P and R branch structure of a molecule with linear equilibrium geometry. The regularly spaced lines and null gap between R(0) and P(1) permit unambiguous numbering of the J assignment. Sample data with typical signal to noise on R(3)-R(1) are shown in Fig. 6. Peak absorption strengths in this figure correspond to 0.2% absorption; the noise background is 10^{-4} absorbance in a 5 kHz bandwidth.

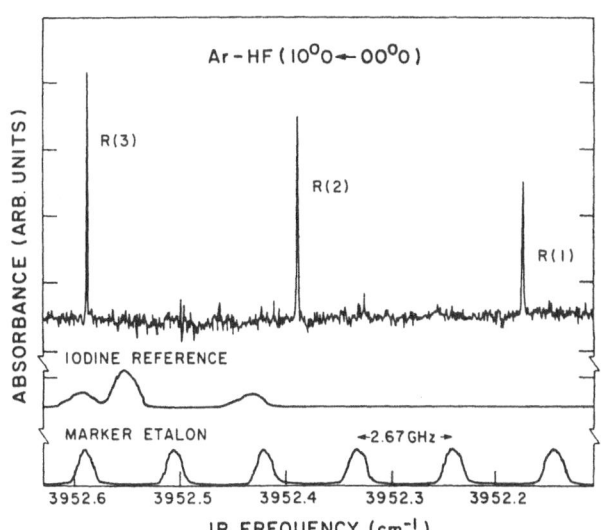

Fig. 6. Sample data scan of R(3)-R(1) of ArHF on the $(10^0 0) \leftarrow (00^0 0)$ fundamental transition. Peak absorption is a few tenths of a percent; baseline noise corresponds to a 10^{-4} absorbance in a 5 kHz bandwidth. The entire scan segment takes roughly 40 seconds. The rotational temperature of the ArHF is approximately 10 K.

   Transition frequencies for R(20) through P(19) of the fundamental
are obtained. In order to decrease systematic and statistical errors,
each transition is observed and measured in two to three separate experimental runs; each individual measurement represents an average of at
least five IR frequency measurements. Combination differences for the
ground and ν_1 excited state are fit to a standard power series expression in J(J+1) to extract the rotational constants; the pure vibrational
frequency is then obtained by a fit to the observed transitions with the
rotational constants held fixed. A summary of the nonlinear least-
squares fits of the molecular constants is shown in Table 1. The ground
state B and D constants obtained for ArHF are in excellent agreement (1
part in 10^5) with the early microwave work.[3,6] This agreement, in
conjunction with the simple mixture of gases, establishes beyond any
reasonable doubt the chemical identity of the complex observed in these
IR experiments.

   Noteworthy is the $3.49(12) \times 10^{-4}$ cm^{-1} increase in the B rotational constant between the (00^00) and (10^00) vibrational states. This
result is in contrast with the negative ΔB observed in rare gas–HCℓ
species excited in the ν_1 vibrational mode.[37] The predicted shift in
B due simply to vibrational elongation of the HF bond is approximately
-1.6×10^{-5} cm^{-1}, i.e., of opposite sign and more than 20 times smaller
than experimental observation. The observed 0.34% increase in B for
ν_1 excited ArHF is more than can be accounted for by rotation of the
HF unit around its center of mass ($\Delta B \leq 0.25\%$) hence some contraction
in the vibrationally averaged van der Waals bond must occur. If θ_{cm}
is held at the vibrationally averaged angle (41.6°) determined for the
ground state, R_{cm} would decrease by 0.006 Å (from 3.510 to 3.504 Å)

Table 1. Molecular constants (in cm^{-1}) obtained for ArHF.

	(00^00)	(10^00)	$(11^10)\Pi^f$	$(11^10)\Pi^e$	(10^02)
B	0.102251(20)	0.102609(18)	0.100325(25)	0.102651(34)	0.08183(14)
	$(0.10226100(3))^a$				$(0.0817)^b$
D	$2.34(11)\times10^{-6}$	$2.07(8)\times10^{-6}$	$3.32(19)\times10^{-6}$	$4.14(13)\times10^{-6}$	---
	$(2.361(3)\times10^{-6})^a$				
H	$2.6(17)\times10^{-10}$	$2.7(13)\times10^{-10}$	$6.0(38)\times10^{-10}$	---	---
E_v-E_o	0	3951.7680(30)	4022.1047(30)	4022.1062(30)	4023.3880(30)

$q_{11^10} = -2.33(6)\times10^{-3}$

Coriolis interaction between $(11^10)\Pi^e$ and $(10^02)\Sigma^+ = 0.01664(20) \sqrt{J(J+1)}$.

^{a}Ref. 3.

^{b}Calculated from a Lennard-Jones potential fit to (10^00) data in the pseudo-diatomic
  approximation.

to induce the observed shift in B. If θ_{cm} decreases upon ν_1 excitation, the decrease in R_{cm} would only need to be ≥ 0.003 Å. In light of the enhanced van der Waals attraction in the excited state, it is likely that the geometric shifts will involve both a shrinking of the van der Waals bond and greater alignment of the HF in the complex. Some recent Stark measurements on ArHF indicate an increase in the molecular dipole moment in the ν_1 excited state which is consistent with a decrease in θ_{cm}.[18]

An increased attraction between Ar and HF in the vibrationally excited complex is also exhibited in the decreased centrifugal distortion constant. Based on the often used pseudo-diatomic approximation,[3] one predicts a harmonic stretch frequency of 45.7 cm^{-1} for the ν_1 excited state i.e., 7.4% higher than the 42.6 cm^{-1} ground state prediction. For a Lennard-Jones 6-12 potential, the binding energy of the complex in this diatomic approximation is readily shown to be

$$ D_e = \frac{R_e^2 \; \mu \; 4\pi^2 \omega_3^2 c^2}{72} , \tag{1} $$

where R_e is the equilibrium Ar-HF center-of-mass separation and μ is the Ar-HF reduced mass. This expression predicts roughly a 15% increase in the vibrationally averaged van der Waals well depth from 117 to 135 cm^{-1} upon ν_1 excitation. Subsequent studies by Fraser and Pine on ArHF complexes in a cooled White cell have provided rotational energies up to the dissociation limit ($J \approx 40$).[23] Line position and predissociation linewidth fits to a four-parameter Maitland-Smith potential yield well depths of 125 and 136 cm^{-1} for the ground and ν_1 excited state, respectively, in surprisingly good agreement with the predictions of the present study.

Although the difference frequency spectrometer cannot excite the low frequency van der Waals modes directly, these modes can be observed in combination, difference and hot bands built on the higher frequency vibrational modes of the van der Waals constituents. In ArHF, the large amplitude zero point motion in the bending vibration modulates the projection of the HF dipole moment derivative along the molecular axis, and hence transfers intensity from the fundamental $(10^00) \leftarrow (00^00)$ into the combination $(11^10) \leftarrow (00^00)$ and difference $(10^00) \leftarrow (01^10)$ bands. In the limit of a free internal rotor, these three "vibrations" correlate with Q(0), R(0) and P(1) of the HF monomer respectively, and the fundamental transition is forbidden. In the jet spectrometer at <10 K, only the ground vibrational state is appreciably populated; we therefore look for the van der Waals bending mode in the $\nu_1 + \nu_2$ combination. Since the degenerate bend has one unit of vibrational angular momentum around the principal axis, one anticipates the P, Q and R branch structure characteristic of a $\Pi \leftarrow \Sigma$ perpendicular transition.

Such a spectrum is indeed observed approximately 70 cm^{-1} to the blue of the ν_1 origin. Signal strengths on the combination band Q branch are 37±4% as strong as the fundamental, which permit R(0)-R(16), Q(1)-Q(17) and P(2)-P(18) to be observed. Since the upper state has a minimum of J=1, P(1) and Q(0) are not anticipated and do not occur. A stick plot of the $\nu_1+\nu_2$ combination band is shown in Fig. 7.

The ν_2 bending vibration in ArHF is doubly degenerate and split by first and second order Coriolis interactions into a π^e and π^f manifold.[38] As a result of the J dependence of parity, the $\Delta J = 0$ (Q branch) transitions access only the π^f component of the doublets, while the $\Delta J = \pm 1$ (R and P branch) transitions access the π^e component of the doublets. Term values for J=1 to 18 in the (11^10) π^f manifold are obtained as the sum of observed Q branch transition frequencies and the measured term values for the lower (00^00) state. These terms have been fitted to standard Hamiltonian[38] expressions for a π state ($|\ell| = 1$); the results are summarized in Table 1.

If one neglects off-diagonal anharmonicity, an approximate bending frequency of 70.3 cm^{-1} can be obtained from the blue shift of the $\nu_1+\nu_2$ versus ν_1 origins. This frequency is remarkably close to the

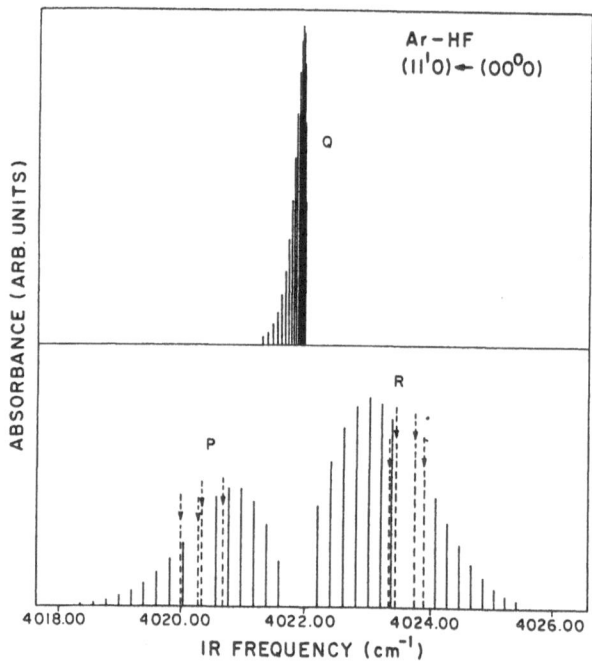

Fig. 7. Stick plot of the observed transition frequencies for the Q(upper) and P/R(lower) branches, respectively. Relative intensities are shown for a 10 K rotational distribution. Transitions to the π^f component of the (11^10) manifold are seen exclusively in the Q branch (displaced for clarity). The P/R branch represents only transitions to the π^e component, which exhibit a strong localized Coriolis perturbation near J'=8 by the (10^02) Σ^+ vibrational manifold. Transitions to states strongly mixed by this interaction are indicated by dotted lines with arrows.

prediction of 67.7 cm^{-1} based on the anisotropic potential surfaces of Hutson and Howard.[12] The harmonic prediction of Klemperer and co-workers[3] based on vibrationally averaged spin-spin HF hyperfine inter-actions in the ground state is also quite close but somewhat higher, 79.8 cm^{-1}. The B constant for the (11^10) π^f bend excited state is lower by 1.9% than the ground state, which is responsible for the ex-treme red shading of the Q branch. This shift is nearly six times larger than, and of opposite sign to, the shift observed in the fun-damental transition. Changes in the vibrationally averaged θ_{cm} of the HF unit can account for at most 13% of the observed shift; hence the major contribution to ΔB must be due to a growth in R_{cm}. If θ_{cm} is held at the ground state value of 41.6°, R_{cm} would have to increase by 0.034 Å (from 3.510 to 3.544 Å) to achieve the observed 1.9% de-crease in B. Again, it is a more likely scenario that both θ_{cm} and R_{cm} will increase an ν_2 excitation, which would be physically rea-sonable since a significant source of the binding is due to aniso-tropic dipole-induced dipole interactions. It would be extremely useful to have Stark data on either $\nu_1 + \nu_2$ or ν_2 excited states to clarify these issues. In any event, this large change in vibra-tionally averaged separation with bending excitation is in sharp contrast with observations for ArHCℓ,[31,32,37] and is indicative of strong bend-stretch couplings in the ArHF potential.

In addition to the Q branch lines, 35 transitions attributed to the P and R branch are observed to the red and blue of the vibrational origin. Fairly regular spacing in the rotational structure is evident, but with dramatic irregularities over a highly localized region of in-termediate J values. Unambiguous J assignment of the transitions is made from measured combination differences in the ground state; agree-ment is consistently within the measurement uncertainty of 0.001 cm^{-1}. These assignments show that P(9) and R(7) are each split into a doublet of transitions, revealing a strong, but localized perturbation in J'=8; such behavior is completely absent in the Q branch measurements. As was discussed above, the P/R branch and Q branch transitions from the ground state terminate on levels in the π^e and π^f manifolds, respec-tively. The localized perturbation in the π^e state can be assigned to Coriolis couplings with the (10^02) Σ^+ overtone of the van der Waals stretch; the "extra" lines observed are transitions directly to the states mixed by the interaction. The π^f state observed in the Q branch, on the other hand, should be unperturbed by the (10^02) state, as is observed to be the case.

B-type Coriolis interaction between a Σ^+ and π^e state requires $\Delta J = 0$, $\Delta \ell = \pm 1$ and scales as[39]

$$\langle J, \ell=0 | \hat{H}_{cor} | J, \ell=1 \rangle = \beta \{J(J+1)\}^{1/2} \quad . \tag{2}$$

Term values for the mixed Σ^+ and π^e state are therefore least-squares fit to a 2×2 Hamiltonian matrix with elements given by

$$H_{11} = \nu_{11^10} + B_{11^10}\{J(J+1) - \ell^2\} - D_{11^10}\{J(J+1) - \ell^2\}^2 \tag{3a}$$

$$H_{22} = \nu_{10^02} + B_{10^02} \, J(J+1) \tag{3b}$$

$$H_{12} = \beta\sqrt{J(J+1)} \tag{3c}$$

where β, ν_{11^10}, ν_{10^02}, B_{11^10}, B_{10^02} and D_{11^10} are adjustable parameters. The results for the deperturbed molecular constants are shown in Table 1.

The B rotational constant for the π^f (11^10) state is significantly higher than for the π^e (11^10) state, which indicates a negative ℓ-doubling constant of $-69.8(18)$ MHz. This is contrary to what is observed in "normal" molecules, where both first and second order vibration-rotation interactions lead to positive ℓ-doubling since the bend excited state is usually the lowest Σ^+ vibration.[38] From the Hutson and Howard surfaces,[12] however, the van der Waals stretch (10^01) and parallel bend (12^00) are predicted to be lower in energy and to have the correct Σ^+ symmetry to perturb the π^e component of the perpendicular bend (11^10). Negative ℓ doublings have also been observed in far IR experiments on ArHCℓ on the (01^10) state,[31,32] although, surprisingly, not in the analogous White cell, near IR studies on the (11^10) excited state.[37] Finally, the centrifugal distortion in the π^e manifold is significantly larger than the ground state value. This is in agreement with the behavior in the π^f manifold and again indicates a weakening of the van der Waals attraction with excitation of the perpendicular bend.

The deperturbed vibrational origin of the (10^02) excited state is 71.6 cm^{-1} above the ν_1 (10^00) excited state. This is in good agreement with numerical integration of the Schrödinger equation for the vibrationally averaged Lennard-Jones 6-12 potential fits for the (10^00) manifold. Using these C_6 and C_{12} parameters, one calculates a vibrational origin of 65 cm^{-1} and a 10^02 rotational constant of 0.0817 cm^{-1}, in very close agreement with the observed values of 71.6 cm^{-1} and 0.0818 cm^{-1}, respectively. Efforts by Fraser and Pine[23] to fit the vibrationally averaged van der Waals well to a one-dimensional Maitland-Smith potential predict 72.25 cm^{-1} for the 10^02 vibrational origin and 0.08264 cm^{-1} for B_{10^02}. These consistency checks leave little doubt that the (10^02) state is responsible for the observed Coriolis perturbations in the (11^10) π^e ← (00^00) combination band.

3.2. HFN$_2$: Stretching and Bending Vibrations in a Nearly Linear Polyatomic Complex

The HFN$_2$ complex offers a nice comparison with ArHF for several reasons. The hydrogen bond is much stronger[11,12] than the van der Waals bond, leading to potentially more anisotropic interactions between the constituents. However, the larger reduced mass of N$_2$ with respect to HF allows much lower frequency bending motions to occur, and may be sufficiently populated in the slit jet to observe directly.

The origin of the ν_1 HF stretch of HFN$_2$ is observed at 3918.2434(2) cm^{-1}, red-shifted by 43.1795(2) cm^{-1} from the rotationless vibration frequency of HF monomer, but blue-shifted by

37 cm^{-1} from observations in a matrix.[15] Initial conditions for
searching for the complex are determined by monitoring the absorption
of the van der Waals complex Ar–HF in an expansion of 1% HF in Ar as
N_2 is added to the mixture. Strong HFN$_2$ transitions are immediately
observed; optimization of expansion conditions shows a broad maximum
at 20% to 50% N_2, 1% HF with intensity monotonically increasing from
100 to 1150 Torr backing pressure. Figure 8 is a stick plot of the
observed transition frequencies. The spectrum displays the P and R
branches, center gap, and regular spacing characteristic of a $\Sigma \leftarrow \Sigma$
transition of a linear molecule; these features provide unambiguous J
assignment. Transition frequencies for R(37) to P(35) (inclusive) are
obtained; all spectral features are isolated and unblended. Unambiguous identification of the absorbing species is made by the excellent
agreement (≤ 0.001 cm^{-1}) between the observed spacing and combination
differences generated using the microwave rotational constants.[7] The
HF P(1) transition at 3920.3119 cm^{-1} is the reference for all frequencies reported.[34]

Analysis of the transition frequencies for the ν_1 fundamental
follows closely the method described for $\nu_1 = 1\leftarrow 0$ in ArHF. Table 2
lists the molecular constant coefficients obtained from our data,
along with the corresponding values for the ground state determined
by microwave spectroscopy. The infrared–determined B" and D" are in

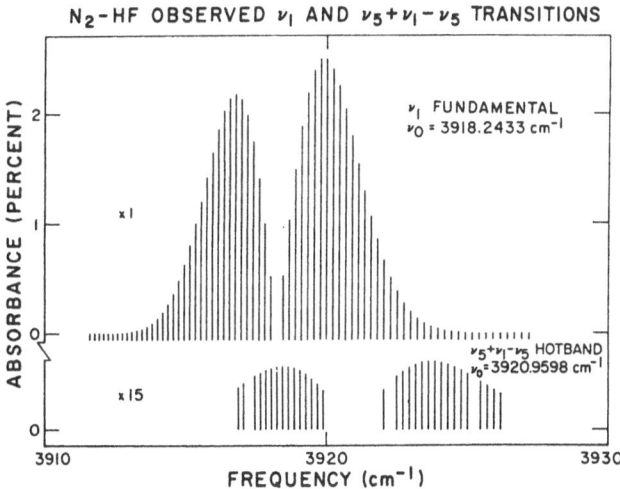

Fig. 8. Stick plot of the observed HFN$_2$ transitions in the ν_1 fundamental (upper) and $\nu_1 + \nu_5 - \nu_5$ bend excited hot band (lower). The ν_1
fundamental transitions from R(37)–P(35) are observed. The hot band
transitions are all split by ℓ-doubling too small to be observed in
this figure. The 60±15 intensity ratio between fundamental and hot
band transitions is consistent with a ν_5 bending vibrational frequency
of 100±25 cm^{-1}.

D. J. NESBITT

Table 2. Molecular constants (in cm^{-1}) obtained for HFN_2.

	Ground State	$\nu_1=1,\nu_5=0$	$\nu_1=0,\nu_5=1$	$\nu_1=1,\nu_5=1$
B	0.1065862(84)	0.1071781(82)	0.107556(18)	0.108123
	(0.10658552)[a]			
D	$5.67(13)\times10^{-7}$	$5.28(12)\times10^{-7}$	$6.64(33)\times10^{-7}$	$6.25(27)\times10^{-7}$
	$(5.737\times10^{-7})^a$			
H	$1.37(59)\times10^{-11}$	$1.42(52)\times10^{-11}$	---	---
E_v-E_o	0	3918.2434(30)	$\nu_5 \cong 100\pm25$ cm^{-1} [c]	$3920.9598(30) + \nu_5$
		(3881)[b]		
			$\|q\| = 4.61(7)\times10^{-4}$	$\|q\| = 4.44(6)\times10^{-4}$
$X_{15} = -2.716(6)$				

[a] Ref. 7.

[b] Ref. 15.

[c] Estimated from relative intensities, based on equilibrated rotations and low
frequency vibrations.

excellent agreement with the microwave results.[7] The contribution
from the sextic terms is 0.039 cm^{-1} at the highest J observed, which
is approximately 50 times the experimental uncertainty.

The rotational constant B increases 0.6% and the centrifugal
distortion constant decreases by 6% on excitation of the ν_1 stretch.
Both changes are indicative of a stronger hydrogen bond in the excited
state. The binding energy calculated for the ground state of HFN_2
is 600 cm^{-1}, using the diatomic approximation and increases by 8% to
650 cm^{-1} in the excited state. To estimate the low frequency stretching vibration, a Lennard–Jones 6–12 potential is assumed and eigenvalues of the pseudodiatomic Hamiltonian are obtained by numerical
integration. The intermolecular stretch frequency is estimated to be
91 cm^{-1} in the ground state, increasing to 93 cm^{-1} in the ν_1 excited
state. These calculations predict a redshift of the ν_1 transition of
48 cm^{-1} from the rotationless frequency of the HF monomer, in fair
agreement with the experimental value of 43 cm^{-1}. With the assumption
of unchanged equilibrium bond angles, the separation between the centers of mass of the subunits is determined to decrease by 0.010 Å upon
ν_1 excitation; again, this value would be slightly smaller if θ_{cm} for
the HF unit decreases with excitation of ν_1.

In addition to the strong ν_1 transitions, the high sensitivity
of the direct absorption/slit expansion technique permits observation
of a much weaker series of doublets that becomes progressively more
intense in the warmer region of the expansion close ($\leq$0.4 cm) to the
nozzle. The spectrum is assigned as a $\Pi \leftarrow \Pi$ hot band of N_2HF, in

which the lower and upper levels each have one unit of vibrational
angular momentum ($|\ell|$ = 1) around the figure axis. Since in a stand-
ard analysis of ℓ doubling, the splitting increases as $J(J+1)$ and is
zero for the rotationless molecule, the ro-vibrational transitions can
be unambiguously assigned by fitting the splittings to

$$\Delta\nu = q'J'(J'+1) - q''J''(J''+1) \tag{4}$$

and varying the J assignment. For only one trial value does the predicted splitting vanish at J = 0, and the standard deviation of the
fit is also observed to reach a sharp minimum at that value. Based on
this assignment, the rotational constants and vibrational frequency
can be extracted as for ArHF; the results are summarized in Table 2.

The band origin is blue-shifted by 2.7160(4) cm^{-1} from the fundamental, indicating that the interaction between the HF and nitrogen
subunits is relatively weaker upon bend excitation, in contrast to the
effect observed for the ν_1 stretch. The average of the rotational
constants for the two states is 0.9% larger than for the ground state,
and increases by 0.5% on ν_1 excitation. The centrifugal distortion
constants for the bend-excited states are also 13% larger than for the
ground and ν_1 = 1 states, again, consistent with a weakening of the
bond upon bend excitation.

There are two bending vibrations in N_2HF that might produce such
a spectrum, the ν_4 HF bend and the ν_5 N_2 bend. The identification
of the hot band as the $\nu_5+\nu_1-\nu_5$ transition is made from two estimates
of the bending vibration frequency obtained from our data that yield
values in general agreement with other estimates of the N_2 bend but
much lower than the 262 cm^{-1} observed for the HF bend.[15] The first
estimate of the bending frequency is obtained from a Boltzmann analysis of the relative intensities (60±15) of the two bands. This yields
an estimate of the bend frequency of 100±25 cm^{-1}, assuming the vibrational degrees of freedom are in equilibrium at the rotational temperature of the fundamental. A confirming estimate of the bend frequency
comes from analysis of the ℓ-doubling parameter. In the absence of a
first-order perturbation of the Π^e state by nearby states of Σ^+ symmetry, the ℓ-doubling parameter q is approximately related to the bend
frequency ω_b by[40]

$$\omega_b = \frac{2B_e^2}{q} \left\{ 1 + 4 \sum_i \xi_{bi} \frac{\omega_b^2}{\omega_i^2 - \omega_b^2} \right\} \tag{5}$$

where B_e is the equilibrium value of the rotational constant, and the
HF subunit is treated as a point mass to make use of the above linear
triatomic expression. In Eq. (5), ξ_{bi} is the Coriolis coupling constant for the ith vibration interacting with the bend, and ω_i are the
corresponding harmonic stretch frequencies. Coriolis coupling constants for both the N_2 and the HF bends are calculated using the experimental values for the bond lengths and for the HF and N_2 stretch
frequencies; the van der Waals stretch frequency is estimated from
the diatomic approximation (see above) as 91 cm^{-1}. Cross coupling
terms are not known and therefore not included. This procedure yields

$\omega_b \approx 80$ cm^{-1} for the N_2 bend and predicts that the ℓ-doubling parameter is positive. In contrast, the ℓ-doubling parameter predicted for the ν_4 HF bend using the known bend frequency[15] determined from matrix isolation studies is 8 times smaller than the observed value. This calculation, in conjunction with harmonic estimates based on observed N_2 quadrupolar couplings in microwave studies,[7] leaves little doubt that the lower state of the hot band transition is the lowest bending frequency, $\nu_5 = 1$, i.e., the N_2 librational mode. Based on these estimates, we hope to be able to locate the $\nu_1 + \nu_5$ combination band in the slit jet spectra, from which a more direct bend frequency measurement can be made.

3.3. HFCO$_2$: Stretching and Bending Vibrations in a Highly Nonrigid Polyatomic Complex

The IR spectroscopy of HFCO$_2$ serves as an intriguing comparison with both ArHF and HFN$_2$. All three complexes have been investigated by microwave studies[3],[5-7]; the equilibrium structures have been determined to be linear. This result was unanticipated for HFCO$_2$ in light of the sp^2 lone pair electron distribution on oxygen and a Lewis acid-base description of the hydrogen bonding. This proved particularly surprising since HFN$_2$O, which is isoelectronic with HFCO$_2$, was subsequently determined to have a bent equilibrium structure. Ab initio calculations further served to verify these structural predictions.[11] If, however, the linear equilibrium geometry in HFCO$_2$ results from a nearly balanced competition between on-axis and off-axis contributions, one would anticipate a shallow bending potential which could be elucidated via hot band absorption spectra in the slit jet apparatus.

A relatively strong series of absorption transitions is observed (peak absorbances $\geq 10^{-3}$) near 3909 cm^{-1} which can be assigned to the $\nu_1 = 1 \leftarrow 0$ HF stretch fundamental spectrum of HFCO$_2$. A stick plot of the observed frequencies is shown in Fig. 9. The clear R and P branch progression and the null gap between R(0) and P(1) permit unambiguous J labeling of the spectral transitions. Combination differences for the ground state consistently agree with values calculated from microwave data to within experimental error.[5] To extract the molecular constants, combination differences for the upper and lower states are fit to standard Hamiltonian expressions[38] for a semirigid linear complex, and analyzed in a similar manner as for ArHF and HFN$_2$. The molecular constants thus obtained are presented in Table 3.

The HFCO$_2$ molecular constants indicate several interesting points. Firstly, there is a significant increase (+1.75%) in the B rotational constant upon ν_1 vibrational excitation. This is immediately evident from the stick plot in Fig. 7, where a compression in the P versus R branch line spacings is quite evident. The sign of the change is identical to what is observed in several other HF containing complexes, and has been interpreted as an indication of stronger bonding in the vibrationally excited state. The magnitude of the change, however, is much greater than what has been observed for any of the linear HF complexes. If one constrains the complex to the nearly linear geometry suggested from the microwave data, the observed change

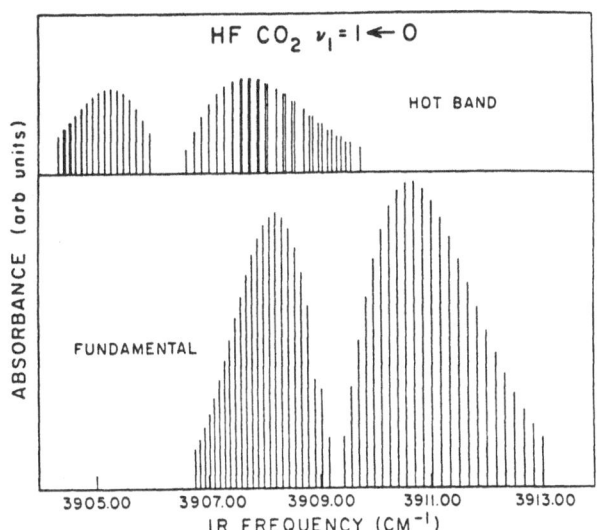

Fig. 9. Stick plot of the observed transitions in the ν_1 fundamental
(lower) + $\nu_1 + \nu_6 - \nu_6$ bend excited hot band (lower) in HFCO$_2$. The bunch-
ing of lines in the P branch is evidence for a large increase in B upon
ν_1 excitation. Doubling in the hot band spectra is clearly visible and
due to rotational splitting of the doubly degenerate bend vibration.
The 2.5±1 intensity ratio between fundamental and hot band transitions
is consistent with a ν_6 bending vibration of 10±5 cm^{-1}, i.e., an order
of magnitude smaller than indicated in HFN$_2$ (see Fig. 8).

in B would require a 0.041 Å decrease (from 1.921 to 1.879 Å) in the
hydrogen bond length. This should be compared to the 4-7 times smaller
change (0.010 and 0.006 Å) in the bond length indicated by the N$_2$HF and
ArHF infrared data, respectively.

Secondly, if the increase in B were in fact reflective of a
greatly enhanced well depth in the upper state, then one would expect
a comparably dramatic reduction in the centrifugal distortion constant
due to tighter binding. What is observed, in fact, is a 55% increase
in D upon ν_1 excitation. Again, this is inconsistent both in sign and
magnitude with the observed behavior for ArHF and N$_2$HF in which D is
observed to decrease by roughly 10%.

It is worth noting that the observed dependences of B and D on
vibrational excitation are contradictory only in the context of a
nearly linear geometry. An increase in both B and D would in fact be
predicted for nonlinear geometries where the CO$_2$ is bent away from the
HF. An increase in B would result from a foreshortened projection of
the nuclei on the A axis. The dramatic increase in D, on the other
hand, could be the consequence of J dependent centrifugal straightening of the complex against a weak bending potential. There would be a

Table 3. Molecular constants (in cm^{-1}) obtained for HFCO$_2$.

	Ground State	$\nu_1=1, \nu_6=0$	$\nu_1=0, \nu_6=1$	$\nu_1=1, \nu_6=1$
B	0.065110(10)	0.066247(11)	0.064709(13)	0.065804(41)
	(0.06508401(4))[a]			
D	3.43(15)×10^{-7}	5.33(16)×10^{-7}	2.7(9)×10^{-7}	4.21(15)×10^{-7}
	(3.56(2)×10^{-7})[a]			
$E_v - E_o$	0	3909.3204(30)	$\nu_6 \cong 10\pm5$ cm^{-1} [c]	$\nu_6 + 3906.3659(30)$
		(3871)[b]		
			$|q| = 4.21(15)\times10^{-4}$	$|q| = 5.78(15)\times10^{-4}$
$X_{16} = 2.954(2)$				

[a]Ref. 5.

[b]Ref. 14.

[c]Estimated from relative intensities in the ν_1 fundamental and hot band, based on equilibrated rotations and low frequency vibrations.

similar effect for geometries where the HF constituent is tilted away from the internuclear axis as well, but would be much smaller due to the light librational mass of the HF bend coordinate, and the relative insensitivity of B to the coordinates of the H atom. The complement of this effect, i.e., rotational saturation, has been observed for high K excitation in HF dimer,[22] in which the molecular geometry evolves into a nearly perpendicular configuration of the two monomers.

The magnitude of this centrifugal straightening effect can be estimated analytically for a semiclassical model of a semirigid complex with harmonic restoring forces in the bend coordinate to be[26]

$$D_{\text{cent. straight.}} \cong \frac{\left(\frac{\partial B}{\partial \theta}\right)^2_{eq}}{k_\theta} . \tag{6}$$

Since $\partial B/\partial \theta = 0$ as $\theta_{eq} = 0$, it is clear from Eq. (6) that significant D enhancement from this model requires both 1) a nonlinear geometry and 2) a very soft bending potential. For example, if one attributes the change in B and D solely to bending of the complex (holding center-of-mass separations of the constituents fixed), the data would be consistent with an upper state bend angle of roughly 15 degrees and a corresponding bending force constant of 0.1 cm^{-1}/degree^2. This translates into a predicted frequency for the CO$_2$ ν_6 librational mode of approximately 10 cm^{-1}, which is low enough to be significantly populated in the jet expansion.

It bears stressing that Eq. (6) applies strictly for a model of the bending vibration with a <u>nonlinear</u> equilibrium geometry. Due to vibrational averaging, however, this centrifugal straightening mechanism should also be important for a <u>linear</u> but highly nonrigid equilibrium geometry. In the latter case the wave function in the bend coordinate would simply be more diffuse at low J, and narrowing toward small θ with increasing J. From these observations alone, therefore, one cannot distinguish between an equilibrium, or simply a vibrationally averaged bent geometry. Both potential surface topologies, however, would require weak restoring forces in the bending coordinate in order to reproduce the observed trends. We are presently pursuing fully quantum close coupled calculations on model potentials to test the accuracy of these classical predictions, and to characterize the correct shape of the energy surface.

We see further evidence for a weak bending potential by a closer inspection of the $HFCO_2$ spectrum. In the P branch of the $\nu_1 = 1\leftarrow0$ fundamental a progression of hot band doublet transitions is observed with a clear P and R branch structure and an origin that is <u>red-shifted</u> from the ν_1 fundamental by 2.954(2) cm^{-1}. The roughly 0.065 cm^{-1} spacing identifies the spectrum as belonging to $HFCO_2$; the doubling of the spectra is characteristic of a parallel transition out of an excited $HFCO_2$ complex with a doubly degenerate bend vibration excited.

The intensity ratio between the fundamental and each component of the corresponding doublets is approximately 2.5±1. This ratio rules out the possibility that the excited vibration correlates to the free ν_2 vibration in CO_2, since even at 300 K the fractional population[38] would be only 6%. The HF bend, ν_4, has been seen in matrix studies at 313 cm^{-1}, and would be negligibly populated at the 16 K jet temperatures. This leaves ν_6, i.e., the CO_2 librational mode, as the only candidate; if one assumes that the vibrational and rotational degrees of freedom are equilibrated, the observed intensity ratio between the fundamental and hot band correspond to a ν_6 frequency of 10±5 cm^{-1}. This value is in qualitative agreement with the predictions made previously which considered only changes in B and D upon ν_1 excitation.

As was performed for the HFN_2 hot band data, J assignment of the transitions is achieved by fitting the splittings to q'J'(J'+1)-q"J"(J +1) and shifting the J labeling until the curve intersects the origin. With the J assignment determined, the average line frequency for each doublet is least-squares fit to standard expressions[38] for a $|\ell| = 1$ Π–Π transition; the molecular constants obtained from these fits are shown in Table 3.

There are several features worth noting about the spectral fits. First as in the case of the fundamental transition, the rotational B constant for the bend excited states <u>increases</u> (+1.7%) with ν_1 vibrational excitation. This is consistent with the previous observation in the ν_1 fundamental, and indicates a shortening of the center-of-mass separations between HF and CO_2, which could arise via a greater vibrationally averaged bend CO_2 bend angle or a deeper attractive well. However, bend excitation consistently <u>decreases</u> the fitted B constants over the corresponding values observed in the ν_1 fundamental, which suggests the following simple physical interpretation. Excitation of bending modes in the complex bends the constituents further away from the energy minimum, and places a node in the orientation of strongest interactions. This tends to weaken the intermolecular bond, and hence increase the center-of-mass displacements of the complexed species. A similar effect is evident in both the ArHF and HFN_2 complexes.

Secondly, the vibrational origin for the transition is red-shifted by 2.95 cm^{-1} from the ν_1 fundamental. This implies a positive off diagonal anharmonicity with respect to the low frequency bend. This would be somewhat inconsistent with the previous explanation of the overall decrease in B upon bend excitation, which relied on a weakening of the bonding interactions between HF and CO_2, and which would tend to blue-shift the HF stretch back in the direction of the free HF monomer.

Thirdly, centrifugal distortion for the bend excited complexes increases dramatically upon $\nu_5+\nu_1-\nu_1$ excitation, similar to what is observed on the fundamental transition. Again, this increase in both B and D is in contrast with expectations based on nearly linear complexes and suggests a significant shift in the vibrationally averaged geometry in the ν_1 excited state.

Finally, the magnitude of the centrifugal distortion constants consistently decreases on excitation of the bend. Note that the exact opposite behavior is observed in both ArHF and HFN_2, and was attributed to a weakening of the bond upon bend excitation. However, this behavior in $HFCO_2$ would be completely consistent if a major contribution to D comes from centrifugal straightening of a highly nonrigid complex. For finite angular momentum around the figure axis, there will be a centrifugal barrier which goes to infinity at the linear geometry; this can compete effectively with the tendency to straighten the complex with increasing end-over-end rotation, and hence reduce the centrifugal distortion. It is noteworthy that the decrease in D on excitation of the bend occurs for both ν_1 = 0 and 1. This would suggest a significant contribution to D even for the ground state occurs from centrifugal straightening, which may explain the anomalously large values measured in the early microwave studies.[5]

4. SUMMARY AND CONCLUSION

A general method for direct IR absorption spectra of weakly bound complexes in slit supersonic expansions is described which permits high detection sensitivity ($\lesssim 3 \times 10^9/cm^3$/quantum state) and high spectral resolution ($\Delta\nu \lesssim 0.001$ cm^{-1}). Advantages of the approach are 1) lack of spectral congestion in the cooled jet, 2) control of expansion temperature for observation of low frequency intermolecular cluster vibrations, 3) sub-Doppler resolution from velocity narrowing in the slit expansion, 4) long path length × number density for direct absorption, and 5) applicability of the method with low power high resolution light sources. Fundamental, combination and low frequency intermolecular hot band spectra have been obtained and analyzed for ArHF, HFN_2 and $HFCO_2$. The vibrational state dependence of the rotational constants provides information on structural shifts in the molecules upon excitation. An enhancement in intermolecular bonding upon HF stretch excitation is consistently observed. Excitation of bending nodes in the complex, on the other hand, consistently weakens the bonding. In HFN_2, the observed behavior of the rotational constants and vibrational frequencies is consistent with a nearly linear geometry and a relatively strong restoring potential with respect to N_2 rotation. For $HFCO_2$, however, the behavior suggests a very nonrigid potential for CO_2 libration (ν_6), and hence a significantly bent vibrationally averaged geometry, at least in the ν_1 excited state. The ν_6 bend frequency is comparable to the time scale for end-over-end rotation for high J's in the $HFCO_2$ complex, which calls into question the validity of standard analyses of vibration-rotation spectral structure. Full quantum calculations on trial potentials are presently being performed to fit the data and thereby infer the topology of the true potential energy surface.

ACKNOWLEDGMENTS

Support for this research by the National Science Foundation (PHY82-00805, CHE86-05970) through the University of Colorado, as well as grants from the Petroleum Research Foundation, Research Corporation and the Henry and Camille Dreyfus Foundation are gratefully acknowledged. The efforts of Christopher M. Lovejoy and Michael D. Schuder for help in collecting and analyzing the data presented are also acknowledged.

REFERENCES

1. D. G. Truhlar, Ed., Resonances in Electron-Molecule Scattering, van der Waals Complexes and Reactive Chemical Dynamics (ACS Symposium Series, Vol. 263, 1984).
2. R. J. LeRoy and J. S. Carley, in Potential Energy Surfaces, edited by K. P. Lawley (Wiley, New York, 1980).
3. S. J. Harris, S. E. Novick and W. Klemperer, J. Chem. Phys. 60, 3208 (1974); T. A. Dixon, C. H. Joyner, F. A. Baiocchi and W. Klemperer, J. Chem. Phys. 74, 6539 (1981).
4. D. H. Levy, L. Wharton and R. E. Smalley, Chemical and Biochemical Applications of Lasers, edited by C. B. Moore (Academic, New York, 1977).
5. F. A. Baiocchi, T. A. Dixon, C. H. Joyner and W. Klemperer, J. Chem. Phys. 74, 6544 (1981).
6. M. R. Keenan, L. W. Buxton, E. J. Campbell, A. C. Legon and W. H. Flygare, J. Chem. Phys. 74, 2133 (1981).
7. P. D. Soper, A. C. Legon, W. G. Read and W. H. Flygare, J. Chem. Phys. 76, 292 (1982).
8. K. C. Jackson, P. R. R. Langridge-Smith and B. J. Howard, Mol. Phys. 39, 817 (1980).
9. A. M. Dunker and R. G. Gordon, J. Chem. Phys. 64, 354 (1976).

10. M. A. Benzel and C. E. Dykstra, J. Chem. Phys. **78**, 4052 (1983).
11. A. E. Reed, F. Weinhold, L. A. Curtiss and D. J. Pochatko, J. Chem. Phys. **84**, 5687 (1986).
12. J. M. Hutson and B. J. Howard, Mol. Phys. **45**, 791 (1982).
13. J. A. Beswick and J. Jortner, Adv. Chem. Phys. **47**, 363 (1981).
14. L. Andrews and G. L. Johnson, J. Chem. Phys. **76**, 2875 (1982).
15. L. Andrews, B. J. Kelsall and R. T. Arlinghaus, J. Chem. Phys. **79**, 2488 (1983).
16. D. H. Rank, B. S. Rao and T. A. Wiggins, J. Chem. Phys. **37**, 2511 (1962).
17. K. D. Kolenbrander and J. M. Lisy (private communication).
18. Z. S. Huang, K. W. Jucks and R. E. Miller (private communication).
19. Ph. Marteau, C. Boulet and D. Robert, J. Chem. Phys. **80**, 3632 (1984).
20. J. M. Farrar and Y. T. Lee, Chem. Phys. Lett. **26**, 428 (1974).
21. B. Schramm and U. Leuchs, Ber. Bunsenges. Phys. Chem. **83**, 847 (1979).
22. A. S. Pine, W. J. Lafferty and B. J. Howard, J. Chem. Phys. **81**, 2939 (1984); A. S. Pine and W. J. Lafferty, J. Chem. Phys. **78**, 2154 (1983).
23. G. T. Fraser and A. S. Pine, J. Chem. Phys. **85**, 2502 (1986).
24. C. M. Lovejoy, M. D. Schuder and D. J. Nesbitt, Chem. Phys. Lett. **127**, 374 (1986).
25. C. M. Lovejoy, M. D. Schuder and D. J. Nesbitt, J. Chem. Phys. (in press).
26. C. M. Lovejoy and D. J. Nesbitt, J. Chem. Phys. (submitted).
27. B. A. Wofford, J. W. Bevan, W. B. Olson and W. J. Lafferty, J. Chem. Phys. **83**, 6188 (1985).
28. M. F. Vernon, J. M. Lisy, H.-S. Kwok, D. J. Krajnovich, A. Tramer, Y. R. Shen and Y. T. Lee, J. Phys. Chem. **85**, 3327 (1981).
29. T. E. Gough, R. E. Miller and G. Scoles, Appl. Phys. Lett. **30**, 338 (1977).
30. G. D. Hayman, J. Hodge, B. J. Howard, J. S. Muenter and T. R. Dyke, Chem. Phys. Lett. **118**, 12 (1985).
31. M. D. Marshall, A. Charo, H. O. Leung and W. Klemperer, J. Chem. Phys. **83**, 4924 (1985).
32. D. Ray, R. L. Robinson, D.-H. Gwo and R. J. Saykally, J. Chem. Phys. **84**, 1171 (1986).
33. J. L. Hall and S. A. Lee, Appl. Phys. Lett. **29**, 367 (1976).
34. G. Guelachvili, Opt. Commun. **19**, 150 (1976).
35. K. Veeken and J. Reuss, Appl. Phys. B **34**, 149 (1984).
36. U. Buck (private communication).
37. B. J. Howard and A. S. Pine, Chem. Phys. Lett. **122**, 1 (1985).
38. G. Herzberg, Infrared and Raman Spectra (van Nostrand Reinhold, New York, 1945).
39. P. R. Bunker, Molecular Symmetry and Spectroscopy (Academic, New York, 1979).
40. C. H. Townes and A. L. Schawlow, Microwave Spectroscopy (Dover, New York, 1975).

SUB-DOPPLER RESOLUTION INFRARED SPECTROSCOPY OF BINARY MOLECULAR
COMPLEXES

R.E. MILLER
Department of Chemistry
University of North Carolina
Chapel Hill, N.C. 27514

ABSTRACT. A newly constructed infrared laser - molecular beam
apparatus has been used to investigate the infrared spectroscopy and
vibrational predissociation of a number of HF containing van der Waals
molecules. In all cases, well resolved spectra have been obtained and
assigned, thus giving not only unique species identification but also
accurate molecular constants. The resolution of the opto-thermal
method is sufficient to observe homogeneous broadening of the
transitions in most cases. It is therefore possible to accurately
examine the dependence of the vibrational relaxation rate on various
properties, such as, the complexity of the partner in the binary
complex and the particular vibrational mode excited.

INTRODUCTION

The usefulness of gas phase infrared spectroscopy in the study of
van der- Waals molecules has increased steadily over the past ten
years. In the first experiments, carried out in both gas cells [1]
and molecular beams [2-5], no rotational fine structure was observed
in the spectra which meant that reliable spectroscopic assignments
could not be made. This made it difficult, particularly in molecular
beam experiments, to uniquely assign the observed spectrum to a
specific van der Waals cluster. Indeed, even when mass spectrometric
detection is used in conjunction with molecular beams, care must be
taken to avoid fragmentation of large clusters [6].
    Much of the interest in infrared van der Waals spectroscopy
arises from the fact that vibrational predissociation accompanies
vibrational excitation. However, the rates for this process remain
the subject of considerable controversy for many molecular clusters
[7]. This is a result of the fact that the source of the spectral
broadening, which leads to an unresolved vibrational spectrum, has
always been somewhat unclear. If the major limitation is due to
homogeneous broadening associated with vibrational predissociation of
the excited state clusters, then the width of the spectrum can be used
to determine the lifetime. On the other hand, the rotational band
contour must clearly be responsible for some, if not the majority, of

A. Weber (ed.), Structure and Dynamics of Weakly Bound Molecular Complexes, 131–140.

the spectral width. The uncertainty in the lifetime, therefore, arises from the fact that without rotationally resolved spectra, it is difficult to tell which of these conditions is applicable.

Recently, a number of infrared laser methods have developed to the point where rotationally resolved spectra are now obtainable for a wide range of binary complexes. Pine and co-workers [8] have made use of a difference frequency laser, in conjunction with a long path length gas cell, to observe rotationally resolved spectra for systems such as $(HF)_2$, $(HCl)_2$ and Rg-HF. For the case of more complex clusters, we have shown that the low temperatures obtainable from a free jet source can be used to observe well resolved spectra at sub-Doppler resolution [9,10]. More recently, Howard and co-workers [11] and Nesbitt and co-workers [12] have also observed rotationally resolved spectra by measuring direct absorption in a free jet expansion.

In this report, we will summarize the results of a series of experiments, carried out using the opto-thermal detection method, on a range of HF containing binary complexes. In all of the cases considered here the infrared spectra show well resolved rotational structure which can easily be assigned to obtain accurate molecluar constants. As a result of the high molecular beam collimation used in these experiments, the instrumental resolution is sufficient to observe homogeneous broadening in most of the systems studied to date. These linewidths have been used to calculate vibrational relaxation lifetimes for a large number of binary molecular complexes. Results of this type are timely in view of the fact that it is now possible to carry out accurate closed coupled calculations for systems of the type reported here [13,14].

EXPERIMENTAL METHOD

The details of the opto-thermal detection method have been given in a number of previous publications [15,16]. For this reason we will restrict our present discussion to the new developments made on the present apparatus, namely those which enable it to be used as a general sub-Doppler resolution infrared spectrometer. A major improvement has been made in the present apparatus by developing, with considerable help from the group at Rice University [17], a computerized scanning procedure for the Burleigh F-center laser. Figure 1 shows a schematic of the experimental arrangement. The computer used in this system is a PDP11/73 and all interfaces with the apparatus are made through the Q-bus. Three confocal etalons, with free spectral ranges of 150.1 MHz, 734.4 MHz and 7500 MHz, and a gas cell containing low pressure water vapor are used to monitor and calibrate the wavelength of the laser. Extended single mode tuning of the F-center laser is accomplished by simultaneously scanning the three laser tuning elements, namely the grating, intracavity etalon and the cavity end mirror. Periodically, when the intracavity etalon needs to be reset, operator intervention is required to ensure that the laser returns to the same frequency before scanning is resumed.

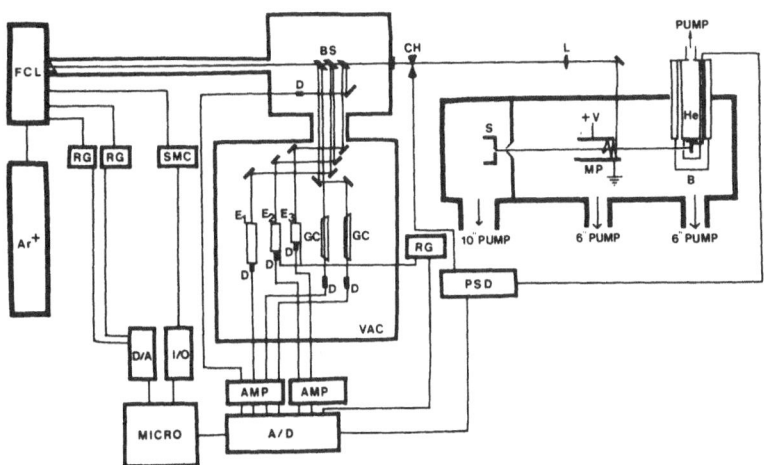

<u>Figure 1</u>: Schematic diagram of the apparatus. RG–ramp generator,
            SMC–stepping motor controller, E–etalons, D–detectors,
            GC–gas cells, BS–beam splitters, CH–chopper, PSD–phase
            sensitive detector, S–beam source, V–Stark voltage,
            MP–multipass cell, B–bolometer, FCL–F–center laser.

Typical scanning times vary from 15 to 45 min. per wavenumber at a
resolution of 10^{-3} cm^{-1}. Considerably slower scanning rates are used
when higher resolution is required.
    In order to improve the sensitivity of the detection method we
have used a parallel mirror multipass arrangement [18] in most cases.
The single orthogonal laser–molecular beam crossing geometry is only
used when very high resolution (approximately 10^{-4} cm^{-1}) is required.
For example, this higher resolution was needed in several cases in
order to determine the homogeneous linewidth associated with
vibrational predissociation. The two gold coated mirrors of the
multipass cell can also be used to apply an electric field to the
laser–molecular beam interaction region in order to carry out Stark
measurements. This geometry ensures that the laser polarization is
perpendicular to the static electric field so that the selection rule
is $\triangle M=\pm 1$ [19].
    The gas mixtures used to supply the free jet source were mixed on
line using a series of neddle valves. In this way, the concentrations
of the gases are easily changed, as is the system being studied. In
the experiments discussed here, HF in helium mixtures, varying in HF
concentration from 0.6% to 1.0%, were passed through one neddle valve
while the flow of the secondary gas (used to supply the partner in the
binary complex) was regulated by a second neddle valve. The final gas
mixture typically contained 10% of the secondary gas. The rather high
concentration of condensible gases in the beam lead to some detector
problems due to the accumulation of material on the liquid helium

cooled bolometer. This leads to an increase in the heat capacity of
the bolometer and hence a reduction in its response time. In
practice, after a beam exposure time of six to eight hours the
sensitivity of the bolometer became too low to be useful. At this
point the bolometer had to be warmed up in order to remove the
condensible material.

INFRARED SPECTROSCOPY

In this section, we will summarize some of our recent results
which demonstrate the capabilities of the opto-thermal molecular beam
technique. N_2-HF serves as a good starting point. It is well known
from the microwave spectrum [20] that N_2-HF is a linear molecule so
that the assignment of the infrared spectrum should be
straightforward. Figure 2 shows a portion of the N_2-HF spectrum
obtained using the multiple crossing geometry. In this case, the
molecular beam was formed by expanding a mixture of 0.6% HF and 10% N_2
in helium at a source pressure of 800 kPa. Under these conditions,
the linewidth of the transitions is approximately 25 MHz. Clearly,
the sensitivity and resolution of the method are both excellent. A
detailed analysis of this spectrum [19] has provided accurate
molecular constants for both the ground and the excited vibrational
states of the complex, namely B_1=0.10719 $\pm$ 0.0001 cm^{-1} and
ν_o =3918.2397 $\pm$ 0.005 cm^{-1} .
    For HF complexes formed from polyatomic molecules it is expected

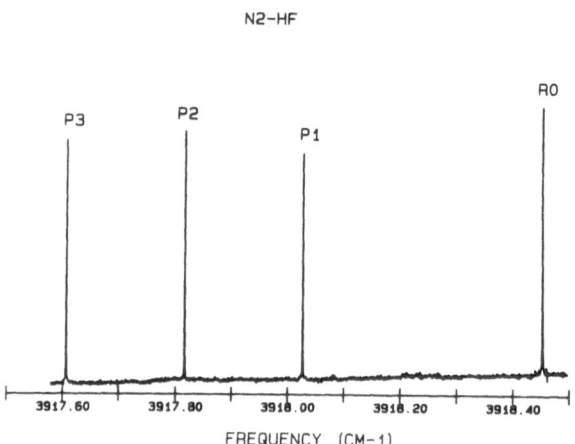

Figure 2: The central portion of the N_2-HF spectrum recorded at a
          resolution of 25 MHz.

that the density of rotational lines in the spectrum will increase,
partially as a result of the smaller rotational constants, but also
due to the fact that the complex will in general be an asymmetric top.
We have investigated several complexes of this type, including C_2H_2-HF
and C_2H_4-HF. Figure 3 shows a portion of the C_2H_2-HF spectrum which
is again well resolved into individual rotational transitions. In
this case, however, the spectrum is considerably more complicated.
The microwave spectrum [21] indicates that the C_2H_2-HF complex is T-
shaped and is therefore a slightly asymmetric rotor. It is the
asymmetry of this molecule which leads to the majority of the
splitting in the Q branch shown in Figure 3. It is interesting to
note that if C_2H_2-HF is indeed T-shaped, then there should be a 3:1
intensity ratio between the odd and even K states resulting from the
associated nuclear spin statistics. This is clearly visible in the P2
transtion in Figure 3 where, in the absence of spin statistics, the
central (K=0) peak would be much larger than the side peaks (K=±1).
Once again, accurate rotational constants have been determined for
this system by assigning and fitting the entire spectrum [22].

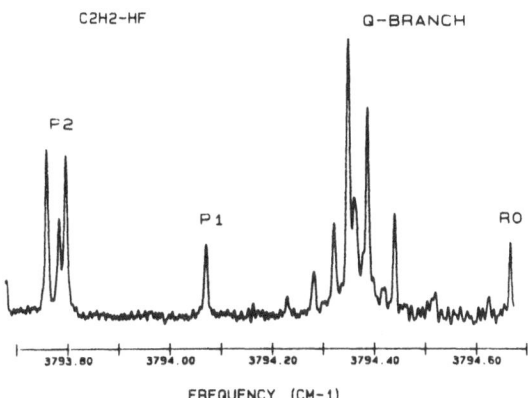

Figure 3: The central portion of the C_2H_2-HF spectrum. In this
spectrum the resolution is limited by the homogeneous
broadening associated with the predissociation of the
complex. Splitting in the Q-branch is primarily due to
the asymmetry of the molecule.

It is obvious from the two spectra discussed above that
rotational line congestion quickly becomes a problem as the complexity
of the molecule is increased. This is not only due to the large
increase in the density of rotational transitions but also to the
increase in the linewidth of the individual transitions. This
homogeneous broadening is a result of vibrational predissociation of
the complex, a subject we will discuss in more detail shortly.

The rotational temperature of binary complexes in molecular beams has been the subject of much speculation, since it is required in order to estimate the rotational contour contribution to the width of unresolved vibrational bands. In most cases, estimates of this type have been made by assuming that the rotational temperature of the complex is essentially the same as that of the monomer. In the systems we have studied to date, we have found that the rotational temperature of the complex is generally considerably less than that of the monomers in the molecular beam. For example, in a 0.6% mixture of HF in helium expanded from a source pressure of 800 kPa, the rotational temperature of HF is approximately 30 K while that of the dimer is 1.75 K. This large difference results from the fact that the B rotational constant of the dimer is considerably smaller than that of the monomer so that R-T relaxation is more efficient in the case of the former. It is clear from these results that, by the time the expansion is complete, the complexes have long forgotten about the condensation energy released at the time of their formation far upstream.

VIBRATIONAL PREDISSOCIATION

As indicated in a previous section, spectra of the type discussed above can be used to determine not only accurate molecular constants, from which molecular structures can be obtained, but also vibrational predissociation lifetimes. This information can be obtained in cases where the linewidth of the observed transitions is dominated by lifetime broadening rather than by instrumental effects, such as doppler broadening or laser jitter. Table 1 gives a summary of the lifetimes we have determined from these high resolution spectra. In several cases, a single laser - molecular beam crossing geometry had to be used since the homogeneous broadening was smaller than the instrumental linewidth obtained using the multipass cell. It should be noted that, in the case of N_2-HF, the linewidth had to be corrected to take into account the nuclear quadrupole splitting [19].

Let us begin our discussion of predissociation lifetimes by examining the Ar-HF and Kr-HF binary complexes. It is clear from Table 1 that the lifetimes of these systems are extremely long. They are, in fact, far too long to be determined from a linewidth measurement since the associated homogeneous broadening would only be 1 kHz. Instead, this lower limit on the lifetime was determined from the molecular time of flight from the laser crossing point to the bolometer. It is possible to place this limit on the lifetime since the laser induced bolometer signal is observed to be positive for these complexes. This indicates that the lifetime of the complex is greater than the flight time of the molecules from the laser crossing region to the bolometer, which, for the present experimental arrangement, is $3*10^{-4}$ s.

In all of the other systems reported in Table 1 the laser induced bolometer signals were negative, indicating that vibrational predissociation occurs on a time scale which is short with respect to the molecular flight time. Indeed, for all of these cases it was

Table 1. Summary of the vibrational predissociation lifetimes determined for various HF containing binary complexes. The uncertainty in the vibrational mode assignment for H_2-HF results
from the fact that we have not yet determined   the structure
for this molecule. In all cases, however, it is the HF stretching vibration which is excited. The lifetimes have been calculated from the linewidths using the expression
$t = (2\pi\Delta\nu_{FWHM})^{-1}$.

Molecule	Vibrational Mode	Lorentzian FWHM (MHz)	Lifetime (s)
Ar-HF	ν_1	–	$>3.\times10^{-4}$
Kr-HF	ν_1	–	$>3.\times10^{-4}$
$(HF)_2$	ν_1	13.4	$12.\times10^{-9}$
$(HF)_2$	ν_2	320.	0.5×10^{-9}
N_2-HF	ν_1	7.2	$22.\times10^{-9}$
OC-HF	ν_1	190.	0.9×10^{-9}
H_2-HF	ν_1 or ν_2	6.0	$27.\times10^{-9}$
C_2H_2-HF	ν_1	200.	0.8×10^{-9}
C_2H_4-HF	ν_1	480.	0.34×10^{-9}

possible to determine the lifetime from the homogeneous broadening of
individual ro-vibrational transitions. It is clear, from Table 1,
that there is a rather strong correlation between the predissociation
lifetime and the complexity of the partner in the binary complex.
Indeed, the lifetime for C_2H_4-HF is more than five orders of magnitude
shorter than that of Ar-HF, while the lifetimes of the other systems
fall somewhere between these two extremes. These results support, at
least in a very qualitative sense, the energy or momentum gap picture
of vibrational predissociation which suggests that the availability of
low translational energy exit channels results in fast predissociation
rates.
    Although this correlation is rather strong, it is clearly not the
only factor controlling the rate of vibrational predissociation. For
example, in $(HF)_2$ the excitation of the two HF stretching vibrations
leads to very different lifetimes despite the fact that the available
exit channels are essentially identical. This can be qualitatively
understood if one considers that one of the HF vibrations is strongly
coupled to the hydrogen bond while the other is essentially free. The
N_2-HF and OC-HF systems provide us with another example where there are
very similar rotational and vibrational states available to the

fragments while the lifetimes are very different, namely 22 ns and 0.8 ns, respectively. It is worth noting that the binding energies of N_2-HF and OC-HF are also rather different, namely 618 cm^{-1} and 987 cm^{-1}, respectively [23]. This suggests that the coupling between the inter- and intra-molecular motions might be larger for the case of OC-HF. In order to obtain a quantitative understanding of these differences it will be necessary to compare these results with complete closed coupled calculations. Calculations of this type are now feasible for most of the binary complexes studied here [13,14].

As indicated in Table 1, we have recently observed transitions associated with the HF stretching vibration of H_2-HF. This binary complex has not been studied previously by any other spectroscopic method. Although the spectrum has not yet been assigned, we have obtained the vibrational predissociation lifetime from the width of the observed transitions. As indicated in Figure 4, isotopic substitution in this system would be very interesting since this would result in a large change in the energetics of the exit channels while keeping both the intra- and inter-molecluar potentials, and thus the molecular geometery, the same. Of particular interest is the fact that for H_2-HF the H_2 v=1 fragment channel is closed, while it is open for the cases D_2-HF and HD-HF. Indeed, for HD-HF the exit channel producing a vibrationally excited HD fragment is very nearly degenerate with the available energy. In this case, near resonant vibrational energy transfer could lead to rapid vibration predissociation. Experiments on the isotopically substituted H_2-HF binary complexes are presently underway.

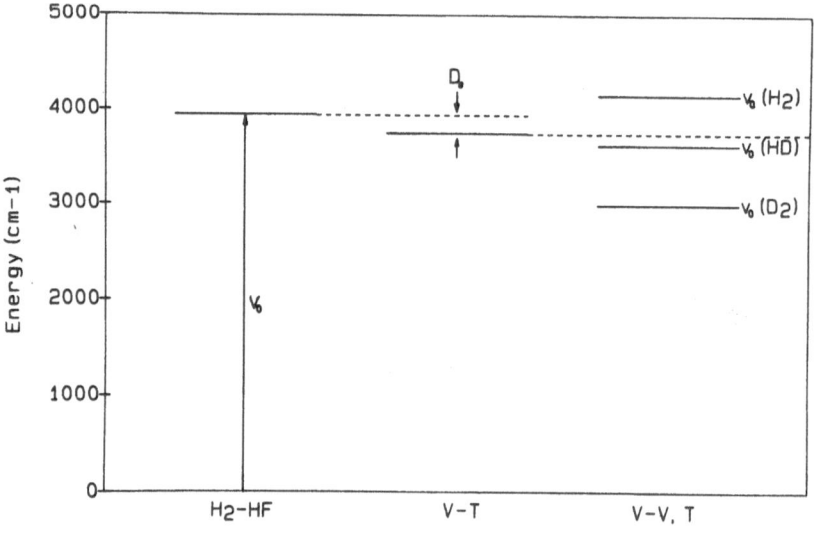

VIBRATIONAL EXIT CHANNELS

Figure 4: The energetics of the V-V,T channels for H_2-HF, HD-HF and D_2-HF.

A major factor which presently limits our ability to obtain quantitative agreement between the experimentally determined lifetimes and those calculated by various theoretical methods is the difficulty associated with determining accurate inter and intramolecular potentials. In particular, the intermolecular stretching dependence of the intramolecular potential, which is clearly the important quantity in vibrational predissociation, is very difficult to determine. However, some information can be obtained directly from the near infrared spectra discussed here. For example, in Ar-HF the rotational constant is observed to increase upon vibrational excitation and the vibrational band is shifted to the red of the monomer. Both of these effects indicate that the intermolecular potential becomes deeper when the molecule is vibrationally excited.

By measuring infrared Stark spectra of these systems, we have been able to obtain information on the stretching dependence of the intermolecular potential anisotropy. This has been done by measuring the dipole moment of the complex in both the ground and excited vibrational states. The major change in the dipole moment comes about because of the change in the amplitude of the intermolecular bending motion upon intramolecular vibrational excitation. This motion tends to average out the dipole moment of the HF monomer and hence decrease the dipole moment of the complex. The stiffer bond associated with the vibrationally excited complex leads to a smaller bending amplitude and hence a larger dipole moment. For example, in Ar-HF, the bending angle decreases from 48.3° to 43° upon vibrational excitation. It is precisely this coupling between the intra and intermolecular motions which is responsible for vibrational predissociation. Clearly, data of this type will help in constructing realistic potentials for use in calculating the vibrational predissociation dynamics of these systems.

SUMMARY

The results summarized in this report demonstrate the fact that the opto-thermal detection method has now become a general spectroscopic technique for the study of van der Waals molecules. Not only do the spectra provide us with accurate molecular constants and structures but also unabiquous determinations of the vibrational predissociation lifetimes. The systems reported here are of great interest since there exist theoretical methods for calculating the detailed dynamics of these binary complexes.

ACKNOWLEDGEMENTS

This work was supported by the National Science Foundation (CHE-86-03604), Research Corporation and the Petroleum Research Fund.

REFERENCES

[1] D.H. Rank, B.S. Rao, and T.A. Wiggins, J. Chem. Phys. 37
     (1962) 2511; D.H. Rank, P. Sitaram, W.A. Glickman, and T.A.
     Wiggins, J. Chem. Phys. 39 (1963) 2673.
[2] T.E. Gough, R.E. Miller, and G. Scoles, J. Chem. Phys.
     69 (1978) 1588.
[3] M.A. Hoffbauer, W.R. Gentry, and C.F. Giese in Laser-
     Induced Processes in Molecules, K. Kompa and S.D. Smith, Eds.,
     Springer-Verlag, West Berlin, 1978, Springer Ser. Chem. Phys.
     Vol. 6, p. 252.
[4] M.P. Casassa, D.S. Bomse, and K.C. Janda, J. Chem. Phys.
     74 (1981) 5044.
[5] M.F. Vernon, D.J. Krajnovich, H.S. Kwok, J.M. Lisy, Y.R.
     Shen, and Y.T. Lee, J. Chem. Phys. 77 (1982) 47.
[6] U. Buck and H. Meyer, J. Chem. Phys. 84 (1986) 4854.
[7] R.E. Miller, J. Phys. Chem. 90 (1986) 3301.
[8] A.S. Pine and W.J. Lafferty, J. Chem. Phys. 78 (1983)
     2154; A.S. Pine, W.J. Lafferty, and B.J. Howard, J. Chem. Phys.
     81 (1984) 2939.
[9] R.E. Miller R.F. Vohralik, and R.O. Watts,
     J. Chem. Phys. 80 (1984) 5453.
[10] R.E. Miller and R.O. Watts, Chem. Phys. Lett. 105 (1984)
     409.
[11] G.D. Hayman, J. Hodge, B.J. Howard, J.S. Muenter, and
     T.R. Dyke, Chem. Phys. Lett. 118 (1985) 12.
[12] C.M. Lovejoy, M.D. Schuder, and D.J. Nesbitt,
     Chem. Phys. Lett. 127 (1986) 374.
[13] N. Halberstadt, Ph. Brechignac, J.A. Beswick, and M.
     Shapiro, J. Chem. Phys. 84 (1986) 170.
[14] A.C. Poet, D.C. Clary and J.M. Hutson, Chem. Phys. Lett.
     125 (1986) 477.
[15] T.E. Gough, R.E. Miller, and G. Scoles,
     Appl. Phys. Lett. 30 (1977) 338.
[16] R.E. Miller, R.O. Watts, and A. Ding, Chem. Phys. 83
     (1984) 155.
[17] J.V.V. Kasper, C.R. Pollock, R.F. Curl, Jr., and F.K.
     Tittel, Appl. Optics 21 (1982) 236.
[18] T.E. Gough, D. Gravel, and R.E. Miller, Rev. Sci. Inst.
     52 (1981) 802.
[19] Z.S. Huang, K.W. Jucks, and R.E. Miller, J. Chem. Phys.,
     in press.
[20] P.D. Soper, A.C. Legon, W.G. Read, and W.H. Flygare,
     J. Chem. Phys. 76 (1982) 292.
[21] W.G. Read and W.H. Flygare, J. Chem. Phys. 76 (1982)
     2238.
[22] Z.S. Huang and R.E. Miller, to be published.
[23] P.D. Soper, A.C. Legon, W.G. Read, and W.H. Flygare,
     J. Chem. Phys. 76 (1982) 292.

THE INFRARED SPECTRUM OF H_2-Ar: NEW OBSERVATIONS IN THE v = 1-0 AND 2-0
BANDS OF H_2

A. R. W. McKellar
Herzberg Institute of Astrophysics
National Research Council of Canada
Ottawa, Ontario   K1A 0R6
Canada

ABSTRACT. With the help of a new Fourier transform infrared
spectrometer and a long-path low-temperature absorption cell, spectra
of the H_2-Ar molecule have been obtained in the 1-0 band region ($\lambda \approx$
2.2 µm) with increased resolution and greatly increased line position
accuracy. The first resolved spectrum of H_2-Ar in the 2-0 band region
($\lambda \approx 1.2$ µm) has also been obtained; this result should improve determinations of the dependence of the hydrogen - argon potential surface
on the bond length of hydrogen.

1.   INTRODUCTION

Among triatomic systems, the intermolecular potential energy surfaces
for the hydrogen - inert gas combinations are perhaps the best characterized [1-4]. Much of the experimental data used to determine these
surfaces have come from infrared spectroscopy of the hydrogen - inert
gas Van der Waals molecules, although molecular-beam scattering,
classical bulk properties, hyperfine spectra, and other types of
experiment are also important, especially for determining the repulsive
part of the potential for small intermolecular distances.
    The infrared spectrum of H_2-Ar was first observed by Kudian et al.
[5] in 1965, who simply recorded the absorption spectrum of mixtures of
hydrogen and argon in the H_2 stretching band using low sample temperatures (99 K), moderate densities (≈10 amagat, where 1 amagat is the
density of an ideal gas at 0 C and 1 atm. pressure), and a long
absorption path of 13 m. Since then, there have been three generations
of improvements in the experiments, achieved by using even longer paths,
lower densities, lower temperatures, and higher spectral resolution and
sensitivity [6-8]. Isotopic molecules formed with D_2 and HD have also
been studied. The basic technique has however remained the same,
namely "classical" infrared absorption spectroscopy of bulk equilibrium
gas samples; it is uncertain whether more modern techniques involving
supersonic expansions and infrared lasers are applicable to these
(essentially nonpolar) hydrogen-containing Van der Waals molecules since
their infrared transition strengths are so low.

A. Weber (ed.), Structure and Dynamics of Weakly Bound Molecular Complexes, 141–147.

The present paper represents the beginning of still another generation of results on the infrared spectra of hydrogen-containing Van der Waals molecules. The physical conditions remain similar to those used in 1982 [8], but by combining the existing long-path absorption cell with a modern Fourier transform spectrometer, significant and worthwhile improvements can be achieved in resolution, sensitivity, measurement accuracy and wavelength coverage. The new spectra of H_2-Ar in the 1-0 band region of H_2 show an improvement in resolution of about a factor of 2 compared to the 1982 results, but the reliability of the line position measurements is improved by almost a factor of ten, for reasons which are discussed below. Furthermore, it has been possible to obtain a spectrum of H_2-Ar in the much weaker 2-0 band region. This new result should further improve the determination of the dependence of the H_2-Ar intermolecular potential surface on the H_2 bond length, an aspect in which the H_2 - inert gas complexes are already the best characterized among weakly bound systems [1].

2.   EXPERIMENTAL RESULTS

The absorption cell used here is that described earlier by McKellar [8]. It has a base path length of 5.5 m and is constructed of stainless steel and fitted with multiple-traversal mirrors. Liquid nitrogen cooling was used to obtain temperatures of 77 K or 91 K (the latter achieved by maintaining a suitable overpressure on the boiling nitrogen).
    Spectra were recorded using a Bomem Model DA3.02 interferometric spectrometer fitted with a quartz-halogen tungsten filament source and either a CaF_2 beamsplitter and InSb detector (for the 1-0 band), or a quartz beamsplitter and Ge detector (for 2-0). The absorption cell was positioned optically between the interferometer and detector, and the entire infrared radiation path outside the cell was evacuated.
    Figure 1 shows an example of the first series of spectra recorded in July and August 1986 with the new spectrometer and the 5.5 m cell, using a temperature of 77 K, a path of 178 m, and a gas mixture of 50% parahydrogen and 50% argon. This is the central portion of the H_2-Ar spectrum which accompanies the $S_1(0)$ transition of H_2 (v = 1-0, J = 2-0, where these quantum numbers refer to H_2). The full resolution of the spectrometer was used, giving an apodized instrumental width of 0.036 cm^{-1} (FWHM). Comparison of Fig. 1 with the same region in Fig. 5 of ref. 8 shows a definite improvement in both resolution and signal to noise ratio. In the lower trace, which required about 7 hours of acquisition time, the rms noise level is about 0.05% of the continuum intensity.
    The most important improvement, however, turns out to be in the measurement of the line positions themselves. The precision for the new results approaches 0.002 cm^{-1}, as would be expected on the basis of the H_2-Ar line widths (≈ 0.08 cm^{-1} at 60 torr) and their signal to noise ratios (≈ 20 to 100). In ref. 8, the much lower precision of 0.02 cm^{-1} was claimed, a value which is approximately confirmed both by the analysis of LeRoy and Hutson [1] and by comparisons with the present results. The reason for the disappointing precision of the 1982

measurements had to do with residual inaccuracies in the signal
averaging system used with the grating spectrometer in that work [8].
This system gave remarkably good-looking spectra, but was not sufficiently reproducible and stable from scan to scan to deliver measurement accuracies of better than about one fifth of a linewidth.

As another example of the new results, the H_2-Ar spectrum in the
$Q_1(0)$ region is shown in Fig. 2. For this transition, the H_2 molecule
remains in the spherically symmetric $J = 0$ state, and H_2-Ar therefore
behaves as a diatomic molecule, giving simple P- ($\Delta\ell = -1$) and R- ($\Delta\ell = +1$)
branches as indicated in Fig. 2 (ℓ is the quantum number for rotation
of the H_2 around the Ar). The weak unlabelled lines in Fig. 2, with
intensities of about 0.3% are real $H_2(J=1)$-Ar transitions due to a
small amount of residual ortho-H_2 in the para-H_2 sample used here (cf.
Fig. 4 of ref. 8).

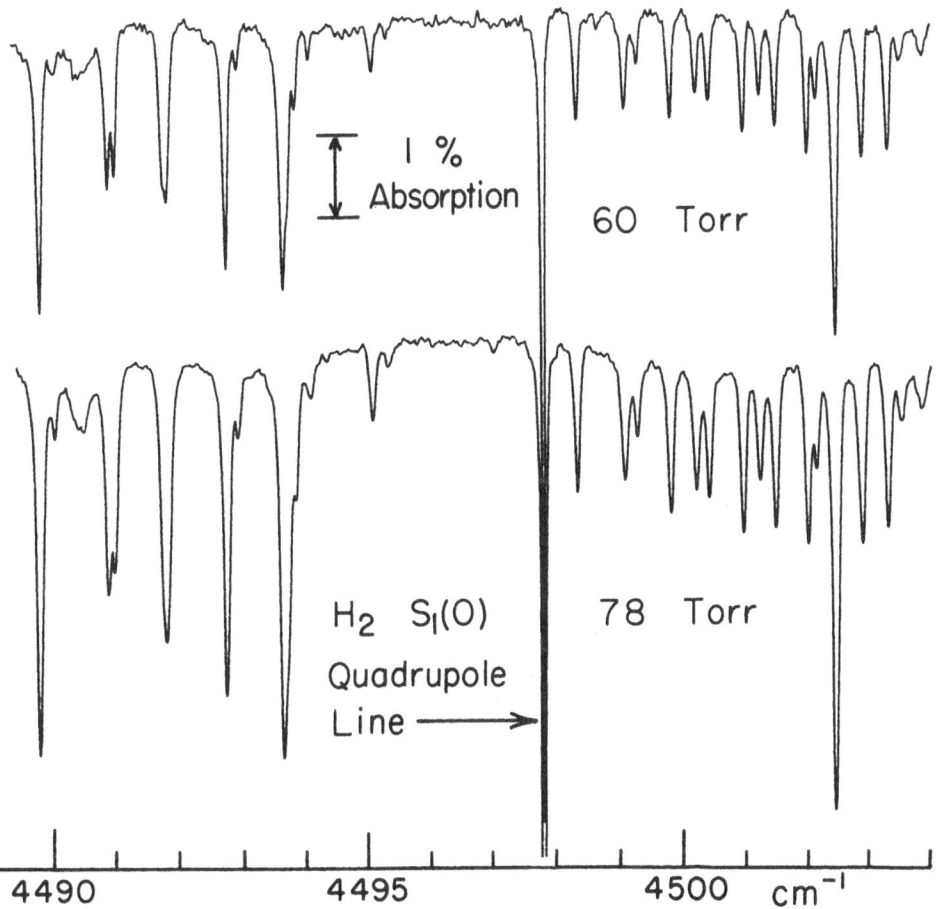

Figure 1. A portion of the infrared spectrum of the H_2-Ar molecule,
recorded with a 50% para-H_2 + 50% Ar mixture at 77 K, 178 m absorption
path, and two densities: 60 torr (0.28 amagat) and 78 torr (0.36 amagat).

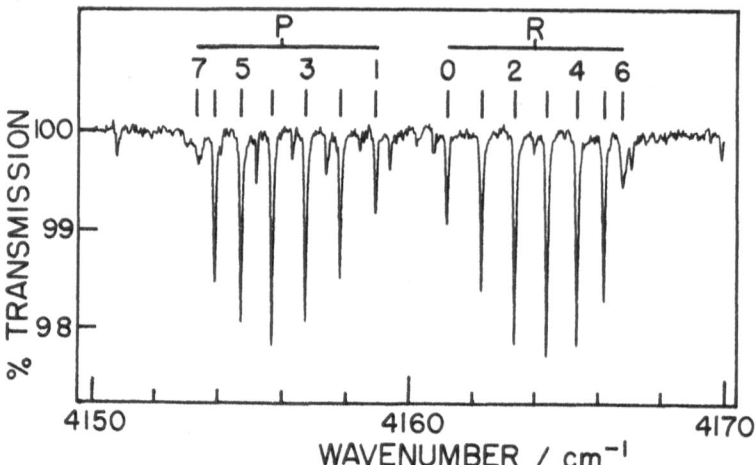

Figure 2. Spectrum of the H_2-Ar molecule accompanying the
$Q_1(0)$ transition of H_2 in a para-H_2 + Ar mixture at 77 K and
78 torr pressure. The spectral resolution is 0.036 cm^{-1} and
the path is 178 m. The P- and R-branches are indicated;
weaker lines not marked are $H_2(J=1)$-Ar transitions arising
from residual ortho-H_2 in the gas sample.

Since the present measurements have not been completed for the
$Q_1(1)$ and $S_1(1)$ transitions in normal H_2 + Ar mixtures, it is premature
to present here a new list of line positions for the 1-0 band region.
Rather, I would like to describe a new result for the H_2-Ar molecule in
the 2-0 overtone region around 8400 cm^{-1}. The v = 2-0 collision-induced
band of H_2 is much weaker than the fundamental band, but one still
expects that "fine" structure due to bound H_2-Ar molecules will
accompany the band in H_2 + Ar mixtures if the observation is made at
densities low enough that the structure is not washed out by pressure-
broadening. Indeed, McKellar and Welsh [9] detected the vestiges of
such structure at moderate densities (17 amagat) while measuring the
overtone band shape at low temperatures. However, subsequent attempts
to observe the H_2-Ar structure in this region have not been successful
until now. The importance of such measurements is easy to understand:
they give considerably extended information about the dependence of the
intermolecular potential on the H_2 bond length, because the hydrogen
molecule shows greater departures from its equilibrium bond length in
the v=2 state than in v=1 or 0.

The structure due to bound H_2-Ar molecules accompanying the $S_2(0)$
(v = 2-0, J = 2-0) transition of H_2 is shown in the lower trace of Fig. 3.
In order to obtain this spectrum, it was necessary to increase the gas
density by a factor of 15 relative to that used for the $S_1(0)$ region in
Fig. 1, and this in turn required a higher temperature (91 K) in order
to keep the argon in a gaseous state. For comparison, a spectrum of
$S_1(0)$ under exactly the same experimental conditions, but with half the

path length, is shown in the upper trace of Fig. 3. Note that at this
density, the P- and R-branch structure between 4490 and 4505 cm⁻¹ is
mostly unresolved (this is exactly the region shown in Fig. 1 in detail),

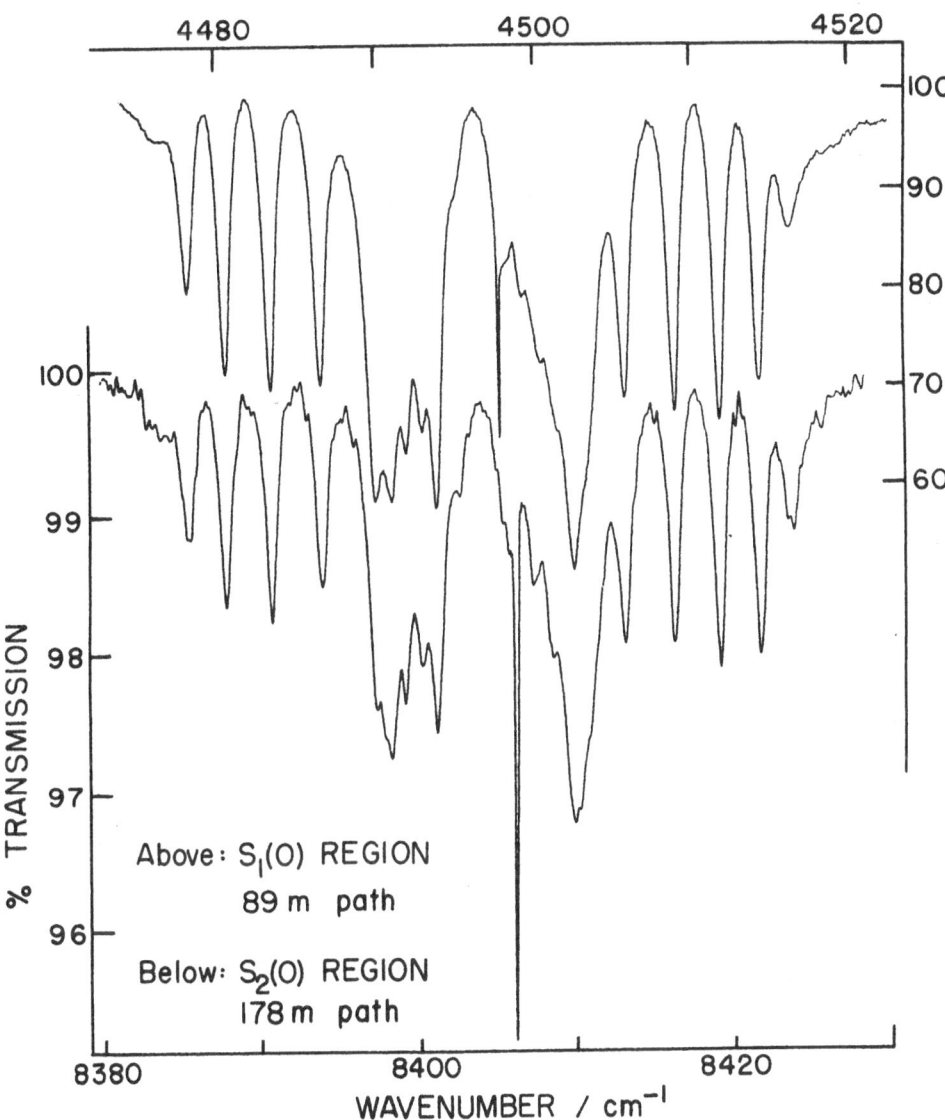

Figure 3. Spectra of the H₂-Ar molecule accompanying the $S_1(0)$ (top)
and $S_2(0)$ (bottom) transitions of H₂ in a para-H₂ + Ar mixture at 91 K
and 1.4 atm. pressure (4.3 amagat density). The path and spectral
resolution was 89 m and 0.10 cm⁻¹ for $S_1(0)$, and 178 m and 0.20 cm⁻¹ for
$S_2(0)$. The sharp lines near the center of each pattern are the quadrupole transitions arising from free H₂ molecules in the gas sample.

and also that there is a very strong similarity between the $S_1(0)$ and
$S_2(0)$ spectra. In fact, the largest apparent difference in shape
between the two spectra in Fig. 3 is not directly related to the H_2-Ar
pattern at all, but rather to the relative position and intensity of the
H_2 quadrupole transition, the sharp line near the center of each pattern.
The shift of the quadrupole line, which marks the true H_2 transition
frequency, relative to the apparent center of the H_2-Ar band is simply
a consequence of the dependence of the intermolecular potential well
depth on the H_2 vibrational state: the well depth increases with
increasing v, shifting the band center to lower frequencies. This is
the most basic of the dependences of the potential surface on the H_2
bond length.

Measured line positions for $S_2(0)$, accurate to about 0.03 cm^{-1},
are given in Table I. Also listed are preliminary predictions corresponding to the latest hydrogen - argon potential surface of LeRoy and

TABLE I. Positions of resolved transitions in the spectrum
of H_2-Ar accompanying the $S_2(0)$ transition of H_2 (in cm^{-1})

Transition	Observed	Predicted [10]
N(7)	8385.55	8385.574
N(6)	8387.96	8387.957
N(5)	8390.86	8390.858
N(4)	8394.04	8394.019
T(1)	8413.25	8413.245
T(2)	8416.32	8416.349
T(3)	8419.24	8419.234
T(4)	8421.78	8421.772
T(5)	8423.8	8423.77

Hutson [1]. The agreement is very good, supporting both the validity
of the approach to parameterization and the accuracy of the parameters
of ref. 1. It is unlikely that much better accuracy can be obtained
for the 2-0 region unless one uses a much longer absorption cell or a
novel experimental approach. What can be added in the near future is
further information on other transitions in the band. No trace of $Q_2(0)$
was detected in these experiments, a result which is not surprising in
view of the complete absence of any "overlap-induced" contribution to
the 2-0 collision-induced band [9]. However, it should be possible
to record the $Q_2(1)$ and $S_2(1)$ regions using normal H_2 + Ar mixtures,
albeit with reduced accuracy since they are inherently weaker than
$S_2(0)$. One attempt has been made already to do this, but the results
are not yet satisfactory. Furthermore, it should be possible to record
the 2-0 band for the D_2-Ar molecule, although here again the spectrum
is inherently weaker than that in Fig. 3.

To a first approximation, one expects the band strengths for H_2-
Ar to vary as $Q^2 \nu$ (where Q is the quadrupole matrix element and ν the
transition frequency), like the collision-induced band [9]. On this
basis, the $S_2(0)$ spectrum should be about 23 times weaker than $S_1(0)$,

whereas the observed difference is about a factor of 43. This weakness helps to explain why the 2-0 band spectra have been difficult to observe in the past. A more complete analysis of the induced dipole moment function for bound H_2-Ar pairs, such as has been performed for the 1-0 band spectra [3,10], may be required to explain the observed 2-0 band strength.

3. CONCLUSIONS

Promising new results on the H_2-Ar Van der Waals molecule have been obtained using a long-path low-temperature cell in conjunction with a modern Fourier transform infrared spectrometer. An extensive range of studies is planned with the Bomem spectrometer in the next few years. Comprehensive spectra of the complexes formed by H_2, HD, or D_2 with Ar, Kr, or Xe should be relatively straightforward to obtain in the 1-0 and possibly 2-0 band regions. Of greater difficulty will be the corresponding Van der Waals spectra expected to accompany the pure rotational lines of hydrogen in the 160 to 600 cm^{-1} region; $S_0(0)$ and $S_0(1)$ for H_2, and $S_0(0)$ for HD and D_2. In addition, the 15 K long-path cell and refrigeration system described by McKellar and Welsh [11,12] are now located in Ottawa, so that very low temperature studies of $(H_2)_2$, $(D_2)_2$, $(HD)_2$, H_2-Ne, etc., are also anticipated both in the mid- and far-infrared regions.

REFERENCES

1. R.J. LeRoy and J.M. Hutson, 'Improved potential energy surfaces for the interaction of H_2 with Ar, Kr, and Xe,' J. Chem. Phys., in press.

2. R.J. LeRoy and J.S. Carley, Adv. Chem. Phys. 42, 353 (1980).

3. A.M. Dunker and R.J. Gordon, J. Chem. Phys. 68, 700 (1978).

4. R.J. LeRoy and J. Van Kranendonk, J. Chem. Phys. 61, 4750 (1974).

5. A.K. Kudian, H.L. Welsh, and A. Watanabe, J. Chem. Phys. 43, 3397 (1965).

6. A.K. Kudian and H.L. Welsh, Can. J. Phys. 49, 230 (1971).

7. A.R.W. McKellar and H.L. Welsh, J. Chem. Phys. 55, 595 (1971).

8. A.R.W. McKellar, Faraday Discuss. Chem. Soc. 73, 89 (1982).

9. A.R.W. McKellar and H.L. Welsh, Proc. Roy. Soc. A322, 421 (1971).

10. R.J. LeRoy and J.M. Hutson, private communication (1986).

11. A.R.W. McKellar and H.L. Welsh, Can. J. Phys. 50, 1458 (1972).

12. A.R.W. McKellar and H.L. Welsh, Can. J. Phys. 52, 1082 (1974).

DYNAMICAL PROPERTIES OF SIMPLE HYDROGEN-BONDED CLUSTERS

J. W. Bevan
Texas A&M University
Chemistry Department
College Station, Texas 77843
USA

ABSTRACT. Observation of rovibrational substructure in a range of vibrational bands occurring in intermolecular hydrogen-bonded complexes
have been obtained.   Static gas phase infrared spectroscopic analysis
of fundamental, overtone, combination and hot bands in common and some
isotopically, substituted species have been recorded using color center laser and Fourier transform infrared spectrometers.   Specifi-
cially, rovibrational analysis of vibrational bands in the linear complexes HCN---HF, HCN---HCl, HCN---HCN and OC---HF have been successfully investigated.   As HCN---HF has been studied in greatest detail, a consideration of its analysis will be emphasized.   Rovibrational data for both intramolecular and intermolecular vibrations, the
latter including low frequency far infrared vibrations have been evaluated.   Accurately determined complex molecular parameters, including
excited state rotational constants, band origin frequencies and information on transition line profiles can provide the basic information
relevant to the characterization of excited vibrational state structures, anharmonic potential energy surfaces, and where appropriate,
vibrational predissociation.   Effects of Coriolis interactions as contributing factors to the observed ν_1 and its $\nu_1 + \nu_7^1 - \nu_7^1$ hot band
profile in HCN---HF are also considered.   Certain analyses of corresponding bands in OC--HF are also presented to further illustrate the
scope of current investigations.
      Progress and limitations in investigations of the specific
molecular complexes are illustrated and reviewed.

INTRODUCTION

It has long been recognized that static gas phase infrared spectroscopy can be a powerful physicochemical technique for the characterization of hydrogen-bonded interactions.[1-3]   In this presentation,
the results of preliminary investigations in some simple hydrogen-
bonded interactions which are linear in their equilibrium configurations are reported.   Specific complexes which have been studied by

A. Weber (ed.), Structure and Dynamics of Weakly Bound Molecular Complexes, 149–169.
© 1987 by D. Reidel Publishing Company.

static gas phase color center laser and Fourier transform infrared spectroscopy by our research group will be reviewed.

Molecular information has been obtained from rovibrational analysis of vibrational bands in four specific systems HCN---HF, OC---HF, HCN---HCN and HCN---HCl. Two of these complexes HCN---HF and OC---HF have been selected to illustrate the type, quality and limitations of information determined from such studies. As our prime objective in these projects is to characterize the excited state structure, potential energy surface, and energy transfer associated with vibrational predissociation, determination of certain molecular parameters has been emphasized. In particular, we have attempted to evaluate excited state rotational and distortion constants, band origin frequencies and vibrationally predissociating transition profiles in fundamental, overtone, combination and hot bands of the complexes. This information has then been used to evaluate anharmonicity and anharmonic cross terms and predissociating excited state lifetimes. In the case of HCN---HF where the analysis is most complete, it has been possible to make a determination of rovibrational parameters for each fundamental vibration represented in Table I. It has also been possible to estimate an equilibrium rotational constant, B_e, for the common isotopic species of this complex.

Table I. Hydrogen-Bonded Vibrations in HCN---HF (in cm^{-1})[a]

H——C ≡≡≡ N ----H —— F	$\nu_1 = 3716\ cm^{-1}$
	$\nu_2 = 3310$
	$\nu_3 = 2121$
	$\nu_4 = 168$
	$\nu_5^1 = 726$
	$\nu_6^1 = 550$
	$\nu_7^1 = 76$

[a]Arrows are not drawn to scale.

Furthermore, the availability of an accurately determined pressure broadening parameter for rovibrational transitions in the ν_2 HCN---HF band enables estimation of the predissociative lifetimes for each predissociating fundamental in its common and perdeuterated isotopic species. These results demonstrate a vibrational state specificity of vibrational relaxation in this complex.

I. Experimental

The concentration of the static gas phase complex in the ground vibrational J=0 state, n_{oo} can be estimated using the relationship[4]

$$K_{oo} = \frac{n_{oo}(Y---H-X)}{n_{oo}(Y)\ n_{oo}(HX)} = \left\{ \frac{h^2}{2\pi\mu kT} \right\}^{3/2} \exp \left[\frac{D_o}{RT} \right] \qquad (1)$$

As pressure broadening parameters are typically in the range 15-30 MHz torr^{-1} for the complexes being investigated, we attempt to restrict the total pressure to $\leq$ 5 torr to reduce the effects of pressure broadening on the resolution. In order to obtain a reasonable S/N ratio for a wide range of transitions and to permit optimization of complex concentration, a temperature controlled White cell was used with absorption pathlength typically in the range 48-160 meters. An isopentane-liquid N_2 cooled refrigeration system was used to temperature control the static gas phase mixture to within 1°K. Gas phase mixtures were prepared at the preferred component ratios and carefully cooled to a temperature which optimized concentration of the complex and signal to noise ratio. The spectrometers used to obtain the recorded spectra have been described previously and will be briefly considered.

A computer-controlled continuously tunable Burleigh FC1-20 color center laser[5] spectrometer with $\leq$ 3MHz free-running instrumental linewidth was used to record spectra in the wavenumber range 3100-3950 cm^{-1}. In addition Bomem DA3-002 Fourier transform interferometers at the National Bureau of Standards and at Texas A&M University were used at 0.01 to 0.004 cm^{-1} instrumental resolution over the frequency range 30-8000 cm^{-1}. In these latter cases, great care was taken to ensure compatibility of the spectrometer with the multireflection White cell. Transfer optics utilized the collimated beam of the spectrometer and simultaneously formed an image of the input aperture in the plane of the field mirror of the temperature-controlled corrosion resistant White cell and an image of the flat mirror of the interferometer on the first back mirror of the cell.[6] This arrangement passed the maximum amount of usable signal radiation without the need of a field lens at the entrance of the White cell. Both interferometer and transfer optics path were evacuated to eliminate extraneous atmospheric absorptions.

II. Results

A.   Rovibrational Analysis of Fundamentals and Associated Hot Bands

Most rovibrational analyses were carred out using a linear regression fit to the familiar expression[7]

$$\nu = \nu_0 + B'[J'(J'+1)-\ell'^2] \pm q_v'/2[J'(J'+1)] - D_j'[J'(J'+1) - \ell'^2]^2$$

$$- (B''[J''(J''+1)-\ell'^2] \pm q_v''/2[J''(J''+1)] - D_j'[J'(J'+1) - \ell''^2]^2) \quad (2)$$

where ν_0 is the band origin frequency, B'', B' ground and excited state rotational constants, D_J'', D_J' corresponding distortion constants and where appropriate q_v'', q_v' the ℓ-type doubling constants. In many cases simultaneous fitting of microwave and infrared transitions were made by weighting each datum by the inverse square of the uncertainty of measurement. Relatively few of the bands currently analyzed showed demonstrable effects of perturbation.

Often the lowest J rovibrational transitions around the origin are difficult to observe in static gas phase spectra. Consequently, identification of the band origin can prove difficult. Alternatively, the presence of relatively highly populated hot bands associated with the low frequency bending or stretching vibrations of the hydrogen-bond lead to pronounced hot band series which interlace with the fun-damental transitions complicating assignment further. While it is relatively easy to identify members of a P(J) and R(J) branch, conclu-sive identification of J assignment in such bands is often not possi-ble based simply on a fit in which the standard deviation is mini-mized. However, use of combination frequency differences based on previously determined microwave rotational and distortion constants, particularly those determined from supersonic jet or molecular beam techniques[8] rapidly permits conclusive assignments to be determined. The lower precision gas phase microwave studies usually provide accu-rate rotational constants whereas distortion constants are not so well-determined.[9] However, these latter investigations prove invalu-able for assignment of rovibrational hot bands as rotational transi-tions in excited vibrational states are usually not observed in single resonance supersonic jet or molecular beam microwave spectra due to the low effective temperature conditions. An example of the assign-ment of a segment of the P(J) branch in ν_2 HCN---HF is shown in Fig. 1. The fundamental transitions are accompanied by an extensive series of hot bands associated with the low frequency bending ν_7^1 and stretching ν_4 vibrations of the hydrogen bond. Table II demonstrates the results of a rovibrational analysis on the fundamental and corresponding hot bands. A corresponding analysis of the ν_1 funda-mental and its ν_5^1, $2\nu_5^2$ and ν_3 hot bands of the OC---HF complex are given in Table III. Results from these and other analyses indicate that at best the rotational constants from gas phase infrared analysis are comparable to those determined from static gas phase microwave analysis. The corresponding distortion constants are determined from the wide range of rovibrational transitions and are quite accurately determined with ≤ 5% standard deviation. The latter constants compare favorably with those evaluated by pulsed Fourier transform microwave spectroscopy.

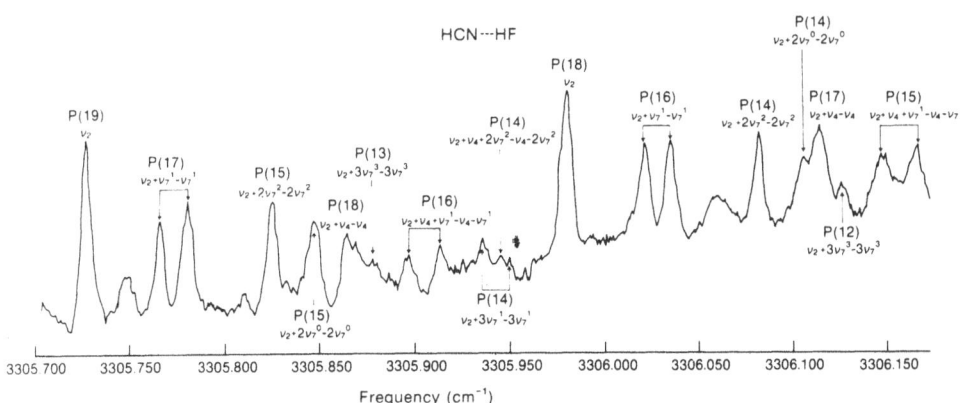

Figure 1. Doppler-limited color center laser static gas phase spectrum of ν_2 HCN---HF.

In HCN---HF, it has been possible to complete evaluation of precise rovibrational data for each fundamental vibration of this complex. These analyses include each intramolecular fundamental so that frequency shifts can be determined from corresponding band origins of vibrations in the experimentally [10] and theoretically [38] determined monomers components (Table IV). As in HCN---HCN[11], the ν_3 (C≡N) stretching band of HCN---HF shows a pronounced increase of the band origin frequency on complex formation. These analyses have also been extended to intermolecular vibrations, including direct observation of the low frequency hydrogen-bond bending vibration in the far infrared by Fourier transform infrared spectroscopy. We have not as yet been able to observe the low frequency intermolecular hydrogen-bond ν_4 stretching vibration directly in the far infrared. This information was, however, rapidly determined from observation of the combination difference band ν_1-ν_4 and the ν_1 fundamental in the near infrared. Rovibrational parameters for each fundamental vibration of the HCN---HF complex are thus available, and are given in Table V. A particularly unusual aspect of these analyses has been the observation of a large positive α_6 value for this high frequency intermolecular

Table II. Molecular constants determined from fit to ν_2 and its ν_4, ν_7 hot band rotational-vibrational transition frequencies in H-C≡N-H-F.[a]

Vibrational Band	B"(MHz)	D_J"(kHz)	q_v"(MHz)	B'(MHz)	D_J'(kHz)	q_v'(MHz)	ν_0(cm^{-1})
ν_2	3591.05(19)	7.003(75)		3586.09(19)	7.003(69)		3310.32836(31)
$\nu_2+\nu_7^1-\nu_7^1$	3622.22(24)	8.10(17)	12.40(19)	3648.85(24)	8.04(16)	12.41(19)	3309.92100(39)
$\nu_2+2\nu_7^2-2\nu_7^2$	3653.92(23)	8.54(17)		3648.85(23)	8.40(17)		3309.50978(37)
$\nu_2+2\nu_7^0-2\nu_7^0$	3642.9(16)	9.17(33)		3642.6(15)	8.99(54)		3309.51101(30)
$\nu_2+3\nu_7^3-3\nu_7^3$	3685.5(18)	9.02(63)		3683.3(19)	8.99(18)		3309.09378(31)
$\nu_2+3\nu_7^1-3\nu_7^1$	3673.0(26)	9.63(33)		3672.4(19)	9.50(36)		3310.02116(45)
$\nu_2+\nu_4-\nu_4$	3529.3(14)	6.02(24)		3523.1(18)	6.00(21)		3310.16439(77)
$\nu_2+\nu_4+\nu_7^1-\nu_4-\nu_7^1$	3560.36(90)	5.753(54)	15.68(42)	3554.88(75)	5.88(26)	15.59(42)	3309.74223(19)
$\nu_2+2\nu_4^2-\nu_4-2\nu_7^2$	3591.39(36)	7.35(28)		3586.09(42)	7.56(24)		3309.32142(61)

[a]One standard deviation.

Table III. CO--HF: Molecular constants of lower and excited states.[a]

	ν_2	$\nu_1+\nu_5^1-\nu_5^1$	$\nu_1+2\nu_5^2-2\nu_5^2$	$\nu_1+\nu_3-\nu_3$
$\nu o/cm^{-1}$	3844.0311(20)	3849.4393(40)	3854.781(43)	3852.664(17)
B"/MHz	3064.84(25)	3087.05(67)	3109.3(17)	3021.50(84)
D"/MHz	10.30(29)	10.75(56)	13.2(19)	10.08(91)
q"/MHz		7.06(46)		
B'/MHz	3126.24(25)	3148.10(70)	3174.6(24)	3081.2(11)
D'/kHz	9.46(42)	9.49(74)	11.6(27)	8.94(97)
q'/MHz		6.61(47)		
σ/cm^{-1}	0.0017	0.0035	0.0020	0.0013
X_{13}/cm^{-1}	8.633(19)	α_1''/MHz	61.50(30)	α_5''/MHz -22.22(75)
X_{15}/cm^{-1}	5.4007(60)	α_3''/MHz	43.3(12)	α_5'/MHz -22.94(83)
		α_3'/MHz	45.0(14)	

[a]From Ref. 23. All errors are one standard deviation.

Table IV. Free Monomer and Intramolecular Fundamental Frequencies in HCN---HF (in cm^{-1}).

ν_o Complex		ν_o Monomer[a]		$\Delta\nu_{obs.}$	$\Delta\nu_{harms}$[b]
ν_1	3716.2012(1)	ν_1	3961.418(3)	-245.206	-127
ν_2	3310.329(6)	ν_1	3311.473(1)	-1.144(7)	-13
ν_3	2120.935(12)	ν_3	2096.855(12	+24.080(24)	+12
ν_5^1	726.5312(3)	ν_2	713.4590(1)	+13.072(2)	+26

[a] C.f. refs. 24-26   [b] C.f. ref. 38

Table V.   Molecular Parameters Determined For HCN---HF

Ground State[c]

B_o = 3591.187(18)MHz D_j^o = 7.066(33)kHz

Excited States

	B'/MHz	D_j/kHz	α_v/MHz	q_v/MHz	ν_o/cm^{-1}	Ref.
ν_1	3660.326(30)	8.556(30)	-69.139(52)	--	3716.2116(15)	27,28
ν_2	3586.09(19)	7.003(75)	5.10	--	3310.32836(31)	29,30
ν_3	3576.01(57)	6.89(21)	15.2(6)	--	2120.935(12)	31
ν_4	3529.319(21)[a]	6.02(24)	61.87(3)	--	168.33(2)	28
ν_5^1	3588.998(33)	7.063(24)	2.189	(1.31)[b]	726.53122(29)	32
ν_6^1	3526.88(27)	8.15(24)	64.31(27)	1.734(24)	550.0285(2)	33
ν_7^1	3622.482(90)[a]	8.100(15)	-31.30(1)	12.91(20)[a] 12.401(19)	76.1713(5)	34

[a]Gas phase microwave Ref. 39.   [b]Estimated $q_v = fB_v^2/\omega_v$

[c]B_e = 0.121182(43)cm^{-1} = 3632.90(1.29)MHz

hydrogen-bonded bending vibration.   This is currently the object of further investigation.

The most important information determined from such analyses is the band origin frequencies which provide basic information for evaluation of the hydrogen bond potential energy surface.   Such band origins can be evaluated to better than 1 x 10^{-3} cm^{-1}.

In the case of $\Pi \leftarrow \Sigma$, $\Pi \leftarrow \Pi$, $\Sigma \leftarrow \Pi$ transitions, it is often possible to evaluate ℓ-type doubling splitting in transitions.   It is thus possible to evaluate the q_v values for such states.   The accuracy of the q_v values are comparable to those determined from gas phase microwave spectroscopy having standard deviations of $\leq$ 2%.   This, in addition, facilitates identification of states in other hot bands of the complex.   Such information can also be quite important particularly for high excited states as it can also be related to Coriolis constants of the complex.[12]

These investigations of common isotopic species have also been extended to analysis of isotopically substituted bands. Table VI illustrates results for ν_1, ν_2, ν_3 and ν_6^1 DCN---DF.

B.   Rovibrational Analysis of Overtones and Combination Bands

Rovibrational analysis of overtones and combination bands have significant implications for quantitative characterization of the anharmonic potential energy surface. Unlike the low frequency hot bands associated with fundamentals which have comparable absorption cross sections, corresponding values for overtones and combination bands can be one or two orders of magnitude smaller,[13] providing a more difficult challenge for observation by gas phase infrared spectroscopy. We have now carried out rovibrational analysis in several overtones of the complex HCN---HF, including $2\nu_5^0$ for which the fit is included in Table VII. Hot band assignments for $2\nu_5^0 + \nu_7^1 - \nu_7^1$ and $2\nu_5^0 + 2\nu_7^2 - 2\nu_7^2$ were also possible in this case.
    Overtones of $2\nu_2$, $2\nu_2 + \nu_7^1 - \nu_7^1$ and the overtone of $2\nu_1$ have also been observed but we were unable to resolve rovibrational substructure in this latter overtone, presumably due to the effects of predissociative linebroadening. Thus far, we have been able to observe three combination bands, two of these, $\nu_2 + 2\nu_4 + \nu_7^1$ and $\nu_2 + 2\nu_4 + 2\nu_7^0$ are observed as perturbations with the ν_1 and $\nu_1 + \nu_7^1$ states and will be discussed later. A combination difference band $\nu_1 - \nu_4$ in HCN---HF was also observed but we were unable to resolve the rovibrational structure in this band. However, as previously mentioned a Voigt simulation of the band profile using alternative but precisely determined rotational and distortion constants provided a relatively accurate band origin for the ν_4 low frequency H-bond stretching vibration ($\nu_4 = 168.344(21)$ cm^{-1}).

C.   Predissociative Lifetimes

It is always preferable to investigate line profiles in molecular beams as there is less necessity to correct for Doppler and pressure broadening effects. However, gas phase studies of predissociative lifetimes can be quite informative for many hydrogen-bonded species as their predissociative lifetimes are often quite significant. Although influenced by Doppler broadening, pressure broadening and in the case of Fourier transform infrared spectroscopy a larger instrumental contribution (~ 120 MHz), it has been possible to effectively simulate line profiles for many species using a Voigt profile[14] and determine accurate predissociative lifetimes.

$$I(\nu) = \frac{1}{\sigma\sqrt{\pi}} \int \frac{\exp\{-[(\nu'-\nu_{if})/\sigma]^2\}}{\{1 + [(\nu - \nu')/\Gamma]^2\}} \, d\nu'$$

$$(3)$$

Table VI. Molecular Parameters Determined For DCN---DF[29,30]

Vibration	ν_0[cm⁻¹]	B"[MHz]	D_J"[kHz]	q_v"[MHz]	B'[MHz]	D_J'[kHz]	q_v'[MHz]	σ[cm⁻¹]b
ν_1 ($\Sigma\leftarrow\Sigma$)	2730.89090(19)	3351.892(30)a (3351.87(4))	5.834(13) (5.5(8))	-- --	3395.3403(71)	5.4896(19)	--	5.5x10⁻⁴
ν_2 ($\Sigma\leftarrow\Sigma$)	2638.13091(16)	3351.892(32)	5.834(13)	--	3342.891(10)	5.8202(36)	--	6.7x10⁻⁴
$\nu_2+\nu_7^1-\nu_7^1$ ($\Pi\leftarrow\Pi$)	2637.53484(19)	3379.765(68) (3379.86(11))	6.266(28) (8(3))	11.34(10) (12.01(22))	3370.718(11)	6.2221(46)	11.378(4)	7.4x10⁻⁴
$\nu_2+2\nu_7^2-2\nu_7^2$ ($\Delta\leftarrow\Delta$)	2636.93156(32)	3407.87(22) (3408.2(3))	6.45(13) (9)	--	3398.969(2)	6.482(14)	--	7.2x10⁻⁴
$\nu_2+2\nu_7^0-2\nu_7^0$ ($\Sigma\leftarrow\Sigma$)	2637.0284(3)	3403.328(96) (3403.46(20))	8.583(54) (13)	--	3394.7358(17)	8.5031(70)	--	6.9x10⁻⁴
ν_3 ($\Sigma\leftarrow\Sigma$)	1943.00460(39)	3351.892(32) (3351.87(4))	5.834(13) (5.5(8))	--	3339.920(32)	5.617(20)	--	7.6x10⁻⁴
$\nu_3+2\nu_7^2-2\nu_7^2$ ($\Delta\leftarrow\Delta$)	1941.88232(44)	3407.87(22) (3408.2(3))	6.45(13) (9)	--	3397.178(25)	6.478(91)	--	1.09x10⁻⁴
$\nu_3+3\nu_7^3-3\nu_7^3$ ($\Phi\leftarrow\Phi$)	1941.20643(27)	3436.07(29) (3436.7(3))	5.86(10) (8)	--	3425.959(10)	5.902(3)	--	7.5x10⁻⁴
ν_6^1	409.16600(18)	3351.89(16)	5.834(14)	--	3310.737(15)	6.512(10)	--	1.8x10⁻⁴

a Parentheses refer to static gas phase microwave data in Ref. 39.
b σ is one-standard deviation in fit.

Table VII. Molecular parameters^{37} for the $2\nu_5^0$, $2\nu_5^0 + \nu_7^1 - \nu_7^1$, and $2\nu_5^0$ $+ 2\nu_7^2 - 2\nu_7^2$ bands of HCN---HF.^{a}

	$2\nu_5^0$	$2\nu_5^0 + \nu_7^1 - \nu_7^1$	$2\nu_5^0 + 2\nu_7^2 - 2\nu_7^2$
ν_0	1437.53991(24)^a	1437.99009(65)	1438.19291(27)
B''(IR)	3591.20(20)	3622.49(24)	3653.66(16)
B''(MW)	3591.15(2)^d	3622.48(9)^b	3653.6(2)^c
D_J''(IR)	7.13(4)	7.96(13)	4.42(5)
D_J''(MW)	6.99(2)^c	9.(3)^b	1.(6)^b
q''(IR)	--	12.44(14)	--
q''(MW)	--	12.9(2)^c	--
B'(IR)	3587.86(12)	3619.23(20)	3650.98(16)
D_j'(IR)	6.96(4)	7.76(11)	4.86(5)
q'(IR)	--	12.27(14)	--

^{a}All units are given in MHz except for the band origins ν_0 which are quoted in cm^{-1} and D_J in KHz. Uncertainties cited are one standard deviation.

^{b}Ref. 39. ^{c}Ref. 40.

$^d 2\nu_2$ HCN monomer $\nu_0 = 1411.416(1) cm^{-1}$ [Ref. 26] resulting in a positive shift of 26.124(1) cm^{-1} on complex formation.

where ν_{if} is the center transition frequency

$$\sigma = [(\Delta\nu_D)^2 + (\Delta\nu_{Inst})^2], \quad \Delta\nu_D = \frac{\nu_{if}}{c} \left\{\frac{2RT}{M}\right\}^{1/2} \tag{4}$$

and Γ is the half-width at $1/e$ intensity of the line profile and M is the molecular weight.

In the case of rovibrational analysis of ν_2 HCN---HF using gas phase color center laser spectroscopy we have been able to evaluate the pressure broadening parameter to be 27(6) MHz torr^{-1} under the experimental mixture conditions.

Employment of relatively low pressures ($\leq$ 1-2 torr where possible) and prior evaluation of the pressure broadening parameter enables a range of estimated excited state predissociative lifetimes to be evaluated (Table VIII). The results indicate a state specificity of excited state predissociative lifetimes which as expected change significantly on deuteration within a given complex species.

Table VIII. Comparison[35] of Predissociative Lifetimes for the ν_1, ν_2 and ν_3 bands of HCN---HF and DCN---DF.

| Band | Lifetime(s) | |
	HCN---HF	DCN---DF
ν_1	$1.1(1)\times10^{-10}$	$5.7(6)\times10^{-10}$
ν_2	$1.3(4)\times10^{-8}$	$5.2(5)\times10^{-9}$
ν_3	$1.0(4)\times10^{-9}$	$3.0(5)\times10^{-9}$

It is pertinent to note that we have not identified significant rotational linewidth dependance within a given vibrationally predissociating band of any complex we have investigated.

As observed[14] in (HF)$_2$, the lifetime of excited states in the bound ν_1 HF stretching vibration in HCN---HF increases on deuteration. This had been predicted theoretically[15] though the observed lifetime is significantly shorter than predicted. The ν_2 excited state deuterated lifetime is greater than the corresponding common isotopic species, an effect which can be correlated with the increased coupling between the C-D and C$\equiv$N stretching vibrations. However, the lifetime of the ν_3 (C$\equiv$N stretching) states in DCN---DF are significantly greater than those of the common isotopic species. Experiments are currently underway to compare these vibrational relaxation rates with directly determined predissociation rates.

D.   Evaluation of Equilibrium Rotational Constants

The determination of excited state rotational constants for each fundamental vibration of a complex permits evaluation of approximate B_e constants. In HCN---HF, sufficient information is available from previous microwave analyses and our current infrared studies to permit such an evaluation for its common isotopic species using the expression

$$B_v = B_e - \sum_i \alpha_i (v_i + d_i/2) + \sum_k \gamma_k \ell_k^2 \qquad (5)$$

and α_i values given in Table 5. B_e is determined to be 3632.90(1.27) MHz. The major contribution to the standard deviation results from the poorly determined B_3' rotational constant. This band is currently under reinvestigation to rectify this situation. The determination of B_e in HCN---HF permits a comparison of the predicted low frequency stretching force constant based on microwave spectroscopic data with that evaluated directly from infrared spectroscopy.

Millen[17] recently derived an expression relating the harmonic force constant K_4 to the equilibrium rotational constants of the complex B_e and its corresponding monomer components $B_e(Y)$ and $B_e(HX)$:

$$K_4 = \frac{16\pi^2 B_e^3 \mu}{D_J^e} \left\{ 1 - \frac{B_e}{B_e(Y)} - \frac{B_e}{B_e(HX)} \right\} = 18.52 \text{ Nm}^{-1} \qquad (6)$$

The determined value compares favorably with that of 19.88 Nm^{-1} evaluated from an anharmonicity corrected stretching force field calculation.[16]

E.   Anharmonicity and Anharmonic Cross Terms

These parameters can be estimated to a first approximation using the expansion[18,19]

$$G_o (v_1, v_2 \cdots) = \sum_i \omega_i^o v_i + \sum_i \sum_{k \geq i} X_{ik}^o v_i v_k + \sum_i \sum_{k \geq i} g_{ik} \ell_i \ell_k \qquad (7)$$

Large amplitude motion of vibrations in hydrogen bonds necessitate inclusion of high order terms in this expansion. The most intensely investigated complex is HCN---HF for which available anharmonicity and anharmonic cross terms are given in Table IX within second order approximation. As more extensive information becomes available these terms may require some revision.

F.   Coriolis Interactions and the Origin of ν_s Hydrogen Band
      Profiles

At the onset of rovibrational analysis of vibrational bands in these hydrogen-bonded complexes, a question arose as to whether such spectra would be so perturbed as to prevent rapid effective analysis. In the complexes we have so far investigated, this has turned out not to be a major problem, at least, at the current level of resolution. In certain bands, however, the effects of perturbation are evident and an analysis of the effects of Coriolis perturbation has been possible in

Table IX. Anharmonicity and Anharmonic Cross-terms in HCN---HF
           (cm^{-1})

X_{11}[a]	X_{12}	X_{13}	X_{14}	X_{15}	X_{16}	X_{17}
-116.9(1)	---	---	8.0252(73)	---	---	4.2162(53)
	X_{22}[b]	X_{23}	X_{24}	X_{25}	X_{26}	X_{27}
	-51.96(1)	-14.61(22)	-0.161(1)	-18.98(2)	---	0.409(2)
		X_{33}	X_{34}	X_{35}	X_{36}	X_{37}
		-10.45(38)	---	-3.61(22)	--	-0.61(3)
			X_{44}	X_{45}	X_{46}	X_{47}
			-2.009(10)	---	---	1.000(49)
				X_{55}	X_{56}	X_{57}
				-2.44(17)	---	0.20(2)
					X_{66}	X_{67}
					---	---
						X_{77}
						-0.06(2)

[a]Ref. 16.   [b]Unpublished results.

the ν_1 and $\nu_1 + \nu_7^1 - \nu_7$ bands of HCN---HF. As can be seen in a segment of the P branch of ν_1 HCN---HF in Fig. 2, the expected monotonic
transition behaviour does not occur in ν_1 and additional transitions
associated with the perturbing states P* and P** are observed. These
perturbations can be deperturbed using the Hamiltonian matrix.[20]

$$\begin{bmatrix} H_{11} & H_{12} & H_{13} \\ H_{21} & H_{22} & H_{23} \\ H_{31} & H_{32} & H_{33} \end{bmatrix}$$

with: $X = J(J + 1)$

$H_{ii} = \nu_{oi} + B_i'x - D_i'x^2$

$H_{ij} = \xi_{ij} + \eta_{ij} \sqrt{X}$

$H_{ji} = H_{ij}$

ν_0: band origin frequency, B: rotational constant, D: centrifugal distortion, η: heterogeneous perturbation parameter and ξ: homogeneous perturbation parameter

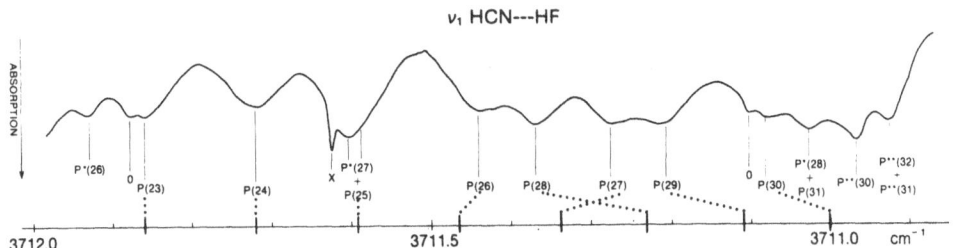

Figure 2.   Perturbed spectrum of ν_1 HCN---HF.

The evaluated parameters determined from a fit to the experimental data are given in Table X.

　　　Observed perturbations are identified in the ν_1 band centered at $J' = 26$ and $J' = 30$. The perturbation centered at $J' = 26$ has been demonstrated to be heterogeneous (Coriolis) type perturbation involving the ν_1 and $\nu_2 + 2\nu_4 + \nu_7^1$ states of the complex. Perturbation corrected transition frequencies for the combination band $\nu_2 + 2\nu_4 + \nu_7^1$ permitted determination of the perturbing state rotational constant and its subsequent identification. A corresponding perturbation between the $\nu_1 + \nu_7^1$ and $\nu_2 + 2\nu_4 + 2\nu_7^0$ states was also demonstrated. We have not, as yet, been able to identify the perturbing state involving the perturbation at $J' = 30$ in ν_1. Furthermore, we are presently unable to conclude whether it has Fermi or Coriolis type character.

　　　With the rovibrational analysis and Voigt profile fits to the observed ν_1 transitions complete it is opportune to attempt to simulate the observed band profile. The linewidth of the free running color center laser spectrometer is ≤3 MHz and thus makes a negligible contribution to the observed spectrum. The Doppler contribution to the linewidth is predicted to be 168 MHz, substantially smaller than the observed 3.1 GHz full width at 1/e maximum intensity of resolved

Table X.   Molecular Parameters From Deperturbation of ν_1 HCN---HF[28]

Band	ν_1	$\nu_1 + \nu_7^1 - \nu_7^1$	$\nu_1 + \nu_4 - \nu_4$
ν_{o1} [cm^{-1}]	3716.2116(20)	3720.4278(33)	3724.2368(53)
B_1' [MHz]	3660.326(30)	3688.537(43)	3598.61(83)
D_1' [kHz]	8.556(35)	6.59(14)	5.24(90)
ν_{o2} [cm^{-1}]	3719.6541(40)	3721.663(44)	
B_2' [MHz]	3525.129(60)	3557.28(31)	
η_{12} [cm^{-1}]	0.005905(22)	0.003528(58)	
ν_{o3} [cm^{-1}]	3715.661(95)		
B_3' [MHz]	3682.33(15)		
η_{13} [cm^{-1}]	0.001935(35)		
σ [cm^{-1}]	0.0058	0.0041	0.0047

rovibrational transitions. Furthermore, the pressure broadening parameter has been determined to be 27 (6) MHz torr^{-1} for rovibrational
transitions in ν_2 HCN---HF.
      If this parameter is assumed to be band independent, pressure
broadening contributes 42(9)MHz to ν_1 transitions under the experimental conditions used to record ν_1 HCN---HF. The major contribution
to the linewidth of rovibrationally resolved transitions comes from
the pressure corrected Lorentzian contributions. The latter are vibrationally predissociative linewidth determined lifetimes corresponding to excited state lifetimes of 1.06 (10) x 10^{-10}s. The associated relaxation rate could be a result of vibrational dephasing or
may be a direct measure of the rate of predissociation. Pump-probe
experiments are currently underway to ascertain this process. The
band origin is predicted to occur at 3716.2116 (20) cm^{-1} and is overlapped with a pronounced P-bandhead of the $\nu_1 + \nu_7^1 - \nu_7^1$ hot band
with origin frequency at 3622.321(80) cm^{-1}. Low P(J) and R(J) tran-

sitions are thus not observed in static gas phase spectra. Other hot
bands including $\nu_1 + 2\nu_7^0 - 2\nu_7^0$, $\nu_1 + 2\nu_7^2 - 2\nu_7^2$ and $\nu_1 + \nu_4 - \nu_4$
are displaced from the ν_1 band origin and consequently only overlap
part of the high R(J) transitions in this band. Combination bands
involving such states as $\nu_1 - \nu_7$, $\nu_1 + \nu_7$ and their hot bands have not
been observed but are expected to have small absorption cross-sections
and are additionally expected in a different frequency range, so they
do not make significant contributions to the observed ν_1 band profile.
Similarly $\nu_1 - \nu_4$ (which has been observed) and $\nu_1 + \nu_4$ (currently not
detected) have large frequency differences and anharmonicity constants
which displace transitions from the ν_1 frequency. As initially concluded by Thomas[21] the primary structure of the ν_1 band is a consequence of a large and negative $\alpha_1 = (B_0 - B_1)$ value of -69.175(52) MHz
which is the dominant term leading to a pronounced P(J) bandhead with
a turnround point at P(51) and a transition maximum at P(24). This
phenomena was initially predicted by Sheppard[22] and is intepreted in
this case as a shortening of r(N---F) distance in the excited state of
the complex. In reality it is a consequence of anharmonicity re-
flecting an increased D_1 dissociation energy as required by the
relationship

$$D_1 - D_0 = \nu_1 \text{ (free HF)} - \nu_1 \text{ (HCN---HF)} = +245.206 \text{ cm}^{-1}. \tag{8}$$

Near the band origin, consecutive low J transitions are separated by
~(B'+B"). However, at higher P(J) transitions as B'>B", (B'-B") J^2
and other terms become increasingly dominant as the separation between
consecutive transitions decreases leading to a turnround point at
~P(51). If the resolution is defined by Rayleigh criteria, transitions will be unresolvable at ~3GHz separation. Thus for P(J) in
the range J = 34 to J = 69 we would expect spectral congestion to prevent rovibrational resolution. Furthermore as the most intense
rovibrational transitions are expected to occur at approximately
P(24), we expect significantly less intensity contribution from P(J)
transitions J > 51. In the case of ν_1 HCN---HF, however, heterogeneous (Coriolis) perturbations are detected centered at J' = 26 and
another smaller perturbation at J' = 30 which we are unable to conclusively demonstrate currently as being either a Coriolis
(heterogeneous) or Fermi (homogeneous) type interaction. As we have
demonstrated these interactions further complicate the spectrum not
only by causing frequency shifts in ν_1 transitions, but in this case
by the presence of additional transitions. Certain of these transitions are associated with normally forbidden (0, 1, 0, 2, 0^0, 0^0, 1^1)
$\leftarrow$ (0, 0, 0, 0, 0^0, 0^0, 0^0) transitions. Similar perturbations occur
in the $\nu_1 + \nu_7^1 - \nu_7^1$ subband involving the equally forbidden (0, 1,
0, 2, 0^1, 0^0, 2^0) $\leftarrow$ (0, 0, 0, 0, 0^0, 0^0, 1^1) transitions. Interestingly these latter transitions involve perturbation of one ℓ component
and intensity stealing. Furthermore, the excited state lifetime of
the perturbing state is significantly different from the ν_1 state. Of
additional interest is the observation that although these states mix
there appears to be no significantly decreased lifetime in the upper

component of these almost degenerate states as might be expected if
the lower state was acting as a channel for intramolecular vibrational
relaxation.

CONCLUSIONS

In this presentation we hope to have illustrated how static gas phase
color center laser and Fourier transform infrared spectroscopy can
contribute, particularly to the rovibrational analysis of not only
fundamental but also overtones, hot bands and combination bands in
hydrogen-bonded complexes.   Such analysis of hydrogen-bonded vibrations have permitted evaluation of excited state rotational constants,
band origin frequencies and in certain cases determination of excited
state lifetimes in predissociating states. This basic information can
contribute to the characterization of excited state structure and
quantitative modeling of hydrogen-bond potential energy surfaces.   In
one case, the common isotopic species of HCN---HF, it has been possible to estimate B_e.   Further extension of such analyses could offer
the opportunity of direct determination of effective r_e structures for
such weakly bound species.   Most importantly, however, the precisely
determined band origin frequencies associated with fundamentals, overtones, hot bands and combination bands have permitted determination of
many anharmonicity and anharmonic cross terms within second order approximation for several of these complexes.   In the case of HCN---HF,
where such an analysis is most advanced the feasibility of modelling
an approximate anharmonic potential energy surface appears realistic,
provided certain critical combination bands can be observed and
analyzed.
        It must be emphasized that the reported results suffer from lim-
itations imposed by spectral resolution.   Whereas almost Doppler-
limited static gas phase spectroscopy is possible using the color center laser system, the present Fourier transform spectrometer has a
lower resolution capability influenced by larger instrumental contributions.   However, the latter has a broadband frequency capability
which we have used between 8000 cm^{-1} to 30 cm^{-1}.   When interfaced with
long path temperature controlled White cells the Fourier transform infrared spectrometer provides a very sensitive technique with the
unique capacity of providing a rapid spectral overview of vibrational
bands in the complex. This has made a wide range of fundamental, hot,
combination and overtone vibrational bands accessible to the investigator.   It has also permitted investigation of intramolecular and intermolecular hydrogen-bonded vibrational bands.   The latter include
the low frequency infrared vibrations so important in characterization
and modeling of the intermolecular potential.   In the near future
Fourier transform supersonic jet infrared absorption spectroscopy
should serve to complement and significantly enhance current static
gas phase studies.
        Finally, it must be emphasized that the current kind of inves-
tigations of hydrogen-bonded interactions are necessarily preliminary.

However, it is expected that they can contribute significantly to facilitate subsequent and more sophisticated spectroscopic studies which should provide new insights into the fundamental nature of one of the most important of molecular interactions, the hydrogen bond.

ACKNOWLEDGEMENTS

Appreciation is expressed for support received from the National Science Foundation, the Robert A. Welch Foundation and CEMR, TAMU. Fourier transform infrared spectra were recorded on Bomem DA2-003 interferometers at the National Bureau of Standards (Gaithersburg) and Texas A&M University, the latter acquired under NSF grant CHE-85-00363. I am deeply grateful to my collaborators referred to in the quoted references and to members of the National Bureau of Standards for their hospitality.

REFERENCES

1.   J. Arnold and D. J. Millen, J. Chem. Soc. 503, 1965.

2.   P. Schuster, G. Zundel and C. Sandorfy, Eds., The Hydrogen Bond (North Holland) 1976.

3.   J. C. Lassegue and J. Lascombe, p. 51, Ch. 2 in Vibrational Spectra and Structure Vol. 11, J. R. Durig, Ed., Elsevier, Amsterdam, 1982.

4.   A. C. Legon, D. J. Millen, D. J. Mjöberg, S. C. Rogers, Chem. Phys. Letts., 55, 157 (1978).

5.   E. K. Kyrö, P. Shoja-Chaghervand, M. Eliades, D. Danzeiser, S. G. Lieb and J. W. Bevan, 63, 1870 (1985).

6.   W. B. Olson, submitted for publication.

7.   D. Papousek and M. R. Aliev, "Molecular Vibrational-Rotational Spectra," Elsevier, New York (1982).

8.   a)   A. C. Legon, Ann. Rev. of Phys. Chem. 34 275 (1983) and references therein.

      b)   T. R. Dyke, Top. Curr. Chem. 120 85 (1984).

9.   A. C. Legon and D. J. Millen, Chem. Rev. 86 635 (1986).

10.   c.f. references 24-26

11. G. A. Hopkins, M. Maroncelli, J. W. Nibler and T. R. Dyke, Chem. Phys. Letts. $\underline{114}$ 97 (1985).

12. H. H. Nielson, Rev. Mod. Phys. $\underline{23}$ 115 (1951).

13. a) S. Liu and C. E. Dykstra, J. Phys. Chem. $\underline{90}$ 3817 (1986).

    b) P. Botschwina (personal communication).

14. A. S. Pine, W. J. Lafferty and B. J. Howard, J. Chem. Phys. $\underline{81}$ 2939 (1984).

15. S. G. Lieb and J. W. Bevan, Chem. Phys. Letts. $\underline{122}$ 284 (1985).

16. S. G. Lieb, A. M. Gallegos, B. A. Wofford and J. W. Bevan (unpublished results).

17. D. J. Millen, Can. J. Chem. $\underline{63}$ 1477 (1985).

18. G. Herzberg, Molecular Spectra and Molecular Structure Vol. 1, D. Van Nostrand, New York (1950).

19. C. Sandorfy, Topics in Current Chemistry, $\underline{120}$ 41 (1984).

20. a) J. T. Hougen, Monograph 115, N.B.S. Washington, D.C. (1970)

    b) M. Lefbvre-Brion in "Atoms, Molecules and Lasers," pp. 411, International Atomic Energy Agency, Vienna (1974).

21. R. K. Thomas, Proc. Roy. Soc. $\underline{A325}$ 133 (1971).

22. N. Sheppard, Proceedings of the First International Conference on Hydrogen Bonding, p. 59, D. Hadzi, Ed., Pergammon London (1959).

23. D. Bender, M. Eliades and J. W. Bevan, submitted for publication.

24. a) D. U. Webb and K. N. Rao, J. Mol. Spect. $\underline{28}$ 121 (1958).

    b) D. H. Rank, D. P. Eastman, B. S. Rao and T. A. Wiggins, J. Opt. Soc. Am. $\underline{51}$ 929 (1961).

25. A. G. Maki and L. R. Blain, J. Mol. Spect. $\underline{12}$ 45 (1964).

26. V. K. Wang and J. Overend, Spectrochim. Acta. $\underline{29A}$ 687 (1983).

27. E. K. Kyrö, R. Warren, K. McMillan, M. Eliades, D. Danzeiser, P. Shoja-Chaghervand, S. G. Lieb and J. W. Bevan, J. Chem. Phys. 78 5881 (1983).

28. D. Bender, M. Eliades, D. A. Danzeiser, M. W. Jackson and J. W. Bevan, J. Chem. Phys. (in press).

29. E. K. Kyrö, M. Eliades, A. M. Gallegos, P. Shoja-Chaghervand and J. W. Bevan, J. Chem. Phys. 85 1283 (1986).

30. A. M. Gallegos and J. W. Bevan (submitted for publication).

31. B. A. Wofford, J. W. Bevan, W. B. Olson and W. J. Lafferty, J. Chem. Phys. 83 6188 (1985).

32. B. A. Wofford, J. W. Bevan, W. B. Olson and W. J. Lafferty, Chem. Phys. Lett. 124 579 (1986).

33. B. A. Wofford, M. W. Jackson, J. W. Bevan, W. B. Olson and W. J. Lafferty, J. Chem. Phys. 84 6115 (1986).

34. B. A. Wofford, M. W. Jackson, J. W. Bevan, W. B. Olson and W. J. Lafferty (unpublished results).

35. M. W. Jackson, B. A. Wofford and J. W. Bevan, J. Chem. Phys. (in press).

36. M. W. Jackson, B. A. Wofford and J. W. Bevan (submitted for publication).

37. M. W. Jackson, B. A. Wofford, J. W. Bevan, W. B. Olson and W. J. Lafferty, J. Chem. Phys. 85 2401 (1986).

38. L. A. Curtis and J. A. Pople, J. Mol. Spectr. 48 413 (1973).

39. A. C. Legon, D. J. Millen, and S. C. Rogers, Proc. Roy. Soc. London Ser. A 370 213 (1980).

40. A. C. Legon, D. J. Millen and L. C. Willoughby, Proc. Roy. Soc. London Ser. A401 327 (1985).

INFRARED SPECTROSCOPY OF HYDROGEN-BONDED AND VAN DER WAALS COMPLEXES

James M. Lisy, Kirk D. Kolenbrander and Daniel W. Michael
Department of Chemistry
University of Illinois at Urbana-Champaign
Urbana, Illinois 61801
U.S.A.

ABSTRACT. The assignment of vibrational spectra to a specific cluster is best accomplished by rotational analysis and the subsequent structural determination. Unfortunately, rotational line congestion, rapid vibrational relaxation and/or incomplete high resolution scans have frequently prevented the resolution and assignment of rotational lines. It is often desirable, even under these circumstances, to assign the vibrational band to a given cluster. Through the application of isotopic substitution, we have developed an approach which utilizes the positive features of mass spectroscopic detection of vibrational predissociation spectra. Applications to hydrogen fluoride trimer, benzene dimer and the argon-benzene binary complex are discussed.

1. INTRODUCTION

A major goal of cluster research is to understand the transition between isolated molecules and the condensed phase. This change is monitored by the determination of one or more molecular properties while observing the dependence of the properties as a function of cluster size. One important facet of such studies is to gauge the magnitude of non-pairwise interactions, i.e., three-body and higher order effects. This knowledge is necessary if an accurate model for the condensed phase is to be obtained. For these efforts to be successful, the monitored properties must exhibit a size-dependence, and an unambiguous identification of a particular sized cluster must be possible.

We have chosen to study the vibrational spectroscopy of molecular clusters as a probe of intermolecular interactions on the intramolecular force field. H-X (X = F, O, C, Cl) stretching frequencies are particularly sensitive to intermolecular hydrogen-bonding, and thus act as excellent probes. In clusters bound by van der Waals forces, the intermolecular interactions are weaker and the frequency shifts are therefore much smaller. For both types of systems, methods are needed to isolate and/or identify the vibrational bands associated with a specific sized cluster. Of the three principal methods for observing

171

A. Weber (ed.), Structure and Dynamics of Weakly Bound Molecular Complexes, 171–180.

the infrared spectra of molecular clusters, mass spectrometric [1], bolometric [2] and direct adsorption [3], only the mass spectrometric method offers an additional observable, i.e., the mass of the cluster. However, electron impact ionization of a van der Waals or hydrogen-bonded cluster often leads to fragmentation in the mass spectrometer. We have developed techniques which use isotopic substitution to eliminate the complications which result form fragmentation. The result is an unambiguous identification of an infrared absorption feature to a specific sized cluster.

2. EXPERIMENTAL METHODS

Details of the molecular beam apparatus [4] and the tunable infrared laser source, an optical parametric oscillator [5], can be found in the literature. Briefly, cluster beams are formed in a continuous supersonic nozzle expansion and detected downstream by a quadrupole mass spectrometer equipped with an electron impact ionizer. Infrared absorption spectra are obtained via vibrational predissociation. Since the binding energy of the cluster is less than the energy of a single infrared photon in the 2500-4000 cm^{-1} frequency range of the laser, the absorption of a single photon results in the dissociation of the complex. By tuning the mass spectrometer to a value corresponding to the cluster of interest, absorption of a photon is thus detected by the resulting decrease in the signal from the mass spectrometer. After a preset number of laser pulses, the laser is adjusted to a new frequency and the process is repeated. In this manner a frequency scan through the region of interest is obtained. The spectra are corrected for fluctuations in the laser energy and care is taken to avoid power-broadening.

The technique required for the successful use of isotopic substitution in the unambiguous assignment of vibrational bands to a given cluster is somewhat specific to the system under study. For larger hydrogen-bonded clusters such as $(HF)_3$, the significant fragmentation of the cluster and of the molecular subunits themselves, which results from electron impact, requires that the larger cluster be studied at a mass at which smaller clusters are also detected. The technique of isotopic substituiton as applied to this system must therefore provide a means of assigning the observed aosorption to one of the possible clusters known to be detected. In the case of van der Waals clusters such as $(C_6H_6)_2$, there is negligible fragmentation of the molecular subunits. Thus spectra can generally be recorded at the parent mass of the cluster of interest, ruling out contamination from smaller clusters. The significant fragmentation of larger clusters to this mass, however, remains a problem. These larger clusters usually have absorption regions which overlap with the cluster of interest. The technique for this system must therefore include a means of identifying and eliminating the presence of the larger clusters. Details of the techniques required for these two distinct systems are contained in the following section.

3. RESULTS

3.1. Hydrogen Fluoride Trimer

The experimental studies on $(HF)_3$ were initiated because of the spectrum shown in Figure 1, which was recorded at m/e=21, H_2F^+. The two

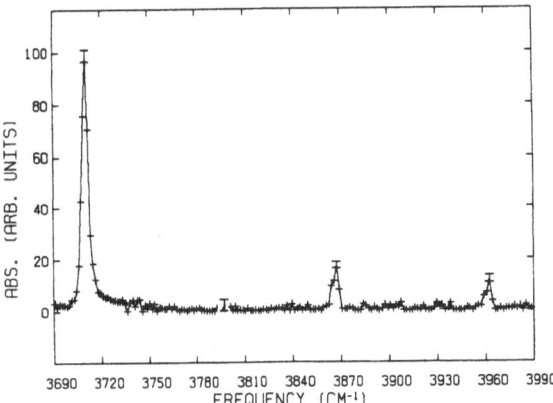

Figure 1. Vibrational predissociation spectrum recorded at m/e=21, using a (1:50) mixture of HF:Ar with a backing pressure of 650 torr through a 70 µ nozzle. Error bars indicate ±σ

bands at 3868 and 3963 cm^{-1} are due to $(HF)_2$ as shown by Pine and co-workers [6]. The third band at 3712 cm^{-1} was independent of carrier gas and its intensity relative to the two dimer bands was dependent on the nozzle backing pressure. There was no evidence in the mass spectra for any major contaminants such as H_2O, which led to the conclusion that this band must be due to a cluster of HF. However, there was no absorption in the 3690-3990 cm^{-1} region at m/e=41, $HF \cdot H_2F^+$, where one might expect larger clusters of HF to be observed. The dependence of the 3712 cm^{-1} band on nozzle backing pressure indicated this band was due to a cluster larger than the dimer. However, without additional information a definitive assignment could not be made.
    To test the hypothesis that the 3712 cm^{-1} band was due to a larger cluster which fragmented to m/e=21, isotopic substitution studies using DF were initiated. For a cluster of $(HF)_{m-n}(DF)_n$ where m>2, the following cracking patterns would be expected:

$$(HF)_m(DF)_0 \xrightarrow{e^-} H_2F^+ \text{ m/e=21}$$

$$(HF)_{m-1}(DF)_1 \xrightarrow{e^-} H_2F^+, HDF^+ \text{ m/e=22}$$

$$(HF)_{m-2}(DF)_2 \xrightarrow{e^-} H_2F^+ (m>3), HDF^+, D_2F^+, \text{ m/e=23}$$

In Figure 2, scans at m/e=22 are shown for two gas mixtures with different ratios of HF to DF. Of the three bands observed, the

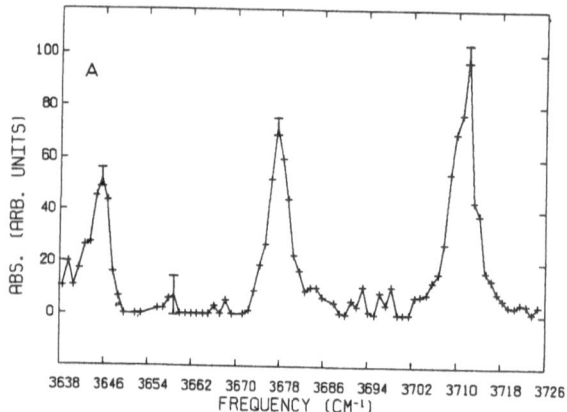

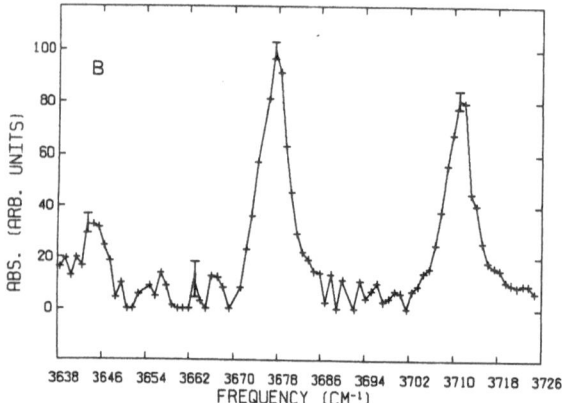

Figure 2. Vibrational predissociation spectra recorded at m/e=22 using a 70 μ nozzle. (A) (1:2:200) mixture of DF:HF:Ar with a backing pressure of 650 torr. (B) (1:1:100) mixture of DF:HF:Ar with a backing pressure of 450 torr. Error bars indicate ±σ.

relative intensity of the 3678 cm^{-1} band increases with the relative concentration of DF, while the other two bands at 3712 and 3645 cm^{-1} maintain a ~2:1 intensity ratio and thus have the same composition. At m/e=22, all three bands must come from clusters with at least one DF and the 3678 cm^{-1} band must have a different composition with more DF. A scan at m/e=23 shown in Figure 3 has only the 3678 cm^{-1} which indicates the 3712 and 3645 cm^{-1} bands have one DF and the 3678 cm^{-1} band has at least two DF units. The simplest cluster consistent with all of the experimental data is the trimer.

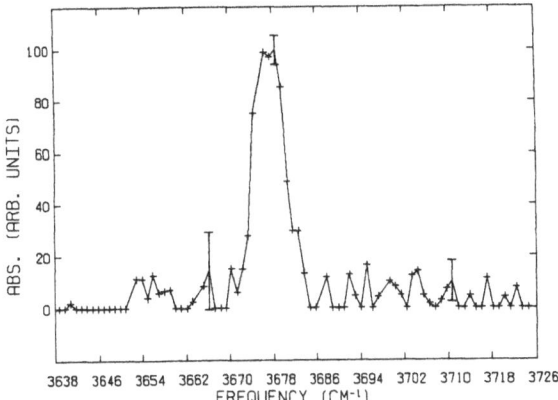

Figure 3. Vibrational predissociation spectra recorded at m/e = 23, using a (1:1:100) mixture of DF:HF:Ar with a backing pressure of 450 torr through a 70 μ nozzle. Error bars indicate ±σ.

Consideration of possible structures for $(HF)_3$ must include the observed spectra for the fully protonated and partially deuterated species. For $(HF)_3$ to have only one infrared active band, there must either be sufficient symmetry to reduce the number of infrared active bands or extremely weak coupling between the HF monomers. As the bands are shifted ~250 cm^{-1} from the HF monomer fundamental and in $(HF)_2DF$ the two HF stretching frequencies are split by 67 cm^{-1}, the weak coupling scenario obviously fails. The only structure which has sufficient symmetry is cyclic with each HF participating in a hydrogen bond. Since each HF would be identical, this structure would have C_{3h} symmetry. $(HF)_3$ under this point group would have an infrared active double degenerate HF stretch and an infrared inactive totally symmetric HF stretch. Cyclic $DF(HF)_2$ and $HF(DF)_2$ have C_{2h} symmetry with two and one infrared active HF stretches respectively, fully consistent with the experimental results.

A force field analysis offers a useful test for the cyclic trimer structure. Since the HF and DF stretching frequencies are much larger than the hydrogen-bond stretches and bends, the analysis can be performed in the harmonic approximation using only two parameters: the H(D)F force constant and the force constant representing the interaction between the H(D)F units. Using values of 762.5 N/m for the diagonal H(D)F force constant and -14.5 N/m for the off-diagonal interactions, the agreement with the four HF stretching frequencies, as shown in Table I, is excellent.

TABLE 1.   Trimer Stretching Frequencies (cm^{-1})

Species	Observed	Calculated
$(HF)_3)$	3712(1)	3712
$(HF)_2DF$	3712(1)	3712
	3645(1)	3644
$HF(DF)_2$	3678(1)	3679

Additional experimental information dealing with the dynamics of the predissociation process: the vibrational relaxation lifetime, fragmentation channels, and kinetic energy of the fragments can be found in the literature [4].

3.2.   Benzene Dimer, Argon-benzene

The study of van der Waals clusters by vibrational predissociation is somewhat more complicated than that for hydrogen-bonded species. The weak interactions which characterize van der Waals clusters often result in significant overlap of vibrational bands attributable to each of the possible polymers. This overlap makes difficult the assignment of observed features to a single specific cluster. Isotopic substitution, however, permits the simple, unambiguous assignment to a specific cluster.

The principle behind this method relies on the large red shift in the vibrational fundamental frequencies which generally accompanies deuterium substitution in a molecule. Applying this technique to the study of 3000-3100 cm^{-1} C-H stretching bands of benzene dimer is made possible by generating a molecular beam composed of a mixture of C_6H_6 and C_6D_6 in argon. The possible benzene dimers in the beam are $(C_6H_6)_2$, $C_6H_6-C_6D_6$ and $(C_6D_6)_2$, with m/e = 156, 162 and 168, respectively. If the mass spectrometer is set to m/e = 168, the only binary cluster which can be detected is $(C_6D_6)_2$. This cluster cannot absorb in the 3000 cm^{-1} range since it contains no C-H units. It is possible, through fragmentation, for trinary and/or larger clusters to be detected at this mass. Most of these trinary and larger clusters will contain at least one C_6H_6 unit and will hence be susceptible to vibrational predissociation in the C-H stretching region. By reducing the benzene concentration and/or the nozzle backing pressure, the formation of larger clusters can be retarded and eventually eliminated. The lack of any absorption at m/e = 168 indicates the elimination of the trinary and larger contaminants. The mass spectrometer is then tuned to m/e = 156, and a spectrum of $(C_6H_6)_2$ is recorded free of larger cluster contamination. C_6D_6 is used only to identify and eliminate the larger cluster contamination and is not used in the actual recording of the $(C_6H_6)_2$ spectrum. The effects of contamination and the subsequent elimination are shown in Figure 4.

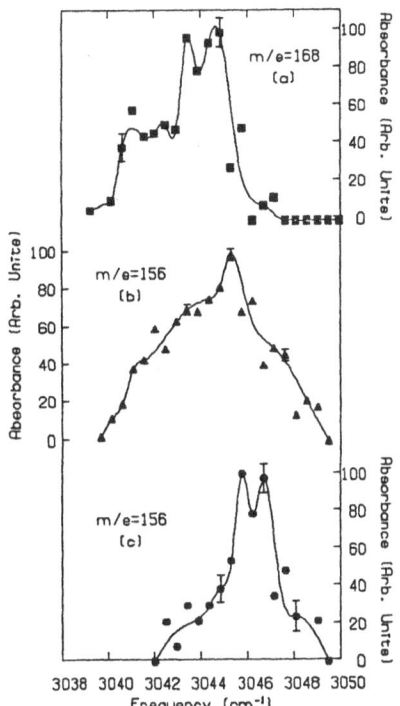

Figure 4. Vibrational predissociation spectra of benzene clusters. (a) m/e=168, (5:5:1000) C_6H_6:C_6D_6:Ar, laser fluence 3 mJ/cm^2. (b) m/e=156, otherwise same as (a). (c) m/e=156, (2:2:1000) C_6H_6:C_6D_6:Ar, laser fluence 1 mJ/cm^2. For (a), (b) and (c) 600 torr backing pressure, 100 μ nozzle, 298 K. A smooth curve is drawn for each. Errors bars represent ±σ.

This technique has also been applied to Ar–C_6H_6 to eliminate the possible contamination of Ar–$(C_6H_6)_2$ to the spectrum at m/e=118. In this case, the mass spectrometer is tuned to m/e=124 (Ar–C_6D_6) and the vibrational predissociation signal in the C-H stretching region is eliminated. The mass spectrometer is then tuned to m/e=118 and the absorption spectrum, as shown in Figure 5, is recorded. There does remain a slight possibility of contamination by $(Ar)_n$–C_6H_6 in this spectrum.

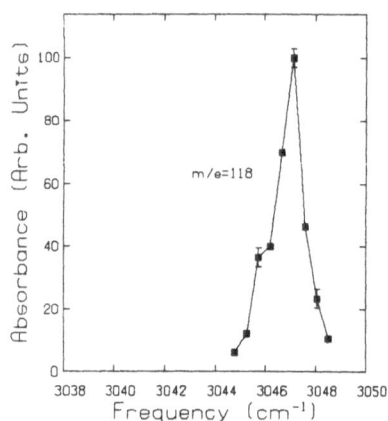

Figure 5. Vibrational predissociation spectrum of Ar-C$_6$H$_6$ at m/e=118. Straight lines connect the data points. (2:1000) Ar-C$_6$H$_6$, 600 torr backing pressure, 100 μ nozzle, 298 K.

4. DISCUSSION

It is possible to assign vibrational bands without full rotational resolution to specific cluster species as we have shown for both hydrogen-bonded and van der Waals complexes. Isotopic substitution is used in both cases. However, the method is applied in a manner which is dependent on the type of cluster. More importantly, the development of these techniques permits us to consider what type of information this size specific vibrational analysis can supply in the understanding of intermolecular forces.

The results from (HF)$_3$ allow us to examine the fundamental question of whether three-body interactions are important in this system. From a theoretical study [7], the stabilization energies per hydrogen-bond are virtually identical: 4.57 and 4.67 kcal/mol for (HF)$_2$ and (HF)$_3$, respectively. The vibrational analysis of (HF)$_3$ and its deuterated analogs gives a different picture. The harmonic interaction force constant for the trimer, ~14.5 N/m is an order of magnitude larger than the dimer. Thus it appears that evidence of non-additivity may not be reflected in all of the cluster properties. Current work in our laboratory is focused on (HF)$_n$, n>3, to see if the larger HF clusters behave in a similar manner or if the trimer behavior is anomalous. It is interesting to note that (HF)$_3$ does not yield an ion at m/e=41, HF·H$_2$F$^+$. Earlier experimental results indicating that (HF)$_3$ was non-polar were based on the assumption that the trimer could be observed at m/e=41 [8]. While this mass peak may be non-polar, it is also non-trimer. Preliminary results from our laboratory [9] indicate the smallest cluster that contributes to the m/e=41 mass peak is (HF)$_4$.

The results on benzene are the most recent from our laboratory. We hope to address a number of questions concerning the nature of benzene dimer, most notably the structure. There have been three different structures proposed on the basis of experimental results: T-shaped [10], offset planar [11] or roof-shaped [12]. Attempts are currently underway to check for polarity using an electrostatic deflection field. This experiment was performed earlier by Klemperer and co-workers [8], but under conditions where larger benzene clusters may have contributed to the dimer mass peak.

The isotopic substitution method clearly provides a number of opportunities to investigate cluster size-dependent behavior. While we have concentrated on vibrational properties, variations of the method are applicable to other types of investigation as demonstrated by Schlag and co-workers on the electronic spectra of benzene dimer [10]. The extension of these methods to larger clusters bodes well for the future understanding of the transition between the isolated molecule and the condensed phase.

5. ACKNOWLEDGMENTS

The authors would like to thank Professors L. Andrews and W. Klemperer for their assistance and advice. Professor C. E. Dykstra and Ms. S. Y. Liu have made collaboration on the theory of HF clusters an enjoyable and educational experience. The authors are very grateful for support from the National Science Foundation, Structural Chemistry and Thermodynamics Division (CHE 8205811 and CHE 8506698), the Research Corporation, and the Dreyfus Foundation. Acknowledgment is made to the donors of the Petroleum Research Fund, administered by the American Chemical Society for partial support of this research.

6. REFERENCES

1. K. Kolenbrander and J. M. Lisy, J. Chem. Phys. 85, 2463 (1986); R. D. Johnson, S. Burdenski, M. A. Hoffbauer, C. F. Giese and W. R. Gentry, ibid. 84, 2624 (1986); C. M. Western, M. P. Casassa and K. C. Janda, ibid. 80, 4781 (1984); J. Geraedts, S. Stolte and J. Reuss, Z. Physik. 304, 167 (1982).

2. G. Fischer, R. E. Miller, P. F. Vohralik and R. O. Watts, J. Chem. Phys. 83, 1471 (1985); D. F. Coker, R. E. Miller and R. O. Watts, ibid. 82, 3554 (1985).

3. B. J. Howard and A. S. Pine, Chem. Phys. Lett. 122, 1 (1985); G. D. Hayman, J. Hodge, B. J. Howard, J. S. Meunter and T. R. Dyke, ibid. 118, 12 (1985); C. M. Lovejoy, M. D. Schuder and D. J. Nesbitt, ibid. 127, 374 (1986); B. A. Wofford, J. W. Bevan, W. B. Olson and W. J. Lafferty, J. Chem. Phys. 85, 105 (1986).

4. D. W. Michael and J. M. Lisy, J. Chem. Phys. **85**, 2528 (1986).

5. D. W. Michael, K. Kolenbrander and J. M. Lisy, Rev. Sci. Instrum. **57**, 1210 (1986).

6. A. S. Pine, W. J. Lafferty and B. J. Howard, J. Chem. Phys. **81**, 2929 (1984).

7. S. Y. Liu, D. W. Michael, C. E. Dykstra and J. M. Lisy, J. Chem. Phys. **84**, 5032 (1986); D. W. Michael, C. E. Dykstra and J. M. Lisy, ibid. **81**, 5998 (1984).

8. T. R. Dyke, B. J. Howard and W. Klemperer, J. Chem. Phys. **56**, 2442 (1972).

9. D. W. Michael and J. M. Lisy, work in progress.

10. K. C. Janda, J. C. Hemminger, J. S. Winn, S. E. Novick, S. J. Harris and W. Klemperer, J. Chem. Phys. **63**, 1419 (1975).

11. K. S. Law, M. Schauer and E. R. Bernstein, J. Chem. Phys. **81**, 4871 (1984); M. Schauer and E. R. Bernstein, ibid. **82**, 3772 (1985).

12. K. O. Bornsen, M. L. Selzle and E. W. Schlag, J. Chem. Phys. **85**, 1726 (1986).

AB INITIO CALCULATIONS OF VIBRATIONAL FREQUENCIES AND INFRARED
INTENSITIES FOR THE HYDROGEN-BONDED COMPLEX HCN---HF

P. Botschwina
FB Chemie der Universität Kaiserslautern
D-6750 Kaiserslautern
West Germany

ABSTRACT. Potential and dipole moment surfaces for the stretching motions of the hydrogen bonded complex HCN---HF were computed by SCF and
CEPA (Coupled Electron Pair Approximation). The calculated anharmonic
stretching vibrational frequencies and the intramolecular frequency
shifts are in good agreement with recent experimental results of Bevan
and coworkers. Vibrational anharmonicity is shown to be of great importance for the HF stretching vibration. The intensity of this vibration is enhanced by a factor of 13 with respect to free HF. The theoretical equilibrium dissociation energy is in close agreement with
the experimental value of Legon and Millen.

1. INTRODUCTION

Over the last 20 years the investigation of hydrogen bonded systems in
the gas phase by means of infrared spectroscopy has developed into an
active field of research. Following earlier work at lower resolution
which is nicely reviewed by Sandorfy [1], high-resolution infrared
spectra are now becoming available from which a wealth of structural
and dynamic information about hydrogen bonded species can be obtained.
Additional and partly complementary information has been gained through
microwave and radiofrequency spectroscopy (see [2] for a review) which
is particularly useful for the determination of structures and electric
properties.
  A system which has attracted much attention and which is probably
the most intensive studied to-date is the complex HCN---HF. Experimental work on this species has been extensively reviewed by Legon and
Millen [3] and morerecent work by Bevan and coworkers [4-9] is discussed in the course of this article. This system appears thus to be an
ideal candidate to examine the accuracy of large-scale ab initio calculations. Although numerous ab initio calculations have been carried out
for HCN---HF [10-26], high-quality calculations of spectroscopic properties such as vibrational frequencies and infrared intensities are
still not available in the literature. Most of the previous ab initio
calculations were carried out at the SCF level using rather small basis

181

A. Weber (ed.), Structure and Dynamics of Weakly Bound Molecular Complexes, 181–190.
© *1987 by D. Reidel Publishing Company.*

sets. A noteworthy exception is the work of Benzel and Dykstra [22] who
calculated the equilibrium geometry, equilibrium dissociation energy,
and electric dipole moment of HCN----HF by means of CI-SD (Configuration
Interaction with single and double substitutions) and ACCD (Approximate
Coupled Cluster Doubles [27]) using fairly large basis sets. These authors gave also a detailed discussion of the nature of the attractive
interactions in this system. Quite recently, Liu and Dykstra [26] reported a theory of vibrational transition frequency shifts due to hydrogen bonding. While quite good agreement with experiment was found for
several complexes such as $(HF)_2$ or HCCH----HF, the shift in the HF stretching vibrational frequency upon forming the complex HCN----HF was severely underestimated by 85 cm^{-1} or 35 %.

This work deals with the calculation of various spectroscopic properties for the hydrogen bonded system HCN----HF. Potential energy and
dipole moment surfaces were calculated by SCF and Meyer's Coupled Electron Pair Approximation (CEPA [28]) using a significantly larger basis
set than hitherto employed. The project is not yet completed and we here
will restrict ourselves to an anharmonic treatment of the stretching
fundamental vibrations (three intramolecular ones and one intermolecular one) and their IR intensities. Numerically accurate work on a variety of combination tones and overtones will require further calculations
at different nuclear configurations and is presently underway.

2. RESULTS FOR THE MONOMERS HCN AND HF

A basis set of 70 contracted Gaussian-type orbitals (C and N: 11s,6p,2d
contracted to [8,4,2] and H: 6s,2p in contraction [4,2]) was employed
in the SCF and CEPA-1 calculations (all valence electrons correlated)
for hydrogen cyanide. The same basis set was used in a recent investigation of the astrophysically important $HCNH^+$ ion [29], where a more
detailed description is given. The CEPA-1 equilibrium bond lengths for
HCN are r_e(CH) = 1.0665 Å and R_e(CN) = 1.1534 Å which are both very
close to the experimental values of 1.0655 and 1.1532 Å [30]. The dependence of the potential energy on the stretching coordinates $\Delta r = r
- r_e$ and $\Delta R = R - R_e$ is represented by a polynomial function of the
form

$$V - V_e = \sum_{i,j} C_{i,j} \Delta r^i \Delta R^j \tag{1}$$

and the non-vanishing terms $C_{i,j}$ included in the least-squares fit are
given in Table I. This table includes also a corrected potential
and the stretching part of the experimental quartic force field of
Strey and Mills [31]. Using the SCF, CEPA-1, and corrected potentials
stretching vibrational energies and wave functions were calculated variationally by diagonalizing an approximate vibrational Hamiltonian
which neglects the anharmonic interaction with the bending mode in a
sufficiently large basis set of harmonic oscillator product functions
[32]. In the calculations with the corrected potential, stretch-bend
interaction was parametrically taken into account [33, 34]; the values
of the two parameters are given in a footnote to Table II. This table

TABLE I. Potential energy functions (stretching coordinates only) for HCN.

PEF term[a]	SCF	CEPA-1	corr.	exp. [30,31]
$r_e(Å)$	1.05728	1.06651	1.06549	1.06549
$R_e(Å)$	1.12401	1.15339	1.15321	1.15321
r^2	0.21819	0.20017	0.20133	0.2007
r^3	-0.21226	-0.20229	-0.20365	-0.2004
r^4	0.14406	0.13639	0.13744	0.1356
r^5	-0.08322	-0.07741	-0.07789	
r^6	0.04541	0.04156	0.04187	
r^7	-0.02499	-0.02321	-0.02339	
r^8	0.01207	0.01144	0.01151	
r^9	-0.00375	-0.00366	-0.00367	
r^{10}	0.00052	0.00052	0.00052	
R^2	0.77418	0.61577	0.59907	0.60064
R^3	-0.84349	-0.72169	-0.70187	-0.7081
R^4	0.60053	0.51261	0.49851	0.4347
R^5	-0.35482	-0.29489	-0.28688	
R^6	0.18395	0.14893	0.14489	
R^7	-0.08230	-0.07437	-0.07232	
R^8	0.01788	0.01788	0.01788	
rR	-0.01061	-0.01214	-0.01214	-0.0128
r^2R	0.00668	0.00323	0.00323	0.0007
rR^2	-0.00709	-0.00068	-0.00068	0.0070

a)
An obvious shorthand notation is employed to designate the expansion coefficients in equ. 1, e.g., R^8 stands for the term $C_{0,8}$. Atomic units are used.

compares the calculated stretching vibrational frequencies with the experimental ones. While the SCF potential severely overestimates all frequencies, the uncorrected CEPA-1 potential yields errors of only 13 and 30 cm^{-1} for ν_1 (∿CH stretch) and ν_3 (∿CN stretch).

The CEPA-1 equilibrium quartic centrifugal distortion constant D_J^e agrees with the experimental value [30] within 3 %. This deviation is mainly due to an overestimation of the harmonic CN stretching vibrational frequency ω_3 by 26 cm^{-1} or 1.2 %.

The dipole moments of HCN at the equilibrium geometries are obtained to be -3.260 D (SCF) and -2.992 D(CEPA-1) where the positive end of the dipole is at the hydrogen site. Throughout this work, CEPA dipole moments were calculated as energy derivatives. The CEPA-1 value for

TABLE II. Calculated and experimental spectroscopic properties for HCN.

	SCF	CEPA-1	corr.[a]	exp.[b]
B_e (MHz)	46595	44488	44512	44512
D_J^e (kHz)	77.12	83.29	85.46	85.53
ω_1 ($\sim$CH)(cm^{-1})	3604.0	3439.0	3445.8	3442.3
ω_3 ($\sim$CN)	2406.9	2154.5	2127.0	2129.1
ν_3	2384.5	2127.8	2096.9	2096.9
ν_1	3505.0	3324.5	3311.5	3311.5
$2\nu_3$	4753.4	4235.5	4173.8	4173.1
$\nu_1+\nu_3$	5875.3	5437.8	5394.7	5393.7
$3\nu_3$	7106.7	6323.0	6230.2	
$2\nu_1$	6921.2	6544.1	6516.2	6519.6
$\Gamma(\omega_1)$(cm^{2}mol^{-1})	1990	1761	1759	
$\Gamma(\omega_3)$	426	7.0	5.1	
$\Gamma(\nu_3)$	472	7.6	5.5	6.9, $\sim$1
$\Gamma(\nu_1)$	2049	1750	1748	1610,1631,1884
$\Gamma(2\nu_3)$	0.0002	0.0029	0.0088	
$\Gamma(\nu_1+\nu_3)$	10.5	5.2	4.9	5.0
$\Gamma(3\nu_3)$	0.001	0.0046	0.0042	
$\Gamma(2\nu_1)$	11.0	19.3	19.7	

a)
  Underlined experimental values have been employed to
  correct the CEPA-1 potential. Stretch-bend interaction
  parameters are $\Delta_{1,2}$ = -18.2 cm^{-1} and $\Delta_{2,3}$ = -3.4 cm^{-1}.

b)
  References to experimental work are given in Ref. 32 and
  36.

μ_e compares well with the experimental value of DeLeon and Muenter [35] of -3.016 D. Integrated molar IR intensities of absorption for several stretching vibrational transitions arising from the vibrational ground state, which are defined as

$$\Gamma_{f0} = (4\pi^2 N_0 / 3\hbar c) \left| \mu_{f0} \right|^2 \qquad (2) ,$$

are given in Table II. The intensities calculated from the CEPA-1(ED) dipole moment surface are in good agreement with available experimental data for both cases of vibrational wavefunctions, generated either from the uncorrected or corrected CEPA-1 potential. As is well known (see, e.g., [36] and references therein), SCF severely overestimates the intensity of the weak ν_3 band. Using the corrected potential and the CEPA-1(ED) dipole moment function even the intensity of the very weak $5\nu_1$ band with calculated origin at 15525 cm^{-1} (exp.: 15552 cm^{-1}) agrees with the recent experimental value of Smith et al. [37] within a factor of 2-3.

The basis set for HF consists of 40 contracted GTOs (11s,6p,2d// 6s,2p in contraction [8,4,2//4,2]. Calculated and experimental properties for this molecule are listed in Table III. The CEPA-1 r_e value is smaller than the experimental value by only 0.0002 Å and the fundamental vibrational frequency is overestimated by 21 cm^{-1} or 0.5 %. The dipole moment in the vibrational ground state μ_0 differs from experiment by 0.022 D or 1.2 %. The CEPA-1 transition dipole moments μ_{01}, μ_{02} and μ_{12} agree with the experimental values of Sileo and Cool [38] within 2-4 %.

TABLE III. Calculated and experimental properties of HF.

	SCF	CEPA-1	exp.[a]
r_e(Å)	0.8983	0.9166	0.9168
B_e(cm^{-1})	21.83	20.97	20.96
D_J^e(MHz)	62.73	63.84	64.49
ω_e(cm^{-1})	4458.4	4160.2	4138.3
ν(cm^{-1})	4294.9	3982.4	3961.4
2ν(cm^{-1})	8431.9	7791.6	7750.8
3ν(cm^{-1})	12417.6	11433.8	11372.8
μ_e(D)	-1.906	-1.819	-1.796
μ_0(D)	-1.936	-1.841	-1.819
μ_1(D)	-1.996	-1.885	-1.865
μ_2(D)	-2.058	-1.926	-1.909
$\left| \mu_{01} \right|$ (D)	0.123	0.0959	0.0985
$\left| \mu_{02} \right|$ (D)	0.010	0.0130	0.0127
$\left| \mu_{12} \right|$ (D)	0.177	0.133	0.138

[a] Experimental data from Ref. 38-40.

3. HCN---HF

Calculated equilibrium geometries for HCN---HF and the geometry changes with respect to the monomers are listed in Table IV. The intermolecular equilibrium bond length R_e(N---F distance) is calculated to be
2.830 Å (CEPA-1) which is slightly smaller than the TZP/ACCD value of
Benzel and Dykstra [22] of 2.848 Å. The CEPA-1 intramolecular equilibrium bond length changes are +0.0006 Å for CH, -0.0032 Å for the CN,
and +0.0092 Å for the HF bond length. Only slightly different values
are obtained from the present SCF calculations which agree with the
previous ones of Curtiss and Pople [12] who used a small 4-31G basis
set. For comparison, the experimental R_0 value is 2.804 A and the lengthening of the HF bond on formation of HCN---HF was estimated from
nuclear hyperfine coupling constants to be 0.014 Å [3].

TABLE IV. Calculated equilibrium geometries for HCN---HF[a].

	SCF[12] 4-31G	SCF[19] 3-21G	SCF[24] FOGO	SCF[25] MINI-1	SCF[b] 110 CGTOs	CEPA-1[b]
R_e(N---F)	2.833	2.848	2.87	3.037	2.888	2.830
r_{1e}(CH)	1.052	1.052	1.058	1.143	1.0582	1.0671
r_{2e}(CN)	1.137	1.135	1.140	1.212	1.1217	1.1502
r_{3e}(HF)	0.929	0.942	0.920	0.983	0.9057	0.9258
Δr_{1e}	0.001			0.003	0.0009	0.0006
Δr_{2e}	-0.003			-0.002	-0.0023	-0.0032
Δr_{3e}	0.007			0.004	0.0074	0.0092

a)
  Bond lengths in Å.
b)
  This work. All 81 valence electron pairs, constructed from
  canonical molecular orbitals, are correlated in the CEPA-1
  calculations which make use of the technique of the Self-
  Consistent Electron Pairs theory (SCEP [41,42]).

The equilibrium dissociation energy D_e of HCN---HF was calculated
to be 2119 cm^{-1} by SCF and 2396 cm^{-1} by CEPA-1. An experimental value
of 2183 ± 134 cm^{-1} was derived by Legon and Millen [3] from absolute
intensities of rotational transitions and an approximate statistical
treatment of the thermodynamic equilibrium. As the present CEPA-1 value
was not corrected for basis set superposition errors, it is probably

somethat too large, although there may be some error compensation between superposition error and missing dispersion energy. The CEPA-1 value for D_e agrees perfectly with the ACCD value of Benzel and Dykstra [22] who used a smaller basis set of TZP quality. This is, however, somewhat fortuitous as their corresponding SCF value of 2222 cm^{-1} is larger than the present one by 103 cm^{-1}.

The equilibrium electric dipole moment of HCN---HF was calculated to be -5.957 D (SCF) and -5.724 D (CEPA-1(ED)). For comparison, Benzel and Dykstra yielded -6.19 D at the SCF and -5.96 at the CI-SD level (TZP basis). The sum of the equilibrium dipole moments of the monomers is -5.166 D (SCF) and -4.811 D (CEPA-1(ED)) so that we obtain dipole moment enhancements of -0.791 D (SCF) and -0.913 D (CEPA-1(ED)). The corresponding values of Benzel and Dykstra are -0.82 D (SCF) and -0.90 D (CI-SD). The experimental $\Delta\mu_a$ value, which is not strictly comparable with the ab initio equilibrium values, is 0.80 D [43].

The analytical four-dimensional potential energy functions which were used for the calculation of vibrational frequencies and electric transition dipole moments are still somewhat preliminary. They allow to describe the fundamentals with a numerical accuracy of about 2 cm^{-1}. Calculated and experimental stretching vibrational frequencies for $H^{12}C^{14}N$---HF are given in Table V. Compared with experiment [9, 44], the intermolecular stretching frequency ν_4 is underestimated by SCF by 22 cm^{-1}, but agrees with the anharmonic CEPA-1 value. The experimental intramolecular frequency shifts [4,6,8] are nicely reproduced by the variational calculations with both the anharmonic SCF and CEPA-1 potential. In particular, all the CEPA-1 shifts agree with experiment to within 3 cm^{-1}. Inclusion of vibrational anharmonicity is crucial to correctly reproduce the shift in the HF stretching vibrational frequency occurring upon complex formation. The harmonic approximation underestimates this shift by as much as 77 cm^{-1} or 31 %.

TABLE V. Equilibrium rotational and quartic centrifugal distortion constants, harmonic and anharmonic stretching vibrational frequencies (in cm^{-1}) and intramolecular frequency shifts (in parentheses) for $H^{12}C^{14}N$---HF.

	SCF [12][a] 4-31G	SCF[b] 110 CGTOs	CEPA-1[b]	exp.
B (MHz)	3544	3456	3529	3591[c]
D_J^e(kHz)		7.17	5.99	6.99[c]
ω_1($\sim$ HF str.)	3990(-127)	4294(-164)	3992(-168)	
ω_2($\sim$ CH str.)	3682(-13)	3602(-2)	3439(-3)	
ω_3($\sim$ CN str.)	2396(+12)	2428(+21)	2182(+28)	
ω_4(inter)	193	154	173	
ν_1		4053(-242)	3738(-245)	3716(-245)[d]
ν_2		3505(0)	3326(+1)	3310(-1)[d]
ν_3		2405(+19)	2155(+27)	2121(+24)[d]
ν_4		147	169	169[e]

Footnotes to TABLE V.

    a) In the calculation of the harmonic frequencies use of an
    approximate separation of high and low frequency modes was
    made.

    b) This work. The frequencies are numerically accurate to
    about 2 cm^{-1}.

    c) Ground state values B_o and D_J^o [3].

                e)
    d) Ref. 6.   Ref. 9 and 44.

The calculated CEPA-1 equilibrium quartic centrifugal distortion
constant (see Table V) is smaller than the experimental ground state
value [3] by 14 %. Such a difference appears to be quite reasonable for
a linear molecule or complex with large-amplitude bending motions. For
example, for HOC^+ with its very shallow bending potential a CEPA-1 potential yields D_J^e = 100.3 kHz [45] while the experimental D_J^o value is
114.9 kHz [46].

Since IR intensities for complexes with low intermolecular vibrational frequencies are rather strongly temperature-dependent we quote
only the electric transition dipole moments, calculated both in the
familiar double harmonic approximation and from the anharmonic treatment. The results for the fundamentals are given in Table VI.

TABLE VI. Calculated electric transition dipole moments (in
D) for stretching vibrations of HCN---HF.

	ν_1	ν_2	ν_3	ν_4
SCF harm. [22]	0.210	0.108	0.057	0.092
harm.	0.316	0.097	0.065	0.076
anharm.	0.348	0.098	0.068	0.088
CEPA-1 harm.	0.315	0.095	0.026	0.054
anharm.	0.347	0.094	0.028	0.067

The transition moments for the strongest band, the HF stretching
vibration, are almost identical at the SCF and CEPA-1 level. This results from a fortuitous cancellation of errors in the potential and
dipole moment surface within the SCF approximation. Compared with free
HF there is an intensity enhancement by a factor of 13. The intensity
of the CH stretching vibration ν_2 increases only by 27 % on formation
of the complex. As for free HCN, the intensity of ν_3 (CN stretch) is
poorly reproduced by SCF. Anharmonicity effects are small for the transition moment of the ν_2 band, but noticably larger (7-24 %) for the
other stretching vibrations.

REFERENCES

1. C. Sandorfy, in "Hydrogen Bonds", Topics in Current Chemistry, Vol. 120 (Springer, Berlin, 1984).
2. T. R. Dyke, in "Hydrogen Bonds", Topics in Current Chemistry, Vol. 120 (Springer, Berlin, 1984).
3. A. C. Legon and D. J. Millen, Chem. Rev. 86, 635(1986) and references therein.
4. E. Kyrö, R. Warren, K. McMillan, M. Eliades, D. Danzeiser, P. Shoja-Chaghervand, S. G. Lieb, and J. W. Bevan, J. Chem. Phys. 78, 5881 (1983).
5. B. A. Wofford, J. W. Bevan, W. B. Olson, and W. J. Lafferty, J. Chem. Phys. 83, 6188(1985).
6. B. A. Wofford, J. W. Bevan, W. B. Olson, and W. J. Lafferty, Chem. Phys. Letters 124, 579(1986).
7. B. A. Wofford, M. W. Jackson, J. W. Bevan, W. B. Olson, and W. J. Lafferty, J. Chem. Phys. 84, 6115(1986).
8. E. K. Kyrö, M. Eliades, A. M. Gallegos, P. Shoja-Chaghervand, and J. W. Bevan, J. Chem. Phys. 85, 1283(1986).
9. M. W. Jackson, B. A. Wofford, J. W. Bevan, W. B. Olson, and W. F. Lafferty, J. Chem. Phys. (in press).
10. A. Johannson, P. Kollman, and S. Rothenberg, Chem. Phys. Letters 16, 123(1972).
11. J. E. Del Bene and F. T. Marchese, J. Chem. Phys. 58, 926(1973).
12. L. A. Curtiss and J. A. Pople, J. Mol. Spectrosc. 48, 413(1973).
13. J. E. Del Bene, Chem. Phys. Letters 24, 203(1974).
14. P. Kollman, J. McKelvey, A. Johansson, and S. Rothenberg, J. Am. Chem. Soc. 97, 955(1975).
15. P. Kollman, J. Am. Chem. Soc. 99, 4875(1977).
16. S. Vishveshwara, Chem. Phys. Letters 59, 26(1978).
17. A. Hinchliffe, Advan. Mol. Relax.Int. Processes 19, 227(1981).
18. E. L. Mehler, J. Chem. Phys. 74, 6298(1981).
19. P. Hobza and R. Zahradnik, Chem. Phys. Letters 82, 473(1981).
20. Y. Bouteiller, M. Allavena, and J. M. Leclercq, Chem. Phys. Letters 84, 91(1981).
21. B. A. Pettit, R. J. Boyd, and K. E. Edgecombe, Chem. Phys. Letters 89, 478(1982).
22. M. A. Benzel and C. E. Dykstra, J. Chem. Phys. 78, 4052(1983).
23. J. M. Leclercq, M. Allavena, and J. Bouteiller, J. Chem. Phys. 78, 4606(1983).
24. H. Huber, P. Hobza, and R. Zahradnik, J. Mol. Struct. 103, 245 (1983).
25. P. Hobza and J. Sauer, Theoret. Chim. Acta 65, 279(1984).
26. S.-Y. Liu and C. E. Dykstra, J. Phys. Chem. 90, 3097(1986).
27. R. A. Chiles and C. E. Dykstra, Chem. Phys. Letters 80, 69(1981).
28. W. Meyer, J. Chem. Phys. 58, 1017(1973).
29. P. Botschwina, Chem. Phys. Letters 124, 382(1986).
30. G. Winnewisser, A. G. Maki, and D. R. Johnson, J. Mol. Spectrosc. 39, 149(1971).
31. G. Strey and I. M. Mills, Mol. Phys. 26, 129(1973).
32. P. Botschwina, Chem. Phys. 68, 41(1982).

33. P. Botschwina, Chem. Phys. Letters 107, 535(1984).
34. P. Botschwina, Habilitationsschrift (Kaiserslautern, 1984).
35. R. L. DeLeon and J. S. Muenter, J. Chem. Phys. 80, 3892(1984).
36. P. Botschwina, Chem. Phys. 81, 73(1983).
37. A. M. Smith, K. K. Lehmann, and W. Klemperer, to be published.
38. R. N. Sileo and T. A. Cool, J. Chem. Phys. 65, 117(1976).
39. K. P. Huber and G. Herzberg, "Molecular Spectra and Molecular Structure. IV. Constants of Diatomic Molecules" (Van Nostrand, New York, 1979).
40. G. Di Lonardo and A. E. Douglas, Can. J. Phys. 51, 434(1973).
41. W. Meyer, J. Chem. Phys. 64, 2901(1976).
42. W. Meyer, R. Ahlrichs, and C. E. Dykstra, in "Advanced theories and computational approaches to the electronic structure of molecules", ed. C. E. Dykstra (Reidel, Dordrecht, 1984).
43. A. C. Legon, D. J. Millen, and S. C. Rogers, J. Mol. Spectrosc. 70, 209(1978).
44. S. G. Lieb and J. W. Bevan, Chem. Phys. Letters 122, 284(1985).
45. P. Botschwina, unpublished result.
46. G. A. Blake, P. Helminger, E. Herbst, and F. C.DeLucia, Astrophys. J. 264, L69(1983).

THE ROLE OF TUNNELING MODELS IN ANALYZING HIGH-RESOLUTION SPECTRA OF
WEAKLY BOUND MOLECULAR COMPLEXES

Jon T. Hougen
Molecular Spectroscopy Division
National Bureau of Standards
Gaithersburg, MD 20899
U.S.A.

ABSTRACT. As the number of large amplitude motions in a weakly bound
molecular complex increases, the size of the vibration-rotation basis
set necessary for carrying out accurate calculations with model
potentials (and consequently the difficulty of the calculation)
increases also. In this paper we shall discuss one alternative to such
full-scale calculations, which can be used for assigning spectra in the
broad class of problems where the large amplitude motions can be
described as "tunneling" motions. The strong and weak points of the
method, which uses group theory to derive phenomenological vibration-
rotation-tunneling Hamiltonians, will be described together with some
successful applications and some possibilities for future work.

1. MEASURES OF DIFFICULTY

It is possible to devise two fairly simple scales of difficulty for
characterizing vibration-rotation problems in weakly bound binary
molecular complexes. In the first scheme we count the number of large
amplitude degrees of freedom in the complex (LAM), since these must in
principle be treated by fairly complicated mathematical techniques.
This number can be found by taking the total number of degrees of
freedom for the complex and subtracting from it all vibrational degrees
of freedom of the monomer units, as well as three translational and two
rotational degrees of freedom for the complex as a whole. (Only two
rotational degrees of freedom can be removed because there are only two
good rotational quantum numbers, J and M_J.) We find after a bit of
thought that the resulting expression takes the form

$$LAM = R_1 + R_2 + 1 , \tag{1}$$

where R_1 and R_2 are the rotational degrees of freedom present in monomer
1 and monomer 2.

If we assume that the remaining rotational degree of freedom in a
nonlinear complex (or its analog in a polyatomic linear complex) does
not contribute significantly to the complexity of the problem because
this degree of freedom does not occur in the potential energy function,

191

A. Weber (ed.), Structure and Dynamics of Weakly Bound Molecular Complexes, 191–199.
© 1987 by D. Reidel Publishing Company.

then we obtain as an alternative measure of difficulty for <u>polyatomic</u>
complexes the number LAMV of "soft" modes occuring in the potential
energy function:

$$LAMV = R_1 + R_2 = LAM - 1 .\qquad\qquad(2)$$

For complexes with only one bond between the monomer units, it is
convenient to consider these soft modes as consisting of one weak-bond
stretching vibration plus a collection of geared or antigeared internal
rotations of one monomer unit against the other. (If one of the monomer
units is an atom, the geared versus antigeared distinction obviously
disappears.)

The number of rotational degrees of freedom is zero for atoms, two
for linear molecules, and three for nonlinear molecules. Thus, for the
cases of atom + atom (Ar-Ar), atom + linear molecule (Ar-HCl), atom +
nonlinear molecule (Ar-CH$_3$Cl), linear molecule + linear molecule (HCN-
HF), linear molecule + nonlinear molecule (NH$_3$-HF), and nonlinear
molecule + nonlinear molecule (H$_2$O-H$_2$O), we find LAM values from Eq. (1)
of 1, 3, 4, 5, 6 and 7, respectively, and LAMV values from Eq. (2) of 1,
2, 3, 4, 5 and 6, respectively.

A problem of difficulty level 1 is the determination of diatomic
molecule potential curves from spectroscopic and scattering data. A
problem of difficulty level 2 (in the LAMV scheme) is illustrated by the
significant body of work presented at this meeting on determining the
potential surface of Ar-HCl. Difficulty levels 3 and 4 have been hinted
at by the experimental data presented on Ar-CH$_3$Cl and HCN-HF, but the
impact of attacking the full potential surface for problems of
difficulty level 3 or 4 has not yet been felt by most workers in the
field. The latter statement is even more true for a problem of
difficulty level 6 such as the water dimer.

2. LIMITING CASES

In view of the difficulty of carrying out a complete numerical
treatment of all large amplitude motions for some complexes, it is only
natural that practicing spectroscopists should be interested in the
question of whether or not the full complexity of the multi-dimensional
potential surface might somehow be reduced to a smaller set of
parameters which could permit assigning and fitting spectra that might
otherwise remain an unidentified jumble of lines. In the microwave
spectroscopy of conventional molecules it has long been known that the
set of all bond angles, all bond lengths, and all atomic masses in the
molecule enters the rotational spectrum (to a rather high precision)
only through the three rotational constants A,B,C. Do any analogs of
this simple parameterization scheme exist for weakly bound molecular
complexes? The answer, as might be expected, is yes if consideration is
restricted to certain limiting cases, and no if intermediate cases must
be treated.

There is a more fundamental reason, of course, for seeking an
approximate model characterized by a few dominant parameters, even if
the spectrum has already been completely identified and precise

numerical calculations have already been carried out, since such descriptions contribute greatly to our qualitative understanding of the important physical and chemical effects in the problem.

We shall assume for the rest of this discussion that the van der Waals or hydrogen-bond <u>stretching</u> problem has more than one bound level, and that the dissociation energy for this weak-bond stretching vibration is high enough to support excitation of various internal rotation motions. This assumption allows us to treat the stretching motion as a vibrational mode with somewhat larger excursions from the equilibrium position than is normal in ordinary chemical bonds. With this restriction on the stretching vibration, two limiting cases for the internal rotations come immediately to mind.

2.1 Case (1)

In case (1), the barriers to internal rotation are very low, so that the problem resembles two almost freely rotating monomer units held together by a weak spring connecting their centers of mass. The rotational motions of the monomer units will be slightly modified by the anisotropy of the intermolecular potential. This effect can be treated by perturbation theory, using a free rotor basis set for each monomer unit, and a Hamiltonian written in terms of a short series in spherical harmonics describing the relative orientations of the monomers, with coefficients for the spherical harmonics which are short series in the weak-bond stretching coordinate.

If the free rotational motions are so severely modified that perturbation theory breaks down, a numerical matrix diagonalization procedure can be used, where a reasonable number of free rotor functions are included in the basis set. This is essentially the approach used in most of the work on atom-diatom and diatom-diatom complexes reported at this meeting.

2.2 Case (2)

In case (2), the barriers to internal rotation are very high, so that the problem resembles, to some extent, a normal vibration-rotation problem, characterized by an equilibrium configuration, equilibrium rotational constants, root-mean-square amplitudes of vibrational excursions, and the like.

A particularly interesting subset of problems falling under case (2), are those with a high degree of symmetry present. This symmetry can arise either from symmetries in the individual monomer units (Ar-NH_3), or from the fact that the complex is formed from two identical monomer units (HF-HF). Under these circumstances, even in the presence of relatively high barriers to the internal rotation motions, the spectrum of the complex is likely to exhibit easily observed splittings resulting ultimately from tunneling motions between symmetrically equivalent conformations. It is this latter type of problem which we shall consider in more detail for the remainder of this paper.

3. CHARACTERIZATION OF THE TUNNELING PROBLEM

The use of a tunneling model involves the implicit assumption that tunneling splittings are small compared to vibrational frequencies. This in turn implies that the molecular complex will execute many small amplitude oscillations about a particular equilibrium framework before it tunnels through some potential barrier to another framework. The first question to answer in a tunneling problem, therefore, is: What is the equilibrium configuration of the complex? The answer to this question is important, because the equilibrium geometry of the complex will determine the vibrational frequencies and rotational constants of the molecule, which in turn will determine the coarse structure (i.e. the vibrational and rotational structure in the absence of tunneling effects) of any observed spectrum.

A second question follows immediately: How many distinct symmetrically equivalent equilibrium configurations can be reached via one or more sequential tunneling motions, starting from a given initial configuration? The answer to this question determines the additional "tunneling degeneracy" of any vibration-rotation state determined from the vibrational frequencies and rotational constants of the previous paragraph. It is the splitting of these tunneling degeneracies which gives rise to the splittings of the lines observed in the spectrum. It is important to remember, when counting frameworks which can be reached via tunneling processes, that one must include only those tunneling motions which occur within the duration of the experimental observation, and exclude those which do not. This is essentially equivalent to stating that one must include only those tunneling motions which lead to splittings observable with the experimental resolution available, and exclude all others.

The third question concerns the number of tunneling paths involved in the problem. It has two parts. First, how many tunneling paths exist which point away from any given equilibrium configuration (e.g., internal rotation of a monomer unit about a two or three-fold axis, inversion of a monomer unit, etc.)? Second, how many different tunneling paths connect any given pair of equilibrium configurations (e.g., in-plane rotation of the monomer units, out-of-plane rotation, clockwise rotation, counterclockwise rotation, etc.)? The answers to these questions determine the number of independent variables (degrees of freedom) which must be considered in any mathematical treatment of the tunneling problem, and thus determine also the complexity of the mathematical description of the tunneling patterns.

4. DECOMPOSITION OF THE TUNNELING SPLITTING INTO A VIBRATIONAL AND A ROTATIONAL PART

A mathematical description of tunneling splittings can be formulated (<u>1</u>) in terms of a tunneling Hamiltonian matrix whose elements are expressed as a sum of terms, each of which is factored into a vibrational part and a rotational part. A given term in a given tunneling matrix element then takes the form:

$$\nu_{tun} \cdot f(J,K_a,K_c) \ . \tag{3}$$

The vibrational factor ν_{tun} can be thought of as a true tunneling frequency, depending on vibrational wavefunction penetration of the classically forbidden region under a potential hump. In tunneling problems which can be described using only one independent coordinate, this tunneling frequency can be determined by application of the well-known Wentzel-Kramers-Brillouin procedure. In tunneling problems which require more than one independent coordinate for their description, one can hope that some multi-dimensional modification of the WKB procedure will be useful, but such methods have not yet been fully worked out. In a given tunneling problem, there will be one tunneling frequency for each path pointing away from a given configuration. Because some pairs of paths will be related in the sense that one points "forward", the other "backward" (e.g., rotation of an ammonia monomer by $+2\pi/3$ or $-2\pi/3$ about its symmetry axis), not all of these tunneling frequencies need be distinct. In any case, since a vibrational tunneling frequency is a familiar concept, ν_{tun} will not be discussed further in this paper.

The rotational factor $f(J,K_a,K_c)$ in Eq. (3) can be thought of as a kind of rotational Franck-Condon factor, in the sense that it is determined by the overlap of a rotational wavefunction describing the orientation in the laboratory of an axis system fixed in the complex in one way with a rotational wavefunction describing the orientation of an axis system fixed in the complex in another way ($\underline{1}$). This rotational Franck-Condon factor modulates the tunneling frequency from the previous paragraph as a function of the asymmetric top rotational quantum numbers J, K_a, K_c. The rotational factor $f(J,K_a,K_c)$ is relatively unfamiliar, since it is identically equal to unity for all J and K in the ammonia inversion problem (perhaps the most familiar example of tunneling in molecular spectroscopy), and is unity for all purely vibrational tunneling problems, i.e., for $J = 0$. We shall thus focus our attention on $f(J,K_a,K_c)$ in this paper.

The change in the way the axis system is fixed in the complex depends upon the tunneling path, and, if complete mathematical rigor is not demanded, this change can be described rather briefly and in a physically reasonable way: Consider an axis system fixed in the complex in some way. As the desired tunneling motion is carried out for this complex, an angular momentum will in general be generated. This angular momentum can be cancelled by rotating the whole complex backwards in some way, keeping the axis system originally fixed to the complex stationary. The orientation of the backward-rotated complex in this stationary axis system after the tunneling motion, and its original orientation before the tunneling motion, represent the two ways mentioned above for fixing the axis system in the complex. The Eulerian angles relating one of these orientations to the other occur in appropriate mathematical expressions for the rotational Franck-Condon factors ($\underline{1}$).

5. PROTONATED ACETYLENE

The protonated acetylene molecule $C_2H_3^+$ can be thought of as an ionic complex between acetylene and a proton (though the bonding may turn out to be too strong to fully qualify for the weakly bound complexes under discussion at this meeting). On the basis of ab initio calculations ($\underline{2},\underline{3}$), it seems possible that the equilibrium configuration of this ion may be either Y shaped (the classical structure) or T shaped (the bridged, or non-classical structure). If we assume that one of these two shapes is the equilibrium configuration, say the T shape for definiteness, we find that six equivalent T-shaped molecular frameworks can be drawn with labeled atoms, and all six of these configurations can be reached from any given framework by allowing the hydrogen atoms to migrate about freely in the molecule. One possible migration path is illustrated in Fig. 1(a) below.

(a)   (b)

Fig. 1   (a) A T-shaped protonated acetylene complex, showing the
         presumed migration path of the protons. (b) A circle with six
         equally spaced points around the circumference representing the
         six configurations generated by successive migrations of the
         protons in (a). Allowed tunneling paths ───→; forbidden
         tunneling path – – →.

For convenience we can represent the six equivalent T-shaped configurations by six equally spaced points around the circumference of a circle, as shown in Fig. 1(b). If we make the further assumption that the hydrogen atoms travel, without the possibility of overtaking one another, around a planar elliptical orbit enclosing the two carbon atoms, we find that there are only two allowed tunneling paths pointing away from each configuration point. These can be represented by one clockwise and one counterclockwise arrow on the circumference of the circle. Furthermore, there is only one allowed tunneling path between two configuration points, namely the arc of the circle connecting them.

The description of the previous paragraph ($\underline{4}$) is remarkably similar to the description which would arise ($\underline{1}$) for the van der Waals complex $H_3N \cdot CO_2$, whose equilibrium geometry has the linear CO_2 molecule balanced on top of the NH_3 pyramid ($\underline{5}$). This similarity arises because the H_3N-CO_2 problem can be formally converted to the $C_2H_3^+$ problem by discarding the C and N atoms and moving the three H atoms and two O atoms into the same plane. In both of these problems the tunneling matrix elements can be factored into a vibrational part and a rotational part, and in both of these problems the rotational Franck-Condon factors depend on a single angle θ describing the rotation of the axis system during the tunneling motion. A dramatic difference arises, however, because in the H_3N-CO_2 problem, this rotation takes place about the a-axis, i.e., about the molecule-fixed axis of quantization of the total angular momentum of the near symmetric top, while in the $C_2H_3^+$ problem this rotation takes place about the c-axis, i.e., perpendicular to the molecule-fixed axis of quantization of the total angular momentum.

In mathematical terms, using the notation of Wigner rotation matrices (6), tunneling matrix elements diagonal in $K \equiv K_a$ for the H_3N-CO_2 problem have the form (1)

$$\nu_{tun} \; D^{(J)}{}_{KK}\{\theta,0,0\} = \nu_{tun} \; e^{+iK\theta} \; , \tag{4}$$

where ν_{tun} represents the tunneling frequency, and the D function represents the rotational Franck-Condon factor. Because the angle θ describes rotation about the axis of quantization of K in the near symmetric top, it occurs in the first position in the D function. The value of the D function then turns out to be independent of the rotational quantum number J. Instead, as indicated above, it is a rather simple, but complex (i.e., non-real) function of the quantum number K.

Tunneling matrix elements diagonal in $K \equiv K_a$ for the $C_2H_3^+$ problem, in the same mathematical terms have the form (4)

$$\nu_{tun} \; D^{(J)}{}_{KK}\{0,\theta,0\} \; . \tag{5}$$

Because the angle θ now describes a rotation about an axis perpendicular to the axis of quantization of K in the near symmetric top, it occurs in the second position in the D function. Therefore, unlike the D function in Eq. (4), the D function in Eq. (5) is a complicated function of both J and K, but it is real.

Without going into the actual details of the energy level patterns, we see that the tunneling splittings in both problems contain only two numerical pieces of information, one is the tunneling frequency ν_{tun}, the other is an angle θ of axis-system rotation which depends on the angular momentum generated by the tunneling motion. Of course, the significant mathematical differences in the D functions for the two problems, and hence the significantly different energy level splitting patterns which arise when the tunneling Hamiltonian matrix is diagonalized, also contain the qualitative piece of information of whether angular momentum is generated parallel or perpendicular to the near symmetric top axis during the tunneling motion.

6. THE HF DIMER

The $\overset{H}{\diagdown}$ F$\cdots$H$-$F complex (7-10), which is bent with the F$\cdot$HF atoms lying essentially on a straight line, could have been an even more interesting example of tunneling splittings than it actually turned out to be. All experimental observations to date are consistent with the existence of two equivalent equilibrium configurations, in which the integrity of the monomer HF molecules is maintained, but the H atom participating in the hydrogen bond is changed. Furthermore, all experimental observations are consistent with the existence of only one tunneling path between these two configurations, a path which passes through a planar trans intermediate structure of symmetry C_{2h}, and which therefore involves a geared in-plane internal rotation of one HF with respect to the other. This tunneling problem is quite similar to the tunneling problem in ammonia, in the sense that $f(J,K_a,K_c) \equiv 1$, and in that sense it is relatively well understood.

Two quite different hypothetical tunneling models can be imagined, however. One of these involves breaking chemical bonds and interchanging the H atoms in the two HF monomer units. This would lead to four equivalent equilibrium configurations and four-fold degeneracies in the vibration-rotation levels (before tunneling splittings are considered). This tunneling model has not been investigated in detail, but it seems probable that two different tunneling frequencies would be involved, one for the geared internal rotation of the HF monomer units described in the preceding paragraph, the other for the transfer of H atoms from one HF monomer to the other. If one were to guess, one might think that one of these tunneling frequencies would lead to a large splitting of each four-fold degenerate level into two two-fold degenerate levels, while the other tunneling frequency would lead to a small splitting within each two-fold degenerate level.

A second possible model retains the integrity of the two HF monomer units, thus restricting consideration to two equivalent equilibrium configurations and a two-fold degeneracy in the unsplit vibration-rotation levels, but it considers two possible tunneling paths between these two equilibrium configurations (10). The first path is, as above, characterized by a planar geared internal rotation and a trans C_{2h} intermediate configuration. The second path, however, is characterized by a planar anti-geared internal rotation and a cis C_{2v} intermediate configuration. The existence of the second path can obviously not lead to a further splitting of the levels, since the two-fold "framework" degeneracy has already been split by the trans tunneling motion. This second tunneling path can therefore only contribute an additional tunneling splitting to the already split levels. The treatment of Ref. (10) indicates that when the two tunneling paths between the one pair of frameworks are considered simultaneously for HF-HF, the two distinct contributions to the tunneling splitting interfere constructively for even values of the near prolate symmetric top quantum number K, but interfere destructively for odd values of K. Thus, if the two tunneling frequencies were identical, no tunneling splitting would be observed at all for odd K in HF-HF. Unfortunately this constructive and destructive interference of two tunneling frequencies as a function of even and odd K was not observed in HF-HF (nor in the mathematically similar HO-OH molecule), so it must be labeled a theoretical prediction, rather than a theoretical fact, at this point.

7. THE WATER DIMER

Molecular beam electric resonance studies of the water dimer $(H_2O)_2$ by Dyke and coworkers (11-13) indicate that it has a reasonably well defined structure of symmetry C_s, in which one hydrogen of one water monomer is hydrogen bonded to the oxygen of the other water monomer. Tunneling splittings are, however, observed in the spectrum, and it was pointed out that, even without transferring hydrogen atoms from one water molecule to the other, a total of eight different equilibrium configurations could be reached by a succession of one or more chemically reasonable tunneling motions, starting from some particular configuration. A given vibration-rotation energy level can thus in principle split into eight components. More careful group theoretical analysis (11) indicates that in fact this splitting will lead to only six components, two of which are still doubly degenerate.

As Dyke has reported at this meeting, considerable progress has recently been made in assigning the water dimer spectrum, and a relatively complete assignment may soon be available. The J and K dependence of the splitting patterns is not yet completely understood, however, and it seems to the author that a global fit of the water dimer spectrum might be facilitated by some sort of treatment in which the tunneling Hamiltonian matrix elements are factored into a tunneling frequency times a rotational Franck-Condon factor, as in Eq. (3) above. Further work is necessary to test this presumption, however, and the water dimer spectrum may well prove to be one of the best tests of the tunneling approach presented in this paper.

8. REFERENCES

1. J. T. Hougen, J. Mol. Spectrosc. **114**, 395-426 (1985).
2. J. Weber, M. Yoshimine, and A. D. McLean, J. Chem. Phys. **64**, 4159-4164 (1976).
3. G. P. Raine and H. F. Schaefer III, J. Chem. Phys. **81**, 4034-4037 (1984).
4. J. T. Hougen, J. Mol. Spectrosc. to be submitted.
5. G. T. Fraser, K. R. Leopold, and W. Klemperer, J. Chem. Phys. **81**, 2577-2584 (1984).
6. E. P. Wigner, "Group Theory," Academic Press, New York, 1959.
7. T. R. Dyke, B. J. Howard, and W. Klemperer, J. Chem. Phys. **56**, 2442-2454 (1972).
8. A. S. Pine and W. J. Lafferty, J. Chem. Phys. **78**, 2154-2162 (1983).
9. I. M. Mills, J. Phys. Chem. **88**, 532-536 (1984).
10. J. T. Hougen and N. Ohashi, J. Mol. Spectrosc. **109**, 134-165 (1985).
11. T. R. Dyke, J. Chem. Phys. **66**, 492-497 (1977).
12. T. R. Dyke, K. M. Mack, and J. S. Muenter, J. Chem. Phys. **66**, 498-510 (1977).
13. J. A. Odutola and T. R. Dyke, J. Chem. Phys. **72**, 5062-5070 (1980).

BIBLIOGRAPHY OF ROTATIONAL SPECTRA OF WEAKLY BOUND COMPLEXES

Stewart E. Novick
Department of Chemistry
Wesleyan University
Middletown, Connecticut 06457
USA

ABSTRACT. The following bibliography contains references to high resolution experimental studies of weakly bound complexes. The list was originally compiled to include only microwave experiments, but has recently grown to include other experiments done to rotational resolution. In addition some experimentally oriented theoretical calculations have been included. For ease of use, references to some complexes appear twice; Ar HF, for example, will appear in the Ar listings and in the HF listings. This bibliography is included in these Proceedings as a service to my colleagues. I make no claims for completeness and apologize in advance for the inevitable omissions. I would appreciate learning about any errors and oversights in the compilation.

Latest update: 12/18/86

He	Cl2	JCP	84	1165*1986	JI CLINE,DD EVARD,F THOMMEN KC JANDA
He	I2	JCP	68	671*1978	RE SMALLEY,L WHARTON,DH LEVY
He	Br2	JCP	81	5514*1984	LJ VAN DE BURGT,J-P NICOLAI MC HEAVEN
Ne	DCl	MP	41	619 1980	AE BARTON,DJB HOWLETT,BJ HOWARD
Ne	Cl2	JCP	84	3630*1986	DD EVARD,F THOMMEN,KC JANDA
Ne	Cl2	JPC		*	DD EVARD,F THOMMEN,JI CLINE KC JANDA
Ne	Br2	JCP	82	5295*1985	F THOMMEN,DD EVARD,KC JANDA
Ne	OCS	CPL	118	12*1985	GD HAYMAN,J HODGE,BJ HOWARD JS MUENTER,TR DYKE
Ne	OCS	JCP	86	1987	GD HAYMAN,J HODGE,BJ HOWARD JS MUENTER,TR DYKE
Ne	OCS				FJ LOVAS,RD SUENRAM
Ar	HF	JCP	60	3208 1974	SJ HARRIS,SE NOVICK,W KLEMPERER
Ar	DF	JCP	74	2133 1981	MR KEENAN,LW BUXTON,EJ CAMPBELL

A. Weber (ed.), Structure and Dynamics of Weakly Bound Molecular Complexes, 201–212.
© *1987 by D. Reidel Publishing Company.*

Ar	HF	JCP	74	6539 1981	AC LEGON,WH FLYGARE TA DIXON,CH JOYNER,FA BAIOCCHI W KLEMPERER
Ar	HF	JCP	85	4890*1986	CM LOVEJOY,MD SCHUDER,DJ NESBITT
Ar	HF	JCP	85	2502*1986	GT FRASER,AS PINE
Ar	HF	JCP	85	6905*1986	ZS HUANG,KW JUCKS,RE MILLER
Ar	HF theory	MP	45	791 1982	JM HUTSON,BJ HOWARD
Ar2	HF	JCP	83	4817 1985	HS GUTOWSKY,TD KLOTS,C CHUANG CA SCHMUTTENMAER,T EMILSSON
Ar2	HF	JCP	86	1987	HS GUTOWSKY,TD KLOTS,C CHUANG CA SCHMUTTENMAER,T EMILSSON
Ar3	HF	JACS	107	7174 1985	HS GUTOWSKY,TD KLOTS,C CHUANG JD KEEN,CA SCHMUTTENMAER T EMILSSON
Ar	HCl	JCP	59	2273 1973	SE NOVICK,P DAVIES,SJ HARRIS W KLEMPERER
Ar	HCl	JCP	65	1114 1976	SE NOVICK,KC JANDA,SL HOLMGREN M WALDMAN,W KLEMPERER
Ar	HCl	JCP	66	1826*1977	EW BOOM,D FRENKEL,J VAN DER ELSKEN
Ar	HCl	JCP	74	6520 1981	JM HUTSON,BJ HOWARD
Ar	HCl	JCP	84	1171*1986	D RAY,RL ROBINSON,D GWO RJ SAYKALLY
Ar	HCl	CPL	122	1*1985	BJ HOWARD, AS PINE
Ar	HCl	JCP	83	4924*1985	MD MARSHALL,A CHARO,HO LEUNG W KLEMPERER
Ar	HCltheory	MP	41	1123 1980	JM HUTSON,BJ HOWARD
Ar	HCltheory	MP	43	493 1981	JM HUTSON,BJ HOWARD
Ar	HCltheory	MP	45	769 1982	JM HUTSON,BJ HOWARD
Ar	HCltheory	JCP	78	4025 1983	CJ ASHTON,MS CHILD,JM HUTSON
Ar	HCltheory	JCP	81	2357 1984	JM HUTSON
Ar	HCltheory	JCSFT2	82	1163 1986	JM HUTSON
Ar	HCl calc	JCP	80	4630 1984	LS BERNSTEIN,J WORMHOUDT
Ar	HBr	MP	39	817 1980	KC JACKSON,PRR LANGRIDGE-SMITH BJ HOWARD
Ar	HBr	JCP	72	3070 1980	MR KEENAN,EJ CAMPBELL,TJ BALLE LW BUXTON,TK MINTON,PD SOPER WH FLYGARE
Ar	BF3	JACS	100	8074 1978	KC JANDA,LS BERNSTEIN,JM STEED SE NOVICK,W KLEMPERER
Ar	C2H2	JCP	72	6020 1980	RL DELEON,JS MUENTER
Ar	HCN	JCP	81	4922 1984	KR LEOPOLD,GT FRASER,FJ LIN DD NELSON,W KLEMPERER
Ar	NH3	JCP	82	2535#1985	GT FRASER,DD NELSON,A CHARO W KLEMPERER
Ar	NH3	JCP	85	5512 1986	DD NELSON,GT FRASER,KI PETERSON K ZHAO,W KLEMPERER,FJ LOVAS RD SUENRAM
Ar	ClCN	JCP	75	631 1981	MR KEENAN,DB WOZNIAK,WH FLYGARE
Ar	ClF	JCP	61	193 1974	SJ HARRIS,SE NOVICK,W KLEMPERER WE FALCONER
Ar	OCS	JCP	63	881 1975	SJ HARRIS,KC JANDA,SE NOVICK

					KLEMPERER
Ar	OCS	CPL	118	12*1985	GD HAYMAN,J HODGE,BJ HOWARD
					JS MUENTER,TR DYKE
Ar	OCS				FJ LOVAS,RD SUENRAM
Ar	F2CO	JCP	79	4724 1983	JA SHEA,EJ CAMPBELL
Ar	(CH2)4O	JCP	77	5242 1982	SG KUKOLICH,JA SHEA
Ar	CH3Cl	JCP	71	4189#1979	JM STEED,LS BERNSTEIN,TA DIXON
					KC JANDA,W KLEMPERER
Ar	CH3Cl	JCP	75	1113 1981	RL DELEON,JS MUENTER
Ar	CH3Cl				GT FRASER,RD SUENRAM,FJ LOVAS
Ar	NO theory	JPC	90	3331 1986	PDA MILLS,CM WESTERN,BJ HOWARD
Ar	NO	JPC	90	4961 1986	PDA MILLS,CM WESTERN,BJ HOWARD
Ar&Ne	NO ph dis	JCP	85	1418 1986	K SATO,Y ACHIBA,H NAKAMURA
					K KIMURA
Ar	N2O	JCP	75	5285 1981	CH JOYNER,TA DIXON,FA BAIOCCHI
					W KLEMPERER
Ar	N2O	JCSFT2	82	1137*1986	J HODGE,GD HAYMAN,TR DYKE
					BJ HOWARD
Ar	O2 calc	JCP	79	1170*1983	A VAN DER AVOIRD
Ar	O2	CP	92	9 1985	J METTES,B HEYMEN,P VERHOEVE
					J REUSS
Ar	O3	JCP	71	4487 1979	RL DELEON,KM MACK,JS MUENTER
Ar	O3	FDCS	73	63 1982	JS MUENTER,RL DELEON,A YOKOZEKI
Ar	SO3	JCP	73	137 1980	KH BOWEN,KR LEOPOLD,KV CHANCE
					W KLEMPERER
Ar	SO2	JCP	73	2044 1980	RL DELEON,A YOKOZEKI,JS MUENTER
Ar	SO2	FDCS	73	63 1982	JS MUENTER,RL DELEON,A YOKOZEKI
Ar	CO2	JCP	70	4095 1979	JM STEED,TA DIXON,W KLEMPERER
Ar	H2S	JCP	82	1674 1985	R VISWANATHAN,TR DYKE
Ar	NCCN	JCP	80	1417 1984	WL EBENSTEIN,JS MUENTER
Ar	(CH2)2O	JMSt	135	435 1986	RA COLLINS,AC LEGON,DJ MILLEN
Ar	NH2CHO	Baltimore		1986	CW GILLIES,J ZOZOM,GT FRASER
					RD SUENRAM,FJ LOVAS
Ar	CH3OH	Baltimore		1986	CW.GILLIES,J ZOZOM,GT FRASER
					RD SUENRAM,FJ LOVAS
Ar	CH2CHCN	Maratea		1986	RD SUENRAM,FJ LOVAS,GT FRASER
					J ZOZOM,CW GILLIES
Kr	HF	CPL	70	420 1980	EJ CAMPBELL,MR KEENAN,LW BUXTON
					TJ BALLE,PD SOPER,AC LEGON
					WH FLYGARE
Kr	HF	CP	54	173 1981	LW BUXTON,EJ CAMPBELL,MR KEENAN
					TJ BALLE,WH FLYGARE
Kr	HF	JCP	85	2502*1986	GT FRASER,AS PINE
Kr	HF	Maratea		*1986	ZS HUANG,KW JUCKS,RE MILLER
Kr	HCl	JCP	72	922 1980	TJ BALLE,EJ CAMPBELL,MR KEENAN
					WH FLYGARE
Kr	HBr	JCP	72	3070 1980	MR KEENAN,EJ CAMPBELL,TJ BALLE
					LW BUXTON,TK MINTON,PD SOPER
					WH FLYGARE
Kr	HCN	JCP	78	3483 1983	EJ CAMPBELL,LW BUXTON,AC LEGON

Kr	ClF	CJP	53	2007 1975	SE NOVICK,SJ HARRIS,KC JANDA
					W KLEMPERER
Kr	SO3	JCP	74	4211 1981	KR LEOPOLD,KH BOWEN,W KLEMPERER
Kr	OCS	CPL	118	12*1985	GD HAYMAN,J HODGE,BJ HOWARD
					JS MUENTER,TR DYKE
Kr	OCS				FJ LOVAS,RD SUENRAM
Xe	HF	JCP	75	2041 1981	FA BAIOCCHI,TA DIXON,CH JOYNER
					W KLEMPERER
Xe	HF	JCP	85	2502*1986	GT FRASER,AS PINE
Xe	HCl	JCP	70	5157 1979	KV CHANCE,KH BOWEN,JS WINN
					W KLEMPERER
HF	Ar	JCP	60	3208 1974	SJ HARRIS,SE NOVICK, W KLEMPERER
DF	Ar	JCP	74	2133 1981	MR KEENAN,LW BUXTON,EJ CAMPBELL
					AC LEGON,WH FLYGARE
HF	Ar	JCP	74	6539 1981	TA DIXON,CH JOYNER,FA BAIOCCHI
					W KLEMPERER
HF	Ar	JCP	85	4890*1986	CM LOVEJOY,MD SCHUDER,DJ NESBITT
HF	Ar	JCP	85	2502*1986	GT FRASER,AS PINE
HF	Ar	JCP	85	6905*1986	ZS HUANG,KW JUCKS,RE MILLER
HF	Ar theory	MP	45	791 1982	JM HUTSON,BJ HOWARD
HF	Ar2	JCP	83	4817 1985	HS GUTOWSKY,TD KLOTS,C CHUANG
					CA SCHMUTTENMAER,T EMILSSON
HF	Ar2	JCP	86	1987	HS GUTOWSKY,TD KLOTS,C CHUANG
					CA SCHMUTTENMAER,T EMILSSON
HF	Ar3	JACS	107	7174 1985	HS GUTOWSKY,TD KLOTS,C CHUANG
					JD KEEN,CA SCHMUTTENMAER
					T EMILSSON
HF	Kr	CPL	70	420 1980	EJ CAMPBELL,MR KEENAN,LW BUXTON
					TJ BALLE,PD SOPER,AC LEGON
					WH FLYGARE
HF	Kr	CP	54	173 1981	LW BUXTON,EJ CAMPBELL,MR KEENAN
					TJ BALLE,WH FLYGARE
HF	Kr	JCP	85	2502*1986	GT FRASER,AS PINE
HF	Kr	Maratea		*1986	ZS HUANG,KW ZUCKS,RE MILLER
HF	Xe	JCP	75	2041 1981	FA BAIOCCHI,TA DIXON,CH JOYNER
					W KLEMPERER
HF	Xe	JCP	85	2502*1986	GT FRASER,AS PINE
HF	N2	JCP	76	292 1982	PD SOPER,AC LEGON,WG READ
					WH FLYGARE
HF	N2	JCP	86	*1987	KW ZUCKS,ZS HUANG,RE MILLER
HF	CO	JCP	73	583 1980	AC LEGON,PD SOPER,MR KEENAN
					TK MINTON,TJ BALLE,WH FLYGARE
HF	CO	JCP	74	4944 1981	AC LEGON,PD SOPER,WH FLYGARE
HF	CO	Maratea		*1986	JW BEVAN
HF	CO2	JCP	74	6544 1981	FA BAIOCCHI,TA DIXON,CH JOYNER
					W KLEMPERER
HF	N2O	JCP	74	6550 1981	CH JOYNER,TA DIXON,FA BAIOCCHI
					W KLEMPERER
HF	H2CO	JCP	78	3509 1982	FA BAIOCCHI,W KLEMPERER

HF	H2CO	JMSp				FJ LOVAS,RD SUENRAM,S ROSS
						M KLOBUKOWSKI
HF	C6H6	JPC	87	2079	1983	FA BAIOCCHI,JH WILLIAMS
						W KLEMPERER
HF	C2H6	JCP	75	2681	1981	LW BUXTON,PD ALDRICH,JA SHEA
						AC LEGON,WH FLYGARE
HF	NCCN	JCP	74	4936	1981	AC LEGON,PD SOPER,WH FLYGARE
HF	(CH2)3	JCP	75	2681	1981	LW BUXTON,PD ALDRICH,JA SHEA
						AC LEGON,WH FLYGARE
HF	C2H4	JCP	76	4857	1982	JA SHEA,WH FLYGARE
HF	C2H4	Maratea		*1986		ZS HUANG,KW ZUCKS,RE MILLER
HF	CH3CCH	JCP	80	4605	1984	JA SHEA,RE BUMGARNER,G HENDERSON
HF	H2O	JCSCC	130	341	1975	JW BEVAN,AC LEGON,DJ MILLEN
						SC ROGERS
HF	H2O	PRSL	372	441	1980	JW BEVAN,Z KISIEL,AC LEGON
						DJ MILLEN,SC ROGERS
HF	H2O	PRSL	381	419	1982	Z KISIEL,AC LEGON,DJ MILLEN
HF	H2O	CPL	92	333	1982	AC LEGON,LC WILLOUGHBY
HF	H2O	JCP	78	2910	1983	Z KISIEL,AC LEGON,DJ MILLEN
HF	H2O	CPL	117	543	1985	G CAZZOLI,PG FAVERO,DG LISTER
						AC LEGON,DJ MILLEN,Z KISIEL
DF	D2O	JMSt	131	201	1985	Z KISIEL,AC LEGON,DJ MILLEN
HF	H2S	JCP	77	1166	1982	R VISWANATHAN,TR DYKE
HF	H2S	JCP	81	20	1984	LC WILLOUGHBY,AJ FILLERY-TRAVIS
						AC LEGON
HF	ClF	JCP	65	5115	1976	SE NOVICK,KC JANDA,W KLEMPERER
HF	HCN	CPL	41	137	1976	AC LEGON,DJ MILLEN,SC ROGERS
HF	HCN	JMSt	70	209	1978	AC LEGON,DJ MILLEN,SC ROGERS
HF	HCN	PRSL	370	213	1980	AC LEGON,DJ MILLEN,SC ROGERS
HF	HCN reviewFDCS		73	71	1982	AC LEGON,DJ MILLEN
HF	HCN	PRSL	401	327	1985	AC LEGON,DJ MILLEN,LC WILLOUGHBY
HF	HCN	CPL	124	579*1986		BA WOFFORD,JW BEVAN,WB OLSON
						WJ LAFFERTY
HF	HCN	JCP	85	1283*1986		EK KYRO,M ELIADES,AM GALLEGOS
						P SHOJA-CHAGERVAND,JW BEVAN
HF	HCN	CPL	129	489	1986	Z KISIEL,AC LEGON,DJ MILLEN
						HM NORTH
HF	HF	JCP	56	2442	1972	TR DYKE,BJ HOWARD,W KLEMPERER
HF	HF	JCP	81	5417	1984	BJ HOWARD,TR DYKE,W KLEMPERER
HF	HF	JMSp				RD SUENRAM,FJ LOVAS,WJ LAFFERTY
HF	HF	JCP	78	2154	1983	AS PINE,WJ LAFFERTY
HF	HF	JCP	81	2939	1984	AS PINE,WJ LAFFERTY,BJ HOWARD
HF	HF	JCP	84	590	1986	AS PINE,BJ HOWARD
HF	HF	JCP	83	2070	1985	HS GUTOWSKY,C CHUANG,JD KEEN
						TD KLOTS,T EMILSSON
HF	HF	Maratea		*1986		ZS HUANG,KW ZUCKS,RE MILLER
HF	HF calc	FDCS	73	45	1982	AE BARTON,BJ HOWARD
HF	Cl2	JCP	77	1632	1982	FA BAIOCCHI,TA DIXON,W KLEMPERER
HF	HCl	JCP	67	5162	1977	KC JANDA,JM STEED,SE NOVICK
						W KLEMPERER
HF	C2H2	JCP	76	2238	1982	WG READ, WH FLYGARE

HF	C2H2	Maratea			*1986	ZS HUANG,KW ZUCKS,RE MILLER
HF	CH3CN	PRSL	370	239	1980	JW BEVAN,AC LEGON,DJ MILLEN SC ROGERS
HF	CH3CN	JMSt	67	29	1980	AC LEGON,DJ MILLEN,SC ROGERS
HF	CH3CN	JPC	85	3440	1981	PD SOPER,AC LEGON,WG READ WH FLYGARE
HF	CH3CN calc	JCSFT2	82	1189	1986	P COPE,DJ MILLEN,AC LEGON
HF	CH3CN	JCSFT2	82	1197	1986	P COPE,DJ MILLEN,LC WILLOUGHBY AC LEGON
HF	(CH3)3CCN	PRSL	370	257	1980	AS GEORGIOU,AC LEGON,DJ MILLEN
HF	Kr	CP	54	173	1981	LW BUXTON,EJ CAMPBELL,MR KEENAN TJ BALLE,WH FLYGARE
HF	PH3	CP	74	127	1983	AC LEGON,LC WILLOUGHBY
HF	PH3 review	JPC	87	2064	1983	AC LEGON
HF	OCS	JCP	79	614	1983	JA SHEA,WG READ,EJ CAMPBELL
HF	OCS	JMSt	131	159	1985	AC LEGON,LC WILLOUGHBY
HF	NH3					BJ HOWARD
HF	(CH2)3O	JMSt	69	69	1980	AS GEORGIOU,AC LEGON,DJ MILLEN
HF	(CH2)2O	PRSL	373	511	1981	AS GEORGIOU,AC LEGON,DJ MILLEN
HF	H2S	JCP	81	20	1984	AJ FILLERY-TRAVIS,AC LEGON
HF	SO2	CPL	123	4	1986	AJ FILLERY-TRAVIS,AC LEGON
HF	SO2	JCP	85	3180	1986	AJ FILLERY-TRAVIS,AC LEGON
HF	HI	JCP	86		1987	RE BUMGARNER,SG KUKOLICH
HF	B2H6	JCP	85	683	1986	HS GUTOWSKY,T EMILSSON,JD KEEN TD KLOTS,C CHUANG
HF	H2	Maratea			*1986	ZS HUANG,KW JUCKS,RE MILLER
HF	HCCCN	PRSL	394	387	1984	K GEORGIOU,AC LEGON,DJ MILLEN HM NORTH,LC WILLOUGHBY
HCl	Ar	JCP	59	2273	1973	SE NOVICK,P DAVIES,SJ HARRIS W KLEMPERER
HCl	Ar	JCP	65	1114	1976	SE NOVICK,KC JANDA,SL HOLMGREN M WALDMAN,W KLEMPERER
HCl	Ar	JCP	66	1826	*1977	EW BOOM,D FRENKEL,J VAN DER ELSKEN
HCl	Ar	JCP	74	6520	1981	JM HUTSON,BJ HOWARD
HCl	Ar	JCP	84	1171	*1986	D RAY,RL ROBINSON,D GWO RJ SAYKALLY
HCl	Ar	CPL	122	1	*1985	BJ HOWARD, AS PINE
HCl	Ar	JCP	83	4924	*1985	MD MARSHALL,A CHARO,HO LEUNG W KLEMPERER
HCl	Kr	JCP	72	922	1980	TJ BALLE,EJ CAMPBELL,MR KEENAN WH FLYGARE
HCl	Xe	JCP	70	5157	1979	KV CHANCE,KH BOWEN,JS WINN W KLEMPERER
HCl	HF	JCP	67	5162	1977	KC JANDA,JM STEED,SE NOVICK W KLEMPERER
HCl	N2	JCP	79	57	1983	RS ALTMAN,MD MARSHALL,W KLEMPERER
HCl	CO	JCP	73	583	1980	AC LEGON,PD SOPER,MR KEENAN TK MINTON,TJ BALLE,WH FLYGARE
HCl	CO	JCP	74	2138	1981	PD SOPER,AC LEGON,WH FLYGARE
HCl	CO2	JCP	77	4344	1982	RS ALTMAN,MD MARSHALL,W KLEMPERER

HCl	OCS	JCSFT2	81	1709	1985	EJ GOODWIN, AC LEGON
HCl	HCN	JCP	76	2267	1982	AC LEGON, EJ CAMPBELL, WH FLYGARE
HCl	C2H2	JCP	75	625	1981	AC LEGON, PD ALDRICH, WH FLYGARE
HCl	C2H4	JCP	75	2126	1981	PD ALDRICH, AC LEGON, WH FLYGARE
HCl	C2H4	JCP	79	1105	1983	SG KUKOLICH, PD ALDRICH, WG READ EJ CAMPBELL
HCl	C6H6	JCP	78	3501	1983	WG READ, EJ CAMPBELL, G HENDERSON
HCl	C6H6	JACS	103	7670	1981	WG READ, EJ CAMPBELL, G HENDERSON WH FLYGARE
HCl	(CH)4O	JCP	78	3545	1983	JA SHEA, SG KUKOLICH
HCl	Hg	JCP	79	4082	1983	EJ CAMPBELL, JA SHEA
HCl	Hg	JCP	81	5326	1984	JA SHEA, EJ CAMPBELL
HCl	(CH2)3	JACS	102	7584	1980	AC LEGON, PD ALDRICH, WH FLYGARE
HCl	(CH2)3	JACS	104	1486	1982	AC LEGON, PD ALDRICH, WH FLYGARE
HCl	PH3	JCSCC		997	1982	AC LEGON, LC WILLOUGHBY
HCl	H2O	CPL	95	449	1983	AC LEGON, LC WILLOUGHBY
HCl	CH3OH	CPL	112	59	1984	P COPE, AC LEGON, DJ MILLEN
HCl	HCl	JCP	81	73*	1984	N OHASHI, AS PINE
HCl	HCl	JCP	84	590	1986	AS PINE, BJ HOWARD
HCl	DCl					BJ HOWARD
HCl	H2S	JCSFT2	80	51	1984	EJ GOODWIN, AC LEGON
HCl	SO2	CPL	123	4	1986	AJ FILLERY-TRAVIS, AC LEGON
HCl	BF3	JCP	85	4261	1986	JM LOBUE, JK RICE, TA BLAKE SE NOVICK
HCl	H2CO	Maratea			1986	RD SUENRAM, FJ LOVAS, GT FRASER J ZOZOM, CW GILLIES
HCl	HCN	JCP	76	2267	1982	AC LEGON, EJ CAMPBELL, WH FLYGARE
HCl	HCN	Maratea		*1986		JW BEVAN
HCl	NH3	CPL			1986	EJ GOODWIN, NW HOWARD, AC LEGON
DCl	Ne	MP	41	619	1980	AE BARTON, DJB HOWLETT, BJ HOWARD
HBr	Ar	MP	39	817	1980	KC JACKSON, PRR LANGRIDGE-SMITH BJ HOWARD
HBr	Ar	JCP	72	3070	1980	MR KEENAN, EJ CAMPBELL, TJ BALLE LW BUXTON, TK MINTON, PD SOPER WH FLYGARE
HBr	Kr	JCP	72	3070	1980	MR KEENAN, EJ CAMPBELL, TJ BALLE LW BUXTON, TK MINTON, PD SOPER WH FLYGARE
HBr	CO	JCP	73	583	1980	AC LEGON, PD SOPER, MR KEENAN TK MINTON, TJ BALLE, WH FLYGARE
HBr	CO	PNAS	77	5583	1980	MR KEENAN, TK MINTON, AC LEGON TJ BALLE, WH FLYGARE
HBr	PH3	JCSCC		997	1982	AC LEGON, LC WILLOUGHBY
HBr	PH3	JPC	87	2085	1983	LC WILLOUGHBY, AC LEGON
HBr	HCN	JCP	78	3494	1982	EJ CAMPBELL, AC LEGON, WH FLYGARE
HBr	H2S	JMSt	145	261	1986	AI JAMAN, AC LEGON
HBr	NH3					NW HOWARD, AC LEGON
N2	HF	JCP	86	*1987		KW ZUCKS, ZS HUANG, RE MILLER
N2	SO3	JCP	73	137	1980	KH BOWEN, KR LEOPOLD, KV CHANCE

						W KLEMPERER
N2	HCN	JCP	82	4434	1985	EJ GOODWIN,AC LEGON
N2	NH3	JCP	84	2472	1986	GT FRASER,DD NELSON,KI PETERSON
						W KLEMPERER
N2	BF3	JACS	100	8070	1978	KC JANDA,LS BERNSTEIN,JM STEED
						SE NOVICK,W KLEMPERER
NO	NO	MP	44	145	1981	CM WESTERN,PRR LANGRIDGE-SMITH
						BJ HOWARD,SE NOVICK
NO	NO	JCP	83	2064*	1985	P BRECHIGNAC,SD BENEDICITIS
						N HALBERSTADT, BJ WHITAKER
						S AVRILLIER
NO	NO	JACS	104	4715	1982	SG KUKOLICH
NO	NO ph dis	JCP	85	2333*	1986	MP CASASSA,JC STEPHENSON,DS KING
NO	NO ph dis	JCP	85	6235*	1986	MP CASASSA,AM WOODWARD
						JC STEPHENSON,DS KING
NO	NO2	JACS	104	6927	1982	SG KUKOLICH
NO	Ar	JPC	90	4961	1986	PDA MILLS,CM WESTERN,BJ HOWARD
HCN	Kr	JCP	78	3483	1983	EJ CAMPBELL,LW BUXTON,AC LEGON
HCN	NH3	JCP	80	3073	1984	GT FRASER,KR LEOPOLD,DD NELSON
						A TUNG,W KLEMPERER
HCN	C2H2	JCP	78	3521	1983	PD ALDRICH,SG KUKOLICH,EJ CAMPBELL
HCN	C2H4	JCP	78	3552	1983	SG KUKOLICH,WG READ,PD ALDRICH
HCN	CO2	JCP	80	1039	1984	KR LEOPOLD,GT FRASER,W KLEMPERER
HCN	(CH2)3	JCP	78	4832	1983	SG KUKOLICH
HCN	C6H6					WA KLEMPERER
HCN	PH3	CP	85	443	1984	AC LEGON,LC WILLOUGHBY
HCN	PH3	CPL	111	566	1984	AC LEGON,LC WILLOUGHBY
HCN	HCN	CPL	47	589	1977	AC LEGON,DJ MILLEN,PJ MJOBERG
HCN	HCN	JMSp	89	352	1981	RD BROWN,PD GODFREY,DA WINKLER
HCN	HCN	CPL	102	126	1983	AC LEGON,LC WILLOUGHBY
						AD BUCKINGHAM
HCN	HCN	PRSL	399	377	1985	K GEORGIOU,AC LEGON,DJ MILLEN
						PJ MJOBERG
HCN	HCN	JCP	83	2129*	1985	M MARONCELLI,GA HOPKINS,JW NIBLER
						TR DYKE
HCN	HCN	JCP	85	105*	1986	BA WOFFORD,JW BEVAN,WB OLSON
						WJ LAFFERTY
HCN	(HCN)2	Columbus			1986	RS RUOFF,T EMILSSON,TD KLOTS
						C CHAUNG,HS GUTOWSKI
HCN	HBr	JCP	78	3494	1982	EJ CAMPBELL,AC LEGON,WH FLYGARE
HCN	N2	JCP	82	4434	1985	EJ GOODWIN,AC LEGON
HCN	CO	CP	87	81	1984	EJ GOODWIN,AC LEGON
HCN	HCF3	JCP	84	1988	1986	EJ GOODWIN,AC LEGON
HCN	(CH2)20	JCP	85	676	1986	EJ GOODWIN,AC LEGON,DJ MILLEN
HCN	HF	CPL	41	137	1976	AC LEGON,DJ MILLEN,SC ROGERS
HCN	HF	JMSp	70	209	1978	AC LEGON,DJ MILLEN,SC ROGERS
HCN	HF	PRSL	370	213	1980	AC LEGON,DJ MILLEN,SC ROGERS
HCN	HF	CPL	124	579*	1986	BA WOFFORD,JW BEVAN,WB OLSON
						WJ LAFFERTY

HCN	HF review	FDCS	73	71	1982	AC LEGON,DJ MILLEN
HCN	HF	PRSL	401	327	1985	AC LEGON,DJ MILLEN,LC WILLOUGHBY
HCN	HF	JCP	85	1283*1986		EK KYRO,M ELIADES,AM GALLEGOS P SHOJA-CHAGERVAND,JW BEVAN
HCN	HF	CPL	129	489	1986	Z KISIEL,AC LEGON,DJ MILLEN HM NORTH
HCN	HCl	JCP	76	2267	1982	AC LEGON,EJ CAMPBELL,WH FLYGARE
HCN	HCl	Maratea		*1986		JW BEVAN
HCN	H2S	JCSFT2	80	1669	1984	EJ GOODWIN,AC LEGON
HCN	SO2	Maratea		*1986		JW BEVAN
HCN	SO2	JCP	85	6828	1986	EJ GOODWIN,AC LEGON
HCN	OCS	JMSt				AI JAMAN,AC LEGON
HCN	H2CO					EJ GOODWIN,AC LEGON
HCN	H2O	CPL	98	369	1983	AJ FILLERY-TRAVIS,AC LEGON LC WILLOUGHBY
HCN	H2O	PRSL	396	405	1984	A FILLERY-TRAVIS,AC LEGON LC WILLOUGHBY
HCN	P(CH3)3	JMSt	125	171	1984	HL HIRANI,AC LEGON,DJ MILLEN LC WILLOUGHBY
HCN	CHF3	JCP	84	1988	1986	EJ GOODWIN,AC LEGON
H2O	HF	JCSCC	130	341	1975	JW BEVAN,AC LEGON,DJ MILLEN SC ROGERS
H2O	HF	PRSL	372	441	1980	JW BEVAN,Z KISIEL,AC LFGON DJ MILLEN,SC ROGERS
H2O	HF	PRSL	381	419	1982	Z KISIEL,AC LEGON,DJ MILLEN
H2O	HF	CPL	92	333	1982	AC LEGON,LC WILLOUGHBY
H2O	HF	JCP	78	2910	1983	Z KISEL,AC LEGON,DJ MILLEN
H2O	HF	CPL	117	543	1985	G CAZZOLI,PG FAVERO,DG LISTER AC LEGON,DJ MILLEN,Z KISIEL
D2O	DF	JMSt	131	201	1985	Z KISIEL,AC LEGON,DJ MILLEN
H2O	HCl	CPL	95	449	1983	AC LEGON,LC WILLOUGHBY
H2O	H2O	JCP	60	2929	1974	TR DYKE,JS MUENTER
H2O	H2O	JCP	66	492	1977	TR DYKE
H2O	H2O	JCP	66	498	1977	TR DYKE,KM MACK,JS MUENTER
H2O	H2O	JCP	72	5062	1980	JA ODUTOLA,TR DYKE
H2O	CO2	JCP	80	2439	1984	KI PETERSON,W KLEMPERER
H2O	C2H2	JCP	81	3842	1984	KI PETERSON,W KLEMPERER
H2O	C2H4	JCP	85	725	1986	KI PETERSON,W KLEMPERER
H2O	CO	Baltimore			1986	KI PETERSON,DJ YARON,W KLEMPERER
H2O	N2O	Columbus			1985	KI PETERSON,DJ YARON,TA FISHER W KLEMPERER
H2O	NH3	JCP	83	3768	1985	P HERBINE,TR DYKE
H2O	NH2CHO	Baltimore			1986	RD SUENRAM,FJ LOVAS,GT FRASER J ZOZOM,CW GILLIES
H2O	HCN	CPL	98	369	1983	AJ FILLERY-TRAVIS,AC LEGON LC WILLOUGHBY
H2O	HCN	PRSL	396	405	1984	A FILLERY-TRAVIS,AC LEGON LC WILLOUGHBY
NH3	HCN	JCP	80	3073	1984	GT FRASER,KR LEOPOLD,DD NELSON

					A TUNG,W KLEMPERER
NH3	HCN	JCP	82	2535*1985	GT FRASER,DD NELSON,A CHARO
					W KLEMPERER
NH3	C2H2	JCP	80	1423 1984	GT FRASER,KR LEOPOLD,W KLEMPERER
NH3	CO2	JCP	81	2577 1984	GT FRASER,KR LEOPOLD,W KLEMPERER
NH3	CO2	JCP	82	2535*1985	GT FRASER,DD NELSON,A CHARO
					W KLEMPERER
NH3	NH3	JCP	82	2535#1985	GT FRASER,DD NELSON,A CHARO
					W KLEMPERER
NH3	NH3	JCP	83	6201 1985	DD NELSON,GT FRASER,W KLEMPERER
NH3	NH3 calc	JCP	85	2077 1986	SY LIU,CE DYKSTRA,K KOLENBRANDER
					JM LISY
NH3	NH3 calc	Columbus		1986	DD NELSON,W KLEMPERER
NH3	Ar	JCP	82	2535#1985	GT FRASER,DD NELSON,A CHARO
					W KLEMPERER
NH3	Ar	JCP	85	5512 1986	DD NELSON,GT FRASER,KI PETERSON
					K ZHAO,W KLEMPERER,FJ LOVAS
					RD SUENRAM
NH3	HCCH	JCP	82	2535*1985	GT FRASER,DD NELSON,A CHARO
					W KLEMPERER
NH3	OCS	JCP	82	2535*1985	GT FRASER,DD NELSON,A CHARO
					W KLEMPERER
NH3	N2O	JCP	82	2535*1985	GT FRASER,DD NELSON,A CHARO
					W KLEMPERER
NH3	N2O	JCP	83	5442 1985	GT FRASER,DD NELSON,GJ GERFEN
					W KLEMPERER
NH3	CO	JCP	84	2472 1986	GT FRASER,DD NELSON,KI PETERSON
					W KLEMPERER
NH3	N2	JCP	84	2472 1986	GT FRASER,DD NELSON,KI PETERSON
					W KLEMPERER
NH3	CF3H	JCP	84	5983 1986	GT FRASER,FJ LOVAS,RD SUENRAM
					DD NELSON,W KLEMPERER
NH3	H2O	JCP	83	3768 1985	P HERBINE,TR DYKE
NH3	HF				BJ HOWARD
NH3	CH3OH				GT FRASER,FJ LOVAS,RD SUENRAM
NH3	H2S	Maratea		1986	TR DYKE
NH3	HCl	CPL		1986	EJ GOODWIN,NW HOWARD,AC LEGON
NH3	HBr				NW HOWARD,AC LEGON
BF3	Ar	JACS	100	8070 1978	KC JANDA,LS BERNSTEIN,JM STEED
					SE NOVICK,W KLEMPERER
BF3	N2	JACS	100	8070 1978	KC JANDA,LS BERNSTEIN,JM STEED
					SE NOVICK,W KLEMPERER
BF3	CO	JACS	100	8070 1978	KC JANDA,LS BERNSTEIN,JM STEED
					SE NOVICK,W KLEMPERER
BF3	CO2	JACS	106	897#1983	KR LEOPOLD,GT FRASER,W KLEMPERER
BF3	N2O	JACS	106	897 1983	KR LEOPOLD,GT FRASER,W KLEMPERER
BF3	NCCN	JACS	106	897 1983	KR LEOPOLD,GT FRASER,W KLEMPERER
BF3	HCl	JCP	85	4261 1986	JM LOBUE,TA BLAKE,JK RICE
					SE NOVICK

CO2	CO2	CPL	120	313*1985	GA PUBANZ,M MARONCELLI,JW NIBLER
CO2	CO2	Baltimore		*1986	KW JUCKS,ZS HUANG,RE MILLER
					WJ LAFFERTY
CO2	(CO2)2			*	AS PINE,GT FRASER
SO2	SO2	JCP	83	945 1985	DD NELSON,GT FRASER,W KLEMPERER
SO2	HF	CPL	123	4 1986	AJ FILLERY-TRAVIS,AC LEGON
SO2	HF	JCP	85	3180 1986	AJ FILLERY-TRAVIS,AC LEGON
SO2	HCl	CPL	123	4 1986	AJ FILLERY-TRAVIS,AC LEGON
SO2	HCN	Maratea		*1986	JW BEVAN
SO2	HCN	JCP	85	6828 1986	EJ GOODWIN,AC LEGON
CH3OH	NH2CHO	Baltimore		1986	FJ LOVAS,RD SUENRAM,GT FRASER
					J ZOZOM,CW GILLIES
N2O	N2O	Maratea		1986	BJ HOWARD
CH3CN	HCCH	JCP	85	6898 1986	NW HOWARD,AC LEGON
HCCH	(HCCH)2			*	D PRICHARD,JS MUENTER,BJ HOWARD

REVIEWS

A Systematic Look at Weakly Bound Diatomics.
J.S. Winn, Acc. Chem. Res. 14, 341 (1981).

Pulsed-Nozzle, Fourier-Transform Microwave Spectrocopy of Weakly Bound
Dimers. A.C. Legon, Ann. Rev. Phys. Chem. 34, 275 (1983).

Microwave and Radiofrequency Spectra of Hydrogen Bonded Complexes in the
Vapor Phase. T.R. Dyke, Topics Current Chem. 120, 86 (1984).

Intermolecular Interaction Involving First Row Hydrides: Spectroscopic
Studies of Complexes of HF, H2O, NH3, and HCN. K.I. Peterson, G.T.
Fraser, D.D. Nelson, Jr., and W. Klemperer in Comparison of Ab Initio
Quantum Chemistry with Experiment for Small Molecules. The State of the
Art. pg. 217. Edited by R.J. Bartlett. Reidel Holland. Kluwer Academic
(1985).

Vibrational Spectroscopy, Photochemistry, and Photophysics of Molecular
Clusters. F.G. Celii and K.C. Janda, Chem. Rev. 86, 507 (1986).

Gas-Phase Spectroscopy and the Properties of Hydrogen-Bonded Dimers:
HCN...HF as the Spectroscopic Prototype.
A.C. Legon and D.J. Millen, Chem. Rev. 86, 635 (1986).

Infrared Laser Photodissociation and Spectroscopy of van der Waals
Molecules. R.E. Miller, J. Phys. Chem. 90, 3301 (1986).

- Not totally solved.

* - Not microwave work.

```
JCP        - J. CHEM. PHYS.
JPC        - J. PHYS. CHEM.
CPL        - CHEM. PHYS. LETT.
CP         - CHEM. PHYS.
CJP        - CAN. J. PHYS.
MP         - MOL. PHYS.
PNAS       - PROC. NAT. ACAD. SCI.
TFS        - TRANS. FARAD. SOC.
JMSp       - J. MOL. SPECT.
JMSt       - J. MOL. STRUCT.
JCSCC      - J. CHEM. SOC. CHEM. COMM.
PRSL       - PROC. ROY. SOC. (LONDON) A
FDCS       - FARADAY DISCUSS. CHEM. SOC.
JCSFT2     - J. CHEM. SOC., FARADAY TRANS. 2
Columbus   - OHIO STATE SPEC SYMPOSIUM, June 1986
Baltimore  - SYMPOSIUM AT REGIONAL ACS, Sept 1986
Maratea    - NATO ADVANCED RESEARCH WORKSHOP, Sept 1986
```

STRUCTURE AND DYNAMICS OF MERCURY VAN DER WAALS COMPLEXES

W. H. Breckenridge, M. C. Duval
C. Jouvet, B. Soep.
C. N. R. S.
Bât. 213 – Laboratoire de Photophysique Moléculaire
91405 – ORSAY Cedex – France.

ABSTRACT.   The electronic structure of 6p Hg-Ar complexes has been investigated and described by a simple electrostatic model using the 6p average orientation with respect to the internuclear axis. The vibrational structure of the triatomic complex Hg-N_2 was also described in terms of two vibrational modes, a stretching and a bending vibration. This description is essential as it is observed that the torsional modes with vibrational angular momentum induce strong fine structure relaxation $^3P_1 - {}^3P_0$.
xxx

We have developed a method for the investigation of collisional processes through the dissociation of van der Waals complexes. In this method the relevant potential energy surface is accessed through the spectroscopy of the complex and the dynamics through its predissociation. We describe here the electronic structure of the ^3P mercury argon complexes in simple electrostatic terms and derive a relationship to determine all the spin orbit potentials arising from the Hg (6s 6p)-Ar interaction.

Furthermore for the other example, Hg-N_2, we describe the molecular structure and vibrational modes in the excited A state correlated to $^3P_1$ mercury. We show how the molecular movements can be separated in stretching and librational (Torsional) motion and how this latter motion influences the fine structure of the predissociation :
$$Hg(^3P_1) - N_2 \rightarrow Hg(^3P_0) + N_2$$

THE MERCURY-ARGON SYSTEM

Diatomic metal-rare gas complexes are especially attractive systems, since model potential calculations can be performed[1, 4-8] and laser spectroscopic observations of their excited states can provide information on the potentials over a very large range of internuclear distance and energy [9-13]. The first optical detection of a metal-rare gas complex was that of the sodium-argon system,[9] and yielded interesting comparisons with model potentials.

A. Weber (ed.), Structure and Dynamics of Weakly Bound Molecular Complexes, 213–229.

In this study of the excited mercury-argon complexes, our goal has been to investigate how the nature of the excited mercury orbital and its average orientation with regard to the internuclear axis influences the binding in the complex. We have observed the Hg–Ar states correlating with the 6^3P mercury atomic levels $^3P_2$, $^3P_1$ and $^3P_0$. Under the molecular field due to the van der Waals attraction each of the mercury 6^3P levels with J=0, 1 or 2 splits into J+1 molecular states. All these states have different average orientations of the 6p mercury atomic orbital with respect to the internuclear axis. We provide a simple model which describes the potential of these six states in terms of two spin free potentials V_π and V_Σ, following Baylis[1]. This simple model accounts for our observations and allows predictions to be made[2].

Fluorescence excitation spectra :

The state correlating with 7^3S_1 can be populated by optical excitation of vibrational levels of the intermediate states A and B correlating with 6^3P_1 (fig. 1). The C and D minima of the 7^3S_1 state are then observed through laser induced double resonance fluorescence spectra where a pump laser is tuned to a transition of the A or B state anf a probe laser is scanned through the transitions to the C or D minimum (fig. 1).

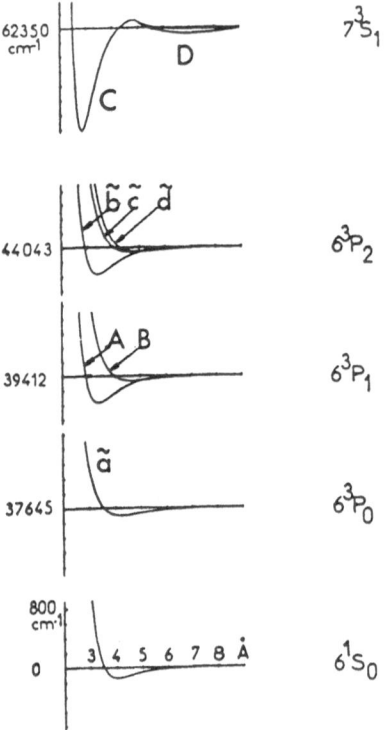

Fig. 1 – Diagram of the Hg–Ar potential curves correlating with the mercury levels 6^1S_0, 6^3P_0, 6^3P_1, 6^3P_2 and 7^3S_3.

Emission spectra :

We observed by emission from the upper C or D minimum the complete set of electronic states involving the 6p orbitals of mercury, i.e. the states correlating with the $^3P_2$, $^3P_1$ and $^3P_0$ atomic levels in a manner similar to that of Nieffer, Atkinson and Kraus in their study of the higher states of Hg_2[3].

The metastable states correlating with 6^3P_0 and 6^3P_2, which cannot be accessed directly from the ground state, were observed by resolved emission from the upper 7^3S_1 states : the C and D minima (fig. 1). The large region of R which is spanned through the emission from the different levels of C and D (fig. 1) allows extensive exploration of the various $\tilde{a}$, $\tilde{b}$, $\tilde{c}$ and $\tilde{d}$, potentials correlating with 6^3P_0 and 6^3P_2 and permits determination of the vibrational structure of the $\tilde{a}(^3P_0)$ and $\tilde{b}(^3P_2)$ states. The relevant bound-free emission of these metastable states were also observed, yielding an accurate description of the repulsive parts of their potentials [2].

The 7^3S_1 C state :

The C minimum was investigated in detail [2] since a precise knowledge of its characteristics (R_e, ω_e, D_e) was necessary to simulate the Franck-Condon intensities of the emission spectra to the states correlating with 6^3P_2 and 6^3P_0. Extensive A→C fluorescence excitation spectra were recorded for different 6^3P_1 A intermediate levels (v"=3 to 6). All these spectra consists of a long vibrational progression of 21 members and the 0-0 position of the A→C transition was conclusively assigned as is displayed in fig. 2 where the lowest wavelength band, by comparison with the simulated Franck-Condon intensities, corresponds to the origin. The C minimum potential can be accurately represented, up to v'=16, by a Morse potential as the Birge-Sponer plot is linear. The C state spectroscopic constants from the linear plot are :

$$\omega_e = 112 \pm 1 \text{ cm}^{-1}, \quad \omega_e x_e = 2.01 \pm 0.04 \text{ cm}^{-1}, \quad D_e = 1560 \text{ cm}^{-1}.$$

Franck-Condon simulations of the excitation spectra (A→C transition) were performed. As seen for level A(v"=6) in the bottom part of figure 2, the simulations describe correctly the spectral minima up to v'=18 but less accurately the absolute intensities because intensities are subjected to large fluctuations of about 20% owing to the double resonance excitation method. Under these conditions we consider the agreement good and the origin of the transition well assigned.

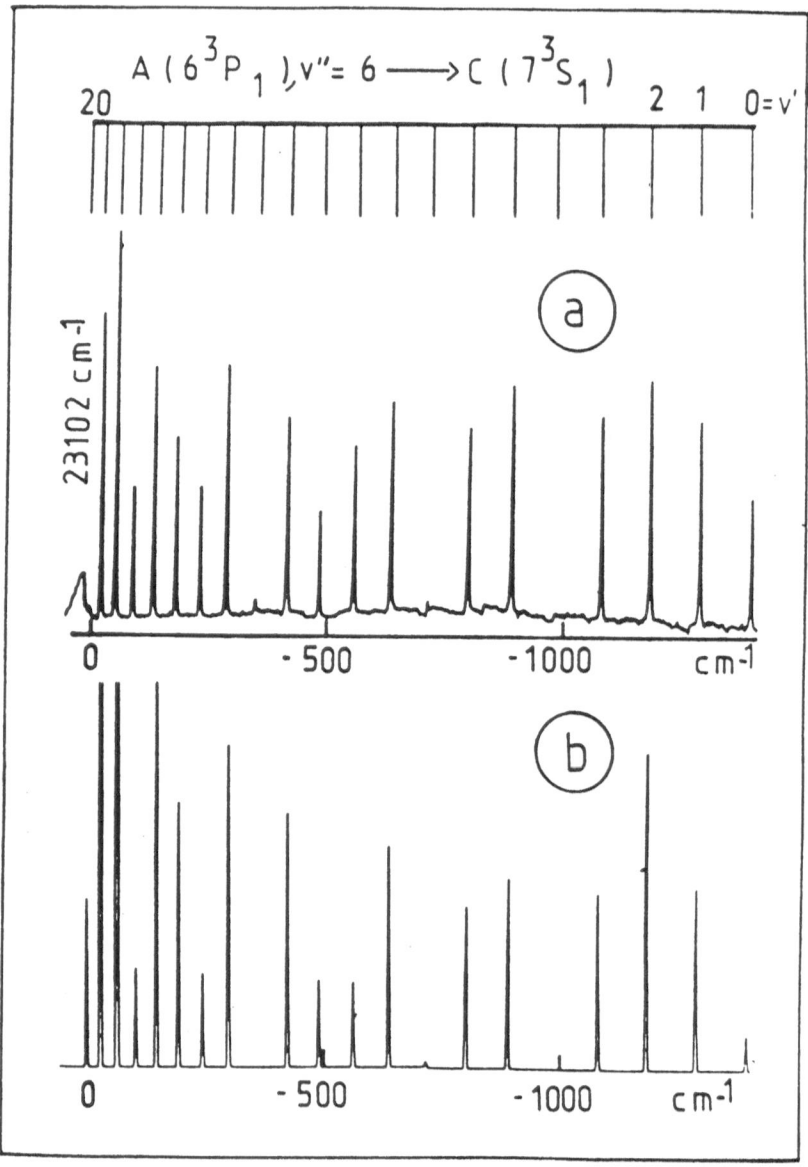

Fig. 2 – Fluorescence excitation spectrum of the $A(^3P_1) \to C(7^3S_1)$
transition in Hg–Ar.   a) The experimental spectrum.
b) The simulated spectrum where the A state is prepared
in the v"=6 vibrational level by the pump laser.
A spectroscopic constants from ref. (13) and for the C state :
R_e = 2.81 Å, ω_e = 112 cm^{-1}, $\omega_e x_e$ = 2.01 cm^{-1}
D_e = 1560 cm^{-1}.

Emission to the $\tilde{a}$ (6^3P_0) state.

The emission spectra to the $\tilde{a}$ state obtained from different vibrational levels of the upper C potential are represented in fig. 3 for v'=19 and 14. Only one state was expected and observed. We observed a short progression which converges to the dissociation limit. By inspecting the emission spectra from different C vibrational levels, the $\tilde{a}$ (v=0) level can be assigned in the spectrum, yielding : $D_0 = 103 \pm 6$ cm^{-1}.

The determination of the vibrational frequency was difficult, however, because of the frequency resolution limits of the system. Taking Morse functions for the upper and lower states, we used extensive Franck-Condon simulations of the various vibrational progressions to match the emission spectra. The resultant spectroscopic constants are then :

$$\omega_e = 13 \pm 2 \text{ cm}^{-1}, \quad \omega_e x_e = 0.38 \pm 0.1 \text{ cm}^{-1}, \quad D_e = 110 \pm 7 \text{ cm}^{-1}.$$

The Franck-Condon simulated spectra in fig. 3 indicate an equilibrium distance of 4.33 Å for the $\tilde{a}$ state. These data were verified through the observation of the repulsive branch of the Hg-Ar potential in bound free C$\rightarrow\tilde{a}$. spectra[2].

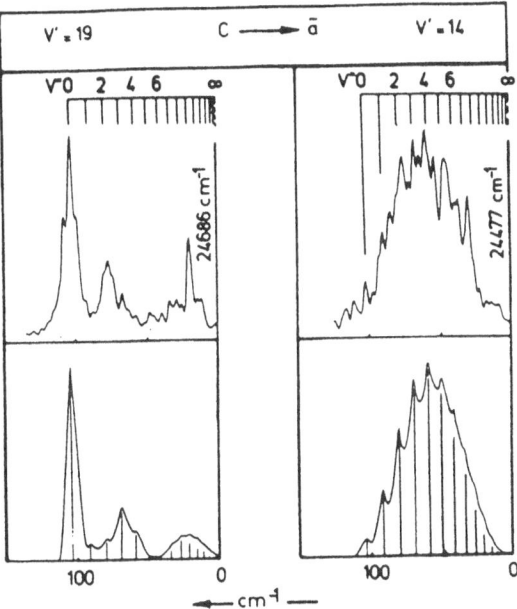

Fig. 3 – C(7^3S_1)$\rightarrow\tilde{a}$(6^3P_0) emission spectra, taken from v'=19 and 14 into the bound region of the $\tilde{a}$ potential. At the bottom of each figure, convolutions of the simulated Franck-Condon intensities with a 10 cm^{-1} slit function are displayed. C spectroscopic constants as in fig. 2 and resultant $\tilde{a}$ Morse potential constants of ω_e=13 cm^{-1}, $\omega_e x_e$=0.38 cm^{-1}, D_e=110 cm^{-1} and R_e=4.33 Å.

Emission to the $\tilde{b}$, $\tilde{c}$ and $\tilde{d}$ (6^3P_2) states.

The bound–bound emission spectra observed from the upper C potential to the states correlating with the 6^3P_2 mercury level are represented in fig. 4. The spectrum observed from the C (v'=2) level exhibits only one long progression. An extensive search for the origin of this $\tilde{b}$ state was necessary using different Franck-Condon windows from vibrational levels of the upper C state. The $\tilde{b}$ (v"=0) level, assigned as seen in fig. 4, gives D_0= 414 cm^{-1}.

A Birge-Sponer plot shows for the first values good agreement with a Morse potential. From the linear part of this plot, we deduce for the $\tilde{b}$ state:

$$\omega_e = 46.5 \pm 1 \text{ cm}^{-1}, \quad \omega_e x_e = 1.37 \pm 0.02 \text{ cm}^{-1}, \quad D_e = 397 \text{ cm}^{-1}.$$

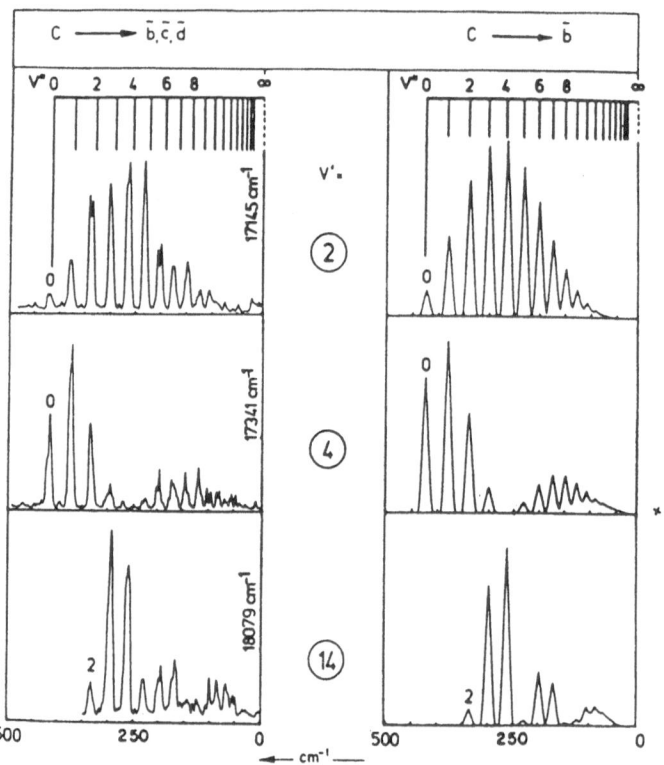

A – Observed spectra   B – Simulated spectra

Fig. 4 – C(7^3S_1)→$\tilde{b}$, $\tilde{c}$, $\tilde{d}$(6^3P_2) emission spectra, taken from v'=2,4, and 14 into the bound regions of the $\tilde{b}$, $\tilde{c}$, $\tilde{d}$ potentials. The emission consists mainly of the C→$\tilde{b}$ transition, as shown on the right hand of the figure, where the spectra are well simulated using the $\tilde{b}$ lower state constants.

The C→$\tilde{b}$ vibrational progression intensities were simulated by Franck-Condon factor calculations, assuming the upper and the lower states can be represented by Morse potentials. The simulations reproduced satisfactorily the observed emission for all levels of thg lower $\tilde{b}$ state as shown on the right side of figure 4, giving credence to the vibrational assignments and yielding a $\tilde{b}$ state equilibrium distance of R_e = 3.31 Å.

We expected three vibrational progressions corresponding to the $6^3P_2(\Omega=2,1,0)$ states and as we shall see (table 2) the observed $\tilde{b}$. progression is that of the deepest $6^3P_2(\Omega=2)$ state. The $\tilde{c}$ and $\tilde{d}$ states should have less binding energy and greater equilibrium distances than the $\tilde{b}$ state, hence the bound-bound emission to the $\tilde{c}$ and $\tilde{d}$ states would only appear in the spectra from the highest vibrational levels of C, where Franck-Condon overlap would be greatest. No evidence of such transitions was observed in bound-bound spectra, but in bound-free emission more than one transition can be observed[2].

A SIMPLE MODEL FOR THE ELECTRONIC STRUCTURE OF THE Hg(6^3P)-Ar STATES

The potentials of the Hg(6^3P)-Ar states, although widely different, can be described approximately as depending only upon the average orientation of the mercury 6p orbital with respect to the internuclear axis. In Hg(6^3P)-Ar, the mercury atom, excited to the 6s6p configuration, is perturbed by the polarizable argon atom in its 3^1S_0 ground state. The wavefunction describing the system can be approximated by the $\mid J, M_J \rangle$ atomic wavefunction of mercury. The van der Waals interaction can be considered as a perturbation, since the separation between van der Waals states is very much smaller than the separation between the J states of the atom. The Hamiltonian of the Hg-Ar system can therefore be represented as :

$$H = H_{Hg} + H_{Ar} + H_{int}$$

where H_{Hg} and H_{Ar} are the isolated mercury and argon atom Hamiltonians and H_{int} is the Hamiltonian corresponding to the electrostatic interaction between the atoms. We assume also the H_{int} does not include any spin-orbit terms dependent on the internuclear distance. This should be a valid first approximation, given the large spin orbit interaction in the free Hg states and the large equilibrium distances of the Hg-Ar complexes.

The first order energy of a van der Waals state described in the $\mid J, M_J \rangle$ basis is :

$$V = \langle\, J, M_J \mid H_{int} \mid J, M_J \,\rangle.$$

We shall consider a non-rotating molecule which we denote by M_J, the projection of J on the internuclear axis (i.e. $M_J=\Omega$). Since we have chosen the symmetry axis to be the internuclear axis, we also have $M_L=\Lambda$ and $M_S=\Sigma$.

We assume H_{int} is purely electrostatic, and change the $|J, M_j >$ basis to a more relevant basis, i.e. $|L, S, M_L, M_S >$ where we can formally separate the spin and space variables :

$$|J, M_j > = \Sigma_i C_i \ |L^i, S^i, M_L^i, M_S^i >$$

It is easily shown that each potential using the latter expansion is the sum of two potentials V_π and V_Σ

where $V_\pi = < L = 1, \ |M_L| = 1 \ |H_{int}| \ L = 1, \ |M_L| = 1 >$
and $\quad V_\Sigma = < L = 1, \quad M_L = 0 \ |H_{int}| \ L = 1, \quad M_L = 0 >$,
as is represented in table 1.

Hg atomic levels	Hg-Ar states	Hg-Ar potential energies
$6 \ ^3P_0$	$6 \ ^3P_0(0) \ \tilde{a}$	$V_{00} = 1/3(2V_\pi + V_\Sigma) = 1/3(2V_{11} + V_{10})$
$6 \ ^3P_1$	$6 \ ^3P_1(0) \ A$	$V_{10} = V_\pi$
	$6 \ ^3P_1(1) \ B$	$V_{11} = 1/2(V_\pi + V_\Sigma)$
$6 \ ^3P_2$	$6 \ ^3P_2(2) \ \tilde{b}$	$V_{22} = V_\pi \quad = V_{10}$
	$6 \ ^3P_2(1) \ \tilde{c}$	$V_{21} = 1/2(V_\pi + V_\Sigma) \quad = V_{11}$
	$6 \ ^3P_2(0) \ \tilde{d}$	$V_{20} = 1/3(V_\pi + 2V_\Sigma) = 1/3(4V_{11} - V_{10})$

Table 1 : Potentials of the $Hg(6^3P)$-Ar states : expressed as functions of V_π and V_Σ potentials accounting for the $|M_L|=1,0$ values, and as functions of the V_{10} and V_{11} potentials of the $Hg(^3P_1)$-Ar states. (The $M_J=\Omega$ values are noted in parenthesis after the atomic level to which the state correlates.)

Using the potentials measured for the $^3P_1(1, 0)$ states we can predict all the potentials of the states correlating with $^3P_2$ and $^3P_0$, and fit them to the nearest Morse potential. These values are listed in table 2 and compared with the experimental values. For the two states experimentally observed and characterized, $^3P_2(\Omega=2)$ and $^3P_0(\Omega=0)$, the fit of the spectroscopic constants (ω_e, D_e, R_e) is surprisingly good for such a simple hypothesis.

	$^3P_0(0)$ $\tilde{a}$		$^3P_2(2)$ $\tilde{b}$		$^3P_2(1)$ $\tilde{c}$	$^3P_2(0)$ $\tilde{d}$
	calculated	measured	calculated	measured	calculated	calculated
ω_e (cm^{-1})	15	13±2	39.6	46.5±1	12.1	10.8
$\omega_e x_e$ (cm^{-1})	0.56	0.38±0.1	1.5	1.37±0.02	0.53	0.49
R_e $(\mathring{A})$	4.21	4.33±0.02	3.38	3.31±0.02	4.66	4.95
D_e (cm^{-1})	101	110±7	369	437±10	67	59

Table 2 : Spectroscopic constants of the $\tilde{a}$, $\tilde{b}$, $\tilde{c}$ and $\tilde{d}$ metastable states, experimentally observed in this work and/or calculated from the $A^3P_1(\Omega=0)$ and $B^3P_1(\Omega=1)$ potentials whose constants are (ref. 13): A state: $\omega_e=39.6$ cm^{-1}, $\omega_e x_e=1.5$ cm^{-1}, $D_e=369$ cm^{-1}, $R_e=3.38$ cm^{-1}; B state : $\omega_e=12.1$ cm^{-1}, $\omega_e x_e=0.53$ cm^{-1}, $D_e=67cm^{-1}$, $R_e=4.66$ cm^{-1}.

This description of the van der Waals potential in electrostatic terms therefore allows a prediction of the $^3P_0(\Omega=0)$ and $^3P_2(\Omega=2,1,0)$ states, and a comparison between the various Hg-Ar states correlating with the 6 ^3p mercury levels as it describes the energy of the spin orbit states in terms only of the average orientation of the 6p mercury orbital with respect to the internuclear axis.

The emission to the metastable $\tilde{a}$ $(^3P_0)$ and $\tilde{b}$, $\tilde{c}$, $\tilde{d}(^3P_2)$ states of Hg-Ar providing a wealth of spectroscopic informations difficult to obtain experimentally by other methods, since these states are optically inaccessible from the ground state, allows the comparison with the already known A and B $(^3P_1)$ states [10-13]. A model was developed which accounts for the structure of these spin orbit states.

The relevant potentials can be related simply by electrostatic interactions between ground state argon and the excited 6^3p mercury levels, wherein the average orientation of the 6p mercury orbital with respect to the complex internuclear axis accounts for the binding.

THE Hg - N₂ COMPLEX

At room temperature collisions of Mercury in ^3P states with rare gas atoms are inefficient in quenching the ^3P states, as appears in Table 3. However collisions with diatomics (N_2) induce electronic relaxation from $^3P_1$ to $^3P_0$ with small cross sections and much greater cross section from $^3P_2$ to $^3P_1$ and $^3P_0$. It is clear that the molecular movements added by the new degrees of freedom of the atom-diatom complex (as compared to the atom-atom case) are effective in relaxing the $^3P_1$ state. It appeared extremely interesting to study the structure and decay of complexes formed by Hg^3P_1 and N_2, as the channel leading to N_2 (v=1) is closed energetically, in opposite to collisions [15].

TABLE 3

	$\sigma \ \overset{\circ}{A}^2$	réf.
$Hg \ (^3P_2) + N_2 \rightarrow Hg \ (^3P_{1,0})$	20	3
$Hg \ (^3P_2) + He \rightarrow Hg \ (^3P_1)$	$<10^{-2}$	4
$Hg \ (^3P_2) + Xe \rightarrow Hg \ (^3P_1)$	$<10^{-3}$	4
$Hg \ (^3P_1) + N_2 \rightarrow Hg \ (^3P_0)$	$0.72; 0.97; 1.1$	5.6.7
$Hg \ (^3P_1) + Ar \rightarrow Hg \ (^3P_0)$	$<2 \times 10^{-2}$	8
$Hg \ (^3P_1) + CO \rightarrow Hg \ (^3P_0)$	30	9
$Hg \ (^3P_0) + N_2 \rightarrow Hg \ (^1S_0)$	$<10^{-5}$	9
$Hg \ (^3P_0) + Ar \rightarrow Hg \ (^1S_0)$	$<10^{-5}$	4
$Hg \ (^3P_0) + He \rightarrow Hg \ (^1S_0)$	$<10^{-5}$	4

When excited in a state correlated to $^3P_1$ the Hg-N₂ complex exhibits both fluorescence to the ground state and dissociation into $^3P_0$ and N_2. The fluorescence excitation spectra and the action spectra, probing $^3P_0$ for variable excitation of the complex, reveal the efficiency of each excited level to induce the electronic relaxation to $^3P_0$. The relaxation efficiency of each vibrational level or group of rotational levels of the complex can be deduced quantitatively by measuring the time decay of those levels. The observed rate is given by $1/\tau$ = Fluorescence rate (same as mercury) + dissociation rate. On the other hand, the sole observed decay for $^3P_1$ mercury rare gas complexes (He, Ar, Ne) is fluorescence with the same lifetime as free mercury (120 ns) and no relaxation to $^3P_0$ has been detected. The specific movements characteristic of the Hg-N₂ system will be studied in the following through the spectroscopy of the complex.

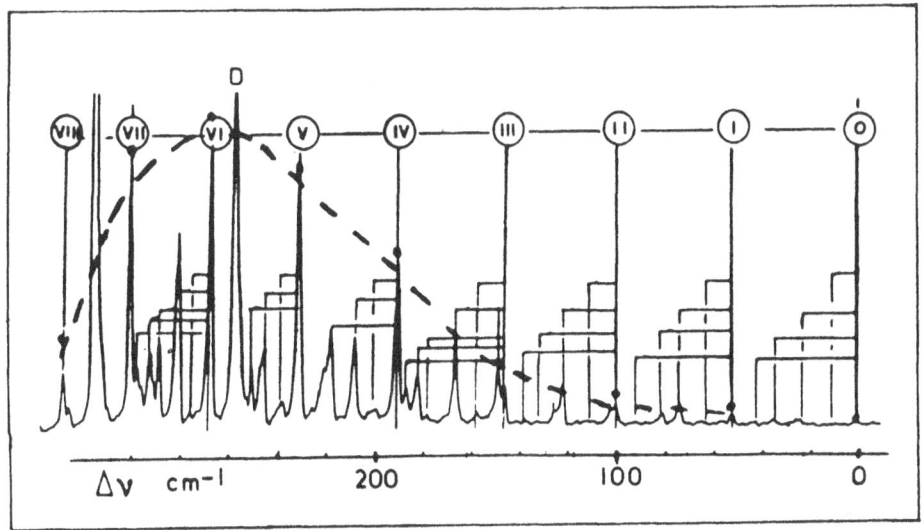

Fig. 5 – Fluorescence excitation spectrum of Hg-N$_2$.

The fluorescence excitation spectrum is displayed in Fig. 5. There are two sets of lines as is usual in complexes involving a p atomic orbital[10, 14] corresponding to the average orientations of this 6p orbital with respect to the Hg-N$_2$ axis. The blue shifted shallower state converges to a continuum due to the direct dissociation into Hg $^3P_1$ + N$_2$ which we have identified by double resonance of Hg ($^3P_1 \rightarrow \, ^3S_1$). The red shifted state exhibits two types of progressions. One is labeled with roman numerals in Fig. 5 and can be identified as a stretching progression of the Hg-N$_2$ bond. This identification comes from the similarity in frequency and appearance of the equivalent Hg-Ar progression. Therefore one is led to distinguish between two types of vibrational movements in the Hg-N$_2$ complex, the stretching vibration and the bending vibrational movement of N$_2$ with respect to mercury. We shall assume, in the simplest approximation[16] that the corresponding radial and angular potentials are independent, the actual potential being the sum of a radial and angular dependence.

THE STRETCHING MOVEMENT

The intensity distribution of the pure stretching vibrations indicates a strong displacement in the radial equilibrium distances between the X ground state, and the red shifted excited state. A one-dimensional Franck Condon simulation of the fluorescence excitation spectrum yields a separation $\Delta R = -1 \text{\AA}$ between the X and A equilibrium distances. Moreover we can infer that the X state equilibrium distance should be very close to the Hg-Ar one and as the binding energies of both ground state complexes are close : $D_0 = 125$ cm^{-1} for Hg-Ar and $D_0 = 100$ cm^{-1} for Hg-N$_2$, (X) (from the separation in Fig. 5 of the Hg line and the dissociation continuum).

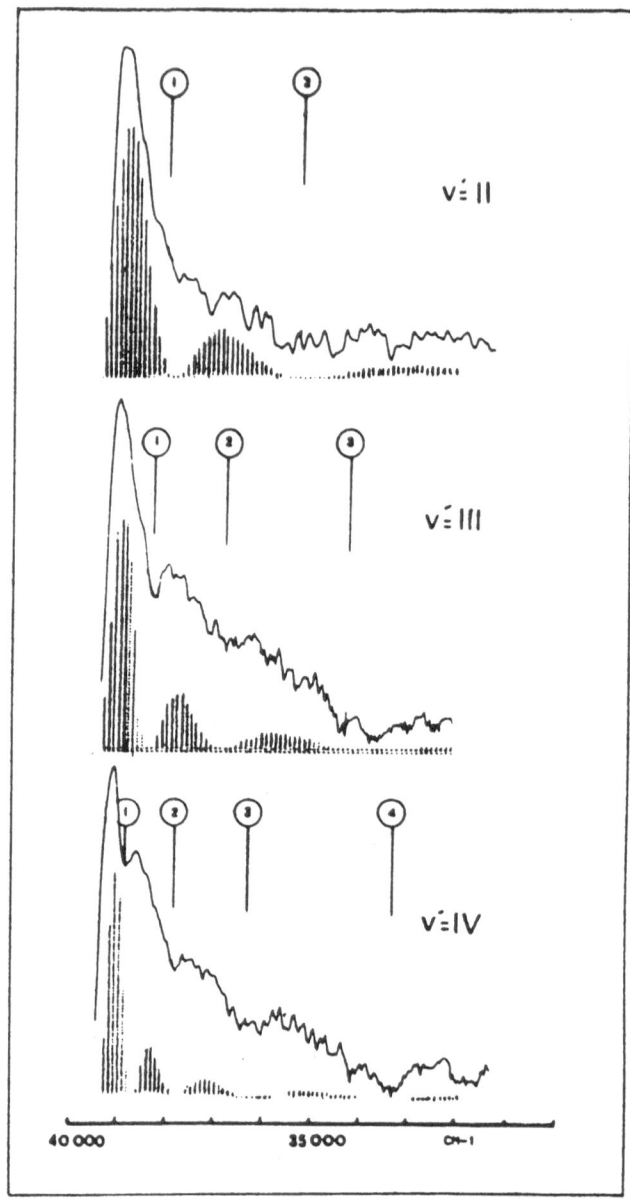

Fig. 6 – Fluorescence emission spectra originating from pure stretching quanta : from top to bottom v' = II, III, IV in Log Scale. F.C. simulations are displayed under each relevant spectrum in linear scale.

The number of the stretch quanta has been assigned, as the intensity of the O-O band is very small, through the emission spectra in the bound free and bound bound regions, after excitation of the most prominent stretching peaks. The number of oscillations in these spectra will reveal the oscillations of the excited state streching wavefunction (when no torsion is excited) as was done for the Na-Ar [9,17] or Hg-Ar [2,13] diatomics. The peaks labeled II and III assigned to 2 and 3 quanta of stretching as they display respectively 2 and 3 minima corresponding to the nodes of the excited state stretching wave function Fig. 6. A Birge-Sponer ΔG plot can be constructed with this labeling and yields for the A state the following constants listed in Table 4. The ground state constants were deduced from the onset of the blue continuum (14) : (D_0) and from the bound bound emission spectra revealing a stretching frequency of $\nu_s \simeq 15$ cm^{-1} (Table 4 : X"STATE 90°).

Using the preceding constants (table 4) to fit the ground state with a Morse one-dimensional potential, the resulting bound free Franck-Condon simulations were completely out of range from the experimental spectra. We used instead of a larger equilibrium distance for the potential, listed in Table 4 : X STATE 0°, to fit the experimental spectra from v'= 2,3,4. The fact that this potential fits the different spectra obtained for different initial levels v=2 to 4 gives reasonable confidence in it, but we have to explain why this latter potential is so different from the one deduced from the bound bound transitions. We are led to the conclusion that there is a drastic geometry change between the ground X state and the excited A state. We are now going to observe the evidence of this geometry change in the prominent torsional progressions.

TABLE 4

DIMENSIONAL Hg-N$_2$ MORSE CONSTANTS				
	ωe cm^{-1}	ωe xe cm^{-1}	De $\overset{o}{A}$	De cm^{-1}
A STATE	55	1,6	3,3	460
X STATE 90°	15	0,5	4,3	107
X STATE 0°	14	0,5	5,2	90

THE LIBRATIONAL MOVEMENT

These modes are most interesting as they yield information on the anisotropy of the potential surface and have up to now received in optical spectra little attention in contrast to ground state spectroscopy. The fluorescence excitation spectrum in Fig. 5 and 7a displays some small lines blue shifted from the stretching peaks whose lifetime is much shorter (c. a. 15ns) as compared to the other peaks (c. a. 70 ns at the maximum). The observed efficient relaxation into $^3P_0$ of those bands is observed in the action spectrum(7b) where these bands are prominent and belong to a progression starting from the stretching band onto the blue. For each stretching quantum there is a progression which seems to stop close to the next stretching quantum, we thus inferred the barrier hight to free rotation to be of the order of the stretching maxima separation ($\sim$50 cm^{-1}).

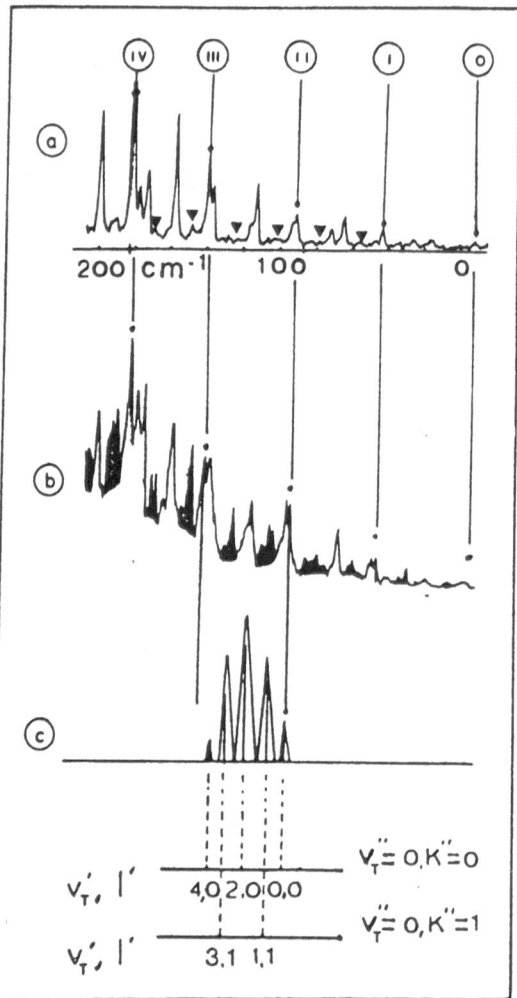

Fig. 7 – a) Fluorescence excitation spectrum 0→IV, the triangles indicate the $^3P_0$ most active modes.
   b) $^3P_0$ action spectrum, the shaded bands correspond to the most active modes.
   c) Simulation of the torsional transitions originating for a T shaped ground state (V_0=25) where the lowest populated states are K"=0,1 (as in Nitrogen, owing to nuclear spin statistics). The excited state is linear with V_0=41 cm^{-1}.

Using a treatment developed by Ewing for $Ar-O_2$ complexes [18] the libration is described in terms of a hindered rotor by the anisotropic potential $V_0/2$ $(Cos\ 2\theta)$, where V_0 may be parametrized as a function of the stretching quantum to account for the decrease of the rotational barrier with the stretch quanta. We produced good simulations of the band positions with $V_0 \approx 40$ cm^{-1} for the lower v' in Fig. 7c.

The intensity in the torsional modes is clearly the result of a geometry change between the A and X states as shows the former simulation where A is linear and X, T shaped, as far the similar $Ar-N_2$[18] and $Ar-O_2$[19] complexes. The result of this geometry change is the discrepancy observed earlier in the bound free emission where in fact the potential observed is the one attained by vertical emission to the ground state surface (Fig. 9). Owing to the T shaped equilibrium geometry in the ground state the linear geometry accessed in the emission is thus likely to be much more repulsive than the T shape in the ground state at short distances, this corresponds to the fitted potential listed in table 4 : X STATE 0°.

Bound free spectra from combination bands were also used to check the previous assignments of the torsional bands. The spectrum of the III-2 peak emission is shown in Fig. 8 together with the III peak emission and a double modulation is manifest. The observed structure in the III-2 spectrum is the result of the excitation of $v(Torsion) = 2$ ($l = 0$, l being the vibrational angular momentum) where the wave function has maxima for $\theta = 0°$ and $\alpha \stackrel{\circ}{-}$.

Fig. 8 – Fluorescence emission from peak III (top) compared to III$_{-2}$ (bottom). The triangles indicate the double modulation observed in this combination band.

Thus schematically the emission is the sum of two contributions from the 0°
and the $\overset{\bullet}{\alpha}$ potentials as is described in Fig. 9. This assignment of the
torsional bands describes consistently the general features of the excited A
state spectrum but does stress the fact that all the torsional bands with
efficient $^3P_0$ relaxation correspond to I ≠ 0. In these conditions for I ≠ 0,
the vibrational angular momentum is non zero corresponding to a rotation
along the Hg-N$_2$ axis. We also observed by time decay measurements within
the rotational contour of the peaks assigned to 0 quantum of torsion that
overall rotation was extremely efficient to induce the electronic relaxation to
$^3P_0$; the decay time varies between 70 ns at the peak to 15 ns at the blue
edge.

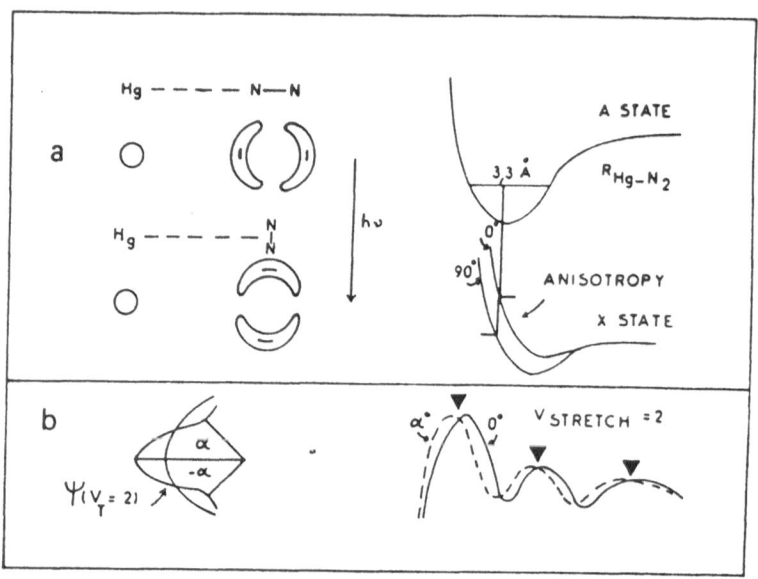

Fig. 9 – a) Schematics of the A→X bound free emission showing the influence
of the anisotropy on the emission.
    b) Bound free emission from $V_{stretch}$ = 2, V_T = 2 showing that the
actual spectrum should be a superposition of both the O° and $\overset{\bullet}{\alpha}$
contributions. The wave function ψ_T (V=2) is plotted versus the torsion angle.

We have provided through the various spectra (fluorescence excitation,
action, emission) a wealth of consistent information on the vibrational
structure of the HgN$_2$ complex and its decay dynamics into $^3P_0$. The spectra
are described by stretching and bending combination bands. The bending
modes correspond to hindered rotation of N$_2$ in a linear equilibrium geometry
with a 40 cm^{-1} barrier in the excited state. The assignment of the bending
motions shows the efficiency of vibrational angular momentum upon the $^3P_0$
relaxation. These findings are in agreement with a model developed by Jouvet
and Beswick [20].

REFERENCES

[1] - W.E. Baylis, *J. Chem. Phys.*, **51**, 2665 (1969)

[2] - M.C. Duval, O. Benoist d'Azy, W.H. Breckenridge,
C. Jouvet, B. Soep, *J. Chem. Phys.* To appear.

[3] - R.J. Niefer, J.B. Atkinson and L. Kraus,
J. Phys. B : At. Mol. Phys. **16**, 3767 (1983).

[4] - E. Czuchaj and J. Sienkiewicz
J. Phys. B : At. Mol. Phys. **17**, 2251 (1984).

[5] - E. Czuchaj, H. Stoll and H. Preuss,
J. Phys. B : At. Mol. Phys. In press.

[6] - M. Krauss and F.H. Mies
Topics in Applied Physics 30 (Rhodes, C.K.-H.K.V. Lotz,
Berlin 1984) p. 5.

[7] - P. Valiron, A.L. Roche, F. Masnou-Seeuws and M.E. Dolan
J. Phys. B : At. Mol. Phys. **17**, 2803 (1984).

[8] - J. Pascale and J. Vandeplanque
J. Chem. Phys. **60**, 2278 (1974).

[9] - J. Tellinghuisen, A. Ragone, M.S. Kim, D.J. Auerbach,
R.E. Smalley, L. Wharton and D.H. Levy
J. Chem. Phys. **71**, 1283 (1979).

[10]- C. Jouvet, Thèse, Université de Paris-Sud, Centre Orsay (March
1985).

[11]- W.H. Breckenridge, M-C. Duval, C. Jouvet and B. Soep
Chem. Phys. Lett. **122**, 181 (1985).

[12]- M.C. Duval, C. Jouvet and B. Soep
Chem. Phys. Letters **119**, 317 (1985).

[13]- K. Fuke, T. Saito and K. Kaya
J. Chem. Phys. **81**, 2591 (1984).

[14]- C. Jouvet and B. Soep
J. Chem. Phys. **80**, 2229 (1984).

[15]- a) E.E. Nikitin, *Theory of elementary atomic and molecular
processes in gases.* Clarendon press. Oxford 1974.
b) H. Horiguchi and S. Tsuchiya
Jour. of Chem. Soc. Faraday Trans. II, **71** 1164 (1975).

[16]- S. Holmgren, M. Waldman, W. Klemperer
J. Chem. Phys. **67** 4414 (1977) and ref. therein.

[17]- R.E. Smalley, D.A. Auerbach. P.S.H. Fitch, D.H. Levy and
L. Wharton
J. Chem. Phys. **66** 3778 (1977).

[18]- G. Henderson, G.E. Ewing, $Ar-O_2$
J. Chem. Phys. **59** 2280 (1973).

[19]- G. Henderson, G.E. Ewing, $Ar-N_2$
Mol. Phys. **27** 903 (1973).

[20]- C. Jouvet, A. Beswick
To be published.

THE STRUCTURE OF VAN DER WAALS MOLECULES OF s-TETRAZINE

Donald H. Levy
James Franck Institute and Department of Chemistry
University of Chicago, Chicago, Illinois 60637
U.S.A.

ABSTRACT. The structural behavior of van der Waals molecules of s-tetrazine is reviewed. Three types of van der Waals molecules are considered: complexes with rare gas atoms, complexes with hydrogen bonding partners such as HCl and water, and dimers of tetrazine. Differences between the dynamical properties of tetrazine dimers and dimethyl-tetrazine dimers are discussed.

1. INTRODUCTION

Over the last several years, my research group has spent a great deal of effort in the study of van der Waals molecules of the aromatic heterocycle s-tetrazine $(C_2H_2N_4)$.[1] Because of the nitrogens in the aromatic ring, the transition from the ground electronic state to the first excited singlet state is a $\pi^* \leftarrow n$ transition whose origin is at 18128 cm^{-1}, a convenient frequency in the visible where high resolution dye lasers are readily available. The availability of such lasers means that it is relatively easy to resolve rotational structure not only in the electronic spectrum of tetrazine itself but also in the spectra of fairly large van der Waals complexes of tetrazine. Therefore, tetrazine is a convenient chromophore on which to base a study of the structure of a variety of complexes.
    In addition to this technical advantage, the study of tetrazine van der Waals molecules benefits from the extensive spectroscopic work that has been done on uncomplexed tetrazine.[2] The rotational structure in the electronic spectra of several isotopic varieties of tetrazine has been resolved in both the static gas[3] and in a supersonic jet.[4] Analysis of this rotational structure has provided rotational constants for various isotopic species and an accurate determination of the geometry of the molecule in both its ground and excited electronic states.[5] Studies of the vibrational structure in

231

A. Weber (ed.), Structure and Dynamics of Weakly Bound Molecular Complexes, 231–250.

both absorption[6] and single vibronic level fluorescence
spectra[7] have provided vibrational frequencies for most of
the vibrational modes that are observed in the electronic
spectrum. Even with the spectral simplification provided by
supersonic cooling, the spectra produced by a mixture of
tetrazine and several of its complexes is congested and con-
fusing, and without the roadmap provided by the previous
spectroscopic work on uncomplexed tetrazine, the interpretation of the van der Waals spectra would have been very much
more difficult.

   In this review I discuss the stucture of a number of van
der Waals complexes of tetrazine and its derivatives. The
object of the review is to illustrate the diversity of structure possible in van der Waals molecules even when attention
is restricted to a single chromophore. First I discuss the
simplest complexes between tetrazine and one or more rare gas
atoms; next, complexes between tetrazine and small hydrogen
bonding partners, and finally dimers of tetrazine and of
dimethyl-tetrazine.

2.   COMPLEXES OF TETRAZINE WITH RARE GAS ATOMS

Figure 1 shows the fluorescence excitation spectra of the
complexes Ar-tetrazine and Ar_2-tetrazine, these spectra being
qualitatively similar to those of all rare gas-tetrazine
complexes.[8] The rotational structure seen in Fig. 1 consists
of a strong central Q-branch and evenly spaced P- and
R-branches. This pattern is characteristic of a parallel
transition of a near symmetric top (transition moment parallel to the near symmetric top axis) in which the rotational
constants of the ground and excited electronic states are
nearly the same. The uncomplexed tetrazine molecule is a
near symmetric oblate top with the unique top axis being the
out-of-plane axis. The $\pi^* \leftarrow n$ transition to the first excited
singlet state of tetrazine is polarized out-of-plane, and
therefore in uncomplexed tetrazine the rotational structure
is parallel polarized and looks similar to the structure
shown in Fig. 1. The fact that the addition of one or two
argon atoms to the tetrazine chromophore preserves the
parallel polarized rotational structure indicates that the
argon atoms attach to tetrazine above and below the tetrazine
ring on the near symmetric top axis. Had the argon atoms
attached to the sides of the ring, the two in-plane rotational constants would not have been almost identical as they
are in uncomplexed tetrazine, and the complex would no longer
be a near symmetric top. Addition of mass to the out-of-
plane axis affect the two in-plane moments of inertia in the
same way, and leaves the complex a near symmetric top.

   Once two rare gas atoms have been complexed to tetrazine,

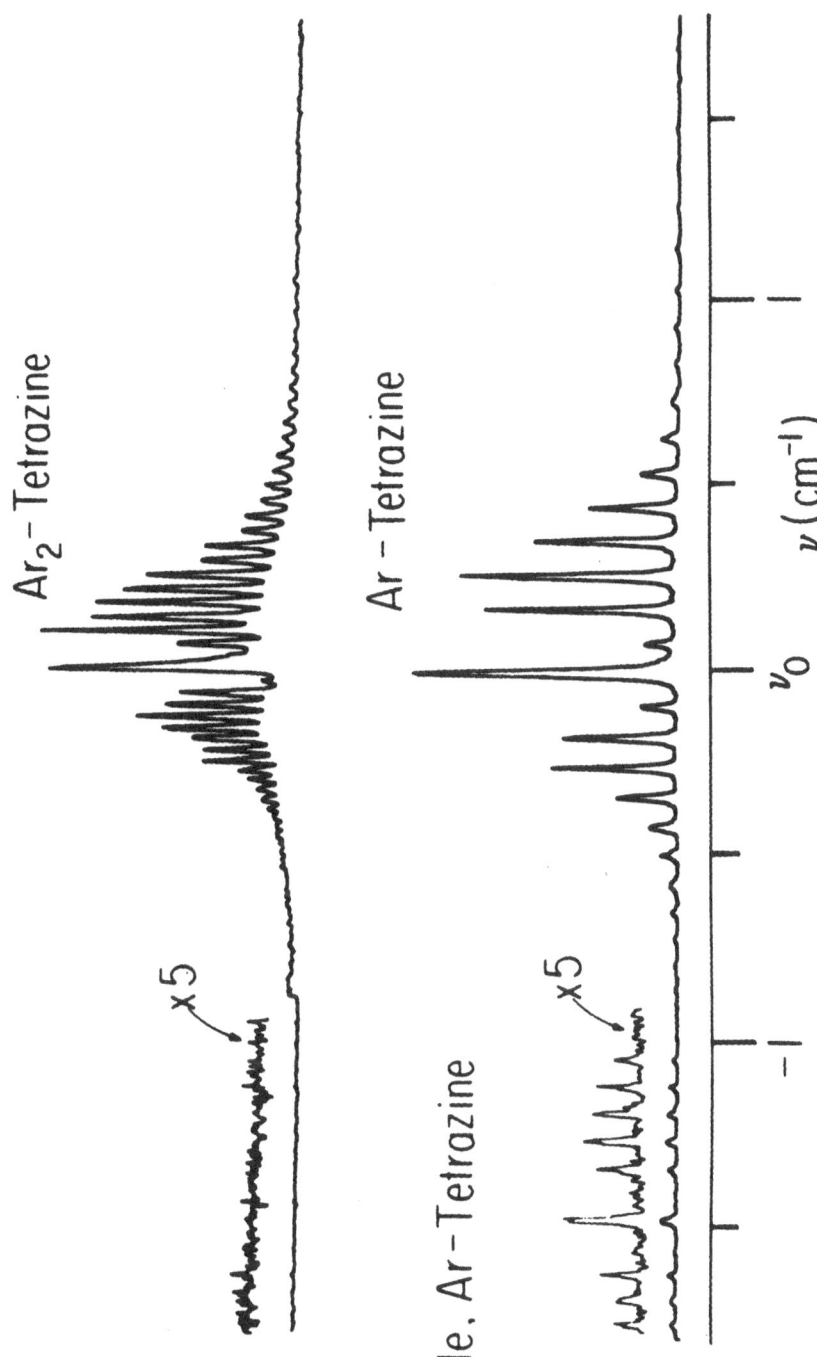

Figure 1.--Rotationally resolved electronic spectrum of the van der Waals molecules Ar-tetrazine, Ar_2-tetrazine, and He,Ar-tetrazine.

the favored out-of-plane binding sites are filled, and there
was some question as to where additional rare gas atoms would
bind. We have studied[8] a series of mixed helium and argon
complexes of tetrazine containing up to six rare gas atoms,
and in all cases we found that the rare gas atoms bind
out-of-plane. When there are more than two rare gas atoms in
the complex, the additional atoms attach to the rare gas
atoms occupying the out-of-plane sites, but do not attach
directly to the open positions at the side of the ring. For
example, Ar_3-tetrazine has an argon atom attached to one side
of the ring, and a diatomic argon molecule attached by one
end to the other side of the ring.

3. COMPLEXES OF TETRAZINE WITH POLAR MOLECULES

In contrast to the case of rare gas complexes, the polar
molecules HCl and water attach to the side of the tetrazine
ring forming hydrogen bonds with the lone pair of the ring
nitrogen atoms. As mentioned above, addition of mass in the
plane of the ring breaks the near equivalence of the two in-
plane inertial axes, and the effect of this on the spectrum
may be seen in Fig. 2. The spectrum in this figure is qualitatively different from that shown in Fig. 1 and is typical
of a perpendicular transition of a near symmetric top. In
this case the near symmetric top axis is near a line drawn
between the centers of mass of the tetrazine ring and the HCl
molecule, and since it lies in the plane of the ring, it is
perpendicular to the out-of-plane transition moment.

The geometry of the molecule was determined[9] by analyzing
the rotational structure in the spectra of three isotopic
species: $H_2C_2N_4$-HCl, $D_2C_2N_4$-HCl, and $H_2C_2N_4$-DCl. In this
analysis, the tetrazine and HCl moieties were assumed not to
change structure upon complexation, and only the geometric
parameters that specify the postion and orientation of the
HCl relative to the tetrazine were varied to fit the observed
spectra. When tetrazine was in its ground electronic state,
the HCl was found to be bound by a linear N...H-Cl hydrogen
bond with a distance of 3.237 ± 0.009Å between the nitrogen and
chlorine atoms. The line between the nitrogen and chlorine
atoms was at an angle of $46°\pm1°$ with respect to the line which
bisects the two N-N bonds in the tetrazine ring. When tetrazine was electronically excited, the hydrogen bond shortened
slightly ($\Delta R = -0.022\pm0.005$Å) and the bond angle increased
slightly ($\Delta\theta = 2.4\pm0.9°$). The slight decrease of bond length
produced by electronic excitation appears to be real and is
surprising inasmuch as the spectrum of the complex is shifted
104 cm^{-1} to the blue of the spectrum of uncomplexed tetrazine. This blue shift indicates that the van der Waals bond

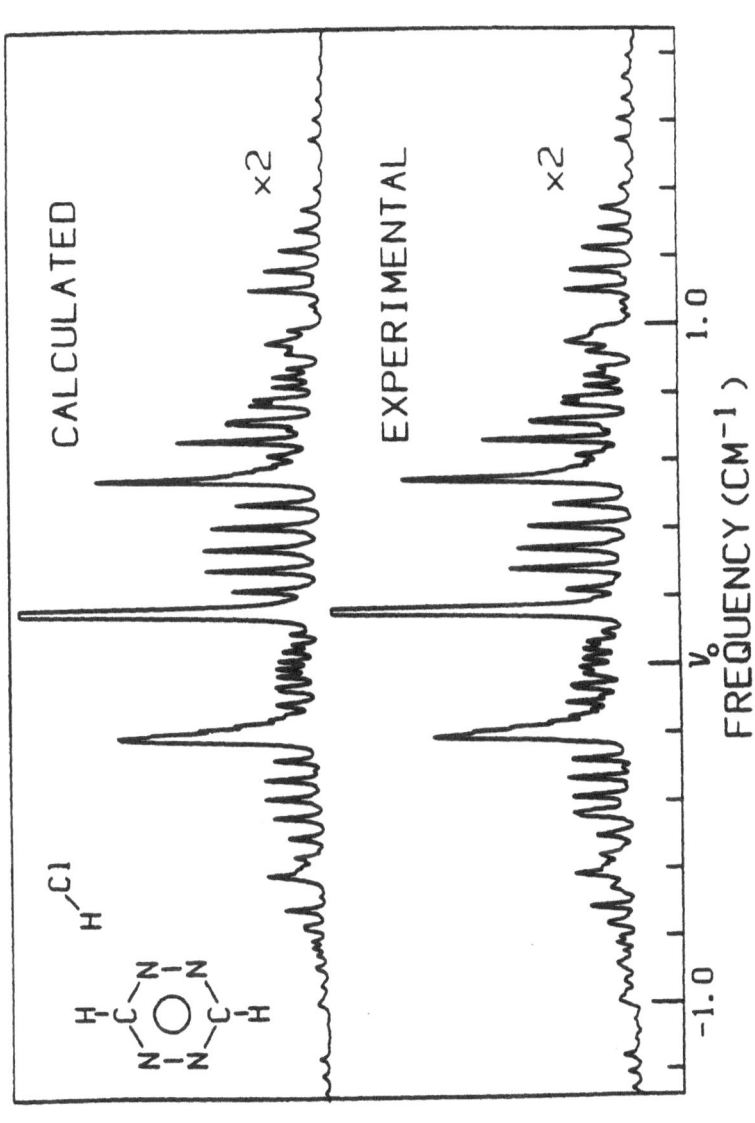

Figure 2.--Rotationally resolved electronic spectrum of tetrazine-HCl. The lower trace is the experimental spectrum, and the upper trace is a spectrum synthesized from the best rotational constants extracted from the experimental spectrum.

weakens upon electronic excitation of the chromophore.

In addition to the band at +104 cm^{-1} which has been assigned to tetrazine-HCl (described above), the addition of HCl to the supersonic expansion produces a second prominent feature at +99 cm^{-1} (see Fig. 3) which we have assigned to the species tetrazine-(HCl)$_2$. Note that while this feature is still blue-shifted with respect to uncomplexed tetrazine, it is red-shifted with respect to tetrazine-HCl. Thus the addition of a second HCl produces an incremental spectral shift that is both much smaller and in the opposite direction from that produced by the first HCl.

The rotational structure of the +99 cm^{-1} feature can also be resolved and gives some information on the structure of the tetrazine-(HCl)$_2$ complex. There are several geometries which are consistent with the experimental spectrum, but with one exception which will be discussed below, they all are extended structures which have one HCl molecule bound directly to the nitrogen of the tetrazine ring, and the other HCl bound to the first HCl. Of this class of structures, the most reasonable would seem to be a geometry which has the extended HCl dimer arranged so as to bring the chlorine of the second HCl into proximity with the tetrazine ring hydrogen thereby producing a second (weak) hydrogen bonding interaction.

Geometries which have the two HCl molecules both bound to the ring in either meta or para positions predict a rotational structure which is totally inconsistent with the observed spectrum. There is one geometry with the two HCl molecules bound to the tetrazine ring in ortho positions which does reproduce the observed rotational spectrum, but we are inclined to reject this stucture on other grounds. First, this ortho structure would have the two HCl molecules bound in equivalent postitions and it is difficult to understand why they would not then produce similar spectral shifts. Second, the ortho geometry which fits the rotational structure requires the angle between the N...H-Cl line and the line bisecting the two N-N bonds in the tetrazine ring to be 20° rather than the 45° observed in tetrazine-HCl. This would require an attractive interaction between the two HCl molecules which would tend to bring them closer together. In this geometry the dipole moments of the two HCl molecules are nearly aligned producing a *repulsive* dipole-dipole interaction of 146-331 cm^{-1} depending on where the point dipole is assumed to be located on the HCl bond. Thus, if the second HCl were to bond directly to the ring, the ortho configuration would not seem to be the most stable possibility.

As pointed out by Klemperer,[10] if the structure of tetrazine-(HCl)$_2$ is an HCl dimer bound to tetrazine rather than two HCl monomers bound to tetrazine, the energetics of

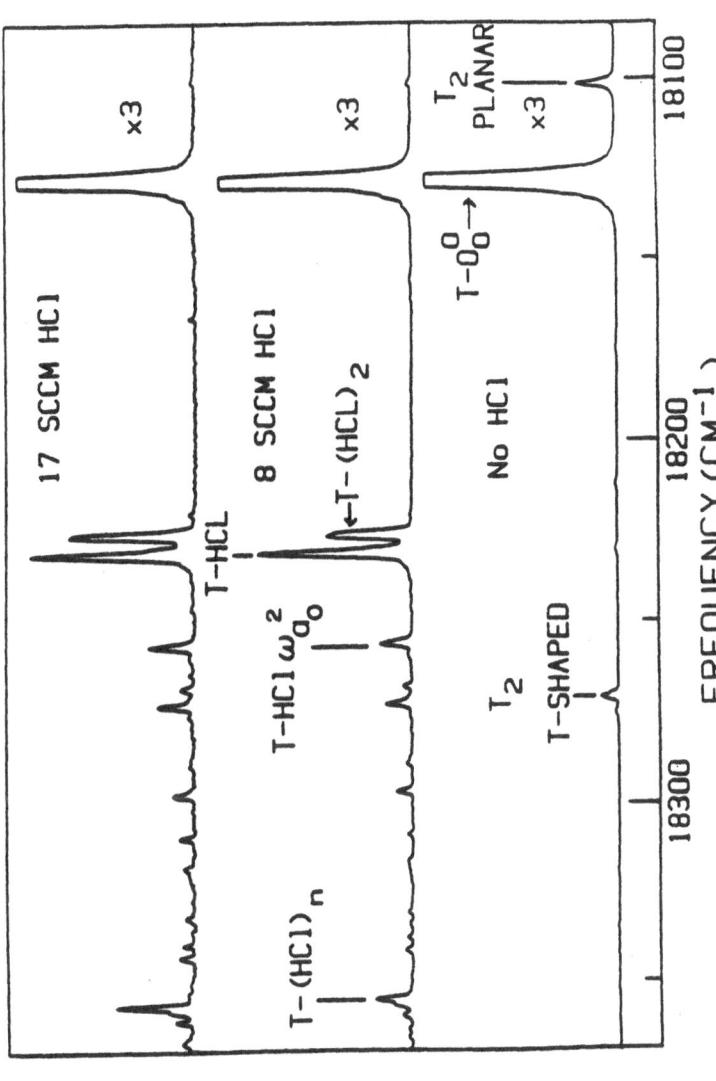

Figure 3.--The electronic spectrum produced by mixtures of tetrazine and HCl in a helium carrier gas. The upper two spectra show the new features that appear when HCl is added to the mixture and indicate the dependence of these features on HCl concentration. Rotational structure is not resolved in these spectra.

the binding are difficult to understand. The binding energy
of the HCl dimer has been measured by Pine and Howard[11] to be
431 ± 22 cm^{-1}. We have been able to bracket the binding energy
of tetrazine-HCl between 800 and 1500 cm^{-1}, and therefore
binding the second HCl directly to the tetrazine ring rather
than to the first HCl would be expected to produce a stronger
bond. It is possible that an additional attractive interaction between the chlorine atom of the second HCl and the
tetrazine hydrogen stabilizes this extended structure. It is
also possible that the structure that we observe is in fact
not the lowest energy structure but is observed because its
formation is kinetically favored in the supersonic expansion.

A third feature of interest in the spectrum produced by
mixtures of tetrazine and HCl is seen at +26 cm^{-1} with respect
to the tetrazine-HCl band (+130 cm^{-1} with respect to uncomplexed tetrazine). The relative intensity of this feature
with respect to the tetrazine-HCl band remains constant under
a variety of expansion conditions, and it has therefore been
assigned as the first member of a vibrational progression in
one of the low frequency intermolecular vibrational modes.
Our first assumption was that this vibration was a stretching
motion between the tetrazine and HCl molecules but this
turned out not to be the case. Analysis of the rotational
structure of this band indicated that, unlike the 0-0 band,
the HCl had a large average out-of-plane displacement, the
rms value of the angle between the HCl bond and the tetrazine
plane being 32°. Therefore the vibrational motion associated
with the +26 cm^{-1} feature is an out-of-plane bend of the HCl.
This vibration is non-totally symmetric since it breaks the
planar symmetry, and therefore only transitions to even overtones are allowed. This means that the fundamental frequency
for this motion is 13 cm^{-1}, half the observed frequency.

The spectrum produced by a mixture of tetrazine and water
in a helium carrier gas (shown in Fig. 4) is qualitatively
similar to the tetrazine + HCl spectrum shown in Fig. 3. The
addition of water to the expansion produces a prominent feature at +86 cm^{-1} with respect to the uncomplexed tetrazine 0-0
band, and this has been assigned to the complex tetrazine-
H_2O.[12] Increasing the concentration of water in the expansion
produces two additional features which are at +75 cm^{-1} and +62
cm^{-1} with respect to uncomplexed tetrazine. On the basis of
their concentration dependence and their rotational structure, these are assigned to the complexes tetrazine-$(H_2O)_2$ and
tetrazine-$(H_2O)_3$, respectively. As in the case of tetrazine +
HCl, we find that the addition of a single water molecule
causes a relatively large blue shift, while the incremental
effect of adding additional water molecules is to produce
smaller red shifts with respect to tetrazine-H_2O.

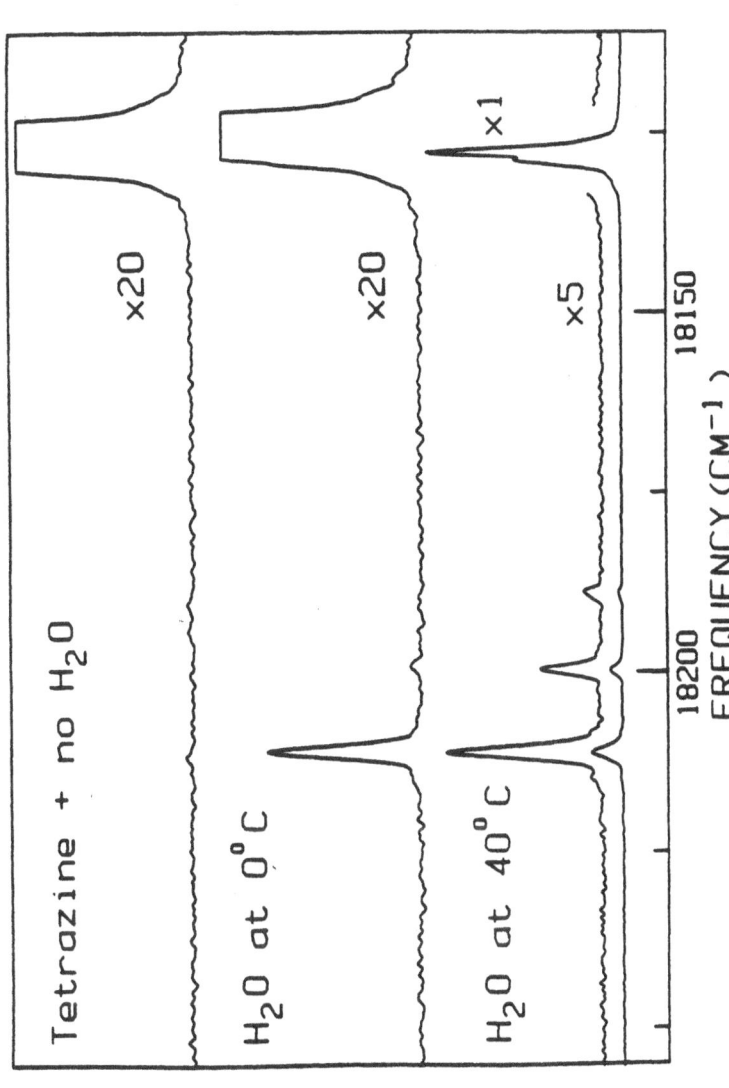

Figure 4.--The electronic spectrum produced by mixtures of tetrazine and water in a helium carrier gas. The lower two spectra show the new features that appear when water is added to the mixture and indicate the dependence of these features on water concentration. Rotational structure is not resolved in these spectra.

The geometry of tetrazine-H_2O was measured by analyzing[12]
the rotational structure in the spectra of the isotopic
species $H_2C_2N_4$-H_2O, $D_2C_2N_4$-H_2O, and $H_2C_2N_4$-D_2O. Two geometries
were found to place the center of mass of the water molecule
in essentially the same postion with respect to the inertial
axes of the complex as a whole, and therefore both of these
structures predicted nearly the same rotational structure for
all isotopic species. The position of the oxygen atom of the
water molecule did not differ very much in the two geometries, and these geometries were distinguished almost entirely by the placement of the water hydrogen atoms with
respect to the tetrazine ring. Although there was no experimental reason for choosing one of these geometries over the
other, one of the geometries was much more favorable for
hydrogen bonding between the water and the lone pair of the
ring nitrogen, and we therefore favor this hydrogen bonding
geometry. In this geometry the water molecule was found to
be hydrogen bonded to the side of the ring with a nitrogen-
oxygen distance of 2.97Å in the ground electronic state. The
angle between a line bisecting the two N-N bonds and a line
between the ring nitrogen and the water oxygen was 71.3°. The
N...H separation was 2.1Å as compared with an N...H separation of 1.95Å in tetrazine-HCl. The vibrationally averaged
geometry of the hydrogen bond was not linear, the O-H bond
making an angle of 22±6° with a line drawn between the N and
O atoms.

The rotational structure of the feature assigned to
tetrazine-$(H_2O)_2$ can be reproduced by several geometries.
Structures having two water molecules bound to the tetrazine
ring in either para or meta positions cannot reproduce the
observed spectrum, but two water molecules symmetrically
bound to the ring in ortho positions will reproduce the rotational spectrum of all three isotopic species. The fit is
somewhat worse than that produced by the extended structure
described below, and in addition the different spectral
shifts produced by the two water molecules suggests that the
two waters are not bound to equivalent sites.

Two extended structures consisting of a water dimer bound
to one of the ring nitrogens will also reproduce the rotational structure in the spectrum. Attaching an $(H_2O)_2$ subunit
which has the same geometry as the free water dimer[13] to the
side of the tetrazine ring gives a reasonable fit to the
spectrum, but a better fit is achieved by using a water dimer
where the internal hydrogen bond is distorted. The details
of these structures have been given elsewhere.[12]

Rotational structure has been resolved in a feature
assigned to tetrazine-$(H_2O)_3$, and the rotational constants of
this cluster have been measured. Because there are so many
parameters needed to define the geometry of such a large

cluster, the measured rotational constants give very little information about the geometry.

4. DIMERS OF TETRAZINE AND DIMETHYL-TETRAZINE

Dimers of tetrazine have been prepared in a supersonic free jet and their structure has been determined by analysis of the rotational structure in their electronic spectra.[14] In the region of the origin band of the tetrazine monomer, three features have been assigned to the tetrazine dimer as shown in Fig. 5. Analysis of the rotational structure of these three bands has shown that the tetrazine dimer exists in two geometric isomers. One of these isomers has the two monomer rings in a common plane with the hydrogen atoms of one ring pointing toward the lone pair electrons on the nitrogen atom of the other ring. In this arrangement the two rings are symmetrically equivalent and there are two excited electronic states which are the symmetric and anti-symmetric linear combinations of an excited state localized on one or the other ring. The transition from the ground electronic state to one of these states is forbidden, and therefore the planar geometric isomer gives rise to a single spectroscopic feature shifted 28 cm^{-1} to the low frequency side of the tetrazine monomer origin at 18128 cm^{-1}. The planar dimer is a near symmetric top with the top axis lying in plane and passing through the centers of the two rings. Therefore the out-of-plane transition moment of the $\pi^* \leftarrow n$ transition is perpendicular to the symmetric top axis, and the rotational structure is characteristic of a perpendicularly polarized transition.

The second geometric isomer has the two rings arranged in a T-shaped structure with the two rings perpendicular to each other. In this isomer the rings are not symmetrically equivalent, and since the interaction between them is small, the two excited electronic states are described as having the electronic excitation localized on one or the other ring. Transitions from the ground state are allowed to both excited electronic states giving rise to two features in the spectrum. Once again, the dimer is a near symmetric top with top axis passing through the centers of the two rings, but in this geometric isomer, the rotational selection rules are different for the two different features. The feature at 144 cm^{-1} to the high frequency side of the tetrazine monomer origin is due to excitation to the top bar of the T. In this case the out-of-plane transition moment is parallel to the top axis giving rise to the typical rotational structure of a parallel transition, and it is from this pattern that this feature is assigned as an excitation of the top ring. The feature 45 cm^{-1} to the low frequency side of of the tetrazine monomer origin has the rotational pattern of a perpendicular transition. This would be expected when the upright ring of

242

D. H. LEVY

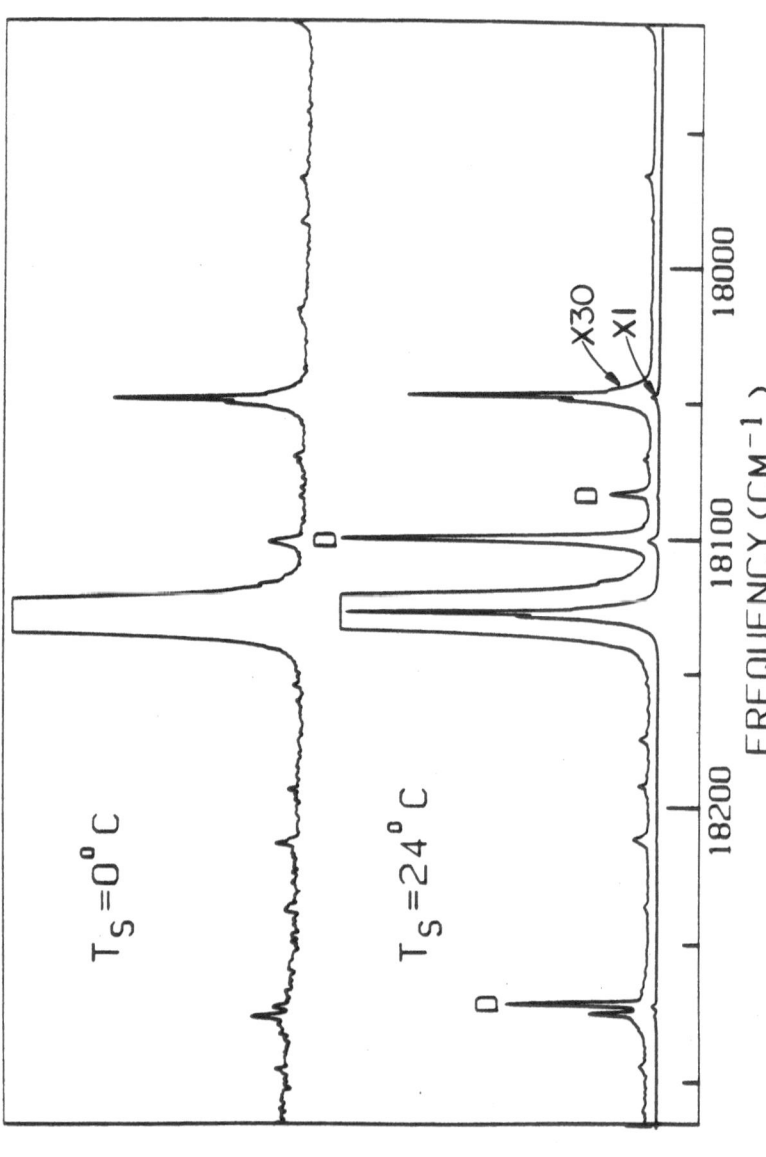

Figure 5.--The low resolution electronic spectrum of tetrazine. The lower trace was taken with a higher concentration of tetrazine in the supersonic expansion relative to the upper trace. Features assigned to the tetrazine dimer are marked D.

the T is excited since in this case the out-of-plane transition moment is perpendicular to the top axis.

The main point that I would like to make about the tetrazine dimer does not involve the structure, but has to do with the dynamics of vibrational and electronic energy flow in the molecule. In Fig. 6 we see the dispersed fluorescence spectrum that is produced when the $6a^2_0$ transition is excited in the upright ring of the T, that is the $6a^2_0$ perpendicularly polarized transition. The mode 6a is an in-plane elongation of the ring having a frequency of 703 cm^{-1} in the excited electronic state, and therefore in the $6a^2$ level the dimer has 1406 cm^{-1} of excess vibrational energy. The emission spectrum shows two types of features, sharp and broad. Analysis of the frequencies of these features has allowed us to assign all of the sharp features as resonance fluorescence; that is, transitions from the originally excited $6a^2$ level to various vibrational levels of the ground electronic state.[15] The broad features are all due to vibrational relaxation where vibrational energy initially in mode 6a has relaxed to excited levels of the six low frequency intermolecular modes that describe the motion of the two halves of the dimer against each other. The broad transitions are composites of $\Delta v=0$ transitions from the several nearly isoenergetic excited levels of the intermolecular modes that have accepted the vibrational energy that was orignally in ring mode 6a. It is important to note that all of the emission, broad and sharp, can be assigned to the dimer and there is no evidence in Fig. 6 of monomer emission.

The emission spectrum shown in Fig. 7 is qualitatively different from that shown in Fig. 6. Fig. 7 was produced by exciting the $6a^2_0$ transition of the top ring of the T, the parallel polarized transition. Note that the top ring excited state is 189 cm^{-1} higher in energy than the upright ring excited state, and that the $6a^2$ level of the top ring has 1595 cm^{-1} combined electronic and vibrational excess energy. The spectrum shown in Fig. 7 has only sharp features of different intensities. The weak features can be assigned as resonance fluorescence of the dimer, but the strong features must be assigned as transtions from the zero-point level of the excited electronic state of the monomer.[15] Thus, excitation to $6a^2$ of the top ring leads to dissociation of the dimer producing one electronically excited monomer fragment and one ground electronic state monomer fragment. The fact that the monomer emission is so much stronger than the dimer emission indicates that the dissociation lifetime is much shorter that the fluorescence lifetime. Quantitative measurement of the strengths of the monomer and dimer fluorescence along with the known fluorescence lifetime allows us to measure the photodissociation lifetime as 38 psec.

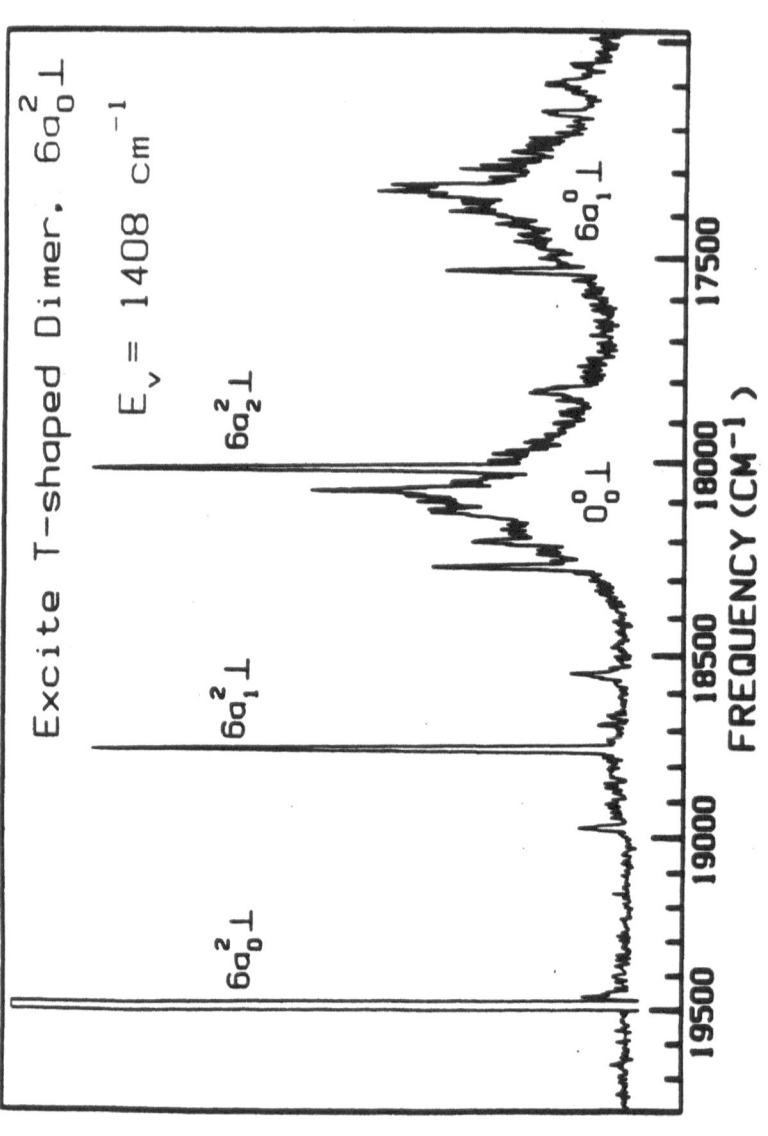

Figure 6.--The emission spectrum produced by exciting the $6a^2{}_0$ transition localized in the upright ring of the T-shaped dimer of tetrazine. Assignments of the broad and sharp features are indicated in the figure. All emission is from the dimer.

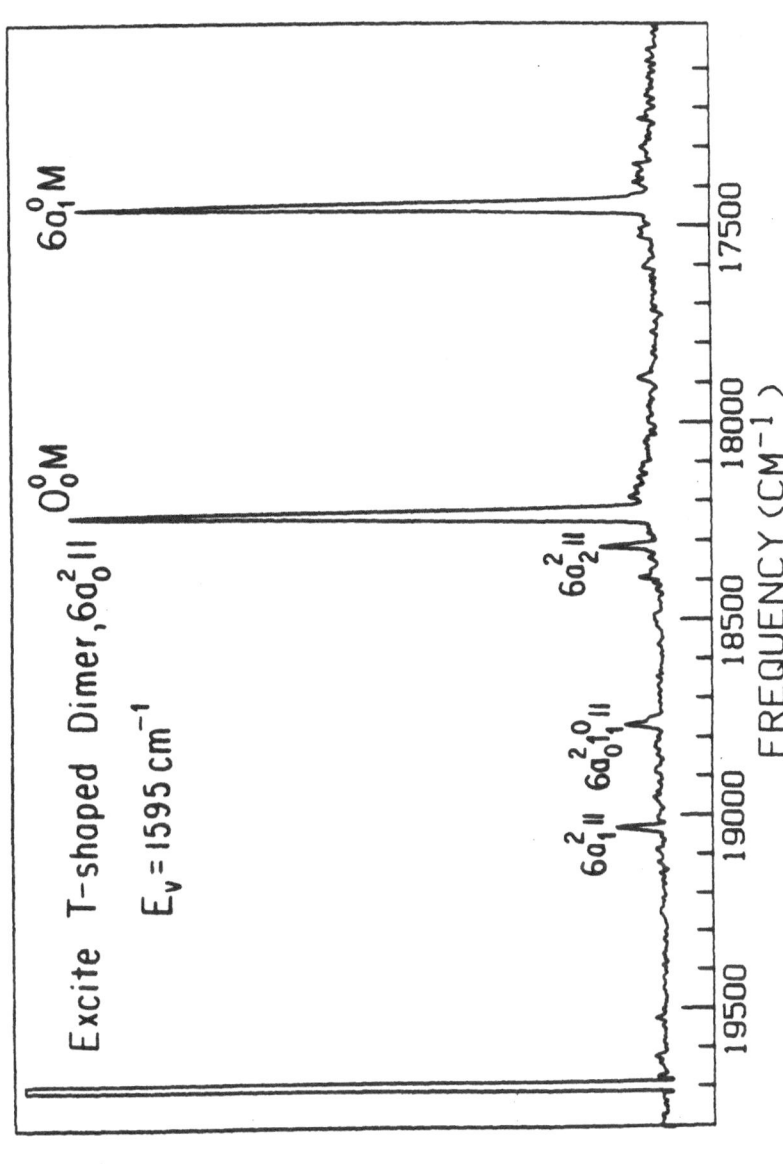

Figure 7.--The emission spectrum produced by exciting the $6a^2_0$ transition localized in the top ring of the T-shaped dimer of tetrazine. Features marked M are due to emission from the monomer fragment produced by dissociation of the dimer.

An important feature of Fig. 7 is the fact that there is no evidence for relaxed emission from the dimer. The fact that we observe no photodissociation when exciting $6a^2$ of the upright ring while we observe almost complete dissociation when exciting the slightly higher $6a^2$ level of the top ring strongly suggests that top ring excitation is just above the dissociation threshold while the upright ring excitation is slightly below the dissociation threshold. The absence of any relaxed dimer emission in Fig. 7 indicates that once the dissociation channel is open, dissociation is the dominant relaxation process and internal vibrational relaxation cannot compete with dissociation in the tetrazine dimer.

The structural and dynamical properties of the tetrazine dimer are in sharp contrast to the dimer of dimethyl-tetrazine.[16] In the dimethyl-tetrazine dimer only a single structure is observed in which the two rings are stacked and parallel with the rings slipped in a direction perpendicular to the out-of-plane axis as shown in Fig. 8.

This major change in structure is accompanied by an equally dramatic change in dynamical behavior. Fig. 9 shows the emission spectrum that is produced when the state $6a^1_0 1^2_0 w_a{}^1_0$ is excited. Mode 1 is the symmetric stretch of the ring and mode w_a is the stretch of the two halves of the dimer against each other. The level $6a^1 1^2 w_a{}^1$ has 2269 cm^{-1} of excess vibrational energy. Almost all of the emission shown in Fig. 9 is broad and can be assigned to relaxed emission from the dimer. The small, sharp feature at 17500 cm^{-1} is significant in spite of its low intensity since it can be assigned as emission from the zero-point vibrational level of the monomer. The fact that monomer emission is observed, however weak, indicates that the excitation is above the dissociation threshold and that the dissociation channel is open at this energy. In contrast to the case of the tetra-zine dimer, internal vibrational relaxation is the dominant process in the dimethyl-tetrazine dimer even when dissocia-tion is energetically allowed.

5. SUMMARY

In this paper I have reviewed the spectroscopy of a number of van der Waals complexes of the heteroaromatic s-tetrazine and have pointed out the diversity of structures and dynamical behavior that is observed even when attention is confined to van der Waals molecules of a single chromophore. Complexes of rare gases with tetrazine all have geometries where the rare gas atoms are located above and below the plane of the tetrazine ring. This is true even when there are more than two rare gas atoms in the complex and the out-of-plane geom-etry requires stacking of the rare gas atoms on top of each

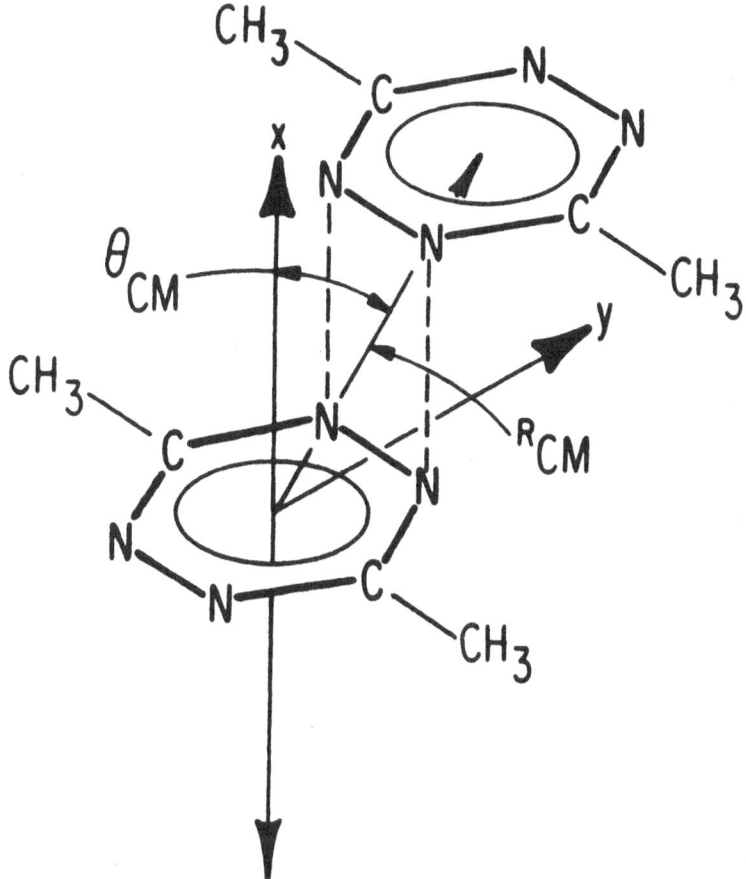

Figure 8.--The structure of the dimethyl-tetrazine dimer.

other. In contrast, hydrogen bonding molecules such as HCl and water attach to the side of the tetrazine ring forming hydrogen bonds to the lone pair electrons on the tetrazine nitrogen atoms. Complexes having two hydrogen bonding molecules appear to have extended geometries with one molecule attached to the tetrazine ring and the second molecule attached to the first.

Dimers of tetrazine have two geometric isomers, the first having the two tetrazine molecules in a common plane and the second having the two tetrazine molecules perpendicular to each other in a T-shaped geometry. When either the planar or the T-shaped dimer is vibrationally excited above the zero-point level but below the dissociation threshold, internal

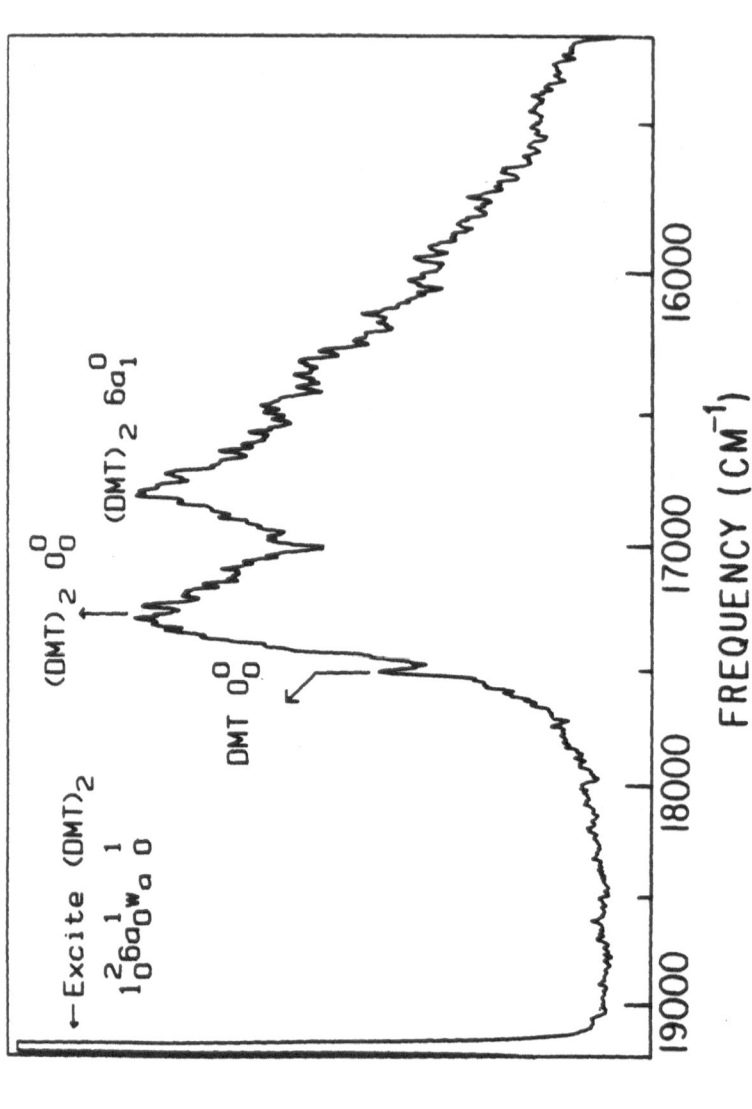

Figure 9.--The emission spectrum produced by exciting the $1^2_0 6a^1_0 w_a^1{}_0$ level of the dimethyl-tetrazine dimer. The sharp feature marked DMT 0^0_0 is emission from the monomer fragment produced by dissociation of the dimer. The broad features are emission from the vibrationally relaxed dimer.

vibrational relaxation from the intramolecular ring vibrational modes to the low frequency intermolecular modes occurs on a time-scale that is competitive with fluorescence. When the T-shaped dimer is excited just slightly over the dissociation threshold, the molecule is almost completely dissociated, and no vibrational relaxation is observed.

The dimer of dimethyl-tetrazine has a geometry that is different from either of the tetrazine dimer geometries, the two rings being stacked parallel and displaced. In contrast to the dynamical behavior of the tetrazine dimer, the dimethyl-tetrazine dimer shows effecient internal relaxation even when excited well above the dissociation threshold. In this case relaxation is much faster than dissociation, and the branching ratio for dissociation is quite small.

This work was supported by the National Science Foundation.

REFERENCES

1. R. E. Smalley, L. Wharton, D. H. Levy, and D. W. Chandler, J. Chem. Phys. **68**, 2487 (1978); J. E. Kenny, D. V. Brumbaugh, and D. H. Levy, J. Chem. Phys. **71**, 4757 (1979); D. V. Brumbaugh, C. A. Haynam, and D. H. Levy, J. Chem. Phys. **73**, 5380 (1980); ibid., J. Mol. Spectrosc. **94**, 316 (1982); D. H. Levy, C. A. Haynam, and D. V. Brumbaugh, Faraday Discuss. Chem. Soc. **73**, 137 (1982); D. V. Brumbaugh, J. E. Kenny, and D. H. Levy, J. Chem. Phys. **78**, 3415 (1983); C. A. Haynam and D. H. Levy, J. Phys. Chem. **87**, 2091 (1983); C. A. Haynam, D. V. Brumbaugh, and D. H. Levy, J. Chem. Phys. **81**, 2270 (1984); C. A. Haynam, L. Young, C. Morter, and D. H. Levy, J. Chem. Phys. **81**, 5216 (1984); Y. D. Park and D. H. Levy, J. Chem. Phys. **81**, 5527 (1984); and D. H. Levy, Faraday Trans. 2, **82**, 1107 (1986).

2. R. R. Karl and K. K. Innes, Chem. Phys. Lett. **36**, 275 (1975); J. H. Meyling, R. P. Van Der Werf, and D. A. Wiersma, Chem. Phys. Lett. **28**, 364 (1974); K. K. Innes, A. H. Kalantar, A. Y. Khan, and T. J. Durnick, J. Mol. Spectrosc. **43**, 477 (1972); K. K. Innes, A. Y. Khan, and D. T. Livak, J. Mol. Spectrosc. **40**, 177 (1971); S. W. Thakur and K. K. Innes, J. Mol. Spectrosc. **52**, 130 (1974); L. A. Franks and K. K. Innes, J. Chem. Phys. **47**, 863 (1967); D. T. Livak and K. K. Innes, J. Mol. Spectrosc. **39**, 115 (1971); K. K. Innes, J. Chem. Phys. **76**, 2100 (1982); and G. G. Asmuth, K. K. Innes, and V. A. Job, J. Mol. Spectrosc. **83**, 266 (1980).

3. A. J. Merer and K. K. Innes, Proc. Roy. Soc. London, Ser. A, **302**, 271 (1968).

4. R. E. Smalley, L. Wharton, D. H. Levy, and D. W. Chandler, J. Mol. Spectrosc. **66**, 375 (1977).

5. V. A. Job and K. K. Innes, J. Mol. Spectrosc. **71**, 299 (1978).

6. K. K. Innes, L. A. Franks, H. A. Merer, G. K. Vemulapalli, T. Cassen, and J. Lowry, J. Mol. Spectrosc. **66**, 465 (1977).

7. D. V. Brumbaugh and K. K. Innes, Chem. Phys. **59**, 413 (1981).

8. C. A. Haynam, D. V. Brumbaugh, and D. H. Levy, J. Chem. Phys. **80**, 2256 (1984).

9. C. A. Haynam, C. Morter, L. Young, and D. H. Levy, J. Phys. Chem. (in press).

10. William Klemperer, discussion at this meeting.

11. A. S. Pine and B. J. Howard, J. Chem. Phys. **84**, 590 (1986).

12. C. A. Haynam, C. Morter, L. Young, and D. H. Levy, J. Phys. Chem. (in press).

13. J. A. Odutola and T. R. Dyke, J. Chem. Phys. **72**, 5062 (1980).

14. C. A. Haynam, D. V. Brumbaugh, and D. H. Levy, J. Chem. Phys. **79**, 1581 (1983).

15. L. Young, C. A. Haynam, and D. H. Levy, J. Chem. Phys. **79**, 1592 (1983).

16. C. A. Haynam, D. V. Brumbaugh, and D. H. Levy, J. Chem. Phys. **81**, 2282 (1984).

SPECTRA OF MIXED DIMERS

E. W. Schlag, H. L. Selzle, and K. U. Boernsen
Institut für Physikalische Chemie
Technische Universität München
Lichtenbergstr. 4
D-8046 Garching
Germany

ABSTRACT. Interesting new information about the spectroscopy of van der Waals complexes is obtained from the studies
of mixed isotopomers in a supersonic jet in coincidence
with high resolution mass spectroscopy. This allows to
separate the different overlapping spectra of the isotopic
labeled dimers even for the very weak ^{13}C substituted complexes. In particularly it could be shown, that the absorption in one of the two halves of the benzene dimer is essentially independent of the other half and only in the homodimer a very weak excition interaction is found. The structure of the benzene dimer does not appear to be T-shaped,
but is now found to be rather V-shaped.

1. INTRODUCTION

The multiphoton ionization mass spectroscopy is a highly
selective method for preparing molecular ions /1/. The
ionization can be directly performed in the source of a
mass spectrometer and allows for a simultaneous detection
of the bare molecules and the fragments generated also in
the ionization process. The multiphoton ionization is only
efficient, if there are real states in resonance with the
photon energy. When the wavelength of the exciting laser
light is scanned the variety of real intermediate molecular
states produce a typical molecular spectrum and this spectrum then appears when the ionisation current is measured
as a function of the laser wavelength. Since in general
molecules differ in their visible/UV absorption spectra,
different components of molecular gas mixtures can be selectively ionized by multiphoton ionization of different wavelength. This adds a new degree of freedom in mass spectrometry and makes the method two dimensional /2/.

A. Weber (ed.), Structure and Dynamics of Weakly Bound Molecular Complexes, 251–261.
© 1987 by D. Reidel Publishing Company.

With this method one can analyze gas mixtures with even spurious contributions of the different components without a previous separation by ordinary chemical preparation techniques. A new and important application of this method is given in the explanation of molecular complexes which are only weakly bound by van der Waals interaction.

Interactions between non-bonded aromatic molecules are of fundamental importance in a number of chemical interactions, the most typical example of which being the interaction of two non-bonded porphyrins in the primary reaction center of photosynthesis /3/. Such non-chemical forces, however, also play an important role in liquids, though here also many body effects come into play. Little is understood about these interactions, particularly as an elementary step.

The benzene dimer is perhaps the simplest prototype system in which the interaction between two aromatic molecules can be studied, this being of further importance in that the dimer can be prepared pure by a supersonic expansion in the isolated gas phase, hence in the absence of any interfering many body effects.

Although many dimers have been studied in supersonic jets a detailed analysis has been made difficult by the paucity of data for each system, data which usually just involve one simple low resolution spectrum of the dimer. Fung et al. /4/ have demonstrated, however, that considerably more information can be obtained if various isotopic modifications of the dimer are prepared and spectrosopically analysed. Although this is an old lesson from molecular spectroscopy, it has been hard to achieve for dimers since it is difficult, if not impossible, to prepare isotopically pure dimers. As an example, there is no way of producing the pure $C_6H_6 - C_6D_6$ dimer without simultaneously producing the two homodimers $(C_6H_6)_2$ and $(C_6D_6)_2$. Fung et al. introduced a method by which they produced all three dimers simultaneously and then extracted three separated absorption (excitation) spectra by measuring the spectra in coincidence with their respective masses, thus producing three identifiably different spectra from the mixture.

The benzene dimer $(C_6H_6)_2$ was observed in low resolution by Hopkins et al. /5/. Fung et al. have identified the mass coincidence spectra of the three isotopomers $(C_6H_6)_2$, $(C_6D_6)_2$ and $C_6D_6 - C_6H_6$ by studying a mixed jet containing 50 : 50 C_6H_6 and C_6D_6 seeded in an Ar jet. This work involved the strong ν_6 absorption which is also the first absorption in the normal benzene spectrum, the 0-0 origin

here being forbidden in both one- and two-photon absorption.

By this isotopic mixture technique /4/ it was possible to
show for the first time that in the heterodimer C_6H_6-
C_6D_6 excitation in the S_1 excited state is a localized
process resident in either half of the dimer, the excitation residing either on the C_6H_6 or the C_6D_6 moiety. Furthermore, it could be seen that the ν_6-dimer transitions
split up in very characteristics ways /4/.

The 0-0 origin is forbidden in the benzene monomer spectrum
but it is slightly allowed in the dimer. This transition
has been very interesting, however, in that the vibrationless 0-0 absorption in this first excited singlet state
showed no splitting in the albeit very weak spectrum. One
would have expected such splitting in the C_6H_6-C_6H_6 homodimer in which two identical molecules of benzene are in
juxtaposition. Such a splitting would give a direct answer
to the question of the most elementary type of band splitting, i. e. of two isolated molecules in the absence of
the usual many body effects in organic crystals /6/. More
recent work by Law et al. /7/ using the same isotopic mixing
technique again failed to observe this effect. This result
led us to reexamine the data with a new high resolution
jet experiment /8/. In this work employing some improved
skimming techniques we were now at last successful in observing the absorption spectrum of the homodimer with a split
of 1,7 cm^{-1}, with a 7 : 5 intensity ratio in the protonated
benzene dimer. This clearly gave us the first measure for
the interaction of two benzene molecules in the S_1 state
in the gas phase.

An important new additional feature which is now observed
is the discovery of a red shift of the homodimers in relation to both of the respective heterodimers, a result which
we first found in the ν_6-absorption band /4/ and now also
in the 0-0 transition /8/. This is true for both homodimers, and hence cannot be simply correlated to the mass of
the dimers, as in typical isotope effects. In order to
study this effect in detail we decided to introduce a small
perturbation. As the perdeuteration in one of the partners
in the heterodimer already presents a rather large perturbation in the vibrational states of the dimer, better
information should be obtainable from a weaker perturbation
of the homodimer with a smaller change in vibrational frequencies. We therefore decided to measure the mass resolved
excitation spectra for ^{13}C isotopic labeled benzene molecules. This should produce an even weaker detuning of vibrational frequencies in the homodimers. We again produced
these dimers from a mixture of isotopically labed species,

but really such a mixture is already present in any normal benzene sample having a natural abundance of ^{13}C isomers. By now tuning to this ^{13}C isomer mass in coincidence with the absorption spectrum, it is possible to measure the relevant mass selected absorption (excitation) spectra directly.

To further investigate the nature of the interaction it was decided to investigate a dimer with similar mass but no aromatic character. Here a benzene-cyclohexane dimer was investigated, and also its ^{13}C substituted isomer.

In 1983 Fung et al. demonstrated with benzene dimers that the information from the spectroscopy of van der Waals complexes could be greatly increased of isotopomers of these complexes could be investigated and separate spectra obtained for each species. This is reasonably straight forward if dimers of one isotopic monomer A are desired, to form A_2, it only being required to suppress higher order aggregates A_n. If mixed dimers AB are required one produces however ipso facto also A_2 and B_2 and AB separatly in the conventional way by scanning a laser, but then adding a second photon to produce the corresponding ion. Hence the absorption spectrum is carried out always with a mass label and this allows absorption spectra to be measured as if each of these three species were present separately, which is particularly important if the bands overlap.

2. EXPERIMENTAL

The experiment is performed in a supersonic jet apparatus where the jet after proper skimming is crossed by two laser beams with a common focus. The first laser is used to achieve resonant excitation in the region of the first electronic excited state, and the second laser is used for ionization out of this excited intermediate state. These ions are detected in a reflectron type time-of-flight mas-spectrometer with a resolution of $M/\Delta M$ of 2000. This allows one to separate the excitation spectra of isotopic compounds even for a very large ratio of partial contributions and unit mass differences as in the ^{13}C studies.

Figure 1 shows a time of flight spectrum at a somewhat higher intensity of the ionizing laser, and the resonant laser being tuned to the region of the absorption of the dimer of benzene h_6 with the natural abundance of ^{13}C. This demonstrates that if the windows of a dual channel boxcar averager are set at the arrival time of specific masses one can readily record the spectra simultaneously. In this way one can separate overlapping bands of different

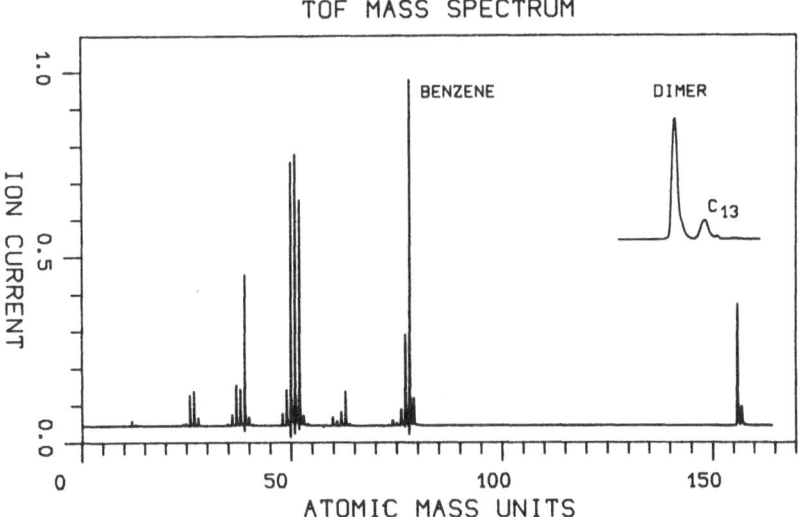

Figure 1. Time-of-flight mass spectrum of benzene dimers with laser excitation near the dimer absorption band.

isotopic species and at the same time measure the relative spectral position of the bands without the need of absolute calibration of every scan.

3. RESULTS

In fig. 2 - 4 the excitation spectra of various isotopic mixed dimers of benzene and the complex with cyclohexane are shown.In the benzene complex the 0-0 transition which is strictly forbidden in the monomer is found. The position is shifted by about 45 cm^{-1} to the red of the not observable monomer transition.

Fig. 2a now shows the absorption spectra at the 0-0 absorption region. In the homodimer at high resolution and very low temperature two peaks can be observed with different height. When a ^{13}C atom is introduced the observed splitting is larger and the two peaks are now of equal height. The ^{13}C-band of the homodimer also lies further to the blue. The observations in the heterodimer which are shown in fig. 2b are quite different. There is only one peak for the ^{12}C-dimer even under high resolution, but

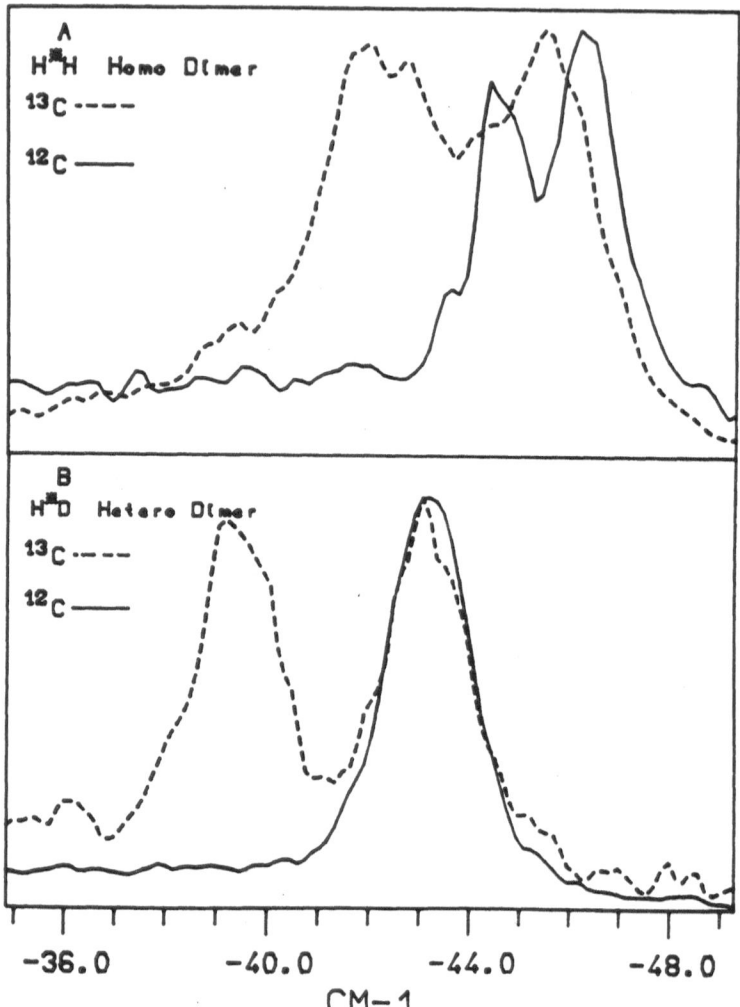

Figure 2.   Excitation spectra of benzene dimers near the d_0-benzene absorption. a) Excitation via the S_1 0-0 transition and detection of d_0-d_0 and ^{13}C-d_0-d_0 ions. b) Excitation via the S_1 0-0 transition and detection of d_0-d_6 and ^{13}C-d_0-d_6 ions.

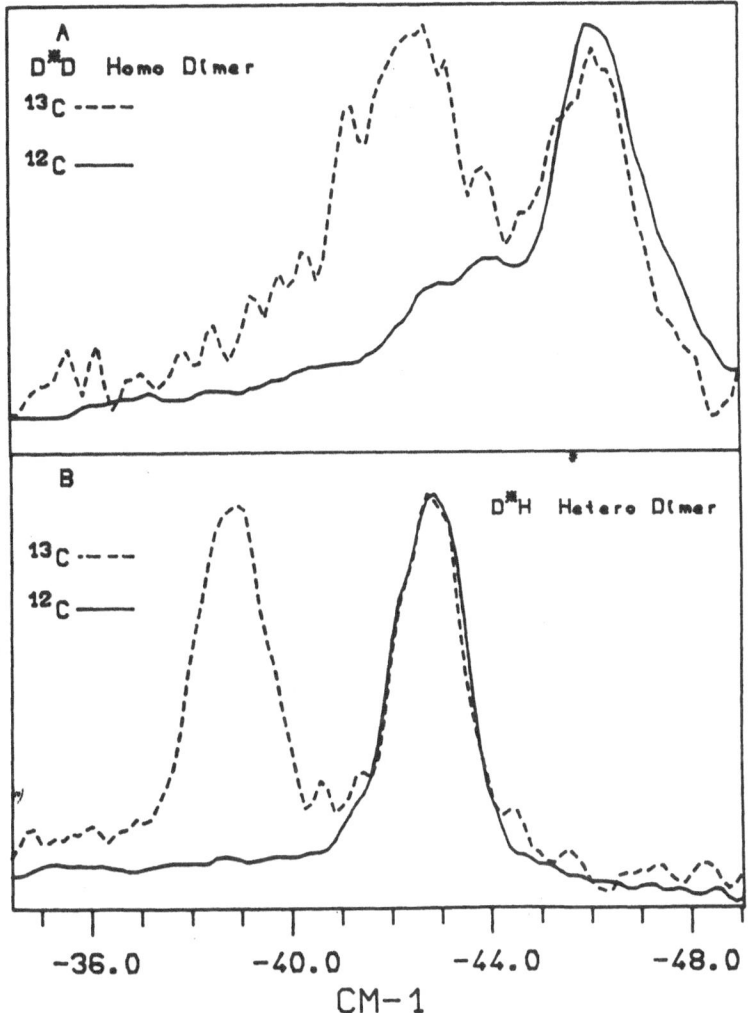

ISOTOPIC SHIFT OF THE HOMO AND HETERO BENZENE DIMER
0-0 TRANSITION

Figure 3.   Excitation spectra of benzene dimers near the
d_6-benzene absorption. a) Excitation via the S_1 0-0 transition and detection of d_6-d_6 and ^{13}C-d_6-d_6 ions. b) Excitation via the S_1 0-0 transition and detection of d_6-d_0
and ^{13}C-d_6-d_0 ions.

there are two peaks in the heterodimer case. The two peaks
also show equal height.

For the case of the simple heterodimer d_0-d_6 this single
peak shows that there is little interaction between the
two halves of the dimer. The same heterodimer with ^{13}C
labeling shows two peaks of equal height, one of which is
not only coincident with the unlabeled dimer but virtually
superposable. This indicates that this peak corresponds to
the excitation of the ^{12}C half of the heterodimer. The
other peak then is the excitation into the ^{13}C half of the
dimer. The fact that the two peaks of d_0-d_6 and ^{13}C d_0-d_6
are superposable shows that the excitation is essentially
unaffected by the isotopic labelling in the other nonexcited
part of the dimer. This again demonstrates the extreme
case of independence of the S_1 excitations in the heterodimer.

The behaviour at the region of the benzene d_6 absorption
is similar to the benzene d_0 case, the only difference is
that the ratio of the two peaks in the ^{12}C homodimer case
is larger. This is shown in fig. 3a and b.

The effect that the heterodimer transition in d^*_0-d_6 is
not influenced at all when the ^{13}C atom is in the non excited half of the dimer lead us to test this for another
heterodimer of equal molecular weight but a different internal force field for the vibration. For the comparison we
chose the dimer with d_0-cyclohexane, a dimer which is equal
in weight to the d_0-d_6 heterodimer, and where therefore
the mass window for the detection is the same.

Fig. 4 shows the experimental results when the benzene
half of the dimer is excited. In the ^{12}C complex only one
peak is observed and in the ^{13}C case two peaks are found,
where on stays exactly at the ^{12}C absorption and the other
is shifted by 3,2 cm^{-1} to the blue which is nearly the same
as in the d_0-d_6 heterodimer case.

4. DISCUSSION

In the benzene dimers the isotopic shift is due to the
internal zeropoint energies of the individual dimer partners and an isotope effect on the vdW bond. The zeropoint
energy difference for the d_0 and d_6 benzene is about 200
cm^{-1}, therefore the heterodimer bands are also separated
by this amount. The observation of only one peak for the
d_0-d_6 heterodimer and the d_0-d_0 heterodimer demonstrates
that the excitation is localized in only one half of the
dimer. Now if one of the heterodimer partners contains a

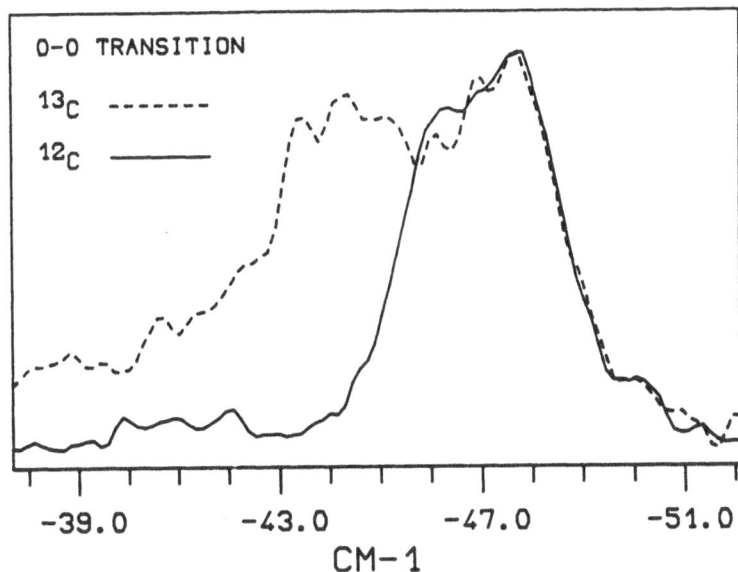

Figure 4.   Excitation spectra of the benzene-cyclohexane
dimer. Excitation via the S_1 0-0 transition and detection
of benzene-cyclohexane and ^{13}C-benzene-cyclohexane ions.

^{13}C atom the zeropoint energies are slightly modified but
the excitation is still localized. One then can determine
the zeropoint energy difference for the benzene monomer
from the shift of one of the peaks relative to the unshifted peak. The unshifted peak corresponds to the dimer where
the ^{13}C is in the non-excited half and the blue shift is
due to the zeropoint energy difference if the partner with
the ^{13}C atom is excited.

These measurements also confirm that there is no site splitting in the heterodimer which means that the two halves
are equivalent. Therefore the splitting in the homodimer
has to be due to exciton interaction and from the observed
splitting one finds an interaction of about 1 cm^{-1}. The
small difference of the ^{13}C transition in respect to normal
d_0 or d_6-benzene can be used to confirm this. If one applies
first order perturbation theory the observed splitting in
the homodimer is twice the interaction matrix element β,

whereas in the case of energy detuning the splitting is given by $\sqrt{\delta^2 + 4\beta^2}$ where δ is the detuning which is about 4 cm^{-1}.

This detuning is already large enough to decouple the two halves of the dimer and the excitation is again localized in one half of the dimer. From the perturbation theory the two peaks have to be of the same height, which is also confirmed from the measurement.

The different peak height in the homodimer now gives information about the structure of the dimer. From the measurements of Janda, et al. /9/ it is known that the benzene dimer is a polar complex, but from the heterodimer measurements one finds that the two halves are equivalent. The highest possible symmetry to show a dipole moment with equivalent benzene molecules is C_{2v}. A reasonable structure for this would be a dimer with a V-type configuration. The dipole moment arises from the lowered symmetry in the charge distribution and gives rise to a transition dipole moment in either half. If one assumes that the transition dipole moments lie in the plane of the molecules one now can find an angle of about 70° from the ratio of the peak height, if one applies a simple vector addition of the transition dipole moments.

5. CONCLUSION

High resolution excitation spectra in combination with high resolution mass spectra gives important new information on the nature of the interaction in weakly bound complexes. In the case of the benzene dimer from isotopic substitution one can determine the interaction matrix element for the exciton interaction, which leads to a small splitting of the transition of 1.7 cm^{-1} in d_0-d_0 and 2.3 cm^{-1} in d_6-d_6. Introducing a single ^{13}C atom mostly decouples the two halves and the excitation is localized in one half. From this high resolution dimer-measurements one can also determine directly the zeropoint energy difference in the benzene monomer when a ^{13}C atom is introduced, a value which cannot be found otherwise. From the intensity ratio of the two homodimer peaks it is also possible to determine the structure of the complex, which now can be best described as a V-shape configuration.

SPECTRA OF MIXED DIMERS

Literature:

/1/   a) U. Boesl, H. J. Neusser, E. W. Schlag, Z.
       Naturforsch. 33a (1978) 1546

       b)L. Zandee and R. B. Bernstein, D. A. Licht in J.
       Chem. Phys. 69 (1978) 3427

/2/   U. Boesl, H. J. Neusser, E. W. Schlag, Chem. Phys.
       55 (1981) 193

/3/   J. Deisenhofer, O. Epp, K. Miki, R. Huber, H. Michel
       J. Mol. Biol. 180 (1984) 385

/4/   K. H. Fung, H. L. Selzle, E. W. Schlag, J. Chem.
       Phys. 87 (1983) 5113

/5/   J. Hopkins, D. E. Powers, R. E. Smalley, J. Phys.
       Chem. 85 (1981) 3739

/6/   E. R. Bernstein, S. D. Colson, R. Kopelman, G.
       W.Robinson, J. Chem. Phys. 48 (1968) 5596

/7/   K. S. Law, M. Schauer, E. R. Bernstein, J. Chem.
       Phys. 81 (1984) 4871

/8/   K. O. Börnsen, H. L. Selzle, E. W. Schlag, Z.
       Naturforsch. 39a (1984) 1255

/9/   K. C. Janda, J. C. Hemminger, J. S. Winn, S. E.Novick,
       S. J. Harris, W. Klemperer, J. Chem. Phys. 63 (1975)
       1419

SOLVENT SHIFTS, SPECTROSCOPY AND STRUCTURE IN VAN DER WAALS COMPLEXES OF PERYLENE

A T Amos*, S M Cohen**, J C Kettley, T F Palmer and J P Simons
Departments of Chemistry and (*) Mathematics
University of Nottingham
University Park
Nottingham NG7 2RD

ABSTRACT   Laser induced fluorescence has been observed from weakly bound complexes formed in supersonic free jet expansions of perylene with rare gas atoms and several organic "solvent" molecules using helium as carrier gas. The fluorescence excitation spectra of the perylene-solvent complexes displayed a considerable variety of intermolecular vibrational structures which were strongly sensitive to the nature of the solvent molecule. Model calculations of the potential surfaces for these van der Waals complexes (vdW) expressing the interaction energy as the sum of pair potentials between the atoms in the molecules, have been used to determine equilibrium geometries and binding energies.

Experimental values for the red spectral shifts observed in the fluorescence excitation spectra of the 1:1 vdW complexes have been used in conjunction with the calculated dispersion energy for the respective ground state to estimate the difference between the Unsöld average energies for the ground and excited states of perylene.

INTRODUCTION

The experimental study and theoretical analysis of large van der Waals (vdW) complexes of aromatic molecules such as s-tetrazine[1], anthracene[2,3], tetracene[4,5], pentacene[6] and perylene[7-11] with rare gas atoms and organic "solvent" molecules has attracted considerable interest. The vdW bonding provides a microscopic laboratory which allows resolution of the influence of solvation on the electronic and molecular structures of the aromatic solute and molecular complexes. They can readily be prepared as isolated clusters under controlled experimental conditions by jet-cooling of the solute-solvent vapour and simply probed by laser induced fluorescence excitation or emission spectroscopy. A helpful framework for their structural analysis can be provided by model calculations of the inter-molecular potentials using, for example, the semi-empirical approaches developed by Ondrechen and Jortner[12] or Topp[13] and their co-workers. These allow predictions of binding energies, force constants, equilibrium conformations, alternative near iso-energetic conformers and barrier heights.

A. Weber (ed.), Structure and Dynamics of Weakly Bound Molecular Complexes, 263–278.

The present work reviews a continuing investigation of the solvation and structure of vdW complexes with perylene. The solvents are selected from a representative range of organic molecules, linear, spherical, cyclic, polar, non-polar - and they are probed and analysed by combining laser induced fluorescence spectroscopy with approximate summed pair-potential calculations. Particular attention has been paid to a theoretical examination of the spectral shifts promoted by complex formation and a revision of the way in which the Longuet-Higgins and Pople[14] theory has been applied to isolated vdW complexes[9]. The analysis of the spectral shifts assumes (i) that they arise solely through the change in the dispersion energies on electronic excitation and (ii) that this difference can be estimated using average excitation energies in the Unsöld[15] approximation

EXPERIMENTAL

The experimental arrangement for investigation of van der Waals complexes between perylene and solvent molecules has been described elsewhere[8] but was modified to operate with helium as the carrier gas by incorporation of a pulsed nozzle valve (General Valve Ltd.). Replacement of the seals of the valve by an 'O' ring ("Kalrez" Du Pont Ltd.) and polyamide fibre plunger ("Vespel" Du Pont Ltd.) and the solenoid windings by copper wire (0.14 mm diameter) coated with a high temperature lacquer (Lewcos F D Sims Ltd.) allowed prolonged operation at high temperatures. A chromel-alumel thermocouple was located close to the nozzle orifice and temperature was monitored using a direct-reading meter (Noronix). A double pulse generator incorporating a delay unit (Advanced Instruments PGS2A) was used to trigger both the pulsed valve and laser and allowed synchronisation of the crossing of the gas jet by the laser beam a few mms downstream from the nozzle. The pulsed valve had a characteristic open-shut time of ~1 ms and was operated at 20 Hz.

Perylene vapour (Aldrich 99+% grade) generated at 200°C in a stainless steel sample oven, diffused into the helium-solvent gas flow prior to expansion through an 800 μm diameter nozzle. Added gases such as argon, krypton and methane were introduced into the helium by a needle valve, while the partial pressures of the other organic solvents were maintained using conventional freezing baths through which the carrier gas was bubbled. For all experiments a nitrogen-pumped tunable dye laser (Molectron UV22, DL200) was used as the excitation light source.

THEORY

Various methods of calculating ground-state geometries, binding energies and spectral shifts of vdW complexes are available ranging from accurate but time-consuming ab initio procedures to semi-empirical or empirical approaches. However, many different solvent molecules have to be considered and in order to find equilibrium geometries, calculations for

many different positions and orientations are required. In addition the
complexity of the organic molecular solutes and solvents conspire to
rule out the ab initio methods, except possibly for a few bench-mark
calculations.

The more practical, though less accurate, methods assume, in effect,
that the interaction energy between the molecules in a vdW complex can
be expressed as the sum of two terms[16-18]

$$\Delta E^s = E_D^s + E_1^s \tag{1}$$

where s refers to the state of the solute i.e. s = g for the
ground-state and s = e for an excited state. The term E_D^s arises in
second-order Rayleigh-Schrödinger perturbation theory and consists of
dispersion and polarisation energies. We shall refer to it as the
dispersion energy since, for the systems we consider here, the
polarisation components are expected to be negligible. E_D^s gives rise to
attractive intermolecular forces. E_1^s is the first order term, including
exchange and therefore it contains terms which arise through the
operation of the Pauli exclusion principle and involve the overlap of
the charge distributions on the interacting molecules. This gives the
repulsive parts of the potential and is supplemented by the electrostatic interaction between the charge distributions, nuclear as well as
electronic.

Because E_1^s is difficult to calculate, especially when the
interacting molecules are large, it is often approximated by a sum of
pair potentials, (W_{uv}^s),[19,20] between the atoms in the molecules, i.e.

$$E_1^s = \sum_u \sum_v W_{uv}^s (r_{uv}) \tag{2}$$

where r_{uv} is the distance between the atom u in the solute molecule
(perylene in our case) and the atom v in the solvent molecule. For the
pair potential we use

$$W_{uv}^s = \frac{B_{uv}^s}{r_{uv}^{12}} + \frac{q_u^s q_v}{r_{uv}} \tag{3}$$

where B_{uv}^s is a constant depending only on the nature of atoms u and v,
and q_u^{s}, q_v are the net charges on the perylene atom (u) in state s
and the solvent atom (v) respectively. The first term in (3) is the
repulsive part of the Lennard-Jones potential and corresponds to charge-
overlap terms. The Coulomb term allows for charge separation effects:
the importance of its inclusion has been stressed by Topp et al[13].
The second order term for the interaction between a solute molecule in
state s and a solvent molecule in its ground state is

$$E_D^s = -\sum \sum{}' \frac{|\langle \phi_s \Psi_0 |H'| \phi_j \Psi_k \rangle|^2}{E_j + F_k - E_s - F_0} \tag{4}$$

where Φ_i, E_i and Ψ_k, F_k are the complete sets of wave functions and energies for the solute and solvent molecules respectively (k = 0 referring to the ground state); H' is the interaction potential. The prime on the summations indicates that the zero order state $\Phi_s \Psi_o$ is omitted from the sum. A standard way to simplify (4) is to introduce Unsold average excitation energies $\bar{E}_s$ and $\bar{F}_{solv}$ for perylene in state s and the particular solvent and apply closure so as to obtain[14]

$$E_D^s = \frac{-1}{\bar{E}_s + \bar{F}_{solv}} (\langle \Phi_s \Psi_o | H'^2 | \Phi_s \Psi_o \rangle - \langle \Phi_s \Psi_o | H' | \Phi_s \Psi_o \rangle^2) \qquad (5)$$

For atoms, distance r apart, a bipolar expansion of H' leads to the well-known r^{-6} dependence of E_D^s. For large molecules, however, where intramolecular distances are comparable with the intermolecular separation, a simple bipolar expansion about two centres, one in each molecule, is very inaccurate. Therefore, it is more satisfactory to divide the charge densities $|\Phi_s|^2$ and $|\Psi_o|^2$ up into regions centered about the various nuclei and make a series of bipolar expansions in terms of the set of inter-atomic distances r_{uv}, along the lines suggested by Claverie (ref 17 section VE2). In this way we can obtain

$$E_D^s = \frac{-1}{\bar{E}_s + \bar{F}_{solv}} \sum_u \sum_v \frac{D_{uv}^s}{r_{uv}^6} \qquad (6)$$

or, as it is more usually written,

$$E_D^s = -\sum_u \sum_v \frac{C_{uv}^s}{r_{uv}^6} \qquad (7)$$

where $C_{uv}^s = D_{uv}^s (\bar{E}_s + \bar{F}_{solv})$

Equations (1), (3) and (7) with s = g the ground state and with empirically determined values for the constants B_{uv}^g and C_{uv}^g can be used to find equilibrium geometries and binding energies. The difference $E_D^g - E_D^e$ between the dispersion energy with perylene in its ground state and the dispersion energy with perylene in an excited state (e) can be used to estimate spectral shifts.

RESULTS AND DISCUSSION

(1) Fluorescence Excitation Spectra

The laser induced fluorescence excitation spectra of a representative range of organic molecular solvent-perylene vdW complexes are collected in Figures 1-2.
        The assignment of the overlapping band systems has been guided by
(i)   the dependence of their relative contributions on the solvent
        partial vapour pressure in the free jet expansion
(ii)   analysis of their vibrational structures and spectral shifts
        relative to that of the bare, uncomplexed perylene, and
(iii)   the assumption that the perylene molecule is only weakly perturbed

by the intermolecular field.

Systems labelled 0, I, II or III are assigned to perylene associated with zero, one, two or three solvent molecules. A cursory inspection of the figures reveals several trends. All the vdW complexes absorb to the red of the free solute molecule and the spectral shifts increase with the solvation number. The increases are generally harmonic. Many of the shifted features have associated satellite bands, assignable either to low frequency intermolecular modes or to a variety of alternative intermolecular conformers (e.g. in the case of the two polar solvents ethanol and bromobenzene). We have only monitored a limited spectral range, for simplicity: the principal features are based upon the band origin and the excited perylene vibrational mode lying at 355 cm^{-1}, probably associated with out-of-plane motion of the ring systems.

Benzene The spectrum shown in figure 1(a) closely resembles that reported by Topp and his co-workers[13], and displays two new band systems, displaced by 390 cm^{-1} and 720 cm^{-1} to the red and assigned to the 1:1 and 1:2 vdW complexes of perylene with benzene. The low frequency mode lying at 95 cm^{-1} in the isolated solute and attributed to an out-of-plane 'butterfly' motion[2] is displaced to 70 cm^{-1}. The near additivity of the spectral shifts for the 1:1 and 1:2 complexes suggests adsorption of the benzene at near equivalent sites on the perylene molecule. A 'sandwich' structure with the benzene rings located above and below the perylene molecular plane would satisfy this constraint and also provide a qualitative rationale for the 'dampening' of the out-of-plane motion. The absence of any very low frequency progressions or spacings implies little intermolecular structural changes on excitation and the existence of a single stable conformer in each of the two complexes.

Cyclohexane Three overlapping band systems can be identified displaced by 230 cm^{-1}, 440 cm^{-1} and 545 cm^{-1} to lower frequencies (Figure 1(b)). The first two are clearly identifiable with 1:1 and 1:2 vdW complexes with the solvent and likely involve near equivalent adsorption sites. Their principal features present doublet structures with a spacing ~12 cm^{-1}, and a second, very much weaker doublet component can also be discerned, displaced by ~30 cm^{-1} to the violet. The irregular spacings suggest contributions from alternative conformers and/or from overlapping low frequency intermolecular progressions. There is little evidence for any contribution from the 'butterfly' mode in the complexes, though it could well be lost among the complicated doublet structures. The third system, displaced by 545 cm^{-1}, is probably attributable to a 1:3 complex where the adsorption sites are no longer near equivalent.

Methylcyclohexane Figure 1(c) indicates the two systems assigned to the 1:1 and 1:2 vdW complexes with methylcyclohexane, which are displaced respectively by 290 cm^{-1} and 560 cm^{-1} to the red - surprisingly large displacements in comparison with cyclohexane itself (but see table 2 and figure 4). Their detailed structures closely resemble those of the

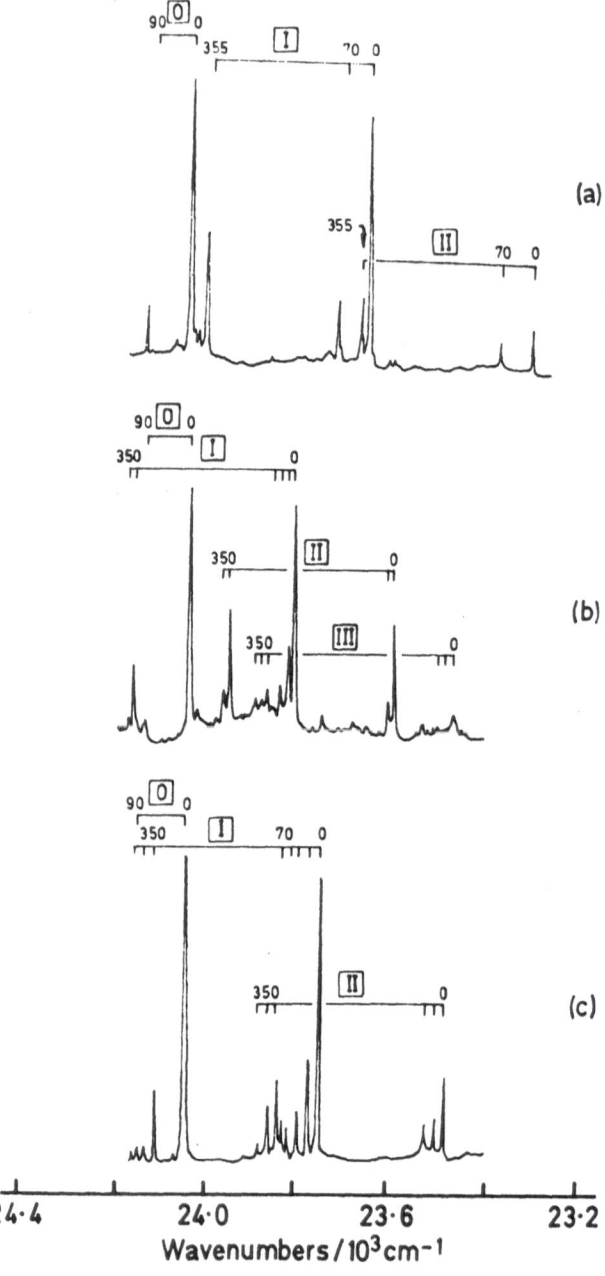

Figure 1   Fluorescence Excitation Spectra of Perylene vdW
            complexes with (a) benzene, (b) cyclohexane and
            (c) methylcyclohexane

cyclohexane complexes however, with satellite features again suggesting overlapping low frequency vibrational progressions and/or alternative conformers. The spacings are larger than in cyclohexane i.e. ~20 cm^{-1} (c.f. 12 cm^{-1}) and ~45 cm^{-1} (cf 30 cm^{-1}). The near harmonic displacements for the 1:1 and 1:2 complexes parallels the behaviour in benzene and cyclohexane, encouraging the conclusion that adsorption leads to a 'sandwich' structure.

Carbon tetrachloride The complexes generated by jet expansion of CCl$_4$-perylene give rise to a remarkably 'clean' electronic excitation spectrum, see fig. 2(a). Displaced systems associated with 1:1, 1:2 and 1:3 complexes can readily be assigned, and remarkably the displacements increase harmonically in all three cases, 260 cm^{-1} (1:1), 500 cm^{-1} (1:2) and 730 cm^{-1} (1:3). Regularly spaced low frequency progressions, ~20 cm^{-1} can be assigned to intermolecular vibrational motion, but there is no evidence of any contributions from alternative conformers. Evidently the perylene structure is able to offer a number of equivalent adsorption sites to the near spherical carbon tetrachloride molecules. Furthermore, their adsorption leads to unique structures despite the possibility of axial, bridged or tripod conformations over the perylene plane.

Bromobenzene Complexing with the dipolar solvent bromobenzene contrasts sharply with carbon tetrachloride. At least three alternative conformers are needed to account for the closely spaced structure in the red-shifted band systems ($\delta\nu$ = 315, 290 and 265 cm^{-1}); they are associated with the features labelled IA, IB and IC on figure 2(b). Their relative intensities were independent of the solvent partial pressure in the nozzle gas flow and are inconsistent with the alternative assignment to a low frequency vibrational progression. However the appearance of additional weak features may suggest slight structural changes in the vdW conformers on excitation. Fragments of other systems are also present but their assignment is currently uncertain.

Ethanol Complexing with ethanol results in much the same behaviour as with bromobenzene; three distinct conformers contribute to the 1:1 electronic excitation spectrum, red-shifted by 170, 150 and 130 cm^{-1} (figure 2(c). They are assigned to the systems IA, IB and IC. Further discussion is deferred, pending the structural calculations discussed in the following section.

(2) Binding Energies and Geometries of the vdW Complexes

Combining equations (1), (2), (3) and (7) allows us to write the following expression for the binding energy ΔE^g of the ground state molecular complex

$$\Delta E^g = \sum_u \sum_v \left(-\frac{C^g_{uv}}{r^6_{uv}} + \frac{B^g_{uv}}{r^{12}_{uv}} + \frac{q^g_u q_v}{r_{uv}} \right) \qquad (8)$$

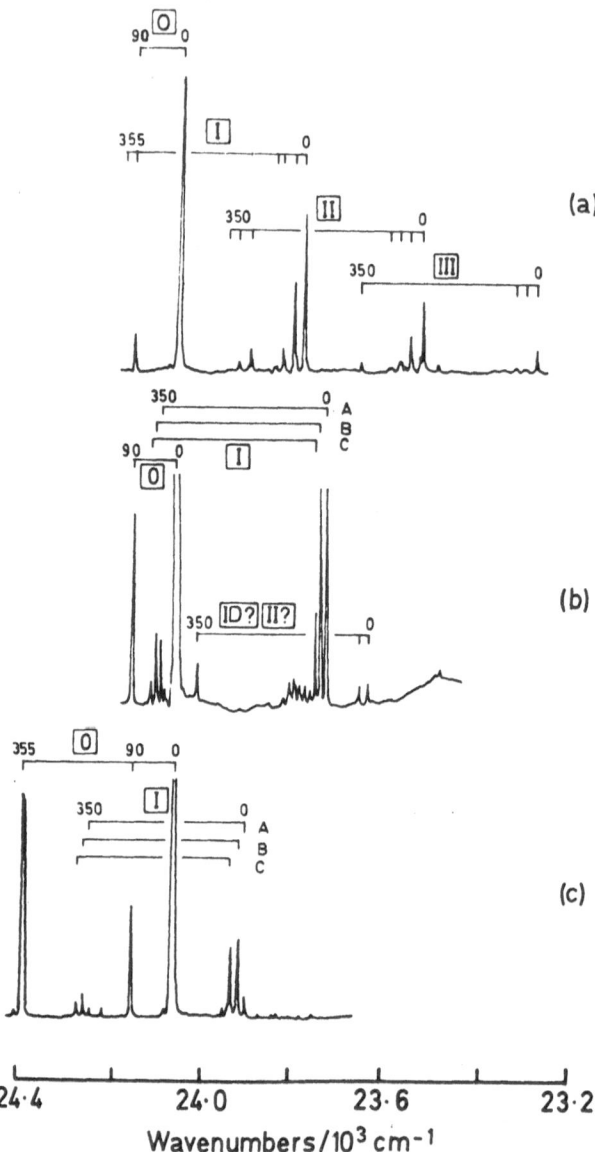

Figure 2   Fluorescence Excitation Spectra of Perylene vdW
            complexes with (a) carbon tetrachloride,
            (b) bromobenzene and (c) ethanol

TABLE I

Parameters for Atom-Atom Potentials

Atom-atom pair	C^e	$B^e/10^3$
C-Ne[a]	928.8	839.2
C-Ar[a]	3100.0	4969.0
C-Kr[a]	4071.0	7273.0
C-Xe[a]	6569.0	16540.0
H-Ne[a]	702.9	406.6
H-Ar[a]	2343.0	2550.0
H-Kr[a]	3071.0	3814.0
H-Xe[a]	4937.0	8553.0
C-C[b]	1558.0	2220.0
C-H[b]	5600.0	355.0
H-H[b]	201.0	56.8
C-O[c]	962.2	787.0
H-O[c]	337.0	120.0
C-Cl[d]	2210.0	2340.0
H-Cl[d]	804.0	3840.0

data taken from a reference 12, b reference 22
c reference 23 d reference 24
e units are kJ mol^{-1}

The atom-atom parameters C^g_{uv} and B^g_{uv} were taken from literature values,[12, 22-24]   applying combination rules where necessary.   Their values are listed in table I.   The charge densities for the Coulomb interaction terms were taken from calculated values given in the literature[13,25-28].   Where no literature values were available estimates were used.   In fact the Coulomb terms were found to contribute only ~5% of the value of ΔE^g so that any errors in the values of the net changes should not be significant.

Perylene was assumed to be planar, with bond lengths and angles taken from Dallinga[29] et al.   Solvent molecular geometries were taken from the literature[30-36].   Ethane was assumed to adopt the more stable staggered configuration and a chair conformation was assumed for cyclohexane and its methyl derivative.   In the latter case, the equatorial form with the methyl H atoms eclipsed to the ring C-H bonds was adopted[32].   Both the staggered and eclipsed forms of ethanol were used since there appears to be some uncertainty on their relative stabilities[33].

Cartesian axes were centred on the perylene molecule with x and y lying in plane and directed along the short and long axes respectively. Symmetry axes in the solvent were generally assumed to lie parallel to the cartesian axes though this constraint was relaxed in some trial calculations.   In the case of benzene, configured with its plane parallel to that of the perylene, rotation of the benzene molecule about the vertical (z) axis minimised ΔE^g when the two stacked hexagonal rings

TABLE II

Solvent Shifts, Binding Energies and Equilibrium Geometries for the 1:1 perylene vdW complexes

Solvent	Red spectral shifts/cm^{-1}	Ionisation Energy/eV[a]	Co-ordinates of Centre of Mass/Å[b] x	y	z	Dispersion Energy E_D^g/kJ mol^{-1}	Binding Energy $-\Delta E^g$/kJ mol^{-1}
Ne	11.1[c]	21.56	0.0	0.0	3.05	-5.75	3.23
Ar	52.1[c]	15.76	0.0	0.0	3.39	-12.31	6.88
Kr	76.4[c]	14.00	0.0	0.0	3.46	-14.81	8.30
Xe	116.7[c]	12.13	0.0	0.0	3.68	-18.31	10.21
CH$_4$	82.0[c]	12.70	0.0	0.0	3.31	-18.83	10.90
Cyclohexane	230.0	9.88	-0.1	0.0	3.93	-42.11	25.24
methyl-cyclohexane	290.0	9.85	-0.3	0.0	4.04	-45.07	28.37
benzene	380.0	9.25	0.0	0.0	3.32	-50.44	26.97
CCl$_4$	260.0	11.47	2.0	0.2	3.79	-32.18	19.38
ethanol							
staggered	150.0	10.48	-0.8	-0.3	3.51	-28.53	17.23
eclipsed	170.0[d]	10.48	-1.3	+0.7	3.52	-29.31	20.70
cyclopropane	156.0[d]	10.09	0.0	-1.1	3.61	-26.04	14.00
ethane	133.0[d]	11.52	0.0	-0.7	3.32	-26.71	12.81

a   Values taken from ref.37
b   Perylene lies in the z=0 plane with its short and long axes along the x axis and y axis respectively and with its centre of mass at the origin. Ionisation energy of perylene is 7.03 eV
c   Values taken from ref.9
d   Values taken from ref.38

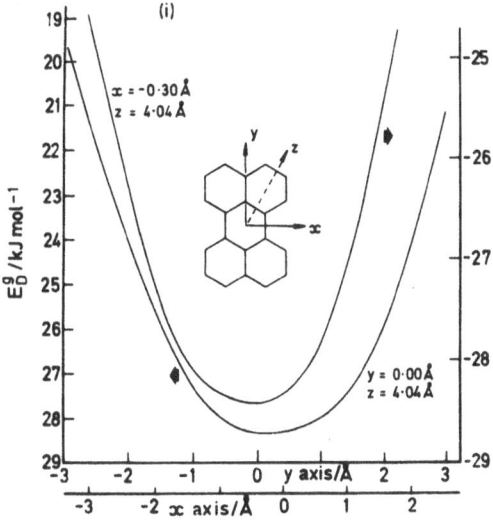

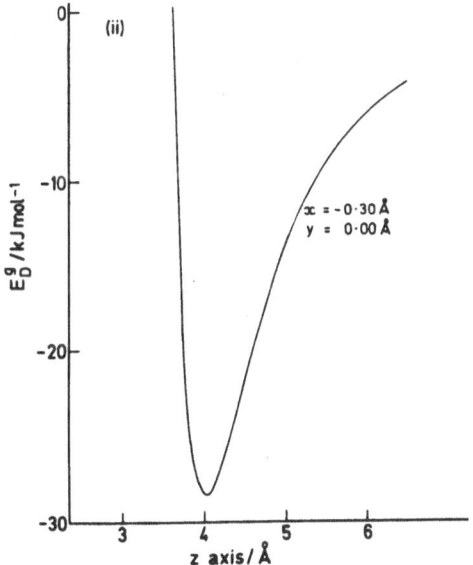

Figure 3   Potential Interaction of Perylene with
            Methylcyclohexane, Potential Energy as a
            function of (i) x and y coordinates,
            (ii) z coordinate

were staggered, i.e. rotated by 30° from the most symmetric (eclipsed)
conformation.

Table II shows the calculated equilibrium geometries and binding
energies for several 1:1 complexes. Despite the major assumptions made
in the potential calculations, and although the parameters used in the
computations are subject to significant uncertainty, the calculated
geometries and binding energies do provide useful qualitative
predictions. The potential-energy curves for methylcyclohexane (figure
3), cyclohexane and benzene show only one minimum for displacement along
the x, y or z directions. In the preferred geometry the ·centre of mass
of the solvent molecule is located almost directly over the central ring
of the perylene molecule. In these cases only one conformation is
predicted to be energetically stable, largely consistent with the
experimental observations on the fluorescence excitation spectra.
(Relaxation of the 'chair' constraint could accommodate an alternative
conformer in the cyclohexane systems). Potential-energy calculations
for the 1:1 perylene-carbon tetrachloride complex predict the most
stable configuration to be one where the solvent molecule is not bound
centrally but is displaced along the x (short) molecular axis with a
single carbon-chlorine bond directed vertically down toward the perylene
molecule. The nearly harmonic shifts observed in the fluorescence
excitation spectra for the higher perylene $-(CCl_4)_n$ complexes and the
lack of any evidence for interaction between the solvent molecules,
strongly suggests that this preference is maintained when further
solvent molecules are bound, with the possibility of two equivalent
sites on each side of the perylene ring system. This may also account
for the apparent ease with which the higher complexes are formed. The
potential energy calculations for ethanol predict the eclipsed to be
marginally more stable than the staggered conformer, with the centre of
mass for ethanol displaced along the y (long) molecular axis.

(3) Spectral Shifts

Theories of solvent spectral shifts, for example, that put forward by
Longuet-Higgins and Pople[14] (LHP), were originally developed for liquid
solvents where, for non-polar solutes, it was assumed that the shift was
caused by the difference in the dispersion term. Some authors[2,7,9],
including ourselves, have modified the LHP method to produce theories
applicable to vdW spectral shifts, making a similar assumption so that
the change, $E_1^g - E_1^s$, in the repulsive and electrostatic part of the
potential is neglected in the belief that it is likely to be small.
Here, we shall continue to neglect this change in E_1, although we feel
that some numerical estimate of its magnitude is called for.

Considering the change in E_D only, the red spectral shift $\delta\nu_{ge}$ will
be

$$\delta\nu = \frac{1}{\bar{E}_e + \bar{F}_{solv}} \sum_u \sum_v \frac{D_{uv}^e}{r_{uv}^6} - \frac{1}{\bar{E}_g + \bar{F}_{solv}} \sum_u \sum_v \frac{D_{uv}^g}{r_{uv}^6} \qquad (9)$$

where $\bar{E}_g$ and $\bar{E}_e$ are the Unsöld average energies for the ground (g) and excited (e) states of perylene. To make progress, it is necessary to estimate the atom-atom interaction constants D_{uv}^e when perylene is in its excited state. However, unlike the ground-state constants D_{uv}^g, there is insufficient experimental information to determine them empirically. Therefore, we make the assumption that they are all unchanged from the ground-state values, equivalent to using the same value for the term in brackets in equation (5), for the two cases $s = g$ and $s = e$. The same assumption is also employed in theories based upon LHP formalism. After manipulation we find

$$\delta\nu = \frac{|E_D^g|}{\bar{E}_g + \bar{F}_{solv}} \, \Delta_e = \frac{\Delta_e}{\bar{E}_g + \bar{F}_{solv}} \sum_u \sum_v \frac{C_{uv}^g}{r_{uv}^6} \tag{10}$$

where $\Delta_e = \bar{E}_g - \bar{E}_e$ is assumed small compared with $\bar{E}_g$ so that terms in Δ_e^2 can be ignored. Equation (10) relates the solvent shift to the ground state dispersion energy as calculated from equation (7). Previous theoretical formulae based on the LHP method can be recovered from (10) by replacing each atom-atom distance r_{uv} by an average intermolecular distance r and invoking the London approximation for E_D^g. However, we believe that equation (10) is more satisfactory for two reasons. It is more consistent to use the same procedure for determining E_D^g in both the potential energy and the solvent shift calculations. When large molecules interact there are wide variations in the values of the atom-atom distances and it is a poor approximation to replace all of them by an average value. To put this another way, the London formula for E_D^g in the LHP method holds only for molecules where intermolecular distances are large compared with intra-atomic distances and this condition is certainly not satisfied for the perylene vdW complexes we consider here.

We have discussed elsewhere[9] the problem of choosing values for the average energies. Here we adopt the usual choice of ionization energies for $\bar{E}_g$ and $\bar{F}_{solv}$ but regard Δ_e as an empirical parameter. To test equation (10), the observed microscopic red shifts for 1:1 perylene-solvent complexes (see Table 2) are plotted against $E_D^g/(\bar{E}_g + \bar{F}_{solv})$ in figure 4. The data fall on a reasonably straight line. The slope of this line gives $\Delta_e = 1.42$ eV which is about 47% of the perylene transition energy and corresponds to an average energy $\bar{E}_e = 5.61$ eV for the perylene excited state. This is close to the average of the ionization energies of the ground and excited state but marginally closer to the former. Note that the best fit straight-line does not quite pass through the origin, as is predicted by (10), but the intercept, $\delta\nu_{ge} = -16$ cm^{-1} is very small.

We conclude from this that a simple extension of the atom-atom potential model leads to a theoretical treatment of spectral shifts in vdW molecules, which, in spite of obvious limitations, can be a useful tool for evaluating and correlating the experimental data.

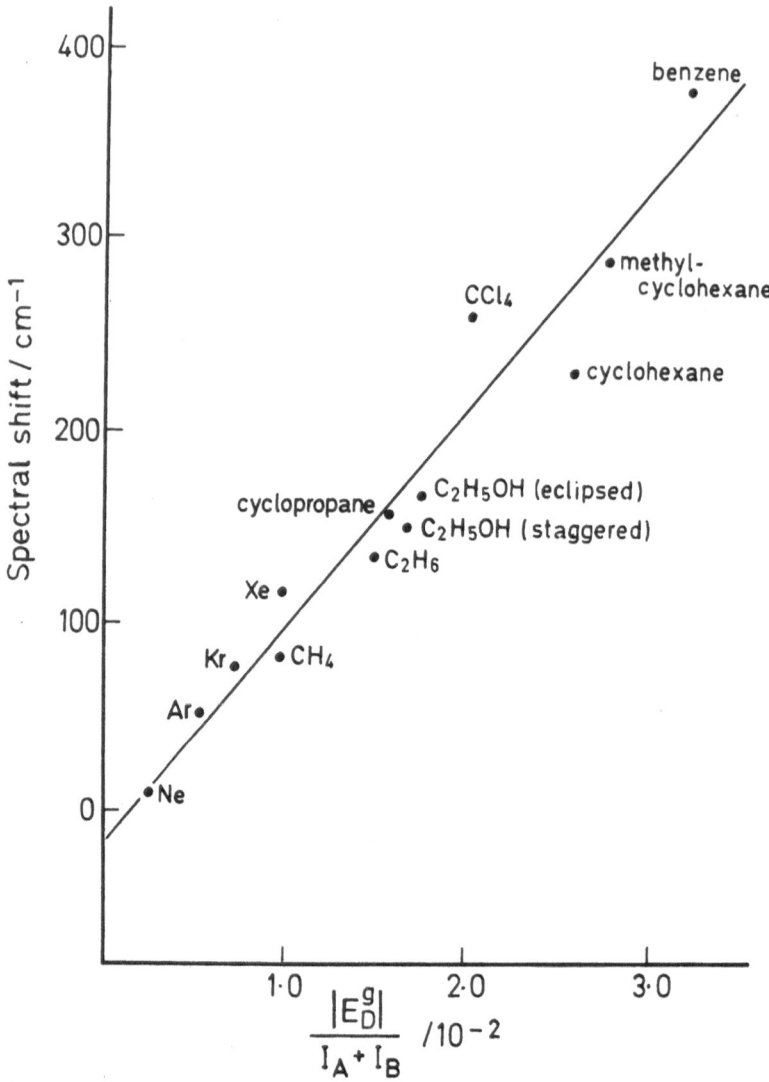

Figure 4   Plot of Observed Red Shift against $|E_D^g|/(I_A + I_B)$
where E_D^g is the calculated dispersion energy
for the ground state of the vdW complex and I_A and
I_B are respectively the ionisation energies of
perylene and the solvent molecule

Acknowledgment
We thank the SERC for an equipment grant and an award to JCK, Mrs M
Krause for preparation of the typescript and Mr F Whetstone for
preparation of the diagrams
** Present address for S M Cohen: Chemistry Department
                                    Rice University
                                    P O Box 1892
                                    Houston, TX 77251
                                    U.S.A.

References

1. R Smalley, L Wharton, D H Levy and D W Chandler, J. Chem.Phys. 68
   (1978) 2487
2. W E Henke, W Yu, H L Selzle, E W Schlag, D Wutz and S H Lin,
   Chem.Phys. 92 (1985) 187
3. A Amirav, U Even and J Jortner, J.Phys.Chem. 86 (1982) 3345
4. A Amirav, U Even and J Jortner, J. Chem.Phys. 75 (1981) 2489
5. I Raitt, A M Grifiths and P A Freedman, Chem.Phys. 63 (1981) 469
6. A Amirav, U Even and J Jortner, J.Phys.Chem. 85 (1981) 309
7. S Leutwyler, J. Chem.Phys. 81 (1984) 5480
8. J C Kettley, T F Palmer and J P Simons, Chem.Phys.Letters 115 (1985)
   40
9. A T Amos, J C Kettley, T F Palmer and J P Simons, Chem.Phys.Letters
   126 (1986) 107
10. M M Doxtader, I M Gulis, S A Schwartz and M R Topp,
    Chem.Phys.Letters 112 (1984) 483
11. S A Schwartz and M R Topp, J.Phys.Chem. 88 (1984) 5673
12. M J Ondrechen, Z Berkovitch-Yellin and J Jortner, J.Am.Chem.Soc. 103
    (1981) 6586
13. M M Doxtader, E A Margle, A К Bhattacharya, S M Cohen and M R Topp,
    Chem. Phys. 101 (1986) 413.
14. H C Longuet-Higgins and J A Pople, J.Chem.Phys. 27 (1957) 192
15. A Unsöld, Z Phys. 43 (1927) 563
16. A T Amos and R J Crispin, Theoretical Chemistry, Advances and
    Perspectives, Vol.2, Eds. H Eyring and D Henderson, Academic Press,
    New York (1976) 1
17. P Claverie, Intermolecular Interactions from Diatomic to Biopolymer
    Ed. B Pullman, Wiley, New York (1978)
18. A van der Avoird, P E S Wormer, F Mulder and R M Berns, Topics in
    Current Chemistry, 'Van der Waals' Systems, Ed. R Zahtadnik,
    Springer, Berlin (1980)
19. A D Crowell and R B Steele, J.Chem.Phys. 34 (1961) 1347
20. J O Hirschfelder, C F Curtiss and R B Bird, Molecular Theory of Gas
    and Liquids, Wiley, New York, (1954)
21. C Bouzou, C Jouvet, J B Leblond, P L Millie, A Tramer and M Sulkes,
    Chem.Phys.Lett. 97 (1983) 161
22. L Battezzati, C Pisani and F Ricca, J.Chem.Soc. Faraday 2, 71 (1975)
    1629
23. F H Stillinger and A Rahman, J.Chem.Phys. 60 (1974) 1545

24. F Serrano Adan, A Banon and J Santamaria, Chem.Phys.Lett. **107** (1984) 475

25. J H Miller, W G Mullaard and K C Smyth, J.Phys.Chem. **88** (1984) 4963

26. K B Wiberg and J J Wendoloski, J.Am.Chem.Soc. **100** (1978) 723

27. N Cyr, A S Perkin and M A Whitehead, Can.J.Chem. **50** (1972) 814

28. S Melberg and K Rasmussen, J.Mol.Structure **57** (1979) 215

29. C Dallinga, L H Toneman and M M Tretteberg, Rec.Trav.Chim., **86** (1967) 795

30. H Huber, Theoret. Chim Acta **55** (1980) 117

31. H J Geise, H R Buys and F C Mijlhoff, J.Mol.Structure **9** (1971) 447

32. S Fitzwater and L S Bartell, J.Am.Chem.Soc. **98** (1976) 5107

33. E R Talaty and G Sinions, Theoret.Chim.Acta **48** (1978) 331

34. U Burket, Tetrahedron **35** (1979) 209

35. Tables of Interatomic Distances, Special Publ. # 11, The Chemical Society, London (1958) M104

36. CRC Handbook of Chemistry and Physics 60th Edition Ed. R C Weast. CRC Press Inc. Boca Raton, Fl. U.S.A. (1979)

37. J G Dillard, K Draxl, J L Franklin, F H Field, J T Herron and H H Rosenstock NSRDS - NBS **26** (1969)

38. M M Doxtader and M R Topp, J. Phys.Chem. **89** (1985) 4291

SPECTROSCOPY IN THE VISIBLE AND NEAR ULTRAVIOLET REGION OF SOME
ORGANIC MOLECULES AND THEIR VAN DER WAALS COMPLEXES

W.M. van Herpen, W.A. Majewski[+], D.W. Pratt[+] and W.L. Meerts
Physics Laboratory, University of Nijmegen,
Toernooiveld, 6525 ED Nijmegen, The Netherlands

[+] Present address: Department of Chemistry, University of
Pittsburgh, Pittsburgh, PA 15260, USA

ABSTRACT. By using a molecular beam apparatus with a single frequency
dye laser we were able to resolve several rovibronic transitions of
some large molecules and their Van der Waals complexes with noble gas
atoms. The rotational constants of tetracene, fluorene and the fluorene-
argon complex have been determined. The structure of the complex was
derived. In the rotational spectra of the tetracene-noble gas complexes
perturbations were shown to be present, increasing with the size and
number of the attached atoms. Fluorescence excitation spectra of trans-
stilbene and the stilbene-argon complex are reported.

1. INTRODUCTION

Detailed spectroscopic information is indispensible in the study of
Van der Waals (VdW) molecules. In recent years, much attention has
been focussed on the understanding of structure, dynamics and energetics
of small VdW complexes. Experimental data allow for realistic description of potential energy surfaces and relaxation phenomena. For larger
complexes the density of states increases rapidly and so does the complexity of its spectra. Less information is therefore available on
these molecules and model calculations lack of experimental verifica-
tion. The insight on potential surfaces and solvent effects is rather
limited.
     To obtain experimental information on the structure of the com-
plexes one needs rotationally resolved spectra. If single-frequency
lasers are used, spectral resolution is limited by the Doppler-width
of the spectral lines. One can either use Doppler-free techniques in
cell experiments [1] or use a free jet expansion. With such an expansion and by strongly collimating the molecular beam the experimental
linewidth can be reduced to a few MHz in the visible region. The seeded
beam technique adds the advantage of a considerable cooling of internal
degrees of freedom of the molecule. Moreover, the high density of VdW
complexes in the beam of the parent molecule with the seeding gas is
very convenient.

A. Weber (ed.), Structure and Dynamics of Weakly Bound Molecular Complexes, 279–290.
© 1987 by D. Reidel Publishing Company.

We report the high resolution spectroscopy of three large molecules and their VdW complexes with rare gas (R) atoms. Fluorescence excitation spectra have been obtained of fluorene (F), tetracene (T) and trans-stilbene (tS). The rotational constants of both F and the F-Ar complex have been determined in ground and excited electronic state. From this the structure of the F-Ar complex has been deduced and compared with model calculations. Rotational bands of T were also assigned. For the T-Ar, T-Kr and T-Xe VdW complexes perturbations in the excited state were shown to exist, depending on energy and size of the attached rare gas atom. The tS molecule shows complicated rovibronic spectra. Isomerisation in the excited state and internal rotation will be heavily influenced in a VdW complex. The observed bands of the tS-Ar complex are quite different from those of the tS parent molecule. This is attributed to the steric effect of the Ar-atom in the complex.

2. EXPERIMENTAL

In our studies on large molecules and their VdW complexes we applied the seeded beam technique. A detailed description of the apparatus can be found elsewhere [2,3]. The source (figure 1) of the molecular beam is made of quartz. It is wrapped with heating wires. Together with a

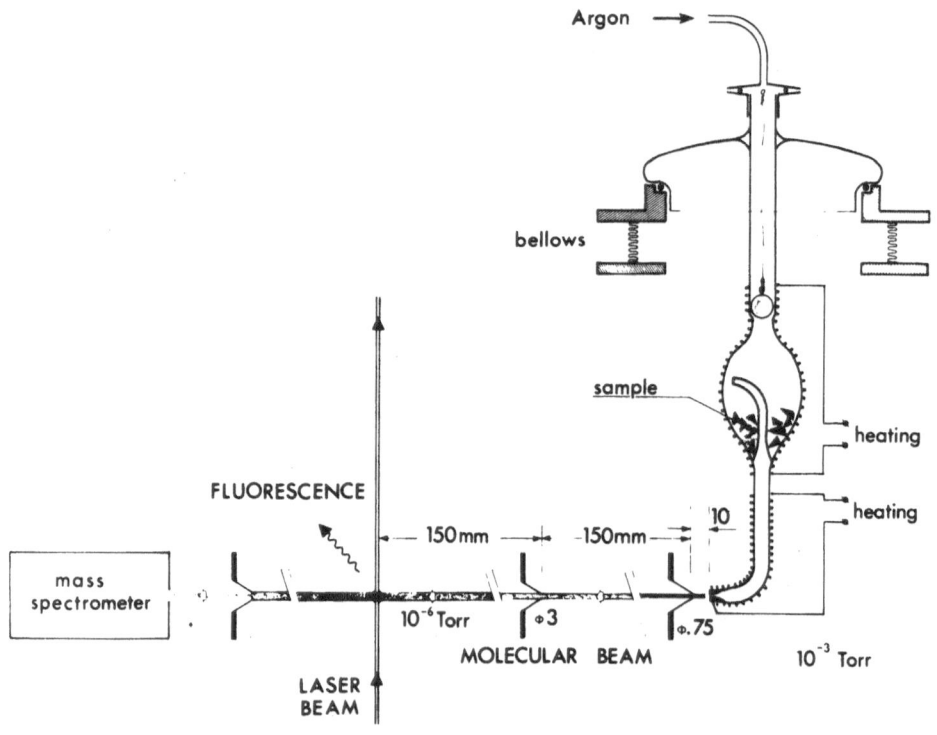

Figure 1. Schematic view of the source and molecular beam.

few thermocouples this allows temperature control of the source. The sample is heated in a reservoir compartment to raise its vapour pressure to typically 1 Torr. A mixture of the vapour and the seeding gas is expanded (at a pressure ranging up to 4 bar) by a circular nozzle of 100 μm diameter. Two conical skimmers in a two-step pumping system are used to collimate the molecular beam. This results in a residual Doppler width of about 15 MHz at UV wavelengths. The interaction zone with the laser is at 30 cm from the beam orifice. Here, the undispersed LIF is imaged to the photocathode of a photomultiplier tube (EMI 9863QA). To suppress background signal from scattered laser light, the molecular beam is chopped and phase sensitive detection is applied. We used a standard photon counting system, interfaced with a computer (PDP11/23 plus). The computer largely expands the dynamic range of the stored data. In the analysis accurate corrections can be made for non-linearity in the scan. Moreover, the computer facilitates the tedious determination of position and intensities of the numerous spectral lines.

For the experiments described in this paper, we applied two different laser systems. The measurements on tetracene were performed with a standing wave linear dye laser (Coherent Radiation 591) operating in the blue spectral region. For fluorene and trans-stilbene we used a frequency doubled modified ring dye laser (Spectra Physics 380D). The doubling crystal, made of $LiIO_3$, was placed intracavity and angle tuned [4]. Both lasers operated single frequency with a bandwidth below 3 MHz. For relative frequency measurements temperature stabilized, sealed-off Fabry-Perot interferometers were used with accurately gauged free spectral ranges. The absolute frequency calibration in the UV is most easily performed by the fundamental wavelength of the dye laser. During a scan a small fraction of the fundamental laser power is used to measure the absorption spectrum in an iodine cell as a reference [5]. For the measurements on T we employed a home-built wavelength meter of the Michelson interferometer type [3]. This apparatus compares the laser wavelength with the accurately known wavelength of a stabilized He-Ne laser.

3. RESULTS AND DISCUSSION

3.1. Fluorene

We studied [6] the $S_1(^1B_2) \leftarrow S_0(^1A_1)$ transition in F and the related transition in the F-Ar VdW complex under rotational resolution. The 0^0_0 band in F exhibits a strong spectrum around 296 nm. The observed linewidth amounted to 15 MHz, due to the residual Doppler width in our spectrometer. Even most of the Q-branch transitions were resolved. The band consists of an a-type transition, corresponding to a transition dipole moment along the long molecular axis. A total of 225 lines in a central region of 20 GHz were assigned and fitted to an asymmetric rotor model. The fit proved excellent, with a standard deviation of 7.5 MHz. The rotationless transition frequency ν_0 was obtained and the rotational constants both in ground and excited state (table 2). All

TABLE 1   Rotation free transition frequencies (ν_0) and relative shifts ($\Delta\nu = \nu - \nu_0$) for fluorene (F), tetracene (T),
trans-stilbene (tS) and their Van der Waals complexes.

	ν_0 (cm^{-1})	$\Delta\nu_0$ (cm^{-1})
F (0-0)	33 775.547(5)	
F-Ar	33 731.595(5)	-43.952(3)
T (0-0)	22 396.53(2)	
T-Ar		-41.67(5)
T-Kr		-66.9(1)
T-Xe		-110.0(1)
T-Ar$_2$		-80.6(1)
T-Kr$_2$		-124(1)
T (311 cm^{-1})	22 707.84(2)	
T-Ar		-41.42(5)
T-Kr		-66.5(1)
T-Ar$_2$		-79.9(1)
T-Kr$_2$		-123.4(1)
T (471 cm^{-1})	22 867.62(2)	
T-Ar		-41.52(5)
T-Kr		-66.9(1)
T-Ar$_2$		-80.0(1)
tS (0-0)	32 234.05(5)	
tS-Ar		-40.03(5)
tS-Ar		-63.04(5)

TABLE 2   Rotational constants (MHz) of fluorene, the fluorene-
argon complex and tetracene in the ground state S_0
and the first excited electronic state S_1 ($\Delta A = A' - A''$,
etc.)

		fluorene	fluorene-argon	tetracene
S_0	A"	2 183.2(33)	811.1(29)	1 630(1)
	B"	586.520(69)	468.58(14)	213.4(2)
	C"	463.239(65)	401.58(13)	188.8(2)
S_1	ΔA	-73.387(14)	-1.402(27)	17.4(12)
	ΔB	6.716(38)	1.437(31)	-1.81(8)
	ΔC	0.734(41)	4.961(26)	-1.19(8)

lines, even the weak ones, could be accounted for. From a fit of the
line intensities with the rotational temperature as adjustable parameter
we obtained a temperature T_{rot}=2.3(3) K for the molecules in the beam.
    The observed transition, shifted -44 cm^{-1} (table 1) with respect
to the F 0_0^0 band was assigned to the F-Ar VdW complex (figure 2). Since
this complex has smaller rotational constants than the parent molecule,
its spectrum is more dense. Nevertheless the P and R branch are still
well resolved. The linewidth was again 15 MHz. In the central 23 GHz
region of the band we assigned 150 lines. A fit to the asymmetric rotor
Hamiltonian yielded ν_0 and the rotational constants as given in table 2.
Again the fit proved to be very satisfactory, with a 7.6 MHz standard
deviation. Analysis of the spectral intensities showed a single
Boltzmann distribution with a rotational temperature of T_{rot}=2.1(4) K.
Within experimental accuracy the rotational temperatures of the parent
molecule and the complex are thus the same.

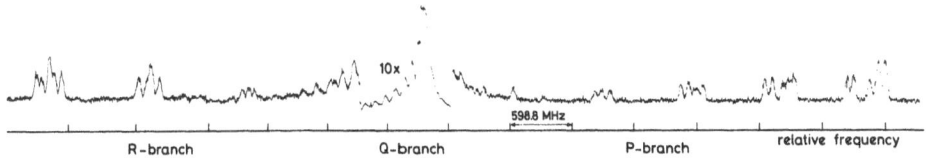

R-branch   Q-branch   P-branch   relative frequency

Figure 2. Part of the fluorene-argon rotational spectrum around
33 731.595(5) cm^{-1}. The frequency markers are spaced 598.64 MHz.

    The position of the argon in the complex can be determined [6]
from a comparison of the moments of inertia along the principal axis of
the parent molecule and the complex. It follows that the argon is
located in the bisecting plane of the molecule at a distance from the
center of mass of fluorene of r=(3.46±0.03) Å with θ=±(8.8±1.0)°,
where θ is the angle with the z-axis. The ambiguous sign of θ arises
from the fact that the moment of inertia tensor depends quadratic on
the relative argon coordinates. This problem can be solved by studying
the rovibronic spectrum of a deuterated F-Ar complex. Such a study is
presently carried out.
    We performed a model calculation of the F-Ar complex using a
Lennard-Jones 6-12 potential with pairwise interactions. Three-body

interactions were neglected. This method was described by Ondrechen et al. [7]. The calculated absolute minimum is above the center ring of the molecule at the argon coordinates $(x_0, y_0, z_0) = (0.13, 0, 3.48)$ Å, with respect to the center-of-mass coordinate system of F. Figure 3 shows two cuts through the coordinates of the potential minimum. The experimentally determined positions are very near this minimum.

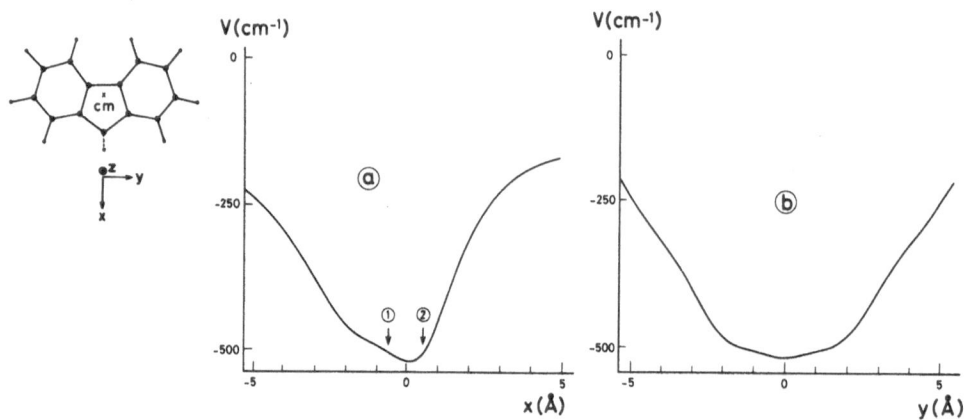

Figure 3.   Rotational energy curves for a single argon with fluorene. The position of the argon is given in the center-of-mass coordinate system of fluorene. (a) Minimal potential energy as a function of x (y=0 and z is adjusted); the two possible experimental positions are indicated. (b) Energy as a function of y (x=0.13, z=3.48).

3.2. Tetracene

We obtained fluorescence excitation spectra of the $S_1(^1B_{2u}) \leftarrow S_0(^1A_g)$ electronic transition of tetracene [8,9]. The rotational band of the 0-0 vibrationless transition was resolved as well as vibronic bands belonging to the 311 cm^{-1} and 471 cm^{-1} vibrational modes in the excited S_1 state. The spectra of the 0-0 and 311 cm^{-1} bands are almost identical. They were both assigned as a b-type transition, corresponding with a transition dipole moment along the short molecular axis. The spectra consist of hundreds of strong spectral lines, spreading over a wide range. The 471 cm^{-1} band is connected to a not totally symmetric vibrational mode. The transition to this mode is symmetry forbidden but becomes weakly allowed by a coupling with the $S_2(^1B_{1u})$ state. The band becomes long axis polarized and shows an a-type structure. A total of 65 lines was assigned in the central part of the b-type perpendicular transitions and 160 lines in the a-type parallel band. A fit to the asymmetric rotor model was excellent. The obtained rotational constants for the different transitions did not differ within their statistical undertainty. The values given in table 2 stem from combined data of the

various bands. To our knowledge T is the largest molecule reported so
far with a fully resolved and assigned rotational spectrum. From the
intensity distribution in the 471 cm^{-1} band, where a number of single
lines were observed, we determined the rotational temperature in the
molecular beam. A single Boltzmann distribution was found with $T_{rot}=$
2.3(3) K. All strong spectral features in the different bands could be
accounted for. The linewidth in the spectra is 15 MHz, and is determined by the residual Doppler width of the spectrometer.

The assignment of transitions in tetracene-noble gas VdW complexes
has been reported by Amirav et al. [10,11]. The complexes have transitions, more or less regularly red shifted with respect to the related
transition in the parent molecule. We explored VdW molecules containing
argon, krypton and xenon. These are readily formed in the molecular
beam by using the noble gas as seeding gas. We performed model calculations of the T-R (R=Ar,Kr,Xe) complex, using a Lennard-Jones potential
[7]. The distance between the R-atom and T was adjusted for minimum

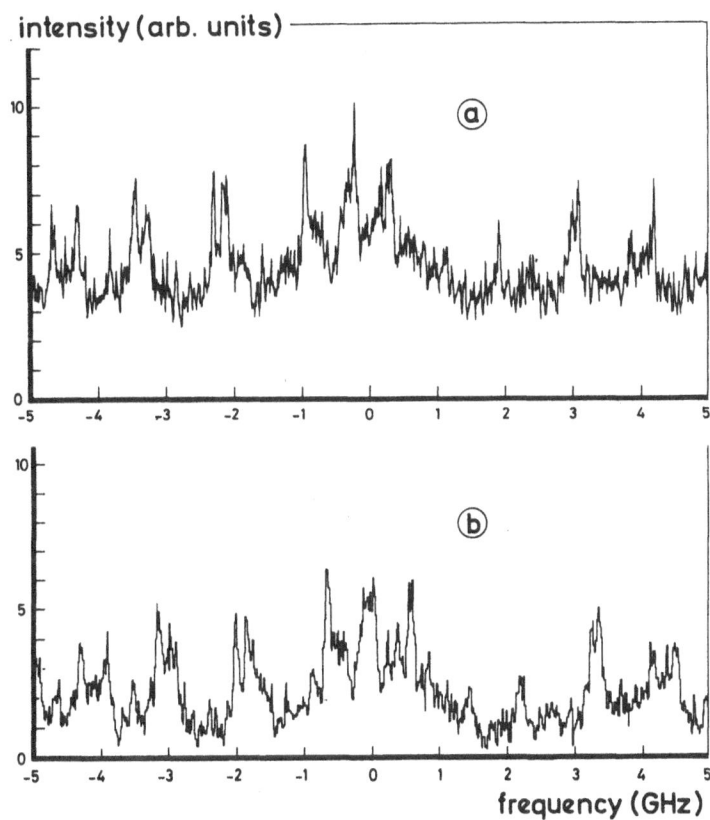

Figure 4. Rotationally resolved spectra of tetracene-argon. (a) The
311 cm^{-1} band; (b) the 0_0^0 transition.

energy. The absolute minima are found above the inner rings of T. The
potential depicts a double well structure along the long molecular axis.
Consequently a tunneling motion may exist of the R atom along the molecular plane, resulting in doubling of the rotational bands. The potential barrier is most pronounced for argon and very shallow for the other
considered noble gases. Other potential minima were shown to be absent
and the existence of chemical isomers of the complex is thus unlikely.
A study of the rotational spectra provides a test of the calculated
potential surface.

High resolution excitation spectra were observed of T-R (R=Ar,Kr,
Xe) complexes. In figure 4 and 5 rotationally resolved spectra are
shown of T-Ar and T-Kr. The 0-0 band of T-Xe is not indicated. It consists of some small humps on a broad background. Other bands of T-Xe
are of very low intensity. The red shifts of the complexes (table 1)
differs for argon, krypton and xenon, but is almost constant for different bands in a complex. This strongly supports the indentification. The
0^0_0 and 311 cm^{-1} band in T are identical. In T-Ar they still look much
alike (figure 4) with a clear perpendicular structure. In T-Kr these
bands, however, are quite different, with an almost unrecognizable
structure (figure 5). The 471 cm^{-1} band in T-Ar is of parallel shape
like it is in the parent molecule. The orientation of the a-axis in
both molecules is thus very much alike. The weak 471 cm^{-1} band in T-Kr
has no recognizable pattern. It contains a small hump on a broad structureless background. The linewidth in the T-Ar spectra is estimated at
15 MHz and limited by our spectrometer. For T-Kr and T-Xe complexes the
fluorescent lifetime has diminished to 11 ns and 1.5 ns respectively
[12] and no individual lines were observed. It is clear that the T-Kr
and T-Xe spectra show no ordinary rotational shape but are heavily perturbed. This anomalous structure is too pronounced to be caused by the
different isotopes of krypton or xenon. For T-Ar perturbations are not
so clear. However, despite much effort we could not assign the spectra.
There seems to be an excess of lines in the parallel band, while the
sequence of Q-branches in the perpendicular band looks distorted. All
bands are observed as single transitions. We did not find evidence for
a splitting due to a tunneling type motion of the inert gas atom. Such
splitting either does not exist or it is very small. Also spectra of
the T-Ar$_2$ and T-Kr$_2$ complexes have been observed. The red shift of a
T-R$_2$ complex with respect to T is not twice the shift of a T-R molecule
(table 1). The transitions in the T-R$_2$ complex qualitatively behave in
the same way. They show hardly any structure and consist of a broad
flat band. No individual lines were observed.

We conclude that the perturbations in the complexes increase with
excited state energy, with the number and with the mass of the attached
noble gas atom. The effects may be caused by S-T intersystem crossing,
S_1-S_0 state mixing or even with a coupling of the S_1 state with higher
energy levels. The latter interaction is present in the T molecule as
the 471 cm^{-1} vibrational band gains its intensity by a coupling of the
S_2 state.

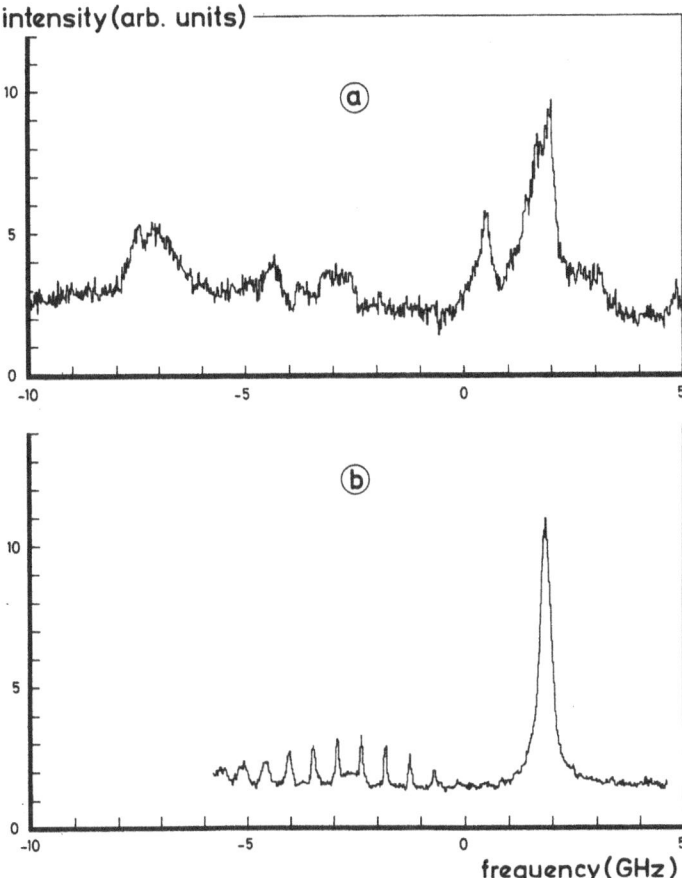

Figure 5. Comparison between the 311 cm^{-1} (a) and 0_0^0 (b) rovibronic spectra of tetracene-krypton.

3.3. Trans-stilbene

The geometry and photo-isomerization of stilbene (diphenylethylene) forms an intriguing problem in physical chemistry. The energy of the electronic states strongly depends on the angle θ of rotation about the ethylenic bond. The first excited S_1 electronic state with minima at the trans ($\theta=0°$) and cis ($\theta=180°$) configuration is crossed by the S_2 state, which has an absolute minimum at the perpendicular ($\theta=90°$) geometry. In the S_0 ground state the trans and cis configuration are separated by a large energy barrier and a perpendicular configuration does not exist. It was shown [13] that the S_0 ground state of tS has phenyl groups rotated by about 30° with respect to the ethylene plane and is thus non-planar.

An experimental assignment of the vibrational modes in the S_0 and S_1 states was presented by Syage et al. [14] and by Zwier et al. [15]. The fluorescent lifetimes [16] strongly decrease at excited state vibrational energies above 1200 cm^{-1}. We measured high resolution fluorescence spectra of tS and the tS-Ar VdW complex to obtain more detailed information on the geometry and internal motion of both molecules. A total of 8 vibronic transitions in tS has been studied under rotational resolution in a molecular beam experiment. The excited state vibrational energy ranges up to 1447 cm^{-1}. The tS 0-0 vibrationless band is depicted in figure 6. It appeared that the observed linewidth in all considered bands of tS exceeded the instrumental linewidth and amounted about 80 MHz in the lower vibrational states. These observed widths conform with lifetime measurements [16]. At higher energies (i.e. above 1200 cm^{-1}) the spectra become congested, probably due to an increased linewidth. The bands considered all show very similar spectra with only detailed differences.

One can estimate the moments of inertia and consequently the rotational constants of tS from the theoretical geometry. The molecule with phenyl groups rotated by about 30° forms an almost prolate symmetric top. However, the observed spectrum of for example the 0-0 band (figure 6) does not resemble a symmetric top transition. The experimental density of lines is higher than expected. Although assignment of the spectrum

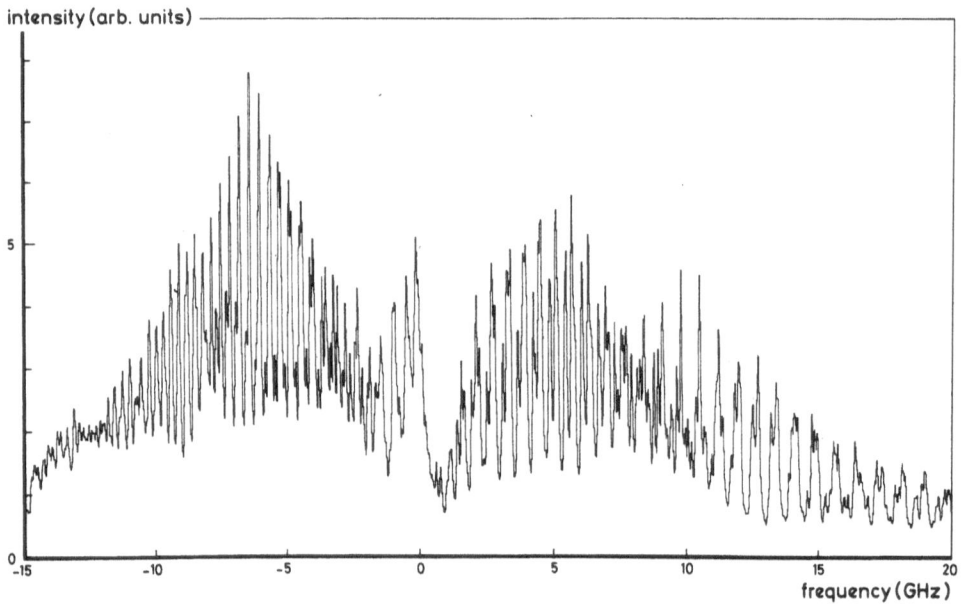

Figure 6. Fluorescence excitation spectrum of trans-stilbene 0_0^0, around 32 234.05 cm^{-1}.

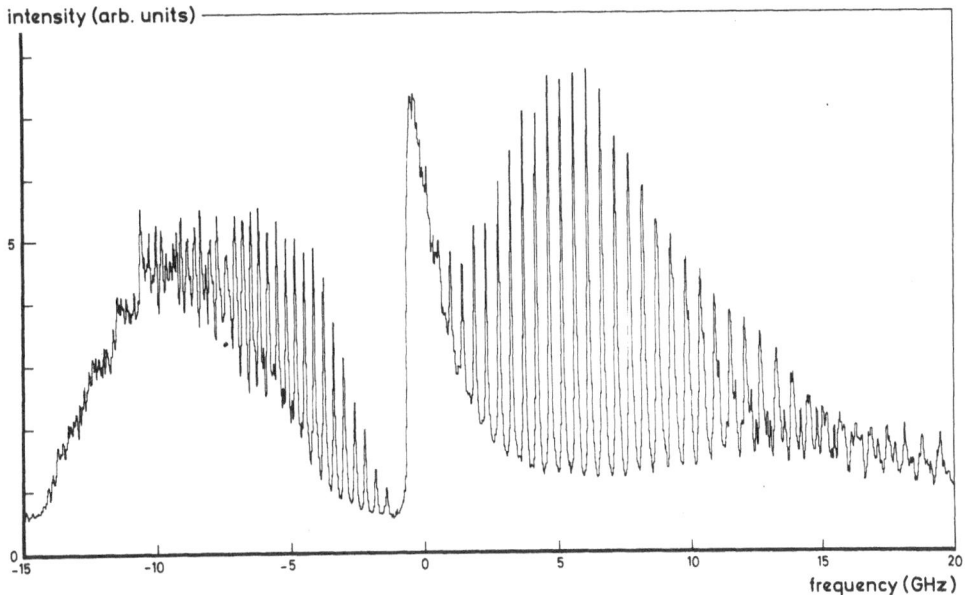

Figure 7. Fluorescence excitation spectrum of trans-stilbene-argon. The band is shifted -63.04(5) cm^{-1} with respect to stilbene 0^0_0.

has not yet been accomplished, we tend to the conclusion that the tS spectra show evidence of an internal motion of the phenyl groups. Such an internal motion would increase the number of observed spectral lines. The process of trans-perpendicular-cis isomerization is too slow to be observed in these fluorescence excitation spectra.

VdW complexes of tS with noble gas atoms form a very interesting subject. Not only the determination of the complex structure is of interest but it is also clear that an attached atom will hinder isomerization. The internal motion will also be affected. Bands of VdW complexes of tS containing helium and argon have been observed at vibrational resolution [15,17]. Unfortunately the intensity of the bands is low and an unambiguous assignment could not be made. Complexes with helium show regularly shifted bands with respect to the related 0-0 transition in the parent molecule. Complexes with argon, however, are much more irregularly shifted.

We observed two bands of complexes of tS with argon, shifted (table 1) -40 and -63 cm^{-1} with respect to the 0^0_0 band if tS. The -63 cm^{-1} band is depicted in figure 7. The linewidth in both spectra did not deviate from the linewidth observed in the parent molecule. On first sight, the spectrum of figure 7 resembles a near symmetric top band. The -40 cm^{-1} transition shows a somewhat less regular structure but has a comparable spacing of lines. It is remarkable that these bands have a less dense spectrum compared to tS, although the rotational constants

are expected to be smaller. This strongly suggests that the internal
motion of the phenyl groups is diminished in the complex. We have not
yet completed the assignment of these rotationally resolved spectra.
However, we come to the preliminary conclusion that the derived rotational constants of both bands compare very well with a complex geometry
of a single argon atom placed above a phenyl group of the parent molecule. Other configurations or complexes with more argon atoms have
rotational constants, which deviate too much from the experimental
values.

4. ACKNOWLEDGEMENTS

This work is part of the research program of the Stichting voor Funda-
menteel Onderzoek der Materie (FOM) and has been made possible by
financial support from the Nederlandse Organisatie voor Zuiver Weten-
schappelijk Onderzoek (ZWO). Part of the program has been supported by
NSF grant nr. INT-8319313.

REFERENCES

[1] E. Riedle, H.J. Neusser and E.W. Schlag, *J. Chem. Phys.* **75** (1981)
4231

[2] W. Majewski and W.L. Meerts, *J. Mol. Spectrosc.* **104** (1984) 271

[3] J.P. Bekooy, W.L. Meerts and A. Dymanus, *J. Mol. Spectrosc.* **102**
(1983) 320

[4] W. Majewski, *Opt. Comm.* **45** (1983) 201

[5] S. Gerstenkorn and P. Luc, *Atlas du Spectroscopie d'Absorption de
la Molecule d'Iode* (CNRS, Paris, 1978)

[6] W.L. Meerts, W.A. Majewski and W.M. van Herpen, *Can. J. Phys.* **62**
(1984) 1293

[7] M.J. Ondrechen, A. Berkovitch-Yellin and J. Jortner, *J. Am. Chem.
Soc.* **103** (1981) 6586

[8] W.M. van Herpen, W.L. Meerts and A. Dymanus, *Laser Chem.* **6** (1986)
37

[9] W.M. van Herpen, W.L. Meerts and A. Dymanus, submitted.

[10] A. Amirav, U. Even and J. Jortner, *J. Chem. Phys.* **75** (1981) 2489

[11] A. Amirav and J. Jortner, *Chem. Phys.* **85** (1984) 19

[12] A. Amirav, U. Even and J. Jortner, *Faraday Disc. Chem. Soc.* **73**
(1982) 153

[13] M. Tratteberg, E.B. Frantsen, F.C. Mijlhoff and A. Hoekstra,
J. Mol. Struct. **26** (1975) 57

[14] J.A. Syage, P.M. Felker and A.H. Zewail, *J. Chem. Phys.* **81** (1984)
4685

[15] T.S. Zwier, E. Carraquillo M., and D.H. Levy, *J. Chem. Phys.* **78**
(1983) 5493

[16] P.M. Felker and A.H. Zewail, *J. Phys. Chem.* **89** (1985) 5402

[17] D. Bahatt, U. Even and J. Jortner, *Chem. Phys. Lett.* **117** (1985)
527

Supersonic Jet Spectroscopy of Complexes of Carbazole and N-ethyl Carbazole with Alkyl Cyanides.

Anita C. Jones, Alan G.Taylor, Elizabeth M.Gibson and David Phillips.
The Royal Institution of Great Britain
21 Albemarle Street
LONDON W1X 4BS.
United Kingdom.

ABSTRACT

The laser induced fluorescence spectra of jet-cooled carbazole and N-ethyl carbazole and their 1:1 complexes with a series of six alkyl cyanides (CH_3CN, C_2H_5CN, n-C_3H_7CN, i-C_3H_7CN, n-C_4H_9CN and t-C_4H_9CN) are presented. In all cases complexation leads to a bathochromic shift (≥ 316 cm^{-1}) in the excitation energy of the chromophore, indicating a stabilisation of the S_1 state of the complex relative to its ground state. Carbazole is expected to hydrogen bond with the ligands through the N-H proton with the cyanides acting as proton acceptors. That this is the case is confirmed by the good correlation between the observed red shifts and the gas phase proton affinities of the cyanides. The spectra of the complexes contain harmonic progressions in a low frequency intermolecular mode, thought to be an out of plane hydrogen bond bending mode. N-ethylation of carbazole prevents the hydrogen bonding interaction but complexation is still observed and may be attributed to dipole-dipole interaction. The spectra of the N-ethyl carbazole complexes exhibit similar vibrational structure to those of carbazole.

1. INTRODUCTION

The spectroscopic properties of carbazole have been widely investigated, in solution [1-3], solid state [4,5] and the gas phase [6]. A number of theoretical investigations of its electronic properties have also been undertaken [3,7,8] . Much of the interest in the photophysical properties of carbazole arises from its role as a chromophore in polymeric systems such as poly(N-vinylcarbazole) [9].

Solution phase studies have provided evidence for the formation of hydrogen bonded complexes between carbazole and a number of proton acceptors [10,11]. The supersonic jet technique allows spectroscopic investigation of such complexes in isolation and can provide insight into the nature of the hydrogen bonding interaction. This paper is concerned with the laser induced fluorescence spectra of jet-cooled complexes of carbazole and N-ethyl carbazole with a series of six alkyl cyanides: CH_3CN, C_2H_5CN, n-C_3H_7CN, i-C_3H_7CN, n-C_4H_9CN and t-C_4H_9CN. The carbazole complexes are shown to be hydrogen-bonded, but N-ethylation blocks the hydrogen-bonding site of carbazole and the dominant interaction in the latter complexes is assumed to be dipole-dipole in nature.

A. Weber (ed.), Structure and Dynamics of Weakly Bound Molecular Complexes, 291–302.
© *1987 by D. Reidel Publishing Company.*

As an introduction to these results, a brief account is given of the fluorescence excitation spectra of the jet-cooled carbazole and N-ethyl carbazole bare molecules. An investigation of the spectroscopy and decay dynamics of these molecules and their homocyclic analogues under supersonic jet conditions has been reported more fully elsewhere [12].

2. EXPERIMENTAL

Carbazole (CAR), N-ethylcarbazole (NEC) and the alkyl cyanides (RCN) were obtained from Aldrich Ltd and used without further purification. The supersonic jet apparatus used in these experiments has been described fully elsewhere [13]. Sample vapour was mixed with helium or argon carrier gas and allowed to expand through a circular or slit-shaped continous flow nozzle to produce a supersonic molecular beam. Details of expansion conditions are given in the appropriate figure captions. The complexing agent was introduced into the expansion by diverting a controlled fraction of the carrier gas flow through a stainless steel reservoir containing the alkyl cyanide at room temperature.

The fluorescence excitation spectra of the bare molecules were produced using, as the excitation source, a 450 W high pressure xenon arc lamp (continuous output) in conjunction with a Monospek 1000 1m monochromator. The excitation spectra of the complexes were produced using a Lambda Physik excimer-pumped dye laser (EMG 103 MSC/FL2002) with bandwidth <0.2 cm^{-1}. Total unfiltered fluorescence was collected mutually at right angles to the molecular and excitation beams and focussed onto an EMI XP2020Q photomultiplier tube. When using the arc lamp excitation source the PM tube was operated in single photon counting mode and the output was recorded on a Canberra model 30 multichannel scaler which was synchronised with the monochromator stepping motor. When using the laser excitation source, the analogue output of the PM tube was measured by a microvoltmeter and recorded on a chart recorder.

3. RESULTS AND DISCUSSION

3.1 The bare molecules.

The fluorescence excitation spectra of carbazole and N-ethyl carbazole are shown in Figure 1 .

The most intense feature in each spectrum is the 0^0_0 transition, which occurs at 30800±20 cm^{-1} for carbazole and at 29650±20 cm^{-1} for N-ethylcarbazole, indicating that the $S_1 \leftarrow S_0$ transition is symmetry allowed. The S_1 state of carbazole has A_1 symmetry (in the C_{2v} point group) and the electronic transition is short axis polarised [4]. The $S_1 \leftarrow S_0$ transition of carbazole has substantial charge transfer character and results in an increase in dipole moment (directed along the C_2 axis) from 2±0.2D in the ground state[14-16] to 3.1D in the excited state[16]. N-ethylation of carbazole lowers the $S_1 \leftarrow S_0$ transition energy ; the red shift of 1150 cm^{-1} has been shown to be due to the selective destabilisation of the highest occupied molecular orbital on N-ethylation as a result of an increase in the electron density localised on the nitrogen atom in this orbital [3]. The spectra are dominated by a_1 modes, as expected for an allowed transition, but some b_2 modes are present as a result of vibronic coupling with the S_2 state of B_2 symmetry. The most active fundamental is the 211 cm^{-1} (CAR)/203 cm^{-1} (NEC) a_1 mode which appears in combination with a number of other modes; for both molecules, excitation to the S_1 state is accompanied by a change in equilibrium geometry along this normal coordinate.

Figure 1. Fluorescence excitation spectrum of (a) carbazole 0_0^0 at 30,800 cm^{-1} and (b) N-ethyl carbazole 0_0^0 at 29,650 cm^{-1}. Sample vapour was seeded into Ar carrier gas at 20-80 mbar and expanded through a 4 x 0.23 mm slit nozzle.

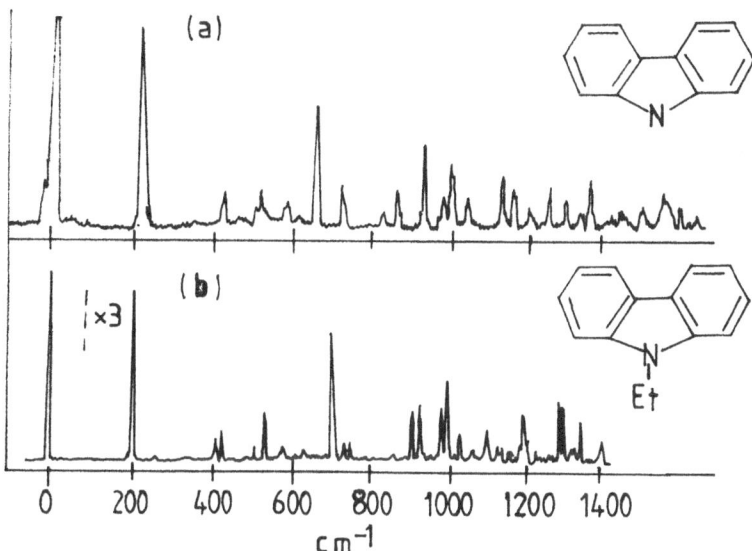

3.2 Complexes with Alkyl Cyanides.

In all cases for both CAR and NEC, complexation caused a decrease in the $S_1 \leftarrow S_0$ excitation energy relative to that of the parent molecule. With the exception of carbazole...NCCH$_3$, (as discussed below) all of the complexes studied exhibited low frequency intermolecular modes in their excitation spectra. Variation of the partial pressure of RCN in the expansion produced no change in the relative intensities of the spectral bands which may thus be attributed to complexes of 1:1 stoichiometry. No additional features to the red of those shown could be produced by increasing the partial pressure of RCN, implying that formation of complexes of higher stoichiometry was unfavourable. Variation of the carrier gas stagnation pressure caused no apparent change in the structure of the spectra, verifying that the transitions observed originate from the zero point vibrational level of the complex. The lowest energy transition discernible in the spectrum was thus assigned as the electronic origin in each case. The red shifts of the complex origins with respect to the corresponding bare molecule origin, $\Delta\bar{\nu}$, are listed in table I.

Table I

Observed red spectral shifts and intermolecular mode frequencies, $\bar{\nu}_a$ and $\bar{\nu}_b$, for CAR...NCR and NEC...NCR.

Ligand	Carbazole			N-ethylcarbazole		
	$-\Delta\bar{\nu}/$ cm^{-1}	$\bar{\nu}_a/$ cm^{-1}	$\bar{\nu}_b/$ cm-1	$-\Delta\bar{\nu}/$ cm^{-1}	$\bar{\nu}_a/$ cm^{-1}	$\bar{\nu}_b/$ cm^{-1}
CH_3CN	316	-a	-a	386	22	235
C_2H_5CN	402	11	112	427	17	230
$n-C_3H_7CN$	466	28	120	-b	-b	-b
$i-C_3H_7CN$	460	28	120	394	21,17	186
$n-C_4H_9CN$	523	25,32	142	395	13	161
$t-C_4H_9CN$	503	24	150	354	16	187

[a] no intermolecular modes observed.

[b] not determined for NEC...n-C_3H_7CN.

3.2.1 Carbazole...NCR Complexes.

The excitation spectrum of Carbazole...NCCH$_3$, (Figure 2) in contrast to those of the other complexes studied, showed only carbazole intramolecular vibronic tranistions, lying 316 cm^{-1} to the red of the corresponding bare molecule bands.

The CAR...NCC$_2$H$_5$ spectrum (Figure 3(a)) is dominated by a progression in 11 cm^{-1}. Two weaker progressions in 20 cm^{-1} appear to the blue, built on modes of 36 cm^{-1} and 45 cm^{-1}.The CAR...NC(CH$_2$)$_2$CH$_3$ and CAR...NCCH(CH$_3$)$_2$ spectra are essentially identical in structure, the former being shown in figure 3(b). Each is dominated by two overlapping progressions in 28 cm^{-1}, one 14 cm^{-1} to the blue of the other.

Figure 2. Fluorescence excitation spectrum of carbazole...NCCH$_3$, The intense feature on the extreme right is the bare molecule 0^0_0 transition at 30,800 cm^{-1}. Sample vapour was seeded into He carrier gas at a stagnation pressure of 3.3 Bar and expanded through a 100 μm nozzle; 20% of carrier gas flow was passed through the cyanide.

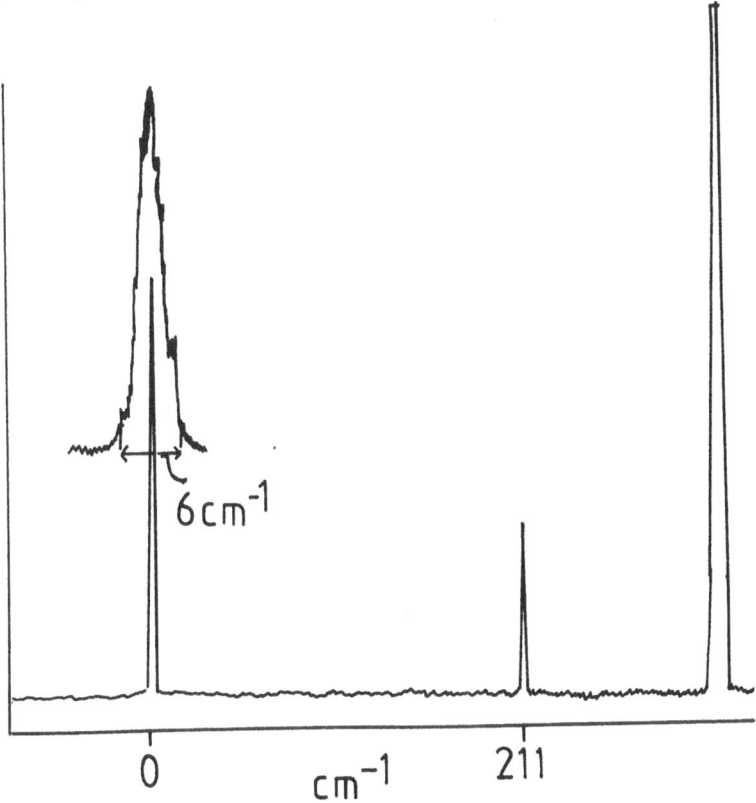

The excitation spectrum of CAR...NCC(CH$_3$)$_3$ also showed two overlapping progressions, in this case the progression frequency was 24 cm^{-1} and the two were separated by 10 cm^{-1}. The CAR...NC(CH$_2$)$_3$CH$_3$ spectrum contained a large number of low-frequency transitions, but unlike the other spectra, there was no prominent regular structure. Two progressions could however be distinguished, one in 25 cm^{-1} and one in 32 cm^{-1}, both commencing at the origin. In all the above spectra, structure identical to that built on the origin occurred a second time in the spectrum, built on a mode of around 100 cm^{-1} listed as $\bar{v}_b$ in table I. The progression frequencies, $\bar{v}_a$, are also summarised in table I.

Carbazole is expected to hydrogen bond with the alkyl cyanides through the

N-H proton, the cyanides acting as proton acceptors. In Figure 4, the observed red shift of the complex origin, relative to that of carbazole, is plotted versus the gas phase proton affinities of the alkyl cyanides [17]. The linear relationship between these two parameters confirms that hydrogen bonding is , indeed, the dominant interaction in these complexes.

Figure 3. Fluorescence excitation spectrum of :
(a) Carbazole...NCC_2H_5, progressions in 11 cm^{-1} (O) and 20 cm^{-1} (X,□) are indicated;
(b) Carbazole...NC $(CH_2)_2CH_3$, two progressions in 28 cm^{-1} are indicated (O,X). Expansion conditions as for Figure 2.

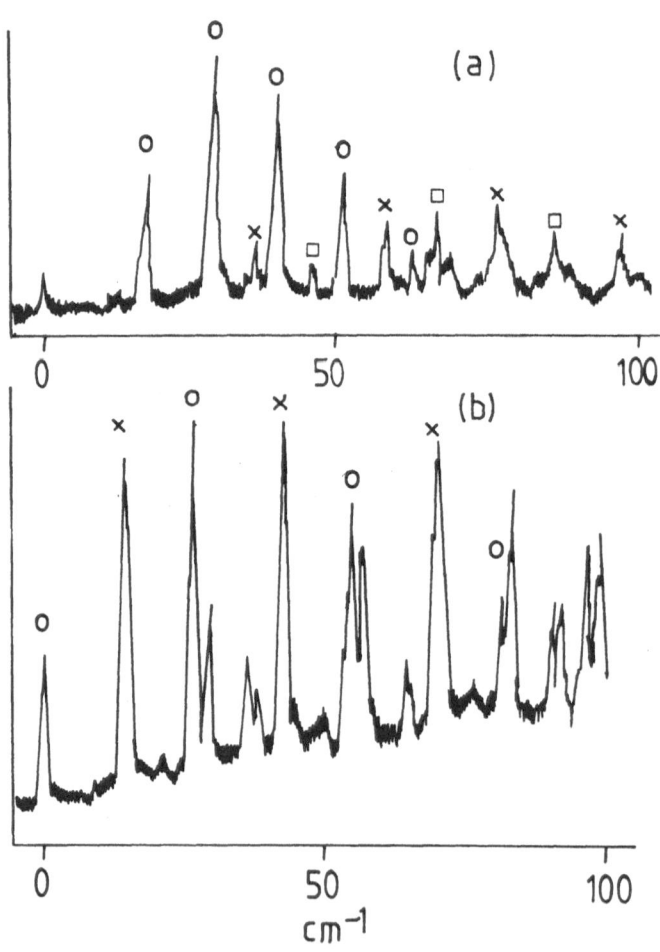

Figure 4. Plot of the electronic spectral shifts ($\Delta\bar{\nu}$) of carbazole...NCR versus the gas phase proton affinity (PA) of the RCN ligands.

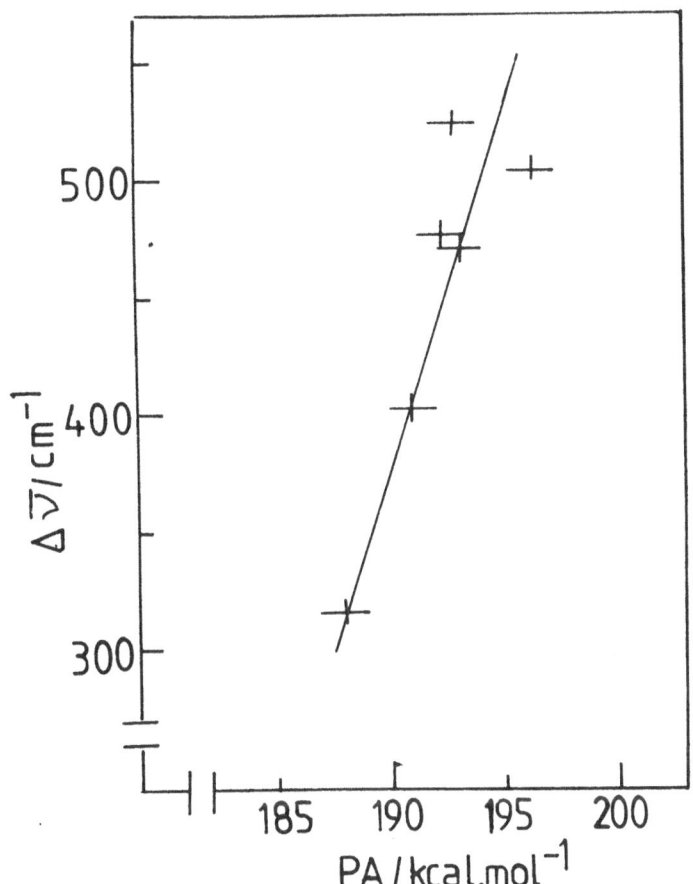

Only complexes of 1:1 stoichiometry were observed since there is only one hydrogen bonding site in carbazole. As discussed above, the $S_1 \leftarrow S_0$ transition of carbazole is accompanied by an increase in dipole moment and a corresponding increase in acidity of the N-H proton which leads to a strengthening in the hydrogen bonding interaction in the CAR...NCR complex in the excited state, resulting in stabilisation of S_1 relative to S_0 and the observed red shift. The magnitude of the red shifts for CAR...NCR is comparable to that observed for CAR...OH$_2$ [18]; however the latter seems anomalously large since the proton affinity of H$_2$O is only 173 kcal.mol^{-1}, suggesting that other factors such as the geometry of the complex are important in determining the strength of the hydrogen bonding interaction.

Hydrogen bonded complexes of indole [19] with ligands of similar proton affinity show red shifts smaller by a factor of 2 to 3 than CAR...NCR, implying that a smaller excitation-induced increase in acidity occurs for indole than for carbazole. Hydrogen bonding can be considered as an interaction mainly electrostatic in nature, but with contributions from polarisation, charge transfer and exchange interactions [19,20]. The alkyl substituent effect on proton affinity, ie. the increase in proton affinity with increasing size of alkyl group, is a polarisation effect in which the protonated base is stabilised as a result of the polarisability of the alkyl group. The calculations of Umeyama and Morokuma [20] predict that although alkylation increases the proton affinity of the base, it has an overall destabilising effect with respect to hydrogen bonding since the stabilisation effects of the increased polarisation and charge transfer interactions are outweighed by the destabilising effect on electrostatic and exchange interactions. Our results contradict this prediction, but agree with those of Hager and Wallace [19] for hydrogen bonded indoles, supporting their conclusion that polarisation forces play an important role in the stabilisation of complexes such as these.

The prominent progressions in a low frequency mode ($\bar{v}_a$ in table 1) in the spectra of CAR...NCR complexes is evidence of a substantial change in geometry between ground and excited states. A simple shortening of the hydrogen bond on excitation would be manifested as a progression in a stretching mode, as observed for CAR...OH$_2$ [8]. The frequency $\bar{v}_a$, however, is too low for a hydrogen bond stretch and is more likely due to a bending mode, being comparable in frequency to bending modes observed in other hydrogen bonded complexes, e.g. 48 cm^{-1} in CAR...OH$_2$ [18]; 23 cm^{-1} in indole...methanol [21]; 45 cm^{-1} in CH$_3$CN...HF; 55 cm^{-1} in (CH$_3$)$_3$CCN...HF [19]. For CAR...OH$_2$ no extended progression was observed in the bending mode which was strongly anharmonic [18], the first overtone occurring at 161 cm^{-1}. The modes denoted $\bar{v}_b$ in table 1 are assigned as hydrogen bond stretches in view of their similarity with stretching frequencies found in other systems, e.g. CAR...OH$_2$ 130 cm^{-1} [18], indole...methanol 146 cm^{-1} [21]; CH$_3$CN...HF 168 cm^{-1}; CH$_3$CN...HCl 100 cm^{-1}; C$_2$H$_5$CN...HCl 95 cm^{-1} [22]. In contrast to CAR...OH$_2$, no progression was apparent in this stretching mode for CAR...NCR. The ground state equilibrium geometries of CAR...OH$_2$ and CAR...NCR are expected to differ: the former having a bent structure [18] while the latter is likely to have a linear N-H...N≡C arrangement [22,23]. The excitation induced geometry change in the CAR...NCR complexes clearly differs from that in CAR...OH$_2$, involving distortion along the bending co-ordinate rather than the stretching co-ordinate of the hydrogen bond. The occurrence of two progressions in $\bar{v}_a$ built on different origins in the spectra of CAR...NCC$_3$H$_7$ and CAR...NCC(CH$_3$)$_3$ may indicate the existence of two different isomeric forms of these complexes.

The absence of bending mode transitions in the CAR...NCCH$_3$ spectrum can be explained by symmetry considerations , as follows. The symmetry of this complex is essentially C$_{2v}$ (if the methyl hydrogens are ignored), the same as the parent carbazole molecule, and thus only a_1 modes will be coupled to the symmetry allowed $S_1 \leftarrow S_0$ transition. Complexes of carbazole with the higher alkyl cyanides have their symmetry lowered to C_s under which a´ modes (correlating with a_1 and b_1 in C_{2v}) are allowed. The activity of the bending mode in the latter cases and its absence in the former is therefore consistent with this vibration having b_1 symmetry, i.e. an out- of-plane motion of the RCN relative to the carbazole.

3.2.2. N-ethyl carbazole...NCR Complexes.

N-ethylation of carbazole precludes the possibility of hydrogen bonding, but does not prevent the formation of complexes with the alkyl cyanides. The vibrational structure of the excitation spectra of these complexes is very similar to that seen for the CAR...NCR complexes.

Figure 5.. Fluorescence excitation spectrum of :
(a) N-ethyl carbazole...$NCCH_3$, the progression in 22 cm^{-1} is marked (O);
(b) N-ethyl carbazole...NCC_2H_5, the progression in 17 cm^{-1} (O) and the 230 cm^{-1} mode (□) are indicated.Expansion conditions as for figure 2.

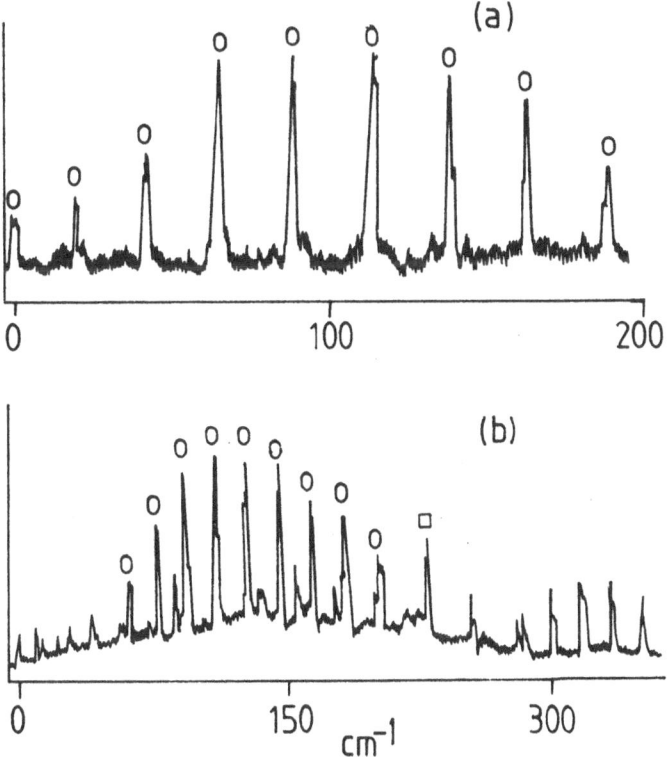

Figure 6. Fluorescence excitation spectrum of N-ethyl carbazole...NCCH(CH$_3$)$_2$. Progressions in 21 cm^{-1} (O) and 17 cm^{-1} (X) are indicated. Expansion conditions as for Figure 2.

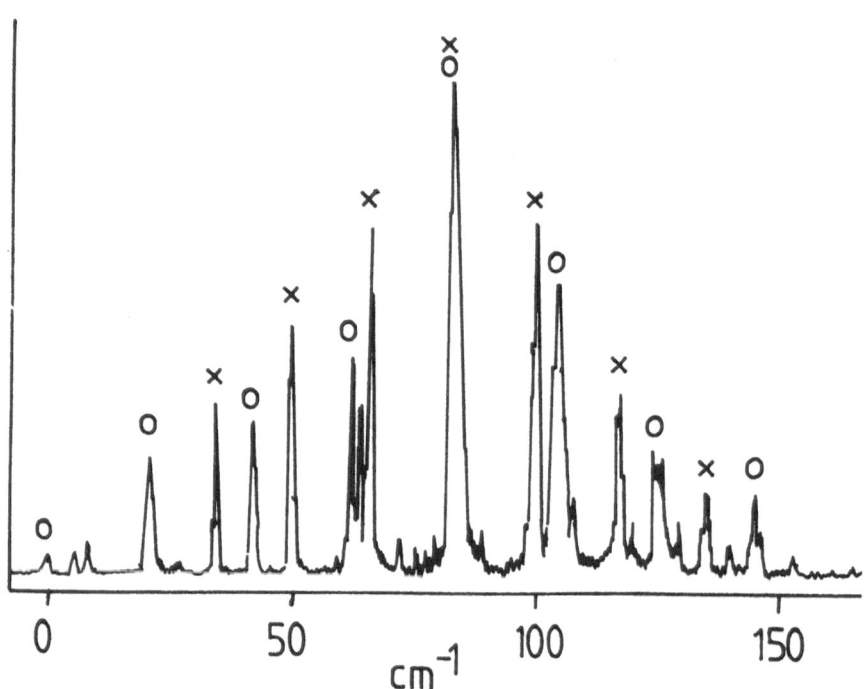

The spectrum of NEC...NCCH$_3$(figure 5(a)), NEC...NCC$_2$H$_5$ (figure 5 (b)) and NEC...NC(CH$_2$)$_3$CH$_3$ is each dominated by a single progression in a mode of 22, 17 and 13 cm^{-1} respectively. The NEC...NCCH(CH$_3$)$_2$ (figure 6) contains two overlapping progressions, one in 21 cm^{-1} and one in 17 cm^{-1}. Similarly the spectrum of NEC...NCC(CH$_3$)$_3$ was dominated by two progressions in 16cm^{-1}, separated by 7 cm^{-1}.

The progressions observed for the NEC...NCR complexes were approximately twice as long as those for the corresponding CAR...NCR, containing typically 9 members compared with 4 or 5 for the latter. In no case, for CAR or NEC complexes, was any appreciable anharmonicity apparent, even in the 9-membered progressions. For all NEC...NCR complexes, a vibronic transition, which was not a member of the observed progressions nor a NEC intramolecular mode, was observed around 200 cm^{-1}; the frequency of this mode, $\bar{\nu}_b$, is given in Table 1 for each complex.

The microscopic solvent shifts for these are comparable in magnitude to those observed for the complexes of carbazole with the more weakly proton accepting cyanides. Since both N-ethylcarbazole and the alkyl cyanides have permanent dipoles, the dominant interaction in the NEC...NCR complexes is assumed to be dipole-dipole in nature. The ground and excited state dipole moments of N-ethylcarbazole are not expected to differ significantly from those of N-methyl carbazole which have been calculated to be 2.2 and 3.3D (along the short molecular axis) respectively [24]. The alkyl cyanides all have approximately equal dipole moments, measured to be 4±0.1D in the gas phase [25]. The increase in stabilisation energy of the complexes in the excited state and the lack of a trend in the observed red shifts of the NEC...NCR series are consistent with this interpretation. The observed red shifts do show some variation, but only over a range of 73 cm^{-1} compared with the 207 cm^{-1} spanned by the CAR...NCR red shifts. This variation can be attributed to the influence of other factors, such as steric interactions, on the binding energy of the complexes.

The most favourable geometry for NEC...NCR will be that which allows closest approach of the opposed NEC and RCN dipoles. There are two obvious alternative structures: (i) with the RCN lying above the NEC plane, aligned with the short molecular axis, i.e. ↑↓ ; (ii) with the RCN and the NEC dipoles aligned end to end (↔ ↔) giving a structure analagous to that of the CAR...NCR complexes. The similarity in the vibrational structure of the NEC...NCR and CAR...NCR spectra implies a similar change in equilibrium geometry on excitation for both types of complex, although the magnitude of the change appears to be greater for NEC...NCR. In view of these similarities, the second alternative, above thus seems more probable. The low-frequency mode $\bar{\nu}_a$ of NEC...NCR can then be attributed to an intermolecular bending motion and the higher frequency $\bar{\nu}_b$ to an intermolecular stretch. The absence of anharmonicity in the progressions in $\bar{\nu}_a$ is indicative of the strength of interaction in these complexes: a population of 9 vibrational quanta in this mode being insufficient to approach the dissociation limit.

Acknowledgements

The authors are grateful to Professor D J Millen for helpful discussions. The authors also wish to thank the SERC for financial support and Coherent (UK) Ltd for provision of a CASE award for A G Taylor.

REFERENCES

[1]   G.E.Johnson, J.Phys.Chem., (1974), **78**, 1512.
[2]   P.D.Harvey, B.Zelant and G.Durocher, Spectros.Intern.J., (1983), **2**, 128.
[3]   R.W.Bigelow and G.P.Ceasar, J.Phys.Chem., (1979), **83**, 1790.
[4]   A.Bree and R.Zwarich, J.Chem.Phys., (1969), **49**. 3355.
[5]   S.C.Chateravorty and S.C.Ganguly, J.Chem.Phys., (1970), **52**, 2760.
[6]   C.A.Pinkham and S.C.Wait, J.Mol.Spectry, (1968), **27**, 326.
[7]   L.E.Nitzsche, C.Chabalowski and R.E.Christoffersen, J.Am.Chem.Soc., (1976), **98**, 4794.
[8]   R.W.Bigelow and G.E.Johnson, J.Chem.Phys., (1977), **66**, 4861.
[9]   H.Hoegl, J.Phys.Chem., (1965), **69**, 755.
[10]   T.S.Spencer and C.M.O'Donnell, J.Am.Chem.Soc., (1972), **94**, 4846.
[11]   A.Ahmad and G.Durocher, Photochem. and Photobiol., (1981), **34**, 573.
[12]   A.R.Auty, A.C.Jones and D.Phillips, Chem.Phys., (1986), **103**, 163.
[13]   A.R.Auty, A.C.Jones and D.Phillips, J.C.S.Faraday, (1986), **82**, 1219.
[14]   E.G.Cowley and J.R.Partington, J.Chem.Soc., (1936), 47.
[15]   W.Liptay, Angew. Chem.Int.Ed. Engl., (1969), **8**, 177.
[16]   W.Liptay, Modern Quantum chemistry, Vol.3, (Academic Press, New York, 1965), p.45.
[17]   R.H.Staley, J.E.Klechner and J.L.Beauchamp, J.Am.Chem.Soc., (1976), **98**, 2081.
[18]   R.Baomback, E.Honegger and S.Leutwyler, Chem.Phys.Lett., (1985), **118**, 449.
[19]   J.Hager and S.C.Wallace, J.Phys.Chem., (1984), **88**, 5513.
[20]   U.Umeyama and K.Morokuma, J.Am.Chem.Soc., (1976), **98**, 4400.
[21]   Y.Nibu, H.Abe, N.Mikami and M.Ito, J.Phys.Chem., (1983), **87**, 3898.
[22]   D.J.Millen, J.Mol.Structure, (1983), **100**, 351.
[23]   A.D.Buckingham and P.W.Fowler, J.Phys.Chem., (1983), **79**, 6426.
[24]   D.Murk, L.E.Nitzsche and R.E.Christoffersen, J.Am.Chem.Soc., (1978), **100**, 1371.
[25]   A.L.McClellan, Tables of Experimental Dipole Moments, (W.H.Freeman and Co. 1963) and references therein.

PHOTOIONIZATION OF HYDROGEN-BONDED MOLECULAR AGGREGATES

P.Bisling, E.Rühl, B.Brutschy, H.Baumgärtel
Institut für Physikalische und Theoretische Chemie
der Freien Universität Berlin
Takustraße 3
D-1000 Berlin 33
Germany

ABSTRACT: The photoionization of hydrogen-bonded molecular aggregates
is dominated by proton transfer in the ionic aggregate and subsequent
fragmentation of the ionic cluster. The ionization potential of aggregates is strongly redshifted in comparison to the isolated molecule.
Vertical transitions are observed and therefore thermochemical data derived from threshold values are lower bounds.

   Aggregates with van der Waals character in the neutral state reveal proton transfer in the ionic state and in addition chemical fragmentation.

1. INTRODUCTION

Weak intermolecular interaction causes association of molecules and at
least the formation of macroscopic systems. Depending on the degree of
aggregation among molecular aggregates microclusters (n < 20) and macro-
clusters are discerned /1/. A rough classification according to the type of intermolecular interaction is possible into true van der Waals
clusters, electron donator-acceptor systems and hydrogen-bonded systems /2/. During the last decade many scientists are increasingly interested in the intrinsic properties of the intermolecular interaction.
This development has been stimulated by improved spectroscopic techniques and molecular beam technology. The combination of these fields
pushed the investigation of the formation and properties of small molecular clusters and cluster ions /3-6/.

   In this paper we report preferably on photoionization studies on
hydrogen-bonded aggregates because none of the different molecular aggregates has received such great attention as this kind. One of the reasons for the great interest in hydrogen bonds is the importance of water, another is the fact that hydrogen bonds strongly affect the geometric structure of biopolymers, e.g. DNA and proteins.

A. Weber (ed.), Structure and Dynamics of Weakly Bound Molecular Complexes, 303–317.
© *1987 by D. Reidel Publishing Company.*

2. EXPERIMENT

In molecular beams molecular aggregates can be produced which are investigated by photoionization mass spectrometry. Only single-photon ionization experiments are discussed in the following. The experimental setup
has been described in detail /7/. It consists of the molecular beam apparatus which is adapted to the Berlin Electron Storage Ring (BESSY).

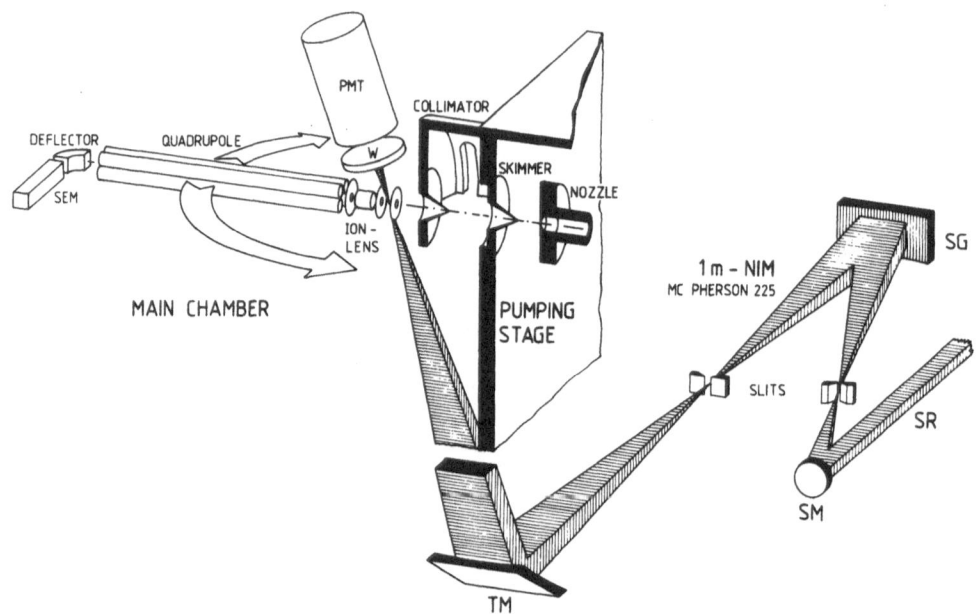

Fig.1: Photoionization mass spectroscopy of molecular clusters using
       Synchrotron radiation

The undispersed synchrotron radiation is focused onto the entrance slit
of a 1m-McPherson 225 monochromator equipped with a holographically ruled , aluminium and MgF_2 coated spherical grating (1800 lines/mm). A
gold coated toroidal mirror refocuses the dispersed synchrotron radiation to a stigmatic target spot (ca.1.5mm^2) onto the cluster beam. A
sodium salicylate coated quartz window is used as a quantum efficiency
converter to monitor the photon flux independently from the photon energy. A photomultiplier feeds this reference count rate into a preset tim-
er to normalize the photoion count rates.
     The photoions are detected with a modified commercial quadrupol
analyzer (BALZERS QMA 160) mounted coaxially to the molecular beam.
The wavelength scale is calibrated better than ±0.01 nm to the argon au-
toionization structure. The mass scale is calibrated with $(Ar)_n^+$ peaks
(n = 1-6). The experimental setup is used to measure photoion yield
curves and photoionization mass spectra.

3. THE HYDROGEN BOND

If a hydrogen atom is bonded to more than one other atom, a hydrogen
bond exists. The properties and description of the hydrogen bond shall
be outlined briefly. An extensive discussion is documented in the literature /8/. The intermolecular hydrogen bond is understood to be a bond
of the type

$$A - H \cdots | B$$

where A-H symbolizes the proton donor molecule and $|B$ the proton accep-
tor molecule. In the proton donor molecule the hydrogen is chemically
bonded to an electronegative atom , e.g. C,N,O,F,P,S etc.; this brings
about some electron deficit at the hydrogen and therefore the A-H-bond
in the proton donator system is polar. On the other hand the proton acceptor molecule possesses regions with high electron density, e.g. atoms
with lone pair electrons ($N\lessgtr$, $|O<$, F); these can be also π-systems
where the π-electron density above and below the σ-skeleton has proton
acceptor properties.
    The formation of the hydrogen bond brings about a change of the
electron density in the molecules involved, chiefly at the hydrogen atom
and the atom $|B$ in the proton acceptor molecule. Usually, the hydrogen
bonding system has linear geometry, the distance between proton donor
and proton acceptor molecule is shorter than the sum of the van der
Waals radii of A-H and B. The bond length A-H is enhanced in the hydrogen bonded system compared to the length of A-H in the isolated proton
donator.
    Schuster /9/ classifies hydrogen bonds into three groups:
a)   Weak assymmetric hydrogen bonds between neutral molecules,
     e.g. $HCN \cdots HF$
b)   Ion molecule complexes with assymmetric hydrogen bonds,
     e.g. $F^- \cdots H_2O$, $NH_4^+ \cdots H_2O$
c)   Strong and nearly symmetric hydrogen bonds, e.g. $F \overset{-}{\cdots} HF$.

    From a quantumchemical point of view the $A-H \cdots |B$ system is a 4-
-electron-3-center bond including the electrons of the 2-electron-2-
-center bond A-H and the lone pair electrons of B. From the linear combination of ψ_{A-H} and ψ_B results in first approximation no bonding effect,
the bonding effect depends on mixing with ψ^*_{A-H}, the antibonding MO of
A-H which is unoccupied in the isolated proton donator /10/.
    The dynamical properties of hydrogen bonds are of course in a first
approximation correlated with the proton dynamics, but as the proton
motion and the motions of the proton donor and proton acceptor are
strong and ineluctably coupled, one is seized to treat the problem in a
two-dimensional space. Details are described by J. Brickmann, E. Weide-
mann and G.L. Hofacker in /8/.

4. RESULTS AND DISCUSSION

Hydrogen-bonded aggregates of ammonia, water, hydrogenfluoride and their
alkyl derivatives have been studied recently by several groups. The results permit a first systematic comparison of their photoionization properties

4.1. Ammonia and Alkylamines

The photoionization mass spectra indicate clusters of the type $(NR_3)_n^+$
(R = H, alkyl) and with higher yield protonated clusters $(NR_3)_nH^+$.
     The ionization potentials of unprotonated clusters and the appear-
ance potentials of the protonated clusters are compiled in table 1 and 2.

Table 1   Ionisation potentials [eV] of $(NR_3)_n$ (R = H,alkyl) cluster

compound	n = 1	n = 2	n = 3	Lit
NH_3	10.16	9.54	-	/11/
CH_3NH_2	8.97	8.1	7.9	/12/
$(CH_3)_2NH$	8.24	7.8	7.7	/12/
$(CH_3)_3N$	7.87	-	-	/12/
$C_2H_5NH_2$	8.82	8.1	7.9	/12/
$(C_2H_5)_2NH$	8.01	7.5	-	/12/
$(C_2H_5)_3N$	7.48	7.57	-	/12/

The ionization potentials of the hydrogen bonded dimers are considerably lower than those of the corresponding monomers; this is a general result which indicates from the thermochemical point of view that the intermolecular bond energy in the ionic dimer is higher than in the neutral dimer (tables 4,5). This tendency proceeds to higher aggregates
where at least a limiting value should appear.
     In ammonia and amines at the first ionization potential one of the
lone pair electrons is ejected /13/. Compared to the isolated subsystems
the binding energy of these lone pair electrons is enhanced in the proton acceptor and lowered in the proton donator subsystem of the dimer.
This implies that the ionization process may be formulated as

$$H_2\bar{N}\text{-}H\cdots|NH_3 \quad \rightarrow \quad H_2\overset{+\bullet}{N}\text{-}H\cdots\cdots|NH_3 + e^- \tag{1}$$

It is very well known from PES spectroscopy and quantumchemical calculations /14/ that ionization of NH_3 is followed by considerable reorienta-
tion of the nuclei; that implies the ionization potentials given in table 1 should be considered to be vertical ones /15/. The bond energy of
the ionic dimer is calculated according to :

$$D(M-M^+) = I(M) + D(M-M) - I_{vert} (M_2)$$

$D(M-M^+)$: Intermolecular bond energy in the ionic dimer
$D(M-M)$: Intermolecular bond energy in the neutral dimer
$I(M)$: First ionization potential of the isolated monomer
$I_{vert}(M_2)$: Vertical ionization potential of the dimer.

The intermolecular dissociation energy obtained in this way is a lower bound.

The ionization potentials of trialkylamine dimers are practically unshifted compared to their monomers. The aggregates of these compounds are more of the van der Waals type because the neutral monomers only reveal proton acceptor functions. The dimer M_2^+ is expected to be of van der Waals type too, but the C-H bond is now polarized by the positive charge at the nitrogen and therefore a rearrangement into a hydrogen-bonded cationic dimer may not be excluded. We assume that this rearrangement has a considerable energy barrier because it is not performed by the motion of the proton in the hydrogen bridge only. The observed appearance potential of M_nH^+ in this case cannot be assigned unambiguously. It may be assigned either to the top of the rearrangement barrier or to the dissociation limit according to equation (2).

The formation of protonated clusters in the threshold region can be assigned to the reaction

$$(NH_3)_n^+ \rightarrow (NH_3)_{n-1}H^+ + NH_2 \qquad (2)$$

The appearance potentials of protonated species are given in table 2.

Table 2 : Appearance potentials AP [eV] $(NR_3)_nH^+$

compound	n = 1	n = 2	n = 3
NH_3			
CH_3NH_2	8.55	8.25	8.2
$(CH_3)_2NH$	8.1	7.85	7.8
$(CH_3)_3N$	7.94	7.75	-
$C_2H_5NH_2$	8.5	8.15	8.0
$(C_2H_5)_2NH$	8.0	7.75	-
$(C_2H_5)_3N$	7.61	7.27	-

The appearance potential of $M_{n-1}H^+$ is blue shifted in comparison to the ionization potential of M_n in the case of ammonia and the primary and secondary amines. This implies that the dissociation limit according to equation (2) is higher than the energy of the M_n^+ state which is reached in the ionization process. As we observe vertical ionization potentials the difference between $AP(M_{n-1})H^+$ and $IP(M_n)$ can only be interpreted as a lower bound of the intermolecular dissociation energy of M_n^+.

The dissociation energy of $(NH_3)_2^+$ into the ammonium ion NH_4^+ and the aminyl radical NH_2 is unexpectedly small $(0,05 \pm 0.05$ eV) compared to that of the $NH_4^+ \cdot NH_3$ complex (1.07 eV) as noted by Cao et al./16/. Quantumchemical calculations /14/ show that the equilibrium geometry of $(NH_3)_2^+$ has to be assigned to $NH_4^+ \cdot NH_2$ instead of $NH_3^+ \cdot NH_3$. Tomoda and Kimura /15/ confirmed this result and calculated the dissociation channels of the dimer cation into $NH_4^+ + NH_2$ and $NH_3^+ + NH_3$ to be 1.0 and 1.7 eV respectively above the ground state of the ionic dimer.

In the ion yield curves of protonated alkylamine clusters one observes a typical inflection at higher energies which indicates that a second decay channel contributes to the ion yield of $(M)_n H^+$. We assign this channel to the reaction

$$(M_{n+2})H^+ \rightarrow M_n H^+ + M + M' \qquad (3)$$

where M is ammonia, a primary or secondary amine and M' is the corresponding aminyl radical. The appearance energies according to equation 3 are compiled in table 3.

Table 3 : Second appearance potentials AP_2 of alkylamine clusters $M_n(H^+)$ according to equation (3)

compound	n = 1	n = 2	n = 3
CH_3NH_2	8.85	8.5	8.4
$C_2H_5NH_2$	8.75	8.4	8.2
$(CH_3)_2NH$	8.35	8.1	8.0
$(C_2H_5)_2NH$	8.3	8.0	-

Referring to the values in tables 1-3 the intermolecular bond energies of ionic clusters, proton solvation energies and bond dissociation energies of neutral clusters can be calculated.

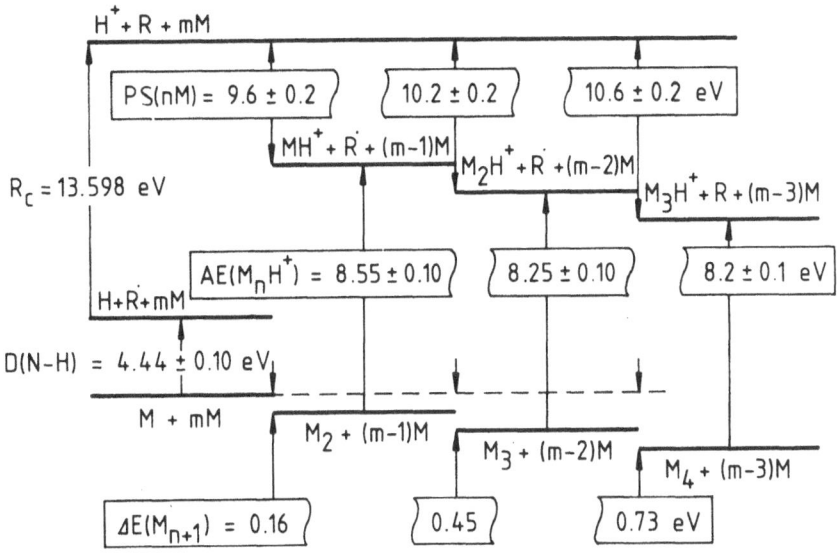

Fig. 2 : Thermochemical scheme used for the evaluation of proton
         solvation energies PS(nM)

To perform these calculations the association energy of the neutral aggregates M_n is necessary. In some cases these values are available from
literature resulting either from experimental investigations or theoretical calculations. In the case of primary and secondary alkylamines we
deduced the intermolecular bond energies $\Delta E(M_{n+2})$ (n = 1.2) from the
threshold values according to equation 4.

$$\Delta E(M_{n+2}) = \Delta E(M_2) + \sum_{i}^{n} AP_2(M_iH^+) - AP_1(M_iH^+) \tag{4}$$

The results are compared with values obtained by EPEN calculations /17/.
The bond energy of dimers are derived from second virial coefficients
/18/. The association energies $\Delta E(M_{n+2})$ are compiled in table 4.

Table 4: Comparison of total intermolecular association energies of alkylamine aggregates obtained from EPEN calculations[a], second virial coefficients [b] and $AP_2(M_{n+2})$[c],[e].

compound	n = 0		n = 1		n = 2	
CH_3NH_2	3.8 [a]	3.4 [b]	10.3 [a]	11.9 [c]	16.9 [a]	16.7 [c]
$C_2H_5NH_2$	-	3.6 [b]	11.0 [d]	9.6 [c]	17.9 [d]	16.7 [c]
$(CH_3)_2NH$	-	3.1 [b]	9.3 [d]	9.6 [c]	15.6 [d]	14.4 [c]
$(C_2H_5)_2NH$	-	3.3 [b]	10.0 [d]	9.6 [c]	16.5 [d]	16.7 [c]

a) /17/   b) /18/   c) /12/
d) extrapolated values obtained by means of EPEN results and second
    virial coefficients; e) values in kcal/mole

The association energy of the trimer and the tetramer reveal that preferably cyclic species exist in the molecular beam because ΔEM_3 and
ΔEM_4 is about threetimes respectively fourtimes the value of ΔEM_2. This
result seems to be surprising from an entropic point of view and the
fact that the intermolecular hydrogen bonds lose their optimal linear
geometry but it has been confirmed by electrical deflection beam experiments /19/ which show that these aggregates have nearly no dipole moment.
      The intermolecular bond dissociation energy of the ionic aggrega-
tes can be calculated if the ionization potentials and the bond energies of M_n are known. In table 5 the bond strength of ionic alkylamine
aggregates are compiled.

Table 5 : Intermolecular bond dissociation energy of ionic
           $D(M_n-M)^+$ (kcal/mole) clusters of alkylamines

compound	n = 1	n = 2	Lit
NH_3	18.2	-	/11/
CH_3NH_2	23.9	12.0	/12/
$C_2H_5NH_2$	21.5	12.0	/12/
$(CH_3)_2NH$	12.0	8.4	/12/
$(C_2H_5)_2NH$	14.4	-	

The values in table 5 document   increasing bond strength of the hydrogen
bond in ionic dimers and trimers compared to the neutral species. Ionization leads to an increased acidity of the N-H bond compared to N-H
and a remarkable shortening of the N····N distance as by Lathan et al.
/14/ pointed by means of quantumchemical ab initio calculation on the
ammonia system. The intermolecular bond strength in the ionic trimer
is clearly decreased in comparison to the dimer which indicates a relative destabilization of the cationic trimer compared to the dimer.
      The absolute proton affinity PA of a molecule M is defined by

$$M + H^+ \rightarrow MH^+ + PA \tag{5}$$

The solvation energy of a proton $(PS(nM))$ may be defined by

$$nM + H^+ \rightarrow M_nH^+ + PS(nM) \tag{6}$$

For $n = 2$ this magnitude is identical with the molecule pair proton affinity defined by Aue and Bowers [20]. The proton affinity of a molecular cluster may be defined by

$$M_n + H^+ \rightarrow M_nH^+ + PA(M_n) \tag{7}$$

The relationship
$$PA(nM) + \Delta EM_n = PA(M_n) \tag{8}$$

connects the proton solvation with the proton affinity of a cluster M_n.

Table 6 shows proton solvation energies of ammonia and amines as following from photoionization-beam experiments. Fig.2 illustrates the thermodynamic cycles used for the calculation of proton solvation energies $PS(nM)$

Table 6 : Proton affinity PA(M) and proton solvation energies PS(nM) (kcal/mole) calculated from threshold energies and compared with literature data

compound	PA(M)		PS(2M)		PS (3M)
NH_3	$202^{[d]}$	$204^{[b]}$	$220^{[d]}$	$230^{[c]}$	-
CH_3NH_2	$222^{[a]}$	$214^{[b]}$	$236^{[c]}$	$236^{[c]}$	$244^{[a]}$
$C_2H_5NH_2$	$225^{[a]}$	$217^{[b]}$	$239^{[c]}$	$238^{[c]}$	$249^{[a]}$
$(CH_3)_2NH$	$228^{[a]}$	$221^{[b]}$	$240^{[c]}$	$241^{[c]}$	$248^{[a]}$
$(C_2H_5)_2NH$	$231^{[a]}$	$226^{[b]}$	$245^{[c]}$	$245^{[c]}$	-

a) [12], b) [21], c) [20], d) [11]

The proton solvation energy increases with increasing degree of association. The values based on threshold measurements are to be considered as lower bounds of the proton solvation energy caused by the fact that vertical ionization potentials are observed in the photoionization experiment. Nevertheless, the agreement with literature data based on ion-molecule reactions is reasonable.

4.2. Water, Alcohol, Ether

The electronegativity of oxygen is increased compared to nitrogen and consequently the polarity of the O-H bond is enhanced in comparison to N-H. This tendency is expressed by different molecular properties, e.g. the dipole moment in the gas phase or the autoprotolytic constant in the liquid phase. As the bond strength of hydrogen bonds depends on the polarity of the proton donor the O-H⋯⋮O bonds usually are stronger than N-H⋯⋮N bonds.

The photoionization of $(H_2O)_2$ has been studied by Ng et al. /22/. From the photoion yield curves they evaluated threshold potentials of $(H_2O)_2^+$ and H_3O^+. The appearance potential of the dimer is 1.4 eV red-shifted versus $IP(H_2O)$.

Methanol, ethanol and trifluorethanol has been investigated by Taylor et al./23/. Their experiment differed slightly from conventional beam experiments. They expanded 20-100 Torr alcohol vapor into the va-cuum and investigated the relative cluster ion intensities at different temperatures of the nozzle (301-345K). From van t'Hoff plots of the equilibrium constant of the reaction

$$CH_3OH^+ + (CH_3OH)_n \rightleftharpoons (CH_3OH)_nH^+ + CH_3O \qquad (9)$$

they obtained the reaction enthalpy. Only protonated species $(CH_3OH)_nH^+$ (n = 1-4) were observed.

Ethers are not pronounced proton donors. The photoionization mass spectra of dimethylether and diethylether reveal molecular aggregates, protonated aggregates and in the case of diethylether some solvated fragment ions. As with tertiary amines the neutral clusters are supposed to be of van der Waals type. The proton transfer indicates typically hy-drogen bond behaviour of the ionic cluster.

Table 7: Ionization potentials [eV] of $(OR_2)_n$ (R = H, alkyl) clusters

compound	n = 1	n = 2	n = 3
H_2O	12.62 [a]	11.21[a]	-
CH_3OH	-	-	-
$(CH_3)_2O$	9.93 [b]	9.44 [b]	-
$(C_2H_5)_2O$	9.65 [b]	9.12 [b]	8.85 [b]

[a] /22/; [b] .this work

Table 8: Appearance potentials [eV] of $(OR_2)_nH^+$ (R=H, alkyl) clusters

compound	n = 1	n = 2	n = 3	n = 4
H_2O	11,73 [a]	-	-	-
CH_3OH	10,2 [b]	9,8 [b]	9.5 [b]	9.3 [b]
$(CH_3)_2O$	9.74 [c]	9.25[c]	9.20[c]	-
$(C_2H_5)_2O$	9.54 [c]	8.99[c]	-	-

[a]/22/; [b]/23/; [c] this work

Nevertheless the experimental result on the photoionization be-haviour of $(OR_2)_n$ (R = H, alkyl) clusters is not as complete, as that with ammonia derivatives. Tables 7 and 8 compile the available results.

The intermolecular bond dissociation energy in $(H_2O)_2^+$ has been calculated to be 36.4 kcal/mol and the proton affinity comes to 165.8 kcal/mol in good agreement with the data given in literature /21/ and obtained by other methods.

The bond dissociation energy in $(CH_3OH)_n^+$ cannot be calculated, as these aggregates have not been observed up to now. The intermolecular bond energy in ionic dimers and trimers of diethylether comes to 14 kcal/mol and 5 kcal/mol respectively.

4.3 Hydrogenfluoride, Methylfluoride

Hydrogenfluoride, like the other hydrogen halides, is an excellent molecule to study the intrinsic properties of hydrogen bonds. It was this molecule; that has been used by Tiedemann et al. /24/ to study the photoionization of hydrogen bonded dimers several years ago.

No $(HF)_2^+$ ion has been observed, but high yields of HF_2^+. The proton affinity of HF evaluated by Tiedemann et al. is substantially lower than that obtained by Forster and Beauchamp /25/ (112 kcal/mol). As mentioned above photoionization delivers lower bounds of absolute proton affinities. In this case the appearance potential of H_2F^+ obviously is lower than IP(HF)$_2$, which means, in the decay

$$(HF_2)_2^+ \rightarrow H_2F^+ + F \qquad (10)$$

may be involved a considerable amount of excess energy.

The fact that no $(HF)_2^+$ was detected has been accounted for by a zero potential energy barrier for reaction (10). In addition this result gives rise for the assumption that the intersection of the Franck-Condon region with the potential surface of $(HF)_2^+$ lies well above the dissociation limit of the ionic dimer into F and H_2F^+.

CH_3F like tertiary amines and ethers is a proton acceptor molecule, but with weaker acceptor properties compared to the amines and ethers. Therefore the neutral aggregates $(CH_3F)_n$ are supposed to be van der Waals complexes, this is confirmed by ab initio calculations /27/. In the ionized species $(CH_3F)_n^+$ increased proton donor properties are expected.

The cluster photoionization mass spectrum shows the signal of CH_3F^+ but no $(CH_3F)_n^+$ signals /26/; the occurence of proton transfer reactions is indicated by the mass peaks of $(CH_3F)_nH^+$ (n = 1,2), in addition one observes $C_2H_4F^+$, $C_2H_6F^+$ and $C_2H_5F_2^+$. The cluster photoionization mass spectrum shows the same main features as the cluster electron impact spectrum reported recently /28,29/, except the relative intensities of some fragmentation channels. The threshold energies of the fragmentation processes are enlisted in table 9.

Table 9: Appearance potentials of fragments from $(CH_3F)_n$ clusters

cation	AP [eV]	cation	AP [eV]
$(CH_3F)H^+$	12,43	$C_2H_6F^+$	12.47
$(CH_3F)_2H^+$	12.08	$C_2F_5F_2^+$	13.00
$C_2H_4F^+$	12.25		

The proton affinity of CH_3F may be calculated according to

$$(CH_3F)_2 + h\nu \quad \rightarrow \quad (CH_3F)H^+ + CH_2F + e^- \qquad (11)$$

The structure of the dimer is predicted to be antiparallel staggered; the intermolecular bond energy comes to -2.5 kcal/mol /30/.Using the APs given in table 9 and thermochemical data from the literature /21,31-34/ one calculates $PA(CH_3F) = 129$ kcal/mol.Measurements of this proton affinity using the bracketing technique deliver $PA(CH_3F) = 151$ kcal/mol /35/.The same behaviour - no dimer ion is detected and the proton affinity is substantially lower than the reference value obtained with bracketing technique - is observed with HF.As the appearance potential corresponds to a vertical transition it is likely that the formation of CH_4F^+ includes excess energy either in the form of internal energy or kinetic energy.

In analogy one obtains the proton affinity of $(CH_3F)_2$ from the AP of the protonated dimer.It results $PA(CH_3F)_2 = 137$ kcal/mol;this value is lower than $PA(CH_3F)$,therefore it is assumed that the formation of the protonated dimer also includes excess energy.

It is remarkable that other fragmentation reactions can compete with the proton transfer process.From the pressure dependence of the cluster electron impact spectra /28/ one may conclude,that these processes start mainly from protonated dimers and trimers.

The formation of $C_2H_6F^+$ can be interpreted as a two step process starting from a trimer

$$(CH_3F)_3 + h\nu \quad \rightarrow \quad C_2H_7F_2^+ + CH_2F + e^- \qquad (12)$$

$$C_2H_7F_2^+ \quad \rightarrow \quad C_2H_6F^+ + HF \qquad (13)$$

The intermediately formed protonated dimer decomposes in a second step into $C_2H_6F^+$.The activation barrier for the second step is 0.39 eV.From the AP one calculates $\Delta H_f(C_2H_6F^+) = 193$ kcal/mol.This value is higher than $\Delta H_f(C_2H_6F^+) = 142$ kcal/mol obtained for protonated fluoroethane /21/.Therefore we favour $C_2H_6F^+$ to be a solvated methyl cation.Following the process

$$C_2H_6F^+ + D(C_2H_6F^+) \quad \rightarrow \quad CH_3^+ + CH_3F \qquad (14)$$

we calculate with $\Delta H_f(C_2H_6F^+) = 193$ kcal/mol the dissociation energy $D(C_2H_6F^+) = 15$ kcal/mol.This is a reasonable value for the intermolecular binding energy in an ionic complex,but does not exclude a dimethyl-fluoronium structure /36-38/,which has been proposed for the product of the ion molecule reaction

$$CH_3F + CH_4F^+ \quad \rightarrow \quad C_2H_6F^+ + HF \qquad (15)$$

The photofragmentation product $C_2H_5F_2^+$ could be a solvation complex of CH_2F^+ and CH_3F,which is formed from a dimer or trimer according to

$$(CH_3F)_2 + h\nu \quad \rightarrow \quad C_2H_5F_2^+ + H + e^- \qquad (16)$$

$$(CH_3F)_3 + h\nu \quad \rightarrow \quad C_2H_7F_2^+ + CH_2F + e^- \qquad (17a)$$

$$C_2H_7F_2^+ \quad \rightarrow \quad C_2H_5F_2^+ + H_2 \qquad (17b)$$

As no $\Delta H_f(C_2H_5F_2^+)$ values are available from the literature the assignment of this ion is somewhat tentative. According to the decay channels (16) and (17) ΔH_f values of 138 kcal/mol and 141 kcal/mol are calculated respectively. To these values correspond a dissociation energy of the ion dipole complex of 8 kcal/mol and 5 kcal/mol respectively

The $C_2H_4F^+$ fragment is proposed to be a protonated fluoroethene ($\Delta H_f(C_2H_4F^{+}) = 159$ kcal/mol), but it should be mentioned, that the heat of formation calculated from the AP given in table 9 and assuming one of the following processes

$$(CH_3F)_2 + h\nu \quad \rightarrow \quad C_2H_4F^+ + HF + H + e^- \quad (18)$$

$$(CH_3F)_3 + h\nu \quad \rightarrow \quad C_2H_4F^+ + HF + H_2 + CH_2F + e^- \quad (19)$$

comes to 185 kcal/mol and 188 kcal/mol respectively.

5. CONCLUSION

The photoionization data obtained from hydrogen bonded clusters reveal strong red shifts of the ionization potentials, which are interpreted to be vertical ones. The fragment ion patterns are dominated by protonated monomers and aggregates $(M_n)H^+$. Proton solvation energies calculated from threshold energies are usually lower bounds of the adiabatic values.

Concerning with the experimental material available at present, it can be assumed that neutral molecular aggregates classified as van der Waals clusters reveal in the mass spectra besides proton transfer in addition other fragmentation processes.

6. REFERENCES

/1/ J.Jortner
     Ber.Bunsenges.Phys.Chem. 88,188 (1984)
/2/ P.Hobza and R.Zahradnik
     "Weak intermolecular Interactions in Chemistry and Biology"
     Elsevier Scientific Publ.Comp. Amsterdam,Oxford,New York 1980
/3/ C.Y.Ng
     Advances in Chem.Phys. 52,264 (1983)
/4/ T.D.Märk and A.W.Castleman jr.
     Adv. in Atomic and Molecular Physics;20,65 (1985)
/5/ A.W.Castleman jr. and R.G.Keesee
     Chem.Rev. 86,589 (1986)
/6/ A.C.Legon and D.J.Millen
     Chem.Rev. 86,635 (1986)
/7/ K.Rademann,B.Brutschy and H.Baumgärtel
     Chem.Phys. 80,129 (1983)
/8/ P.Schuster, G.Zundel and C.Sandorfy eds.
     "The Hydrogen bond"
     North Holland Publ.Comp.,Amsterdam,New York,Oxford 1976
/9/ P.Schuster in
     "Perspectives in Quantum Chemistry and Biochemistry"
     B.Pullman ed., Vol.II; J.Wiley Sons, New York, 1977

/10/ W.Kutzelnigg
      "Einführung in die Theoretische Chemie"
      Vol.2, p.364f; Verlag Chemie, Weinheim, New York 1978
/11/ S.T.Ceyer, P.W.Tiedemann, B.H.Mahan and Y.T.Lee
      J.Chem.Phys. 70, 14 (1979)
/12/ P.Bisling, E.Rühl, B.Brutschy and H.Baumgärtel
      to be published
/13/ A.W.Potts and W.C.Price
      Proc.Roy.Soc.(Lond), A326, 181 (1972)
/14/ W.A.Lathan, L.A.Curtiss, W.J.Hehre, J.B.Lisle and J.A.Pople
      Progr.Phys.Org.Chem., 11, 175 (1974)
/15/ S.Tomoda and K.Kimura
      Chem.Phys.Lett., 121, 159 (1985)
/16/ H.Z.Cao, E.M.Evleth and E.Kassap
      J.Chem.Phys., 81, 1512 (1984)
/17/ G.Brink and L.Glasser
      J.Mol.Struct., 85, 317 (1981)
/18/ J.D.Lambert and E.D.T.Strong
      Proc.Roy.Soc., A200, 566 (1950)
/19/ J.A.Otutola, R.Viswanathan and T.R.Dyke
      J.Am.Chem.Soc., 101, 4787 (1979)
/20/ T.A.Aue and M.T.Bowers
      in"Gas Phase Ion Chemistry", ed.M.T.Bowers
      Vol.II, p 1 ff.; Academ.Press New York 1979
/21/ S.G.Lias, J.F.Liebman and R.D.Levin
      J.Phys.Chem. Ref.Data 13, 695 (1984)
/22/ C.Y.Ng, D.J.Trevor, P.W.Tiedemann, S.T.Ceyer, P.L.Kronebusch,
      B.H.Mahan and Y.T.Lee
      J.Chem.Phys., 67, 4235 (1977)
/23/ K.D.Cook, G.G.Jones and J.W.Taylor
      Int.J.Mass.Spectrom.Ion Phys., 35, 273 (1980)
/24/ P.W.Tiedemann, S.L.Anderson, S.T.Ceyer, T.Hirooka, C.Y.Ng,
      B.H.Mahan and Y.T.Lee
      J.Chem.Phys., 71, 605 (1979)
/25/ M.S.Foster and J.L.Beauchamp
      Inorg.Chem., 14, 1229 (1975)
/26/ E.Rühl, P.Bisling, B.Brutschy and H.Baumgärtel
      J.Electron Spectroscop. Rel.Phen. to be published
/27/ T.Oi, E.Sekreta and T.Ishida
      J.Phys.Chem., 87, 2323 (1983)
/28/ J.F.Garvey and R.B.Bernstein
      Chem.Phys.Lett., 126, 394 (1986)
/29/ J.F.Garvey and R.B.Bernstein
      J.Phys.Chem. , 90, 3577 (1986)
/30/ H.J.Böhm, R.Ahlrichs, P.Scharf and H.Schiffer
      J.Chem.Phys., 81, 1389 (1984)
/31/ D.R.Stull, H.Prophet et al.
      JANAF Thermochemical Tables NSRDS-NBS 37
      Washington D.C. 1971
/32/ H.M.Rosenstock, K.Draxl, B.W.Steiner and J.T.Herron
      J.Phys.Chem., Ref.Data 6, Suppl.1 (1977)

/33/ D.D.Wagman, W.H.Evans, V.B.Parker, R.M.Schumm, I.Halow,
S.M.Bailey, K.L.Churney and R.L.Nuttall
J.Phys.Chem., Ref.Data 11, Suppl.2 (1982)
/34/ D.F.McMillen and D.M.Golden
Ann.Rev.Phys.Chem., 33, 493 (1982)
/35/ J.L.Beauchamp, D.Holtz, S.Woodgate and S.L.Patt
J.Am.Chem.Soc., 94, 2798 (1972)
/36/ N.A.McAskill
Austr.J.Chem., 23, 2301 (1970)
/37/ R.J.Blint, J.B.McMahon and J.L.Beauchamp
J.Am.Chem.Soc., 96, 1269 (1974)
/38/ M.J.K.Pabst, H.S.Tan and J.L.Franklin
Int.J.Mass Spectrom.Ion Phys., 20, 191 (1976)

ACKNOWLEDGEMENTS

Financial support of the Bundesministerium für Forschung und
Technologie and the Fonds der Chemischen Industrie is gratefully
acknowledged.

THE UNAVOIDABLE IMPORTANCE OF ELECTROSTATIC EFFECTS IN THE STRUCTURES
OF WEAKLY BONDED COMPLEXES

Clifford E. Dykstra and Shi-yi Liu
Department of Chemistry
University of Illinois at Urbana-Champaign
Urbana, Illinois 61801
U.S.A.

ABSTRACT.   The electrical interaction between static and polarized
charge distributions of molecules is very important in determining the
approach orientations of monomers making up hydrogen bonded complexes.
A detailed analysis of those interactions is presented. The orientational angles describing the structures of complexes can be predicted
by searching the surface of the electrical interaction energy, and a
large number of binary complexes and several larger complexes have been
studied in this way. The model appears to be generally successful.

1.   INTRODUCTION

Weak intermolecular attraction is ubiquitous in chemical phenomena, and
it is a key feature in connecting isolated, small molecule chemistry
with biological, surface and condensed phase chemistry. Coulson [1]
laid the foundation for much of current thinking about the basis for
weak intermolecular interaction. He partitioned the interaction into
contributions from different physical effects, and he attempted to show
that these effects may be of comparable and competing size. The physical effects that one might consider in an intermolecular interaction
are the electrical interactions of the unperturbed charge distributions, charge polarization, charge transfer, dispersion and exchange
repulsion. All but the last of these are generally attractive, and the
last three are exclusively quantum mechanical in origin. Ab initio
electronic structure tools have been constructed to try to carry out
energy partitioning at the level of the underlying physics [2-4]. Of
course these effects, though conceptualized as independent physical
features, are not strictly independent and are not even separate terms
in the molecular Hamiltonian.

One type of partitioning is unambiguous and it is the separation
between contributions that involve intermolecular quantum phenomena and
those that are classical. The latter contributions are electrostatic,
the sum of the interaction energies of the permanent charge fields and

A. Weber (ed.), Structure and Dynamics of Weakly Bound Molecular Complexes, 319–335.

the induced fields or polarization. One might rightfully regard these
contributions as semi-classical rather than classical, because the
nature of the interacting charge distributions is of quantum mechanical
origin; it is in the intermolecular effects that quantum features are
set aside in this type of partitioning.

Contemporary understanding of intermolecular electrostatics
evolved from the fundamental ideas developed by Buckingham [5]. Among
other things, a handy notion was abundantly clear: Quadrupole moments
are important in molecular systems, and interactions must be thought of
as involving juxtaposition of dipole-dipole, dipole-quadrupole, and
quadrupole-quadrupole interactions. Modern quantum chemistry has provided new tools that make it possible to examine a molecule's electrical properties as never before. With that, rigorous, unambiguous
evaluation of essentially complete electrostatic interaction energies
can be performed. The result is a strengthened conclusion that electrostatic effects are unavoidably important in understanding the nature
of weak molecular complexes. The next step, exploiting this understanding in computational studies of ever larger systems, is an ex-
traordinary opportunity for chemical science to complete the bridge
between microscopic and macroscopic molecular phenomena.

2. INTERMOLECULAR ELECTRICAL INTERACTIONS

Though definitions of molecular electrical multipoles and polarizabilities are available at the textbook level [6], we find that expressing the properties in one   certain manner offers convenience and
facilitates applications. Classical electrostatics dictates that the
energy of interaction for placing a charge, q_i, in an electric potential, $V(x,y,z)$, is the product of the charge and the potential at the
position of the charge, $r_i = (x_i, y_i, z_i)$. For a distribution of N point
charges in an electric potential, the classical interaction energy is a
sum:

$$E^{int} = - \sum_i^N q_i \ V(x_i, y_i, z_i) \tag{1}$$

The potential function V can be expressed as a power series expansion a-
bout some point, which is most conveniently taken to be the origin "o".

$$V(x,y,z) = V_o + x \left.\frac{\partial V}{\partial x}\right|_o + y \left.\frac{\partial V}{\partial y}\right|_o + z \left.\frac{\partial V}{\partial z}\right|_o$$

$$+ \frac{1}{2} x^2 \left.\frac{\partial^2 V}{\partial x^2}\right|_o + \frac{1}{2} y^2 \left.\frac{\partial^2 V}{\partial y^2}\right|_o + \frac{1}{2} z^2 \left.\frac{\partial^2 V}{\partial z^2}\right|_o$$

$$+ xy \left.\frac{\partial^2 V}{\partial x \partial y}\right|_o + xz \left.\frac{\partial^2 V}{\partial x \partial z}\right|_o + yz \left.\frac{\partial^2 V}{\partial y \partial z}\right|_o \tag{2}$$

$$+ \frac{1}{3} \cdot \frac{1}{2} x^3 \frac{\partial^3 V}{\partial x^3}\bigg|_o \quad + \frac{1}{3} \cdot \frac{1}{2} y^3 \frac{\partial^3 V}{\partial y^3}\bigg|_o \quad + \frac{1}{3} \cdot \frac{1}{2} z^3 \frac{\partial^3 V}{\partial z^2}\bigg|_o$$

$$+ \frac{1}{2} x^2 y \frac{\partial^3 V}{\partial x^2 \partial y}\bigg|_o \quad + \frac{1}{2} x^2 z \frac{\partial^3 V}{\partial x^2 \partial z}\bigg|_o \quad + xyz \frac{\partial^3 V}{\partial x \partial y \partial z}\bigg|_o$$

$$+ \dots$$

Using this expansion for the $V(x_i, y_i, z_i)$ values in Eqn. 1, and using the subscript notation where $V_{xy} = \frac{\partial^2 V}{\partial x \partial y}\bigg|_o$ yields the usual multipole expansion.

$$E^{int} = -V_o \sum_i q_i$$

$$- V_x \sum_i q_i x_i - V_y \sum_i q_i y_i - V_z \sum_i q_i z_i$$

$$- \frac{1}{2} V_{xx} \sum_i q_i x_i^2 - \frac{1}{2} V_{yy} \sum_i q_i y_i^2 - \frac{1}{2} V_{zz} \sum_i q_i z_i^2 \qquad (3)$$

$$- V_{xy} \sum_i q_i x_i y_i - V_{xz} \sum_i q_i x_i z_i - V_{yz} \sum_i q_i y_i z_i$$

$$- \frac{1}{6} V_{xxx} \sum_i q_i x_i^3 + \dots$$

$$= -V_o Q - V_x \mu_x - V_y \mu_y - V_z \mu_z - \frac{1}{2} V_{xx} Q_{xx} - \frac{1}{2} V_{xy} Q_{xy} + \dots$$

$$- \frac{1}{6} V_{xxx} R_{xxx} + \dots \qquad (4)$$

The sums appearing in Eqn. 3 correspond to moments of the charge distribution: $Q = \sum_i q_i$ is the total charge or zeroeth moment; $\mu_x = \sum_i q_i x_i$ is the x-component of the first or dipole moment. Next are second moment terms, then third moment terms and so on. In general, the elements of each moment depend on the specific directionality of the coordinate axes. Also, the choice of the evaluation center affects moments beyond the first non-vanishing moment. The nth moment elements can be evaluated at a spatially translated origin if the complete set of moments, through the nth moment, is known at the original origin. This implies that with a truncated moment expansion the interaction energy depends on the evaluation center. Thus, the validity of a <u>truncated</u>

moment expansion depends on there being a physically reasonable
choice of the evaluation center and/or truncation [7]. A physically justified spatially distributed set of moments [8,9] is a
corresponding idea that seems extremely useful for minimizing
truncation error.

Laplace's equation, means that the number of unique elements
needed to evaluate an interaction energy can be reduced. For the
second moment this amounts to a transformation into a traceless form, a
form usually referred to as the quadrupole moment [5]. With modern
computation machinery, such reduction is of lessened benefit, and on
vector machines, it may be less efficient in certain steps. We usually
do not make that transformation and instead use "Cartesian moments."
Interaction results are entirely equivalent. Logan has pointed out the
convenience and utility of the Cartesian form of the multipole polarizabilities [10] and Applequist has set forth a computationally useful
organization for electrical properties in the Cartesian form [11].
Applequist defines a Cartesian polytensor of the first degree as a
sequence of Cartesian tensors, each arranged in a column. The ordering
of tensor elements is

first: x, y, z
second: xx, xy, xz, yx, yy, yz, zx, zy, zz   (5)
third: xxx, xxy, xxz, xyx, xyy, ..., zzz

The first degree polytensor is a stacked list of the (column) moment
tensors in increasing rank. This will be designated M, with M_x being
the x-component of the first moment, M_{xy} being the xy-component of the
second moment, and so on. Notice that in this list, equivalent values
such as M_{xy} and M_{yx} are all included.

$$
M = \begin{pmatrix}
M_o \\
M_x \\
M_y \\
M_z \\
M_{xx} \\
M_{xy} \\
\bullet \\
\bullet \\
\bullet \\
M_{zz} \\
M_{xxx} \\
M_{xxy} \\
\bullet \\
\bullet \\
\bullet
\end{pmatrix}
\tag{6}
$$

The power series in Eqn. 2 has as many partial derivative values
as there are multipole moments. They can also be arranged in a first
degree Cartesian polytensor, V, with row indices in the same anti-
canonical order. The energy of interaction can now be expressed as

$$E_{int} \equiv -V^T \cdot M \tag{7}$$

For quantum mechanical systems, moment operators are used to construct an interaction Hamiltonian.

$$\hat{H} = \hat{H}_0 - V^T \cdot \hat{M} \tag{8}$$

where $\hat{H}_0$ is the free molecule Hamiltonian. V is now a set of parameters in the Schrodinger equation. For any particular choice of V one has a new Schrodinger equation to solve. The Hamiltonian does not include polarization terms but the quantum mechanical charge distributions are certainly polarizable. Polarization develops because of how the wavefunction adjusts to a potential, V. In other words, the wavefunction and likewise the energy eigenvalues, E, can (and do) have non-linear dependencies on the elements of V even though the Hamiltonian is only linear in V. The non-linear dependencies of E on the elements of V are the multipole polarizabilities, hyperpolarizabilities and so on.

Dykstra and Jasien [12] developed a quantum mechanical tool, designated derivative Hartree-Fock (DHF), that analytically yields any order derivative of the SCF energy. It has been used to obtain electrical properties of quite a few molecules [13], and has been carried as far as a sixth dipole hyperpolarizability [14]. We use properties such as these and permanent moments calculated from well-correlated wavefunctions to evaluate electrical interaction energies [15].

Figure 1 shows the contributions of permanent moments and selected polarizabilities to the electrostatic interaction energy of the HCCH-HF complex. At each point of truncation, evaluation has been carried out two ways. The first way accounts for the polarization of one monomer by the other's permanent moments. The interaction energy includes permanent-permanent and permanent-induced moment interactions only. In the second way, an iterative process allows for full mutual polarization and includes the entire induced-induced moment interactions. An interesting point is the exaggerated polarization that occurs when only the dipole polarizability is included. The iterative process in that case is nearly divergent. Physically, this results because α allows for only a shift of charge density. When the dipole-quadrupole and quadrupole-quadrupole polarizabilities are included, the exaggeration is remedied nicely. Formal analysis of convergence limits for charge distributions in non-uniform fields has been done by Larter and Malik [16].

One block in Figure 1, the non-iterative incorporation of μ, Q, R, α, A and C, represents a useful compromise between completeness in the electrical interaction model and economy. A more stringent test of that compromise is an energy difference, and such a case is shown in Figure 2. We have implemented an efficient scheme for calculations on large clusters at this compromise level [17], and have used it for the more-than-three molecule systems discussed herein.

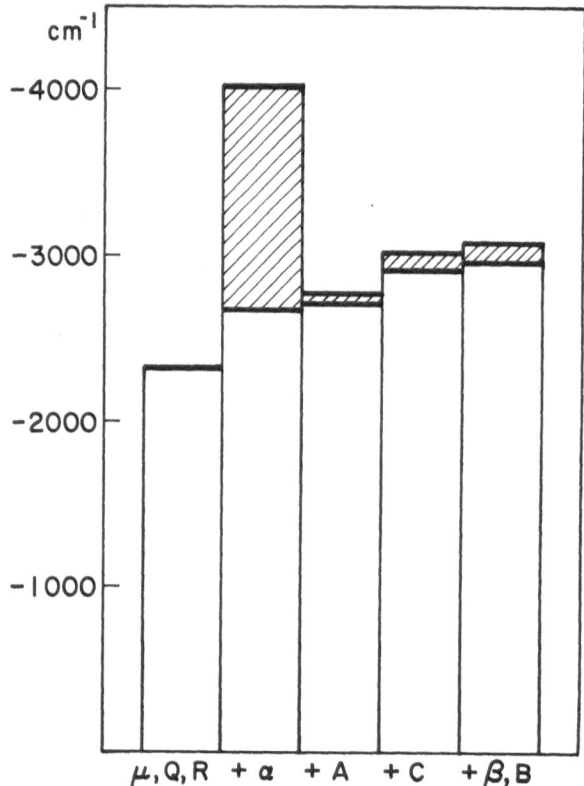

Figure 1.   Contributions to the electrostatic interaction energy of
acetylene and HF.

3.   STRUCTURES OF MOLECULAR COMPLEXES

The structures of numerous hydrogen bonded complexes have been examined
spectroscopically in recent years with molecular beam electric resonance [18-21], Fourier transform microwave spectroscopy [22-24], high
resolution vibrational spectroscopy [25,26], vibrational predissociation spectra [27,28] and matrix isolation IR [29,30]. Theoretical
investigation through high-level ab initio electronic structure calculations has been fairly extensive as well [see, for example: 31-33].
Thus, there is a wealth of structural information or spectroscopic data
from which to glean the fundamental basis for the nature of weak complexes. We compare that information with electrostatic predictions for
a number of complexes.

3.1. Binary Complexes

Exchange repulsion, a quantum mechanical feature, prevents the coales-

cence of two neutral charge distributions. However, it and dispersion
forces depend most on the proximity of the charge distributions and not
so much on their orientations. In the classical electrostatic treatments, both are neglected and instead molecules are frozen translationally. The sensitivity of the optimum orientational angles so
determined to the fixed intermolecular distances is, however, weak [34].
For a number of binary complexes, electrical interaction analysis was
carried out at equilibrium separations with a fully converged iterative
treatment of the mutual polarization, with permanent moments through
the octupole, and with α, A and C polarizabilities, and β and B hyperpolarizabilities.

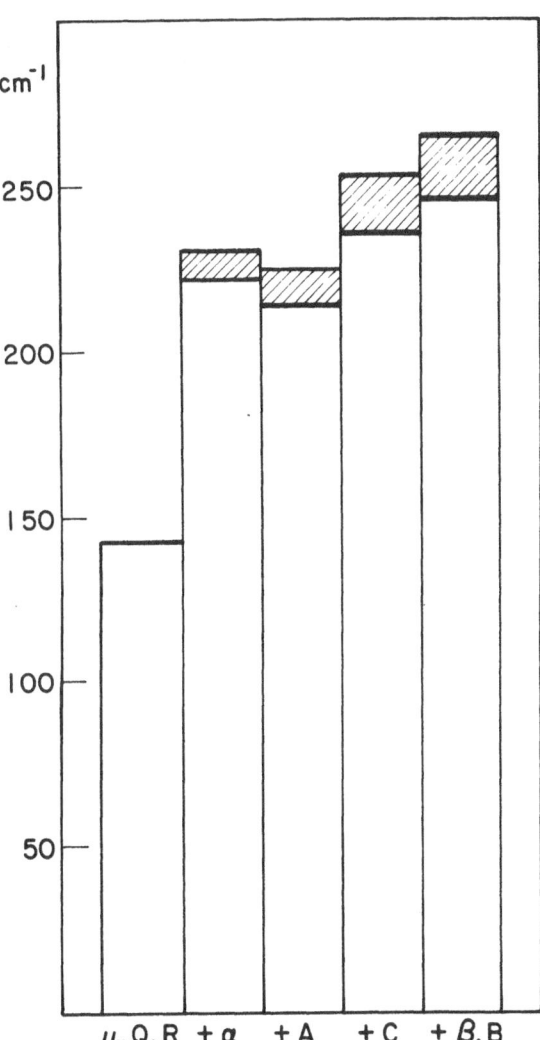

Figure 2.   Electrostatic interaction energy contributions to the inversion barrier of the HF-dimer following Figure 1.

Table I lists the binary complexes that were studied and summarizes
the qualitative orientational information. A number of the complexes
have been experimentally studied and are known to have linear equilibrium structures. In certain cases, where vibrational averaging effects
can not be entirely ascertained, measured data is consistent with the
equilibrium structures being linear. In each case, electrical analysis
yields a linear equilibrium structure with proper left-right orientation
of each monomer. There is no single explanation for linearity in all
the structures. For HCN-HCN and HCN-HF, the strong dipole-dipole interaction dominates. In NN-HF and NN-HCl, it is the dipole of HF or HCl
strongly interacting with the quadrupole of N_2 that leads to a linear
equilibrium structure. The electrical model predicts axially-symmetric
complexes of HCCH and HCN with ammonia, the protons pointing into the
NH_3 lone pair. The NH_3-HCCH complex exhibits considerable vibrational
averaging effects in the microwave spectrum [41], but the spectrum is
consistent with an axially-symmetric structure. Ab initio calculations
[42] have also predicted this type of equilibrium structure.
Spectroscopic studies of NH_3-HCN [40] are indicative of an axially-
symmetric structure as well. The complex of HCCH and HF is T-shaped
[43] as is the HCCH-HCl [44] complex. This is correctly found by the
electrical model and is explained by the dominant interaction of the HF
or HCl dipole with the acetylene quadrupole. The electrical model
predicts the H_2-HF complex to be T-shaped, and well-correlated, large
basis set ab initio results concur [45].

For the HF-dimer, the electrical model's results are in satisfactory agreement with the ab initio surface of Michael et al. [33]. In
these calculations a two-center multipole expansion was used for each
monomer, the center of negative charge and the center of positive
charge. As the values in Table II show, the angle of the free HF, θ_1,
disagrees by 15°. One reason for this is that the electrical model
results were based on properties obtained with a much larger basis than
that used for the ab initio potential surface. The ab initio values
are necessarily in error because they are relatively deficient in
describing the electrical interaction. The correlation effect on θ_1
and θ_2 is found to be no more than 1°, and so SCF level calculations
were sufficient to explore the dependence of θ_1 and θ_2 on the basis.
SCF calculations with a very large 82 function basis, TZ2P, were carried out for the HF-dimer [34], and θ_1 and θ_2 changed to be more in
line with the electrical model. This basis set is nearly as flexible
as that used to obtain the electrical properties, and so we conclude
that the reliability of the electrical model may be as good as an 8°
uncertainty. The transition state for interconversion of $(HF)_2$ appears
well-determined by the electrical properties, too, and the electrical
prediction of the barrier height is 264 cm^{-1}. This is below but not
sharply different from ab initio and other determinations [33, 51, 52].
There is comparable agreement between ab initio and electrostatic
predictions for the interconversion of H_2-HF [45], also. This agreement assures that the electrical model is following in step with the ab
initio potential for angular dependence.

TABLE I. Electrostatic Orientational Predictions for Binary Complexes.

Complex	Electrostatic Prediction	Expt. Agreement		Ab Initio Agreement	
NN-HF	linear	Yes	[35]	Yes	[31]
OC-HF	linear	Yes	[36]	Yes	[31]
HCN-HF	linear	Yes	{37]	Yes	[31]
HCN-HCN	linear	Yes	[38]		
OC-HCl	linear	Yes	[36]		
HCN-HCl	linear	Yes	[39]		
NN-HCl	linear	Yes	[35]		
H_3N-HCCH	axially symmmetric	Yes	[40]		
H_3N-HCN	axially symmetric	Yes	[41]	Yes	[42]
HCCH-HF	T-shaped	Yes	[43]	Yes	[32]
HCCH-HCl	T-shaped	Yes	[44]		
H_2-HF	T-shaped			Yes	[45]
$(HF)_2$	bent	Yes	[1]	Yes	[33]
$(NH_3)_2$	bent, asymmetric	cyclic [46]		[47-49]	
$(HCl)_2$	bent			Yes	[50]

Table II. Comparison of Ab Initio and Electrical Model Predictions for HF and HCl Complexes.

Species	Ab Initio Values		Electrical Model	
	θ_1	θ_2	θ_1	θ_2
(HF)$_2$				
equilibrium	120.1[a]	-6.4[a]	105.5	-10.5
	112.9[c]	-8.6[c]		
transition state	54.4[a]	54.4[a]	56.9	56.9
(HCl)$_2$				
equilibrium	97[b]	-2[b]	92.8	-9.8

[a]Ref. 33.
[b]Ref. 50.
[c]82-function basis calculation.

     The water dimer has been studied with electrostatics. The equilib-
rium structure was determined by searching for the optimal values of θ_1
and θ_2, which are the angles between the line connecting the two centers
of mass and the bonding hydrogen of water molecule 1, and the C_2 symmetry axis of molecule 2, respectively. The electrical potential surface
was generated assuming C_s symmetry for the dimer, the equilibrium angles
were found to be 8.7° and 104.2° for R_{COM} = 3.0 A. This is well within
the range of values determined recently by Frisch and coworkers [47b]
who examined the effects of basis set selection as well as the effect of
correlation. They found values for θ_1 to vary from 0.2° to 9.1° and θ_2
from 100.2° to 143.1°. We also generated certain cuts along the potential surface near the energy minimum to compare with the ab initio
potential of Popkie and coworkers [53]. Where there are enough points
along the potential curve to make a qualitative comparison, the electrical potential agrees reasonably well with the ab initio surface.

     The ammonia dimer is a special problem because as yet there appears
to be greater disagreement with experiment [46] in orientational angles
(~25-30°) than for any other complex. This could indicate a limitation
in the electrostatic treatment, or an incompleteness that does not show
up in the other complexes. However, it may and quite probably does
indicate a mismatch between what a potential surface reveals and the
structural parameters deduced so far from spectroscopic measurements. In
particular, we believe the vibrational motion of (NH$_3$)$_2$ significantly
obscures the usual process of deducing structure from microwave spectra
[34,54]. Also, ab initio calculations [55] have pointed to potential
limitations in using nuclear quadrupolar coupling constants to determine
orientational angles because of hidden approximations. Errors could be
10-20°.

An important result of the quantitative electrostatic predictions of structures is that non-linearity in the hydrogen bonds of species such as $(HF)_2$, $(HCl)_2$ and $(NH_3)_2$ appears surprising only because existing paradigms lead us to expect linearity. When viewed from the standpoint of the juxtaposed multipole moment and induced moment interactions, a linear hydrogen bond is more an accident, occurring mainly when a dipole interaction dominates. Furthermore, weak intermolecular interactions that can't be "hydrogen" bonding, as in N_2-CO, may be considered on exactly the same grounds.

3.2. Larger Complexes

The juxtaposition of multipole moment interactions is clearly evident in the non-linearity of the hydrogen bond of cyclic $(HF)_3$ [56,57]. This might suggest a weakened hydrogen bond between each HF if linearity is a necessary stabilizing feature of the interaction. However, we have shown [56] that to within 5°, the orientational angles of $(HF)_3$ and of $(HF)_4$ are predicted by the electrical model and are the natural consequence of balancing mainly dipole and quadrupole interactions. Furthermore, there is no weakening relative to $(HF)_2$. Applying the electrical model to cyclic $(HCl)_3$, another case, we obtain 22.8° for the orientational parameter compared to a differently calculated value [50], one based on an accurate $(HCl)_2$ potential, of 19°.

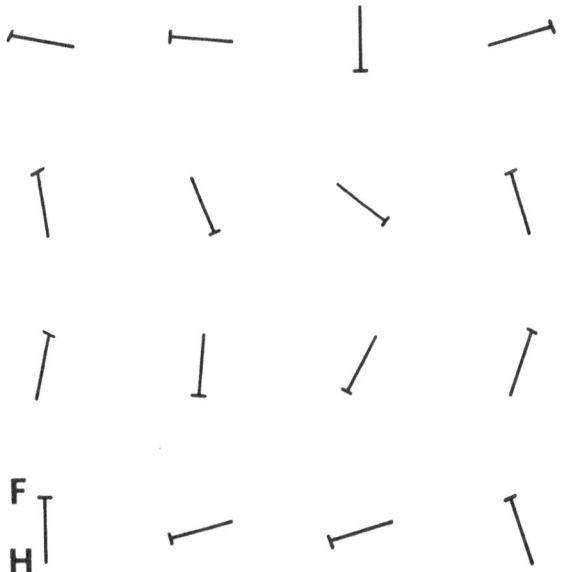

Figure 3a. Optimum electrostatic orientations of HF molecules fixed at center-of-mass separations of 2.8 Å and arranged in a 4x4 2-dimensional array.

Figure 3b. Optimum electrostatic orientations of a 5x5 array of HF molecules.

    As a preliminary demonstration of how electrostatics can be em-
ployed to study aggregations from small size to large size, we have
calculated the structure of a uniform 2-dimensional (idealized surface)
layers of 16, 25 and 36 HF molecules. These are shown in Figures 3a-c.
The fact that the intermolecular interactions can be done classically
instead of quantum mechanically reduces the computing cost by an estimated factor of 10^{10}! Establishing the underlying basis for weak
intermolecular attraction clearly paves the way for more involved
studies of molecular clustering phenomena.

4. OTHER FEATURES OF HYDROGEN BONDING

We have explored the role of electrostatic effects in an important
manifestation hydrogen bonding, the changes in intramolecular vibration. We have found that for a series of binary HF-complexes, the
electrostatic interaction changes the HF stretching potential in a way
that does account rather well for the observed vibrational frequency
shifts [58]. We have also made quantitative predictions of vibrational

transition moment enhancements that arise through electrical polarization [14]. When laboratory measurements of these moments are obtained,
the comparison will test an important aspect of the electrical interaction. We anticipate reasonably good success because we have found that
intermolecular quantum effects on electrical properties of a complex
tend to be unimportant [59], and it is only those effects that have
been neglected in predicting the transition moments.

Figure 3c. Optimum electrostatic orientations of a 6x6 array of HF
molecules. In the central regions of the array, there is a tendency
for molecular pairs to have relative oreintations similar to (HF)$_2$.

ACKNOWLEDGMENTS

For one of us (C.E.D.), theoretical investigation of hydrogen bonding, weak interactions and molecular clustering has been stimulated by experimentalist colleagues, particularly Professors W. H. Flygare and J. M. Lisy. We are grateful for the understanding and ideas they shared with us.
    This work was supported by the Chemistry Division of the National Science Foundation through grant CHE84-19496.

REFERENCES

1. C. A. Coulson, Research 10, 149 (1957).

2. L. C. Allen and P. A. Kollman, Chem. Rev. 72, 283 (1972).

3. K. Kitaura and K. Morokuma, Int. J. Quantum Chem. 10, 325 (1976);
    H. Umeyama and K. Morokuma, J. Am. Chem. Soc. 99, 1316 (1977).

4. P. A. Kollman, J. Am. Chem. Soc. 99, 4875 (1977).

5. A. D. Buckingham, Quart. Rev. (London) 13, 189 (1959); Adv. Chem.
    Phys. 12, 107 (1967).

6. See for example, P.W. Atkins, "Molecular Quantum Mechanics," 2nd
    edition (Oxford University Press, Oxford, 1983).

7. D. E. Stogryn and A. P. Stogryn, Molec. Phys. 11, 371 (1966).

8. A. J. Stone and M. Alderton, Molec. Phys. 56, 1047 (1985).

9. A. D. Buckingham and P. W. Fowler, Can. J. Chem. 63, 2018 (1985).

10. D. E. Logan, Mol. Phys. 46, 271 (1982).

11. J. Applequist, J. Math. Phys. 24, 736 (1983); J. Chem. Phys. 83,
    809 (1985); Chem. Phys. 85, 279 (1984).

12. C. E. Dykstra and P. G. Jasien, Chem. Phys. Lett. 109, 388 (1984).

13. S.-Y. Liu and C. E. Dykstra, J. Phys. Chem., in press.

14. S.-Y. Liu, C. E. Dykstra and D. J. Malik, Chem. Phys. lett. in
    press.

15. C. E. Dykstra, S.-Y. Liu and D. J. Malik, Adv. Chem. Phys. (1987,
    edited by K. D. Sen).

16. R. Larter and D. J. Malik, to be published.

17. C. E. Dykstra, S.-Y. Liu and M. F. Daskalakis, J. Comput. Chem., to be submitted.

18. T. R. Dyke, B. J. Howard and W. Klemperer, J. Chem. Phys. 56, 2442 (1972).

19. S. E. Novick, P. Davies, S. J. Harris and W. Klemperer, J. Chem. Phys. 59, 22373 (1973).

20. T. R. Dyke and J. S. Muenter, J. Chem. Phys. 60, 2929 (1974).

21. W. Klemperer, J. Mol. Structure 59, 161 (1980).

22. T. J. Balle, E. J. Campbell, M. R. Keenan and W. H. Flygare, J. Chem. Phys. 71, 2723 (1979); 72, 922 (1980).

23. T. J. Balle and W. H. Flygare, Rev. Sci. Instrum. 52, 33 (1981).

24. E. J. Campbell, W. G. Read and J. A. Shea, Chem. Phys. Lett. 94, 69 (1983).

25. A. S. Pine and W. J. Lafferty, J. Chem. Phys. 78, 2154 (1983).

26. N. Ohashi and A. S. Pine, J. Chem. Phys. 81, 73 (1984).

27. J. M. Lisy, A. Tramer, M. F. Vernon and Y. T. Lee, J. Chem. Phys. 75, 4733 (1981).

28. M. P. Cassasa, C. M. Western, F. G. Cellii, D. E. Brniza and K. C. Janda, J. Chem. Phys. 79, 3227 (1983).

29. G. L. Johnson and L. Andrews, J. Am. Chem. Soc. 104, 3043 (1982).

30. L. Andrews, J. Phys. Chem. 88, 2940 (1984).

31. M. A. Benzel and C. E. Dykstra, J. Chem. Phys. 78, 4052 (1983); 80, 3510E (1984).

32. M. J. Frisch, J. A. Pople and J. E. DelBene, J. Chem. Phys. 78, 4063 (1983).

33. D. W. Michael, C. E. Dykstra and J. M. Lisy, J. Chem. Phys. 81, 5998 (1984).

34. S.-Y. Liu and C. E. Dykstra, Chem. Phys., in press.

35. P. D. Soper, A. C. legon, W. G. Read and W. H. Flygare, J. Chem. Phys. 76, 292 (1982).

36. A. C. Legon, P. D. Soper, M. R. Keenan, T. K. Minton, T. J. Balle and W. H. Flygare, J. Chem. Phys. 73, 583 (1980); P. D. Soper, A. C. Legon, W. G. Read and W. H. Flygare, J. Chem. Phys. 76, 292 (1982).

37. A. C. Legon, D. J. Millen and S. C. Rogers, Proc. R. Soc. London Ser. A. 370, 213 (1980).

38. L. W. Buxton, E. J. Campbell and W. H. Flygare, Chem. Phys. 56, 399 (1981).

39. A. C. Legon, E. J. Campbell and W. H. Flygare, J. Chem. Phys. 76, 2267 (1982).

40. W. J. Jones, R. M. Seel and N. Sheppard, Spectrochimica Acta A25, 385 (1969).

41. G. T. Fraser, K. R. Leopold and W. Klemperer, J. Chem. Phys. 80, 1423 (1984).

42. M. J. Frisch, J. A. Pople and J. E. Del Bene, J. Chem. Phys. 78, 4063 (1983).

43. W. G. Read and W. H. Flygare, J. Chem. Phys. 76, 2238 (1982).

44. A. C. Legon, E. J. Campbell and W. H. Flygare, J. Chem. Phys. 76, 2267 (1982).

45. D. E. Bernholdt, S.-Y. Liu and C. E. Dykstra, J. Chem. Phys. 85, 0000 (1986).

46. D. D. Nelson, G. T. Fraser and W. Klemperer, J. Chem. Phys. 83, 6201 (1985).

47. (a) M. J. Fisch, J. A. Pople and J. E. DelBene, J. Phys. Chem. 89, 3664 (1985); (b) M. J. Frisch, J. E. DelBene, J. S. Binkley and H. F. Schaefer, J. Chem. Phys. 84, 2279 (1986).

48. Z. Latajka and S. Scheiner, J. Chem. Phys. 84, 341 (1986).

49. K. P. Sagarik, R. Ahlrichs and S. Brode, Molec. Phys., in press.

50. C. Votara, R. Ahlrichs and A. Geiger, J. Chem. Phys. 78, 6841 (1983).

51. J. F. Gaw, Y. Yamaguchi, M. A. Vincent and H. F. Schaefer, J. Am. Chem. Soc. 106, 3133 (1984).

52. A. E. Barton and B. J. Howard, Faraday Discuss. Chem. Soc. 73, 45 (1982).

53. H. Popkie, H. Kistenmacher and E. Clementi, J. Chem. Phys. $\underline{59}$, 1325 (1973).

54. S.-Y. Liu, C. E. Dykstra, K. Kolenbrander and J. M. Lisy, J. Chem. Phys. $\underline{85}$, 2077(1986).

55. S.-Y. Liu and C. E. Dykstra, Chem. Phys., submitted.

56. S.-Y. Liu, D. W. Michael, C. E. Dykstra and J. M. Lisy, J. Chem. Phys. $\underline{84}$, 5032 (1986).

57. A. Karpfen, A. Beyer and P. Schuster, Chem. Phys. lett. $\underline{102}$, 289 (1983).

58. S.-Y. Liu and C. E. Dykstra, J. Phys. Chem. $\underline{90}$, 3097 (1986).

59. C. E. Dykstra, S.-Y. Liu and D. J. Malik, J. Molec. Structure (Theochem) $\underline{135}$, 357 (1986).

INTERMOLECULAR POTENTIALS, INTERNAL MOTIONS AND THE SPECTRA OF VAN DER
WAALS MOLECULES

G. Brocks and A. van der Avoird
Institute of Theoretical Chemistry
University of Nijmegen
Toernooiveld
6525 ED Nijmegen
The Netherlands

ABSTRACT. The infrared spectra of N_2Ar, O_2-Ar, N_2-N_2 and O_2-O_2 seem to
indicate that these Van der Waals molecules possess orientationally
localized (librational) states, as well as (nearly) free internal rotor
states. Using the full anisotropic potential, we have calculated the
rovibrational bound states of N_2-N_2 and N_2-Ar, and also the rotational
predissociation states of the latter complex, and we have evaluated the
contributions of these states to the infrared spectra. Thus the infrared
spectrum of N_2-Ar, and in principle also the spectra of the other dimers,
can be understood. The onset of the regular free internal rotor structure cannot be explained from the bound states; it is due to rather
narrow (0.2 to 3 cm^{-1}) rotational resonances lying in the collision
continuum. The structure in the lower frequency part of the spectra is
caused by transitions between localized librational states; especially
this part will be sensitive to the detailed intermolecular potential.
In order to exploit this sensitivity one has to measure the infrared
spectra at very low temperature.

1. INTRODUCTION

The mobility of the constituents in different Van der Waals molecules varies considerably. Considering, for instance, the series of
molecule-rare gas atom dimers, we observe, as one extreme, the nearly
free internal rotations of H_2 in the H_2-X dimers. A theoretical description of such dimers is best given by using a basis of free rotor functions, both for expanding the potential and for the rovibrational wave
functions [1]. At the other end of the series are the dimers in which
the (rigid) molecule is strongly aspherical, very long or flat, with
one or two of its dimensions larger than the Van der Waals distance
between the molecule and the rare gas atom. Internal rotations are
completely prevented in that case; at specific angles (and distances)
the potential becomes even infinitely repulsive. When the molecular
size is comparable with the Van der Waals distance, as in benzene-argon
or tetrazine-argon [2], dynamical calculations using a free rotor basis
are still possible, with some special precautions, but they converge

337

A. Weber (ed.), Structure and Dynamics of Weakly Bound Molecular Complexes, 337–355.
© *1987 by D. Reidel Publishing Company.*

very slowly. In cases like anthracene-argon or fluorene-argon [3], this
description breaks down completely and one has to use different internal
coordinates and different basis functions. A more natural description
of the internal motion is that of the atom moving along the molecular
plane and vibrating against it [4]. The amplitude of the motion along
the plane can be substantial. For instance, in benzene-argon and
tetrazine-argon the root-mean-square amplitude of the Van der Waals
(stretch) vibration perpendicular to the plane is 0.13 Å, whereas the
amplitude parallel to the plane is about 0.40 Å, in the ground state.
With two quanta of vibrational excitation the latter amplitude goes up
to 1.0 Å. The probability that the atom tunnels to the other side of
the molecule is still negligible, however [2].

An interesting intermediate case are the Van der Waals complexes
containing N_2 or O_2 molecules. According to the first interpretations
of their infrared spectra [5-8], such complexes seem to exhibit both
rotational vibrations (librations) and nearly free internal rotations,
depending on the degree of excitation or the temperature. The homogeneous dimers N_2-N_2 and O_2-O_2 are very important because their dynamical
behaviour and spectra are determined by the same intermolecular
potentials that cause the properties of bulk N_2 and bulk O_2. The
infrared spectra of N_2-Ar and O_2-Ar are strikingly similar to those of
the homogeneous dimers, but these systems are simpler (with just one
instead of three internal angles) and therefore more accessible to
detailed studies. The early interpretations of the infrared spectra of
these dimers [5-8], even of the diatom-diatom complexes, are based on
a one-dimensional model for the hindered internal rotation. The coupling
with the Van der Waals stretch was disregarded, although the excitation
energies corresponding with the higher frequency parts of the infrared
spectra are larger than the Van der Waals dissociation energy.

O_2 complexes are especially interesting because of the electronic
triplet spin momentum in the ground state O_2 molecule. Mainly via
spin-orbit and spin-spin interactions this triplet spin momentum
couples with the rotational states of O_2. The magnetic spectrum of free
O_2 corresponds (nearly) with Hund's coupling case b [9]. The drastic
changes of this spectrum in O_2-Ar are completely understood [10-12];
they form a direct measure of the rotational (im)mobility of O_2. In the
O_2-O_2 dimer various magnetic coupling terms are important: the intermolecular exchange coupling usually represented by the Heisenberg
effective spin hamiltonian, the intramolecular spin-orbit and spin-spin
couplings in each O_2 monomer and the intermolecular spin-spin (magnetic
dipole-dipole) coupling. All these couplings play a role in the
magnetic properties of solid O_2 [13,14]. A detailed study of their
effect on the magnetic spectrum of O_2-O_2 will soon be published [15].
In the present paper we shall concentrate ourselves on the internal
motions of N_2 in the N_2-N_2 and N_2-Ar dimers and their effects on the
infrared spectra of these dimers.

2.   INTERMOLECULAR POTENTIAL, ROVIBRATIONAL STATES AND INFRARED
      SPECTRUM OF N_2-N_2

Besides a number of empirical model potentials, two detailed
anisotropic N_2-N_2 potentials are available from ab initio calculations
in our institute [16,17]. Just as most of the empirical model potentials
which have been parameterized by fitting solid state data, our ab initio
potentials have been used to calculate various properties of solid N_2.
Without any parameter fitting they yield good agreement with experiment
for several solid state properties, provided that the anharmonicity in
the (collective) vibrations of the N_2 molecules in the solid is
correctly taken into account. It may be worth mentioning that also in
the solid, at low temperatures, the librations of N_2 have fairly large
amplitudes (about 16° even at 0 K), so that it was necessary to develop
special lattice dynamics methods [18] in order to deal with the
anharmonicity of these librations. These methods were strongly inspired
by our experience with Van der Waals molecules.
       Among solid state physicists, it is generally believed that
especially the phonon frequencies depend sensitively on the form of the
intermolecular potential. We have found, however, [17] that the two
different ab initio potentials do not yield more than 9% difference
in these frequencies for both the ordered (α and γ) phases of solid N_2,
whereas the discrepancy with the experimental frequencies is about 8%.
(This is considered to be a very good agreement; most empirical model

TABLE I

Energies of the N_2-N_2 dimer with two different ab initio potentials.

Structure	θ_A θ_B ϕ (degrees)	B-vdA potential [16] ΔE (cm^{-1})	R_e (Å)	vdA-W-J potential [17] ΔE (cm^{-1})	R_e (Å)
(structure)	90 0 0	83	4.15	97	4.09
(structure)	50 50 0	94	4.02	104	3.97
(structure)	90 90 0	97	3.65	91	3.65
(structure)	90 90 90	122	3.50	95	3.64

potentials perform much worse, even after optimization of the parameters. Better agreement could not be expected, if only because of the neglect of many-body contributions to the potential.) So it is not possible by looking at the solid state data to discriminate between these two ab initio N_2-N_2 potentials and, thus, to evaluate their quality.

In Table I we have applied the same two ab initio potentials to various geometries of the N_2-N_2 dimer and we observe striking differences. The older potential [16] yields a crossed (D_{2d}) equilibrium structure, whereas the absolute minimum for the newer potential [17] occurs for a shifted parallel geometry. Also the relative energies of other low lying structures are different. The changes in the well depth are not larger than 20%, however, and the variations in the repulsive and attractive contributions which lead to these changes are even more subtle. This situation is typical for a Van der Waals molecule. It illustrates that it will be very difficult by means of ab initio calculations alone, to obtain potential surfaces that are sufficiently accurate to predict equilibrium structures and barriers to internal rotation. The differences in equilibrium structure and in the barriers to internal rotations will have their consequences for the rovibrational transitions in Van der Waals molecules, however. Making dynamical calculations based on various potentials and comparing the results with the rovibrational spectra of Van der Waals molecules is probably the most discriminating method to evaluate these potentials. With the older of the two ab initio potentials such calculations on the N_2-N_2 dimer have been made already [19,20].

The infrared spectrum of N_2-N_2 has been measured [7] with a resolution of about 1 cm^{-1} in the gas phase at 77 K as a discrete structure on top of the broad collision-induced spectrum (see Fig. 1). The spectrum has been observed in the region of the N_2 monomer stretch frequency at 2330 cm^{-1} and the Van der Waals vibrations are visible in

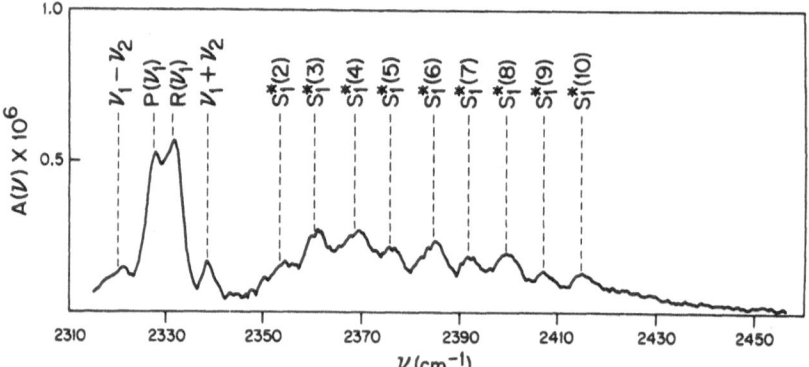

Figure 1. Infrared spectrum of N_2-N_2 at 77 K from ref. [7].

combination with the N_2 stretch. Apart from rotational bands due to the end-over-end rotations of the dimer, the spectrum contains one peak at 9.5 cm^{-1}, which has been assigned as a locked rotational vibration or libration, and a fairly regular series of peaks at higher frequencies, which have been attributed to (slightly perturbed) internal rotations.

In the calculations [19,20] the rovibrational states of N_2-N_2 have been expanded in a basis of free rotor functions, expressed in a dimer-fixed coordinate frame, multiplied by a basis of Laguerre functions for the Van der Waals stretch. The dimer hamiltonian with the anisotropic N_2-N_2 potential that was available from ab initio calculations [16] has been diagonalized in this basis. All bound states

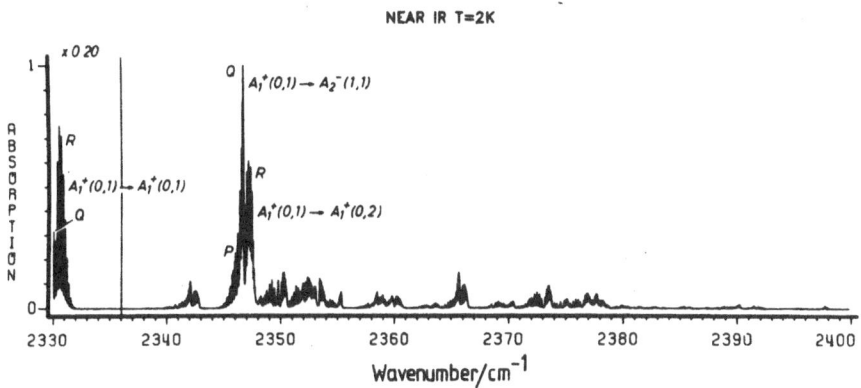

Figure 2. Calculated infrared spectrum of N_2-N_2 at 2 K from ref. [20].

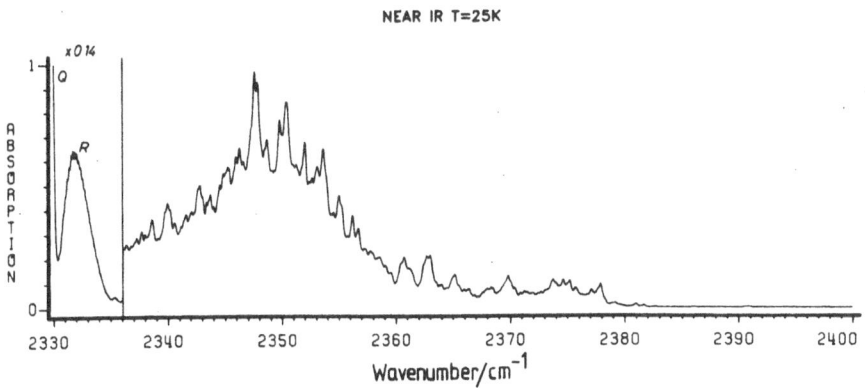

Figure 3. Calculated infrared spectrum of N_2-N_2 at 25 K from ref. [20], using all bound states.

of the N_2-N_2 dimer have thus been calculated. Using an empirical
dipole function obtained from the collision-induced infrared spectrum
[21], we have also calculated the strengths of all the allowed
bound-bound dipole transitions and thus generated the complete infrared
spectra, both in the far- and near-infrared regions. The spectrum in
the region of the N_2 stretch frequency is shown in Fig. 2, at very low
temperature (2 K) and in Fig. 3 at somewhat higher temperature (25 K).

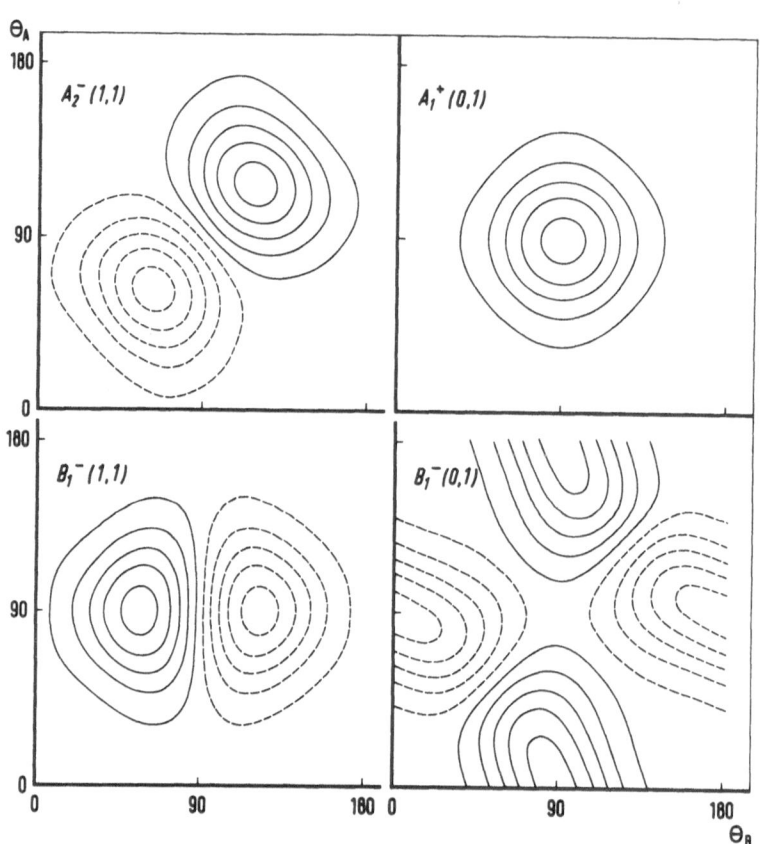

Figure 4. Cuts through ro-vibrational wave functions with R and ϕ
fixed at the values specified below. The states are labelled $\Gamma(k,n)$,
where Γ is the PI(16) symmetry label, k the angular momentum around
the intermolecular axis R and n a sequence number for the levels
belonging to the same Γ and k. Shown are the ground state $A_1^+(0,1)$
(R=7·0 a_0, ϕ=90°) and the excited levels for which transitions from
the ground state are strongest; near-infrared: shifted parallel
(R=7.4 a_0, ϕ=0°) structure $A_2^-(1,1)$; far-infrared: T-structure
(R=7.4 a_0, ϕ=0°) $B_1^-(0,1)$ and θ-excited state (R=7.4 a_0, ϕ=90°) $B_1^-(1,1)$.

In the low temperature spectrum specific transitions from the vibrational ground state to some low lying excited states are clearly distinguishable. These states are more or less localized, see Fig. 4, around the different local minima in the potential. Already at 25 K many states are populated, the energy differences between them do not show a regular pattern as in harmonic vibrations or free rotations, and so the structure in the spectrum is rather blurred. So we think that the lower frequency part of the experimental infrared spectrum [7] at 77 K can certainly not be ascribed to a single librational transition. It will be due to many different transitions which could only be resolved at much lower temperature.

A surprising feature in the calculated spectra, Figs. 2 and 3, is the complete absence of the series of bands at higher frequency which have been assigned [7] to (nearly) free internal rotations. Although all the bound states up to the dissociation limit have been included in the generation of the infrared spectrum, no regular pattern of free internal rotor transitions is visible still. We believed that such a pattern, which has been observed experimentally, might actually be due to rotational (predissociation) states lying in the continuum, i.e. to rotational resonances. Both the fairly high temperature (77 K) of the experiment and the observation that the structure in the spectrum continues beyond the dissociation energy support this assumption. In order to study this possibility we have started calculations on the N_2-Ar dimer, which is simpler than N_2-N_2, but has a very similar infrared spectrum [8], see Fig. 5. The results of these studies are reported in the sequel of this paper.

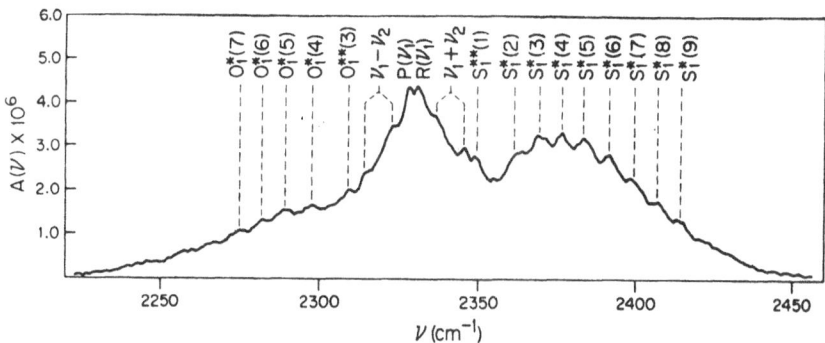

Figure 5. Infrared spectrum of N_2-Ar at 87 K from ref. [8].

3.   BOUND AND ROTATIONAL RESONANCE STATES AND THE INFRARED SPECTRUM
     OF N_2Ar

3.1. Theory

The Hamiltonian which describes the nuclear motions in a dimer
that consists of two rigid molecules, has been discussed by Brocks
et al. [22]. The special case of an atom-diatom system has been subject
to extensive studies [1]. Bound states of this Hamiltonian are usually
calculated by an expansion in a basis set which consists of products
of free rotor angular functions and radial basis functions. The angles
describe the orientations of the molecules and of the complex as a
whole, and the radius is the distance between the centers of mass of
the molecules. The expansion parameters are then variationally optimized;
this method has been called SE [1] or LC RAMP [19]. The details of the
calculation and possible approximations depend on the particular
coordinate reference frame which is chosen [1,22]. However, as our
calculations are exact within the finite space spanned by our basis,
these details are irrelevant here.
The resulting pattern of energy levels is very dependent on the
interaction potential, especially on its anisotropy, and on the
rotational constants of the molecules. For atom-diatom systems, notable
extreme cases are I_2Ar, which behaves like a (an)harmonic oscillator/
rigid rotor molecule [23], and H_2Ar, where H_2 behaves as a free
internal rotor [1]. Approximate labeling schemes for the states, which
are based on the anisotropy of the interaction potential versus the
rotational constant of the diatom, have been discussed by Ewing [24].
In case of a strongly anisotropic potential and/or a small rotational
constant, the labels of the harmonic oscillator/rigid rotor model
can be used. For N_2Ar, which has a triangular equilibrium structure,
this amounts to v_s, the stretch quantum number describing the motion
in the radial direction, v_b, the bend quantum number describing the
orientational motion of the diatom in the complex, and k, the symmetric
top rotational quantum number. (If the atom is relatively heavy
compared to the diatom, and the atom-diatom distance is larger than
the diatom bond length, the asymmetry splitting is small. The complex
is then a near prolate symmetric top.) In case of a nearly isotropic
potential and/or a large rotational constant, the diatom can be
treated as a free internal rotor. The labels of this model are j, the
angular momentum quantum number associated with the rotation of the
diatom, and ℓ, the angular momentum quantum number associated with
the rotation of the vector which connects the center of mass of the
diatom with the atom. In both models, the states are labelled
additionally with the overall rotational quantum number J and with
the parity, which are exact quantum numbers. A correlation diagram for
the two models has been given by Henderson and Ewing [8]. This diagram
is complicated, but can be simplified, if one disregards for the moment
the rotational fine structure caused by J and ℓ. The correlation
between the two models is then simply:

$$v_b + k \Leftrightarrow j \qquad\qquad\qquad (1)$$

In the next section we will try to assign labels to the calculated
states of N_2Ar and to determine their positions in the correlation
diagram.

At the temperatures of bulk gas experiments, such as in refs.
[5-8], continuum states are strongly occupied. Most of these contribute
to the infrared spectrum as a broad structureless band, the so-called
collision induced absorption [21]. Some continuum states, however,
give rise to additional structure. A state, in which the diatom is
excited as an internal rotor, can have a finite life time τ, before
the diatom rotational energy is transformed into translational energy
(between atom and diatom) and the complex dissociates. In the spectrum,
a transition to such a state results in a homogeneously broadened line
having a width $\Gamma \sim 1/\tau$. Such a state is called a rotational resonance.
The calculation of continuum states is costly, especially by numerical
close coupling calculations. In order to establish the possible role
of rotational resonances in the infrared spectrum of N_2Ar we therefore
adopt two procedures. In the first, more approximate procedure, we
consider rotational resonances as if they were bound states, i.e. we
exclude all dissociation channels. This allows a relatively cheap
calculation of the approximate position of a large number of resonances.
Moreover, these (quasi) bound states can be classified along the same
lines as the truly bound states. The second procedure involves more
computational effort as the calculations are repeated while adding
the dissociation channels. The coupling with these channels results in
a level shift of the quasi bound states, and in a broadening, which
is characterized by a certain width. The latter calculations are
performed for a subset of resonance states, from which we hope to
draw conclusions related to all these states. If the coupling to the
dissociation continuum is strong, both width and level shift are large.
Such a state is thus not separately observable in the spectrum. For a
weak coupling the state can be observed due to its small width. But
then the level shift is also small, and the quasi bound level
determined in the first procedure is a good approximation of the
position of the resonance. So, as a result of the first procedure, the
pattern of energy levels for the resonance states and its impact on
the structure of the spectrum can be discussed. As a result of the
second procedure, it can be determined which of the resonances are
actually observable.

The calculation of the resonance position and width in the
second procedure is based on an expansion in the same basis set as
was used for the bound states. The states resulting from the variational calculation which have a positive energy then represent the
continuum in the way explained in ref. [25]. These states can be used
to calculate an approximate resonance phase shift:

$$\delta_i^r(\tfrac{1}{2}[\varepsilon_n + \varepsilon_{n+1}]) = \sum_{m=1}^{n} |\langle i|m\rangle|^2 \tag{2}$$

Here ε_n is the nth positive eigenvalue of the variational calculation
and $|n\rangle$ the associated eigenstate. The state $|i\rangle$ is the closed
channel part of the resonance, i.e. the result of the calculation

where the dissociation channels are excluded. As outlined in ref. [25], this approximate phase shift can be used to obtain the resonance parameters. The true phase shift is a continuous function of the energy E, and a resonance is characterized by a sharp rise of this function around the energy E_i, which is called the position of the resonance. For the discrete function of Eq. (2), the situation where $E_i \approx \varepsilon_n$ would mean that:

$$|<i|n>|^2 \; >> \; |<i|m>|^2 \quad , \quad m \neq n \tag{3}$$

In general, a basis set contains one or more intrinsic parameters; changing these parameters results in a shift of the eigenvalues ε_n of the variational calculation. If the discrete approximation of Eq. (2) resembles the true phase shift, and the differences between the eigenvalues do not vary too drastically in changing the basis set parameters, then the quantity $|<i|n>|^2$ has its maximum in the situation where $E_i = \varepsilon_n$. Reversing the argument, the position of the resonance E_i can be found by maximizing $|<i|n>|^2$, while varying the intrinsic basis set parameters. The width of the resonance can in principle be found by assuming a Breit-Wigner line shape [26] between the points $\frac{1}{2}[\varepsilon_n + \varepsilon_{n-1}]$ and $\frac{1}{2}[\varepsilon_n + \varepsilon_{n+1}]$. The complications which arise if ε_{n-1} and/or ε_{n+1} represent a neighbouring resonance, lead to a somewhat modified procedure. This has been discussed in ref. [25]. From a different starting point Grabenstetter and LeRoy have developed a similar procedure [27].

The calculations will only yield meaningful results when the resonances are isolated, i.e. non-overlapping. Accurate close coupling calculations on ArHCl by Ashton et al. [28] indicate the validity of that assumption for narrow resonances. Those are also the most interesting ones in our case. Our procedure was tested on rotational resonances of H_2Ar [25] and the results were compared to accurate close coupled values [29]. The calculated resonance positions were in excellent agreement; the widths were 25% too high, but their ratios were again in excellent agreement. Considering that these calculations are relatively cheap for resonances at not too high energies, these results are satisfactory.

3.2. Results

For the calculations we have used the semi-empirical potential of Candori et al. [30], which has been constructed using a fair amount of experimental data. It is most convenient if this interaction potential is expressed as a series expansion in Legendre polynomials. For that purpose, the potential has been numerically transformed into such a form. Truncation of the series at the Legendre polynomial of order 8 had an effect of less than 0.005 cm^{-1} on the bound state eigenvalues. With the basis we have used, these eigenvalues were converged within 0.05 cm^{-1}.

A number of calculations for various overall rotational states J (remember that J is the only exact quantum number) have been performed. Approximate labels, as discussed in the previous section, have been

assigned to the resulting states on the basis of energy level separations and the coefficients of the dominant basis functions. If these
states are sorted on their radial stretch quantum number v_s, for each
v_s a correlation diagram can be made with the levels of the harmonic
oscillator/rigid rotor model on one side and the free internal rotor
model on the other side. Such a diagram is given in Fig. 6 for the
levels with v_s=0, v_b, k from 0 to 3 and J=k. In the lower energy range
one observes many crossings in the correlation between the N_2Ar levels
and the free internal rotor levels and few crossings between the N_2Ar
levels and the harmonic oscillator/rigid rotor levels. Especially in
the vibrational ground state N_2Ar has a rigid rotor like spectrum.
Already at the second bending overtone, the rigid rotor character
is diminished, as indicated by the relatively large splitting of the
rotational excitations. A transition range starts here, where the N_2Ar
levels cannot be assigned to either one of the limiting models. In
this range also the v_s and v_b modes are strongly coupled and part of
the assignment becomes rather arbitrary. At higher energy the number
of crossings between the N_2Ar levels and the free internal rotor levels
decreases. With some imagination a j=5 multiplet can be recognised just
below the dissociation limit. For other v_s states, the conclusions are
more or less similar; we will come back to this point later. A general
conclusion is that at temperatures where only bound states are
sufficiently populated, a free internal rotor character is practically
absent. Only at very low temperatures, the rigid rotor character may
be observed; increasing the temperature involves the levels in the
transition region and the resulting spectrum becomes very complicated.
These conclusions are consistent with the $(N_2)_2$ results.

The experimental spectrum of Henderson and Ewing [8] was obtained
for a temperature of 87 K, at which continuum states are strongly
populated. Most of this spectrum was interpreted in terms of a
one-dimensional hindered internal rotor model. Our calculated bound
states cannot explain the observed structure in the spectrum. So, unless
the potential of Candori et al. [30] is substantially wrong, the explanation must invoke the rotational resonance states. If the latter are
treated as bound states (see previous section), the assignment of
approximate labels is similar. Fig. 7 gives the results for the v_s=0
rotational resonances. These states clearly show a free internal rotor
character. Figs. 6 and 7 then give a complete characterisation of the
v_s=0 states of N_2Ar. A distinct free internal rotor pattern is not
observed until the resonance states are reached, i.e. for j=6. Results
for other v_s states are similar. By excitation in the radial stretch
mode, the average distance between atom and diatom increases. The
interaction potential becomes less anisotropic at larger distances.
Therefore, for these radially excited states, a free rotor pattern is
observed for lower v_b and k than for the v_s=0 states. However, the
radial excitation energy compensates for this earlier onset of free
internal rotor character. So on the whole, the free internal rotor
pattern is observed only for resonance states.

The onset of free internal-rotor character can be estimated from
diagrams like Figs. 6 and 7 and it is given in Table II as a function
of the radial (stretch) excitation. In a one-dimensional model [8]

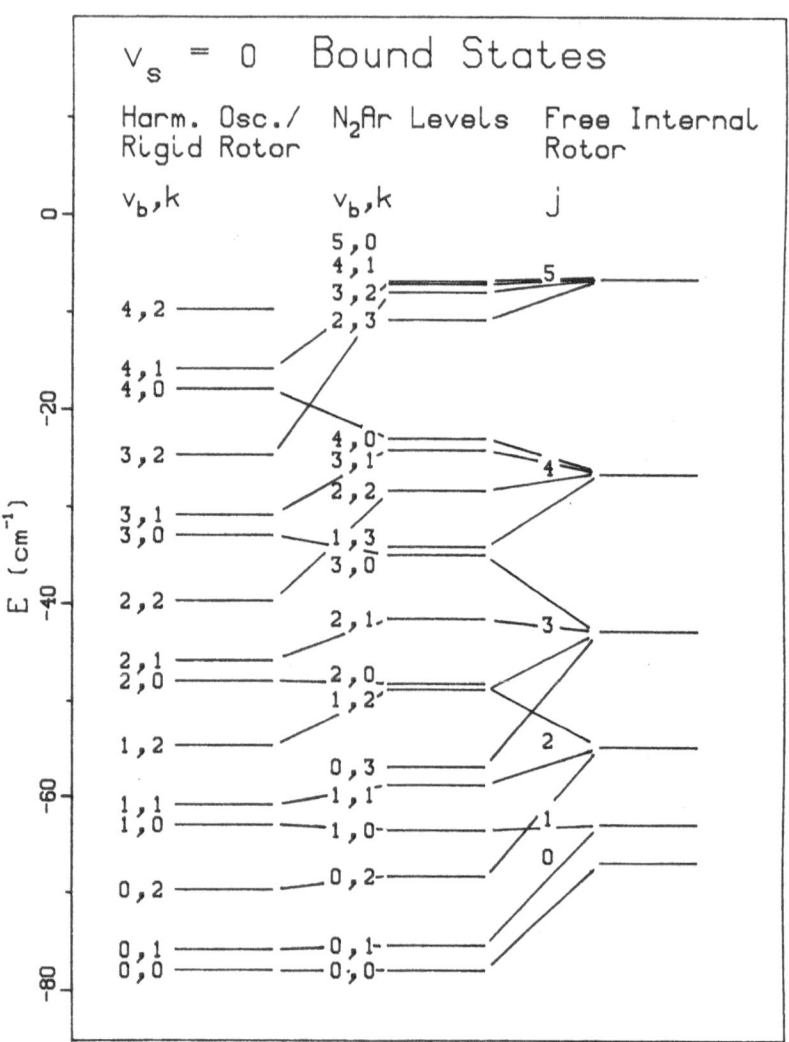

Figure 6. Bound states of N_2Ar with stretch quantum number $v_s=0$ and overall rotational quantum number $J=k$, correlated with the levels of the harmonic oscillator/rigid rotor model and those of the free internal rotor model. The states are labeled with bend quantum number v_b and symmetric top quantum number k or diatom rotational quantum number j.

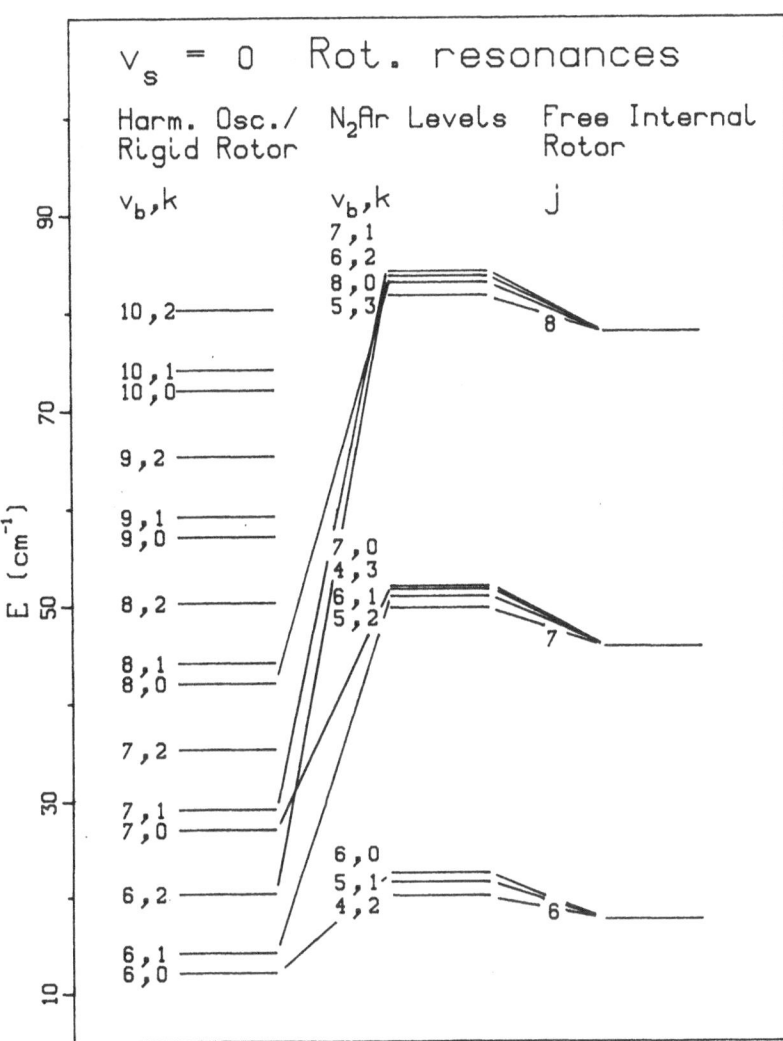

Figure 7. Rotational resonances with $v_s=0$, J=k, labeling see Figure 6.

TABLE II

Radial stretch excited states (v_b=0, k=0), vibrationally averaged
atom-diatom distances and potential values at the T-shaped and
linear geometries of N_2Ar. The onset of the internal rotor
character is estimated from diagrams like Figs. 6 and 7.

v_s	Energy (cm^{-1})	<R> Å	V(<R>) T-shaped	V(<R>) linear	j free internal rotor onset
0	-77.95	3.71	-98.17	-52.30	6
1	-56.17	3.92	-83.00	-70.39	5
2	-37.13	4.12	-67.64	-67.33	4
3	-19.74	4.42	-48.50	-53.50	4
4	-9.53	4.81	-30.58	-36.50	2
5	-2.81	5.83	-9.27	-12.20	0
6	-0.64	7.51	-1.89	-2.49	0

it can be related to the rotational barrier which is by definition the
difference in potential energy between the T-shaped and the linear
structure. In the multi-dimensional case, this definition of the
rotational barrier has little meaning, as is shown in Table II, where
the values of the potential for the vibrationally averaged distances
are given. Instead, one should consider the topology of the potential
surface in all coordinates.

At this stage it is useful to consider whether the calculated
levels can lead to the interpretation of the experimental spectrum [8],
if only qualitatively. For this purpose, a crude procedure for the
construction of a spectrum is chosen. We disregard the overall rotational fine structure associated with different J values. The set of
levels from Figs. 6 and 7 is taken, together with the equivalent levels
for the stretch excited states. Each level is weighted with the
appropriate Boltzmann factor and with its spin statistical weight.
Allowed transitions are determined by a set of selection rules and it
is assumed that they all have equal strengths. For harmonic oscillator
like states the selection rules are

(a) Δv_s=0 ; Δv_b=0,±1 ; Δk=0,±1

(b) Δv_s=±1 ; Δv_b=0   ; Δk=0,±1

If the transition dipole moment is mainly determined by the N_2
quadrupole induced dipole on Ar (and possibly by low order exchange
terms), the selection rules for a free internal rotor are

(a) $\Delta v_S = 0$; $\Delta j = 0, \pm 2$

(b) $\Delta v_S = \pm 1$; $\Delta j = 0$

For each v_S, a threshold is set, below which the harmonic oscillator
selection rules apply, and above which the free rotor selection rules
are valid. These thresholds are given in Table II. The transition lines
that are thus computed are artificially broadened by applying a Lorentz
profile of uniform width (2.5 cm^{-1}). Thus a simulated spectrum is constructed, which is given in Fig. 8 for the experimental temperature.
Fortunately, the gross structure of the spectrum does not depend sensitively on the thresholds or the Lorentz width. The experimental spectrum was measured near the N_2 monomer stretch frequency of 2330 cm^{-1}; in
order to compare the experimental values with Fig. 8 one should thus
subtract this value. Qualitatively the agreement is good, especially in
the higher energy region, where the characteristic structure of a free
internal rotor spectrum is observed. This is direct evidence for our
assertion that the regular structure of the experimental spectrum is
mainly due to rotational resonance states, and not to the bound states.
The peaks in the simulated spectrum are even closer to free internal
rotor transitions than the experimental peaks. However, one should be
careful to draw quantitative conclusions on the basis of such a crude
calculated spectrum. In reality, the transition strengths should be calculated with the use of a reasonable interaction dipole surface; furthermore the overall rotational bands will have to be included. This will
also yield a definite assignment of the structure in the lower frequency
part of the spectrum, which contains contributions from transitions
between localized bending states and from stretch transitions. As calculations on $(N_2)_2$ have shown, however, the required computational
effort is substantial [20].

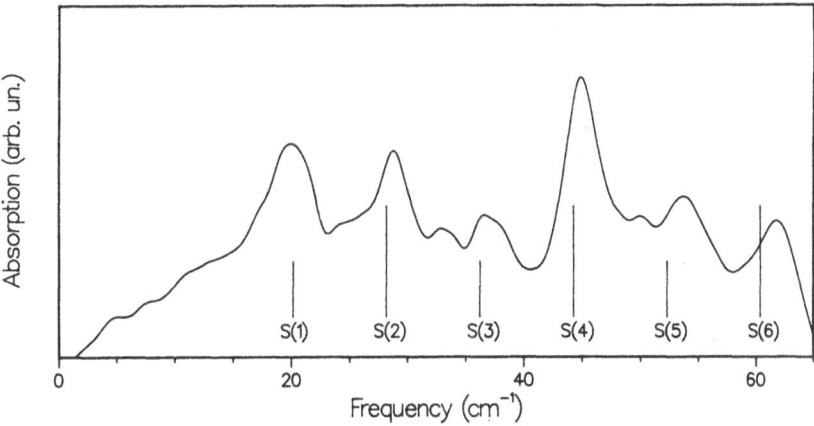

Figure 8. Simulated spectrum of N_2Ar at T=87 K, from simplified model
(see text). The vertical lines indicate the positions of the ortho/para
N_2 free rotor transitions with their spin statistical weights.

TABLE III

Rotational resonance parameters and assignments for the $J = 0$ resonances.

ortho N_2			para N_2		
assignment v_s, j	resonance position (cm^{-1})	width (cm^{-1})	assignment v_s, j	resonance position (cm^{-1})	width (cm^{-1})
3,2 (a)	1.5	0.30	2,3	4.9	1.7
4,2	7.5	0.98	3,3	15.1	1.2
5,2	7.9	1.6	4,3 (a)	20.8	0.46
6,2	11.2	0.33	5,3	23.5	1.8
2,4	12.7	1.7	6,3	24.6	0.65
0,6 (a)	22.6	0.38	2,5	34.6	2.2
3,4	25.9	3.5	3,5	45.9	1.6
4,4	32.3	1.4	0,7 (a)	52.3	0.29
5,4	37.8	0.64	4,5	55.5	1.7
6,4	40.0	0.19	5,5	58.9	0.85
1,6	45.0	1.3	6,5	60.5	0.34
2,6	60.8	1.6	1,7	74.0	1.2
3,6	72.4	2.0	2,7	91.1	2.4
4,6	79.6	1.3	3,7	101.9	1.2
5,6	83.1	0.93	4,7	108.9	0.95
6,6	84.9	0.18	5,7	112.1	0.57
			6,7	113.5	0.33

(a) these states are mixed

The widths of the $J=0$ resonances were obtained by using the procedure outlined in the previous section. The properties of the rotational resonances are intimately connected with the free internal rotor character. Therefore we have changed the labeling of states somewhat, as compared to Figs. 6 and 7. In Table III the states are labeled with the stretch quantum number v_s and the diatom angular momentum j. For $j \geq 4$, this labeling is obvious, apart from some occasional mixing of nearby resonances. For $j=3$ and especially $j=2$ the states are strongly mixed and the labeling is highly approximate. The position of the resonances is shifted, as compared to the quasi bound state calculations. In most cases, however, the shifts are smaller than the widths of the resonances, which confirms the validity of the quasi bound calculations. The only exceptions are the $j=2$ resonances, but these have a large width and are probably overlapping, so the model breaks down anyway.

Despite the strong anisotropy of the potential, the resonances are surprisingly narrow. The calculated widths are in the range of 0.18 to 3.5 cm^{-1}. For H_2Ar, our method overestimated the widths [25], so they might even be smaller. The main reason for this sharpness seems the exclusion of the $j \to j-1$ dissociation route for symmetry reasons. The states with even and odd j belong to ortho and para nitrogen, respectively, and transitions between these two species are excluded. The variation of the widths with the stretch quantum number v_s within the same j series follows the same pattern that has been observed by Ashton et al. in ArHCl [28]. At first the width increases slightly with increasing v_s, but then it drops sharply as the energy reaches the threshold for the dissociation channel, associated with that partic- ular j value. The reason for this behaviour is probably the large increase of the average distance R with stretch excitation, see Table II, whereby the anisotropy of the potential decreases. This anisotropy is the driving factor for the conversion of the internal rotational energy to the translational energy, which leads to the dissociation of the compex. Therefore, it determines the resonance line widths. Exceptions to this pattern (besides the $j=2$ resonances already men- tioned) are the j resonances, which are even below the $j-2$ thresholds, e.g. $(v_s, j) = (0,6)$ and $(0,7)$. Because of the additional exclusion of the $j \to j-2$ and $j \to j-3$ dissociation routes and the energy or momentum gap law [31,32], these resonances are relatively narrow.

Resonances for other J states are expected to behave similarly to the J=0 resonances. Resonances having a width of about 1 cm^{-1} or less can probably be observed in the experimental spectra [5-8]. The states which are more excited in the stretch mode have a smaller width and thus give rise to a sharper structure in the spectrum. As these states have a relatively strong free internal rotor character, as compared to the states that are less excited in the stretch mode, this gives the spectrum an extra free internal rotor accent.

4.   CONCLUSIONS

Using an expansion in a finite basis set, we have been able to calculate both bound and rotational resonance states of N_2Ar. These states have been used to interpret the infrared spectrum of this Van der Waals molecule. At low temperature, the spectrum appears to be determined by transitions in the low frequency range between bound states which represent localized vibrations around a distinct local minimum. At higher temperature, the structure in this spectrum is blurred and it is extended to somewhat higher frequencies. Major con- tributions then result from transitions between bound states which are not well localized. Moreover, these states show appreciable mixing of Van der Waals stretch and hindered rotation modes. The majority of the states up to the dissociation limit have this character. A similar analysis of this frequency range has been made for the infrared spec- trum of N_2-N_2.

At higher frequencies, the experimental spectrum shows the charac- teristic structure of a nearly free internal rotor. This structure appears to be caused by transitions involving states with energies above the dissociation limit, which are called rotational resonances or predissociation states. During the lifetime of such a state, a collision complex is formed. Given a rotational state, the lifetime of the complex is longer the more it is excited in the radial stretch mode (as long as the radial excitation energy itself does not exceed the dissociation energy, of course). This can be understood by noting that the average distance increases with stretch excitation and the effective interaction potential becomes more nearly isotropic. A long lifetime means a small line width and a sharp contribution to the spectrum.

The infrared spectrum of N_2Ar is thus understood. For N_2-N_2, O_2Ar and O_2-O_2 the infrared spectra in the higher frequency range are very similar to that of N_2Ar; these can be interpreted along the same lines. According to the reasoning of the previous paragraph, the states which cause the internal rotor structure in the spectrum, do not depend sensitively on the anisotropy of the potential. This part of the spectrum is therefore not liable to yield a detailed test on the accuracy of the potential. Moreover, the effort which is needed to calculate all the resonances is substantial. The lower frequency part of the spectrum contains information which is more suitable for this purpose. It corresponds with transitions between states which are localized around different local minima. These states are thus very sensitive to the topology of the potential surface. In order to access this information, experiments should be performed at low temperatures, presumably using molecular beams, with a reasonably high resolution.

Finally, we note that, although the present paper refers to the infrared spectra of Van der Waals dimers, the same considerations apply to their Raman spectra.

5.   REFERENCES

1. R.J. Le Roy and J.G. Carley, Advan. Chem. Phys. 42, 353 (1980).
2. G. Brocks and T. Huygen, J. Chem. Phys., September 15 (1986).
3. W.L. Meerts, W.A. Majewski and W.M. van Herpen, Can. J. Phys. 62, 1293 (1984).
4. G. Brocks and D. van Koeven, to be published.
5. C.A. Long and G.E. Ewing, J. Chem. Phys. 58, 4824 (1973).
6. G. Henderson and G.E. Ewing, J. Chem. Phys. 59, 2280 (1973).
7. C.A. Long, G. Henderson and G.E. Ewing, Chem. Phys. 2, 485 (1973).
8. G. Henderson and G.E. Ewing, Mol. Phys. 27, 903 (1974).
9. M. Mizushima, The theory of rotating diatomic molecules, Wiley, New York (1975).
10. J. Tennyson and J. Mettes, Chem. Phys. 76, 195 (1983).
11. A. van der Avoird, J. Chem. Phys. 79, 1170 (1983).
12. J. Mettes, B. Heymen, P. Verhoeve, J. Reuss, D.C. Lainé and G. Brocks, Chem. Phys. 92, 9 (1985).
13. A.P.J. Jansen and A. van der Avoird, Phys. Rev. B 31, 7500 (1985).
14. A.P.J. Jansen and A. van der Avoird, J. Chem. Phys., submitted.
15. G. Brocks and A. van der Avoird, to be published.
16. R.M. Berns and A. van der Avoird, J. Chem. Phys. 72, 6107 (1980).

17. A. van der Avoird, P.E.S. Wormer and A.P.J. Jansen, J. Chem. Phys. 84, 1629 (1986).
18. W.J. Briels, A.P.J. Jansen and A. van der Avoird, Advan. Quantum Chem. 18, 131 (1986).
19. J. Tennyson and A. van der Avoird, J. Chem. Phys. 77, 5664 (1980); 80, 2986 (1984).
20. G. Brocks and A. van der Avoird, Mol. Phys. 55, 11 (1985).
21. J.D. Poll and J.L. Hunt, Can. J. Phys. 59, 1448 (1981).
22. G. Brocks, A. van der Avoird, B.T. Sutcliffe and J. Tennyson, Mol. Phys. 50, 1025 (1983).
23. G. Brocks, unpublished results.
24. G.E. Ewing, Can. J. Phys. 54, 487 (1976).
25. G. Brocks, in preparation.
26. T.Y. Wu and T. Ohmura, Quantum theory of scattering, Prentice-Hall, Englewood Cliffs, N.J., (1962) Ch. 5, sec. T.
27. J.E. Grabenstetter and R.J. Le Roy, Chem. Phys. 42, 41 (1979).
28. L.J. Ashton, M.S. Child and J.M. Hutson, J. Chem. Phys. 78, 4025 (1983).
29. R.J. Le Roy, G.C. Corey and J.M. Hutson, Faraday Discuss. Chem. Soc. 73, 339 (1982).
30. R. Candori, F. Pirani and F. Vecchiocattivi, Chem. Phys. Lett. 102, 412 (1983).
31. J.A. Beswick and J. Jortner, Advan. Chem. Phys. 47, 363 (1981).
32. G.E. Ewing, J. Chem. Phys. 71, 3143 (1979).

NON-BONDING ATOM-DIATOM POTENTIALS *via* A DOUBLE MANY-BODY EXPANSION
METHOD

A. J. C. Varandas
Department of Chemistry
University of Coimbra
3049 Coimbra Codex
Portugal

ABSTRACT. Recent progress on the determination of global functions
for triatomic van der Waals molecules which include a single-valued
potential for NeH_2 and a two-valued potential for H_3 — this standing
as a well known prototype in the theory of reactive and non-reactive
collisions — is reported from the double many-body expansion method.

1. INTRODUCTION

In a recent series of papers starting in 1984[1-6], we have been
stressing the advantages of making a double many-body expansion (DMBE)
for the potential energy functions of small polyatomics. The method has
been shown to be general both for chemically-bound molecules and weakly-
bound complexes, these being frequently called van der Waals molecules.
It consists of dividing the potential energy into an extended-Hartree-
Fock-type energy, which includes the non-dynamical or internal
correlation due to rearrangements of the valence electrons in open-shell
or nearly degenerate orbitals, and the dynamical or external correlation,
which arises from the true dynamic correlation of the electrons[7,8], being
each of these energy contributions written as a cluster expansion usually
referred as the many-body expansion[9] (MBE). In principle, the terms in
the extended-Hartree-Fock-type series expansion are modelled from
accurate *ab initio* calculations while those in the dynamical correlation
series expansion are obtained semiempirically from the dispersion
coefficients for the various separate and united atom limits, and those
also known for the equilibrium geometries of the subsystems. We shall
limit the present discussion to basic justifications, methodological
specifics and recent applications to atom-diatom non-bonding interactions.
  For atom-diatom closed-shell interactions with the diatomic near
the equilibrium geometry (R_m), it is a common approach to use the single-
configuration Hartree-Fock method to compute the repulsive part of the
interaction potential function (it also includes any inductive attraction
due to the presence of permanent multipole moments). However, single-
configuration Hartree-Fock theory suffers from a total neglect of
(dynamical and, if existing, non-dynamical) electron correlation. It

A. Weber (ed.), Structure and Dynamics of Weakly Bound Molecular Complexes, 357–371.

fails therefore to give the long-range attraction at regions of the atom-diatom interaction curve where the dispersion energy plays the dominant role. For large interfragment distances, where orbital overlap can be neglected and the interaction hamiltonian can be expanded in terms of multipole operators of the fragments the dispersion energy can be written as a power series in $1/R$, the R-dependent terms multiplied by an angular expansion in terms of θ_i, the angle between the atom and the internuclear axis of the diatomic. Figure 1 introduces the coordinates used for the present discussion.

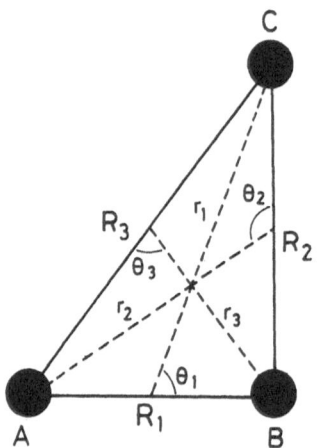

Fig. 1. Coordinate systems (r,R,θ) and (R_1,R_2,R_3) used to represent the triatomic potential energy surface.

Perhaps the most popular approximation[10,11] to the interaction potential is to make a Legendre analysis of the total potential which is written as a sum of the single-configuration Hartree-Fock energy and the dispersion energy, this being suitably damped[12] to mimic charge-overlap and exchange-overlap effects at finite interfragment distances where the R^{-n} asymptotic expansion formally diverges. However, current Legendre-analysis-type functions cannot be used to describe bond-breaking or bond-forming processes. Not only because small values of r are involved for which the Legendre analysis is slowly convergent[13] but also because the single-configuration Hartree-Fock wavefunction often produces incorrect dissociative fragments.

Amongst the best known tools to improve the wavefunction by includ-ing dynamical correlation and the proper asymptotic behaviour in the Hartree-Fock wavefunction are the configuration-interaction (CI) and multiconfiguration self-consistent-field (MC SCF) methods. The state of the art for the calculation of the potential energy surfaces of small molecules combines these two approaches by performing a CI calculation including all singles and doubles excitations (CISD) from a multi-configuration reference (MR) space[8]. Despite the computational effort put into the final result, even for the largest affordable MR CISD calculations and even using orbitals optimized from a MC SCF method, the calculated energies may be several kcal mol^{-1} in error[14]. Brown and Truhlar[14] suggested a semiempirical way to overcome these difficulties. They start by calculating the non-dynamical correlation energy by a MC SCF method and choose as the specific variant of this method the complete-active-space self-consistent-field[15] (CAS SCF) theory; this is equivalent to fully optimized reaction space[16] (FORS) theory. Included are therefore all the configurations obtained by arranging the electrons in core and valence orbitals with the constraint that the core orbitals are double occupied and the active space is defined by the valence orbitals. The semiempiricism of their approach enters in the calculation of the dynamical correlation energy which they first partition into core and valence contributions. The core correlation is assumed to be geometry independent (and thus discarded) while the valence correlation is calculated from a MR CISD calculation, and the rest due to higher excitations and the finite size of the one-electron basis set estimated semiempirically by scaling; the method has therefore been called the SEC[14] (scaled-external-correlation) method. Similarly to Brown and Truhlar, we suggest that the non-dynamical correlation energy should be calculated by a CAS SCF method but adopt a simpler strategy to obtain the dynamical correlation energy which is fully estimated semiempirically in the DMBE method. We also note that the dynamical correlation energy is often divided into interfragment, intrafragment and inter-intra components[17]. Since the interfragment correlation corresponds asymptotically to the dispersion energy, the DMBE method somehow accounts for this fact by allowing a small correction to describe semiempirically the intrafragment and intra-inter correlations at small distances, and indeed other deficiencies related with the method itself.

Even for non-dissociative processes, the need to improve atom-rigid diatomic potential functions obtained from the Legendre analysis of the molecular energy has been pointed out. For example, high-energy scattering at small angles has provided[18] evidence that the dynamics and the intermolecular potential are sensitive to the internal molecular motion. In contrast, the DMBE method may offer an important route to global potentials which are reliable over the complete configuration space of the supermolecule formed by the interacting partners. Once obtained, such functions can hopefully be used both for low and high energy scattering calculations, including reactive transformation.

This paper is organized as follows. In section 2, the DMBE theory is surveyed both for single-valued and multi-valued potentials. Specific examples of atom-diatom non-bonding interactions are presented in section 3, and the results judged against available experimental and theoretical data. Particularly important amongst the atom-diatom non-bonding interactions is H-H_2. The H_3 potential[4,9] is double-valued and stands as a well known prototype for reaction dynamics studies. The importance of describing accurately long-range effects for the hydrogen atom exchange reaction at low and ultralow temperatures where tunnelling dominates is pointed out. There are some concluding remarks in section 4.

2. THE DMBE THEORY

2.1. Single-valued and multi-valued potentials

A potential energy surface which has only a single electronic state of the dissociation products allowed by the Wigner-Witmer spin-spatial correlation rules can be represented by a single analytical function of the interparticle coordinates. For example, the ground-state surface for the interactions involving a rare-gas atom and a hydrogen molecule has asymptotic limits which are determined by the following scheme

$$
RgH_2(^1A') \rightarrow \begin{cases} Rg(^1S) + H_2(\tilde{X}, ^1\Sigma_g^+) \\ H(^2S) + RgH(\tilde{X}, ^2\Sigma^+), \\ Rg(^1S) + 2H(^2S) \end{cases} \tag{1}
$$

and the DMBE energy development assumes the form

$$
V = V_{EHF} + V_{corr} \tag{2}
$$

where

$$
V_{EHF} = \sum_{i=1}^{3} V_{EHF,i}^{(2)}(R_i) + V_{EHF}^{(3)}(\underline{R}) \tag{3}
$$

and

$$
V_{corr} = \sum_{i=1}^{3} V_{corr,i}^{(2)}(R_i) + V_{corr}^{(3)}(\underline{R}) \tag{4}
$$

In Eqs (2-4), EHF and corr denote extended-Hartree-Fock-type and dynamical correlation, respectively, i=1,2,3 are the H_aH_b, RgH_a, and RgH_b diatomic fragments, $\underline{R} = \{R_1, R_2, R_3\}$ is a colective variable, and the superscripts indicate the order of the n-body contribution.

However, in general, dissociation products cannot be formed in their ground-states which implies that two or more surfaces cross in some configurations while showing avoided crossings in others. In such cases, the functional representation of the whole of the ground-state potential including the non-analytic behaviour at the conical intersections can be expressed as the lowest eigenvalue of a mxm matrix[9,19,20]. A typical two-valued surface, which we shall also discuss in the present work, is the ground-state of H_3 which has the following dissociation scheme

$$H_3(^2E') \rightarrow \begin{vmatrix} H(^2S)+H_2(\tilde{X},^1\Sigma_g^+) \\ H(^2S)+H_2(b,^3\Sigma_u^+) \\ 3H(^2S) \end{vmatrix} \tag{5}$$

For this system, the lowest-Riemann sheet of the potential energy surface for the ground-doublet electronic state can be expressed as the lowest eigenvalue of a 2x2 matrix

$$\begin{bmatrix} V_{11} & V_{12} \\ V_{12} & V_{22} \end{bmatrix} \tag{6}$$

with the second eigenvalue representing the upper-Riemann sheet of the two-valued surface. The two Riemann sheets exhibit a conical intersection for D_{3h} geometries[21] and it is known that such intersection extends up to infinity[22]. As a result, both the diagonal (diabatic) and off-diagonal terms in the potential matrix have to include two-body terms in their definitions; if the off-diagonal terms were restricted to a three-body component then they would vanish faster than the diagonal terms and the conical intersection would disappear at large D_{3h} perimeters[20].

The DMBE of the H_3 potential can still be written[23] as in Eq (2) with

$$V_{EHF}=X_{LEPS}^{(2)}+X^{(3)}\pm[Y_{LEPS}^{(2)}+Y^{(3)}]^{1/2} \tag{7}$$

and V_{corr} assuming the form of Eq (3). In Eq (7), $X_{LEPS}^{(2)}$ and $Y_{LEPS}^{(2)}$ are two-body terms defined from simple valence-bond theory, i.e.

$$X_{LEPS}^{(2)}=\sum_{i=1}^{3} Q_i \tag{8}$$

and

$$Y_{LEPS}^{(2)}=\sum_{i>j=1}^{3} (J_i-J_j)^2, \tag{9}$$

where the Coulomb (Q) and exchange (J) integrals are determined semiempirically from the ground-singlet and lowest-triplet extended-Hartree-Fock-type potentials of H_2 according to the London-Eyring-

Polanyi-Sato[24] (LEPS) procedure,

$$Q_i = (\tfrac{1}{2}) [V_{EHF,i}^{(2)}(R_i; X \; ^1\Sigma_g^+) + V_{EHF,i}^{(2)}(R_i; b \; ^3\Sigma_u^+)]. \quad . \tag{10}$$

$$J_i = (\tfrac{1}{2}) [V_{EHF,i}^{(2)}(R_i; X \; ^1\Sigma_g^+) - V_{EHF,i}^{(2)}(R_i; b \; ^3\Sigma_u^+)], \tag{11}$$

and $X^{(3)}$ and $Y^{(3)}$ are three-body terms, i.e., terms which become zero on removing any atom to infinity.

2.2 The functional representation of n-body energy terms

The details of the models used have been surveyed elsewhere[2,25] and will not be repeated here. For the present discussion we need only to outline the basic equations; for situations where the first-order electrostatic energy represents the leading contribution, see Refs. 3 and 25.

A convenient representation for the EHF energy is

$$V_{EHF,i}^{(2)} = R^m P^{(2)}(R_i) T^{(2)}(R_i) \tag{12}$$

$$V_{EHF}^{(3)} = P^{(3)}(\underline{R}) T^{(3)}(\underline{R}) \tag{13}$$

where m is a parameter generally taken as 0 or -1, $P^{(2)}$ and $P^{(3)}$ denote two-body (1D) and three-body (3D) polynomials in the interparticle coordinates, and $T^{(2)}$ and $T^{(3)}$ are 1D and 3D range-determining factors, respectively. The polynomial coefficients together with the non-linear coefficients in the range-determining factors are, in principle, determined from a least squares fit to *ab initio* energies. For bound-state diatomic curves, spectroscopic data offers, if available, the best means to define the diatomic functions according to the procedure described elsewhere[26]. In some cases, alternative forms have been suggested, and we shall refer to some of these in section 3.2.

The two-body terms in the dynamical correlation expansion are approximated by[2,27]

$$V_{corr,i}^{(2)} = \sum_n C_n \chi_n(R_i) R_i^{-n} \qquad n=6,8,10,\ldots \tag{14}$$

where

$$\chi_n(R) = [1 - \exp(-A_n R/\rho - B_n R^2/\rho^2)]^n \tag{15}$$

$$A_n = \alpha_0 n^{-\alpha_1} \tag{16}$$

$$B_n = \beta_0 \exp(-\beta_1 n) \ , \tag{17}$$

and the three-body term is specified by[2,4]

$$V_{corr}^{(3)} = \sum_{i=1}^{3} \sum_n C_n \chi_n(R_i)\{1-(\tfrac{1}{2})[g_n(R_j)h_n(R_k)+$$

$$+g_n(R_k)h_n(R_j)]\}R_i^{-n} \qquad n=6,8,10,\ldots \tag{18}$$

where

$$g_n(R) = 1 + k_n^{(i)} \exp[-k_n'^{(i)}(R-R^0)] \tag{19}$$

and

$$h_n(R) = [\tanh(\eta_n^{(i)}R)]^{\eta_n'^{(i)}} \ . \tag{20}$$

In Eqs (15-19), ρ_i (i=1-3) are system-dependent scaling parameters[27], and α_0, α_1, β_0 and β_1 are parameters set equal to the proposed[2] universal constants for any spherically symmetric interaction, namely $\alpha_0=25.9528$, $\alpha_1=1.1868$, $\beta_0=15.7381$, and $\beta_1=0.09729$. Still in Eq (18), the indices j and k are defined by $j=(i+1)(\mathrm{mod}\ 3)$ and $k=(i+2)(\mathrm{mod}\ 3)$, respectively, where $i,j,k=i-1,j-1,k-1$ in an obvious correspondence. All other symbols have the meaning given previously, with the summations over n being usually truncated at n=10. Note that the C_n dispersion coefficients are known[28] from reliable semiempirical methods for many systems of practical interest and k and k' (R-independent for simplicity[1]) are, given η and η', determined from[25] (e.g., for pair AB)

$$k_n^{(1)} = [2C_n^{C(AB)} - C_n^{AC} - C_n^{BC}]/(C_n^{AC} + C_n^{BC}) - \{\tanh[\eta_n^{(1)}R_1^0]\}^{\eta_n'^{(1)}} \tag{21}$$

and

$$k_n'^{(1)} = (1/R_1^0)\ln\{(2C_n^{C(AB)} - C_n^{AC} - C_n^{BC})/[(C_n^{AC} + C_n^{BC})k_n^{(1)}]\} \tag{22}$$

where (AB) stands for the united atom of AB and (AB) represents the AB molecule fixed at R_1^o. By developing the $C_n^{C(AB)}$ dispersion coefficients as a Legendre analysis in the C-AB centre-of-mass coordinate system it is possible to reproduce the asymptotic anisotropy to any known order, though this does not imply that the anisotropy will be well represented at finite separations without further parameter optimization[5].

3. SELECTED EXAMPLES

3.1. Global potential for NeH$_2$ including anisotropy

This section deals with the NeH$_2$ potential which has been the object of recent work in our group[5]. Quite naturally, this is a good van der Waals system to test the DMBE method since it affords sufficient simplicity to allow accurate SCF calculations and the Ne-H$_2$ van der Waals well has been characterized from a wealth of experimental techniques (Refs.29-31 and therein). Moreover, a comparison can be made with results from the semiempirical Tang-Toennies[10] (TT) and Hartree-Fock-plus-damped-dispersion[11] (HFD) models.

For NeH$_2$ a minimal basis set of 1s, 2s, 2p$_x$, 2p$_y$, and 2p$_z$ centred on the neon atom and 1s centred on the hydrogen atoms leads to 7 molecular orbitals. In C$_{2v}$, 4 of these orbitals transform as a$_1$, 1 as b$_1$ symmetry and 2 as b$_2$ symmetry, thus giving for the Hartree-Fock configuration

$$1a_1^2 \ 2a_1^2 \ 3a_1^2 \ 1b_1^2 \ 1b_2^2 \ 4a_1^2 \ 2b_2^0 \ .$$

By assuming 1a$_1$ as a frozen core orbital and the remaining orbitals as active one gets for the CAS SCF wavefunction a total of six configurations: the Hartree-Fock configuration plus the other five which are obtained by promoting a pair of electrons from the occupied orbitals into the 2b$_2$ orbital (for an infinite atom-diatom separation this orbital corresponds to the H$_2$ antibonding orbital). However, for not too small atom-diatom distances, a good approximation[32] may be to assume all orbitals but 4a$_1$ as frozen which gives the CAS SCF wavefunction formed by two configurations: core-4a$_1^2$ and core-2b$_2^2$. At this level of approximation, the two configurations for collinear geometries of NeH$_2$ would be core-4σ^2 and core-5σ^2.

Since we wish to compare our model with others based on single-configuration Hartree-Fock theory which is believed to give a good representation of the repulsive atom-diatom interaction when the H$_2$ molecule is near the equilibrium geometry, we use for the present discussion the EHF terms calculated at this level of approximation. Thus, the three-body EHF term has been obtained from a fit to single-configuration Hartree-Fock energies for Ne-H$_2$ geometries with H$_2$ near equilibrium, and extrapolated to the other regions of the molecular configuration space by relying entirely on the merits of the functional form we describe below, Eq (23). The actual SCF calculations[5] employed

the extended GTO basis set of Rodwell and Scoles[11], and were carried out
for a total of 72 geometries covering values of R_1 which correspond
to the turning points and expectation value of the radial coordinate for
the ground-vibrational state of H_2 (R_1=1.363, 1.535 and 1.449a_0,
respectively) and values of θ_1=0, 45 and 90°; 3≤r_1≤8a_0. Single-
configuration SCF calculations were also performed[5] for the NeH
fragments (for 1≤R≤8a_0) to define the two-body EHF terms; for H_2 see
Ref.26. By using the functional form of Eq (12) with m=-1 and a cubic
$P^{(2)}$ polynominal, these NeH points were fitted with typical errors of 1%.
Similarly, the three-body SCF points were fitted to

$$V^{(3)}=V^o(\sum_{\substack{i=0 \\ }}^{4}\sum_{\substack{j=0 \\ j+k\leq 2}}^{2}\sum_{k=0}^{2}c_{ijk}Q_1^i S_{2a}^{2j}S_{2b}^{2k})[1-\tanh(\gamma_0+\gamma_1 Q_1)] \qquad (23)$$

where $S_{2a}=Q_2^2+Q_3^2$, $S_{2b}=Q_2^2-Q_3^2$, and Q_1, Q_2, and Q_3 are well known[9] D_{3h}-
symmetry coordinates; V^o, c_{ijk}, γ_0, and γ_1 are adjustable parameters.
Typical errors in this fit were 0.1%.
     A comparison of the van der Waals attributes of the NeH potential
obtained by combining EHF and dynamical correlation [Eq (14)] two-body
energy terms with those from a molecular beam potential[33] suggests that
two-body intra-atomic and intra-inter correlation effects are
destabilizing and cannot be neglected. As in previous work[27], such
effects have been found to be well represented as a fraction (≅12%) of
the two-body EHF curve. Similarly, a comparison for 148≤T≤323K of the
Ne+H_2 mixed second-virial coefficients [B_{12}(T)] calculated using the
Pack's[34] method with the experimental values[35] suggests the addition of a
three-body correction term supposed to account for three-body intra-
atomic and intra-inter correlation effects. By representing this term as
a fraction of the three-body EHF energy, agreement between theory and
experiment for B_{12}(T) is found[5] to be best when such fraction is about
3.5%; r.m.s.d.≅0.18cm³. Clearly, this root mean squared deviation is well
within the error bars of ±1cm³ commonly accepted for the best virial
data measurements[35]. We denote the final potential as DMBE II to allow
a distinction with potential I obtained without the addition of such
three-body correction term.
     Table 1 compares the attributes of the most recent experimental[29-31]
potentials for Ne-H_2 with those obtained from the DMBE method. Although
we believe that the experimental potentials we show are the most
reliable ones presently available, it should be noted that they differ
in significant details from other experimental potentials devised for
this system. These differences have been discussed in tetail previously[10,29-31] and will not be elaborated any further here. Also given for
comparison in Table 1 are the TT[10] and HFD[11] semiempirical potentials.
The DMBE results are shown to be in good agreement with the commonly
accepted values, though the minimum of V_2 is slightly too deep at the
level of no freely adjustable parameters (DMBE I).

TABLE I

Survey of van der Waals parameters for the isotropic (V_0) and leading anisotropic (V_2) components of the Ne-H$_2$ interaction potential with the H$_2$ molecule fixed at its equilibrium geometry; σ=distance of energy zero, r_m=equilibrium geometry, ε=well depth. All quantities are in atomic units.

Potential	V_0			V_2		
	σ	r_m	$10^4\varepsilon$	σ	r_m	$10^5\varepsilon$
Andres et al[29]§	5.52	6.24	1.047	6.03	6.73	0.992
Waaijer-Reuss[30]§	5.52	6.24	1.047	5.98	6.55	1.064
Wagner et al[31]¶	5.53	6.24	1.074	6.12	6.82	0.954
TT[10]	5.52	6.20	1.139	6.02	6.69	1.001
HFD[11]¶	5.53	6.24	1.074	6.09	6.79	0.980
DMBE I[5]	5.50	6.20	1.128	5.80	6.40	1.766
DMBE II[5]	5.57	6.29	1.077	6.10	6.79	1.053

§,¶ Same isotropic part.

3.2. Two-valued potential for H$_3$

Perhaps the most obvious way to make a potential energy surface is to perform a large number of accurate correlated electronic structure calculations and fit the results by least squares. However, practical computational difficulties associated with the calculation of the dynamical correlation energy make this approach only feasible for few electron systems. Of these systems, H$_3$ is by far the most important one for the chemist as it represents the intermediate species of the simplest chemical reaction[9,36], e.g.,

$$D+H_2 \rightarrow DH+H \quad , \tag{24}$$

and provides also the simplest example of an anisotropic atom-diatom van der Waals interaction[9,36].

The DMBE potential described in this section has been obtained from such a fit to 316 ab initio points[4,37,38] which are believed to lie within about 0.1-0.2 kcal mol^{-1}[37-39] of the true Born-Oppenheimer energy. The fit was performed to the CAS SCF-like energy which is obtained after removing the dynamical correlation energy, semiempirically described by Eq (18). Two sets of extended-Hartree-Fock-type terms were involved, the second of which vanishes for all linear geometries. The problem was therefore separated in two steps in a manner similar to that previously suggested in a fit by Truhlar and Horowitz[40] (TH). First, the surface for linear geometries is modelled and then the data for nonlinear geometries is used to calibrate the bend potential. In the notation of section 2, the linear CAS SCF-like energies were fitted by the sum of the LEPS function that reduces to H_2 when one atom is infinitely removed and another term that vanishes for all linear symmetric geometries (V_a in the notation of Ref.40, except for the fact that the sum over odd powers was replaced such that the functional form has the correct analytical structure[41,42] for expansions about the conical intersection). Following previous experience[40], the EHF diatomic triplet state curve of H_2 has been taken as an effective potential (also Ref.43) represented by a four-parameter functional form similar to Eq (12) (m=0), with two parameters determined from the requirement that the height and position of the barrier for the hydrogen atom exchange reaction were reproduced; the remaining two parameters were determined from a least squares fitting procedure to the H_3 CAS SCF-like data. The correct dissociation of the upper-Riemann sheet into $H(^2S)+H_2(b\ ^3\Sigma_u^+)$ was then insured by means of a switching function which smoothly joins the effective triplet potential into the true one as one atom moves away from the strong interaction region of the potential to the atom-diatom dissociation channel. The second term of the linear potential was next adjusted to fit the rest of the collinear points. The bend potential was successfully fitted by using a sum of two additional three-body terms (which vanish smoothly at linear geometries) to represent $X^{(3)}$ in Eq (7) each of which was designed to allow flexibility for a particular aspect of the H_3 surface. One of the bend three-body terms has been taken identical to V_{II} from Ref.40 while the extra term was designed to improve the fit near the conical intersection, and expressed in terms of coordinates previously[41,42] suggested. Thus, in contrast with the TH functional form, the DMBE potential allows analytic continuation of the energy from the lower to the upper-Riemann sheet, having been used[4] for the calculation of the nonadiabatic coupling[41]. Note that no $Y^{(3)}$ term [Eq (7)] has been explicitly considered and no attempt has been made to reproduce the H-H_2 asymptotic anisotropy through Eqs (21,22), though a reliable description of the H-H_2 potential has been obtained at short and intermediate distances which are crucial for dynamics studies. For example, the characteristic parameters for the van der Waals well of the H-H_2 isotropic interaction potential are ε=48 cal mol^{-1} and $r_m \cong 6.5 a_0$ in agreement with the best available estimates (e.g., Ref.9).

Figure 2 compares the classical potential energy barrier for the DMBE and TH potentials. The agreement is shown to be excellent over most of the reaction coordinate, and indeed over most of the points common to the TH and DMBE fits; the final r.m.s.d. at 316 fitted *ab initio* points is 0.242 kcal mol^{-1} — this results from a r.m.s.d. of 0.175 kcal mol^{-1} at 153 linear geometries and a r.m.s.d. of 0.291 kcal mol^{-1} at 163 nonlinear geometries. Note that the height of the barrier of the DMBE potential was imposed to be 9.65 kcal mol^{-1} in accordance with the best *ab initio* estimates[39,44] (that for the TH potential is 9.80 kcal mol^{-1}) while the nearest distance between atoms at the collinear symmetrical saddle point is essentially the same (R_1=1.757a_0) for both potentials. However, at large interatomic separations, the effort put into the DMBE potential to describe long-range effects should pay dividends in providing a more realistic description of the asymptotic channels at the atom-diatom dissociation limits. There is some indication of such an improvement from variational-transition-state-theory

(VTST) calculations[45] for the D(H)+H$_2$ reactions at moderately low
temperatures as the rate constants from the DMBE potential show
slightly better agreement with experiment[46] than those from similar
calculations for the TH potential.

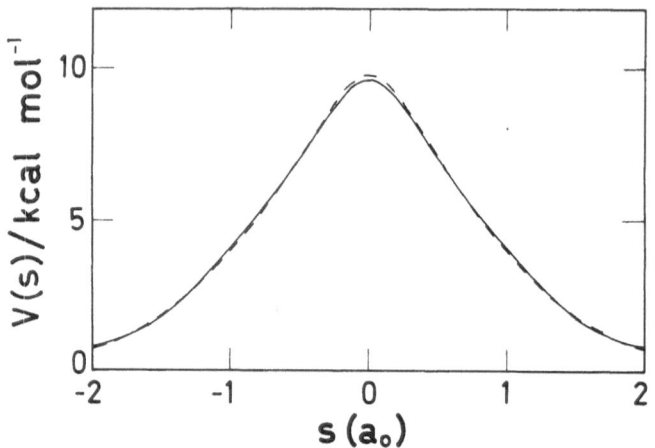

Figure 2. Classical potential energy barrier as a function of the
reaction coordinate s for the H+H$_2$ exchange reaction: ------, TH[40];
———, DMBE[4].

     A final note to comment on the importance of a realistic
description of the width of the minimum energy path at large atom–diatom
separations when modelling chemical reactivity at low and ultralow[47]
temperatures. At such temperatures, the rate constant is essentially
controlled by quantum mechanical tunnelling, and it is known from
quantum mechanics that the probability of barrier penetration is
proportional to an inverse exponential of the product of the barrier
width, the square-root of the barrier height, and the square-root of the
mass of the particle doing the tunnelling. Thus, isotope effects provide
information on the tunnelling. For example, at T=360K, the ratio of the
rate constant D+H$_2$→DH+H to that for H+D$_2$→HD+D is 6.3[48], but the same
ratio at T=4K exceeds 20,000[49]. By using an Eckart[50] potential, Miyazaki
and coworkers were able to rationalize their experiments on production
and decay of D and H atoms in γ-radiolysis of D$_2$-H$_2$ mixtures at 4K and
lower temperatures from a tunnelling migration mechanism. Note that
tunnelling alone can make, e.g., for D+H$_2$, the reaction occur because
the process is slightly exothermic. However, there are approximations in
using an (1D) Eckart potential which led the authors[49] to conclude that
it would be highly desirable to calculate the rate constants more
precisely by use of a more realistic potential energy surface. Thus, it
would be interesting to investigate whether the H$_3$ DMBE potential which
has been useful in interpreting reaction rates in the gas phase[45] can
also provide a consistent interpretation of ultralow-temperature solid-
state results. Such calculations are currently in progress.

4. CONCLUDING REMARKS

Potential energy surfaces have been traditionally divided into two main categories: surfaces with deep minima in the valence region associated with stable molecules, and surfaces which are minimum free except for the shallow van der Waals well arising in the long-range part of the potential due to attractive long-range forces. Strategies for developing model potentials have until recently adopted a similar view by specifying different requirements depending on whether the surface belongs to one category or to other. Rather than attempting a global description of the entire configuration space for the triatomic van der Waals system, such models cover only the space associated with the intermolecular coordinates (r,θ) and rarely deal with the intramolecular motion of the interacting subsystems. If such an approach may have the advantadge of simplicity, it has major limitations since real systems should show the effects of coupling between those motions. Moreover, such potentials are obviously inadequate for studying atomic recombination processes.

Since intermolecular potentials for van der Waals interactions are widely discussed in the literature[9,51], we have emphasized in this paper the DMBE method developed in our group, which aiming at generality by not descriminating between normal molecules and van der Waals molecules still preserves simplicity to an acceptable level for dynamics studies. Although the examples given are on nonbonding interactions, they cover the single-valued potential for the $Ne-H_2$ closed-shell interaction and the two-valued H_3 potential which represents a benchmark system for the study of chemical dynamics as well as for non-reactive collisions[36]. For both systems, the DMBE method has provided accurate potentials when judged from the best available empirical and theoretical data. Thus, along with the results presented elsewhere (Refs.3,25, and therein) for chemically stable molecules, the method offers a unified approach to the potential functions of small polyatomics, irrespective of the type of interaction.

ACKNOWLEDGEMENTS

This work was supported by the "Instituto Nacional de Investigação Científica" (INIC), Lisbon, Portugal.

REFERENCES

[1] A.J.C. Varandas, *Molec. Phys.*, 53, 1303 (1984).
[2] A.J.C. Varandas, *J. Mol. Struct. Theochem*, 120, 401 (1985).
[3] A.J.C. Varandas and J. Brandão, *Molec. Phys.*, 57, 387 (1986).
[4] A.J.C. Varandas, F.B. Brown, C.A. Mead, D.G. Truhlar and N.C. Blais, 'A double many-body expansion of the two lowest-energy potential surfaces and nonadiabatic coupling for H_3', *J. Chem. Phys.*, submitted for publication.
[5] A.J.C. Varandas, C.A. Rocha and M.A. Matias, 'An accurate DMBE energy potential for the ground-state of NeH_2', Abstracts of the

9th Annual Meeting of the Portuguese Chemical Society, held June 2, 1986 in Coimbra; to be published.

[6] M.A. Matias, J. Brandão, L.M. Tel and A.J.C. Varandas, 'Reliable 3D potential for HeLi$_2$ from the DMBE method', Abstracts of the International Symposium Molecules in Physics, Chemistry and Biology, dedicated to Professor R. Daudel, held June 15, 1986 in Paris; to be published.

[7] O. Sinanoglu and K.A. Brueckner, *Three approaches to electron correlation in atoms*, Yale University Press, 1970.

[8] I. Shavitt, in: *Advanced theories and computational approaches to the electronic structure of molecules*, ed. C.E. Dykstra, Reidel, Dordrecht, 1984.

[9] J.N. Murrell, S. Carter, S.C. Farantos, P. Huxley and A.J.C. Varandas, *Molecular Potential Energy Functions*, John Wiley, 1984.

[10] K.T. Tang and J.P. Toennies, *J. Chem. Phys.*, **68**, 5501 (1978); *Idem*, **76**, 2524 (1982).

[11] W.R. Rodwell and G. Scoles, *J. Phys. Chem.*, **86**, 1053 (1982).

[12] M.E. Rosenkrantz and M. Krauss, *Phys. Rev.*, **A32**, 1402 (1985); P.J. Knowles and W.J. Meath, 'Non-expanded dispersion and induction energies, and damping functions, for molecular interactions with application to HF-He', preprint.

[13] C. Douketis, J.M. Hutson, B.J. Orr and G. Scoles, *Molec. Phys.*, **52**, 763 (1984).

[14] F.B. Brown and D.G. Truhlar, *Chem. Phys. Lett.*, **117**, 307 (1985).

[15] B.O. Roos, P.R. Taylor and P.E.M. Siegbahn, *Chem. Phys.* **48**, 157 (1980).

[16] K. Ruedenberg, M.W. Schmidt, M.M. Gilbert and S.T. Elbert, *Chem. Phys.*, **71**, 41 (1982).

[17] G. Das, A.F. Wagner and A.C. Wahl, *J. Chem. Phys.*, **68**, 4917 (1978).

[18] A.P. Kalinin and V.B. Leonas, *Chem. Phys. Lett.*, **114**, 557 (1985).

[19] J.N. Murrell, S. Carter, I.M. Mills and M.F. Guest, *Molec. Phys.*, **42**, 605 (1981).

[20] J.N. Murrell and A.J.C. Varandas, *Molec. Phys.*, **57**, 415 (1986).

[21] R.N. Porter, R.M. Stevens and M. Karplus, *J. Chem. Phys.*, **49**, 5163 (1968).

[22] T. Carrigton, *Accts Chem. Res.*, **1**, 20 (1974).

[23] A.J.C. Varandas, 'The double many-body expansion of potential functions from interacting ^{2}S atoms', to be published.

[24] H. Eyring and S.H. Lin, in: *Physical chemistry and advanced treatise*, eds. H. Eyring, D. Henderson and W. Jost, Academic Press, 1974.

[25] A.J.C. Varandas, 'Intermolecular and intramolecular potentials: topographical aspects, calculation, and functional representation *via* a double many-body expansion method', to be published.

[26] A.J.C. Varandas and J. Dias da Silva, *J. Chem. Soc. Faraday 2*, **82**, 593 (1986).

[27] A.J.C. Varandas and J. Brandão, *Molec. Phys.*, **45**, 857 (1982).

[28] K.T. Tang, J.M. Norbeck and P.R. Certain, *J. Chem. Phys.*, **64**, 3063 (1976); J.M. Standard and P.R. Certain, *J. Chem. Phys.*, **83**, 3002 (1985).

[29] J. Andres, U. Buck, F. Huisken, J. Schleusener and F. Torello, *J. Chem. Phys.*, **73**, 5620 (1980).

[30] M. Waaijer and J. Reuss, *Chem. Phys.*, **63**, 263 (1981).

[31] R.S. Wagner, R.L. Armstrong, C. Lemaire and F.R. McCourt, *J. Chem. Phys.*, 84, 1137 (1986).

[32] J.W. Birks, H.S. Johnston and H.F. Schaefer III, *J. Chem. Phys.*, 63, 1741 (1975).

[33] P.J. Toennies, private communication quoted in Ref.17.

[34] R.T. Pack, *J. Chem. Phys.*, 78, 7217 (1983).

[35] J. Dymond and E.B. Smith, *The virial coefficients of pure gases and liquids*, Clarendon Press, 1980.

[36] D.G. Truhlar and R.E. Wyatt, *Ann. Rev. Phys. Chem.*, 27, 1 (1976).

[37] B. Liu, *J. Chem. Phys.*, 58, 1925 (1973); P. Siegbahn and B. Liu, *Idem*, 68, 2457 (1978).

[38] M.R.A. Blomberg and B. Liu, *J. Chem. Phys.*, 82, 1050 (1985).

[39] B. Liu, *J. Chem Phys.*, 82, 581 (1984).

[40] D.G. Truhlar and C.J. Horowitz, *J. Chem. Phys.*, 68, 2466 (1978); 71, 1514E (1979).

[41] T.C. Thompson and C.A. Mead, *J. Chem. Phys.*, 82, 2408 (1985).

[42] T.C. Thompson, G. Izmirlian, Jr., S.J. Lemon, D.G. Truhlar and C.A. Mead, *J. Chem. Phys.*, 82, 5597 (1985).

[43] A.J.C. Varandas, *J. Chem. Phys.*, 70, 3786 (1979).

[44] D.M. Ceperley and B.J. Alder, *J. Chem. Phys.*, 81, 5833 (1984); R.N. Barnett, P.J. Reynolds and W.A. Lester, Jr., *J. Chem. Phys.*, 82, 2700 (1985).

[45] B.C. Garrett, D.G. Truhlar, A.J.C. Varandas and N.C. Blais, 'Semi-classical transition state calculations for the reactions of H and D with thermal vibrationally excited H_2', *Int. J. Chem. Kinetics*, (Special issue containing the proceedings of the Farkas Memorial Symposium held December 15, 1985 in Jerusalem).

[46] D.N. Mitchell and D.J. Le Roy, *J. Chem. Phys.*, 58, 3449 (1973).

[47] V.I. Goldanskii, *Ann. Rev. Phys. Chem.*, 27, 85 (1976).

[48] D.J. Le Roy, B.A. Ridley and K.A. Quickert, *Discuss. Faraday Soc.*, 44, 92 (1967).

[49] T. Miyasaki and K.P. Lee, *J. Phys. Chem.*, 90, 400 (1986); and references therein.

[50] C. Eckart, *Phys. Rev.*, 35, 1303 (1930); H.S. Johnton, *J. Phys. Chem.*, 62, 532 (1962).

[51] G.C. Maitland, M. Rigby, E.B. Smith and W. Wakeham, *Intermolecular Forces*, Clarendon Press, 1981.

THE DETERMINATION OF INTERMOLECULAR FORCES BY DATA-INVERSION METHODS

E. Brian Smith
Physical Chemistry Laboratory,
South Parks Road,
Oxford OX1 3QZ
U.K.

ABSTRACT.   The   determination   of   intermolecular   forces   from
experimental measurements is often limited by the methods used to
process the data.   Traditional methods which attempt to reconcile
properties to simple analytical representations of intermolecular
potential energy functions (subsequently referred to as 'potentials')
frequently   lead   to   misleading   conclusions.   In   this   paper
data-inversion techniques are described which enable potentials to be
obtained   directly   from   both   thermophysical   and   microscopic
properties.   For spherically symmetric interactions the methods can
lead to a complete definition of the potential over a wide range of
separations.   In the case of anisotropic interactions the inversion
methods can be used to define effective potentials.

1.   INTRODUCTION

The   determination   of   intermolecular   forces   from   experimental
measurements is almost invariably a difficult and uncertain process.[1]
This is as true for spectroscopic and molecular beam data as for the
more obviously difficult thermophysical properties. The traditional
methods used to extract information about intermolecular forces,
which often employ force-fitting to inadequate analytical functions,
can lead to very misleading results.   However in recent years
inversion methods have been developed which enable intermolecular
forces to be obtained from a wide range of properties.   This paper
summarizes the main features of the methods.
    In considering inversion methods it is usually pertinent to ask
two questions about any set of physical measurements.   (i) What
specific information about the potential energy function U(r) does it
contain and (ii) how can that information be extracted.   Often even
the answer to the first question can be valuable.   To illustrate the
possible   pitfalls   it   is   instructive   to   take   as   an   example   the
analysis of the ultra-violet absorption spectrum for argon dimers in
the 780-1080 Å region obtained by Tanaka and Yoshino[2] in 1971.   These
important measurements were first analysed by fitting to a Morse

A. Weber (ed.), Structure and Dynamics of Weakly Bound Molecular Complexes, 373–387.

Curve. The resulting potential predicted second virial coefficients of argon that were too positive by up to 80 cm^3 mol^{-1} at 80K, a discrepancy of about ten times the experimental uncertainty. It is instructive therefore to ask the two questions discussed above. To the first we can answer that, as no rotational fine structure was resolved, it can be shown that the only information about the potential energy function contained in these data is the width of the potential energy well as a function of its depth.[3] The data are not sufficient to define a complete potential curve. Second, a well-established method of inversion, the Rydberg–Klein–Rees method, is available to extract the information that is contained in the data. From this well-width function so obtained it proved possible, using other available data, to establish an accurate Ar–Ar potential energy function.[3] That such problems can occur with spectroscopic data that arises directly from the pair interaction serves as a warning to those concerned with thermophysical properties which are usually related to U(r) by integration over one or more variables. The problem is to unfold the integrals to obtain U(r).

2. THE INVERSION METHODS

The inversion methods seek to relate the intermolecular potential energy at a separation determined by a measured cross-section to the energy at which the cross-section was determined. We describe the basis of the methods drawing our first examples from thermophysical properties.   For these properties the characteristic energy associated with the appropriate cross-section is taken to be kT where T is the temperature at which the property that defines the cross-section is measured.

Thermophysical   properties   determined   as   a   function   of temperature, P(T), are related to the intermolecular potential energy function U(r) by integration over one or more subsidiary variables. The inversion methods we describe provide a method of unfolding these integrals to obtain r as a function of U from P as a function of T. For some properties formal inversion methods are possible, for most the methods proposed can only be formally justified for very simple potentials such as $U(r) = \pm a/r^n$.

The basis of the methods comprises two steps. First the property P is used to define a temperature-dependent length $\bar{r}$ defined in terms of an experimentally determined cross-section. For example in the case of the viscosity of gases we define $\bar{r}$ in terms of the collisions integral $\Omega^{(2,2)}$ so that $\bar{r} = [\Omega^{(2,2)}/\pi]^{\frac{1}{2}}$. The second step is to write the master equation

$$U^{(1)}(\bar{r}) = G^{(0)}(T)kT \qquad\qquad (1)$$

where $G^{(0)}(T)$ is the inversion function for the property under investigation. It can be calculated using an approximate potential $U^{(0)}$ from

$$G^{(0)}(T) = \frac{U^{(0)}(\bar{r})}{kT} \; ;$$

$U^{(1)}(\bar{r})$ is an improved estimate of the potential. The efficiency of the method depends on $G(T)$ being relatively insensitive to the detailed form of the potential. The improved potential $U^{(1)}$ may be used to calculate a more accurate inversion function $G^{(1)}(T)$, which in turn will lead by application of equation (1) to an even better estimate of $U(r)$. Up to three iterations may be carried out if data of sufficient quality are available. These principles are first illustrated by application to some of the more important thermophysical properties.

2.1 Second virial coefficients

The classical second virial coefficient of a gas may be written[4]

$$B(T) = \frac{2\pi N}{3kT} \exp \left[\frac{\varepsilon}{kT}\right] \int_0^\infty \Delta \exp \left[\frac{\Phi}{kT}\right] d\Phi, \qquad (2)$$

where ε is the maximum energy of interaction (well-depth) and Φ is the total intermolecular potential energy measured from the bottom of the well, so that $\Phi = U + \varepsilon$. The quantity Δ is equal to $r_L^3 - r_R^3$, where r_L and r_R are the inner and outer coordinates of the potential energy function at Φ. In the repulsive region where $U(r)$ is single valued, $\Delta = r_L^3$. This expression can be formally inverted since $\frac{3}{2}B(T)TN\pi \exp(\varepsilon/T)$ is the Laplace transform of Δ. Thus the direct inversion of $B(T)T$ should lead to the repulsive branch for $U(r) > 0$ and the well-width as a function of is depth, $\Delta(\Phi)$ for $U(r) < 0$. One method for inverting $B(T)T$ would be to fit it to a function for which the inverse Laplace transform is known. This has only proved useful for the data for helium at very high reduced temperatures.[5] Even for this favourable case it was necessary to correct for the attractive forces.

Numerical inversion procedures based on quadrature transform equation (2) into a set of N linear equations

$$\underline{A} \cdot \underline{\Delta} = \underline{L}$$

where $\underline{A}$ is a NxN matrix and $\underline{\Delta}$ and $\underline{L}$ are Nx1 matrices. Inversion by $\underline{\Delta}^{-1} = \underline{A}^{-1} \cdot \underline{L}$ gives N values of Φ at the appropriate values of Φ obtained from N values of $B(T)$. Unfortunately this formal inversion procedure is inherently unstable since the matrix $\underline{A}$ is ill-conditioned. This means that although L (or $B(T)$) is relatively insensitive to small changes in Φ, the latter is unusually sensitive to small changes in L. The ill-conditioning manifests itself in solutions which oscillate about the true one; the lower the precision of the experimental data the greater the oscillations. Thus, although using a specially devised dynamic programming technique the inversion of $B(T)$ is possible for data accurate to one part in 10^4 it is not practicable for experimental data which have a precision of at best 0.1% and more commonly ±1% or worse.[6]

Although the inversion of B(T) by Laplace transform methods is impracticable, iterative methods based on the procedure outlined earlier in this section have proved very effective.[7] The method defines the characteristic length $\bar{r}$ as

$$\bar{r} = \left\{\frac{3}{2\pi N}\left[B + T\left[\frac{dB}{dT}\right]\right]\right\}^{1/3},$$

a choice that has recently been justified by Maitland et al.[8], who make use of the fact that

$$\left[B + T\left[\frac{dB}{dT}\right]\right]\frac{3kT^2}{2N\pi \ \exp(-\mathcal{E}/T)}$$

is the Laplace transform of $U(r)\Delta$. Thus, unlike B(T) itself, the function contains information about $U(r)$ as well as the well-width function Δ. Following this definition of $\bar{r}$ application of equation (1) leads to $U(r)$. This method has been evaluated on simulated data for the Ar-Ar interaction and is found to be extremely successful (Fig. 1).

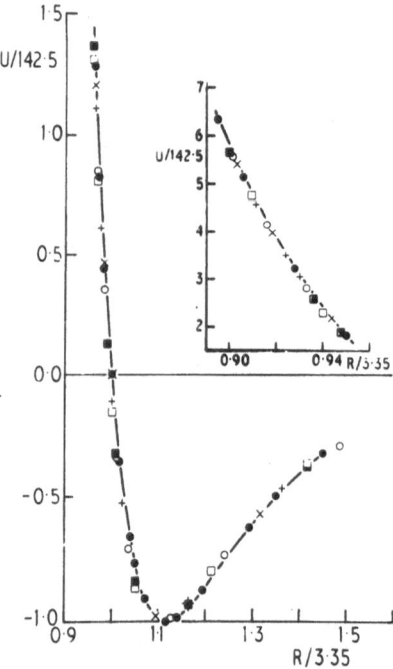

Fig. 1   The inversion of simulated second virial coefficient data. The solid line represents the potential used to calculate the data and the symbols refer to the different potentials used to make the first estimate of the inversion function. The results shown are those obtained after three iterations (Ref. 1, p 143).

2.2 Gas transport properties[9]

Gas transport coefficients can be expressed in terms of collision integrals $\Omega^{(\ell,s)}$ which are linked to U(r) by integration of the separation, impact parameter, and relative kinetics energies. Thus, we can write

$$\Omega^{(\ell,s)} = \left[\frac{kT}{2\pi\mu}\right]^{\frac{1}{2}} \int_0^\infty \exp(-y^2)y^{2s+3}Q^{(\ell)}(g)dy,$$

where

$$Q^{(\ell)}(g) = 2\pi \int_0^\infty (1 - \cos^\ell \chi)bdb,$$

and

$$\chi(g,b) = \pi - 2b \int_{r_c}^\infty \frac{dr/r^2}{\{1-[U(r)/\frac{1}{2}\mu g^2] - (b^2/r^2)\}^{\frac{1}{2}}}$$

where $y^2 = \frac{1}{2}\mu g^2/kT$, μ is the reduced mass of the colliding species, r_c is the turning point separation, b is the impact parameter, and g is the initial relative velocity. Because of this complexity it has not proved possible to specify with any certainty the information about U(r) which is contained in any given set of transport property measurements.

In view of the difficulties that are inherent in the inversion of molecular beam data one would not be surprised to find that for these apparently more complex properties the three integrations would effectively prevent any useful information about U(r) from being extracted. However, this is not the case. The inversion of gas transport properties, particularly viscosity, has proved one of the most powerful methods of obtaining potential energy functions. We will illustrate the principles involved with reference to gas viscosity. This property, η, can be defined as

$$\eta = C \frac{(MT)^{\frac{1}{2}}}{\Omega^{(2,2)}} f_\eta$$

where C is a factor independent of the gas, M the molecular mass and T the temperature, and f_η is a factor close to unity. Defining $\bar{r} = [\Omega^{(2,2)}/\pi]^{\frac{1}{2}}$, we can show for potential functions of the form $U = -a/r^m$ that $U(\bar{r}) = kTG(m)$ where G(m) is determined only by m and the sign of a. This observation was used by Hirschfelder and Eliason[10] to provide an economical way of calculating high-temperature transport properties for known potential energy functions. Dymond[11] used it as the basis of an inversion method that was used to establish the repulsive forces of the inert gases at energies, U, greater than 5ϵ. It has since proved possible to extend the method so as to obtain by data inversion not only the repulsive branch but also the well of the potential energy function.[9]

For simple inverse-power potentials G(m) is a constant, but for real potential functions G is a function of temperature. However G(T), though temperature dependent, proves very similar for a wide range of potentials and the inversion method based on the calculation of G(T) from an approximate potential works well. It is obtained as a function of $T^*(= kT/\varepsilon)$ and an estimate of ε is required. The inversion procedure is carried out for a series of estimates of ε. The one resulting in the potential which best fits the original data is accepted as defining the best estimate of ε.[12] The application of the method to invert viscosity data calculated from a known potential is illusrated in Fig. 2. After three iterations the original potential is recovered with very high accuracy.

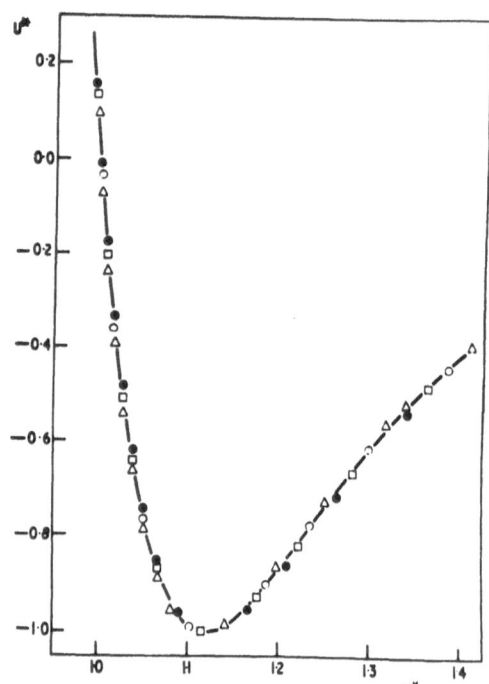

Fig. 2   The attractive portion of the   reduced potential energy function, ($U^X = U/t/k$) obtained by the inversion of simulated gas viscosity data for argon.   The solid line represents the potential used to generate the data and the symbols refer to the different approximate potentials used to make a first estimate of the inversion function.   The results are those obtained after three iterations.[6]

Equivalent inversion methods have been used to invert the thermal conductivity and diffusion data of monatomic species. Each of these properties can be related to a single collision integral in a manner similar to the viscosity. Thermal diffusion is determined largely by the ratio of two collision integrals related by a recursion formula and inversion leads not to $U(r)$ but to its logarithmic derivative with respect to temperature.[13]

2.3 Ion mobilities

In the last twenty years or so it has become possible to measure the mobility of trace amounts of ions through a neutral gas as a function of the ratio of the electric field strength to the neutral gas density.[14] Early attempts to relate these measurements to the ion-neutral gas potential were not successful but the subsequent development of an effective inversion method has enabled the potential to be determined to 5% or better over a wide range of separations. The method[15] is based on the simplifying assumption that the information about the ion-neutral potential is contained in the collision integral $\Omega^{(1,1)}$ which is a function not of the true temperature as for normal diffusion processes but of an effective temperature, T_{eff}, which is a function of the kinetic energy associated with the drift velocity, V_d, itself.

$$T_{eff} = T + \frac{M\,V_d^2}{3k}$$

M is the molecular mass of the neutral gas. Thus the inversion method relates the gaseous ion mobility to the interaction potential in essentially the same manner as that developed for normal diffusion coefficients differing only in the fact that the characteristic energy is related to the effective temperature rather than the true temperature. In practice higher order correction terms may be incorporated into the calculation.

Then $U^{(1)}(\bar{r}) = G^{(0)}(T_{eff})\,k\,T_{eff}$

$\pi\bar{r}^2 = \Omega^{(1,1)}(T_{eff})$.

The procedure can be applied iteratively as for the other applications discussed above.

The value of the method hs been evaluated by applying it to data simulated using assumed potentials for Li^+-He based on *ab initio* quantal calculations. After two iterations excellent agreement is obtained between the potential recovered by the inversion procedure and the true potential (Fig. 3).

2.4 Total cross-sections

For total cross-section data obtained from the scattering of spherically symmetric molecules a data-inversion method has been developed which permits the potential to be determined from <u>relative</u>

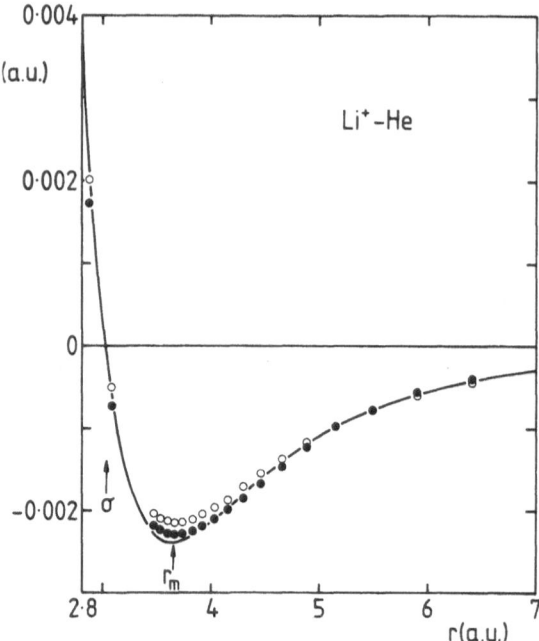

Fig. 3   The potential for Li⁺-He obtained from the inversion of
simulated ion mobility data.   The solid line represents the
potential used to calculate the data and the open and closed
circles represent the results from one and two iterations
respectively.   The results from the third iteration cannot be
distinguished from the solid line.   This diagram is based on that
of Viehland, Harrington & Mason, Ref. 15.

measurements only without recourse to specific functional models.
The method[16,17] can be formally justified by assuming that the total
cross-section, Q, at a given relative velocity, V, is determined by
the potential at separations close to the distance of closest
approach, r_0, so that $Q = 2\pi r_0^2$.   It is further assumed that the
magnitude of the total cross-section is determined by collisions for
which the deflection angle is just too small to measure due to the
uncertainty principle.   For this critical angle $\theta_c = \lambda/2r_0$ where λ is
the de Broglie wavelength.   For such small-angle scattering we may
write

$$\theta_c \approx \frac{|U(r_0)|}{E}$$

where E is the relative energy defined by $E = \frac{1}{2}\mu V^2$ where μ is the reduced mass. Equating these two expressions for θ_C and noting $\lambda = h/\mu V$ we obtain

$$|U(r_0)| \approx \frac{hV}{4r_0}$$

We can write in general

$$|U(r_0)| = G\left[\frac{V}{S_0}\right] \cdot \frac{hV}{r_0}$$

where $G(V/S_0)$ is only a weak function of the potential, and S_0 is a parameter determined by the spacing of the glory undulations.[16]

By the procedure outlined above we can relate cross-sections as a function of V to points $(r_0, |U|)$ on the potential. The procedure is first to determine S_0 from an analysis of the glory undulations. Then the inversion function $G(V/S_0)$ is calculated for an approximate potential function which is used to obtain an improved potential function. The method was tested on simulated data for the Ar-Ar interaction. Three iterations produced a potential function indistinguishable from the potential used to generate the data.[17] (Fig. 4).

2.5 Other properties

The methods described can be applied to a wide range of properties. The appropriate definitions of cross-section for the different properties and the master inversion equations are given in the Table.

3. APPLICATIONS TO ANISOTROPIC MOLECULES

3.1 Low and moderate anisotropy

The inversion methods described above may be applied to the data for anisotropic molecules. We have investigated the second virial coefficient, gas viscosity, and self-diffusion data calculated for molecules interacting with the di-Lennard-Jones potential[19] which can be expressed as

$$U_{ij} = 4\varepsilon \sum_{\ell=1}^{2} \sum_{m=1}^{2} \left[\left[\frac{\sigma}{r_{\ell m}}\right]^{12} - \left[\frac{\sigma}{r_{\ell m}}\right]^{6} \right]$$

We have considered the two sites in each molecule to be separated by a distance ℓ and to interact with parameters $\varepsilon/k = 33.46$ K and $\sigma = 0.3266$ nm to facilitate comparison with previous work.[18]

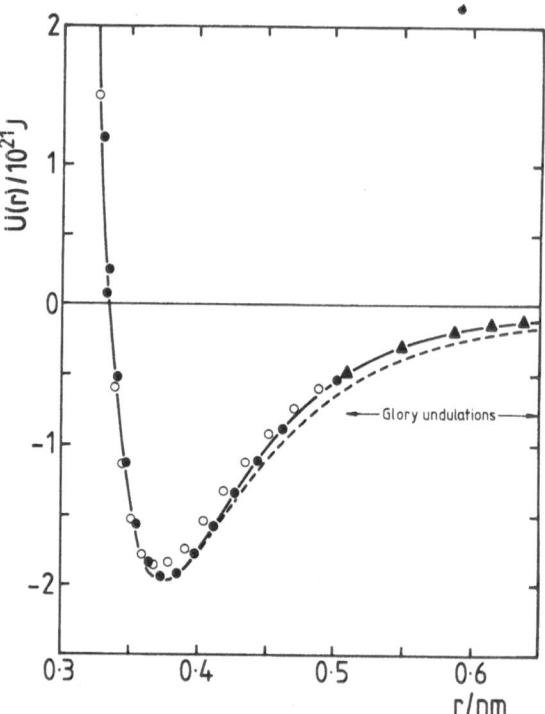

Fig. 4 Ar—Ar potential obtained by the inversion of simulated total
        cross-section data. The solid line is the true potential and the
        open circles are the results with the first iteration using a LJ
        (12-6) initial approximation (shown as a broken line). The
        filled circles are the second approximation and the solid
        triangles are points obtained from analysis of the glory region.
        (This diagram is based on that of Mason, Hermans and van den
        Meijdenberg, Ref. 17).

    The classical second virial coefficients were calculated from the
anisotropic potential by means of the expression

$$B(T) = -\frac{N}{4}\int_0^\infty r_{12}^2 dr_{12}\int_0^\pi \sin\theta_1 d\theta_1\int_0^\pi \sin\theta_2 d\theta_2\int_0^\pi f_{12}(r_{12},\theta_1,\theta_2,\Phi_{12})d\Phi_{12}$$

which is valid for linear molecules; $\Phi_{12} = \Phi_1 - \Phi_2$, and $f_{12} = \exp(-U_{12}/kT)-1$ is the Mayer f function.
    The transport properties were calculated using the Monchick—Mason
approximation[20] which provides a simple scheme whereby the collision
integrals may be calculated. The principal assumptions made in their

approach are that first, the energy tranfer between translation and
internal modes does not affect the transport properties, and second,
the relative angular orientation of the molecules remains unchanged
throughout the collisional process.   They suggested therefore that
the calculation should be performed by computing the fixed–angle
collision integrals $\Omega^{(\ell,s)}(T,\omega)$ for the collision of two molecules at
the fixed orientation.   The collision integral for the nonspherical
potential is then obtained quite simply by averaging the fixed
orientation collision integrals with equal weight over all space;
thus

$$(\Omega^{(\ell,s)}(T) = \frac{1}{8\pi} \int_{-1}^{1} \int_{-1}^{1} \int_{0}^{2\pi} \Omega^{(\ell,s)}(T,\omega)d\phi_{12}d(\cos\theta_1)d(\cos\theta_2).$$

The   inversion   methods   developed   for   spherically   symmetric
interactions have been applied to the properties of anisotropic
molecules calculated as described, for values of ℓ/σ up to 0.6584.[19]
The results are illustrated for ℓ/σ = 0.3292 and 0.6584 in Figures 5
and 6.

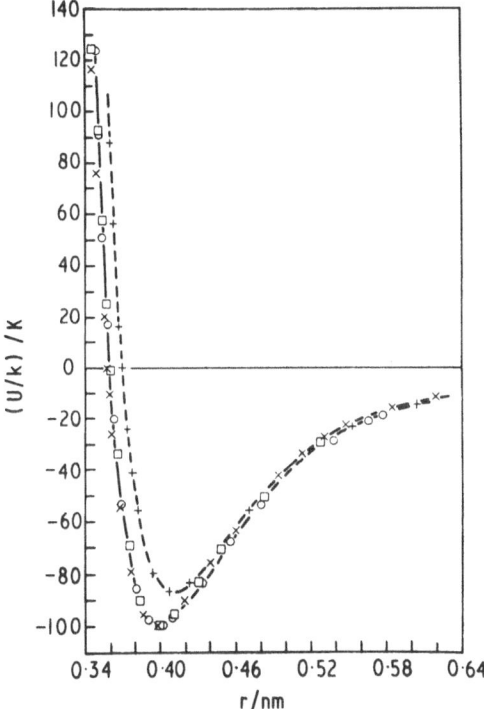

Fig. 5   Results from the inversion of properties calculated for the
      di–Lennard–Jones potential with ℓ/σ = 0.3292. (Ref. 19)

The potentials obtained by inversion are, except at the lowest value of anisotropy, very different from the spherical average, the differences becoming more marked at higher anisotropies. For all but the highest value of anisotropy investigated the inversion potentials obtained from different properties are remarkably similar.

We can conclude that: (i) for diatomic molecules with a degree of anisotropy comparable to that which has been proposed for fluorine (for which ℓ/σ is estimated as approximately 0.5) an effective spherically symmetric potential can be defined which can be used to calculate to high accuracy both second virial coefficients and transport properties; (ii) these potential functions can be obtained directly by the application of established inversion techniques.

The conclusions are quite unexpected, though as far as they concern transport properties they are subject to the validity of the Monchick–Mason approximation. To the extent that they offer the hope of correlating the thermophysical properties of an anisotropic substance by using only a simple potential function they may be of practical value. They could provide an economical method of estimating accurately a wide range of thermophysical data when only limited measurements are available.

3.2 Effective spherical potentials

The use of temperature–independent spherically symmetric potential functions to represent the interactions between anisotropic molecules has been common practice for many years, largely because more sophisticated approaches were not practicable. However, recently the possibility that the use of such effective potentials could be theoretically justified has been raised. Shaw et al[21] used thermodynamic perturbation theory to obtain from the di–Lennard–Jones potential an effective potential which adequately reproduces the equilibrium liquid state properties. Lebowitz and Percus[22] have shown that under certain conditions this approach can be simplified and that the effective spherical potential can be approximated by the median average $U^m(r_{12})$ of the anisotropic potential $U(r_{12},\omega_1,\omega_2)$ defined by

$$\iint \text{sign}\{U(r_{12},\omega_1.\omega_2) - U^m(r_{12})\}d\omega_1 d\omega_2 = 0$$

where ω_1 and ω_2 define the orientation of the molecules. This median average was shown to be superior to the unweighted spherical average or mean potential (obtained by averaging U over all orientations) and to give a good account of the di–Lennard–Jones fluid. At low anisotropies the effective potentials obtained by inversion are close to the median potential, though very different from the mean[23]. However, at higher anisotropies ($\ell/\sigma \approx 0.6$) the median approximation breaks down and its predictions diverge from those obtained by inversion procedures (Figure 6). The median approximation also breaks down for polar interactions where the contribution to median potential from the multipolar interactions can be zero.

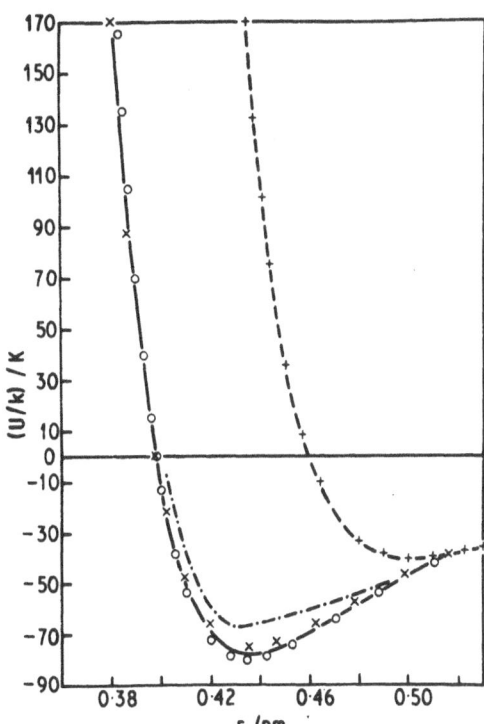

Fig. 6 Results from the inversion of properties calculated from the di-Lennard-Jones potential with $\ell/\sigma = 0.6584$. (Ref. 26)

3.3 Molecules of high anisotropy

The application of the inversion methods to systems of high anisotropy has been investigated by Maitland et al.[24] They selected a realistic potential function to model the system $Ar-CO_2$ in which the maximum energy of the collinear configuration is less than one sixth that of the broadside configuration. Under these conditions of very high anisotropy it was found that the potentials obtained by the inversion of transport properties differed significantly from that obtained from the inversion of second virial coefficient data. The authors propose a 'locus average' potential which is found to be significantly superior to the median and other proposed effective potentials.

In an analysis of the application of the inversion methods to the second virial coefficients of polyatomic molecules Trusler[25] has shown that a single temperature–independent well width function can be found that exactly specifies the second virial coefficients arising from anisotropic potentials. This may be represented by an effective potential of spherical symmetry but the potential may exhibit non–analytic turning points. He showed that the potential obtained by inversion provided an accurate representation of the second virial coefficients.

TABLE: Summary of Inversion Procedures

Property	Cross-section $Q = \pi \, \tilde{r}^2$	Master eqn
Viscosity η	$\Omega^{2,2}(T)$	$U^1(\tilde{r}) = G_\eta^0(T^*).kT$
Diffusion D	$\Omega^{1,1}(T)$	$U^1(\tilde{r}) = G_D^0(T^*).kT$
Second–Virial Coefficient B	$\left[B + \dfrac{TdB}{dT}\right]^{2/3}$	$U^1(\tilde{r}) = G_B^0(T^*)/kT$
Total Cross Section Q	$Q(V)$	$U^1(\tilde{r}) = G_Q^0 \left[\dfrac{V}{S^0}\right] \cdot \dfrac{hV}{\tilde{r}}$
Drift velocity V_d	$\Omega^{1,1}(T^1)$	$U^1(\tilde{r}) = G_{V_d}^0 \left[\dfrac{3}{2}kT + \dfrac{1}{2}mV_d^2\right]$
Adsorption Isotherm B_{AS}	$\left[B_{AS} + T\dfrac{dB_{AS}}{dT}\right]^{2/3}$	$U^1(\tilde{s}) = G_{B_{AS}}^0(T^*).kT$

Also Thermal Diffusion $\longrightarrow$ $\left[\dfrac{\partial \, \ell n \, r}{\partial \, \ell n \, U}\right]$

References

1. G.C. Maitland, M. Rigby, E.B. Smith and W.A. Wakeham, '*Intermolecular Forces : Their Origin and Determination*', Oxford. Clarendon, 1981.
2. Y. Tanaka and K. Yoshimo, *J. Chem. Phys.*, 1970 53 2012.
3. G.C. Maitland and E.B. Smith, *Mol. Phys.*, 1971 22 861.
4. H.L. Frisch and E. Helfand, *J. Chem. Phys.*, 1960 32 269.
5. D.A. Jonah and J.S. Rowlinson, *Trans. Faraday Soc.*, 1966 62 1067.
6. G.C. Maitland and E.B. Smith, *Mol. Phys.*, 1972 24 1185.
7. H.E. Cox, F.W. Crawford, E.B. Smith and A.R. Tindell, *Mol. Phys.*, 1980 40 705; E.B. Smith, A.R. Tindell, B.H. Wells and F.W. Crawford, *Mol. Phys.*, 1981 42 937.
8. G.C. Maitland, V. Vesovic and W.A. Wakeham, *Mol. Phys.*, 1985.

9. D.W. Gough, G.C. Maitland and E.B. Smith, *Mol. Phys.*, 1972 24 151.
10. J.O. Hirschfelder and M.A. Eliason, *Ann. N.Y. Acad. Sci.*, 1957 67 451.
11. J.H. Dymond, *J. Chem. Phys.*, 1968 49 3673.
12. G.C. Maitland and E.B. Smith, *7th Symp. Thermophys. Props N.Y. A.S.Med Eng.*, 1977 412.
13. E.B. Smith, A.R. Tindell, B.H. Wells and J.L. Brun, *Physica*, 1981 C106 117.
14. E.W. McDaniel and E.A. Mason, '*The Mobility and Diffusion of Gaseous Ions*', New York, Wiley 1973.
15. L.A. Vieland, M.M. Harrington and E.A. Mason, *Chem Phys.*, 1976 17 433.
16. E.A. Mason and C.J.N. van den Meijdenberg, *Physica*, 1983 117A 139.
17. E.A. Mason, R.M. Hermans and C.J.N. van den Meijdenberg, *Physica*, 1983 117A 160.
18. P.S.Y. Cheung and J.G. Powles, *Mol. Phys.*, 1975 30 921.
19. E.B. Smith and A.R. Tindell, *Faraday Disc. Chem. Soc.*, 1982 73 221.
20 L. Monchick and E.A. Mason, *J. Chem. Phys.*, 1961 35 1176.
21. M.S.Shaw, J.D. Johnson and B.L. Holian, *J. Chem. Phys.*, 1983 80 1141.
22. J.L. Lebowitz and J.K. Percus, *J. Chem. Phys.*, 1983 79 443.
23. C.G. Gray and C.G. Jarlin, *Chem. Phys. Letts.*, 1983 101 248.
24. G.C. Maitland, M. Mustafa, V. Vesovic and W.A. Wakeham, *Mol. Phys.*, 1986 57 1015.
25. J.P.M. Trusler, *Mol. Phys.*, 1986 57 1075.
26. E.B. Smith, A.R. Tindell and B.H. Wells, *High Temp.-High Press.* 1985 17 53.

TRANSPORT PROPERTIES OF VAN DER WAALS MOLECULES COMPUTED FROM ACCURATE
INTERACTIONS

F.A. Gianturco and M. Venanzi
Department of Chemistry
The University of Rome
Città Universitaria, 00185 Rome
Italy

ABSTRACT. Fully anisotropic atom-molecule interaction potentials have
been employed to study viscosity and diffusion in gaseous mixtures of
He and Ar with Nitrogen and Oxygen molecules over a wide range of temperatures. The rather accurate forms of potential energy surfaces employed
here were previously obtained from multiproperty analysis of each of the
examined systems and are further tested via the calculation of interaction viscosities and binary diffusion coefficients. It is found that,
for the He-N_2 and He-O_2 cases, the agreement between experiments and
calculated quantities is very satisfactory while the Ar-N_2 and Ar-O_2 interactions should still undergo some modification of their form in the
region of relative distance sampled by transport properties.

1. INTRODUCTION

Recent years have witnessed quite a remarkable increase in the quality
of the experimental procedures which measure the shear viscosity and
thermal conductivity coefficients for pure gases and gas mixtures, hence
in the corresponding reliability of the data obtained [1,2] about them.
The same has happened for the measurements of diffusion and thermal diffusion coefficients for binary gas mixtures [3] and therefore the combined use of the above gathering of very precise data have proven to be
essential aids in the accurate determination of the intermolecular potential functions of atomic gases, the simplest systems available [4] .
For polyatomic gases and for their mixtures with either atomic or molecular systems these effects still carry valuable information and provide good indicators of the quality of intermolecular potentials, but the
orientational nature of the interaction forces gives rise to an additional number of contributions to transport processes which are not present
in noble gases and which must also be taken into account when develop-

389

A. Weber (ed.), Structure and Dynamics of Weakly Bound Molecular Complexes, 389–404.
© 1987 by D. Reidel Publishing Company.

ping theoretical models.

In particular, the existence of additional degrees of freedom in molecular partners (internal states) leads both to inelastic contributions
to the effective cross sections which control transport phenomena and to
entirely new types of cross sections appearing through the collisional
coupling between translational space (relative velocity space) and rotational angular momentum space. These cross sections, in fact, give
rise to polarisation contributions in transport processes, contributions
which are present through the existence of tensorial angular momentum
polarisations in the non equilibrium distribution function for the gas
and can be related to a further set of experimental data [5] that study
the influence of externally applied electric and magnetic fields on
transport phenomena.

In addition, another group of relaxation processes which are associated with the internal state structure of the molecular systems occur
during transport events involving polyatomic partners. Thus, the non-
classical sound absorption, due to rotational and vibrational relaxation, and the pressure broadening of microwave and infrared electric dipolar lines provide another ensemble of accurate data again related to
the above phenomena [6] and therefore experiments carried out for the
pressure broadening of depolarised Rayleigh (DPR) and pure rotational
Raman lines have been obtained for a number of molecular systems [7]
and can provide a better understanding of anisotropic potential energy
surfaces (PES).

On the other hand, one of the reasons for relatively slow progress
that has been made in this area, on the theoretical side, for the clari-
fication of the relative importance of various features of these surfaces in determining the behaviour of the effective cross sections is due
to the still scant knowledge that we have of accurate PES for multielec-
tron systems, i.e. for molecular and atomic partners beyond H_2 and He
[8,9] . This essentially means that when dynamical models are put together to treat the above list of collision-based phenomena over the same
range of temperatures observed experimentally, a direct comparison
between computed observables and measured quantities is often obscured
by the insufficient reliability of the employed forms of PES over the
whole range of geometries that are presumably sampled by the experiments.

This is particularly true when the systems under study are molecules like N_2, O_2, CO and CO_2, to cite a few of the most commonly studied
cases, and the corresponding rare gases are Ne, Ar and the heavier terms
of the series. For the above molecular targets, on the other hand, even
their interaction with Helium atoms is still not fully understood.

In the present paper we therefore carry out a study of the behaviour
of computed diffusion coefficients and interaction viscosity for systems

where the corresponding PES are already known   with a good degree of
reliability from our previous studies of scattering experiments and of
other measured bulk data [10-13] . It will be shown that calculations
of transport properties can indeed provide both a further check of the
quality of the chosen interaction and an indication of the regions where
improvements are needed to yield more realistic shapes and size for the
full potential energy surfaces.

2. THEORETICAL DERIVATIONS

In the study described here we are limiting our work to the case of collision integrals arising from the interaction of an atom A with a rigid
rotor BC. Such integrals obviously enter the expressions for the transport properties of a mixture of a monoatomic gas with a molecular gas
[4,14,15] when one can assume that there is little or no vibrational
excitation of the molecule, as is frequently the case near room temperature and below.

One usually starts by defining the average of a quantity F as given
by:

$$\langle F \rangle = (2Z)^{-1} (k_B T/2\mu\pi)^{1/2} (k_B T)^{-3} \int_0^\infty d\varepsilon \ \varepsilon^2 \ \exp(-\varepsilon/k_B T)$$

$$\times \sum_j (2j+1)\exp(-\varepsilon_j/k_B T) \sum_{j'} \int d\bar{r} I(j \to j' | \theta) F \tag{1}$$

where $\quad Z = \sum_j (2j+1)\exp(-\varepsilon_j/k_B T)$ \hfill (2)

μ is the reduced mass of the system and ε is the relative kinetic energy
of particles before a collision and when the rotational state of the ro-
tor is labelled by $|j\rangle$. $I(j \to j'|\theta)$ is the degeneracy averaged, sta-
te-to-state differential cross section for the process and θ provides
the scattering angle in the c. of m. system. The additional energy ε_j
denotes the internal energy of the rotor and $d\underline{r}$ is the volume element
in the same c. of m. system.

The classical, or rather semi-classical definition of the binary
diffusion coefficient of the gas mixture is thus given by [4] :

$$D_{12}^{(0)} = 3k_B T/16n\mu\Omega^{(1,1)} \tag{3}$$

where the collision integral $\Omega^{(1,1)}$ is

$$\Omega^{(1,1)} = \langle 1 - (\varepsilon'/\varepsilon)^{1/2} \cos\theta \rangle \tag{4}$$

Here ε' is the relative kinetic energy of partners after collision:

$$\varepsilon' = \varepsilon - (\varepsilon_{j'} - \varepsilon_j) \qquad (5)$$

and n is the gas density in amagat units.

The viscosity curvature or interaction viscosity, η_{12}, is in turn given via another collision integral [4,15] :

$$\eta_{12} = 5k_BT/8\Omega^{(2,2)} \qquad (6)$$

where

$$\Omega^{(2,2)} = \langle(\varepsilon/k_BT)\{1-(\varepsilon'/\varepsilon)\cos^2\theta - (1-\varepsilon'/\varepsilon)^2/6\}\rangle$$

$$= \langle(\varepsilon/k_BT)\{\sin^2\theta + (1-\varepsilon'/\varepsilon)^2/3 - (1-\varepsilon'/\varepsilon)^2(\sin^2\theta)/2\}\rangle \qquad (7)$$

One of the most common approximations that has been used to calculate the above collision integrals is due to Monchick and Mason (MM) [16] and proceeds in a number of stages. As we shall see later, it turns out to be exactly equivalent to the dynamical treatment of quantal coupled equations known as Infinite Order Sudden (IOS) Approximation [17] .

In a first approximation, one assumes in the semiclassical model that the dominant inelastic collisions are those for which $j'-j = \pm 1$, hence:

$$\Delta\varepsilon = \varepsilon_{j'} - \varepsilon_j \ll k_BT \sim \varepsilon \qquad (8)$$

on average and therefore one can set $\varepsilon' = \varepsilon$ in the previous equations. This assumption goes under the name of energy sudden (ES) approximation. One consequence of this simplification is that all contributing trajectories are only slightly distorted during encounters and therefore $I(j \to j'|\theta)$ could be replaced by the corresponding differential cross section for elastic scattering, I_{el}. The corresponding collision integrals could then be rewritten as formally identical to those pertaining to monoatomic gases with spherically symmetric pair potentials [18] :

$$\Omega^{(1,1)} = (1/2)(k_BT/2\mu\pi)^{1/2}(k_BT)^{-3}\int\varepsilon^2\exp(-\varepsilon/k_BT)I_{el}.(1-\cos\theta)d\bar{r}d\varepsilon \qquad (9)$$

and:

$$\Omega^{(2,2)} = (1/2)(k_BT/2\mu\pi)^{1/2}(k_BT)^{-4}\int\varepsilon^3\exp(-\varepsilon/k_BT)I_{el}.\sin^2\theta d\bar{r}d\varepsilon \qquad (10)$$

The elastic differential cross section is however computed here via the fully anisotropic PES of the problem at hand. It should also be noted that the viscosity collision integral of eq. (7) can also be simplified to its form of eq. (10) by employing the average value of $\sin^2\theta = 2/3$ that applies to a rigid sphere collision [16] .

Finally, the classical evaluation of I_{el} was further simplified within the MM scheme of calculation by computing the collision integrals between partners holding fixed during each encounter the relative angle of orientation, ω . This is then exactly the same calculation as that for a central potential relative to each chosen ω value and it allows one to compute the final collision integral for the full anisotropic PES by averaging over all ω values taken with equal weight, i.e. by writing:

$$\Omega^{(n,s)}(T) = \int \Omega^{(n,s)}(T,\omega)d\omega \Big/ \int d\omega \qquad (11)$$

2.1 The IOS quantal approximation

For the earlier derivation of the anisotropic interactions from several, different properties we have essentially employed a simplified form of the coupled scattering equations that is now well-known as IOS approximation [17] and we have also shown, in a comparison with rigorous close-coupled (CC) calculations [19] , that it is a physically accurate assumption to make for the present systems interacting with He while it is probably less likely to hold when stronger interactions are present, as is the case for the interactions with Ar [20] .

In essence, one combines two different approximations, the energy sudden (ES) and the centrifugal sudden (CS) approximation. The first consists of replacing the set of wavenumbers for the translational motion K_j, in the usual coupled equations in the SF frame [21] , by the single constant value $\bar{K}$. It is obviously equivalent to the weak-excitation approximation of the MM scheme (see previous eq. (8)) and to considering negligible the distorsion of each contributing classical trajectory that is brought about by the inelastic collisions.

Furthermore, in the CS approximation all the centrifugal potentials of the collision are assumed to be degenerate so that the orbital angular momentum quantum number l is replaced by a single value $\bar{l}$ that is held constant during collisions. The conservation of $\hat{j}_z$ implied by this approach is therefore the equivalent to what the MM model assumes during the resolution of the classical equations of motion, i.e. that the relative orientation of the colliding partners is being held fixed during encounters.

It is thus easy to show [22] that the corresponding degeneracy-averaged state-to-state differential cross sections may be written as:

$$I(j\rightarrow j'|\theta) = (k_0/k_j)^2 \sum_{j''} C^2(j,j'',j'; 0,0,0)I(0\rightarrow j''|\theta) \qquad (12)$$

where the C's are the usual Clebsch-Gordan coefficients.

The differential cross section from the ground state $|0\rangle$ is in turn given by:

$$I(0 \to j''|\theta) = (\bar{k}^2/k_0^2)|f_j^{\bar{k}''}(\theta)|^2(2j''+1)^{-1} \tag{13}$$

where the coefficients on the r.h.s. of (12) come from an angle-dependent partial wave expansion of the scattering amplitude [13] :

$$f^{\bar{k}}(\theta;\gamma) = \sum_L f_L^{\bar{k}}(\theta)P_L(\hat{R} \cdot \hat{r}) \tag{14}$$

where γ is now the internal orientation angle, and is equal to arcos $(\hat{R} \cdot \hat{r})$, with $\underline{R}$ and $\underline{r}$ being the collision coordinate and the molecular coordinate, respectively [22] .

The IOS scattering amplitude are in turn obtained as [13] :

$$f^{\bar{k}}(\theta;\gamma) = (i/2\bar{k}) \sum_\ell (2\ell+1) \{ 1 - \exp[2i\eta_\ell^{\bar{k}}(\gamma)] \} P_\ell(\cos\theta) \tag{15}$$

The last equation is therefore the quantal equivalent of elastic scattering by a central potential which causes, at each chosen γ and for each partial wave, the phaseshift $\eta_\ell^{\bar{k}}(\gamma)$.

The classical approximations discussed above within the MM model are thus clearly exactly the same as those introduced here for the quantal IOS.

The total differential cross section can now be written via the simple form:

$$I(\theta) = \sum_{j'} I(j \to j'|\theta) \tag{16}$$

$$= 1/2 \int_{-1}^{+1} I(\theta;\gamma) \, d\cos\gamma \tag{17}$$

where:

$$I(\theta;\gamma) = |f^{\bar{k}}(\theta;\gamma)|^2 \tag{18}$$

hence the total cross section is simply the equally weighted average of the elastic cross sections for all fixed orientations γ . The corresponding collision integrals of eqs.(4) and (7) are now given by quantal, simplified expressions obtained by assuming that $\epsilon' = \epsilon$ and that j and j' can take any positive integer values. Thus:

$$\Omega^{(n,s)} = \langle (\varepsilon/k_B T)^{s-1} | 1-\cos^n \theta | \rangle$$

$$= 1/2(k_B T/2\mu\pi)^{1/2} (1/k_B T)^{s+2} \int d\varepsilon \cdot \varepsilon^{s+1} \exp(-\varepsilon/k_B T) Q^{(n)}(\varepsilon) \quad (19)$$

where

$$Q^{(n)}(\varepsilon) = \int Z^{-1} \sum_j (2j+1)\exp(-\varepsilon_j/k_B T) d\phi \cdot d(\cos\theta) (1-\cos^n\theta)$$

$$\times \sum_{j''} |f_{j''}^{\bar{k}}|^2 (2j''+1)^{-1} \sum_{j'} C^2(j,j'',j';0,0,0) \quad (20)$$

By using the closure property of the Clebsh-Gordan coefficients in eq. (20), and by the use of eq. (14) for the expansion coefficients needed in (20), one can then obtain the $Q^{(n)}$ from the following quadratures:

$$Q^{(n)}(\varepsilon) = 1/2 \int_{-1}^{+1} Q^{(n)}(\varepsilon;\gamma) d(\cos\gamma) \quad (21)$$

where

$$Q^{(n)}(\varepsilon;\gamma) = 2\pi \int_{-1}^{+1} (1-\cos^n\theta) I(\theta;\gamma) d(\cos\theta) \quad (22)$$

which is formally identical to the MM result of eq. (11).

One therefore sees from the above derivations that the full PES is always needed to compute the scattering amplitudes before the quadrature over internal orientations, either in the classical or quantal scheme. On the other hand, the effect of inelastic collisions can be included either by summing over all possible final states as in eq. (20) or by the use of eq. (12) and by including only those $(j \rightarrow j'|\theta)$ which are allowed by the energy balance of eq. (5). This is equivalent to the use of the IOS approximation to generate state-to-state differential cross sections in a quantal scheme and then to using them in the exact kinetic theory formulae for the transport collision integrals [23] .

3. THE CALCULATIONS

Preliminary calculations have been carried out for N_2 and O_2 molecules interacting with He and using full PES that have been obtained by us [10,11] via the analysis of scattering observables, i.e. of total differential cross sections and state-to-state differential cross sections.

The angle-dependent phaseshifts of eq. (15) were obtained in all cases by a 10-point Gauss-Mehler quadrature of the JWKB formula [24] .

The necessary scattering amplitudes of eq. (18) were in turn obtained
for up to 25 different γ values and saturation of the quadrature procedure was tested for a smaller range of angles, to make sure that the
latest value corresponded to convergence. The partial-wave summation in
eq. (15) was also tested at the highest energies and found to converge
with about 250 terms, a value which was then also kept at the lower collision energies. For the Ar projectile the number of terms in the l-
summation was increased to 400.

The energy averaging needed in eq.s (19) and (20) was carried out
by dividing the range of integration into several sectors of ~20 meV
each and by integrating within each sector with a 64-point Gauss-Laguer-
re quadrature. The threshold behaviour of the cross sections was obtained down to ξ = 0 by extrapolating from the lowest energy that yielded
stable cross section values upon integrating the scattering equations.
For all systems different extrapolation procedures were tested and found
to affect very negligibly the final values of the corresponding collision integrals.

The diffusion coefficients and the interaction viscosity values are
shown, in Table 1a and 1b respectively, for the He-N$_2$ system where two
of the most recent interaction potentials were employed: the one proposed by Mc Court and collaborators [25] and used recently in classical
calculations of collision integrals [26] , which we have labelled as
HFD1, and the more recent PES that we obtained from the analysis of
scattering experiments [19] and of further comparison with several other
properties [13] , which we have labelled M3SV mod.

It is worth noting, from the results of the Tables, that one can attempt a separation of the effects which originate from using different
approximations in computing the collision integrals from the possible
effects which can arise from differences in the features of the full PES
employed to represent the interaction.

Thus, the quantum calculations implied by the IOS scheme described
here turn out to be exactly the same as the calculations produced by
the classical solution of the dynamics as developped within the MM scheme once the same potential and the same treatment of inelastic processes
is carried out, as shown by the third and fourth column in each Table.

On the other hand, the differences between shapes and sizes of each
PES do indeed show up in the different behaviour of the computed coefficients: the HFD1 potential, in fact, exhibits always smaller collision integrals, hence presents a smaller 'molecular size' than that of
both the present M3SV mod surface and of previous surfaces [25] , an effect which it tries to compensate with a larger angular anisotropy. The
latter feature, however, is not so strongly felt by the present viscosity calculations (which are mostly sensitive to changes of size) and

TABLE 1a. Comparison of computed and measured binary diffusion coefficients ($cm^2 s^{-1}$ at 1 atm) for the He-N_2 system, $[D_{12}]_1$ The computed values refer to various PES forms and different schemes of calculation
that are described in the text

T (K)	HFD1		M3SV mod		Expt
	MM^a	IOS^b	MM^c	IOS^d	
77.3	0.0679	0.0692	0.0669	0.0680	0.0725($\pm$0.001)
195.10	0.341	–	0.338	0.338	0.34444($\pm$0.001)
218.07	0.412	–	0.408	0.408	0.4159($\pm$0.0012)
234.20	0.465	–	0.461	0.461	0.4689($\pm$0.0014)
255.50	0.539	–	0.534	0.534	0.5427($\pm$0.0016)
300.00	0.708	0.714	0.700	0.700	0.7068($\pm$0.0021)
320.00	0.789	–	0.780	0.780	0.787($\pm$0.0024)
337.90	0.865	–	0.855	0.855	0.8638($\pm$0.0026)
373.09	1.023	1.029	1.010	1.011	1.018($\pm$0.0031)
463.15	1.473	–	1.454	1.454	1.464($\pm$0.029)
563.1	2.049	–	2.020	2.021	2.047($\pm$0.041)

a) from ref. [26] ; b) as quoted in ref. [26] ; c) from ref. [13];
d) this work; e) from several authors, as quoted in ref. [13] .

is also appearing to influence only marginally the diffusion coefficients. In other words, the present comparison seems to indicate that collision integrals are sensitive tests of the onset of the repulsive region of the full PES, which qualitatively depicts the 'size' of the molecular target as seen by the impinging, smaller atomic projectile. Because of this effect, one can therefore unequivocally say that the new
M3SV mod potential yields a larger N_2 target than the HFD1 model and
requires a much weaker anisotropy to reproduce the experimental phase
differences seen between total and partial differential cross sections
of this pair of gases [19] .

Similar results for the He-O_2 mixture are reported in Tables 2a
and 2b, where comparison between measured and computed diffusion coefficients (Table 2a) and interaction viscosity (Table 2b) are shown.

One sees from the above calculations that the KSK potential surface obviously produces, incorrectly, transport cross sections which are
smaller than those needed to match the experimental results: both the
η_{12} and $[D_{12}]_1$ values are in fact larger, at all examined tempera-

TABLE 1b. Comparison of computed and measured mixture viscosity coefficients, η_{12} (μ P), for the He-N$_2$ system. The computational methods
are the same as those for Table 1a.

T (K)	HFD1		M3SV mod		Expt
	MM[a]	IOS[b]	MM[c]	IOS[d]	
77.3	56.6	56.5	56.1	56.5	
298.05	148.1	146.8	146.8	146.9	147.5($\pm$1.5)
373.09	1,72.0	170.4	170.4	170.5	171.5($\pm$1.7)
463.15	198.7	-	196.7	196.8	197.8($\pm$2.0)
563.15	226.4	-	224.0	224.1	225.5($\pm$2.3)
673.15	255.2	-	252.3	252.4	254.0($\pm$2.5)
763.15	277.1	-	274.4	274.5	276.1($\pm$2.8)
873.16	304.1	-	304.1	300.1	300.0($\pm$3.0)
1,000.00	333.5	-	329.0	329.2	-

a) from ref. [26] ; b) as quoted in ref. [26] ; c) from ref. [13];
d) present work; e) as quoted in ref. [26] .

ture, than the experimental values, this being especially so for diffusion coefficients. This is not the case, however, when the more realistic full surface obtained from scattering data [10] is employed within
the MM and IOS approximations. The values yielded by both classical (MM)
and quantal (IOS) treatments with the same physical approximations are
in fact no more than 1% smaller than experiments, hence well within
their quoted error bars [31] . The overall 'size' of our M3SV potential
and its shape around the turning point region at low T values seem to
be here within the range suggested by transport experiments. Moreover,
both the MM and the IOS approximate schemes produce numerical results
which are identical with each other.

It is useful to compare the relative behaviour of η_{12} and $[D_{12}]_1$
within the above two systems, since N$_2$ and O$_2$ are very similar targets
with comparable anisotropies [36] . One sees from Tables 1a and 2a, for
instance, that binary diffusion coefficients are larger, at the same T
values, for He-O$_2$ than for He-N$_2$, a fact which indicates smaller cross
sections for the former when compared to the latter. If one keeps in
mind that transport properties tell us about the effective 'size' of a
molecular target then the above result is in line with the found featu-

TABLE 2a. Comparison and measured binary diffusion coefficients $(cm^2s^{-1}$ at 1 atm) for the He-O_2 system, $\left[D_{12}\right]_1$. The computed values refer to different interactions and different schemes for the collision integrals, as discussed in the text.

T (K)	KSK[a]	M3SV[b]		Expt[e]
	MM[c]	MM[c]	IOS[d]	
195.15	0.3740	0.3547	0.3553	0.3644($\pm$0.001)
218.15	0.4538	0.4285	0.4292	0.4378($\pm$0.0012)
255.50	0.5968	0.5604	0.5610	0.5710($\pm$0.0016)
300.00	0.7871	0.7350	0.7356	0.7448($\pm$0.0021)
320.00	0.8797	0.8195	0.8201	0.8305($\pm$0.0024)
373.15	1.1457	1.0615	1.0620	1.072($\pm$0.0031)
400.20	1.2926	1.1948	1.1950	1.208($\pm$0.0034)
678.15	3.207	2.907	2.908	2.88($\pm$0.055)

a) as given in ref. [27]; b) from ref. [10] ; c) from ref. [29] ;
d) present work; e) from ref. [28] .

res of the two interaction potentials: the spherical averages for He-N_2 and He-O_2 show [10,13] $\overline{R}_m$ and $\overline{\sigma}$ values which are about 3% larger for the former than for the latter, hence one expects larger cross sections in the Nitrogen mixtures as opposed to the Oxygen mixtures.

This behaviour is even more marked when one observes the interaction viscosity data (Tables 1b and 2b), where the computed and measured values for O_2 are often 5% larger than those for N_2. Since it is the scattering at angles around $\vartheta = \pi/2$ which dominates the viscosity collision integral $\Omega^{(2,2)}$, while those around $\vartheta = \pi$ are the ones which contribute the most to the diffusion collision integral $\Omega^{(1,1)}$, it is usually assumed that inelastic contributions are even more important for diffusion than they are for the η_{12} data. Hence, the latter data are even more clearly dominated by 'size' effects while the former property is also affected by the anisotropic features of the surfaces involved. In other words, the fact that the He-N_2 PES is more anisotropic in the R_m and σ regions than the He-O_2 surface is reflected more markedly in the size of the $\Omega^{(1,1)}$ integrals, which in turn control the magnitude of $\left[D_{12}\right]_1$. For both systems, on the other hand, we find that the computed values of the collision integrals yield η_{12} and diffusion coefficients that are in good accord with measurements. The use of the

TABLE 2b. Computed and measured interaction viscosity coefficients, $\eta_{12}(\mu P)$, for the He-O_2 system. The potential forms and computational schemes are the same as in Table 2a.

T (K)	KSK[a]	M3SV[b]		Expt[e]
	MM[c]	MM[c]	IOS[d]	
298.1	165.0	156.4	156.5	154.2($\pm$1.5)
378.1	194.6	183.3	183.3	185.5($\pm$1.7)
478.1	229.2	214.3	214.3	218.2($\pm$2.0)
573.1	260.1	241.8	241.8	243.7($\pm$2.3)
678.1	292.7	270.6	270.6	273.0($\pm$2.5)

a) as given in ref. [27] ; b) from ref. [10] ; c) from ref. [29] ;
d) this work; e) as given by ref. [30] from measurements in ref.
[31].

IOS prescription to generate transport properties for these mildly anisotropic causes is therefore confirmed to be an effective and inexpen-
sive strategy of calculation, as indicated already in earlier work with
less accurate potentials [37] .

The extension of this work to heavier rare gases has recently involved interactions with Argon and we therefore present below a similar
analysis for Ar-N_2 and Ar-O_2.

In the case of Ar-N_2 (Table 3a), several surfaces have been proposed in recent years. Pattengill et al. [32] (PLBC) used a simple model
potential consisting of a Lennard-Jones (12,6) central potential with
P_2-type repulsive and attractive anisotropies. Later on, Kim [38] calculated a partial PES by using the electron gas model which was further
modified by Lee and Kim [39] in a subsequent calculation of the same
surface.

Van der Biesen et al. [40] proposed a trial potential whose spherical part was produced by the Ar-Ar interaction and where the anisotropic part was determined by fitting to their measured total cross sections over the energy range of 66-910 meV. More recently Brunetti et
al. [41] obtained spherically averaged potentials from the analysis of
the measured absolute total cross sections and then proceeded to assess the angular dependence by a combined analysis of integral and differential cross sections and of some spectroscopic data [10] . A very
recent study of partial, state-to-state differential cross sections
measured with high degree of accuracy [20] has cast however some strong

TABLE 3a. Computed and measured diffusion coefficients, $\begin{bmatrix} D_{12} \end{bmatrix}_1$ ($cm^2 s^{-1}$ at 1 atm), for the Ar-N_2 system as function of temperature (K).

T (K)	IOS-PLBC[a]	MM-PBLC[d]	IOS-M3SV[b]	Expt[c]
77.3	0.0131	0.0139	0.0166	-
195.1	-	-	0.0964	0.0928
205.2	-	-	0.1061	0.1020
218.1	-	-	0.1190	0.1143
237.1	-	-	0.1392	0.1333
252.7	-	-	0.1567	0.1496
280.0	-	-	0.1892	0.1793
300.0	0.1923	0.1944	0.2146	0.2033

a) potential from ref. [32] and calculations from ref. [33] ;
b) potential from ref. [11] and present calculations;
c) from ref. [28] ; d) from ref. [33] .

doubts on the reliability of the anisotropic nature of the above estimate, although it was not possible to disentangle fully the found
shortcomings of the PES from the inadequacies of the dynamical approximations used to perform the comparison with the measured data.

We have employed the proposed PES of ref. [10] and compared its
preliminary results with those quoted by Dickinson and Lee [33] for the
diffusion coefficients. The latter IOS calculations had employed the
PLBC surface and turned out to be rather close to the classical MM calculations reported in ref. [33] for the same surface.

It is interesting to note that the present computations use quantal IOS with a unitary S-matrix, i.e. by taking advantage of the closure properties of the Clebsch-Gordan coefficients of eq. (12) and therefore generating the total differential cross section via the quadrature
(17). If one were to use individual DCS'S in the correct averaging of
eq. (1), then the problem of correcting for higher j-values to attain
unitarity would have appeared [37] . The IOS calculations of column 1
af Table 3a contain such a correction while our calculations of column
3 in the same Table directly keep unitarity by using the sum rule   of
eq. (16). One sees that the classical MM calculations with the same surface [33] are rather different than the earlier IOS data, a result which
is at variance with our previous findings for He-N_2 and He-O_2. Moreover,

TABLE 3b. Computed and measured diffusion coefficients, $[D_{12}]_1$ (cm^2 s^{-1} at 1 atm), for the Ar-O$_2$ system as function of temperature (K)

T (K)	M3SV[a]		ESMSV[b]	Expt[c]
	IOS[d]	MM[e]	IOS[b]	
195.1	0.0915	0.0917	0.0907	0.0922
205.2	0.1009	0.1011	0.1001	0.1015
218.1	0.1137	0.1136	0.1122	0.1135
234.2	0.1301	0.1303	0.1286	0.1297
246.0	0.1429	0.1430	0.1411	0.1421
264.8	0.1643	-	0.1616	0.1627
300.0	0.2074	0.2076	0.2017	0.2035
400.2	0.3516	0.3518	0.3371	0.3398

a) present fit of potential of ref. [34] ; b) potential of ref. [34] and calculations from ref. [28] ; c) from ref. [28] ; d) present calculations; e) from ref. [35] .

the reduced dimension of the repulsive core of the M3SV potential, where $\overline{\sigma}$ is 3.31 Å as opposed to 3.5 Å for the PBLC surface, produces as before smaller transport cross sections, hence larger diffusion coefficients. Comparison with experiments [28] seems to suggest that the correct 'size' value should be somewhere in between the values of above, although further studies on the validity of the dynamical approximation need to be done and are presently under way in our group [42] .

Finally, Table 3b compares measured and computed diffusion coefficients for the Ar-O$_2$ gaseous mixture. The two potentials employed are two different fits of an anisotropic surface recently proposed in the literature [34] . The one termed M3SV uses a Morse function to describe the repulsive region of the interaction and parametrizes the anisotropy as previously described by us [19] , while the one termed ESMSV contains an exponential repulsive core and only one Morse function for the well region.

One sees there that the onset of the repulsive region does affect the final results on $[D_{12}]_1$, while once more the classical MM approximation calculations [35] are indistinguishible from the quantal IOS calculations with unitary S-matrix as carried out by us. In both cases agreement with experiments is rather satisfactory and certainly better than in the case of the Ar-N$_2$ surface.

In conclusion, the present preliminary study demonstrates once mo-re that computing transport properties for which accurate experimental data are available is a very illuminating way of further testing the quality of potential energy surfaces in regions often outside the ran-ge of collision experiments, thus complementing the latter and even suggesting directions which changes and modifications should be follo-wing in order to increase overall agreement with as many properties as possible.

ACKNOWLEDGEMENTS

The financial support of the Italian Nat. Research Council (CNR) and of the Italian Ministry of Education (MPI) is gratefully acknowledged. We also want to thank Dr. Alan Dickinson for several helpful discus-sions on various aspects of the present calculations.

REFERENCES

1   J. Kestin, S.T. Ro and W.A. Wakeham, J. Chem. Phys. 56, 4036 (1972)
2   J. Kestin and E.A. Mason in Transport Phenomena, ed. by J. Kestin (AIP, New York, 1973); pg 137
3   e.g. see: P.S. Arora and P.J. Dunlop, J. Chem. Phys. 71, 2430 (1979)
4   G.C. Maitland, M. Rigby, E.B. Smith and W.A. Wakeham, Intermole-cular forces. their origin and determination (Clarendon, Oxford 1981)
5   J.J.M. Beenakker in Transport Phenomena, G. Kirezenow and J. Marro ed.s (Springer, Berlin, 1974); J.J.M.Beenakker and F.R. Mc Court, Ann. Rev. Phys. Chem. 21, 47 (1970)
6   H. Rabitz, Ann. Rev. Phys. Chem. 25, 155 (1974)
7   H.F.P. Knaap and P. Lallemand, Ann. Rev. Phys. Chem. 26, 59 (1975)
8   W. Meyer, P.C. Hariharan and W. Kutzelnigg, J. Chem. Phys. 73, 1880 (1980)
9   J. Schaefer and W. Meyer, J. Chem. Phys. 73, 6153 (1980)
10   M. Faubel, K.M. Kohl, J.P. Toennies and F.A. Gianturco, J. Chem. Phys. 78, 5629 (1983)
11   R. Candori, F. Pirani and F. Vecchiocattivi, Chem. Phys. Lett. 102, 412 (1983)
12   R. Candori, F. Pirani, F. Vecchiocattivi and F.A. Gianturco, Faraday Disc. 73, 289 (1981)

13   F.A. Gianturco, M. Venanzi, R. Candori, F. Pirani, F. Vecchio-
     cattivi, A.S. Dickinson and M.S. Lee, Chem. Phys., 1986 (in
     press)

14   L. Monchick, K.S. Yun and E.A. Mason, J. Chem. Phys. 39, 654
     (1963)

15   L. Monchick, A.N.G. Pereira and E.A. Mason, J. Chem. Phys. 42,
     3241 (1965)

16   E.A. Mason and L. Monchick, J. Chem. Phys. 36, 1622 (1962)

17   G.A. Parker and R.T. Pack, J. Chem. Phys. 68, 1585 (1978)

18   L. Monchick and E.A. Mason, J. Chem. Phys. 35, 1676 (1961)

19   F.A. Gianturco and A. Palma, J. Phys. B, 18, L519 (1985)

20   M. Faubel and G. Kraft, J. Chem. Phys. 1986 (in press)

21   A.M. Arthurs and A. Dalgarno, Proc. R. Soc. A256 , 540 (1960)

22   e;g. see: F.A. Gianturco, The transfer of molecular energy by
     collisions (Springer, Btrlin, 1979)

23   G.C. Maitland, V. Vesovic and W.A. Wakeham, Mol. Phys. 42, 803
     (1981)

24   R.T. Pack, J. Chem. Phys. 60, 633 (1974)

25   F.R. Mc Court, R.R. Fuchs and A.J. Thakkar, J. Chem. Phys. 80,
     5561 (1984)

26   A.S. Dickinson and M.S. Lee, J. Phys. B 18, 4177 (1985)

27   G.C. Corey and F.R. Mc Court, J. Chem. Phys. 81, 3892 (1984)

28   R.D. Trengove, K.R. Harris, H.L. Robjohns and P.J. Dunlop,
     Physica 131A, 506 (1985)

29   A.S. Dickinson, private communication

30   W.A. Wakeham, private communication

31   J. Kestin, H.E. Khalifa, S.T. Ro and W.A. Wakeham, Physica 88A,
     242 (1977)

32   M.D. Pattengill, R.A. La Budde, R.B. Bernstein and C.F. Curtiss,
     J. Chem. Phys. 55, 5517 (1971)

33   A.S. Dickinson and M.S. Lee, J. Phys. B 18, 3987 (1985)

34   F. Pirani and F. Vecchiocattivi, Chem. Phys. 59, 387 (1981)

35   A.S. Dickinson, private communication

36   F.A. Gianturco, A. Palma and M. Venanzi, Mol. Phys. 56, 399
     (1985)

37   A.S. Dickinson and D. Richards, J. Phys. B16, 2801 (1983)

38   Y.S. Kim, J. Chem. Phys. 68, 5001 (1978)

39   S. Lee and Y.S. Kim, J. Chem. Phys. 70, 4856 (1979)

40   J.J.H. Van der Biesen et al., Physica 116A, 101 (1982)

41   B. Brunetti et al., J. Chem. Phys. 79, 273 (1983)

42   F.A. Gianturco and M. Venanzi, in preparation.

DETERMINATION OF ANISOTROPIC INTERACTION POTENTIALS FROM ROTATIONAL
STATE RESOLVED SCATTERING DATA

M. Faubel
Max-Planck-Institut für Strömungsforschung
Bunsenstr. 10
3400 Göttingen / W.-Germany

ABSTRACT: Rotationally inelastic collisions of Ar-O_2, Ar-N_2, Ar-CO
and Ar-HCl are studied in a crossed molecular beams time of flight
experiment.
The Ar-O_2 and Ar-N_2 scattering shows the excitation of very large
rotational quantum transitions. In Ar-HCl collisions, in contrast, the
scattering is prevailingly elastic, or restricted to small $\Delta j \lesssim 2$ transitions.
The Ar-O_2 and the Ar-HCl molecular beam scattering results are used
for a comparison with the best available spectroscopically determined
interaction potentials for these molecules.

1. INTRODUCTION

Scattering studies of inelastic molecular collisions and the spectroscopy of bound states of van der Waals molecules provide two basically
different experimental methods for a direct investigation of intermolecular forces. The two methods yield often complementary pieces of information on the molecular interaction potential.
    In the molecular beam scattering experiments, to be discussed sub-
sequently, we investigated the partially state resolved rotationally inelastic scattering in the homologous Ar-O_2, Ar-N_2 and Ar-CO systems
/1/2/ and, also, in Ar-HCl collisions /3/. The experimental results for
the hydrogen halide collision differ dramatically from the results for
Ar-O_2, N_2 and CO by an almost complete absence of large rotationally
inelastic transitions. Both, the Ar-O_2 /4/ and the Ar-HCl /5/6/ van
der Waals complex molecules have been well characterized by spectroscopy
and are, thus, suitable for a direct comparison of the two principle investigation methods for anisotropic intermolecular potentials.

2. THE SCATTERING APPARATUS

The apparatus employed in the present experiment is shown in Fig.1.
This crossed molecular beams time of flight machine is discussed else-

405

A. Weber (ed.), Structure and Dynamics of Weakly Bound Molecular Complexes, 405–421.

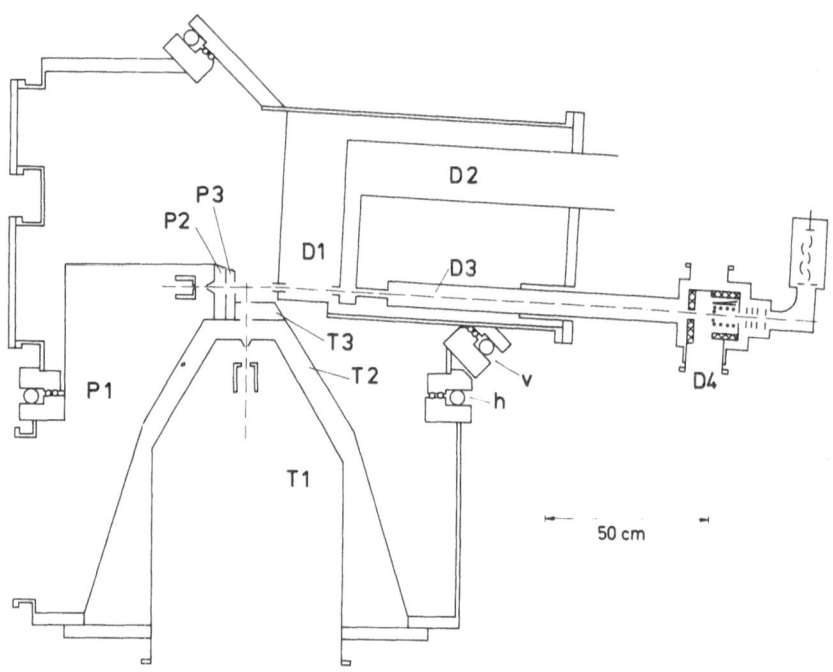

Fig.1: Cross section of the molecular beam scattering apparatus. The
horizontal primary beam and the upwards directed target beam originate
from nozzles in the vacuum chambers "P1" and "T1". They intersect at
right angles in the main vacuum chamber. Scattered molecules are detected by the ultrahigh vacuum mass spectrometer in chamber "D4". Both, in
plane and out of plane scattering angles are accessible through rotation
of the bearings "v" and "h".

where in detail /1/7/. Briefly, its principal components are the two
nozzle beam sources in the vacuum chambers "P1" and "T1" and the time of
flight detector, a sensitive mass spectrometer with the ionizer located
in the ultrahigh vacuum chamber "D4" of Fig.1. With two large 12 000
ℓ/sec diffusion pumps on the beam source chambers "P1" and "T1" the two
nozzles can operate at a maximum gas flux of 2 mol/hour while, at full
through-put, a pressure of lower than 10^{-3} mbar is sustained in "P1"
and "T1". A skimmer and additional differentially pumped collimators reduce the initial molecular beam spread by more than a factor of hundred
to the experimentally required angular divergence of less than one degree
before the horizontal primary beam and the vertical target beam intersect
in the center of the main chamber H1. Additional pumping stages, "D1"
through "D4", on the 145 cm length time of flight drift path buffer the
$1.5 \cdot 10^{-11}$ mbar vacuum of the detector from the 10^{-6} mbar high vacuum in
the main chamber. For most gases, but H_2 and CO, the residual gas
partial pressure in the detector region is 10^{-15} mbars, and, as low as

10^{-16} mbar in favorite cases as, for example, with helium. For the time
of flight measurements the primary beam is intercepted in the differential pumping chamber "P2" by a 12 cm diameter slotted wheel chopper. Its
typical speed of rotation is 400 cycles per second, allowing for a beam
pulse time of 7 μsec with 1 mm wide chopper slits. Because of the background of 10^{-15} mbar on the argon mass 40 a pseudo random chopping sequence with 255 elements is employed for improving on the noise in the
time of flight measurements with argon. The setting of the horizontal
(Θ_{lab}) and of the vertical (Φ_{lab}) elevation angle of the detector with
respect to the intersecting molecular beams is accomplished through the
rotation of the two ball bearing supported vacuum joints "h" and "v", respectively. This particular feature allows for an alternative of 'in
plane' or of 'out of plane' measurements when this is indicated by experimental reasons, for example, of optimum energy resolution for a given
combination of collision partners.

The actual energy resolution and angular resolution of the apparatus
depends on the nozzle beam performance of the involved collision partners and on kinematical factors. Thus far a highest energy resolution of
0.75 meV f.w.h.m. has been obtained for the scattering of He on O_2,
N_2, CO and CH_4. Fig.2a shows a typical time of flight spectrum from
this previous work with He /7/ for the scattering of He on N_2 at a
'perpendicular plane' angle Θ=39.5° and a collision energy of 27.3meV.
This spectrum has been accumulated in 30 hours. It shows three distinct
peaks with a half width of 0.75meV (~6cm^{-1}) for the elastic and for the
j_i=0 → j_f=2 and the j_i=1 → j_f=3 rotational transitions of the N_2 molecule with an energy loss of 1.5 and of 2.5meV , respectively. Because
the N_2 rotational states are cooled in the nozzle expansion to T_{rot}~5K
the majority of the target molecules is in the respective ground states
of the N_2's para and ortho modification. I.e. almost 2/3 of the N_2 molecules are in j_i=0 and 1/3 in j_i=1 . Thus the tof spectrum is not complicated by deexcitation peaks or excitation from excited rotational
levels.

The beam collimation requirements for high resolution tof spectroscopy imply also a high angular resolution of the scattering experiment
as is illustrated in Fig.2b by the total differential cross section for
Ar-Ar and for Ar-O_2 collisions at E~97meV /1/7/. The effective experimental angular resolution is here approximately 0.3° fwhm. This allows for the resolution of the fast diffraction oscillations which have
here a spacing of 1 degree, only. In comparison with the Ar-Ar scattering these diffraction structures show a comparably strong damping in
Ar-O_2. This is to be attributed to rotationally inelastic transitions
as is to be discussed subsequently.

3. ROTATIONAL EXCITATION BY ARGON COLLISIONS

The energy resolution obtainable in argon time of flight scattering experiments, Fig.3., is roughly ten times lower than in the helium scattering because of the earlier onset of condensation in nozzle beams of
argon. Also, for this reason the collision energy in the argon scattering can not be as readily reduced by cooling the nozzle as with the he-

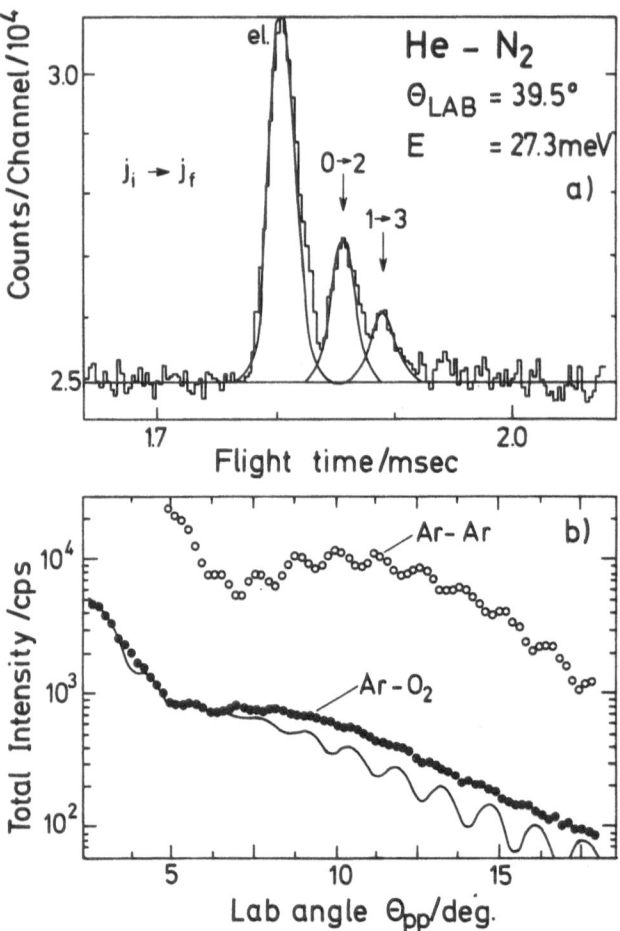

<u>Fig.2:</u> Examples for the ultimate resolution limits of the Fig.1 scat-
tering apparatus: (a) Time of flight (tof) spectrum of the He-N₂ scattering showing individually resolved N_2 rotational transitions. The
(fwhm) energy resolution is here 0.75 meV (≈6cm⁻¹). (b) Total differential scattering cross section for Ar-Ar and Ar-O₂ showing the fast
diffraction oscillations. The effective angular resolution is 0.3 degrees.

lium. This results in a further degrading of the time of flight energy
resolution with Ar beams. However, in compensation for the reduction of
resolution the Ar is found to be more efficient in causing rotational
excitation. In the examples of argon scattering time of flight spectra
in Fig.3 strong rotational excitation up to $\Delta j \cong 10$ transitions is present in the spectra for Ar-O₂ and Ar-N₂ , Fig.3a and Fig.3b, and, to
a somewhat lesser extend also for Ar-CO , in Fig.3c. A spectrum for
Ar-HCl in Fig.3d, in contrast, shows only a small amount of rotational

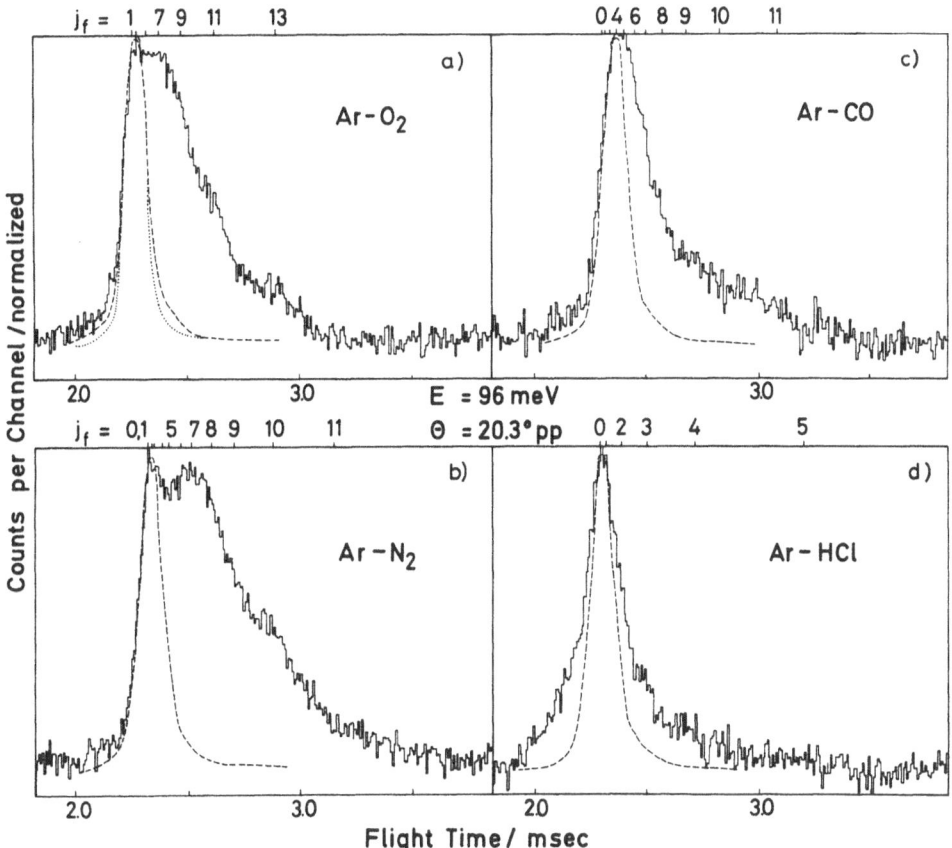

Fig.3: Argon time of flight spectra showing unresolved rotationally inelastic excitation structures for the scattering by: O_2 (a), by N_2 (b), by CO (c) and by HCl (d). The shape of the isolated elastic peak is shown by the (broken line) experimental Ar-Ne tof-spectrum. Flight times for the excitation of individual final states are indicated in the upper scales of the respective spectra. All spectra are shown for the perpendicular plane scattering angle $\Theta=20.3°$. The collision energy of $E\approx96$meV is varying by less than 2meV with the different target molecules.

excitation with a moderate broadening of the wings of the -dashed line - elastic scattering peak. These spectra are all shown here for the per- pendicular plane scattering angle $\Theta=20.3°$. The collision energies are $E=97.0$ meV for $Ar-O_2$, $E=95.9$ meV for $Ar-N_2$, $E=96.2$ meV for $Ar-CO$, and $E=98$ meV for $Ar-HCl$. They differ slightly as a conse- quence of the different molecular weights of the target gases. All other experimental variables are kept constant. The primary beam nozzle tempe-

rature is 518 K , the target nozzle is at 290 K , and the argon primary beam velocity is 735 m/sec . The arrival times of the elastic peaks
for the different targets vary slightly between 2.2 and 2.3 msec. The
exact kinematical values for the flight times for the elastic and the rotationally inelastic transitions are indicated in the upper scales on the
respective time of flight spectra. The experimental shape of the isolated
elastic peak has been determined from the Ar-Ne scattering at this angle.
For the comparison its position is arbitrarily shifted to the location
of the intensity maximum in each inelastic spectrum. The half width and
the shape of the elastic scattering peak depend here only slightly on the
target mass as is confirmed by an additional (dotted line) Ar-Ar time of
flight spectrum in Fig.3a.

In comparison with the isolated elastic peak the half width of the
rotationally inelastic Ar-O_2 and the Ar-N_2 tof-spectra of Fig.3 has
increased by roughly a factor of four . A narrower main peak in the
Ar-CO scattering, Fig.3c, is followed by a long slowly decreasing inelastic tail extending up to the 0 → 11 rotational transition. The
energy loss associated, for example, with the in Fig.3a highest observable 1 → 13 rotational transition of the O_2 is 32 meV. Thus, a substantial fraction of up to one third of the available collision energy
is transferred into internal molecular motion by a single collision of
Ar with O_2 , N_2 and with CO at this comparably small scattering
angle of Θ=20.3° , already. Quite in a contrast, however, hardly one
quarter of this energy transfer occurs in the collision of Ar with the
strongly polar HCl molecule. These principal differences in the rotational excitation between the four different molecules persist for all experimentally investigated scattering angles.

An example for the angular dependence of the rotational excitation
in one collision system is shown in Fig.4 for Ar-O_2 . The range of the
perpendicular plane scattering angles is here investigated from Θ=17°
to Θ=38° . At the most favorable angles for experimental energy resolution, between Θ=22.8 and Θ=31.3° , individual rotational peaks are
discernible for the widely spaced rotational transitions 1 → 9 , 1 → 11
and 1 → 13 . This part of the time of flight spectra can be deconvoluted
for the individual rotational transitions as shown by the Gaussian shaped
dotted lines. The smooth broken line represents the envelope of this fit.
The relative amplitude of excitation for higher rotational levels increases for larger scattering angles as would be expected. However, beyond a kinematically determined maximum scattering angle for a given
transition (see Fig.5) the respective inelastic peak disappears completely. Thus, for example, the 1 → 13 transition is strongly excited at
Θ=27.1° , but, has been vanished in the tof spectrum at Θ=29.1° and for
all larger perpendicular plane scattering angles shown in Fig.4.

For illustration of this kinematical effect a Newton diagram for
Ar-O_2 is drawn to scale in Fig.5 for the present collision energy
E=97 meV . The upper part of this composite diagram shows the in plane
and the lower part the perpendicular plane kinematics circles for the
end points of the laboratory velocity of scattered argon atoms. The perpendicular plane experimental scattering angles of the tof spectra of
Fig.4 are indicated by radial lines in the lower part of Fig.5. From the
diagram it is obvious that the just discussed 1 → 13 transition disap-

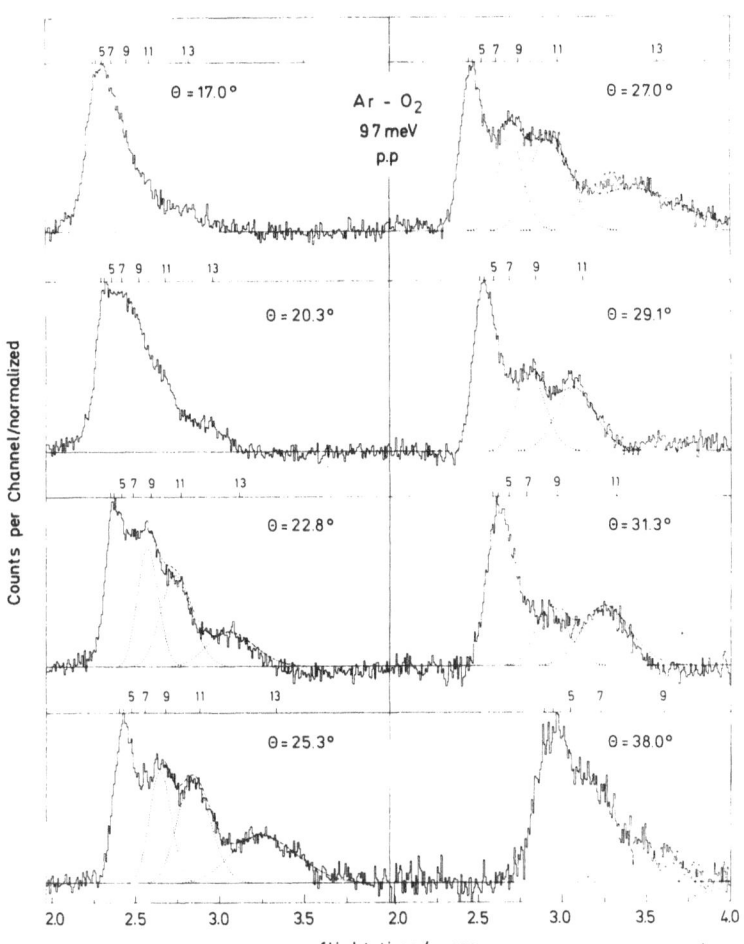

Fig.4: A series of Ar–O_2 time of flight spectra shows the progression of rotational excitation with increasing perpendicular plane scattering angles. Individual rotational state to state transitions appear as shoulders and more separated tof peaks at the scattering angles from 22.8 to 38.0 degrees. The rotationally inelastic excitation is remarkably effective. The energy loss associated with the here notably excited 1 → 13 transition is 32 meV . This is one third of the total available collision energy of E=97meV .

pears at Θ_{pp} > 27.5° .
   The contribution of the 1 → 13 and smaller rotationally inelastic transitions to the total scattering intensity is strong enough to produce in the total differential cross section noticeable structures at the ki-nematical cutoff angles. Measurements of total differential cross sect-

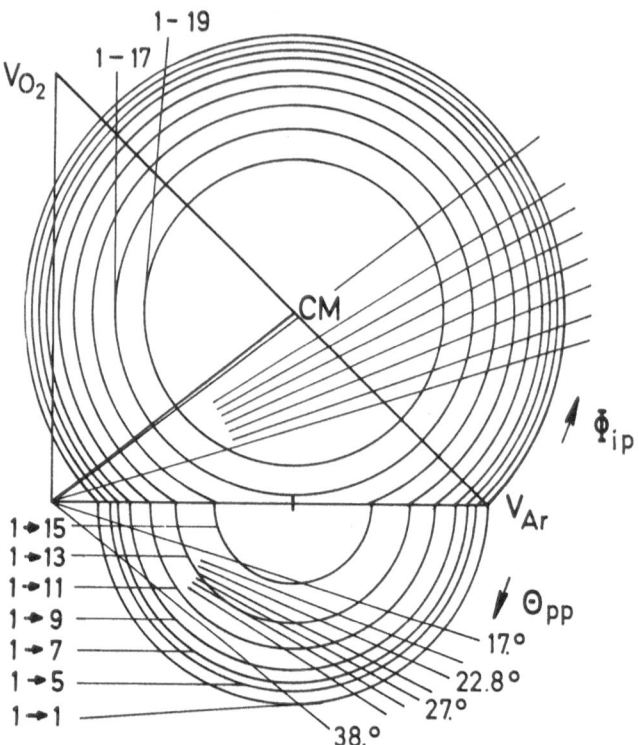

<u>Fig.5:</u> Velocity diagram for the scattering of Ar from O$_2$ at the
present experimental conditions for E=97meV . The upper part of this
composite diagram shows the Newton circles for individual O$_2$ rotational
state transitions from the elastic 1 → 1 up to the j$_i$=1 → j$_f$=19 rotational transition for 'in plane' scattering angles: ϕ^i. The lower part
is for perpendicular plane scattering angles: Θ .

ions, only, are much simpler than the measurement of tof spectra and are
shown for the Ar-O$_2$, the Ar-N$_2$ and the Ar-CO scattering in Fig.6.
The range of perpendicular plane scattering angles extends here from
Θ=5° to Θ=55° . Also shown is an in plane measurement of the Ar-O$_2$
total differential cross section in the lowest angular distribution in
Fig.6. At small scattering angles from 5° to 15° the three collision
systems show similar rainbow structures with a rapid decrease of intensity by more than an order of magnitude. Fast diffraction oscillations,
as shown before in Fig.2b, are not resolved in the Fig.6 measurements
because the angular resolution is here reduced to 1° for intensity reasons. In the region of large scattering angles usually no particular
structure is expected in the center of mass cross sections and no structure is observed in the Ar-O$_2$ in plane cross section in Fig.6. All three
perpendicular plane total differential cross sections in Fig.6, however,

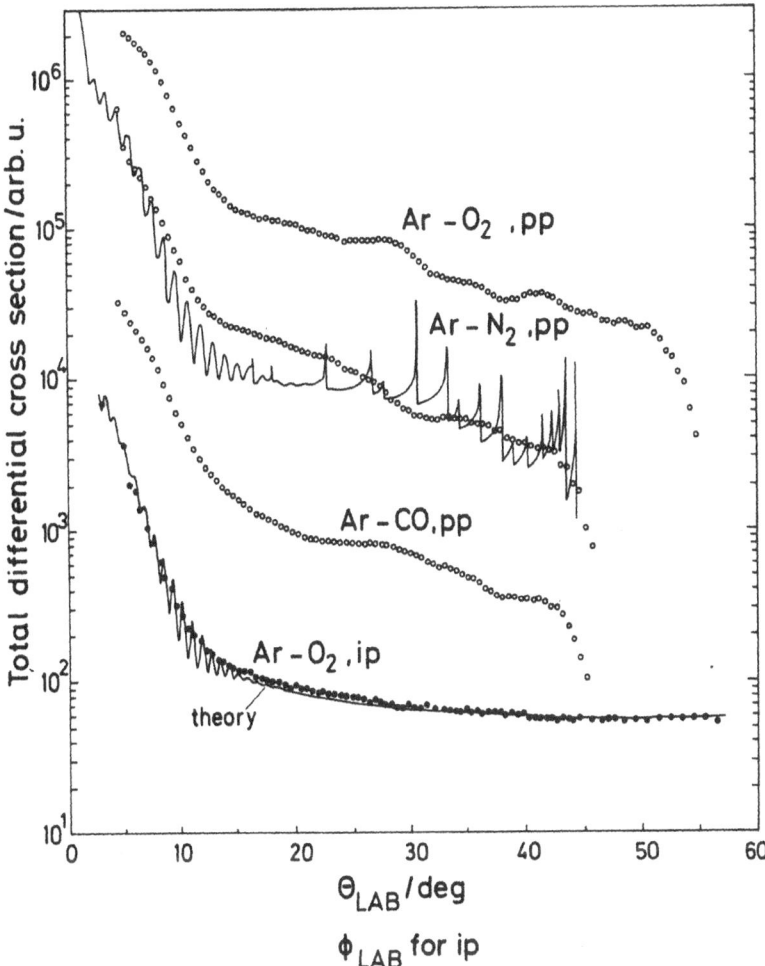

Fig.6: Experimental total differential cross sections (o) of Ar-O_2, Ar-N_2 and Ar-CO for E=96meV, measured for perpendicular plane scattering angles Θ. Note the intensity steps and irregularities associated with the kinematical cut off angles for individual rotational transitions. The lowest intensity distribution curve shows the in plane experimental total cross section for Ar-O_2 (•) smoothly decreasing for large scattering angles ϕ. For the thin line theoretical cross sections, see text.

show irregularities or several rapid changes of intensity in the angular range from $\Theta > 15°$ to $\Theta < 55°$ which are to be associated with the

kinematical cutoff angles for individual rotationally inelastic transitions. In particular the Ar-O_2 perpendicular plane cross section shows a
rapid decrease of intensity in the vicinity of the previously discussed
angle $\Theta=27.5°$ for the kinematical cutoff of the $1 \to 13$ rotational
transition at $E=97$ meV .

It appears, thus, that detailed conclusions on the magnitude of individual inelastic transitions can also be drawn from a comparison of
appropriate perpendicular plane versus in plane total differential cross
section measurements, only.

4. DISCUSSION

The interaction potentials for the here investigated collision systems
have been repeatedly discussed in literature and are reasonably well
known /8/6/. They have been derived from a number of quite diverse experimental sources of information. The more important ones are the transport properties of bulk gases, virial coefficients, relaxation times and
line broadening data, and, also, total integral and total differential
scattering cross sections from molecular beam experiments. In the case
of Ar-O_2 and Ar-HCl , in addition, the spectroscopic data of the bound
van der Waals complex are available. For the subsequent comparison and
discussion of our present inelastic scattering results we used the most
recent interaction potentials for the two spectroscopically accessible
systems Ar-O_2 /8/4/ and Ar-HCl /6/. These are shown in Fig.7 for
the principal collinear and perpendicular alignment angles $\gamma=0°$ and
$\gamma=90°$. For the heteronuclear Ar-HCl system also the $\gamma=180°$ orientation
is shown. Superficially, the location and the steepness of the repulsive
barrier, the well depth of 10 to 15 meV and the anisotropy appear to
be quite similar for the two potentials.

For an approximate qualitative estimate of the magnitude of the rotational excitation the 'rotational rainbow' model provides a very useful simple concept /9/10/. Here the maximum of excitation is estimated
for the collision of a hard sphere atom with a hard wall ellipsoid molecule. The maximum of rotational excitation is obtained from the maximum
angular momentum transfer $|\vec{L}| = |\vec{r} \times \vec{q}|$ with $\vec{r}$ describing the point
of impact on the ellipsoid and $\vec{q}$ being the recoil momentum of the scattered atom. The adiabatic recoil depends on the scattering angle ϑ ,
the incident momentum $p_0=hk$ and on the ellipsoid's surface normal vec-
tor $\vec{n}$ as: $\vec{q} = \vec{n} \cdot 2p_0 \cdot \sin \vartheta/2$. Thus, the actual amount of angular momentum transfer is limited by the maximum length of the 'handle'
$|\vec{r} \times \vec{n}|$ which is the maximum geometrical distance at which a surface
normal of any point $\vec{r}$ on the ellipsoid passes by the center of mass of
the ellipsoid. For an ellipsoid with the major axes $2 \cdot A$ and $2 \cdot B$ and with
an offset δ of the center of mass from the center of symmetry the maximum excited rotational state is then:

$$j_R = |2 k \sin \frac{\vartheta}{2} \cdot \{(A-B) \mp \delta(1+B/A)^{1/2}\}| . \qquad (1)$$

This means one has usually two different maxima of the rotational excitation like, for example, with the CO and the HCl. These coincide,

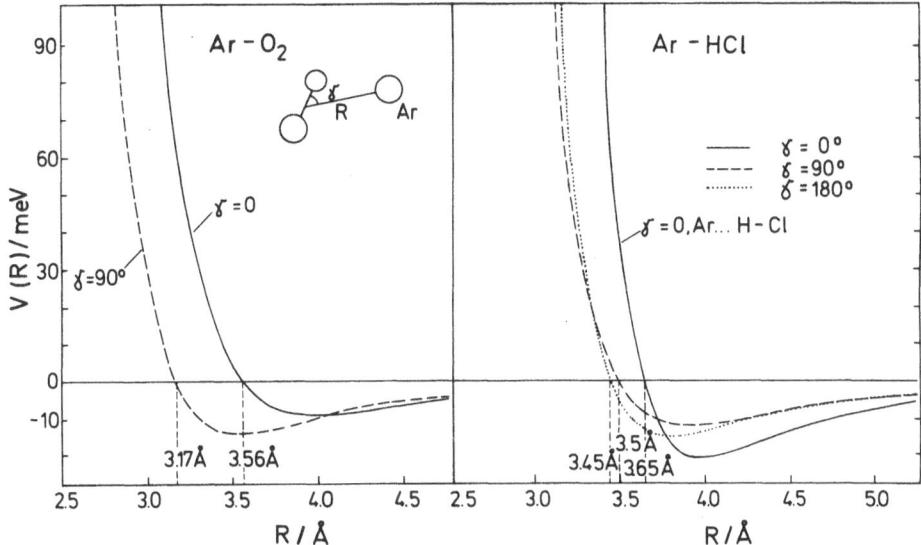

Fig.7: The best available van der Waals interaction potentials for
Ar-O$_2$ (Ref.8) and for Ar-HCl (Ref.6) shown for the orientations
γ=0 , 90° and 180° .

however, for homonuclear symmetric molecules as the N$_2$ and the O$_2$.
Thus, eq.(1) is providing a straightforward explanation of the observed
structural differences in the envelope of the rotational excitation tof
spectra for Ar-N$_2$ and Ar-CO in Fig.3.
  For the given interaction potentials of Fig.7 the principal parameters of an equivalent ellipsoid may be estimated, for example, from the
location of the zero crossings of the potential at γ=0°, 90° and 180°.
For the symmetric Ar-O$_2$ case one finds, thus, (A-B) ~ 0.38 Å with
δ=0 . For Ar-HCl the parameters are A ~ 3.55 Å, B ~ 3.50 Å and
δ=0.1 Å . The experimental collision wave numbers for both systems at
E=97 and 98 meV, respectively, are almost identical with k=28.9 Å^{-1}
and 29.8 Å^{-1}. Then eq.(1) predicts a maximum rotational excitation
j$_R$ ~ 6 at ϑ_{cm} ~ 32° (Θ_{LAB}=20°) for the Ar-O$_2$ scattering. The two rotational rainbow values for Ar-HCl at the same scattering angle are
j$_R^{(1)}$ ~ 1.6 and j$_R^{(2)}$ ~ 3 . This is the experimentally observed order
of magnitude in Fig.3, although, quantitatively, the value for Ar-O$_2$
seems to be some 50% too small in comparison with the experiment.
  In a more rigorous approach for a comparison of the available potential and the present inelastic scattering data, quantum scattering cross
sections were calculated within the 'infinite order sudden approximation'.
This approximation has the benefit of being quick and cheap. By comparison with exact 'close coupling' calculations this approximation had been
found to yield, within some 10%, reasonably accurate results for the scattering of light atoms as, for example, for He-N$_2$ collisions /7/.
  In the Argon scattering, now, in contrast to the He-N$_2$ tof spectra,

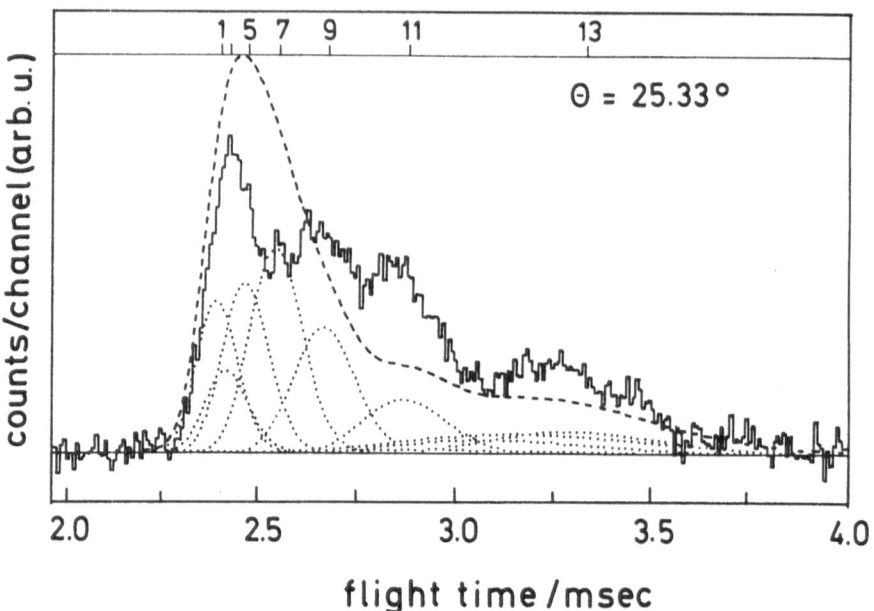

Fig.8: Comparison of an experimental Ar-O₂ tof-spectrum for Θ=25.3°
and at E=97meV versus the IOS theory prediction (---,and...) for the
Ar-O₂ potential of Ref.8.

large rotationally inelastic energy losses are experimentally observed.
The 'IOS' approximation is, thus, possibly not very accurate with its implicit assumption of energy sudden collisions. However, exact or alternative approximate calculations for checking·its reliability have thus
far been prevented by the very large number of some 132 coupled channels
and of 100 to 400 partial waves which have to be dealt with in an exact
treatment of the Ar-O₂ scattering at E=97 meV .

    A comparison of an experimental result with theoretical IOS cross
sections is shown in Fig.8 for an Ar-O₂ time of flight spectrum at Θ=25.3°.
The bell shaped dotted line individual rotational transitions are derived
from the theoretical cross sections by convolution with the respective
experimental tof resolution functions of the apparatus. The broken line
envelope of these individual peaks is normalized to the area of the experimental time of flight spectrum. The comparison shows that the higher
rotational transitions 1 → 13 , 1 → 11 and 1 → 9 are here underestimated by the theory by factors between two and four. The lower transitions in the range from 1 → 3 to 1 → 7 come out much stronger than observed in the experiment.

    The strongest individual theoretical transition is found to be the
1 → 7 peak, in agreement with the previous rough estimate from the classical rotational rainbow model. Thus, the anisotropy of the available literature potential of Fig.7a will need to be increased by the equivalent

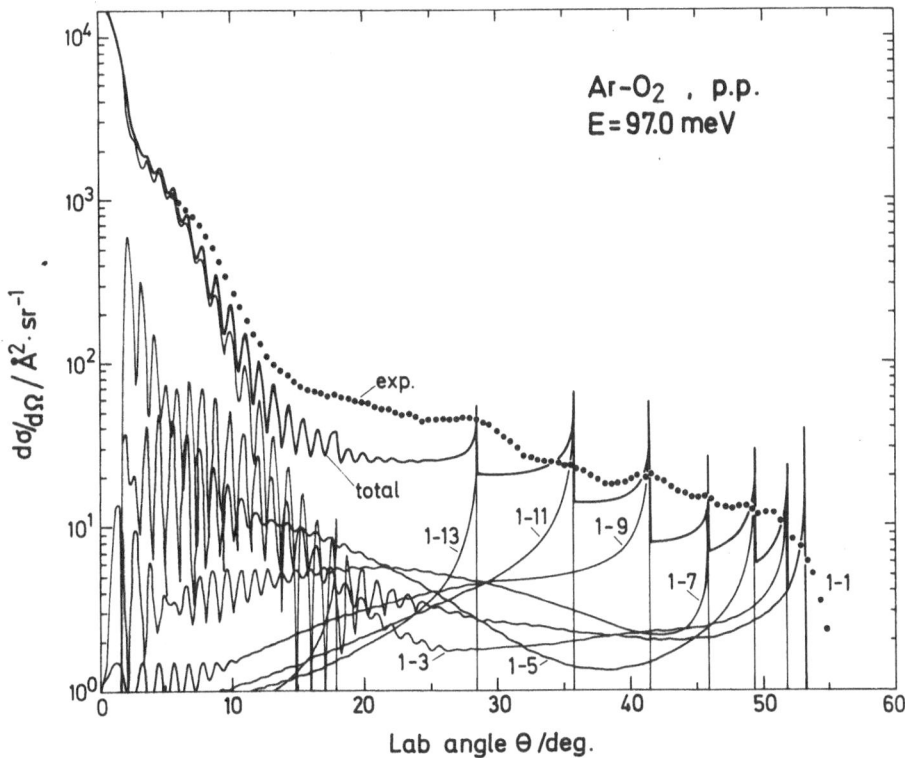

Fig.9: Comparison of theory and experiment for the Ar-O_2 total differential cross section in perpendicular plane scattering geometry. Singularities in the theoretical and steps in the experimental cross section appear for kinematical reasons at the cut off angles of individual rotational transitions. The differences in the amplitude of the theoretical versus the experimental cross section show the incorrect theoretical description of rotationally inelastic excitation. Experiment and theory agree for in plane scattering where the Ar-O_2 potential of Ref.8 had originally been fitted (see Fig.6).

of roughly 0.2 to o.3 Å in the parameter (A-B) of the ellipsoid model. This is a surprisingly large correction in view of the satisfactory agreement of this potential with the recent radio frequency spectroscopy data for the Ar-O_2 van der Waals molecule obtained by the Nijmegen group /4/.

The inadequacy of the available potential /8/ for the description of Ar-O_2 rotationally inelastic collisions is, also, readily observed in a comparison of the theory with the experimental perpendicular plane total differential cross sections shown in Fig.9. The individual theoretical state to state cross sections show here the intensity singularities at the previously discussed cutoff angles for the subsequent rotational

states (cf.Fig.5). In the experimental total differential cross section
these sharp spikes are smoothed by the angular averaging of the apparatus, and appear then as finite intensity steps at the cutoff angles. The
amplitudes of these steps allow for a direct comparison of the experimental versus the theoretical intensity contribution of the at a particular cutoff angle vanishing rotational state transition. Thus, as noticed
and discussed before with the tof spectrum in Fig.9, now also in the angular distribution the $1 \to 13$ transition appears largely underestimated
by the theory while the lower transitions are too strong. The overall
discrepancy between the total experimental and the total theoretical differential cross sections is as large as a factor of two .

This discrepancy, however, is not observed in the comparison of the
theory with the experimental $Ar-O_2$ in plane total differential cross
section in Fig.6. It is, thus, obvious that the earlier in plane measurements /11/ which entered the determination of the $Ar-O_2$ literature potential /8/ did not represent a particularly sensitive choice of experimental data points. They should, rather, have also included an additional out of plane total intensity measurement.

A quite similar degree of discrepancy exists between the theoretical
prediction from the best available $Ar-N_2$ interaction potential /8/ and
the experimental perpendicular plane total differential scattering cross
section as is illustrated by the comparison with the respective thin continuous line in Fig.6. The $Ar-N_2$ system, at present, seems not to be
particularly well suited for a spectroscopic investigation of the van der
Waals potential and is not to be further commented here (see, however,
ref.12).

For $Ar-HCl$ the Figs. 10a and 10b show a comparison of a high angular resolution experimental perpendicular plane total differential cross
section with preliminary theoretical cross sections. The apparatus parameters determining the angular resolution of $\delta\Theta \sim 0.3°$ are identical with
the values used previously for the $Ar-Ar$ and the $Ar-O_2$ measurements
shown in Fig.2b. Thus, for the first time also, the fast oscillation pattern in the $Ar-HCl$ scattering are here experimentally resolved. The experimental $Ar-HCl$ total cross section in Fig.10 appears to be quenched
by out of phase diffraction oscillations of inelastic transitions, similarly as in Fig.2b the oscillations in the $Ar-O_2$ are considerably
damped in comparison to the Ar-Ar scattering. In the investigated range
from $\Theta \sim 3$ degrees to $\Theta < 20$ degrees the experimentally observed rainbow
structure in the $Ar-HCl$ scattering is closely resembling to the shape
of the $Ar-O_2$ and the $Ar-Ar$ (as well as to the $Ar-N_2$ and the Ar-CO)
total differential cross sections. The noticeable shift of all rainbow
features toward not quite a factor of two larger scattering angles
corresponds to the deeper average potential well of the Ar-HCl system.

With the calculation of exact theoretical cross sections for this
collision system similar problems arise and have still to be solved as
with the $Ar-O_2$. For an initial test of the Ar-HCl potential of ref.6
the IOS approximation was used and is shown in the comparison in Fig.10a.
The most pronounced discrepancy between the IOS theory and the measured
total differential cross section is the appearance of the theoretical
rainbow structures at appreciably smaller angles, almost 30% below the
experimental angles. The spacings of the fast diffraction oscillations

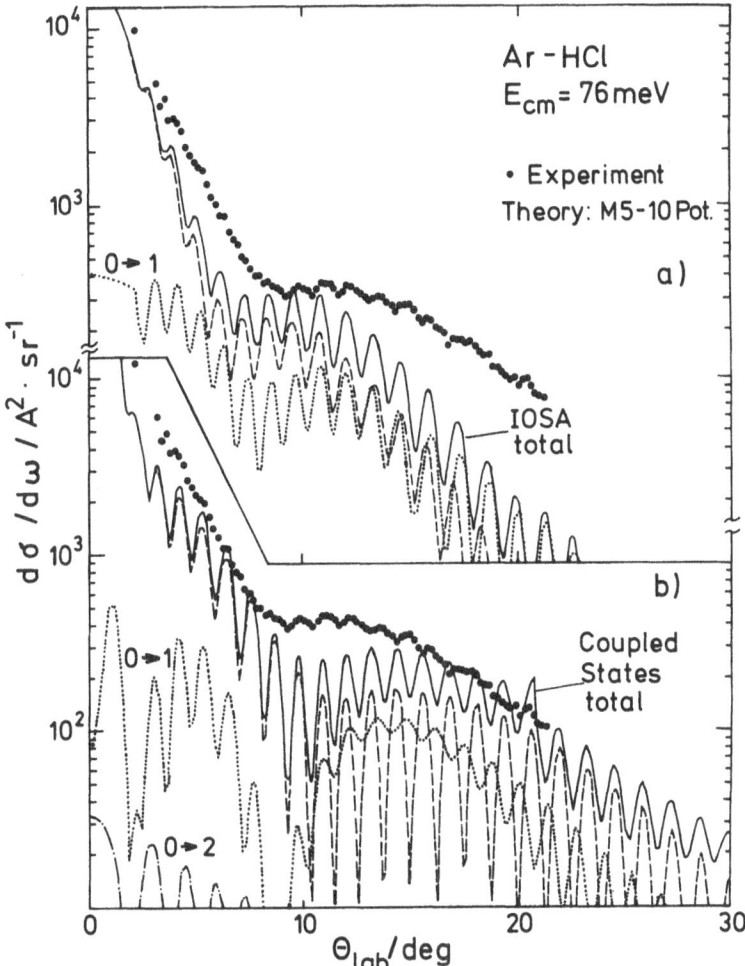

Fig.10: A comparison between the 'infinite order sudden' (a) and the 'coupled states' (b) approximation theoretical cross sections for Ar-HCl shows the inadequacy of the iosa for this particular collision system. The positions of the fast diffraction oscillations in the experimental Ar-HCl angular distribution, in particular, are not correctly reproduced by the csa, either. Likely, the strength of $\Delta j=1$ rotational transitions is underestimated by the "M5" potential of Ref.6.

is approximately correct. The positions of the IOS diffraction maxima agree with the experiment at small scattering angles. However, the oscillations run out of phase at angles larger than 10 degrees .

The discrepancy in the position of the theoretical rainbow is of a different kind than the previous kinematical caustics discrepancies in

the Ar-O$_2$ and the Ar-N$_2$ total differential cross section. Here, the
infinite order sudden approximation is not able to reproduce even the
elastic scattering with reasonable accuracy. This is demonstrated by the
coupled states cross sections in Fig.10b for the same Ar-HCl potential
and the same collision energy of E=76 meV /13/. This phenomenon in the
Ar-HCl scattering had been noticed earlier in comparisons of coupled
states cross sections with one·channel elastic calculations of the total
cross section /14/. The failure of the sudden approximation is particularly surprising because experimentally, as shown by the tof spectrum in
Fig.3d, the Ar-HCl scattering appears to be dominated by the elastic or
the nearly elastic scattering. The general shape of the coupled states
approximation total differential cross section in Fig.10b compares quite
favorably with the experimental Ar-HCl scattering distribution. The
saddle near 10°, only, seems to be slightly overemphasized by the theory.
Just in this narrow region the CS approximation can not really be trusted
when considering available comparisons of CS versus exact CC cross
sections. But, this structure could possibly also be smoothed out when
all three populated initial states j=0 , 1 and 2 of the HCl target
beam will properly be included in this preliminary comparison /14/.

Another disagreement is observed in Fig.10b in the positions of the
fast diffraction oscillations. They are in acceptable agreement for very
small angles below 7°, only. In the remaining range only the diffraction
maxima near Θ=10° and Θ=21 degrees seem to occur at the correct experimental location. When the CS approximation is assumed to produce essentially correct results this can only be attributed to an inadequacy
of the employed interaction potential /6/ in reproducing strong enough
rotational excitation cross sections which might shift the diffraction
structure of the total cross section.

5. CONCLUSIONS

The here presented new detailed measurements of inelastic scattering and
of total differential cross sections allow to subject to new scrutiny
the available description of typical heavy partner van der Waals collision systems.

The experimental investigation of the inelastic scattering reveals
large disagreements by as much as a factor of two in the average rotational energy transfer, and, even in the total experimental laboratory
cross section. Thus, the best available potentials for e.g. Ar-N$_2$,
Ar-O$_2$ and, also, for Ar-HCl may have to be considered as giving, at
best, an educated rough guess for the actual rotationally inelastic collision dynamics of these van der Waals molecule systems. Interestingly,
it seems that these potential features responsible for the magnitude of
the inelastic scattering are hardly noticeable in the analysis of the
van der Waals spectroscopy data for our Ar-O$_2$ and Ar-HCl test cases.

These statements are, thus far, correct within the framework of approximations used in all previous analyses of the respective interaction
potentials. On the basis of evidence of the rotationally inelastic scattering measurements these approximations, also, appear likely to be incorrect. Thus, redetermination of "improved" anisotropic interaction po-

tentials from the more detailed experimental data on the basis of the
same inadequate theory is probably not an advisable recipe for a true
improvement of the potentials. In order to further clarify these questions very expensive exact quantum calculations of the respective inelastic cross sections will be required. These appear just feasible now on
the very largest available computers and, thus, could be and should be
done, soon.

REFERENCES

/1/ M. Faubel and G. Kraft; J.Chem.Phys. 85, 2671 (1986).
/2/ G. Kraft; Ph.D. Thesis, Göttingen 1985.
/3/ H. Brämer; Diplom Thesis, Göttingen 1985.
/4/ J. Mettes, B. Heijmen, P. Verhoeve, J. Reuss, D.C. Laine, and
       G. Brooks; Chem.Phys. 92, 9 (1985).
/5/ S.E. Novick, P. Davis, S.J. Harris, and W. Klemperer; J.Chem.Phys.
       59, 2273 (1976).
/6/ J.M. Hutson and B.J. Howard; Molec.Phys. 45, 769 (1982).
/7/ M. Faubel; Adv.At.Mol.Phys. 19, 345 (1983).
/8/ R. Candori, F. Pirani, and F. Vecchiocattivi; Chem.Phys.Lett. 102,
       412 (1983).
/9/ D. Beck, in: "Physics of Electronic and Atomic Collisions", p. 331,
       S. Datz, ed., North Holland Publ., Amsterdam 1982.
/10/ J.M. Bowman and R. Schinke; Top.Curr.Phys. 33, 61 (1982).
/11a/F.P. Tully and Y.T. Lee; J.Chem.Phys. 57, 866 (1972),
   b/G. Rotzoll; Chem.Phys.Letters 88, 179 (1982).
/12/ G. Brocks; this conference, TP-6.
/13/ We are grateful to R. Schinke and to U. Buck for providing these
       CS cross sections.
/14/ U. Buck and J. Schleusener; J.Chem.Phys. 75, 2470 (1981).

VAN DER WAALS INTERACTIONS FROM GLORY AND DIFFRACTION SCATTERING

Vincenzo Aquilanti, Fernando Pirani
and Franco Vecchiocattivi
Dipartimento di Chimica dell'Universita'
06100 Perugia, Italy

ABSTRACT

Recent molecular beam measurements of integral and total differential
cross sections are reviewed with particular emphasis on the information
content which the glory and diffraction patterns bear on interatomic and
intermolecular forces. High accuracy potentials are derived for rare
gas-rare gas dimers. The anisotropic interactions of rare gases with
ground state open shell atoms (N, O and F) and diatomic molecules (N_2,
O_2 and NO) are also investigated. A discussion is given of new advances
on the extraction of details of the interaction from cross section data:
An adiabatic representation with nonadiabatic corrections within the
Coupled States scheme is proposed and tested.

1. INTRODUCTION

A decisive contribution to the present knowledge of van der Waals
interactions comes from atomic and molecular beam measurements of
scattering cross sections [1-4]. In these experiments detailed
information is obtained by the study of collisional interference and
diffraction effects [5] which are sensitive to specific characteristics
of the interaction between the colliding partners. A direct inversion
procedure, which leads from scattering results to interatomic potential
energy curves, has been attempted only in a few particularly favorable
cases [6], where isotropic atom-atom interactions are involved and high
resolution conditions in angle or energy are obtained in the
experiments. However in most cases the scattering results are fitted by
assumed potential energy functions in a trial and error procedure.
   When for a given system in addition to scattering results other
properties sensitive to the interaction are available, such as
spectroscopic and thermophysical results, a combined analysis of all
these experimental data can extend the range of validity of the
potential energy function and can improve its reliability: Several
examples are available in the literature [7-12].

A. Weber (ed.), Structure and Dynamics of Weakly Bound Molecular Complexes, 423–439.
© 1987 by D. Reidel Publishing Company.

Considering the high level of accuracy for the potentials which can be obtained by the analysis of experimental data, an important aspect is the theoretical scheme which is used to connect the experimental observables with the potential. Accurate computational procedures and powerful approximation schemes have been developed. However accurate calculations are often impractical and approximations are used, sometimes in a not well controlled way. This is the case of many systems with anisotropy of interaction.

In Fig.1 one can see that the potential for the interaction between a closed shell atom with a P-state atom or a diatomic molecule can be represented in both cases by formally using the same potential expansion. We show below that this fact can lead to a unified treatment of the collision dynamics of such systems. This approach is very useful for the study of the anisotropy in atom–atom systems and results to be very promising in the atom–diatom case.

This paper is a progress report on recent results from our laboratory about atom–atom and atom–diatom integral and differential cross sections measurements, where the glory and diffraction structure is well resolved. In atom–atom systems involving an open shell partner the anisotropy of interaction is studied by using a magnetic selection technique in the glory scattering experiments. An emphasis is given to interpretative problems for anisotropic systems. An adiabatic approach is developed for such systems, where the non–adiabaticity can be easily controlled due to the relatively low collision energy in these experiments involving thermal beams.

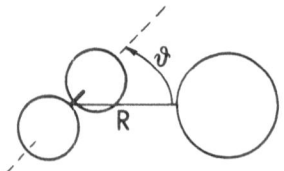   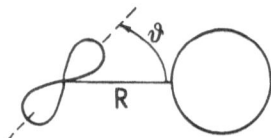

$$V(R,\vartheta) = V_0(R) + \sum_\lambda V_\lambda(R) \cdot P_\lambda(\cos\vartheta) \qquad V(R,\vartheta) = V_0(R) + V_2(R) \cdot P_2(\cos\vartheta)$$

Fig.1 – Showing that the electrostatic interaction of a closed shell atom with a diatomic molecule and a P-state atom can both be represented by a Legendre expansion, assuming the molecule as a rigid rotor and describing the atom by the symmetry of its electronic cloud. Note that in the latter case only two terms in the expansion need be included (for further details, see Ref. 13–15): The relationship of V_0 and V_2 with the more familiar electrostatic interaction curves $V_\Sigma = V(\vartheta = 0)$ and $V_\pi = V(\vartheta = \pi/2)$ are

$$V_0 = (V_\Sigma + 2V_\pi)/3 \ ; \ V_2 = 5(V_\Sigma - V_\pi)/3.$$

For the atom–diatom case discussed in Sec.4, $V(\vartheta = 0)$ and $V(\vartheta = \pi/2)$ will be identified respectively as the ridge and valley bottom lines of the potential energy surface (See also Fig.8).

2. ATOM—ATOM: CENTRAL FORCES

When two closed shell atoms interact their potential is described by a single function depending only on the internuclear distance, $V(R)$. Experimental results of very high accuracy are today obtainable for such systems, moreover in these cases one can use reliable computational procedures to connect the interatomic potential and the experimental observables. This is true not only for scattering properties, but also for spectroscopic results and gasous properties. It follows that for several systems with isotropic interaction high precision potential energy curves are available.

     An example of the high level of accuracy obtainable for the experimental scattering results in these cases and of their high sensitivity to the relevant features of the potential energy curves is taken from recent work from our laboratory.

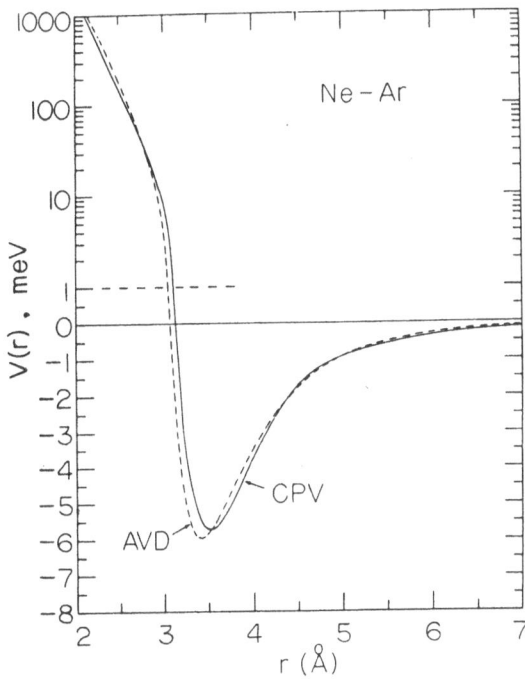

Fig.2 — The potential energy curves which had been proposed for the Ne—Ar interaction. The CPV curve is the one proposed by Candori et al. (Ref.16) while the AVD curve was proposed by Aziz and van Dalen (Ref.17).

        In Fig.2 two potential energy curves, V(R), are reported which had
been recently proposed [16,17] for the Ne-Ar interaction on the basis of
an analysis of several experimental data. The two interactions are very
similar, although not identical: Small differences of the order of a
few percents exist in the depth of the well and in its location.
However the high quality of the present state-of-the-art scattering
experiments allows discriminating between the two proposed interactions.

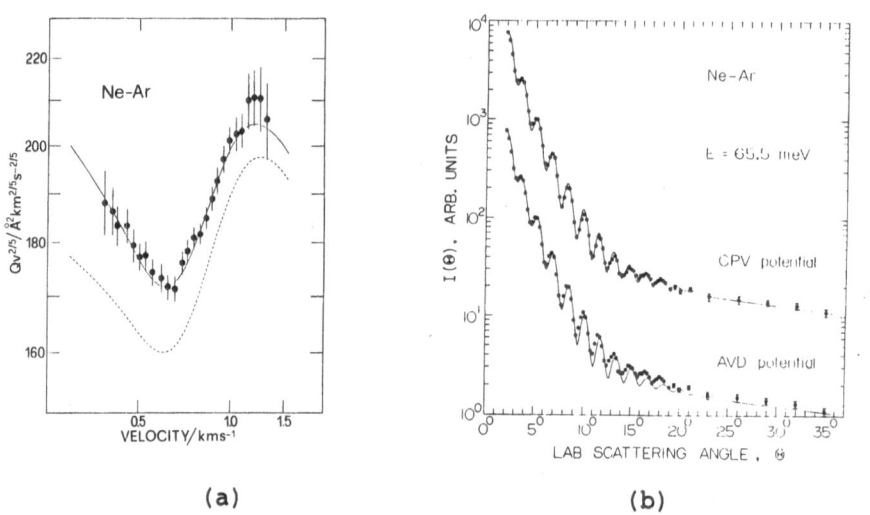

(a)   (b)

Fig.3 - Experimental results for Ne-Ar scattering. a) Integral cross
sections as a function of the collision velocity compared with
calculations performed assuming the CPV (full line) and AVD (dashed
line) interactions of Fig.2. b) Differential elastic cross sections
(Ref.18) and comparison with the CPV and AVD potentials.

        In Fig.3 the integral [16] and differential [18] cross sections for
Ne-Ar collisions as measured in our laboratory are reported and compared
with calculations performed using the two potentials of Fig.2. Well
established computational schemes are available for solving the
Schroedinger equation for scattering by V(R), at a given reduced mass, μ,
at total energy E, and at total angular momentum J:

$$[-(\hbar^2/2\mu)\ d^2/dR^2 + V^J(R)]\ \psi = E\ \psi \qquad (1)$$

where $V^J(R)$ is obtained by adding to V(R) the centrifugal barrier at
angular momentum J. Also, well established is the origin of structure
in cross sections, and how it is related to features of the interaction
[5]: The glory undulations in integral cross sections (Fig.3a) are a

manifestation of the van der Waals well and are sensitive to its capacity; the diffraction pattern (Fig.3b) in differential cross section is a measure of the range of the interaction. It is evident that the quality of both experimental sets of data removes every uncertainty: The CPV potential appears to be the most adequate. Very recently, that potential has been refined by a multiproperty analysis [11] which has made it adequate for the second virial coefficients and the transport properties of gaseous mixtures and has extended its range of validity between 2.5 and 6.5 Å. Similar analysis of experimental data have provided potential energy curves for other rare gas-rare gas systems [7].

3. ATOM-ATOM: INTERACTIONS OF MAGNETICALLY ANALYZED OPEN SHELL ATOMS

The interaction between two atoms is anisotropic when at least one of the two partners is an open shell atom not in S-state (See Fig.1 for an illustration of the interaction of a P-state atom with a closed shell). In this case at each internuclear distance several states are possible due to the presence of an internal angular momentum. To study the interatomic potential energy in such cases in our laboratory integral scattering cross sections are measured as a function of the relative collision velocity for some light open shell atoms, such as $F(^2P)$ and $O(^3P)$. The main problems in these experiments are the production of intense, stable and velocity selected beams of open shell atomic species and the control of their electronic states and of the relative population of the involved magnetic sublevels. In our experiment open shell atom beams are obtained by a microwave discharge source and velocity selected by a slotted disk selector. An inhomogeneous magnetic field is used to defocus from the beam direction those atoms which show a non-zero effective magnetic moment. In this way one can characterize the electronic state and vary the relative population of their magnetic sublevels. The details of this experimental procedure have been reported elsewhere [19-22].

The theory needed to interpret the experimental results and to obtain information on the potential from them has been given and reviewed elsewhere [13-15,22]. Basically, one has to solve a multichannel Schroedinger equation which can be written as a matrix generalization of (1)

$$[-(\hbar^2/2\mu)\ \underline{1}\ d^2/dR^2 + \underline{V}^J(R)]\vec{\psi} = E\ \underline{1}\ \vec{\psi} \qquad (2)$$

where the potential energy matrix V^J is not diagonal (diabatic representation) and is the sum of three contributions:

$$\underline{V}^J = \underline{V}_{el}(R) + \underline{V}_{rot}(R) + \underline{V}_{so}(R) \qquad (3)$$

The dimension of the basis is restricted to the fine structure components, which are splitted asymptotically by the spin-orbit interaction of the open shell atom, $\underline{V}_{so}$, the collision effects being described by the electrostatic interaction between atoms which is relatively short ranged (Fig.1), and by centrifugal contributions described by the therm V_{rot}, which decays as R^{-2}. According to the relative importance of these terms, five alternative representations are

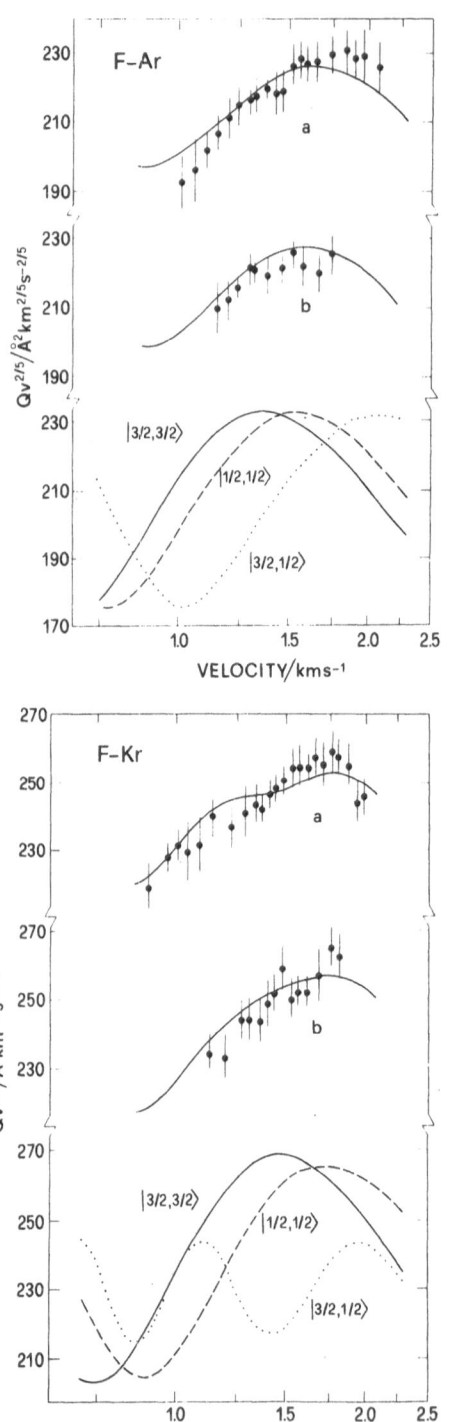

Fig.4 — Integral scattering cross sections for F–Ar and F–Kr as a function of the collision velocity and for two different populations of the magnetic sublevels of the fluorine atoms: a) the population corresponding to a zero magnetic field; b) the population corresponding to a ratio between the magnetic field and the square velocity equal to 0.1 T $Km^{-2} s^2$. The two populations lead to cross sections which are different combinations of those corresponding to the three adiabatic energy curves which represent the interaction in these two systems. These cross sections are plotted at the bottom of each set of experimental results.

possible corresponding to different coupling schemes for the angular momenta involved (Hund's cases). The choice of representations and recipes for simplifications based on decoupling schemes have been given elsewhere [13,14].

The simplification considered in the following is justified by the large spin orbit splittings of these atoms with respect to centrifugal effects. This allows to restrict our attention to only two cases, (a) and (c), valid at short and large range respectively, when the electrostatic interaction is stronger or respectively weaker than spin–orbit splitting. A proper label for scattering states is then $|j\Omega>$ where j is the atomic angular momentum and Ω its projection along the R axis. In this Centrifugal Sudden or Coupled States (CS) or Ω–conserving approximation, the scattering, which is considered as taking place along adiabatic effective potential curves, $\mathcal{E}(R)$, is described by the equivalent of (1) in the adiabatic representation:

$$[-(\hbar^2/2\mu)(\underline{1}\ d/dR + \underline{P}(R))^2\vec{\psi} + \underline{\mathcal{E}}(R)]\vec{\psi} = E\ \underline{1}\ \vec{\psi} \qquad (4)$$

The diagonal matrix $\underline{\mathcal{E}}$ contains the eigenvalues of $\underline{V}_{el}$ and $\underline{V}_{so}$ (of course independent of the diabatic representation chosen):

$$T(R)[\underline{V}_{el}(R) + \underline{V}_{so}(R)]\ \widetilde{T}(R) = \underline{\mathcal{E}}(R) \qquad (5)$$

These eigenvalues are effective potential energy curves in the adiabatic representation. The coupling between them is represented by the matrix P, which is related to the orthogonal diagonalizing matrix $\underline{T}$ appearing in (5) by

$$\underline{P}(R) = dT(R)/dR \cdot \widetilde{T}(R) \qquad (6)$$

This approach has been used for studying the interaction of fluorine atoms and oxygen atoms with rare gases from the measurements described above: scattering is described by effective adiabatic curves, and the nonadiabatic coupling between them will be shown to be negligible.

In Fig.4 the integral scattering cross sections for fluorine atoms colliding with Ar and Kr as recently measured in our laboratory [23,24] are reported as a function of the collision velocity and for different populations of the magnetic sublevels of the open shell partner determined by magnetic analysis. The glory pattern is seen to be the combination of the individual contributions from the three adiabatic potential energy curves $|3/2,1/2>$, $|3/2,3/2>$ and $|1/2,1/2>$. These curves, as obtained from the analysis of our data (Fig.5), were also tested on other observables: spectroscopic for the ground state of F-Xe [25], total differential cross sections [26] and relative integral cross sections at low energy [27] (none of these scattering experiments however included an analysis of the magnetic sublevels involved). Also shown in Fig.5 are the non adiabatic coupling terms, P(R), which connect the $|3/2,1/2>$ and $|1/2,1/2>$ states: Since non adiabatic corrections are of the order of $\hbar^2 P^2/2\mu$, they are clearly very small for the present systems. The maxima of P(R) functions mark the separation between coupling schemes, the molecular case (a) at short distances and the diatomic case (c) at long range: The figure clearly shows that the

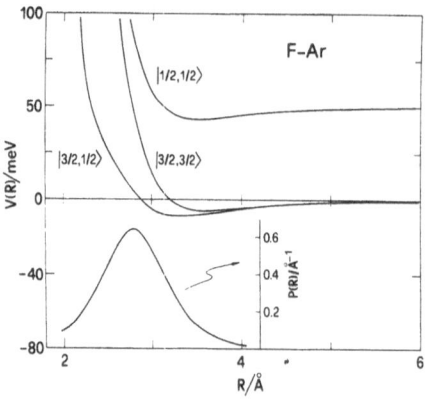

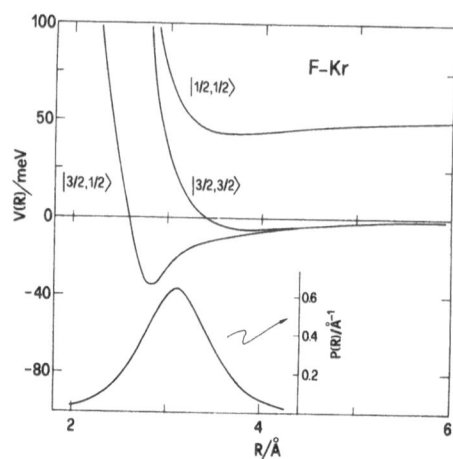

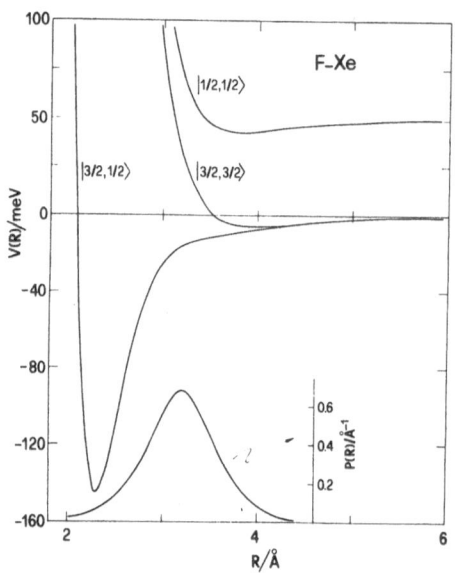

Fig.5 - Adiabatic potential energy curves for the $|3/2,3/2\rangle$, $|3/2,1/2\rangle$ and $|1/2,1/2\rangle$ states of F-Ar, F-Kr and F-Xe as obtained from the analysis of the experimental integral scattering cross sections measured in Perugia (Ref.21-24). Also shown in the figure are the non adiabatic coupling terms, P(R), which connect the $|3/2,1/2\rangle$ and $|1/2,1/2\rangle$ states. The maxima of P(R) functions mark the separation between coupling schemes, the molecular case (a) at short distances and the diatomic case (c) at long range: The change in character, evident also in the change of the potential well shape, varies systematically from F-Ar to F-Xe.

character of the van der Waals wells changes systematically along the series.

Similar experimental results have been obtained for the interaction of O(^{3}P) atoms with all the five rare gases. They are being analyzed using the same theoretical approach. In particular, for the O–He system, for which a comparison is possible with accurate quantum chemical calculations of open shell interactions, it is found that van der Waals forces are still underestimated by theory [28].

4. ATOM–DIATOM INTERACTIONS

4.1. Effect of anisotropy on glory and diffraction structure

As discussed above for the anisotropic atom–atom systems, also for the atom–diatom case the anisotropy of interaction can produce a modification of the interference structure in the integral and differential cross sections.

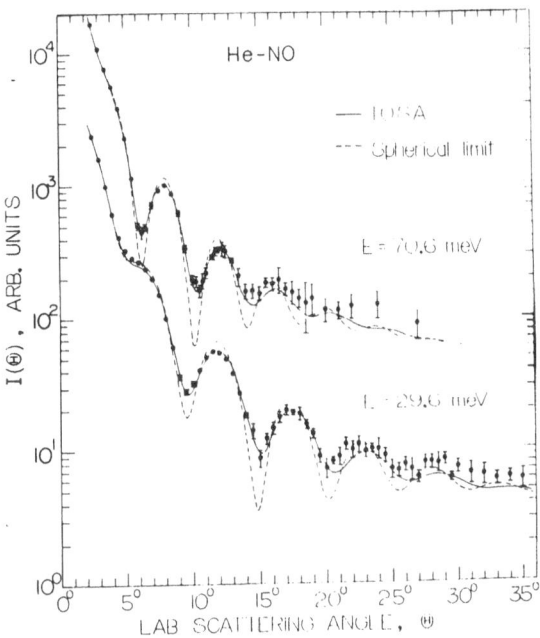

Fig.6 – Total differential cross sections for He–NO collisions as measured in Perugia at two different collision energies (Ref.29). The experimental results are compared with IOS calculation performed assuming an anisotropic interaction and with a calculation performed with an optimized spherical potential. A partial damping of the diffraction structure due to the effect of the anisotropy is evident.

As an example, in Fig.6 the differential total cross sections for He-NO collisions as measured in our laboratory at two collision energies [29] are reported. In the figure the experimental results are compared with calculation for a spherically symmetric potential. A damping of the experimental diffraction undulations respect to those calculated for such potential is well evident. Similar behavior has been observed also in other atom-diatom systems [29,30]: However it is also interesting to observe that the damping of the interference structure in the scattering cross sections can be different when different rotational states of the diatom are involved.

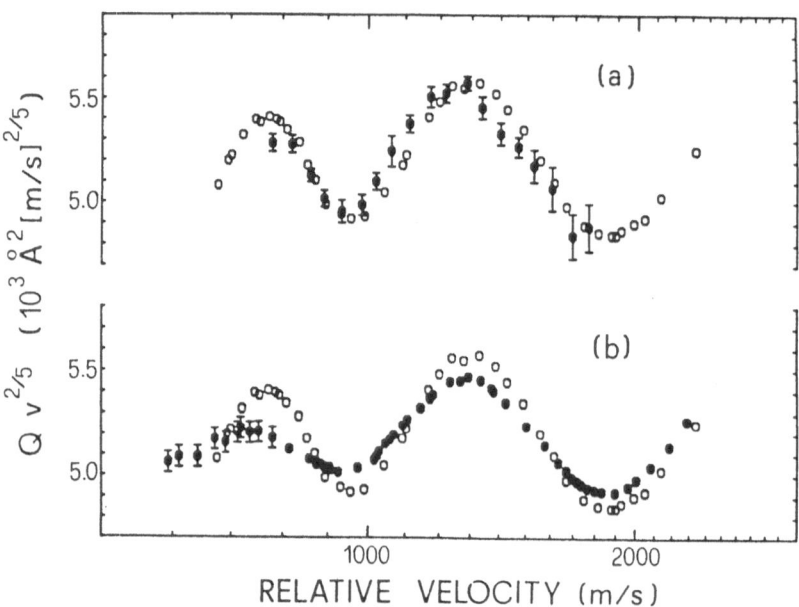

Fig.7 - Integral cross sections for Kr-O_2 (full dots) as a function of the collision energy and for two different rotational temperatures of the diatom (Ref.31): a) 1000 K; b) 10 K. The atom-diatom cross sections are compared with those for Ar-Kr (open circles) which is the isotropic system with an interaction similar to that one for Kr-O_2. These experimental data clearly show that the rotational level of the diatom affects the glory structure: At low rotational temperature the glory undulations appear partially damped when compared with those of the isotropic Ar-Kr system, while at high rotational temperature the undulations show an amplitude much closer to that of the isotropic system.

In Fig.7 the integral cross sections for $Kr-O_2$ collisions as a function of the relative velocity are reported as measured for two different rotational temperatures of oxygen molecule [31]. The cross sections are compared with those of the Ar-Kr system. It can be seen that while at low rotational temperature a damping of the glory interference is observed compared with that of the atom-atom system, at high rotational temperature the amplitude of the undulatory structure becomes similar for the two systems. Such an effect has also been documented for $Ar-O_2$[8], $Ar-N_2$[9] and $Kr-N_2$[32]. It is the purpose of the following section to show that these modifications in the structure can be exploited to extract valuable information about some characteristics of the potential energy surfaces, in particular their anisotropy.

4.2. The adiabatic route from scattering data to the potential

Although formally the scattering of an atom by a rigid rotor is described by a multichannel Schroedinger equation of the same form as the one considered above for the atom-atom case (equation 2), an additional complication arises in this case from the fact that the dimension of the basis is here infinite, being extended to all the rotor states labeled by the quantum number j. These states are the eigenvalues of the operator V_{mol} of molecular rotation, which enters into the potential energy matrix (3) in place of V_{so} .

The most popular simplification is the so called Infinite Order Sudden (IOS) approximation, which assumes essential degeneration of rotational and centrifugal channels, leading to cross section formulas which are independent of the rotational temperature of the diatomic molecule: this approximation, which has proven to be useful for the estimate of the magnitude of the interactions involved, is not adequate when a dependence on rotational temperature is actually observed, as in the cases mentioned above.

The failure of the IOS approximation in these cases can be shown to be mainly due to the assumption of degenerate eigenvalues for $\underline{V}_{mol}$, while the assumption of the degeneracy for $\underline{V}_{rot}$ can still be assumed to be valid. The approximation which exploits this assumption is then formally identical to the CS approximation discussed in Sec.3, in the present context also denoted as j_z-conserving. The accuracy of such an approximation is well established, but the computational labor involved is still exceedingly heavy if an initial guess is made for the potentials and then computations are carried out in the usual diabatic representation, until successive refinements lead to agreement with experiments. This is due to the fact that for the typical experiments described above usually hundreds of total angular momentum waves have to be included.

Recently an alternative approach to this problem has been suggested in our laboratory [33]: It exploits the adiabatic representations within the CS scheme, and therefore is essentially an extension to this case of the same approach which had been proven useful in the atom-atom case (Sec.3). Accordingly, adiabatic eigenenergies are generated by simultaneous numerical diagonalization of $\underline{V}_{el}$ and V_{mol} . This states can be labelled $|j\Omega>$ as before. Examples of the behavior of such eigenvalues are shown in Fig.8 for the $He-N_2$ and $Ar-N_2$ systems. For the

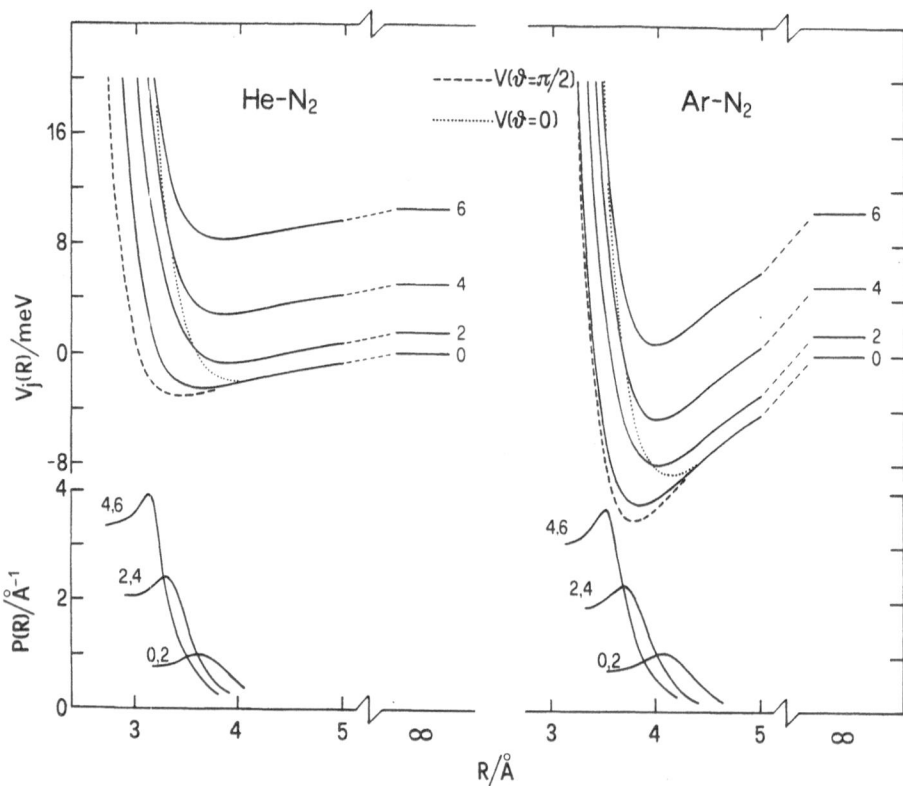

Fig.8 – In the upper parts, adiabatic energy curves are shown, obtained as described in the text, for $\Omega = 0$. Also shown are the $V(\vartheta = 0)$ and $V(\vartheta = \pi/2)$ curves, which, as anticipated in Fig.1, represent the ridge and valley bottom lines for these potential energy surfaces: States have essentially the character of librating modes below the ridge and of hindered rotator above it. In the lower parts, nonadiabatic coupling matrix elements (equation 6) are shown: As expected from experience on a variety of other physical situations [37], such couplings peak where the adiabatic curves cross the ridge.

He–N_2 curves the potential energy surface obtained by a recent multiproperty analysis [12] has been used, while for the Ar–N_2 case rather than using a potential of similar reliability, a model interaction [34] has been used instead, because for this interaction extensive numerical experience exists. In particular for this Ar–N_2 model system an assessment has been recently presented of the validity of CS and IOS approximations [35].

From plots such as those shown in Fig.8, it is possible to individuate at once which channels are effective in a given scattering experiment , which of these channels can be considered effectively decoupled, which nonadiabatic coupling between states has to be considered explicitly. It is the latter feature which dominates inelastic effects. Consider for example the glory structure in integral cross sections, which, as stressed before, is mainly sensitive to features of the wells [36]: For its description, it is clear from Fig.8 that most adiabatic wells are unaffected by nonadiabatic couplings, but that coupling has to introduced explicitly between the two lowest states for both systems.

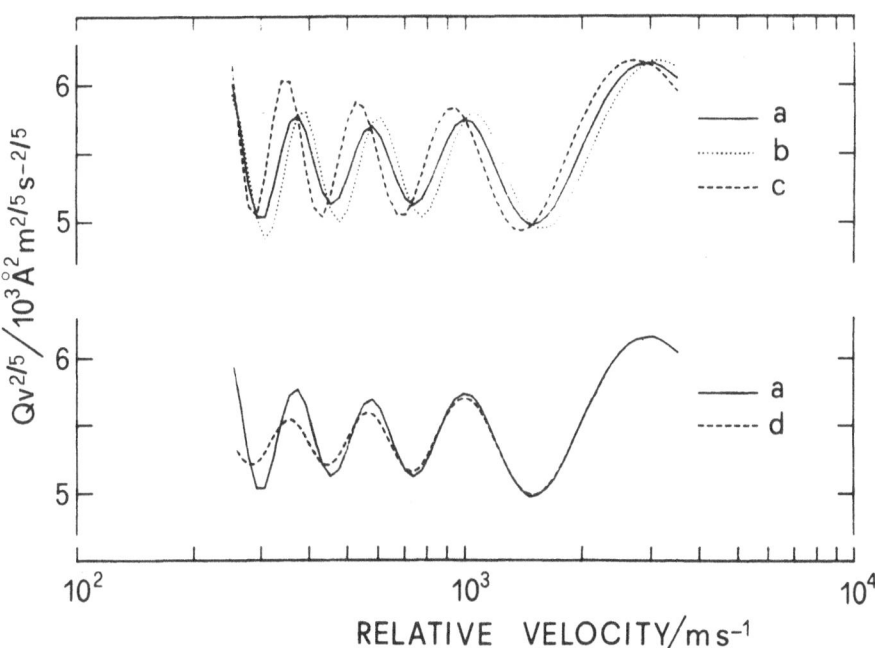

Fig.9 – The glory structure for the model Ar–N_2 system obtained by the present two state adiabatic approach with nonadiabatic coupling (curves a). Curves b and c show the glory scattering from the adiabatic curves correlating with the rotor states j=0 and j=2, respectively: They give curve a when combined as described in the text. Curve d is the IOS result: The perfect agreement at higher velocities deteriorates at lower velocities where the IOS assumption of degenerate rotor states becomes incorrect.

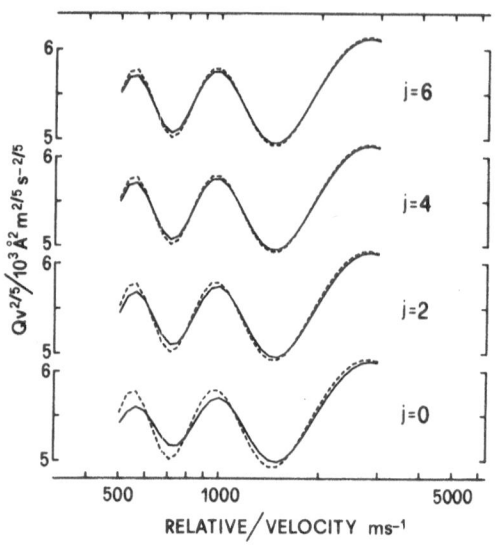

(a)

Fig.10 – Effect of initial
rotational states, (a), and of
the initial rotational
temperature, (b), of the N_2
molecule on glory structure
for the model Ar–N_2 system,
obtained as described in the
text: The dashed curve,
representing the glory
structure from the isotropic
potential V_0 (Fig.1), is
repeated on each graph for
comparison and shows that the
effects of the anisotropy
decrease the higher is the
rotational angular momentum of
the diatomic molecule.

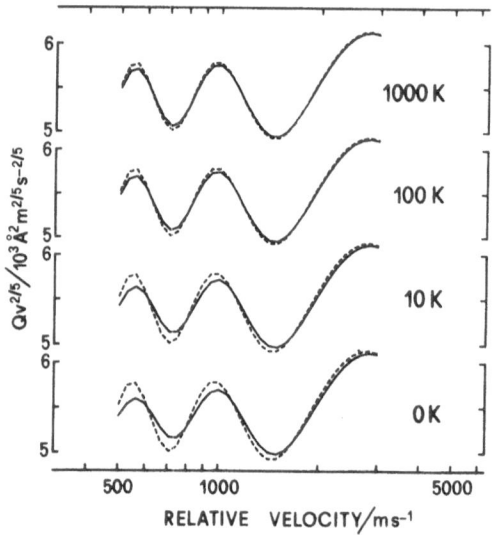

(b)

Connor et al. [35], in their calculation for the glory effect have shown that CS and IOS both agree with exact results in the 500 to 3500 m/s velocity range, introducing j=0 as the only entrance channel. We succeeded [33] in reproducing their results within mutual numerical errors by describing the scattering as due only to lowest j=0 and 2 states, but properly combining them by explicitly computing the effect of the nonadiabatic coupling between them. It is interesting to note that the type of coupling between states can be analyzed to be of the perturbed resonance type first described by Rosen and Zener, and therefore it has been found that the so called Demkov formula is appropriate in this case to describe the nonadiabatic transitions. The results of such computation are exhibited in Fig.9: The comparison with IOS shows perfect agreement at high collision energy, the discrepancy at low energy being due to the failure of the IOS assumption of degenerate rotor states, which is not made in our approach.

The present approach makes feasible the computation of integral cross sections as a function of relative velocity at different values for the initial rotational angular momentum of the diatomic molecule (Fig.10a); these results can in turn be combined to give finally the dependence of the glory pattern as a function of the rotational temperature (Fig.10b). The states included in the computation for the highest rotational temperature are the even ones from j=0 to 44, and all their projections from Ω =0 to 44.

The results show that the effect of the rotational temperature exists and is in the same direction as indicated by the experimental observations: They also show that the present computational approach represents an accurate computational route for extracting information on the potentials from the modification of the glory structure because of anisotropy effects. Further computational and analytical work confirms this: In particular semianalitycal formulas based on the theory of perturbation of librators below the ridge and of rotors above the ridge are available for the construction of adiabatic curves [33]. The extension of the present approach to differential total and inelastic cross sections and to the computation of transport properties is also being pursued.

REFERENCES

1 – U.Buck, Adv. Chem. Phys. 30 (1975) 313.
2 – G.Scoles, Ann. Rev. Phys. Chem. 31 (1980) 81.
3 – U.Buck, Comm. At. Mol. Phys. 17 (1986) 143.
4 – F.Vecchiocattivi, Comm. At. Phys. 17 (1986) 163.
5 – H.Pauly, in: Atom–Molecule Collision Theory: a Guide for The Experimentalist, ed. R.B.Bernstein (Plenum Press, New York, 1979) p.111.
6 – U.Lackschewitz, J.Maier and H.Pauly, J. Chem. Phys. 84 (1986) 181, and references therein.
7 – R.A.Aziz, in: Inert Gases, ed. M.L.Klein (Springer, Berlin, 1984) Chap.2 and references therein.

8 - F.Pirani and F.Vecchiocattivi, Chem. Phys. 59 (1981) 387.
9 - R.Candori, F.Pirani and F.Vecchiocattivi, Chem. Phys. Letters 102 (1983) 412.
10 - M.Keil and J.A.Parker, J. Chem. Phys. 82 (1985) 1947.
11 - R.Candori, F.Pirani and F.Vecchiocattivi, J. Chem. Phys. 84 (1986) 4833.
12 - F.A.Gianturco, M.Venanzi, R.Candori, F.Pirani, F.Vecchiocattivi, A.S.Dickinson and M.S.Lee, Chem. Phys. (1986), in press.
13 - V.Aquilanti and G.Grossi, J. Chem. Phys. 73 (1980) 1165.
14 - V.Aquilanti, P.Casavecchia, G.Grossi and A.Lagana', J. Chem. Phys. 73 (1980) 1173.
15 - V.Aquilanti and G.Grossi, Lett. Nuovo Cim. 42 (1985) 157.
16 - R.Candori, F.Pirani and F.Vecchiocattivi, Chem. Phys. Letters 90 (1982) 202.
17 - R.A.Aziz and A.van Dalen, J. Chem. Phys. 81 (1984) 779.
18 - L.Beneventi, P.Casavecchia and G.G.Volpi, J. Chem. Phys. 84 (1986) 4229.
19 - V.Aquilanti, G.Liuti, F.Pirani, F.Vecchiocattivi and G.G.Volpi, J. Chem. Phys. 65 (1976) 4751.
20 - V.Aquilanti, E.Luzzatti, F.Pirani and G.G.Volpi, J. Chem. Phys. 73 (1980) 1181.
21 - V.Aquilanti, E.Luzzatti, F.Pirani and G.G.Volpi, Chem. Phys. Letters 90 (1982) 382.
22 - V.Aquilanti, G.Grossi and F.Pirani, in: Electronic and Atomic Collisions, invited papers at XIII ICPEAC, eds. J.Eichler, I.V. Hertel and N.Stolterfoht (Berlin, 1983), p.441.
23 - V.Aquilanti, E.Luzzatti, F.Pirani and G.G.Volpi, in: Electronic and Atomic Collisions, abstracts of XIV ICPEAC, eds. M.J.Coggiola, T.L.Heustis and R.P.Saxton (Palo Alto, CA, 1985), p.350.
24 - V.Aquilanti, E.Luzzatti, F.Pirani and G.G.Volpi, to be published.
25 - P.C.Tellinghuisen, J.Tellinghuisen, J.A.Coxon, J.E.Velazco and D.W.Setser, J. Chem. Phys. 68 (1978) 5187.
26 - Ch.Becker, P.Casavecchia and Y.T.Lee, J. Chem. Phys. 69 (1978) 2377; J. Chem. Phys. 70 (1979) 2986.
27 - K.Muller, Dissertation, Max Planck Institut fur Stromungsforschung, Bericht 1/1984, Gottingen (1984).
28 - V.Aquilanti, R.Candori, E.Luzzatti, F.Pirani and G.G.Volpi, J. Chem. Phys. (1986), in press.
29 - L.Beneventi, P.Casavecchia anf G.G.Volpi, J. Chem. Phys. (1986), in press.
30 - P.Casavecchia, A.Lagana' and G.G.Volpi, Chem. Phys. Letters 112 (1984) 445.
31 - F.Pirani, F.Vecchiocattivi, J.J.H.van den Biesen and C.J.N.van den Meijdenberg, J. Chem. Phys. 75 (1981) 1042.
32 - G.Liuti, E.Luzzatti, F.Pirani and G.G.Volpi, Chem. Phys. Letters 121 (1985) 559.
33 - V.Aquilanti, L.Beneventi, G.Grossi and F.Vecchiocattivi, to be published; see L.Beneventi, Tesi di Laurea in Chimica, Universita' di Perugia (1985).
34 - M.D.Pattengil, R.A.LaBudde, R.B.Brenstein and C.F.Curtis, J. Chem. Phys. 55 (1971) 5517.

35 – J.N.Connor, D.C.Clary and H.Sun, Mol.Phys. 49 (1983) 1139.
36 – F.Pirani and F.Vecchiocattivi, Mol. Phys. 45 (1982) 1003.
37 – V.Aquilanti, S.Cavalli and G.Grossi, in: Chaotic Behaviour in Quantum Systems, ed. G.Casati (Plenum Press, New York, 1985), p.299; Chem. Phys. Letters 110 (1984) 43; V.Aquilanti, in: The Theory of Chemical Reaction Dynamics, ed. D.C.Clary (Reidel, Dordrecht, 1986) p.383.

HIGH RESOLUTION CROSSED MOLECULAR BEAM STUDIES OF VAN DER WAALS FORCES

Laura Beneventi, Piergiorgio Casavecchia
and Gian Gualberto Volpi
Dipartimento di Chimica - Università di Perugia
06100 Perugia, Italy

ABSTRACT. High resolution crossed molecular beam measurements of diffraction oscillations for some atom-atom and atom-diatom systems in the
thermal energy range are reported. From the observation of a well resolved diffraction structure in the Ne-Ne, Ar, Kr and Xe systems a refinement or assessment of the corresponding interatomic potentials to better
than 1% has been obtained. From the simultaneous best-fit, within the
infinite-order-sudden approximation, of the diffraction data, and of
absolute total integral cross sections and second virial coefficients
from literature, rather accurate potential energy surfaces for the
He-N_2, O_2 and NO systems have been derived.

1. INTRODUCTION

The present knowledge of the potential energy curves for the interaction
between rare gases has recently been reviewed. From the whole picture
it emerges that the interatomic potentials for the symmetric pairs are
known to within 1%-2%, while those for the asymmetric pairs are known
with less precision. This fact is due to the existence of a smaller
amount and quality of macroscopic and microscopic experimental data for
the mixed interactions. Then, a call was made for new precise measurements to aid in constructing potentials for some of the systems.
    We have, therefore, carried out elastic differential cross section
measurements in high resolution conditions for Ne-Ne, Ne-Ar, Ne-Kr and
Ne-Xe, in which the diffraction quantum oscillations superimposed on the
fall off of the main rainbow structure are, for the first time, clearly
resolved at collision energies corresponding to room temperature beams.
In Ne-Ne the symmetry and the diffraction oscillatory patterns have been
clearly distinguished by performing isotope resolved angular distribution measurements. These new high quality data have allowed us to determine the location of the low repulsive wall of the interaction with an
accuracy of better than 0.6% and, with the help of absolute integral
cross section and second virial coefficient data available in literature,
to refine or assess, with high precision, the interatomic potentials for
these systems.
    The extension of such precision to nonspherical systems has repre-

441

A. Weber (ed.), Structure and Dynamics of Weakly Bound Molecular Complexes, 441–454.

sented the challenge of recent years. But, in spite of considerable[2] improvement during the past decade in both experimental techniques and theoretical methods[3] for treating low energy atom-molecule collisions, as we go beyond the hydrogen-rare gas systems[4] much work has still to be done to reach in the description of the interactions the accuracy achieved for the atom-atom spherical systems.

We have extended to some simple atom-diatom system the same approach used for the rare gas systems, and found that precise diffraction scattering measurements, coupled to absolute total integral cross sections, second virial coefficients, and semiempirical long-range coefficients, also lead for systems as He interacting with N_2, O_2 and NO to quite precise determinations of the intermolecular potentials. However, while for spherical systems the determinations can be very accurate and conclusive,[5,6] for nonspherical systems their validity remains confined within the limits of the approximate scheme of analysis employed.

In both atom-atom and atom-diatom cases we exploit very detailed[8] information content of diffraction scattering. The origin of the diffraction oscillations in the differential cross section is well understood.[9] Their angular positions give a direct measure of the diameter, σ, of the repulsive wall to better than 1%, and this represents the most detailed information provided by a scattering experiment.

2. EXPERIMENTAL

The experiments were performed on a high resolution crossed molecular beam apparatus recently built in our laboratory.[5,10] Briefly, two well collimated differentially pumped supersonic nozzle beams are crossed at 90° in a large scattering chamber kept below $3 \cdot 10^{-7}$ torr in operating conditions, and the in-plane scattered lighter particle is detected by a rotatable triply differentially pumped ultra-high-vacuum quadrupole mass spectrometer detector kept below $1 \cdot 10^{-10}$ torr in the ionization region by extensive ion-, turbo-, and cryo-pumping.

Helium and Neon gases were expanded at high pressures (18-50 Atm), while the Ar, Kr, Xe, NO, N_2 and O_2 pressures were kept much lower (0.5-3 Atm) to avoid condensation. Under our expansion conditions the molecular species are expected to be in their lowest rotational levels,[11,7] with NO also relaxed to the lower $^2\Pi_{\frac{1}{2}}$ electronic state.[7,10,12]

The primary and secondary beams were defined to an angular divergence of 0.4° and 1.8°, respectively. Under our geometrical arrangement, which is the same for all the experiments reported here, the collision volume is always contained in the detector viewing angle, which, for a point collision zone, is 0.50°. The narrow divergence in angle and spread in velocity of the colliding beams as well as the high angular resolution of the detector proved to be critical for the observation of a pronounced oscillatory structure in the differential cross sections.

Although precise values of σ may be extracted from scattering data presenting well resolved diffraction oscillations, its absolute accuracy depends on the calibration of the velocity distributions of the two beams as well as of the angular positions. Therefore, great care was

devoted to all these aspects. The beam velocities were measured by absolute time-of-flight analysis to better than 0.6%. The location of the
primary beam was determined to within 0.03° by carefully comparing the
scattered signals on both sides of the primary beam and making the appropriate laboratory to center-of-mass kinematic transformations.

The laboratory angular distributions, $I(\Theta)$, were obtained by taking
at least four scans of 30-180 s counts at each angle. The secondary
target beam was modulated at 160 Hz for background subtraction.

3. RESULTS AND DISCUSSION

It is useful to separately examine the results obtained for the atom
-atom and atom-diatom systems.

3.1. Atom-Atom

As example of the quality of the data measured for the Ne-rare gas
series, we report in Figs. 1 and 2 the angular dependence of the differential cross section for Ne-Ne and Ne-Kr, respectively.

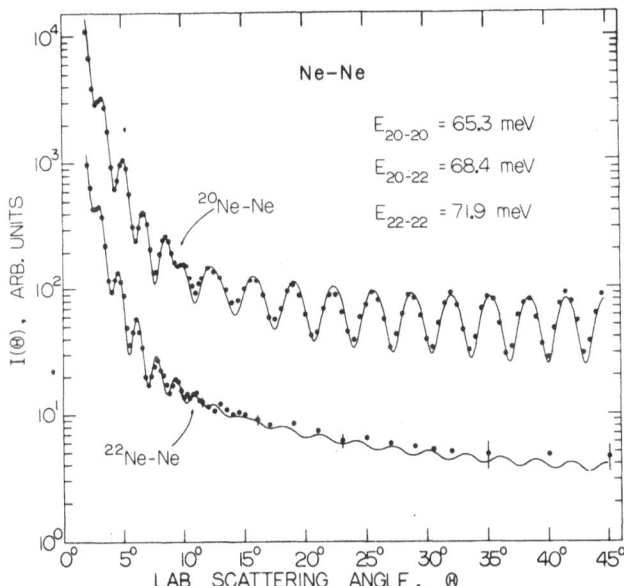

Fig. 1 - Differential cross section data for Ne-Ne obtained by detecting ^{20}Ne (upper plot) and ^{22}Ne (lower plot). The continuous line is
the prediction of the HFD-C2 potential of Aziz, Meath and Allnatt
(Ref. 13). The natural isotopic abundance of the Ne target beam is taken
into account in the calculations of the cross sections.

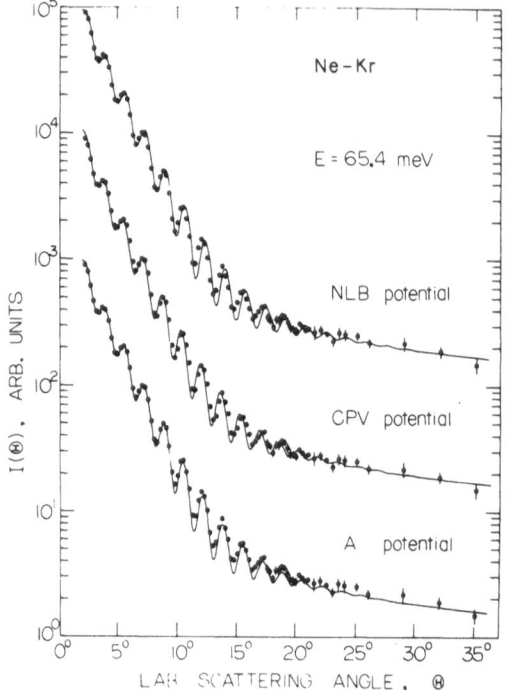

Fig. 2 - Differential cross
section data for Ne-Kr compared with predictions from
different interatomic potentials. Upper plot: potential
of Ng, Lee and Barker (Ref.
17). Middle plot: potential
of Candori, Pirani and
Vecchiocattivi (Ref. 18).
Lower plot: potential of
Aziz (Ref. 1).

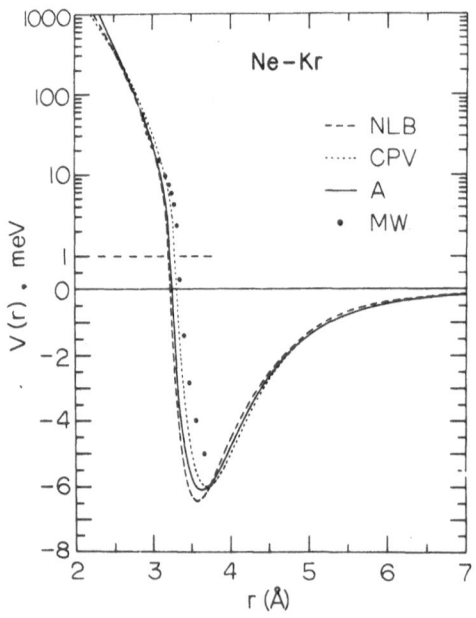

Fig. 3 - Comparison between
the interatomic potentials
quoted in Fig. 2 for the
Ne-Kr system. The dots represent the energies
obtained by Maitland and
Wakeham (Ref. 16) from the
inversion of transport
coefficients.

The data for Ne-Ne obtained by detecting [20]Ne (upper plot in Fig. 1) exhibit well resolved diffraction and symmetry oscillations. The average spacing of the diffraction oscillations is about 1.6°, while that of the oscillations due to the effect of nuclear symmetry is about double. The two patterns coalesce at $\Theta \cong 9°$. The diffraction structure has been clearly distinguished by detecting [22]Ne (lower plot in Fig. 1) in a separate experiment. For Ne-Kr the diffraction oscillations superimposed on the fall off of the main rainbow structure are also clearly resolved (see Fig. 2). Their average spacing is also about 1.6°. In rare gas -rare gas differential scattering experiments quantum oscillations so closely spaced have never been resolved before at the quality level of the present experiments. Differential cross section data with such detailed oscillatory structure provide for the systems in question very detailed information on the interaction potential which is not obtainable by other means and represent a very stringent test of the best available potentials for Ne-Ne, Ar, Kr and Xe.

A comparison with our scattering data of the prediction of the HFD-C2 potential of Aziz, Meath and Allnatt[13] for Ne-Ne, derived from a multiproperty analysis of scattering, bulk and spectroscopic properties, shows good agreement (see Fig. 1). This indicates that for this system the location and slope of the low repulsive wall of the potential up to about 70 meV is correct and therefore the potential of Aziz et al.[13], having also been tested against many other properties which probe different regions of the interaction, is confirmed to represent the most accurate description of the Ne dimer to date.

Recently, by using diffraction scattering we have critically assessed the status of Ne-Ar[5,6]. In particular, we have found[5] that a multiproperty potential, always of the HFD form, proposed by Aziz and van Dalen[14] was in disagreement with our differential cross section data, being the interaction in this system better described by the multiproperty potential put forth by Candori et al.[15]

The present knowledge of the Ne-Kr and Ne-Xe potentials is less satisfactory.[1] Here, as example, we examine the Ne-Kr case. In Fig. 3 we report the three best available Ne-Kr potentials together with the energies obtained by Maitland and Wakeham[16] from direct inversion of transport properties. As can be seen, the differences in the potentials are very small (a few percent). Neverthless, our high resolution differential scattering data are able to easily discriminate between them. None of the available potentials affords a very good fit to the experimental data (see Fig. 2). The NLB potential[17] gives diffraction oscillations which are increasingly shifted towards too large angles with increasing scattering angle with respect to the experiment, while the reverse can be noticed for the CPV potential[18]. This is due to the fact that the low repulsive wall (i.e., the σ, and then r_m, parameter) of the NLB and CPV potentials are too inward (r_m=3.58 Å) and too outward (r_m=3.70 Å), respectively, located on the distance scale. The phase and period of the measured oscillations are found to be consistent with a r_m intermediate value of 3.65±0.02 Å.[19] The prediction of the A potential[1] follows rather closely in phase and period the experimental data, having r_m=3.63 Å (see Fig. 2). However, it predicts the general fall off of the rainbow not as well as , for instance, the CPV potential, indica-

ting that the shape of the outer wall and/or the well depth are not
correct. The inverted energies of Maitland and Wakeham (see Fig. 3)
give a σ value of 3.359±0.008 Å[16] and are not in agreement with our
direct determination of the location of the low repulsive wall
(σ=3.265±0.020 Å).[19] Clearly, the present differential cross section
data permit a sophisticated refinement[19] of the Ne-Kr potential, by
simultaneously analyzing them with other properties, as shown for
Ne-Ar.[5,6] Comparisons similar to those described here for Ne-Kr have
also been carried out for Ne-Xe[19] and previously also for Ne-Ar.[5]

The results of a preliminary analysis can be summarized as follows.[17]

Potentials derived by Ng, Lee and Barker[17] from differential cross
section measurements without resolving the diffraction oscillations are
inaccurate for all the Ne-heavier rare gas systems.

Potentials derived by Candori, Pirani and Vecchiocattivi[18] from a
multiproperty analysis are accurate for Ne-Ar, but less accurate for
Ne-Kr and still less for Ne-Xe.

Potentials derived by Aziz and coworkers[1,13,14] from a multiproperty analysis are accurate for Ne-Ne, less for Ne-Kr and Ne-Xe, and
inaccurate for Ne-Ar.

The locations of the low repulsive wall of the potentials derived
by Maitland and Wakeham[16] from direct inversion of transport coefficients are accurate for Ne-Ar, almost satisfactory for Ne-Xe, but inaccurate for Ne-Kr.

The potential distance parameters derived from the recent combining
rules of Tang and Toennies[20a] are in excellent agreement with our high
resolution differential cross section measurements. Agreement is also
noted with the parameters calculated from the correlations in terms of
polarizabilities of Liuti and Pirani.[20b]

3.2. Atom-Diatom

The same approach used to study the atom-atom interactions has been
extended to atom-diatom systems. High resolution diffraction scattering
measurements have been carried out for He-O_2, N_2 and NO. Rather accurate potential energy surfaces (PES) have been derived by also simultaneously analyzing absolute total integral cross sections and second
virial coefficient data. A remarkable difference with respect to the
atom-atom systems discussed before is that now all the conclusions reached are confined within limits of validity of the approximate scheme
of analysis employed, since it is still impracticable to perform exact
quantum close-coupling calculations for atom-diatom potential fitting
purposes.

In Fig. 4 we report total differential cross section (DCS) data
for He scattered off O_2, NO and N_2, together with the elastic DCS data
for He-Ar, which is taken as standard. The data show well resolved
diffraction oscillations which appear damped with respect to the corresponding isotropic He-Ar case. The location of these oscillations determine to high precision (≅1%) the absolute position of the repulsive wall
of the spherical average potential. Sensitivity of these data to the
well depth instead is somewhat lower. The quenching of the oscillatory
structure is a manifestation of the anisotropy of the potential energy

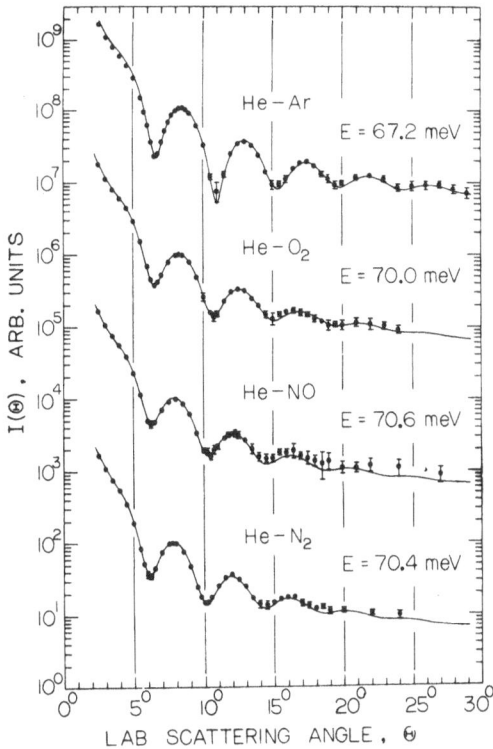

Fig. 4 - Total differential
cross section data for scattering of He off O_2, NO and
N_2 and elastic differential
cross section data for
He-Ar. The curves for the
He-diatom systems are calculated from the best-fit
anisotropic potentials in
the IOS approximation. The
curve for He-Ar is calculated from the accurate po-
tential of Ref. 21.

surface.[22] So far the information content about anisotropy obtained
from this kind of scattering measurements has been limited by the quality of the experimental data.[23,24] We have recently shown[7] that an accurate evaluation of the quenching of the diffraction structure within
the infinite-order-sudden (IOS) approximation brings to estimates of the
potential repulsive anisotropy which, although indirect, appears to be
in agreement with those from more direct determinations, as from state
-to-state rotationally inelastic DCS or differential energy loss spectra.
The IOS approximation[25,3] is expected to work reasonably well for the
systems investigated, under the present experimental conditions. We note
that also the state-to-state inelastic scattering data[26,27] for some of
the systems in question have mainly been analyzed by IOS methods and[28]
comparisons with close-coupling results are limited but encouraging.
We believe that our IOS analysis provides reasonably consistent information for the comparison of proposed interactions.

In the data analysis we take a potential model whose reduced form
is the same for all orientations and the angular dependence is given by
the size parameters, ε and R_m, angle dependent, i.e.:

$$V(R,\gamma) = \varepsilon(\gamma) \cdot f(x) \quad , \quad x = R/R_m(\gamma)$$

where the parameters $\varepsilon(\gamma)$ and $R_m(\gamma)$ are given by:

$$\varepsilon(\gamma) = \overline{\varepsilon} \cdot \{1 + A_2 \cdot P_2(\cos \gamma)\}$$

$$R_m(\gamma) = \overline{R}_m \cdot \{1 + B_2 \cdot P_2(\cos \gamma)\} \tag{1}$$

The reduced potential curve $f(x)$ is chosen to be the piecewise analytic exponential-spline-Morse-spline-van der Waals (ESMSV) form. We are using only an effective P_2 term for He-NO, as previously done by other authors[23,12] and us,[10] because the data do not warrant the use of a more elaborate description of the anisotropy (i.e., a parametrization which includes also a P_1 term). Physically this means disregarding the slight eccentricity of NO.[12] For He-O_2 and N_2 only even terms appear in the expansion and again a P_2 term was sufficient to describe the aniso-tropy effects observed in the present experiments.

It is well known[29] that potentials as those described above converge very rapidly and that the spherical limit, $V_s(R)$, of such potential forms, obtained by setting equal to zero the anisotropy parameters in Eq. (1), is not the same as the $V_0(R)$ term (the spherical average of the PES) in a Legendre series expansion of the PES itself. In particular, $V_s(R)$ is a much better "effective" spherical potential than $V_0(R)$; that is, the scattering and bulk properties of $V_s(R)$ are much closer to those of $V(R,\gamma)$ than are those of $V_0(R)$ (see also Ref. 7).

The procedure of analysis consists in determining firstly the parameters of the spherical limit potential from the simultaneous best-fit of the location and fall off of the diffraction oscillations in our total DCS (which are sensitive to the $\overline{\sigma}$, and then $\overline{R}_m$, parameter), of absolute total integral cross sections[30] (which are sensitive to the long -range attraction and somewhat to the well depth and its location), of the location of the only glory maximum exhibited by these light systems in the velocity dependence of the relative total integral cross section[31] (which is sensitive to the $\overline{\varepsilon} \cdot \overline{R}_m$ product), and of second virial coefficient data over an extended temperature range[32] (which are mainly sensitive to the repulsive wall of the potential and slightly less to the area of the well). Once determined the spherical limit potential, anisotropy A_2 and B_2 parameters were introduced in Eq. (1) until we reproduced, within the experimental error, the quenching of the diffraction oscillations by using the IOS method. This quenching is a measure of the anisotropy of R_m (and then σ) and B_2 is the parameter directly determined from our data. Reliable values of A_2 have also been derived by imposing the constraints that the long range anisotropy had to be consistent with the anisotropy of the diatom polarizability and that the most stable configuration occurs for the T-shaped O_2, N_2 and NO-He complexes.

The best-fit total DCS are shown as solid lines in Fig. 4 for the three systems. As illustrative example we also report for He-N_2 the best-fit curves for the absolute total integral cross section (Fig. 5) and second virial coefficient data (Fig. 6). The characteristic potential and anisotropy parameters, ε, R_m, σ and $\Delta\varepsilon$, ΔR_m and $\Delta\sigma$, are listed in Table I for all systems (A more complete tabulation can be found in Ref. 7). Comparisons have been carried out with previous potentials, both empirical[23,27] and theoretical[33,34], which have also been derived

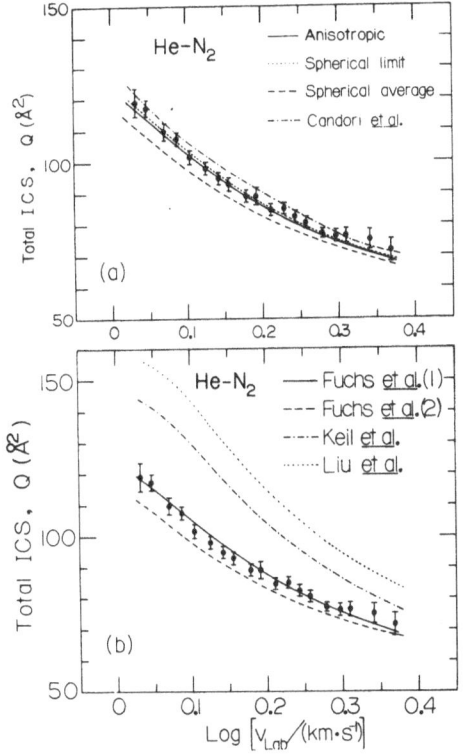

Fig. 5 - Absolute total integral cross sections for He-N$_2$ as a function of laboratory velocity. Experimental points are from Ref. 30.
(a) The continuous, dotted and dashed lines are calculated from the best-fit anisotropic, spherical limit and spherical average potentials, respectively (see Table I and Ref. 7). The dashed-dotted line is the prediction of the effective spherical potential of Candori et al. (Ref. 30).
(b) Predictions of the two anisotropic model potentials of Fuchs et al. (Ref. 34) and of the empirical anisotropic potentials of Keil et al. (Ref. 23) and Liu et al. (J. Chem.Phys. 75, 1496 (1981).

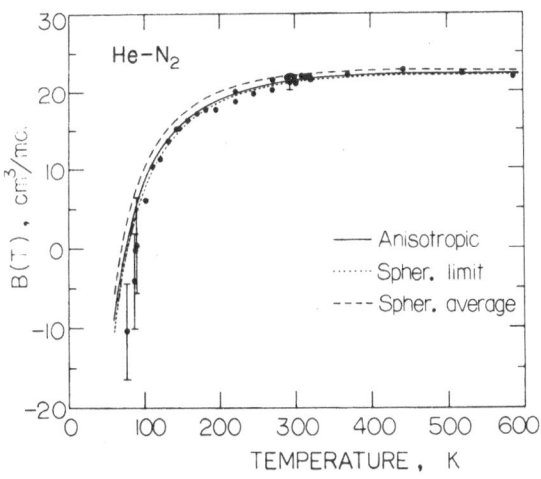

Fig. 6 - Second virial coefficients for He-N$_2$. The experimental points are taken from Ref. 32. The continuous, dotted and dashed lines are calculated from the best-fit anisotropic, spherical limit and spherical average potentials, respectively (see Table I and Ref. 7).

TABLE I.   Characteristic potential parameters.

Systems	He-N_2	He-NO	He-O_2	He-Ar [a]
$\overline{\varepsilon}$ (meV)	2.42±0.12	2.69±0.19	2.69±0.16	2.57±0.13
$\overline{R}^m$ (Å)	3.65±0.04	3.63±0.04	3.49±0.04	3.48±0.04
$\overline{\sigma}^m$ (Å)	3.24±0.03	3.21±0.03	3.09±0.03	3.10±0.02
A_2	-0.456	-0.518	-0.425	
B_2	0.125±0.010	0.140±0.010	0.125±0.010	
$\varepsilon_\perp - \varepsilon_\parallel$	1.65	2.09	1.71	
$R_{m\parallel} - R_{m\perp}$	0.68	0.76	0.64	
$\sigma_\parallel - \sigma_\perp$	0.61±0.05	0.69±0.05	0.58±0.05	
ε_0	2.15±0.11	2.35±0.17	2.40±0.14	
R_{m0}	3.69±0.04	3.67±0.04	3.52±0.04	
σ_0	3.27±0.03	3.25±0.03	3.12±0.03	

[a] Ref. 21

from or tested against experimental bulk and scattering properties within the IOS approximation. As example, we report in Fig. 7 the potentials for parallel and perpendicular configurations for He-N_2 and compare them with previous results. In order to show the sensitivity of the diffraction scattering data to potential features, we report in Fig. 8 for this system the predictions of previous potentials. As can be seen, none of the compared surfaces is in agreement with the experimental data, because of the too weak or too strong anisotropy, or of the incorrect location on the absolute distance scale. A similar comparison with respect to absolute total integral cross section data is also reported in Fig. 5.

The repulsive anisotropy, derived from the quenching of the diffraction oscillations and represented by the $\Delta\sigma$ ($=\sigma_\parallel - \sigma_\perp$) parameter (see Table I),is found to be much larger, for all systems, with respect to previous estimates within the same scheme of analysis from similar experiments performed at a lower resolution by Keil, Slankas and Kuppermann.[23] Agreement is instead found with results obtained from state-to-state rotationally inelastic DCS data,[27,35] which supports the reliability, at least for the systems investigated, of the procedure of analysis used. Similar findings have also been obtained for the He-CO_2 system.[36] It has to be pointed out that the present indirect approach to the repulsive anisotropy is not clearly general and powerful as the inelastic scattering method, but for favorable cases, as He interacting with linear molecules as O_2, N_2 and CO_2, it appears to be able to provide reliable results beacause of the nature of the diffraction oscillations and of

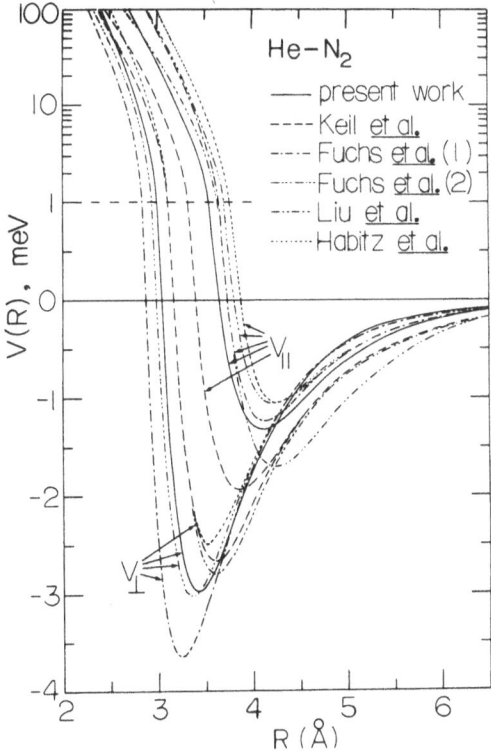

Fig. 7 - Best-fit He-N$_2$ anisotropic potential (continuous curve) for parallel (V$_{\parallel}$) and perpendicular (V$_{\perp}$) configurations compared to the empirical potentials of Keil et al. and Liu et al., and model potentials of Habitz et al. (Ref. 34) and Fuchs et al. (Ref. 34).

the effect on them of the anisotropy of the location of the repulsive wall of the intermolecular potential.

A final comment should be made. Why the earlier total DCS data of Kuppermann and coworkers[23] gave, within the same IOS scheme of analysis, a repulsive anisotropy much lower than our total DCS data at similar collision energies? The explanation could lie in the lower angular and velocity resolution of the earlier experiments, which prevented an accurate evaluation of the damping of the diffraction oscillations. And/or it could be that the previous experiments[23] sampled a more spherical interaction because of the significantly higher rotational state distribution of their thermal molecular beams. The effect of the rotational temperature on scattering cross sections warrants more experimental and theoretical investigation.[37]

The results obtained for the atom-diatom systems can be summarized as follows.

The potential energy surface derived for the He-O$_2$ system is found to be in good agreement with that proposed by Faubel et al..[27]

The surface obtained for He-N$_2$ represents the only potential at the moment able to predict correctly a wide variety of scattering and bulk data.

The surface proposed for He-NO is an improvement with respect to

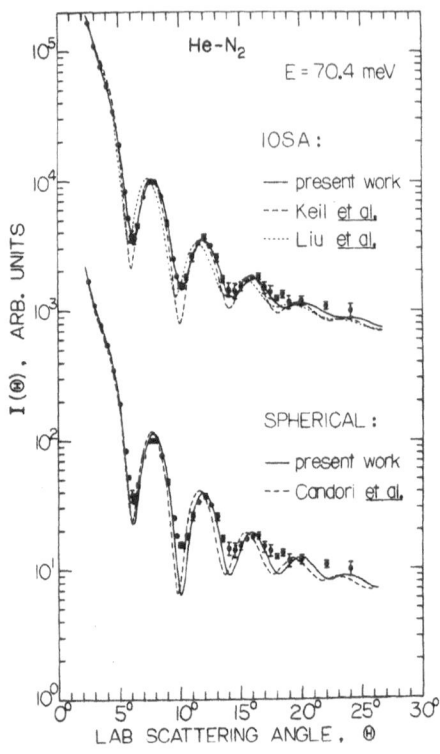

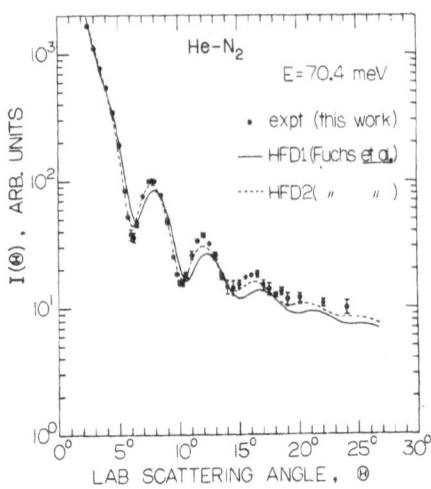

Fig. 8 - Total differential cross section data for He-N$_2$ compared with predictions from different potentials. References are as in Figs. 5 and 7.

that previously available.[23] Comparison of the He-NO surface with those obtained recently[10] in this laboratory for Ar-NO and Kr-NO indicates that both the anisotropy of the repulsive wall and well depth are much stronger when the rare gas is the small He atom.

Theoretical ab initio (for He-O$_2$ and He-N$_2$)[33] and model (for He-N$_2$)[34] potential predictions are not found to be in agreement with the experiment, indicating that more work has still to be done to reach a satisfactory accuracy in the construction of potential energy surfaces for even simple atom-diatom systems.

The location of the repulsive wall of the spherical average interactions, i.e. σ_0 and R_{m0} values (see Table I), in the He-N$_2$, He-NO and He-O$_2$ series is found to follow, as one would expect, the same trend as the molecular polarizabilities and to correlate inversely with the equilibrium bond lengths of the diatomic molecules.

4. CONCLUSIONS

In this work we have reported and discussed some recent high resolution crossed molecular beam measurements of diffraction oscillations in atom-atom and atom-diatom systems. Their high sensitivity to potential characteristics has been assessed by comparison with predictions from only slightly different intermolecular potentials.

For the first time the diffraction structure has been clearly resolved in relatively heavy systems as Ne-Ar, Ne-Kr and Ne-Xe and separated from the symmetry oscillations in Ne-Ne. The high quality of the data has permitted the refinement of the Ne-heavier rare gases interatomic potentials to better than 1%.

Accurate measurements of the diffraction structure in the total differential cross sections for some atom-diatom systems (He-N_2, NO and O_2) have brought, within a multiproperty analysis in the IOS approximation, to rather reliable potential energy surfaces. In particular, for the prototype He-N_2 case the existence of virial coefficient data down to low temperature with small uncertainties, in addition to absolute total integral cross section data with small error, has permitted to derive a potential surface whose spherical part has an accuracy comparable to that achieved for the corresponding isotropic He-Ar system (see Table I).

In the next few years, a fruitful coupling of experimental and theoretical means should allow to extend also to the interactions of nonspherical systems the same precision today attained for the spherical systems.

ACKNOWLEDGMENTS

The financial support of the Italian National Research Council (CNR) and of the Italian Ministry of Education (MPI) is gratefully acknowledged.

REFERENCES

1. R.A.Aziz, in _Inert Gases_, ed. by M.L.Klein (Springer, Berlin, 1984), Chap. 2. See also: G.Scoles, Ann.Rev.Phys.Chem. $\underline{31}$, 81 (1980).
2. U.Buck, F.Huisken, J.Schleusener and H.Pauly, Phys.Rev.Lett. $\underline{38}$, 680 (1977); U.Buck, Comments At.Mol.Phys. $\underline{17}$, 143 (1986), and references therein; M.Faubel and J.P.Toennies, Adv.At.Mol.Phys. $\underline{13}$, 229 (1978); M.Faubel, K.H.Kohl and J.P.Toennies, J.Chem.Phys. $\underline{73}$, 2506 (1980); M.Faubel, Adv.At.Mol.Phys. $\underline{19}$, 345 (1983); H.H.W.Thuis, S.Stolte and J.Reuss, Comments At.Mol.Phys. $\underline{8}$, 123 (1979); J.Mettes, B.Heymen, P.Verhoeve, J.Reuss, D.C.Lainé and J.Brocks, Chem.Phys. $\underline{92}$, 9 (1985).
3. _Atom-Molecule Collision Theory_, ed. by R.B.Bernstein (Plenum, New York, 1979).
4. U.Buck, H.Meyer and R.J.LeRoy, J.Chem.Phys. $\underline{80}$, 5589 (1984).
5. L.Beneventi, P.Casavecchia and G.G.Volpi, J.Chem.Phys. $\underline{84}$, 4828 (1986).
6. R.Candori, F.Pirani and F.Vecchiocattivi, J.Chem.Phys. $\underline{84}$, 4833 (1986).
7. L.Beneventi, P.Casavecchia and G.G.Volpi, J.Chem.Phys. (1986), in press.
8. U.Buck, Adv.Chem.Phys. $\underline{30}$, 313 (1975),
9. G.E.Zahr and W.H.Miller, Mol.Phys. $\underline{30}$, 951 (1975), and references therein.

10. P.Casavecchia, A.Laganà and G.G.Volpi, Chem.Phys.Lett. 112, 445 (1984).

11. M.Faubel, Adv.At.Mol.Phys. 19, 345 (1983).

12. H.H.W.Thuis, S.Stolte, J.Reuss, J.J.H.van den Biesen and C.J.N. van den Meijdenberg, Chem.Phys. 52, 211 (1980).

13. R.A.Aziz, W.J.Meath and A.R.Allnatt, Chem.Phys. 78, 295 (1983).

14. R.A.Aziz and A.van Dalen, J.Chem.Phys. 81, 779 (1984).

15. R.Candori, F.Pirani and F.Vecchiocattivi, Chem.Phys.Lett. 90, 202 (1982).

16. G.C.Maitland and W.A.Wakeham, Mol.Phys. 35, 1429 (1978).

17. C.Y.Ng, Y.T.Lee and J.A.Barker, J.Chem.Phys. 61, 1996 (1974).

18. R.Candori, F.Pirani and F.Vecchiocattivi, Mol.Phys. 49, 551 (1983).

19. L.Beneventi, P.Casavecchia and G.G.Volpi, to be published.

20. (a) K.T.Tang and J.P.Toennies, Z.Phys.D 1, 91 (1986);
    (b) G.Liuti and F.Pirani, Chem.Phys.Lett. 122, 245 (1985).

21. R.A.Aziz, P.W.Riley, U.Buck, G.Maneke, J.Schleusener, G.Scoles and U.Valbusa, J.Chem.Phys. 71, 2637 (1979).

22. R.T.Pack, Chem.Phys.Lett. 55, 197 (1978).

23. M.Keil, J.T.Slankas and A.Kuppermann, J.Chem.Phys. 70, 541 (1979).

24. M.Keil and G.A.Parker, J.Chem.Phys. 82, 1947 (1985).

25. G.A.Parker and R.T.Pack, J.Chem.Phys. 68, 1585 (1978).

26. M.Faubel, K.H.Kohl, J.P.Toennies, K.T.Tang and Y.Y.Yang, Faraday Discuss.Chem.Soc. 73, 205 (1982).

27. M.Faubel, K.H.Kohl, J.P.Toennies and F.A.Gianturco, J.Chem.Phys. 78, 5629 (1983).

28. F.A.Gianturco and A.Palma, J.Phys.B: At.Mol.Phys. 18, L519 (1985).

29. R.T.Pack, J.Chem.Phys. 64, 1659 (1976).

30. E.Luzzatti, F.Pirani and F.Vecchiocattivi, Mol.Phys. 34, 1279 (1977); R.Candori, F.Pirani, F.Vecchiocattivi, F.A.Gianturco, U.T.Lamanna and G.Petrella, Chem.Phys. 92, 345 (1985); ibid. 97, 464 (1985).

31. H.P.Butz, R.Feltgen, H.Pauly and H.Vehmeyer, Z.Physik 247, 70 (1971).

32. B.Schramm,to be published; B.Shramm and A.Buchner, Chem.Phys.Lett. 98, 118 (1983), and references therein.

33. R.Jaquet and V.Staemmler, Chem.Phys. 101, 243 (1986); A.J.Banks, D.C.Clary and H.J.Werner, J.Chem.Phys. 84, 3788 (1986).

34. P.Habitz, K.T.Tang and J.P.Toennies, Chem.Phys.Lett. 85, 461 (1982); R.R.Fuchs, F.R.W.McCourt, A.J.Thakkar and F.Grein, J.Phys.Chem. 88, 2036 (1984).

35. M.Faubel, J.Chem.Phys. 81, 5559 (1984).

36. L.Beneventi, P.Casavecchia and G.G.Volpi, to be published.

37. V.Aquilanti, F.Pirani and F.Vecchiocattivi, this Volume.

WHERE ARE WE IN WEAK INTERMOLECULAR INTERACTIONS?

William Klemperer
Department of Chemistry
Harvard University
Cambridge, MA 02138

This article, written as a recollection of a summarizing lecture of this NATO Workshop, affords the opportunity to portray with broad and probably inaccurate strokes a field of research which has been and continues to be fundamental to many areas of chemistry.

This meeting in my view represents a middle period in the modern study of the nature of intermolecular interactions. It clearly shows that spectroscopic studies in a wide range of energy are highly productive for the precision elucidation of structure and dynamics of molecular complexes. While rotational spectroscopy[1] has provided a solid foundation for this field, this conference points clearly to the rapid growth of high resolution infrared studies which greatly extend our knowledge in a most fundamental manner. This together with the now well established electronic spectroscopy[2] of molecular complexes brings us to the position of being able to have virtually the same information on this class of molecules that we have for chemically bonded stable species. It becomes easy therefore to phrase both questions and discussions of structure in terms similar to those used for qualitative, i.e., conceptual, discussions of stable molecules. Whether this is useful becomes increasingly less clear as our body of structural data grows. Let me first address the question of bonding description in a qualitative manner with an emphasis on what are presently puzzling structures, either by their presence or by their present absence.

In examining systematically the structures of classes of molecular complexes most attention will be paid to bonding angles. This seems always true (Note how much attention is paid to the 104° H-O-H angle in water and the frequently less than 5° changes produced by substitutions). In searching for exemplary systems which provide a rapid assessment of the utility of thinking of intermolecular forces in qualitative chemical bonding terms it would appear useful to find systems for which a continuous range of bond lengths from the 1.5A distance typical of chemical bonding to the 3.0A typical of van der Waals bonding is found. To date no such system has been found. We had

A. Weber (ed.), Structure and Dynamics of Weakly Bound Molecular Complexes, 455–463.

thought an attractive system for this purpose would be the B--N bond. Strong BF_3-NR bonds with lengths at 1.6A are well known for amines. The system BF_3-N_2 shows a 3A bond[3]. There do not appear to be any gas phase complexes with B-N bond lengths near 2A. Thus the question of whether there are classes of bonding ranging continuously over the whole bond length range at this point appears to be answered negatively. This negative result may be intrinsic or it may be just be that the subject is still new enough that the full range of examples has not yet been explored.

Closely related to this question is that of the relation of bond length and bond strength. One extremely interesting example here is given by the XeF species. Here the bond length is 2.2A (R_e=2.293 A) while the bond strength, D_0=3 kcal/mole (D_e=1175 cm^{-1})[4]. There is virtually no change in Xe-F bond length in going to the completely stable molecule XeF_2. Thus, perhaps as an aside, we may ask whether there exist a number of weak but short bonds. The nature of bonding in XeF remains to be more fully explored both experimentally and theoretically, especially with respect to the question of charge and spin distribution.

The discussion by Tony Legon[5] on how to think of angular geometry in acid-base complexes is extremely main stream in the chemical sense. In some manner our discussion of bonding has at its origin an elementary teaching desire. We wish to explain, essentially perturbatively, the building up of a molecule in terms of its constituent parts. Molecular complexes appear to provide us with simple experimental test probes for revealing precisely quantities such as lone pair or frontier orbital distributions. This idea is certainly one that has been used traditionally, as shown by the crystallographic analyses of Dunitz[6]. The structures of free binary complexes show pair interactions free of the extraneous complexity of an (inert/interactive?) imbedding medium. Using the hydrogen halides as probes of "lone pairs" in the binding partner appears to be safe usually. (The structures of ClF and Cl_2 with HF would appear to contradict this. It would appear useful to extend the halogen interactions with HCl to see if structural changes occur). Thus the exploration of SO_2 interactions with these simple proton donor probes is clearly important. The role of weak interactions in modifying reaction rates is a topic of current concern. The importance of building up a total knowledge of the chemistry of SO_2 and other sulfur oxides is obvious. The structures of relatively few SO_2 and SO_3 complexes have been structurally characterized. The numerous gas phase transformations leading from SO_2 eventually to H_2SO_4 (or other acid sulfates) make it likely that a systematic examination of weak bonding in relevant sulfur oxide species is necessary.

HCN complexes are frequently a source of surprise, when compared with hydrogen halide complexes. The observation that HCN forms weaker hydrogen bonds than does HCl is general and points out clearly the very local nature of these weak interactions. We comment upon the structural similarities and differences for HCN and HF complexes. With the Lewis bases NH_3, H_2O, CO, and the π electron systems HCCN and C_2H_4 both HF and HCN form hydrogen bonds[7]. With the acidic species CO_2 and SO_2 HF forms hydrogen bonds. In distinction HCN bonds to these through the nitrogen, acting structually as a Lewis base. Finally with the inert gases Kr and Ar the hydrogen halides form reasonably rigid hydrogen bonded species while the structures of the complexes with HCN are best characterized as highly non-rigid[8].

The structure of liquid water is a classic research area. A tremendous advance in this was achieved by Muenter and Dyke[9] with the production of the rotational spectrum and structural characterization of the water dimer. $(H_2O)_2$ has been the subject of a countless body of theoretical analyses. Its spectroscopy serves as a benchmark against which to test intermolecular potentials of water before going to liquid water simulation. The nature of non-rigid motions in $(H_2O)_2$ and $(D_2O)_2$ has previously been theoretically characterized. The facility with which the hydrogen bonds are broken and reformed or better interchanged is intimately connected to the question of how we view aqueous structures. That the dipole reversing interchange motion occurs at a frequency near that previously observed for $(HF)_2$ is amusing[10]. There are a number of H_2O interchange motions that occur in the water dimer. Their complete analysis in the $(H_2O)_2$ and $(D_2O)_2$ system is likely to be difficult. It appears useful, therefore, to point to a simpler system in which H_2O displays proton interchange. Peterson and Yaron[11] have completed a study of the rotational spectrum of H_2O-CO. The structure is hydrogen bonded but in contrast to $(H_2O)_2$ the hydrogen bond angle O-H--C has an equilibrium non-linearity $18°$. The barrier to proton interchange is high, estimated from hyperfine structure splittings in the radiofrequency Stark spectrum of CO-D_2O to be $V_2 = 1.8$ kcal/mole. The barrier is high enough so that no splittings of rotational transitions are observed in the microwave spectrum of CO-H_2O. It is reasonable to assume a similar high barrier exists for the related proton interchange in $(H_2O)_2$. The reported work of Dyke provides the essential information for unravelling the important internal motions of the water dimer.

The importance of rotational spectroscopy in establishing a detailed structural data base for molecular complexes is well documented at this conference. The analysis of zero point oscillations and of tunneling processes and paths in these complexes remains an active frontier topic. It is convenient to use this occassion to point out that the tunneling motions in low symmetry situations is presently more poorly understood than in the higher symmetry, threefold barrier, problem. There are several facets to this problem. The theory for the general asymmetric two fold barrier problem is not really well developed in a convenient computational form. Moreover, in the absence of clear analogs of highly directed chemical bonds, it is generally unclear what is the geometric path to be considered. This problem, so closely tied to the general vibrational problem, will perhaps become clarified as reliable theoretical potential energy surfaces are developed for the wide variety of systems now considered.

The dramatic increase in infrared spectroscopy of complexes is likely to provide quantitative, reliable answers to a variety of traditional, fundamental questions. The inert gas-hydrogen system, which in the hands of Harry Welsh and his students (primarily Bob McKellar) provided the beginning of van der Waals molecule spectroscopy, continues to provide the model for the very isotropic atom-diatom intermolecular potential. These systems show that the intermolecular potential does indeed depend upon the H_2 bond length. The new results presented here for the overtone spectrum of Ar-H_2 are promising both for providing a new data set which validates predictions and as an encouragement for the experimental exploration of overtone spectra of molecular complexes generally[12]. In this respect the increasingly extensive study of HCN-HF by John Bevan[13] is providing a most complete "intermolecular potential." The coupling of this very extensive experimental study to the accurate electron structure calculations of Botschwina[14] provides a nice example of the interplay of theory and experiment. It is difficult not to carried away with enthusiasm by the manifold results of vibrational spectroscopy presented at this conference. The use of the Stark effect by Roger Miller[15] in his study of the vibrational spectrum of ArHF to establish the dramatic change in average angle makes those of us who routinely use hydride/deuteride structural parameter ratios ignoring the likely dependence of intermolecular potentials upon valence coordinates extremely uncomfortable. (This problem has been discussed by Bob Leroy). The ability to resolve hyperfine structure, including of course Stark effects, in rotational spectroscopy has been of singular importance in establishing angular orientations of binding partners in molecular complexes. Stark effect measurements in vibrational spectra are likely to be of great importance, since it appears unlikely that the typically 10 kHz to 10 Mhz hyperfine structure will be observable in predissociating states. (The demonstration by Miller of the long predissociation time in the inert gas-hydrogen halide excited valence vibrations illustrates the lack of generality of the remark.)

There are many aspects of vibrational spectroscopy of molecular complexes that merit attention, especially when contrasted to pure rotational spectroscopy. The ability to observe the totality of a particular oscillator essentially independent of its bonding to substrates is a clear difference and most likely a great advantage of vibrational spectroscopy over both pure rotational and even electronic fluorescence detected absorption. The question of species distribution, including isomeric forms, continues to be of central concern. The high resolution absorption measurements of David Nesbitt[16] on HF complexes, the low spectral resolution but highly complete with respect to isotopic species studies of Lisy[17] do display directly the ease of disappearance of monomeric and dimeric forms of HF in expansion. It is clear that pure rotational spectroscopy suffers to some extent for seeing trees rather than the forest. While it is difficult to summarize the experiences of all observers in this area, from our own it is usual to try as rapidly as possible to establish a model that fits observation. This is always a quite frequency limited range. Thus, it is noteworthy that isomeric forms have not, as far as the writer is aware, been observed in pure rotational spectroscopy on jets. While at first sight it might seem that this is an intrinsic weakness of low frequency spectroscopy, it seems essential to point out that the manifold consequences of nonrigidity that occur in weakly bonded systems have been understood primarily through analysis of rotational spectra. As an example of this we use the ammonia dimer. David Nelson[18] has recently provided an analysis of this system in terms of the permutation inversion symmetry classifications delineated so clearly by Bunker[19]. A brief physical picture of this system is obtained by noting that NH_3 upon cooling in a jet expansion has two low temperature forms, the proton spin state $I=3/2$ with rotational state $K=J=1$ and the other $I=1/2$, $J=0$. The resultant dimer $(NH_3)_2$ then has a number of distinct forms in view of the observations that the NH_3 units are inequivalent in the dimer. These forms are then given using the labels of 1) proton nuclear spin, 2) angle of the NH_3 C_3 axis with respect to the a inertial axis (the measured angles are 48° and 65°) for each subunit:

$$
\begin{array}{llll}
1.a & (1/2,48,1) & (1/2,65,r) \\
 b & (1/2,65,1) & (1/2,48,r) \\
\\
2.a & (3/2,48,1) & (3/2,65,r) \\
 b & (3/2,65,1) & (3/2,48,r) \\
\\
3.a & (3/2,48,1) & (1/2,65,r) \\
 b & (1/2,65,1) & (3/2,48,r) \\
\\
4.a & (1/2,48,1) & (3/2,65,r) \\
 b & (3/2,65,1) & (1/2,48,r).
\end{array}
$$

These four sets of isoenergetic partners are zeroeth order descriptions of the lowest energy vibrational(?) states of NH_3. The electric dipole moment component μ_a is essentially determined by the angles made by the C_3 axes of the NH_3 pairs. The isoenergetic partners 1a and 1b can facilely interconvert as in $(HF)_2$ or as in the states of $(H_2O)_2$ discussed presently by Dyke. The same is true for the partners 2a and 2b where again the proton spin state is the same in both monomeric units. For the isoenergetic states 3a and 3b interchange or tunneling is not likely since it involves a change in nuclear spin during the motion. This is clearly also the case for the partners 4a and 4b. These states are the rigid rotor states observed in the pure rotational spectrum of $(NH_3)_2$ by Fraser and Nelson[20]. This type of spectrum was seen by Dyke and Muenter[9] in their classic work on the $(H_2O)_2$ and $(D_2O)_2$ pure rotational spectrum. There have been few high resolution vibrational spectra of dimeric species (clearly here we mean dimer in the traditional sense rather than the general binary complex). The work of Alan Pine and coworkers Walt Lafferty and Brian Howard[21] on the hydrogen fluoride dimer is clearly an important example to be discussed below. The infrared spectrum of $(NH_3)_2$ dimer was examined by Charo, Nelson, and Fraser[22] using microwave double resoncence detection of line tuneable CO_2 laser radiation. Only those vibrational (?) states seen in the pure rotational spectrum (those that we have labelled above 3 and 4) were studied. A splitting of about 5 cm^{-1} is observed between states 3 and 4.   The spectrum is best explained presently as the ν_2 vibration of one angle NH_3 but it is not known whether it belongs to the 48° or the 65° unit.   Meerts and coworkers[23] assign a second transition near 1010 cm^{-1} to the other tilt.   We interject these qualitative remarks to indicate the liklyhood of considerable complexity in the spectra of symmetrical dimers in which several states are present in the cooled monomer.   The spectral complexity due to inequivalence and nonrigidity might well not be accounted for in elementary analyses.

The dynamics of vibrational levels of van der Waals molecules lying above the dissociation limit has been a rich exciting area.   The classic work started by Levy, Wharton and Smalley[24] on He-I$_2$ B state is well known.   This certainly has addressed major questions such as time scales for dissociation and propensity rules for vibrational changes. The extension of these questions to the fully polyatomic many valence mode complex is an entire area of research.   The use of vibrational spectroscopy allows these questions to be answered for the ground electronic state potential energy surface. This is likely to be useful since in general it is the ground potential energy surface that attracts most physical interest.   There is generally concern in dynamics on excited electronic states surfaces whether non-adiabatic interactions with lower surfaces occur.   These doubts decrease the crispness of the results from fluorescence studies.

The (HF)$_2$ results of Roger Miller[15] confirm and quantitatively extend the early work of Pine on providing a nice non-statistical time scale for dissociation dynamics.   That there exists a factor of 25 between the dissociation time induced by ν_1 excitation and that of ν_2, would appear to argue strongly against a picture of the universality of strong intermode coupling, at least at this level of excitation energy. The extensive results of John Bevan on the HCN-HF system show a qualitatively similar behaviour in that the linewidth of the HF stretching vibration (ν_1) is broader than that of the CN stretching vibration (ν_3) or the CH stretching vibration (ν_2).   Of the three, ν_3 is closest to the dissociation energy of the complex and might naively be expected on the basis of energy (or momentum) gap law arguments to couple most effectively with states of the dissociation continuum. That this is not the case is further evidence that simple, unmodified Franck-Condon type projections are not adequate and that the cross-modification of the intramolecular and intermolecular potential must be explicitly taken into account.   A considerable body of detailed study does exist for the weakly interacting ArH$_2$ system; relatively few detailed studies exist for the more strongly bound complexes.   The question of what time scales and what linewidths should be expexted for these essentially predissociative processes appears to remain a most active current topic.

This recitation ends on perhaps the most central concern. This conference has been effectively dedicated to pair interactions. Furthermore, most of the spectroscopic studies are concerned with the geometric region near the (intermolecular) potential energy minimum. The question of how to utilize these studies for understanding condensed phases is a broad and poorly defined question. It appears that a complete (at least in angles) intermolecular potential is required. There exist relatively few direct spectroscopic studies of the relevant soft modes even for pair interactions. It is therefore likely that models of bonding are required for the exploration of condensed phases. Thus, bonding models capable of extension to the many molecule system are of considerable importance. The work discussed by Dykstra[25], of the extensions of electrostatic modelling of Buckingham and Fowler[26] clearly is an example of this. It is of importance to test and understand the adequacy and shortcomings of these models. While this testing can be done purely computationally by comparison of increasingly refined electron structure calculations there will always be amongst conservative spectroscopists the desire to see direct comparison with observation. The comparison to spectroscopic observations here is, unfortunately, not elementary. Efforts such as those of Jeremy Hutson in developing effective codes for treating the multicoordinate large amplitude system are essential to effect this comparison.

This conference surely illustrates the great progress under way for the understanding of intermolecular forces in polyatomic systems. The wide variety of techniques that have successfully been applied are encouraging in demonstrating the importance of attempting new methods on what might appear to be difficult systems. That rotational spectroscopy has provided a good structural basis is established. This conference marks a nice turning point toward high resolution vibrational (and of course electronic) spectroscopy which has the promise of moving our knowledge of intermolecular forces to a new level of understanding.

REFERENCES

1.   J. S. Muenter, These Proceedings, pp. 3-22.

2.   D. H. Levy, These Proceedings, pp. 231-250; Adv. Chem. Phys. $\underline{47}$, 323 (1981).

3.   K. C. Janda, L. S. Bernstein, J. M. Steed, S. E. Novick, W. Klemperer, Journal of the American Chemical Society, $\underline{100}$, 8074 (1978).

4.   P. C. Tellinghuisen, J. Tellinghuisen, J. A. Coxon, J. E. Velazco, and D. W. Setser, J. Chem. Phys. $\underline{60}$, 5187 (1978).

5.   A. Legon, These Proceedings, pp. 23-42.

6.   J. D. Dunitz, X-Ray Analysis and the Structure of Organic
        Molecules, Cornell University Press, 1979.

7.   For a review of these systems see K. I. Peterson,   G. T.
        Fraser, D. D. Nelson, Jr., and W. Klemperer, "Intermolecular
        Interactions Involving First Row Hydrides: Spectroscopic
        Studies of Complexes of HF, H_2O, NH_3, and HCN" in Comparison of
        Ab Initio Quantum Chemistry with Experiment: State of the
        Art, R. J. Bartlett, Editor (D. Reidel Publishing Co.,
        1985).

8.   E. J. Campbell, L. W. Buxton, A. C. Legon, J. Chem. Phys.
        78, 3483, (1983); K. R. Leopold, G. T. Fraser, F. J. Lin, D.
        D. Nelson and W. Klemperer, J. Chem. Phys. 81, 4922 (1984).

9.   T. R. Dyke and J. S. Muenter, J. Chem. Phys. 60, 2929
        (1974).

10.   T. R. Dyke, These Proceedings, pp. 43-56.

11.   K. I. Peterson, D. Yaron and W. Klemperer, (to be
        submitted to J. Chem. Phys.).

12.   A. R. W. McKellar, These Proceedings, pp. 141-148.

13.   J. W. Bevan, These Proceedings, pp. 149-170.

14.   P. Botschwina, These Proceedings, pp. 181-190.

15.   R. Miller, These Proceedings, pp. 131-140.

16.   D. J. Nesbitt, These Proceedings, pp. 107-130.

17.   J. M. Lisy, K. D. Kolenbrander, and D. W. Michael,
        These Proceedings, pp. 171-180.

18.   D. D. Nelson and W. Klemperer (submitted to J. Chem.
        Phys.).

19.   P. R. Bunker, Molecular Symmetry and Spectroscopy,
        (Academic Press, 1979).

20.   D. D. Nelson, G. T. Fraser, and W. Klemperer, J. Chem.
        Phys. 83, 6201 (1985).

21.   A. S. Pine and W. J. Lafferty, J. Chem. Phys. 78, 2154
        (1983); A. S. Pine, W. J. Lafferty, and B. J. Howard,
        J. Chem. Phys. 81, 2939 (1984).

22. G. T. Fraser, D. D. Nelson, A. C. Charo, and W. Klemperer, J. Chem. Phys. 82, 2535 (1985).

23. M. Snels, R. Fantoni, R. Snaders, W. L. Meerts, Chem. Phys. (to be published).

24. R. E. Smalley, L. Wharton, and D. H. Levy, Accts. Chem. Res. 10, 139 (1977).

25. C. E. Dykstra and S. Liu, These Proceedings, pp. 321-338.

26. A. D. Buckingham and P. W. Fowler, Can. J. Chem. 63, 2018, 1985; J. Chem. Phys. 79, 6426 (1983).

PART II

DYNAMICS

INTERPRETATION OF LINEWIDTHS IN THE INFRARED PHOTODISSOCIATION SPECTRA
OF VAN DER WAALS MOLECULES

W. Ronald Gentry
Chemical Dynamics Laboratory
Department of Chemistry
University of Minnesota
207 Pleasant Street, S.E.
Minneapolis, Minnesota 55455
U.S.A.

ABSTRACT. The interpretation of lifetime-broadened linewidths for
bound-free vibrational transitions of polyatomic van der Waals (vdW)
molecules has been a subject of some controversy. One point of view
is that these linewidths correspond to the vdW molecule vibrational
predissociation rates, and the other is that they correspond to the
rates of intramolecular vibrational energy redistribution (IVR) within
the undissociated complex. Measurements of the vdW molecule population
decay in real time could, in principle, distinguish between these two
possibilities, but so far there is no single system and transition
for which both the spectroscopic linewidth and time-domain kinetics
have been accurately determined. The issue may have been resolved,
however, by recent observations of narrow-line structure within the
broad-line absorption previously observed for ν_7 excitation of ethylene
dimers. It appears that a consistent interpretation of all these
results can be made by taking the $10 cm^{-1}$ width of the overall absorption profile to correspond to the IVR lifetime of 0.5ps, while the
< 10MHz widths of the rotationally-resolved structure correspond to
the predissociation lifetimes of $\geq$ 10ns.

1. INTRODUCTION

The infrared absorption spectrum for a weakly bound van der Waals (vdW)
molecule may be measured, for photon energies above the vdW dissociation
limit, by observing the depletion of the vdW molecule population as a
function of frequency.[1-3] Such measurements on a large number of polyatomic vdW species have yielded approximately Lorentzian and apparently
homogeneous lines with widths of 1 to $20 cm^{-1}$, corresponding to lifetimes of the initially prepared excited state in the picosecond regime.
A few measurements, on different systems or different transitions, have
yielded much narrower linewidths, corresponding to lifetimes of nano-
seconds or more.[3] A point of some controversy in the interpretation of
these data has been whether the uncertainty-principle lifetime derived
from the homogeneous linewidth in such a spectrum corresponds to the
lifetime for vibrational predissociation of the vdW molecule, or to the

467

A. Weber (ed.), Structure and Dynamics of Weakly Bound Molecular Complexes, 467–475.

lifetime for intramolecular vibrational energy redistribution (IVR) within the undissociated vdW molecule.[1] Since the homogeneous line-width gives only the rate of decay of the coherent quantum state produced by absorption of the infrared photon, one cannot learn from such a measurement alone whether the products of that decay are the separated molecular constituents of the vdW molecule or merely a set of delocalized and dephased "bath" states of the undissociated vdW molecule. As was pointed out in a recent review,[1] one may think of this distinction in terms of the six coordinates which express the relative positions and orientations of two nonlinear molecules within a vdW dimer. Five of these correspond to low-frequency librations or internal rotations, and one to the radial separation of the two molecules. One can imagine circumstances under which coupling of a monomer vibration to the five rotational/librational coordinates might be much faster than coupling to the separation coordinate. In fact, measurements of the product energy distributions, when available, show a general tendency for excess energy to appear in rotations of the dissociation products rather than in translation.[1]

An unambiguous distinction between IVR and dissociation lifetimes can be made by studying the infrared photodissociation process in the time domain, rather than in the frequency domain. Several pump-probe experiments have been carried out for vdW photodissociations, in either the ground or an excited electronic state.[4-7] However, at the time of this writing there is no experiment known to the author which gives a direct measurement of the vdW population decay in real time for the same vdW system and state for which a homogeneous absorption line-width is also known. This is surely a temporary situation, since there are several groups presently pursuing such data.

In the meantime, there have appeared some additional spectroscopic data from very high-resolution experiments by Snels et al[8] at the University of Nijmegen and by Watts et al at the Australian National University[9] which, when combined with earlier results, may be sufficient to resolve the matter, at least for the ethylene dimer system $(C_2H_4)_2$. Since 1978 when we found that ethylene dimers were dissociated with every single line of a pulsed CO_2 laser over the frequency range 900–1100cm^{-1}, this system has become an extremely popular one for infrared photodissociation experiments. The results from many laboratories have been recently reviewed,[1] and can be briefly summarized for our purposes here as follows.

With pulsed or cw multimode CO_2 lasers, the photodissociation spectrum in the 950cm^{-1} region of the C_2H_4 ν_7 vibration consists of a single, broad, approximately Lorentzian line having a width of 10cm^{-1} or more. In experiments with mass spectrometer detectors, which monitor the disappearance of the dimer signal, it was demonstrated that (1) a fraction close to unity of the dimers can be dissociated with a single CO_2 laser line having an effective bandwidth of about 10^{-2}cm^{-1}, and (2) no hole-burning is observed when one laser is used to dissociate a large fraction of the dimer population, and the spectrum of the remaining dimers is then measured with a second laser.[10] Both experiments demonstrate the population of ground-state dimers to be essentially homogeneous in a cold supersonic beam (translational temperature

of order 1K). Subsequent experiments by Buck et al,[11] who used
crossed-beam scattering to separate $(C_2H_4)_2$ from other species $(C_2H_4)_n$
kinematically, and thus to eliminate the concern that larger clusters
might contribute to the dimer ion signal in the mass spectrum, confirmed the single-beam spectra for cold dimers and also demonstrated
that additional broadening and structure appears in the spectrum when
the $(C_2H_4)_2$ species are collisionally excited.

In contrast to the earlier results obtained with multimode lasers,
the experiments at Nijmegen and ANL revealed structure in the $(C_2H_4)_2$
photodissociation spectrum around $950 cm^{-1}$, having instrument-limited
linewidths of about 10MHz, or $3 \times 10^{-4} cm^{-1}$. These data were obtained
with pulsed single-mode CO_2 waveguide lasers and bolometer detectors.
The differences in the two types of experiment are significant, the
most important being that the low-resolution mass spectrometer experiments are more sensitive to broad features of the spectrum and the
high-resolution bolometer experiments are more sensitive to fine structure. While experiments with mass spectrometer detectors must incor-
porate careful studies of the dimer ion signal as a function of expansion conditions in order to verify that the signal comes from the
neutral dimer, experiments with bolometer detectors must either identify the dimer by means of an assignable spectrum or must include aux-
iliary tests to rule out not only contributions from higher clusters
but also vdW molecules which might be formed with impurities or carrier
gas molecules.[12]

In the case of $(C_2H_4)_2$ at least, the tests necessary to attribute
both the broad-line and narrow-line spectra to the ethylene dimer
appear to be definitive. We are therefore faced with the question of
whether a dynamical mechanism can be found which is consistent with the
entire set of observations. It is the thesis of this article that a
model which incorporates rapid IVR followed by slower dissociation is
such a mechanism. The remainder of the article deals with the spectral
consequences of that model. A briefer discussion of the same mechanism was presented previously.[1]

2.   A MODEL FOR SIMULTANEOUS IVR AND DISSOCIATION OF vdW MOLECULES

The simplest model which incorporates the essential qualitative features of the proposed dynamical mechanism is shown in Fig. 1. Here
$|g\rangle$ is the ground state of the system and $|s\rangle$ is a zeroth-order vibrationally excited state carrying oscillator strength for a single-
quantum transition from $|g\rangle$. For a nonsymmetric vdW dimer, e.g.
$C_2H_4 \cdot C_2D_4$, $|s\rangle$ will generally be a slightly perturbed monomer vibra-
tion, and for a symmetric vdW dimer $|s\rangle$ will exhibit the usual pertur-
bation splitting and symmetry selection rules. Nothing important for
the present discussion is lost if we restrict our attention to the
simpler nonsymmetric case. The excited state $|s\rangle$ is coupled to each of
a set of "bath" states $|\ell\rangle$ which have no oscillator strength connecting
them with $|g\rangle$, except through $|s\rangle$. The states $|\ell\rangle$ in this scheme
represent vibrations in the six vdW coordinates (or five, if we factor
out the radial separation coordinate), which are mixed with the

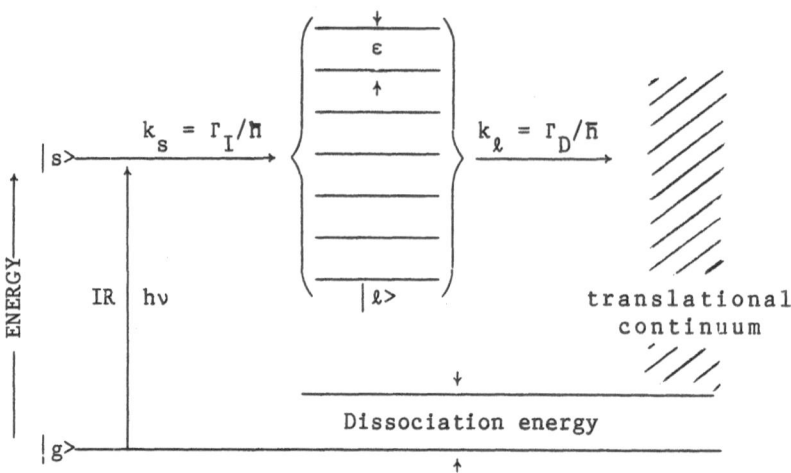

Figure 1. Dynamical coupling scheme.

zeroth-order state $|s\rangle$. In the absence of other couplings, the $|\ell\rangle$
states would be easily recognized in the excitation spectrum as "combination bands" of the low-frequency vdW modes with the much higher-
frequency mode in which $|s\rangle$ is the first excited state. Each mixed
state of $|s\rangle$ and $|\ell\rangle$ is, however, also coupled to the translational
continuum, since $|s\rangle$ lies above the vdW dissociation energy.

    An accurate description of the vdW vibrations and their couplings
to $|s\rangle$ is of course an extremely complicated matter, which has so far
been attempted only for systems much simpler than ethylene dimers.[2,3]
The natural coordinates for describing vdW vibrations will usually
resemble monomer rotations, and the potentials in those coordinates, at
the excitation energies of greatest interest here, will generally be
very anharmonic, or even periodic. The qualitative features will be
clear, however, if we make the simplifying assumption that the states
$|\ell\rangle$ are evenly separated in energy, with spacing ε, and that their
coupling to $|s\rangle$ is characterized by a constant matrix element V. This
model is essentially the same as that used by Bixon and Jortner in 1968
to describe radiationless transitions in polyatomic molecules,[13] and
expounded more recently by Stannard and Gelbart,[14] and others. The
difference from these earlier models of IVR within a covalently bonded
molecule is just that, in the case of vdW molecule predissociation, the
states $|\ell\rangle$ couple to a true translational continuum rather than to a
vibrational quasicontinuum. A characteristically lucid set of notes
incorporating this feature has been written by W. H. Miller and circu-
lated privately.[15] The results are quoted below.

    The complex hamiltonian matrix coupling states $|s\rangle$ and $|\ell\rangle$ gives
rise to the time-dependent survival probability for the state $|s\rangle$

$$P_{s \leftarrow s}(t) = \left| \langle s | e^{-iHt/\hbar} | s \rangle \right|^2 \tag{1}$$

and to the survival probability of the vdW molecule

$$P_M(t) = P_{s \leftarrow s}(t) + \sum_\ell P_{\ell \leftarrow s}(t) \quad . \tag{2}$$

Setting the zero of the energy scale to coincide with that of the state $|s>$, and specifying all energies E and energy widths in units of the level spacing ϵ, gives the absorption spectrum

$$I(E) = |<s|\mu|g>|^2 \pi^{-1} \frac{\Gamma/2}{(E-\Delta)^2 + (\Gamma/2)^2} \quad , \tag{3}$$

where

$$\Gamma = \Gamma_I \frac{A(1+a^2)}{A^2 + a^2} \tag{4}$$

$$\Delta = \frac{\Gamma_I}{2} \frac{a(1-A^2)}{A^2 + a^2} \tag{5}$$

$$a = \tan(\pi E) \tag{6}$$

$$A = \tanh(\frac{\pi}{2}\Gamma_D) \quad . \tag{7}$$

Here, Γ_D is the (assumed constant) decay width of each mixed state $|\ell>$, which is proportional to the state-specific decay rate coefficient $k_\ell = \Gamma_D/\hbar$. Γ_I is given by

$$\Gamma_I = \frac{2\pi V^2}{\epsilon} \tag{8}$$

and is proportional to the rate coefficient for decay of $|s>$, $k_s = \Gamma_I/\hbar$. Γ_I and k_s can therefore always be identified with the IVR process which destroys state $|s>$. The rate of vibrational predissociation, however, may be governed either by k_s or by k_ℓ, whichever is slower.

Now let us consider the various possibilities for the relative magnitudes of Γ_I, Γ_D and ϵ.

Case I: $\Gamma_D \gg \Gamma_I$. In this case, the IVR and dissociation rates are the same, and the observable dynamics correspond to direct coupling of $|s>$ to the continuum, i.e. the spectrum and decay of $|s>$ are independent of the levels $|\ell>$. This condition is expected for a low density of states $|\ell>$, or large spacing ϵ, corresponding to the small-molecule limit. At high resolution, the spectrum of a Case I system will show completely resolved rotational transitions. It may or may not display combination bands involving monomer and vdW modes, depending on the magnitude of V and the spacing ϵ.

Case II: $\Gamma_D \ll \Gamma_I$, $\Gamma_D \gg 1$. If k_ℓ is rate-limiting for dissociation, then IVR and dissociation are distinctly different dynamical phenomena. One generally associates rapid IVR with a large density of

states, so we will first consider the case where the width $\Gamma_D \gg 1$ (re-
calling that Γ_D in eqn. (3) is expressed in units of ϵ). This limit is
effectively reached for $\Gamma_D \gtrsim 2$. We have in this limit $T \to 1$, $\Gamma \to \Gamma_I$ and $\Delta \to 0$.
The resulting spectrum is Lorentzian, having a width Γ_I which is
determined by the IVR rate, not the dissociation rate. Under these
conditions the states $|\ell\rangle$ are unresolvable no matter how high the
spectroscopic resolution might be. For the survival probabilities one
obtains

$$P_{s \leftarrow s}(t) = e^{-\Gamma_I t/\hbar} \tag{9}$$

$$P_M(t) = \frac{\Gamma_I e^{-\Gamma_D t/\hbar} - \Gamma_D e^{-\Gamma_I t/\hbar}}{\Gamma_I - \Gamma_D} \to e^{-\Gamma_D t/\hbar} \tag{10}$$

The dissociation rate $k_\ell = \Gamma_D/\hbar$ is therefore accessible in the time
domain through a measurement of $P_M(t)$, even though it is not accessible
(even in principle) from the spectrum $I(E)$.

Case III: $\Gamma_D \ll \Gamma_I$, $\Gamma_D \ll 1$. If, on the other hand, the energy width
of the states $|\ell\rangle$ is less than their spacing, then the spectrum from
eqn. (3) will be sensitive to both Γ_I and Γ_D. $I(E)$ then has the form
of a set of narrow peaks with width Γ_D, all contained within a Lorentzian envelope having an overall width Γ_I. Under these conditions both
k_s and k_ℓ may be extracted from a measured spectrum $I(E)$. In the time
domain, $P_{s \leftarrow s}$ shows a rapid and approximately exponential decay with the
time constant $\Gamma_I/\hbar$, followed by damped "recurrences" having the spacing
$\epsilon/\hbar$. The recurrences taken together describe an exponential envelope
with the time constant $\Gamma_D/\hbar$.[14]

3. APPLICATION TO $(C_2H_4)_2$.

Although $(C_2H_4)_2$ is in many details more complicated spectroscopically
than the simple model just discussed, the data gathered to date appear
to be sufficient to identify it as a Case III species having a subpico-
second IVR lifetime and a dissociation lifetime of 10nsec or more. The
detailed complications which one must consider are (1) the uneven
splittings and coupling strengths of the $|\ell\rangle$ states, (2) possible power
broadening in the experimental spectra, and (3) inhomogeneous contributions. We consider these in turn.

3.1. Details of the $|\ell\rangle$ Spectrum

We do not at present have any quantitative method for estimating the
density of $|\ell\rangle$ states in the energy range of interest, but it is
reasonable to expect that individual states might display a range of
coupling strengths both to $|s\rangle$ and to the continuum, and therefore to

have spectral intensities and widths which vary. This variation will however not affect the broad IVR Lorentzian profile, provided that the states $|\ell\rangle$ are distributed with more-or-less equal density over the entire absorption band, and provided that the number of states falling within the effective bandwidth in the low-resolution experiments is sufficient to average out intensity differences between states.

The high-resolution results obtained with CO_2 waveguide lasers show snippets of spectra around each of the CO_2 laser lines. Each snippet shows several peaks, with average spacings of perhaps 30MHz, or $10^{-3}cm^{-1}$, with a reasonably uniform distribution of intensity. As mentioned previously, we do not have enough structural information to make a reliable estimate of the density of states $|\ell\rangle$ for $(C_2H_4)_2$, but we can at least àsk whether $10^3/cm^{-1}$ is a plausible number. The classical density of states for a collection of n oscillators of frequency ν is

$$N(E) = \frac{E^{n-1}}{(n-1)!(h\nu)^n} , \qquad (11)$$

where E is the total energy.[16] Taking n = 5 (i.e. leaving out the radial separation coordinate) and $E = 950cm^{-1}$, a frequency $\nu = 30cm^{-1}$ gives $N = 1400/cm^{-1}$. An average vdW mode frequency of $30cm^{-1}$ certainly seems plausible for $(C_2H_4)_2$.

Furthermore, if we use $\nu = 30cm^{-1}$ and calculate the density of states at $E = 3000cm^{-1}$, we obtain $N = 1.4 \times 10^5/cm^{-1}$. This much higher density of states plausibly accounts for the fact that Miller et al[3,17] did not observe any fine structure for $(C_2H_4)_2$ in the $3000cm^{-1}$ region even though their spectra were measured with an F-center laser having a bandwidth of about 1 MHz or $3 \times 10^{-5}cm^{-1}$.

3.2. Power Broadening

There are two explanations for the fact that early experiments with either pulsed or cw multimode lasers did not observe any fine structure in the broad IVR envelope. One is that many lines fall within the effective bandwidth of the laser, averaged over the data accumulation time. Another is that the individual lines are power-broadened enough to overlap with each other. Careful checks were done to ensure that the overall IVR linewidth was not determined by power broadening,[10] but that of course says nothing about the individual $|\ell\rangle$ states which were not observed in those experiments. Taking the transition moment to be that of C_2H_4 (0.188D), the power density required to broaden a line by $10^{-3}cm^{-1}$ is only $100w/cm^2$. The combination of power broadening and laser mode averaging was clearly enough to remove any fine structure from the results of the early experiments with both pulsed and cw lasers.

3.3. Inhomogeneous Contributions

In the sense in which we have consistently used the term,[1] the spectrum

represented by eqn. (3) is entirely homogeneous, since we have assumed
in the model that all transitions originate in the ground state $|s\rangle$.
The operational definition of homogeneity in the experimental spectrum
was that the entire population of vdW dimers could be dissociated
linearly with a single laser frequency, and that no hole-burning was
observed in the two-laser experiments. Neither the unresolvable fine
structure in Case II dynamics nor the resolvable fine structure in Case
III dynamics affects the interpretation of the overall homogeneous
linewidth so defined as being related to the uncertainty principle
lifetime for IVR.

At the same time, it must be recognized that with rotational state
resolution spectroscopically available, in the Nijmegen and ANL experiments, there will certainly be inhomogeneous contributions to the spectrum from excited rotational states in the supersonic beam, even at a
translational temperature less than 1K. It is not possible at this
point to say what fraction of the narrow lines observed in the high-
resolution spectra originate in rotationally excited states of the
ground vibrational state $|g\rangle$, but it seems likely that many do. How-
ever, each of these rotationally excited states $|g\rangle_j$ will connect to
its own manifold of states $|\ell\rangle_j$, distributed over approximately the
same range of energies Γ_I as those originating in the ground state.
Provided that the range of populated rotational energy levels is much
smaller than the width Γ_I, the IVR linewidth will not be affected
significantly by inhomogeneous contributions. Furthermore, as long as
the density of $|\ell\rangle_j$ levels for different populated j states is high
enough for them to be overlapped by power broadening, then the entire
population will behave homogeneously. Based on the fine structure
which has now been observed for $(C_2H_4)_2$, this was certainly the case in
the pulsed laser experiments in our laboratory.[10]

4.   SUMMARY AND CONCLUSIONS

A simple model for the infrared photodissociation of vdW molecules,
represented by Fig. 1 and eqn. (3), appears to be capable of reconcil-
ing qualitatively the low-resolution and recent high-resolution spectroscopic results for $(C_2H_4)_2$ in both the $1000cm^{-1}$ and $3000cm^{-1}$
regions. It is a consequence of the model that a broad homogeneous
linewidth in such an experiment can always be interpreted in terms of a
small uncertainty-principle lifetime for the initially excited state
$|s\rangle$ with respect to decay by IVR. When rotationally-resolved fine
structure is present, the intrinsic linewidth of a single peak is
determined by the dissociation rate for that particular state $|\ell\rangle_j$. In
cases where individual states $|\ell\rangle$ are not resolvable even in the limit
of perfect resolution, the implication is only that the dissociation
width Γ_D is greater than the $|\ell\rangle$ spacing ε, not necessarily that it is
equal to the overall width Γ_I. In those cases the dissociation rate
cannot be deduced from the absorption spectrum, though it may be determined from the population decay in real time.

It appears that Case I, Case II and Case III dynamics are all
represented in examples which can be found in the recent literature.

Case I dynamics and spectroscopy are typical of small vdW systems, many beautiful examples of which will be found elsewhere in this volume. $(C_2H_4)_2$ is apparently Case III at $1000cm^{-1}$, but Case II at $3000cm^{-1}$.

ACKNOWLEDGMENT

This research was supported by the National Science Foundation through grant number CHE-8205769 from the Chemical Physics Program.

REFERENCES

1. W. R. Gentry, ACS Symposium Series 263, 289 (1984). (REVIEW).
2. K. C. Janda, Adv. Chem. Phys. 60, 201 (1985); F. G. Celli and K. C. Janda, Chem. Rev. (REVIEWS).
3. R. E. Miller, J. Phys. Chem. 90, 3301 (1986). (REVIEW).
4. A. Mitchell, M. J. McAuliffe, C. F. Giese and W. R. Gentry, J. Chem. Phys. 83, 4271 (1985).
5. J. Knee, L. Khundakar and A. Zewail, J. Chem. Phys. 83, 1996 (1985).
6. M. P. Casassa, J. C. Stephenson and D. S. King, J. Chem. Phys. 85, 2333 (1986).
7. M. Heppener, A. G. M. Kunst, D. Bebelaar and R. P. H. Rettschnick, J. Chem. Phys. 83, 5341 (1985).
8. M. Snels, R. Fantoni, M. Zen, S. Stolte and J. Reuss, Chem. Phys. Lett. 124, 13 (1986).
9. R. O. Watts, private communication.
10. M. A. Hoffbauer, K. Liu, C. F. Giese and W. R. Gentry, J. Chem. Phys. 78, 5567 (1983).
11. F. Huisken, H. Meyer, C. Lauenstein, R. Sroka and U. Buck, J. Chem. Phys. 84, 1042 (1986); U. Buck, private communication.
12. R. D. Johnson, S. Burdenski, M. A. Hoffbauer, C. F. Giese and W. R. Gentry, J. Chem. Phys. 84, 2624 (1986).
13. M. Bixon and J. Jortner, J. Chem. Phys. 48, 715 (1968).
14. P. R. Stannard and W. M. Gelbart, J. Phys. Chem. 85, 3592 (1981).
15. W. H. Miller, private communication.
16. P. J. Robinson and K. A. Holbrook, Unimolecular Reactions (Wiley-Interscience, 1971).
17. G. Fischer, R. E. Miller, R. O. Watts, Chem. Phys. 80, 147 (1983).

INFRARED PHOTODISSOCIATION OF VAN DER WAALS COMPLEXES SELECTIVELY
PREPARED BY MOLECULAR BEAM SCATTERING

U. Buck, F. Huisken, Ch. Lauenstein, T. Pertsch,
and R. Sroka
Max-Planck-Institut für Strömungsforschung
Bunsenstraße 10
D3400 Göttingen
Fed. Rep. Germany

ABSTRACT. Small ethylene clusters $(C_2H_4)_n$ are dissociated upon
absorption of a CO_2-laser photon by exciting the ν_7-mode of the
monomer. The clusters are selected according to their size by scattering them from a helium beam. Since an appreciable amount of energy
is transferred to the cluster during the collision, the laser photons
interact with internally cold or hot clusters depending on whether
the interaction is before or after the collision center. Measurements
of the frequency and fluence dependences of the photodissociation
cross section are carried out for cold and hot clusters from $n=2$ to
$n=6$. The spectra of the cold clusters show the same shape and linewidth for all cluster sizes investigated, while those of the hot
clusters are characterized by large variations in shape and linewidth. The very structured dimer spectrum is explained by the excitation of hot bands.

1. INTRODUCTION

The infrared photodissociation of weakly bound van der Waals clusters
has attracted much interest in recent years and a variety of systems
has been studied [1-3]. Since the photon energy is usually larger
than the binding energy of the van der Waals bond, the complex predissociates. Typically, this process is measured by monitoring the
depletion of the cluster beam as a function of the infrared laser
frequency via a liquid He bolometer [4] or mass spectrometer [5]. The
interpretation of the frequency bandwidth as predissociation lifetime
has been subject of a lively debate. For many polyatomic molecules
featureless and broad spectra have been observed, which correspond to
lifetimes of the order of picoseconds [1-3]. These results have led
to the suggestion that the observed linewidth is due to a fast intramolecular relaxation and is not caused by the predissociation rate
[2]. On the other hand, the broad widths could also be the result of
a complicated rotational band structure, which was indeed observed in

477

A. Weber (ed.), Structure and Dynamics of Weakly Bound Molecular Complexes, 477–487.

some high resolution experiments [6]. In addition, there is the possibility that the spectra are composed of various contributions from higher clusters, which are usually present if the clusters are generated by an adiabatic expansion at not too low pressures. This can certainly not be measured, if a bolometer is used as detector. But even with a mass spectrometer detector, heavy fragmentation, which occurs when the clusters are ionized by electron bombardment, can obscure the neutral cluster distribution in the beam [7]. In the present contribution we essentially address this last question.

Recently we have introduced a new technique to selectively pre- pare small neutral clusters by taking advantage of the kinematic constraints in the scattering process of clusters with rare gas atoms [7,8]. Depending on their masses, different cluster species are scattered into different angular ranges. In this way we are able to select clusters of specific size by adjusting both the scattering angle and the tuning of the mass spectrometer.

As an example we have chosen pure C_2H_4-clusters, which have been thoroughly investigated by several groups. In the spectral range of the CO_2 laser around 1000 cm^{-1} dimer linewidths between 12 cm^{-1} and 17 cm^{-1} have been reported [9-11] using pulsed and continuous wave lasers. These results were attributed to homogeneous broadening corresponding to lifetimes between 0.3 and 0.5 ps. Very recently structured spectra have been observed in high resolution experiments showing linewidths in the order of some MHz [12,13]. In the 3000 cm^{-1} spectral range only unstructured spectra have been measured using colour-center lasers [14,6] or optical parametric oscillators [15]. This result is even valid for experiments with very narrow bandwidth lasers [6]. In the time domain only one experiment has been carried out giving an upper limit of the lifetime of 10 ns for the dimer upon excitation of the ν_7-mode [16].

In the present experiments the ethylene clusters are prepared according to their size by scattering them from a helium atom beam. Two different experimental arrangements are used. In the first series of experiments the clusters are scattered before the interaction with a pulsed CO_2 laser takes place, in which the ν_7-mode of C_2H_4 is ex- cited. Preliminary results for the dimer have already been published [17]. The measurements for clusters up to n=6 exhibit more or less pronounced structures [18]. Due to the scattering process with he- lium, a certain amount of translational energy is transferred into internal energy of the cluster (e.g. 30 meV for the dimer [19]) so that the photodissociation takes place for internally exited or "hot" clusters. In contrast, the second series of experiments is carried out by dissociating the clusters with the laser before separating them by the scattering process [20]. In this case the clusters do not carry much internal excitation. For these "cold" clusters no struc- ture in the dissociation spectra is observed.

We start with a short description of the experimental methods and present the results of the dissociation curves as a function of laser frequency and fluence. The implications of these results on the structure and lifetime of ethylene clusters are discussed.

2. EXPERIMENTAL

The experiments were carried out in crossed molecular beam machines
with continuously operated supersonic beam sources. For the detection
of the scattered clusters a quadrupole mass spectrometer with electron impact ionization is used. The experimental set up I, in which
the laser crosses the scattered beam after the scattering center and
in which hot clusters are photodissociated, is described in detail
elsewhere [21]. The sources are mounted on a rotatable platform,
while the detector unit which is operated under ultra-high-vacuum
conditions is fixed. The pulsed infrared laser radiation is generated
by a line tunable CO_2 laser (Lambda Physik, model EMG 201E). The
maximum repetition rate is 25 Hz. The laser beam is slightly focussed
onto the axis of the fixed detector by means of a cylindrical ZnSe
lens. The dimensions of the laser beam in the intersection zone are
35mm long and 3mm wide. The laser energy is measured by a power meter
and adjusted by an absorption cell filled with propylene.

The experimental set up II, in which the laser crosses the cluster beam before the scattering center in order to photodissociate
cold clusters is described in detail in Ref. [20]. In this machine
the two sources are fixed, while the detector unit can be moved
around the scattering center. The laser intersects the cluster beam
between the skimmer and the scattering center. Otherwise the conditions are similar to those of set up I.

The measurement of the angular and velocity distributions of the
scattered $(C_2H_4)_n$-beam from a He beam provides unambiguous and
detailed information on 1) the composition of the cluster beam [22],
2) the fragmentation of the $(C_2H_4)_n$ clusters by electron impact
ionisation [22], and 3) the collisional energy transfer [19]. The
kinematic diagram for the two intersecting beams is shown in Fig. 1.
It is constructed from the beam parameters obtained in the experimental set up I and listed in Table I. The limiting laboratory
angles for the scattering of monomers up to hexamers, are given by
23.3°, 11.7°, 7.8°, 5.9°, 4.7°, and 3.9°, respectively. At a stagnation pressure of 4 bar, the 10% mixture of C_2H_4 in He contains,

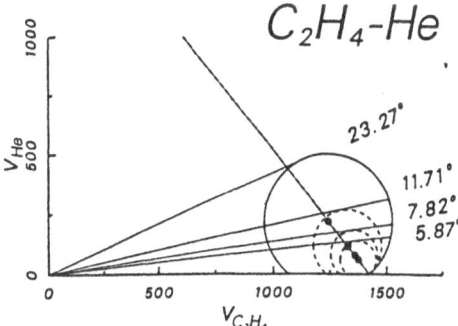

Figure 1. Kinematical diagram for $(C_2H_4)_n$ + He scattering using the
data of Table I. The circles correspond to elastically scattered
clusters of the size n=1 (solid), n=2, n=3, and n=4 (dashed).

TABLE   I. Beam parameters

Gas	10 % C_2H_4/He	He
nozzle diameter/μm	55	30
source temperature/K	310	310
source pressure/bar	4	40
peak velocity/ms^{-1}	1448	1823
speed ratio	26.7	62.1

aside from monomers, essentially trimers and tetramers. Even for
pressures as low as 1 bar and for mixtures of 1% C_2H_4 in He the ratio
of dimers to trimers was found to be less than 0.8. The fragmentation
turned out to be quite complicated. Except for the loss of single
H-atoms or CH_3-radicals, which is known from the study of ion-molecule reactions, the additional loss of one or more C_2H_4-units was
found. Thus mass m=41, which is usually considered to be a typical
dimer mass, is the main fragmentation channel of trimers. Finally,
from the measurements of the time-of-flight distributions, we
determine an average value of about 30 meV which is transferred to
the dimer by collision with He. This is about half of the binding
energy of $(C_2H_4)_2$ [23].

3. RESULTS

The photodissociation spectra are obtained by firing the laser and
measuring the fraction of the non-dissociated clusters with the
detector fixed at a certain scattering angle and the mass spectrometer tuned to a certain mass. According to what has been said in the
last section, the detector adjustment determines the size of the
clusters investigated. Therefore, at mass 41 amu and a scattering
angle of θ=9.5° only dissociation of dimers is detected, since monomers are excluded by the mass selection and higher clusters by the
choice of the deflection angle [17,18,20]. In this way dissociation
spectra are measured as a function of the laser frequency and the
laser fluence. The experimental conditions and results for some
selected cluster sizes are summarized in Table II. Except for the
dimer spectrum, more than one cluster size may contribute. Since one
cluster size is dominating, for some spectra only the most important
contribution is indicated. We will discuss the results separately
for internally excited (hot) and non-excited (cold) clusters.
     The results of the dissociation spectra for hot dimers, trimers,
pentamers, and hexamers are displayed in Fig. 2. Nearly all spectra
exhibit certain structures, which are most pronounced for the dimer
and are only marginally seen for hexamers. The overall maximum

TABLE II. Photodissociation parameters

Θ deg	m amu	fraction of clusters [%]					n	F mJcm^{-2}	P_{diss} %	ν_0 cm^{-1}	Γ cm^{-1}
		2	3	4	5	6					
hot clusters											
9.5	41	100	–	–	–	–	2	34.1	0.64	951.8	31.2
6.0	41	20	80	–	–	–	3	34.1	0.76	951.2	18.5
3.5	41	–	15	23	45	17	5	34.1	0.80	951.3	13.6
3.5	69	–	–	6	29	65	6	11.9	0.61	951.9	12.2
cold clusters											
9.5	41	100	–	–	–	–	2	46.9	0.53	952.9	10.4
6.6	56	96	4	–	–	–	3	26.8	0.83	953.4	11.7
5.0	56	–	23	77	–	–	4	26.8	0.88	953.7	12.2

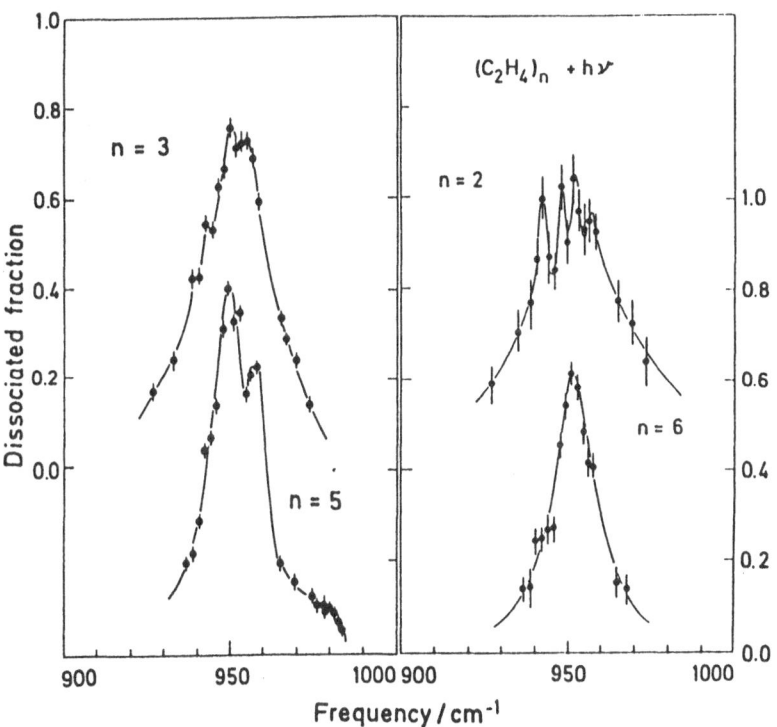

Figure 2. Measured frequency dependence of the photodissociation cross section for hot ethylene dimers, hexamers, trimers and penta-mers.

position is about the same for all curves, $\nu_0 = 952$ cm^{-1}. This corresponds to a blue shift of 3 cm^{-1} from the monomer absorption frequency of the ν_7-mode. The overall full width at half maximum (FWHM) of the curves decreases from $\Gamma = 31$ cm^{-1} for the dimer to $\Gamma = 12$ cm^{-1} for the hexamer. The most pronounced sub-structures for the dimer are at 942 cm^{-1} and for the pentamer at 958 cm^{-1}. The fluence dependences obtained in the maximum of the frequency curves are shown in Fig. 3. Here the undissociated or remaining fraction P_r is plotted with a logarithmic scale against the laser fluence. This quantity P_r can be related directly to the laser fluence by a simple Beer-type law [10]

$$P_r = 1 - P_{diss} = \exp[-\sigma(\nu) \cdot F] \tag{1}$$

where $\sigma(\nu)$ is the absorption cross section and F the laser fluence measured in photons per pulse and unit area. The results clearly show that for small laser fluences the exponential law is fulfilled, while for larger fluences the curve of the dimer and, less pronounced, also that of the trimer level off. As indicated by the slope, the absorption cross sections increase with increasing cluster size.

The results for the frequency dependence of the cold clusters are displayed in Fig. 4. Assuming the validity of (1) $\sigma(\nu)$ is plotted against the laser frequency. The solid lines are Lorentz distributions which are fitted to the data points. The dashed lines are the same Lorentzians but normalized to equal height. The comparison shows that, regardless of cluster size, the curves peak around $\nu_0 = 953$ cm^{-1} and have full widths at half maximum of $\Gamma = 12$ cm^{-1}. The absorption cross sections increase with increasing cluster size as one expects because of the increasing number of oscillators. Therefore the curves of the fluence dependence look similar to the results obtained for the hot clusters except that the absolute values of the cross sections are larger by about 50% for the cold clusters.

4. DISCUSSION

The most striking result of the present investigation is the large difference found in the frequency dependence of the dissociation spectra between hot and cold ethylene clusters if measured as a function of cluster size. Going from the dimer to the hexamer, for the cold clusters there is nearly no difference in the maximum position ($\nu_0 = 953$ cm^{-1}) and FWHM ($\Gamma = 12$ cm^{-1}). This has obviously something to do with the structure of these clusters. It is likely that the parallel shifted structure found for the dimer [23] is continued for the higher clusters, so that the excitation of the ν_7-mode, which is an out-of-plane bending mode, is not very much influenced when another ethylene molecule is added on top of the cluster. These results are in good agreement with all previous findings [9-11], where, aside from small deviations which might be due to

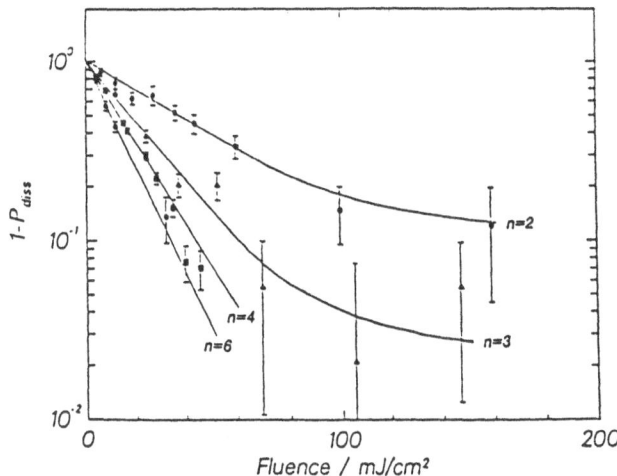

Figure 3. Measured fluence dependence of the photodissociation of hot ethylene clusters: o dimers, Δ trimers, □ tetramers, • mainly hexamers.

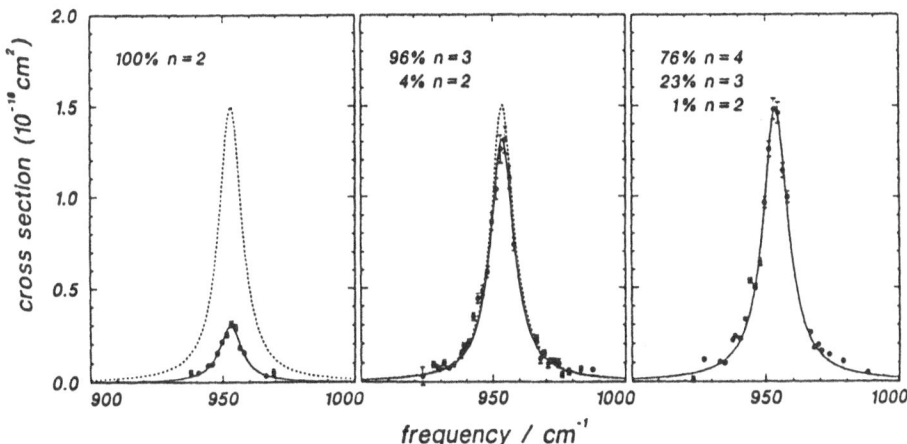

Figure 4. Measured frequency dependence of the photodissociation of cold ethylene clusters. The dashed lines are normalized to the same height.

orientational hole burning [24,25], a homogeneously broadened line
was observed regardless of the different beam compositions. There is,
however, a remarkable difference concerning the magnitude of the
absorption cross sections. Hoffbauer et al. [10] report that they can
dissociate 93.3% of the dimers at 40 mJ/cm^2, while we only get 70% of
the cold dimers dissociated at the same fluence. The comparison with
the results of Casassa et al. [9] shows even larger discrepancies
[20]. We conclude that in most of the earlier experiments contributions of higher clusters were present. This is also supported by
our scattering analysis of different ethylene mixtures reported in
section 3.

The frequency dependence of the hot clusters behaves quite differently. The dimer spectrum is quite broad and exhibits a rich
structure which gradually diminishes going to larger clusters. For
the hexamer the results of the cold clusters are reached. The differences are obviously due to the internal excitation by the collision with He, which effects most strongly the dimer. For the higher
clusters the energy transferred by collisions can be easier distributed among the various degrees of freedom of the internal motion so
that, finally, for the hexamer, the internal excitation does not
influence anymore the results.

The difference between hot and cold clusters are largest for the
dimer. The two spectra are compared directly in Fig. 5. Although some
of the features of the hot dimer spectrum require confirmation by
additional measurements, we see two distinct peaks shifted to the red
by 4 cm^{-1} and 10 cm^{-1}. The peaks are attributed to the excitation of
hot bands. Indeed, two additional peaks at 916 cm^{-1} and 986.4 cm^{-1}
were recently observed by Snels et al. [12]. They attributed these
peaks to the sum and difference bands of a van der Waals mode with
the ν_7 monomer vibration. In a recent calculation this mode was
correlated with combination bands involving bending modes of the
ethylene monomers about their C-C-axes [26]. What is more important
for the interpretation of the present results is the fact that Snels
et al. [12] also found a slightly negative anharmonicity of -1.7 cm^{-1}
when comparing the positions of the two side bands. This should give
rise to hot bands shifted to the red as it is observed in our spectrum. Using their numbers we calculate a red shift of 10 cm^{-1} if the
dimer is excited from an energy level corresponding to the half of
the dissociation energy. This precisely corresponds to the most
probably internal energy transfer in the collisions with He as
measured for the dimer. It is quite convincing that the observed
structure is due to the excitation of hot bands of the bending van
der Waals motion. Therefore the measured spectrum should give detailed information on the structure of the ethylene dimer.

The comparison of the absorption cross sections of hot and
cold clusters gives smaller values for the hot clusters. This can be
easily explained by the fact that, in the case of hot clusters,
more initial states are populated, while the number of oscillators is
the same. Therefore the overall broadening of the hot dimer
spectrum is inhomogeneous, but the single features with FWHM of about
3-5 cm^{-1} have to be considered as homogeneous. According to the

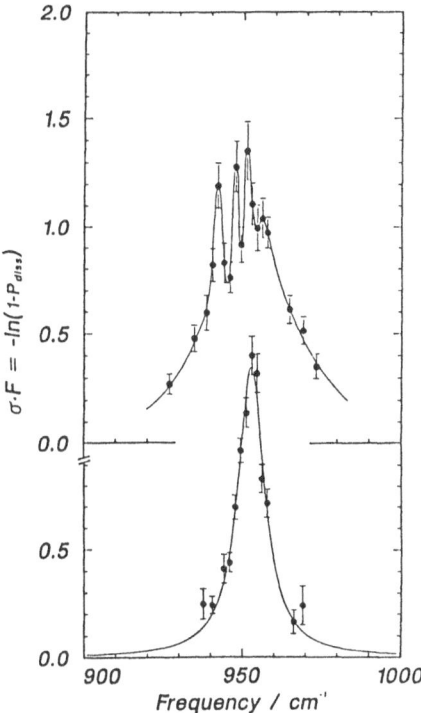

Figure 5. Comparison of the photodissociation spectrum of hot (upper part) and cold (lower part) ethylene dimers.

results of Fig. 3 90% of the dimers can be dissociated at fluences around 100 mJ/cm². If, on the other hand, the absorption band which we observe were entirely inhomogeneous because of unresolved rotational structure, a density of rotational states of at least 10^2/cm^{-1} would be required in order to achieve coincidence with one of the CO_2-laser lines which are about 2 cm^{-1} apart [27]. In addition, the fraction of the population which could be attenuated by the laser would be less than 10^{-2} and not 90%. Thus, in spite of a lot of new results on ethylene clusters, we are left with the problem of interpreting the broad lines (3-12 cm^{-1}) obtained in this experiment and many others carried out previously and the narrow lines (3-15 MHz) observed recently [12,13,28]. Since the vibrational predissociation of ethylene is expected and calculated to be a slow process due to the pure vibrational-rotational coupling [29], the lifetimes (6-10 ns) obtained from the high-resolution experiments might be interpreted as predissociation lifetimes [30]. In contrast, the fast times of the order of ps derived here and from other experiments might be due to strong intramolecular coupling to a set of van der Waals modes of the dimer, which are available upon excitation of the ν_7 vibrational mode.

5. SUMMARY

The present results demonstrate that it is possible to carry out
infrared photodissociation experiments on $(C_2H_4)_n$ clusters, which are
prepared according to their size by scattering them from a helium
beam, and measuring the spectra at different deflection angles and
masses. In addition, the degree of internal excitation can be varied
by crossing the laser before or after the scattering center. For the
cold clusters the measured absorption profiles show no dependence on
cluster size as far as shape and linewidth are concerned. The spectra
of the hot clusters have pronounced structures, which gradually
diminish going from the dimer to the hexamer.

REFERENCES

[1]   K.C. Janda, Adv.Chem.Phys. **60**, 201 (1985)
[2]   W.R. Gentry, Am.Chem.Soc.Symp.Ser. **263**, 289 (1984)
[3]   R.E. Miller, J.Phys.Chem. **90**, 3301 (1986)
[4]   T.E. Gough, R.E. Miller, and G. Scoles, J.Chem.Phys. **69**, 1588
      (1978)
[5]   M.A. Hoffbauer, W.R. Gentry, and C.F. Giese, in Laser Induced
      Processes in Molecules, edited by K. Kompa and S.D. Smith
      (Springer, Berlin, 1978) Vol. **6**, p. 252
[6]   G. Fischer, R.E. Miller, P.F. Vohralik, and R.O. Watts,
      J.Chem.Phys. **83**, 1471 (1985)
[7]   U. Buck and H. Meyer, Phys.Rev.Lett. **52**, 109 (1984)
[8]   U. Buck and H. Meyer, J.Chem.Phys. **84**, 4854 (1986)
[9]   M.P. Casassa, D.S. Bomse, and K.C. Janda, J.Chem.Phys. **74**, 5044
      (1981)
[10] M.A. Hoffbauer, K. Liu, C.F. Giese, and W.R. Gentry,
      J.Chem.Phys. **78**, 5567 (1983)
[11] J. Gereadts, Ph.D. Thesis, Katholieke University of Nijmegen
      (1983)
[12] M. Snels, R. Fantoni, M. Zen, S. Stolte, and J. Reuss,
      Chem.Phys.Lett. **124**, 1 (1986)
[13] K.G.H. Baldwin and R.O. Watts, Chem.Phys.Lett. **129**, 237 (1986)
[14] G. Fischer, R.E. Miller, and R.O. Watts, Chem.Phys. **80**, 147
      (1983)
[15] W.L. Liu, K. Kolenbrander, and J.M. Lisy, Chem.Phys.Lett. **112**,
      585 (1984)
[16] A. Mitchell, M.J. McAuliffe, C.F. Giese, and W.R. Gentry,
      J.Chem.Phys. **83**, 4271 (1985)
[17] F. Huisken, H. Meyer, Ch. Lauenstein, R. Sroka, and U. Buck,
      J.Chem.Phys. **84**, 1042 (1986)
[18] U. Buck, F. Huisken, Ch. Lauenstein, H. Meyer, and R. Sroka,
      J.Chem.Phys. (to be published)
[19] U. Buck, Ch. Lauenstein, H. Meyer, R. Sroka, and M. Tolle,
      Chem.Phys. (to be published)
[20] F. Huisken and T. Pertsch, J.Chem.Phys. (1986), accepted for
      publiation

[21] U. Buck, F. Huisken, J. Schleusener, and J. Schaefer,
      J.Chem.Phys. **72**, 1512 (1980)
[22] U. Buck, Ch. Lauenstein, and R. Sroka, J.Chem.Phys. (to be
      published)
[23] A. van der Avoird, P.E.S. Wormer, F. Mulder, and R.M. Berns, in
      Topics in Current Chemistry, **93**, 1 (1980)
[24] M. Snels, J. Gereadts, S. Stolte and J. Reuss, Chem.Phys. **94**, 1
      (1985)
[25] M.P. Casassa, C.M. Western, and K.C. Janda, J.Chem.Phys. **81**,
      4950 (1984)
[26] A.C. Peet, Chem.Phys.Lett. (to be published)
[27] R.D. Johnson, S. Burdenski, M.A. Hoffbauer, C.F. Giese, and W.R.
      Gentry, J.Chem.Phys. **84**, 2624 (1986)
[28] J. Reuss, this volume
[29] A.C. Peet, D.C. Clary, and J.M. Hutson, Chem.Phys.Lett. **125**, 477
      (1986)
[30] W.R. Gentry, this volume

IR DISSOCIATION OF WEAKLY BOUND VAN DER WAALS COMPLEXES: $(SF_6)_n$, $(SiF_4)_n$, $(SiH_4)_n$, $(C_2H_4)_n$, (n=2,3) IN THE 9-11 μm RANGE

M. Snels[+], R. Fantoni, and J. Reuss[+]
ENEA, Dip. TIB, Divisione Fisica Applicata, C.R.E. Frascati,
P.O. Box 65
00044 Frascati (Roma)
Italy

+ Physics Laboratory, University of Nijmegen
Toernooiveld,
6525 ED Nijmegen
The Netherlands

ABSTRACT. Small $(SF_6)_n$, $(SiF_4)_n$, $(SiH_4)_n$, $(C_2H_4)_n$ (n=2, 3) clusters have been produced in a molecular beam. SF_6 and SiF_4 dimer dissociation spectra were measured and spectral features due to clusters containing different S and Si isotopes could be resolved. The contribution of trimers could be estimated in case of SF_6. Dissociation spectra of small SiH_4 clusters were measured for the first time in the frequency range 880 cm^{-1} - 940 cm^{-1}. In the dissociation spectra of ethylene clusters van der Waals modes coupled with the out of plane ν_7 C_2H_4 vibration have been observed for the dimer, and a long predissociation lifetime (15 ns) has been inferred from the observed narrow (FWHM = 10 MHz) rotational lines.

1. INTRODUCTION

IR predissociation experiments on small clusters can provide information about the formation of the clusters, about the dissociation process itself and sometimes about the structure of the studied complexes. Clusters are usually produced in a nozzle expansion by properly selecting the nozzle diameter, the source pressure and temperature. The cluster formation process can be manipulated by using various seeding gases with different concentration. In order to produce small clusters, a combination of small nozzle diameter, high source pressure and very diluted gas mixtures is required.

The formed complex can be excited by IR (laser) radiation of suitable frequency, which is usually close to the frequency of a vibrational mode of the monomer species. For weakly bound systems the energy of the excited complex exceeds by far their binding energy, thus the IR excitation originates the cluster bond rupture after a fast intramolecular relaxation. From the spectral linewidth of the IR absorption

489

A. Weber (ed.), Structure and Dynamics of Weakly Bound Molecular Complexes, 489–494.
© *1987 by D. Reidel Publishing Company.*

490   M. SNELS ET AL.

the lifetime of the excited complex can be calculated.

In the dissociation-proces the cluster fragments are scattered out
of the supersonic beam. The fragments can be detected in order to
obtain information about the energy partition in the dissociation. The
IR cluster dissociation cross-section can be obtained by measuring the
attenuation of the beam signal. The IR dissociation spectrum is recorded monitoring the beam attenuation as a function of the laser
frequency. This spectrum can give some insight in the structure of the
complex. We measured the dissociation spectra for clusters of SF_6,
SiF_4, SiH_4 and C_2H_4.

2. EXPERIMENTAL

2.1 The apparatus

The molecular beam apparatus consisting of three differentially pumped
sections, is sketched in Fig. 1. In the first chamber the molecular
beam is produced, in the second chamber it is crossed by the radiation
from a CW CO_2 laser. The laser with a cavity length of 1.95 m can be
operated single-mode with $^{12}CO_2$, $^{13}CO_2$ and N_2O, providing more than 250
laserlines between 880 and 1100 cm^{-1}. The long time power stability is
better than 1%, the frequency stability better than 1 MHz. A piezo-
-electric translator, mounted on the laser grating, allows a 75 MHz
fine tuning of the laser frequency within each laserline. The laser
beam is gently focussed to a spot of 0.8 mm diameter on the molecular
beam axis. The distance between laserspot and the detector
(bolometer) is 400 mm. As bolometer serves - on a substrate of
sapphire (2 mm x 5 mm) - a doped Ge detector of 1 mm x 1 mm (Infrared
Laboratories) typically operated at 4.2 K. The N.E.P. at 4.2 K is
$0.5 \cdot 10^{-13}$ WHz$^{-1/2}$, the responsivity $5 \cdot 10^4$ V/W and the response time
2.5 ms. During the measurements the response time increases due to
cryofrost, (condensation of molecules on the cold detector surface).
The bolometer is well shielded from external sources of heat radiation
by screens at temperatures of 77 K and 4.2 K and from the laser
straylight by several screens between laser and detector.

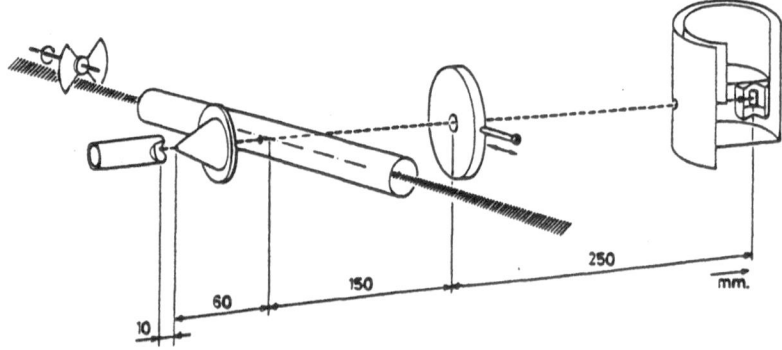

Figure 1. Experimental set-up.

2.2 The measurements

The dissociation spectra were obtained working with a continuous molecular beam and modulating the radiation of the laser. The modulated bolometer signal was amplified by a preamplifier (Infrared Laboratories LN-6C) and fed into a lock-in amplifier. The lock-in signal has been averaged by a microcomputer (Rockwell Aim 65), after subtracting the contribution of the straylight of the laser. A straightforward correction procedure has been followed in order to account for the linear decrease of the bolometer sensitivity during each run of measurements. This decrease of sensitivity was caused by an increase in the response time during the experiment. Full spectra obtained at different days were reproduced within 5%, and in most of the cases even within 2%.

3. RESULTS

3.1 C_2H_4

Figure 2 shows the cluster dissociation spectrum recorded for a 5% C_2H_4 in He mixture (p_{source} = 5 atm, T_{source} = 323 K). Three lorentzian line-shapes are fitted through the experimental points. The main feature in this spectrum centered at 952.1 cm^{-1}, (FWHM =13.8 cm^{-1}) corresponds to the slightly perturbed out-of-plane vibration ν_7 (949 cm^{-1}) of C_2H_4. The two small satellites at 916 cm^{-1} and 986.4 cm^{-1} (FWHM = 4.5 cm^{-1}) are due to transitions involving a low frequency ($\cong$ 36 cm^{-1}) van der Waals mode. This low frequency mode couples with

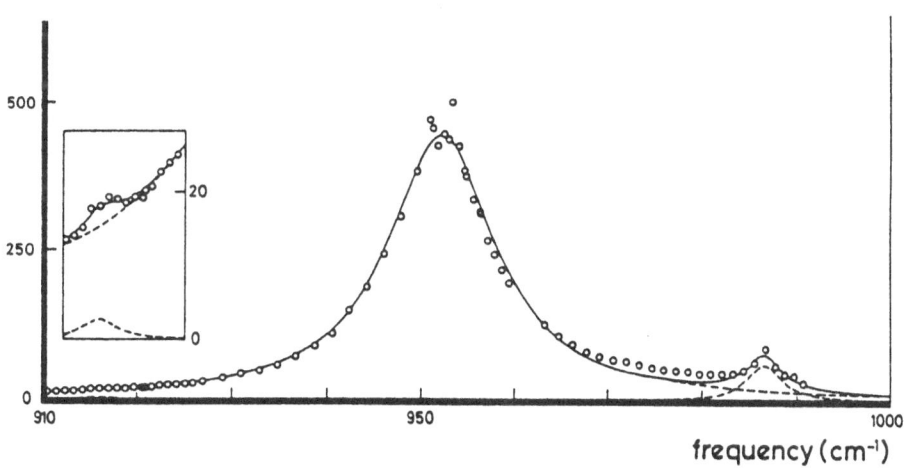

Figure 2. Cluster dissociation spectrum for an expansion of a 5% C_2H_4 in He mixture through a 30 μm diameter nozzle. Source pressure is 5 atm, source temperature 323 K.

the dimer mode v_D at 952.1 cm^{-1}, so gives a difference $(v_D - v_{vdw})$ or a sum frequency $(v_D + v_{vdw})$. From the relative intensity of the two satellites a vibrational temperature of about 16 K can be calculated.

For laser lines around 952.1 cm^{-1} and 986.4 cm^{-1} rotational fine structure (FWHM $\cong$ 10 MHz) was observed [1].

Recently Watts and coworkers [2] confirmed this observation, using different beam conditions and a waveguide laser with a larger free spectral range. They established that the fine structure must be due to C_2H_4 dimers.

3.2 SF$_6$, SiF$_4$, SiH$_4$

In Fig. 3 the dissociation spectrum for SF$_6$ clusters around the v_3 vibration of SF$_6$ (949.0 cm^{-1}) is displayed. The two main peaks at 934.4 cm^{-1} and 954.8 cm^{-1} correspond to dimers consisting of two ^{32}SF$_6$ molecules. Two smaller features, at 923.3 cm^{-1} and 950.4 cm^{-1} are due to dissociation of ^{32}SF$_6$ - ^{34}SF$_6$ dimers.

As shown in Table I the peak positions and intensities agree very

TABLE I   SF$_6$ cluster dissociation

Calculated spectrum (λ=6.8 cm^{-1}, Δv= 0.0)				measured spectrum (1% SF$_6$ in He, 30 μm nozzle, P_o=5 bar, T_o= 294 K)	
peak position (cm^{-1})	cluster species	isotopic abundance (%)	transition intensity (%)	peak position (cm^{-1})	relative intensity* (%)
Dimers					
923.3	^{32}SF$_6$-^{34}SF$_6$	8.0	2.5	922.8	2.3
928.4	^{32}SF$_6$-^{34}SF$_6$	8.0	1.0		
934.4	^{32}SF$_6$-^{32}SF$_6$	90.3	30.1	934.7	35.2
950.4	^{32}SF$_6$-^{34}SF$_6$	8.0	4.4	950.4	4.1
954.8	^{32}SF$_6$-^{32}SF$_6$	90.3	60.2	955.1	60.2
955.5	^{32}SF$_6$-^{34}SF$_6$	8.0	0.2		
Trimers					
934.9	$(^{32}$SF$_6)_3$	85.7	41.3		
957.7	$(^{32}$SF$_6)_3$	85.7	15.8		
961.6	$(^{32}$SF$_6)_3$	85.7	28.6		

* Experimental intensities have been normalized on the 954.8 cm^{-1} peak

well with calculations based on a resonant dipole-dipole model [3]. The FWHM of all the lines is 2.7 cm^{-1}. The shoulder around 960 cm^{-1} is caused by dissociation of trimers. We estimated a trimers/dimer ratio of about 5%.

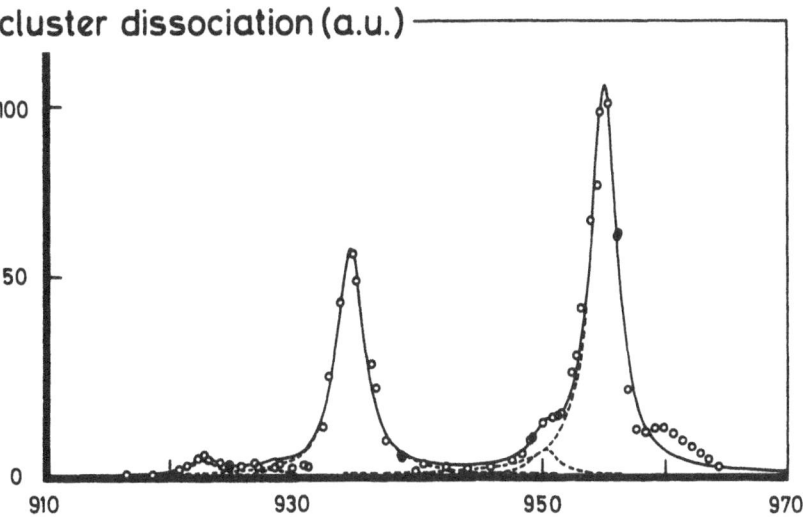

Figure 3. 1% SF_6 in He, P_{source} = 5 atm, T_{source} = 294 K, 30 μm nozzle.

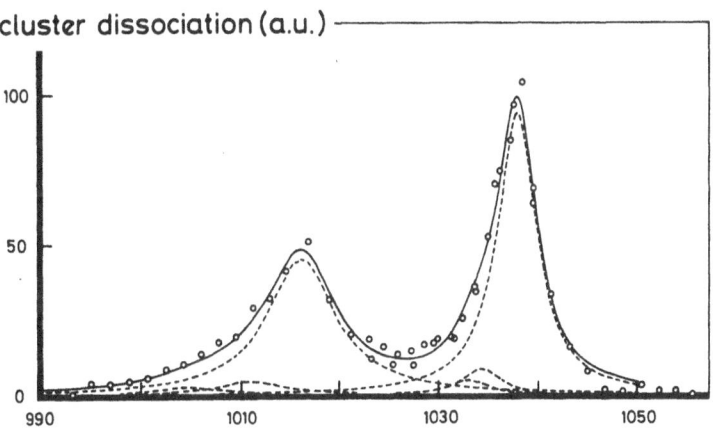

Figure 4. 2% SiF_4 in He, P_{source} = 5 atm, T_{source} = 323 K, 30 μm nozzle.

Figure 4 shows the dissociation of SiF_4 clusters around the v_3 vibration of SiF_4 (1031.4 cm^{-1}). As for SF_6 clusters, we fitted lorentzians through the measured points. The two main peaks are due to $^{28}SiF_4$ - $^{28}SiF_4$ clusters, the smaller features represent the presence of dimers containing an $^{29}SiF_4$ or $^{30}SiF_4$ molecule.

In the case of SiH_4 clusters (Fig. 5) we find again two peaks

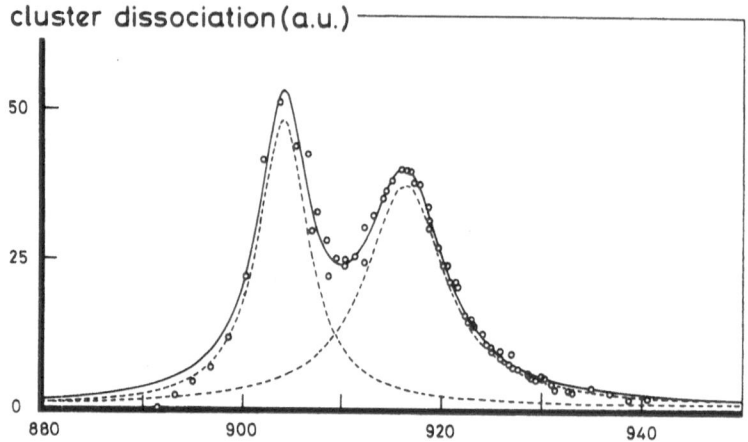

Figure 5. 2% SiH_4 in He, P_{source} = 5 atm, T_{source} = 323 K, 30 μm nozzle.

caused by dimer dissociation. Since the frequency shift for the v_4 vibration (v_4=913.5 cm^{-1} for $^{28}SiH_4$) is very small for the ^{29}Si and ^{30}Si isotopes no attempt was made to fit the contribution of the isotopically mixed dimers. Although the resonant dipole-dipole model predicts a ratio of 1:2 for the peak intensities, we observe here a 1:0.75 ratio.

REFERENCES

[1] M.Snels, R. Fantoni, M. Zen, S. Stolte and J. Reuss, Chem. Phys. Lett., 124 (1986) 1.
[2] K.G.H. Baldwin and R.O. Watts, Chem. Phys. Lett., 129 (1986) 237.
[3] J. Geraedts, M. Waayer, S. Stolte and J. Reuss, Faraday Disc. Chem. Soc., 73 (1982) 375;
     M. Snels, R. Fantoni, Chem. Phys. (1986) in press.

INFRARED (9-11 μm) DISSOCIATION OF THE HYDROGEN BONDED CLUSTERS $(NH_3)_n$
$(n \geq 2)$ DETECTED BY BOLOMETRIC TECHNIQUE

M. Snels[+], R. Fantoni, W. Leo Meerts[+]
Dip. TIB, Divisione Fisica Applicata, C.R.E. Frascati
P.O. Box 65
00044 Frascati (Roma)
Italy

+ Physics Laboratory, University of Nijmegen
Toernooiveld,
6525 ED Nijmegen
The Netherlands

ABSTRACT. Predissociation of NH_3 clusters has been induced by a CW CO_2 laser and detected by a semiconducting bolometer. For the dimer two absorption bands have been found at 979 cm^{-1} and 1004 cm^{-1}, which originate from the excitation of two non-equivalent NH_3 molecules. Heavier NH_3 clusters dissociate at frequencies between 1020 cm^{-1} and 1100 cm^{-1}. A simple electrostatic model can account qualitatively for the observed features.

1. INTRODUCTION

Recently many theoretical and experimental studies on the subject of hydrogen bonded clusters have been carried out.
    Equilibrium structures for some complexes could be established and valuable information on internal motions has been obtained. The IR dissociation of $(NH_3)_n$ complexes has been first investigated by Howard et al. [1], who observed a broad dimer band centered at 977.2 cm^{-1} and dissociation of heavier clusters between 1020 cm^{-1} and 1060 cm^{-1}. A molecular beam electric deflection study [2] investigating small clusters, from dimers to hexamers, has shown that only dimers posses a permanent electric dipole moment larger than 0.3 D. A dipole moment of 0.74 D was indeed measured [3] for $(NH_3)_2$ along the dimer axis.
    Investigations of NH_3 complexes in CO and N_2 matrices [4] demonstrated the existence of two dimer bands of nearly the same intensity separated by 12 cm^{-1} (CO) and 17.5 cm^{-1} (N_2), respectively.
    In the molecular beam machine described in the next section we observed two broad bands (FWHM $\cong$ 14 cm^{-1}) in the $(NH_3)_2$ dissociation spectrum centered at 979 cm^{-1} and 1004 cm^{-1}. A structure richer than previously [1,3] emerged in the first band. Dissociation of heavier

495

A. Weber (ed.), Structure and Dynamics of Weakly Bound Molecular Complexes, 495–500.

clusters occurred at frequencies between 1020 cm^{-1} and 1100 cm^{-1}, which agrees with the results of [2] and with matrix spectra [4].

2. EXPERIMENTAL APPARATUS

The molecular beam is produced by supersonic expansion of a mixture of NH_3 in He through a 30 μm nozzle into a vacuum chamber. For the reported spectra the stagnation pressure was 5 atm. The temperature of the nozzle can be varied between -50°C and 150°C and is stabilized within 0.1°C. A conical skimmer separates the first chamber from the second which is independently pumped. In this chamber the molecular beam is crossed by the radiation from a CW CO_2 laser. This laser can be operated single-mode with $^{12}CO_2$, $^{13}CO_2$ and N_2O gasmixtures, providing more than 250 laserlines between 880 cm^{-1} and 1100 cm^{-1}. The gaussian laser beam is focussed to a spot of 0.8 mm diameter on the molecular beam axis. The laser power in all the present experiments was 5 W. The molecular beam is detected by a Ge bolometer (Infrared Laboratories) operated at 4.2 K. This very sensitive device is located in a third vacuum chamber, 400 mm from the interaction point [5]. The dissociation spectra were obtained working with a continuous molecular beam and modulating the radiation of the laser. The modulated bolometer signal was preamplified, fed into a lock-in amplifier and averaged by a microcomputer. The reproducibility of the observed spectra was better than 5%. A discussion about the use of the bolometer for the cluster dissociation spectra can be found in [5].

3. RESULTS AND DISCUSSION

A mixture of 2% NH_3 in He was expanded from a 294 K nozzle. The dissociation spectrum is displayed in Fig. 1b and shows two absorption bands of comparable intensity centered at 979 cm^{-1} and 1004 cm^{-1}. The two peaks in the first band, at 977.2 cm^{-1} and 980.9 cm^{-1}, correspond to those measured by Fraser et al. [3], the second band has never been observed before. Figure 1a displays a spectrum for a nozzle temperature of 248 K, with the other beam conditions unchanged. The two bands which are due to dimer dissociation remain, but new structures between 1020 cm-1 and 1100 cm-1 appear which correspond to the dissociation of heavier clusters [1].

The cluster dissociation versus the laser power has been measured on all the peaks shown in Fig. 1. The same linear dependence was observed for the investigated temperature range, followed by saturation at about 20 W laser power. Clearly the dissociation of $(NH_3)_2$ is an one-photon process.

Varying the nozzle temperature from room temperature to 150°C (Fig. 1c) we noticed that the dimer signal decreased only by a factor of two. In dimers of SF_6, SiF_4 and SiH_4 a much more drastic dependence has been observed [5]. This observation suggests that the dimer bond in $(NH_3)_2$ is stronger than in the other complexes. A lower limit of 520 cm^{-1} to the binding energy of $(NH_3)_2$ was given by Buck et al. [6], the present results yield the upper limit of about 950 cm^{-1} (Fig. 1c).

The spectrum displayed in Fig. 1c has been recorded fro 5% NH_3 in He at a nozzle temperature of 423 K. This spectrum shows a few extra peaks and a broadening of the two dimer bands. Since in Fig. 1 the two main bands show the same behaviour when we change the beam conditions both must be due to the same cluster species, which is shown to be the NH_3 dimer for the band at 979 cm^{-1} [1,3].

In order to understand the origin of the two dimer bands we have to consider the structure of the $(NH_3)_2$ complex. The structure suggested by Nelson et al. [3] for the NH_3 dimer (θ_1=48.7°, θ_2= 115.8°, χ_1= 0°, χ_2= 180°, ϕ= 0° and R= 3.3374 A) accounts for the experimental

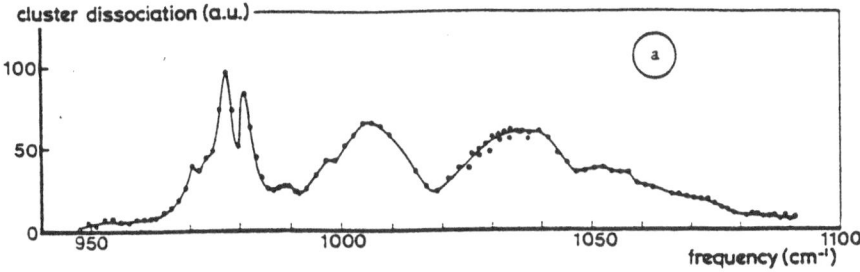

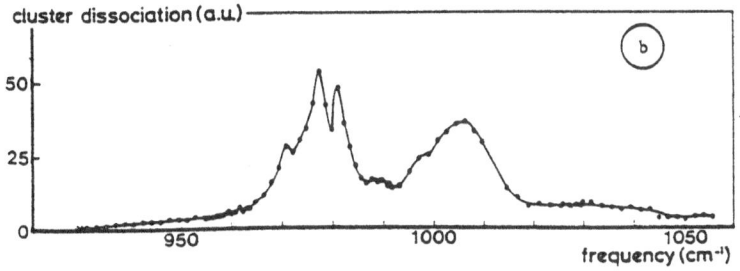

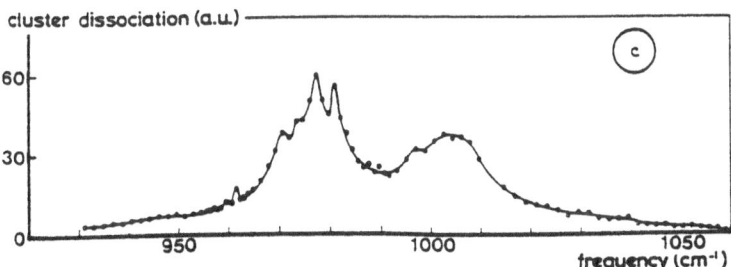

Figure 1.   a,b) Cluster spectrum for 2% NH_3 in He, nozzle temperature 248 K (a), 294 K (b); c) Cluster spectrum for 5% NH_3 in He, nozzle temperature 423 K.

data measured in the microwave region (below 22 GHz). This structure contains two non-equivalent molecules; one of them (M_2) is mainly a hydrogen donor, the other (M_1) a hydrogen acceptor (see Fig. 2a). No evidence for NH_3 molecule inversion in the dimer was obtained [3]. A resonant dipole-dipole interaction, observed in excited dimers of SF_6 [5], could give rise only to a splitting of 0.83 cm^{-1} in the case of $(NH_3)_2$.

It can be concluded that the two bands with a splitting of 25 cm^{-1} in the dimer spectrum originate from the excitation of the two non--equivalent NH_3 molecules in the dimer. Similarly, the two dimer bands observed in matrix experiments were explained as excitation of two non-equivalent molecules, assuming, however, that a complex with a linear N-H--N hydrogen bond was formed.

In order to check this hypothesis for the structure of Ref. [3] we have perfoméd a simple calculation of the electrostatic energy. A charge distribution, yielding the correct dipole moments for the NH_3 monomers both in the ground and in the excited state was assumed.

The total electrostatic energy of the complex is obtained by calculating the energy of one of the NH_3 molecules, in the electric field produced by the other NH_3. The electrostatic energy has been calculated for groundstate (E_o) and for two different excited states of dimer (E_1 and E_2). We find $E_1-E_o = 29.6$ cm^{-1} and $E_2-E_o =19.6$ cm^{-1}. Our simple model yields a splitting of 10 cm^{-1}; the peak corresponding to the excitation of the hydrogen acceptor molecule (M_2), is blue shifted with respect to the peak corresponding to the excitation of the hydrogen donor (M_1). Note that both dimer frequencies are blue-shifted with respect to the frequency of the monomer vibration. We will now try to find an explanation for the different spectral appearance of the two dimer bands. The band at 979 cm^{-1} is structured within 1 cm^{-1} resolution, the other band (at 1004 cm^{-1}) looks rather smooth. Furthermore, Fig. 1 shows that increasing the nozzle temperature leads to a new peak around 961.5 cm^{-1}. In addition, cooling the nozzle narrows the bands. This observation suggests that the width of the two dimer bands is determined by internal degrees of freedom, which become depopulated for colder beams. In order to explain the observed spectral structures we suggest that the NH_3 dimer exhibits a tunneling motion as observed for the HF dimer [7]. Starting from the configuration in Fig. 2a with $\theta_1= 48.7°$, $\theta_2= 115.8°$, $\chi_1= 0°$, $\chi_2= 180°$, $\phi= 0°$ and R= 3.3374 Å, we can obtain an equivalent situation changing slightly the angles θ_1, θ_2, χ_1 and χ_2 to $\theta_1= 64.8°$, $\theta_2= 131.4°$, $\chi_1= 60°$ and $\chi_2= 120°$. The barrier for this tunneling motion is expected to be rather low, since in this motion the roles of the hydrogen-donor and -acceptor is gradually interchanged. A tunneling barrier causes a splitting ΔE_i for every vibrationally excited state v_i. We can classify the 18 dimer vibrations in 6 stretchings and 6 bendings, three of each mainly localized on monomer M_1 and three of each on monomer M_2. More precisely these are symmetric and antisymmetric conbinations of monomer-like vibrations. Furthermore we have six intermolecular (intradimer) vibrations which depend on the dimer coordinates defined in Fig. 2a. A pictorial sketch of the vibrational levels (J=0, K=0) with the tunneling sublevels is shown in Fig. 2b. In general selection rules for tunneling [7] require

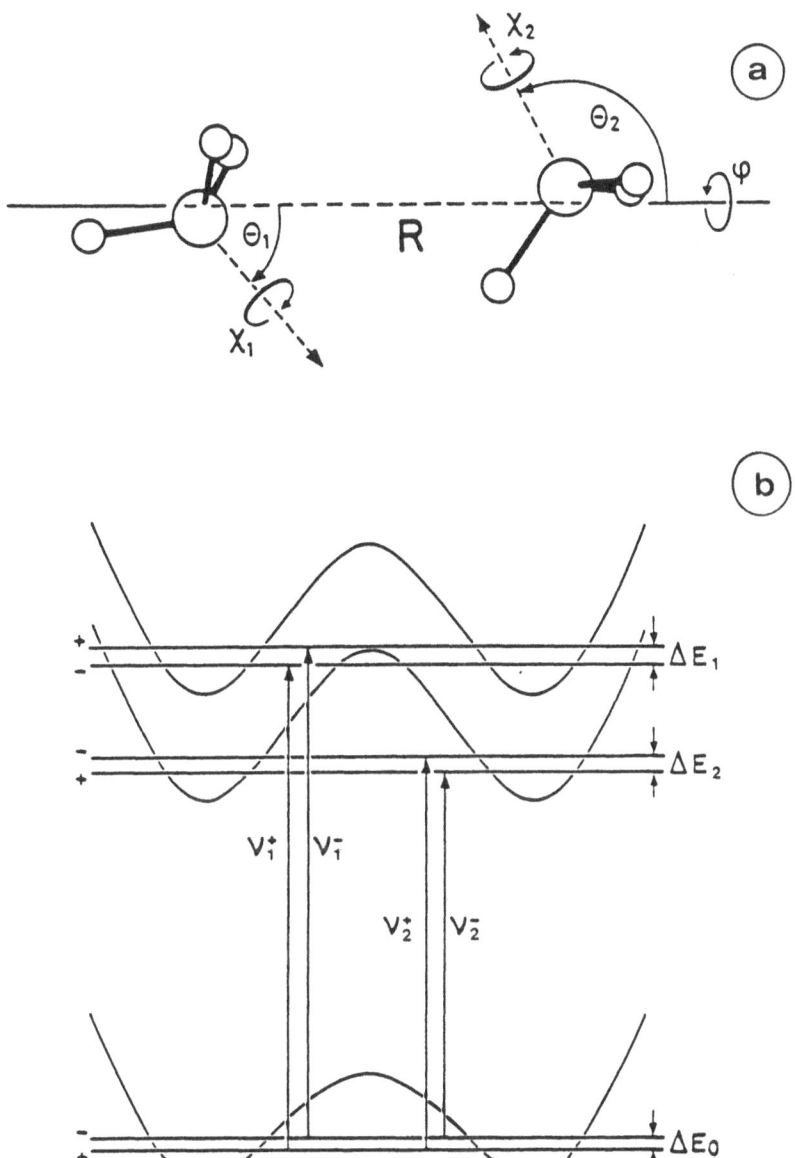

Figure 2.   a) NH_3 dimer, structure according to Nelson et al. /3/;
             b) Tunnelling splitting in the NH_3 dimer ground state (ΔE_0)
             and in the symmetric (ΔE_1) and antisymmetric (ΔE_2) umbrella
             vibrations. Allowed transitions between J=0, K=0, sublevels
             are shown.

a change of symmetry in the tunneling species during the transition. The symmetry group of the NH_3 dimer, including the possibility of tunneling will be presented in a forthcoming paper [8] where a complete discussion of vibro-rotational-tunneling selection rules will be given.

For very cold beam conditions (Fig. 1a) a broad band is observed, which corresponds to heavier clusters. Comparison of spectra measured at different beam temperatures showed that the broad band between 1020 and 1100 cm-1 (Fig. 1a) emerges bit by bit. Initially (at 270 K) only a band around 1045 cm-1 was observed, but reducing the beam temperature produced a dissociation signal at higher laser frequencies. Apparently, the absorption bands shift to the blue for heavier clusters. The constraint of an almost zero permanent dipole moment for all the heavy complexes from the trimers to the hexamers is fulfilled by a symmetric cyclic geometry [2]. In order to obtain the minimum of electrostratic energy for each complex θ and χ angles have been optimized. These angles are defined analogously to the dimer case. θ is the angle between the NH_3 symmetry axis of each monomer and the N--N axis with the right hand adjacent monomer. χ stands for the rotation around the symmetry axis each monomer. The simple electrostatic model, previously applied for the trimer, yields in this case a large blue shift (from 174 cm-1 for the dimer to 195 cm-1 for the hexamer) with respect to the v_2 in NH_3 monomer. The trend is in qualitative agreement with the experiment. The energy per bond calculated in the present model is roughly about 700 cm-1 for all the small clusters, which falls within the lower and upper experimental limits obtained for the dimer.

4. REFERENCES

[1] M.J. Howard, S. Burdenski, C.F. Giese and W.R. Gentry, J. Chem. Phys., 80 (1984) 4137.

[2] J.A. Odutola, T.R. Dyke, B.J. Howard and J.S. Muenter, J. Chem. Phys., 70 (1979) 4884.

[3] D.D. Nelson Jr., G.T. Fraser and W. Klemperer, J. Chem. Phys., 83 (1985) 6201;
     G.T. Fraser, D.D. Nelson Jr., A. Charo and W. Klemperer, J. Chem. Phys., 82 (1985) 2535.

[4] W. Hagen and A.G.G.M. Thielens, Spectrochimica Acta, 38A (1982) 1203;
     G.C. Pimentel, M.O. Bulanin and M. van Thiel, J. Chem. Phys., 36 (1962) 500;
     G. Ribbegard, Chem. Phys., 8 (1975) 185.

[5] M. Snels, R. Fantoni, M. Zen, S. Stolte and J. Reuss, Chem. Phys. Lett., 124 (1986) 1.
     M. Snels and R. Fantoni, Chem. Phys. (1986) in press.

[6] Z. Bacic, U. Buck, H. Meyer and R. Schinke, Chem. Phys. lett., 125 (1986) 47.

[7] A.S. Pine, W.J. Lafferty and B.J. Howard, J. Chem. Phys., 81 (1984) 2939;
     Ian M. Mills, J. Phys. Chem., 88 (1984) 532.

[8] M. Snels, R. Fantoni, R. Sanders and W. Leo Meerts, 'IR dissociation of Ammonia clusters' to be published.

HIGHLY EXCITED SYSTEMS

J. Reuss
Department of Physics
University of Nijmegen
Toernooiveld, 6525 ED Nijmegen
The Netherlands

ABSTRACT.
The structure and dynamics of weakly bound molecular systems are often
characterized by low potential barriers and large tunelling probabilities. Similar effects can be observed for molecules with internal
motions, especially if these molecules are excited to energies near to
or over above the barriers. We discuss this similarity with examples
drawn from Raman overtone spectroscopy (e.g. C_2H_6) and from dimer predissociation spectra (e.g. $(NH_3)_2$). IR overtone spectroscopy often leads
to vibrational transitions which are heavily perturbed by resonances but
still show well defined rotational structure and sharp lines. For dimer
predissociation spectra similarly sharp transitions have been observed
in many cases, notwithstanding the high density of states in the neighbourhood of the observed transitions. In the singular case of $(C_2H_4)_2$
this leads to a predissociative lifetime of about 10 ns, since the minimum value derived from the linewidth practically coincides with the
maximum value derived from time resolved pump-probe laser experiments.

1. INTRODUCTION

The spectroscopic determination of molecular structure has been so
successful that it is very common to immediately associate more or less
symmetric shapes with the chemical formulae which describe molecules,
e.g. a flat pyramid with NH_3, a regular tetraeder with CH_4, and octa-
eder with SF_6, etc.
   The symmetry of these shapes also governs the motion of the con-
stituting N atoms upon vibrational excitation; the 3N-6 so-called normal
modes represent a coherent and perfectly organized oscillation of all N
atoms, at a single frequency, around the equilibrium configuration determined by the potential energy surface of the system.
   The normal modes yield vibrational eigenstates, which have been
probed by IR and Raman spectroscopy to a high degree of sophistication.
For those who strive for general ordering and cultivation there may
hardly be a more satisfying object for scientific endeavour than

A. Weber (ed.), Structure and Dynamics of Weakly Bound Molecular Complexes, 501–512.
© 1987 by D. Reidel Publishing Company.

molecules that permit an extreme condensation of empirical facts – often
thousands and thousands of spectral lines – into simple formulae containing perhaps ten so-called molecular constants. Within the few constants the full and even increasing wealth of spectroscopic information
is contained.

In this beautiful world of molecular models molecules like CH_4
possess such high barriers that the in principle indistinguishable H-
atoms are prevented from visiting each other and changing places. What
happens upon vibrational excitation with respect to tunnelling through
these barriers or even surpassing them? Once barriers are surmounted
where does the ordered world of molecular structure finish? Do large-
amplitude motions of the molecular constituents still resemble the
normal mode pattern? How can large-amplitude motions be excited, after
all? Is the result of such an excitation still an eigen-state? Is there
experimental evidence of stochastic behaviour of which rumours increasingly threaten the orderly world of molecular spectroscopists?

2. THROUGH AND ABOVE THE BARRIERS

Even in classical molecular spectroscopy large-amplitude motion is ad-
mitted, occasionally, if structural barriers appear penetrable; it does
not lead to a complete breakdown of molecular structure, the symmetry
concepts have only to be modified and amplified.

The best known case is formed by NH_3 and its inversion. The N-atom
possess two equivalent positions, symmetrical with respect to the H_3-
plane, with a barrier of about 2076 cm^{-1} in between. Thus, in many regards, NH_3 is better described by a bi-pyramid of D_{3h} symmetry than by
a flat one of C_{3v} symmetry. The double-well potential surface leads to
an inversion splitting of vibrational-rotational states, e.g. 0.8 cm^{-1}
for the ground state and 36 cm^{-1} for the first excited v_2 umbrella-mode
of NH_3. The barrier penetration of ammonia is well studied because
transitions between the split inversion states are infrared-active, i.e.
the electric dipole moment changes if the N-atom moves from one to the
other of its two equivalent positions.

An interesting quantitative question is how much the barrier diminishes between the two equivalent positions, if e.g. the v_1-mode of
NH_3 is excited. According to Herzberg II [1], already $n_1=1$ results in
a significant change of inversion splitting, 1 cm^{-1} as compared to 0.66
cm^{-1} for the ground states. High level excitation attainable with modern
techniques allows to follow this degrading of the barrier (accompanied
by an enhanced splitting of e.g. 2.76 cm^{-1}, for the level $4v_1+v_3$ [2]).
The potential barriers responsible for the classical geometrical shapes
of molecules are lowered and might even be thought to vanish reducing
the ensemble of molecular constituents to a state of rather floppy and
less structured cohabitation, hitherto little studied yet. However, the
example given demonstrates the persistance of barriers upon excitation.

Another example of barrier penetration is found for C_2H_6, with its
staggered equilibrium position of the two CH_3 groups, with respect to
the angular twist around the C-C axis. Here a twisting motion does not
lead to a change of electric dipole moment, i.e. the splitting of states
cannot be observed by direct IR-absorption measurements.

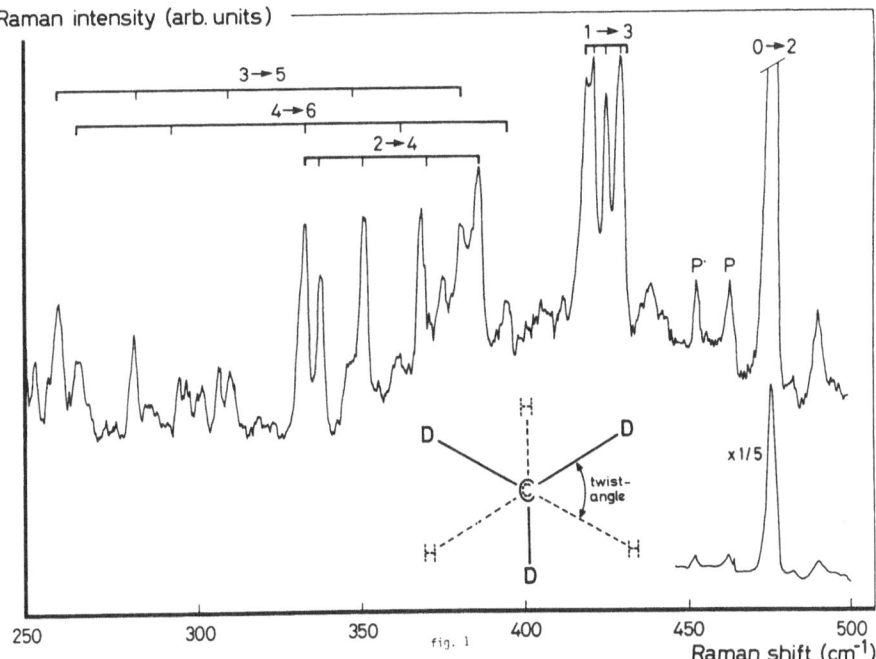

Figure 1. The torsion-Raman-overtone spectrum $n\nu_4 \rightarrow (n+2)\nu_4$ of CH_3CD_3, as function of the Stokes-Raman shift. The insert shows the staggered equilibrium position, head-on. The letter P indicates plasma lines. It is seen that the quintuplet-shifts become smaller for increasing n-values until (for n=4) the torsional motion approaches internal rotation instead of torsional vibration.

In such a case one is inclined to look for Raman-activity which might reveal barrier penetration permitting the CH_3-groups to be twisted into the different equivalent positions. Raman-activity goes along with a change of polarizability α as function of the twist angle Θ. The first attempt leads to a deception; $\Theta=\Theta_{equilibrium}\pm\Delta\Theta$ leads to the same α, that is $\partial\alpha/\partial\Theta=0$ for $\Theta=\Theta_{equilibrium}$. Consequently, excitation of the twisting motion cannot be observed by conventional Raman-Stokes-scattering. However, $\partial^2\alpha/\partial\Theta^2\neq0$ holds for $\Theta=\Theta_{equilibrium}$ and, therefore, Raman-overtone spectroscopy [5,6] leads to a direct observation of the twist-mode excitation with its splitting due to barrier penetration (see Fig. 1).

In general, the splitting reflects how long a CH_3-group dwells at twisting angles where the barrier is high or low, respectively.

The twisting mode ν_4 (275 cm^{-1}) possesses a barrier of about 1000 cm^{-1} and has been observed for $n_4=0 \rightarrow n_4=2$, $n_4=1 \rightarrow n_4=3$, $n_4=2 \rightarrow n_4=4$ and $n_4=3 \rightarrow n_4=5$; the state with $n_4=5$ is definitely above the barrier and C_2H_6 undergoes a nearly free rotation where the motion is faster (slower) at twisting angles corresponding to the potential well (top).

As is evident from Fig. 1, the level-splitting increases for higher levels, as does the red-shift due to anharmonicity of the well, until the character of eigenfunctions changes from dominant vibration to dominant rotation; rotational transitions show a blue shift for increasing n_4 values [3,4,5,6].

3. LINE WIDTH AT HIGH LEVELS OF EXCITATION

In Fig. 2 the set-up is shown by which high vibrational overtone absorption is measured, employing an intracavity absorption cell, photo-acoustical detection and a tunable dye laser. The observed spectra are taken with 0.2 cm^{-1} experimental resolution and may show fully resolved P- and

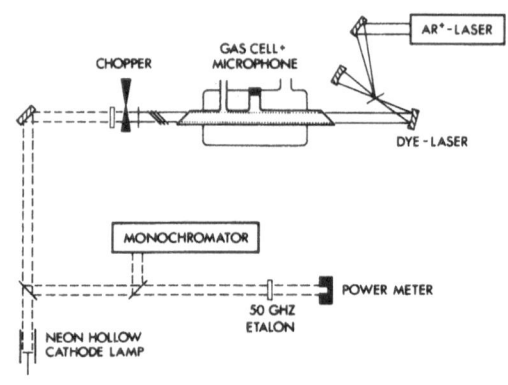

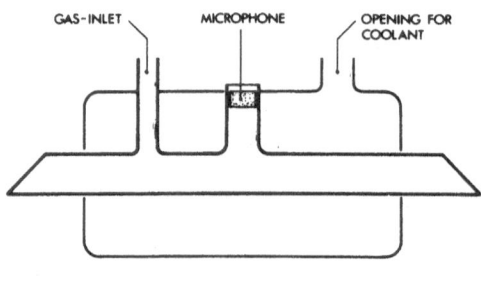

Figure 2. Photo-acoustic (PA) spectrometer of Haenen, Liedenbaum and Stolte, Nijmegen. The cylindrical intracavity PA resonator is cooled to get rid of H_2O overtones. The all-glass design ensures that window reflections do not reach the microphone.

R-branches and hot band structures although the density of vibrational states amounts already to about $100/cm^{-1}$. There is no indication of the onset of what has been called once the quasicontinuum of states; single vibro-rotational states are discernable and possess a "lifetime" of at least 50 ps, otherwise an extra line-broadening would have been observed [7].

At Zürich the group of Martin Quack [8] has investigated systems like HCF_3, where the C-H-stretch is almost resonant with the double C-H bending mode frequency. This anharmonic resonance is not lost for over-tones; on the contrary, the number of participating levels increases and e.g. amounts to 5 for the third C-H-stretch overtone, $4v_s \approx 11300 \ cm^{-1}$. The rest of the molecule apparently participates little in the motion. Recently, the IR measurements [8] were extended into the visible and found in agreement with the Zürich model [7,9,10].

An example of little coupling between vibrational states is fur-nished by vibrationally excited dimers where - due to the low energy of the van der Waals modes - the density of vibrational states is much larger ($10^3/cm^{-1}$) even if the excitation takes place by a single CO_2 laser photon. Also there, sharp absorption lines have been observed, first in pioneering work of the Canberra group [11,12,13]. An example is displayed in Fig. 3 and concerns the dimer $(C_2H_4)_2$; the linewidth is about 3 MHz fwhm and corresponds to a minimum lifetime $\tau=100$ ns, for the vibrationally excited state [14].

4. STRONG FIELD EFFECTS

Linear absorption spectroscopy reveals the position (and transition strength) of the eigenstates which may result from an anharmonic reso-nance between the participating states; eigenstates of free molecules become populated by absorption of quanta from weak radiation fields.

In strong laser fields the situation can be different as we demon-strate by a simple three-level system. The ground state |0> is linked to one excited state |a> by a non-vanishing (off-diagonal) electric-dipole màtrix element. Another nearly degenerate excited state |b> is assumed to be infrared inactive. It interacts, however, with |a> through a Fermi resonance off-diagonal matrix element F. The energy matrix of the isola-ted model system is given by table 1.

TABLE 1   The energy matrix in normal mode representation, for a three level system disregarding radiation coupling

	|0>	|a>	|b>
|0>	0	0	0
|a>	0	E_a	F
|b>	0	F	E_b

E_a and E_b are the deperturbed excited energies of the system: the eigen-values are given by

$$E_{1,2} = \frac{1}{2}(E_a+E_b) \pm \sqrt{\frac{1}{4}(E_a-E_b)^2+F^2}$$

Table 2 shows the relevant energy matrix if we put the model system into a strong single mode radiation field containing nhv-photons with

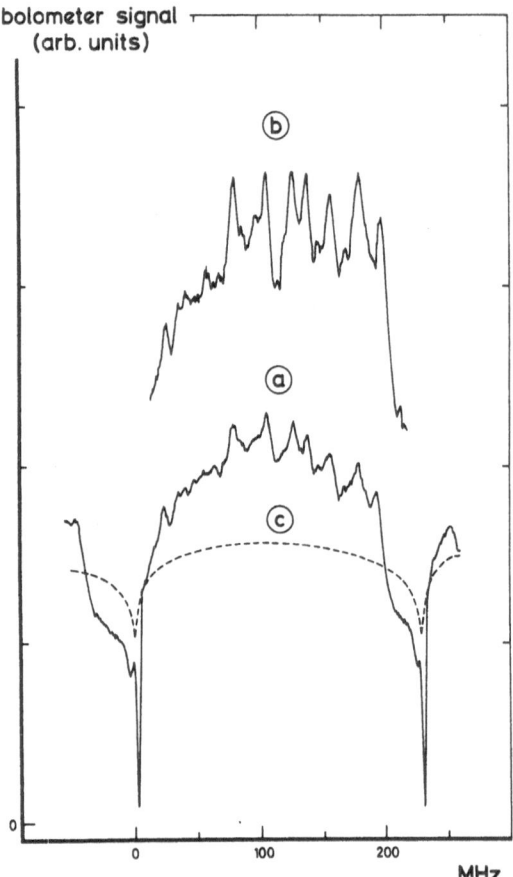

Figure 3. High resolution pre-dissociation spectrum of $(C_2H_4)_2$ of the Nijmegen group around the 10P(10) CO_2 laser line. The line-width yields a lifetime of about 100 ns for the excited dimers. The structure might correspond to $\Delta K_{eoe}'=1$ transitions belonging to different J_{eoe}-states (eoe stands for end-over-end). Double resonance experiments confirm that the sharp peaks are inhomogeneous. (a) and (b) correspond to a 5% mix-ture in He, with P_o=6 atm and 4.5 atm, respectively, (c) to the laser output power signal. The sharp negative peak corresponds to the monomer transition $(J,K_A,K_B)=(4,1,3) \rightarrow (5,0,5)$.

$hv \approx E_a \approx E_b$. The constant α contains the factor μ_{ab}, the electric transition dipole moment. The three diagonal energies are nearly degenerate.

TABLE 2   The energy matrix as in Table 1, with radiative coupling, in a dressed-atom representation.

	$\|0>\|n+1>$	$\|a>\|n>$	$\|b>\|n>$
$\|0>\|n+1>$	$(n+1)hv$	$\alpha\sqrt{n+1}$	0
$\|a>\|n+1>$	$\alpha\sqrt{n+1}$	E_a+nhv	F
$\|b>\|n>$	0	F	E_b+nhv

By increasing the number n of photons we reach the limit $\alpha\sqrt{n+1} >> F$ where the anharmonic resonance between the $|a>$- and $|b>$-state ceases to be important; instead the ground state and the $|a>$-state become strongly coupled. Consequently the system being in $|0>|n+1>$ at t=0 starts to Rabi-oscillate between $|0>$ and $|a>$ with the Rabi-frequency $\alpha\sqrt{n+1}/h$.

For the case that a molecule and a laser field together form a combined system of high excitation (n>>1) the conclusion is that no longer eigenstates may become populated, but one of the deperturbed states.

What happens at the moment when the model system leaves the radiation field? Assume, we find the system with unit probability in $|n>|a>$, at that time, i.e. $|a>$ just a moment later. This state $|a>$ is not an eigenstate for the radiation-free case, i.e. the system starts to oscillate between the two deperturbed states with energy E_a and E_b, with an oscillation frequency of $|E_1-E_2|/h$. Remember, $|a>$ ($|b>$) was taken to be IR active (IR inactive), so that a quantum beat of the fluorescence signal occurs. Such a quantum beat is shown in Fig. 4, taken, however, from an optically excited molecule where the role of the inactive $|b>$-state is played by a dark triplet state $|T>$, which is coupled to the luminescent excited singlet state $|S>$, [15].

5. MULTIPHOTON EXCITATION

Multiphoton dissociation of polyatomic molecules proceeds through the excitation of high vibrational states; therefore, a study of these states with respect to possible stochastization is of great importance. Recent-ly, spontaneous Raman scattering has been employed to investigate the "hot fraction" which occurs after pulsed excitation of an ensemble of molecules absorbing CO_2 laser radiation. This hot fraction showed – as evidenced by anti-Stokes Raman signals – excitation of all vibrational modes, which was interpreted by the authors of [16,17] as possible indication of stochastic processes. In Table 3 we resume the experimentally obtained minimum excitation energy ε_g^{min}, for the onset of stochastic behaviour, for four different molecules. Although the measurements were performed at relatively high pressures (1 torr), collisional effects

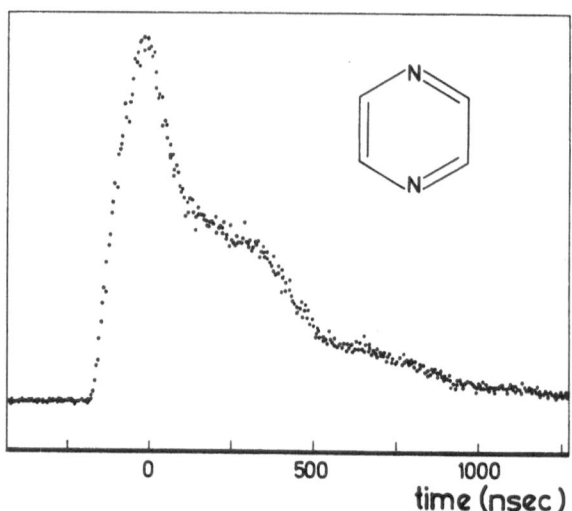

time (nsec)

Figure 4.   Fluorescence quantum beat, after pulsed excitation of pyra-
zine. Demonstration of singlet-triplet state mixing of two levels about
5 MHz apart. Many states are in the neighbourhood of the single rotational-vibrational level - due to singlet-triplet interaction - and can be
reached by shortening the exciting laser pulse.

were shown to be unimportant for the observations. As to the time scale,
stochastization occurs faster than within 5 ns (corresponding to a linewidth of about 30 MHz fwhm).

TABLE 3   Experimental threshold for stochastization

	SF_6	CF_3I	CF_3Br	CF_2ClH
ε_g^{min} [cm^{-1}]	3000	6000	7200	4000÷5000

6. DIMER SPECTROSCOPY

Dimers often represent systems with little hindrance to changes of
orientation and distances of their constituents. Consequently, large
amplitude motions result which are obtainable for conventional molecules
only after heavy excitation of the vibrational degrees of freedom. More
often than not absorption of a single CO_2 laser photon of about 1000 cm^{-1}
already suffices to dissociate a dimer. This easy dissociation indicates
the weakness of even the strongest mode, the van der Waals stretch-mode.

A simple well-studied system of small rigidity is formed by ortho-H_2-Ar, where it is a good approximation to describe the dimer by a virtually freely rotating ortho-H_2-molecule (j=1) bonded to the noble gas atom. The freedom to rotate stems from the angle dependent part of the H_2-Ar potential, the strength of which is small compared to the splitting of the end-over-end rotational levels of a H_2-Ar complex. In contrast to the next example (O_2-Ar), the projection K of the angular momentum on the dimer axis is not a good (approximate) quantum-number; the reason is that a linear configuration (o-o···o) possesses the lowest energy and that there is no effective barrier which produces a preferential orientation of the ortho-H_2 and clear ordering of energy levels according to K [18,19]. The relevant potential parameters have been obtained from the magnetic hyperfine structure of the complex, measured on a magnetic beam resonance apparatus.

For O_2-Ar there exists a barrier of about 30 cm^{-1} which impedes the complex to change from a T-structure ($_o^o$···o) to a linear structure (o-o···o). Consequently, K becomes a "good" quantum-number, and characterizes a zero-order basis to describe rather well the fine structure levels and their Zeeman-effect for O_2-Ar [20,21].

For the investigator the main challenge of these complexes is to find a proper description which reflects the presence of only low

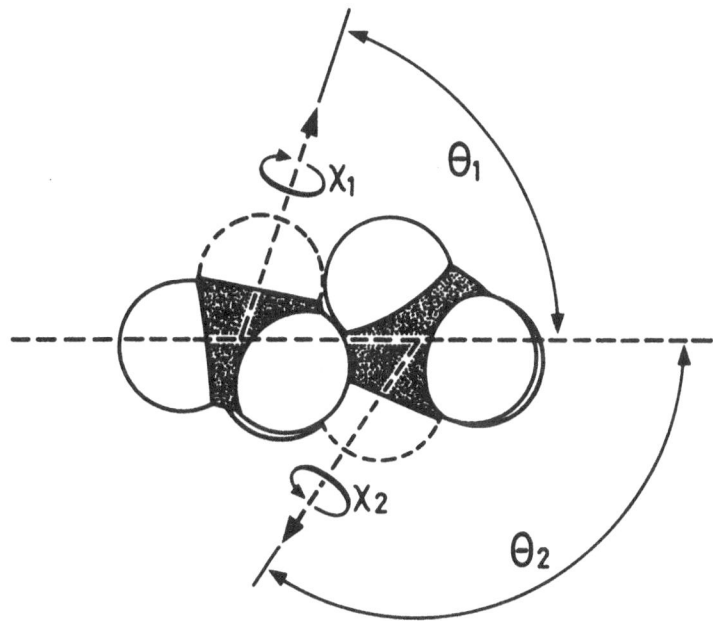

Figure 5. The model of $(NH_3)_2$ based upon close packing of two methane molecules; one H of each CH_4 (broken circle) is thought to be replaced by a filled electron lobe. The drawn angles θ_i and χ_i have values approximately equal to those of [22].

barriers and of liberal large amplitude motions. This task has not been
accomplished for $(NH_3)_2$; however, at least a (surprising) structure
seems to emerge from recent measurements at Harvard [22,23]. Microwave
spectra yield the main component of the inertial moments. In addition
the total permanent dipole moment along the linear axis, $\mu=0.74$ debye,
has been measured. The data are compatible with the model shown in
Fig. 5 [23]. At variance with expectation, this model cannot be based on
the occurrence of a single hydrogen bridge $H_3N--HNH_2$. In agreement with
the model one finds that a single CO_2 laser photon leads to dissociation
of the complex [22,23,24,25,26].

    In Fig. 6 a recent predissociation spectrum of $(NH_3)_2$ is reproduced,
where besides the previously known double-peak structure around 980 cm^{-1}
a further blue shifted broad peak has been measured by Snels et al.,
centered around 1020 cm^{-1} [27]. Note that the v_2-monomer absorption
occurs at about 950 cm^{-1}, i.e. a very pronounced blue shift is observed
in the dimer absorption spectrum. Its origin may partly come from the
static dipole-dipole interaction, which is stronger for ground state
dimers than for $NH_3---NH_3^+$ dimers since NH_3^+ is less bent and possesses a
static dipole moment of $\mu^+=1.24$ debye only, as compared to $\mu=1.47$ debye
for the ground state [28]. If the physical dimension of these dipoles is

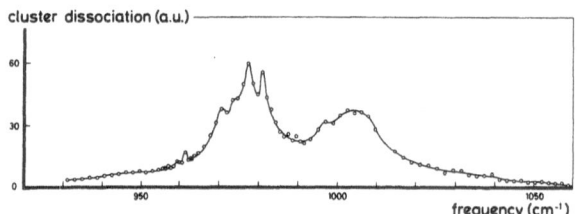

Figure 6. $(NH_3)_2$ predissociation signal (in arbitrary units) vs CO_2
laser frequency. The dimers were produced in an expansion of 5% NH_3 in
He, with $T_0=423$ K and $P_0=5$ atm; the nozzle diameter was 30 μm.

taken into account (the H-atoms bear a positive charge ($q=0.276$ e, $q^+=$
o.245 e), the N-atom a negative one ($3q$, $3q^+$)) the observed blue shift
appears compatible with the model of Fig. 5 [27]. As expected from this
conjecture, the blue shift increases for heavier clusters [24,27];
however, similar blue shifts have been found e.g. for NH_3-CO_2, too [23].
    The two broad peaks around 980 cm^{-1} and 1020 cm^{-1}, see Fig. 6, show
about the same absorption strength. In the model of Fig. 5, the positions
of the two NH_3-constituents are not equivalent; simple electrostatic
dipole forces again yield the right order of magnitude for the splitting
[27]. Due to the particular angles of the two NH_3 axes with respect to
the dimer axis, the dynamic dipole-dipole forces (which produce significant splittings of up to 20 cm^{-1}, for systems like $(SF_6)_2$ and $(SiF_4)_2$,
[29,30]), appear of negligible influence for $(NH_3)_2$ [27].
    In Fig. 6, we shall neglect the spectral structure for frequencies

$\geqq 1020$ cm^{-1}, which is attributed to photodissociation of larger clusters.
Instead, we shall speculate about the finer spectral features in the
left part of Fig. 6. The presented ideas are slightly provocative but
too interesting to be suppressed; they were conceived by M. Snels,
R. Fantoni and W.L. Meerts [27]. A doublet is observed around 980 cm^{-1};
its depth of modulation depends on the beam temperature. Its position
agrees with the findings of [23]. Less convincingly perhaps, a second
doublet appears at 972 cm^{-1}, i.e. 8 cm^{-1} further red shifted. It happens
that the (A-B)-constant of the dimer model (Fig. 5) amounts to 4 cm^{-1},
i.e. the two doublets are produced by $\Delta K_{eoe}=1$ transitions, where K_{eoe}
is defined with respect to the dimer axis. In Fig. 6 appear further
structures which occur at frequencies about 8 cm^{-1} shifted with respect
to each other.

Then why observe $\Delta K_{eoe}=1$ transitions around 970 cm^{-1} and not around
1005 cm^{-1}? The answer is fairly simple; the structure around 970 cm^{-1} is
due to v_2-excitation of the hydrogen donor (molecule 2 in Fig. 5); its
dipole moment is significantly more perpendicular than that of the
hydrogen acceptor. Therefore, $\Delta K_{eoe}=1$ transitions are dominant around
970 cm^{-1}, whereas for the hydrogen acceptor $\Delta K_{eoe}=0$ and 1 transitions
occur with about equal strength and the spectrum is more congested,
therefore.

Why are then doublets observed after all. Again, Fig. 5 shows that
a slight simultaneous change - so called tunnelling - of both inclination
angles θ_1 and θ_2 leads again to an equivalent geometry. Note that X_1 and
X_2 have to be turned around likewise during this tunnelling in order to
let molecule 1 (2) become the hydrogen donor (acceptor). Consequently,
each J_{eoe}, K_{eoe}-level forms a doublet split at the tunnelling frequency
of these donor-acceptor configurations.

The fact that one does not observe doublets for the 1005 cm^{-1} peak
might be attributed again to its congested nature discussed above. But
there may be more to it, as has been discussed by Mills [31] (especially
for the case of (HF)$_2$). If the 970 cm^{-1} peak shows doublet structure it
is because here the ground state doublet splittings add up, for the observed transitions, as consequence of pertaining selection rules. But
necessarily, then, for the 1005 cm^{-1} peak the splittings of ground and
excited level have to be subtracted from each other; assuming about the
same splitting for both levels leads to the unobservability of tunnelling
when the hydrogen acceptor is excited.

This discussion of (NH$_3$)$_2$ spectra demonstrates how narrowly connected the large amplitude motions of section 2 and dimer dynamics really
are.

REFERENCES

[1] G. Herzberg, *Molecular Spectra and Molecular Structure, Vol. 2,
Infrared and Raman Spectra*, Van Nostrand-Reinhold, New York 1945.
[2] T. Kuga, T. Shimizu and Y Ueda, *Jap. J. Appl. Phys.* **24** (1985)
L147.
[3] W. Knippers, K. van Helvoort, S. Stolte and J. Reuss, *Chem. Phys.*
98 (1985) 1.

[4] W. Knippers, K. van Helvoort, M. de Felici, J. Reuss and S. Stolte, *Chem. Phys.*, in print.

[5] R. Fantoni, K. van Helvoort, W. Knippers and J. Reuss, *Chem. Phys.*, to be published.

[6] K. van Helvoort, R. Fantoni, W.L. Meerts and J. Reuss, *Chem. Phys. Lett.*, accepted for publication.

[7] C. Liedenbaum, J.P. Haenen and S. Stolte, private communication.

[8] H.R. Dübal and M. Quack, *J. Chem. Phys.* **81** (1984) 3779.

[9] J.E. Bagott, M.C. Chuang, R.N. Zare, H.R. Dübal and M. Quack, *J. Chem. Phys.* **83** (1985) 1186.

[10] R.N. Zare, private communication.

[11] R.E. Miller and R.O. Watts, *Chem. Phys. Lett.* **105** (1984) 404.

[12] R.E. Miller, P.F. Vohralik and R.O. Watts, *J. Chem. Phys.* **80** (1984) 5453.

[13] G. Fisher, R.E. Miller and R.O. Watts, *J. Phys. Chem.* **88** (1984) 1120.

[14] M. Snels, R. Fantoni, M. Zen, S. Stolte and J. Reuss, *Chem. Phys. Lett.* **124** (1986) 1.

[15] W. van Herpen and W.L. Meerts, private communication.

[16] V.N. Bagratashvili, Yu.G. Vainer, V.S. Doljikov, S.F. Kol'yakov, V.S. Letokhov, A.A. Makarov, L.P. Malyakin, E.A. Ryabov, E.G. Sil'kis and V.D. Titov, *Sov. Phys. JETP* **53** (1981) 512.

[17] V.S. Doljikov, Yn.S. Doljikov, V.S. Letokhov, A.A. Makarov, A.L. Malinovsky and E.A. Ryabov, *Chem. Phys.* **102** (1986) 155.

[18] J. Verberne and J. Reuss, *Chem. Phys.* **50** (1980) 137; *Chem. Phys.* **57** (1980) 189.

[19] M. Waayer, M. Jacobs and J. Reuss, *Chem. Phys.* **63** (1981) 247; 257; 263.

[20] A. van der Avoird, *J. Chem. Phys.* **79** (1983) 1170.

[21] J. Mettes, B. Heijmen, P. Verhoeve, J. Reuss, D.C. Lainé and G. Brocks, *Chem. Phys.* **92** (1985) 9.

[22] G.T. Fraser, D.D. Nelson and W. Klemperer, *J. Chem. Phys.* **83** (1985) 6201.

[23] G.T. Fraser, D.D. Nelson, A. Charo and W. Klemperer, *J. Chem. Phys.* **82** (1985) 2535.

[24] M.J. Howard, S. Burdenski, C.F. Giese and W.R. Gentry, *J. Chem. Phys.* **80** (1984) 4137.

[25] A. Sudbo, P.A. Schultz, Y.T. Lee and Y.R. Shen, *Proceedings of the First International School on Laser Applications to Atoms, Molecules and Nuclear Physics*, Vilhuis, USSR 1978.

[26] J.A. Pople, *Faraday Disc. Chem. Soc.* **73** (1982) 7.

[27] M. Snels, R. Fantoni, R. Sanders and W.L. Meerts, *Chem. Phys.*, to be published.

[28] K. Shimoda, Y. Ueda and J. Iwahori, *Appl. Phys.* **21** (1980) 181.

[29] M. Snels and R. Fantoni, *Chem. Phys.*, in print.

[30] J. Geraedts, S. Stolte and J. Reuss, *Z. Phys.* **A304** (1982) 167.

[31] I.M. Mills, *J. Phys. Chem.* **88** (1984) 532.

VIBRATIONAL PREDISSOCIATION DYNAMICS OF THE NITRIC OXIDE DIMER

Michael P. Casassa, John C. Stephenson, and David S. King
Molecular Spectroscopy Division
National Bureau of Standards
Gaithersburg, MD 20899 USA

ABSTRACT. Details of experimental measurements of the total energy
distribution and time dependence of the vibrational predissociation of
the nitric oxide dimer are presented. Energy disposal measurements
following ν_1 excitation at 1870 cm^{-1} indicated the fragments to be
described by an average rotational energy $\langle E_R \rangle$=75 cm^{-1}, full
equilibration of the lambda doublet species, approximately equal
populations in both spin-orbit states, no significant degree of
alignment, an isotropic flux distribution, and an average kinetic
energy of $\langle E_K \rangle$=400 cm^{-1} per fragment. Although approximately 75% of
the available energy goes into fragment translation, picosecond laser
pump - probe experiments showed that ν_1 decayed exponentially with a
880 ps lifetime. Excitation of ν_4 at 1789 cm^{-1} gave a 39 ps
predissociative lifetime.

I. INTRODUCTION

Vibrational predissociation is observable for van der Waals molecules
(vdW) since a single quantum of vibrational excitation in a constituent
of such a cluster generally exceeds the vdW binding energy of the
complex. Thus, vibrational excitation leads to the rupture of the weak
vdW bond. Herein lies the appeal of these systems as models for
vibrational dynamics: rich and dramatic photochemistry is caused by
excitation to low lying vibrational levels. The resulting vibrational
energy flow and dissociation, proceeding from a well defined state
along a single potential energy surface, can be compared to theoretical
predictions.
   A combination of time-resolved and final-state-resolved
measurements are needed to thoroughly characterize the predissociation
dynamics of these clusters. Experimental observables include product
kinetic, rotational and vibrational energy distributions, vector
velocity and angular momentum distributions, and dissociation
lifetimes. Although there have been many studies of vibrational energy
flow within electronically excited states of vdW molecules, due to
experimental difficulties most studies involving ground electronic
states of vibrationally energized clusters have been limited to the
measurement of infrared absorption and photodissociation spectra.[1]

A. Weber (ed.), Structure and Dynamics of Weakly Bound Molecular Complexes, 513–524.

Fragment velocity distribution have been measured by Hoffbauer et al.[2] and by Bomse et al.[3] for the ground state predissociation of the ethylene dimer following ν_7 excitation. These show that only a small fraction ($\sim$23%) of the excess energy appears as translation. While the observed isotropic flux distributions suggest a long lived complex, such an interpretation is ambiguous for clusters of incompletely characterized structure. Similar isotropic flux distributions have been measured for several other polyatomic clusters, including the water dimer.[4] King and Stephenson have used LEF measurements[5] to characterize the fragments produced in the predissociation of NO-ethylene following ethylene-ν_7 excitation. The product flux resulting from this excited complex was determined by Doppler profile measurements to also be isotropic, with minimal translational energy release (only $\sim$35% of the available energy). Most of the excess energy excited internal (i.e., rotational) degrees of freedom of the ethylene fragment. There have been to date no time-resolved measurements of actual rates of vibrational predissociation of vdW dimers along the ground electronic state potential surface.

In this paper, experiments are discussed which utilize molecular beam and pulsed laser techniques to characterize the rates and mechanisms important in the predissociation of the nitric oxide dimer excited to the v=1 level of either the symmetric (ν_1) or antisymmetric (ν_4) stretch within the electronic ground state.

The structure of the ground state of the nitric oxide dimer has been determined by microwave spectroscopy to have C_{2v} symmetry with a 2.236 Å N---N separation and a 99.6° ONN angle.[6,7] The two NO stretching fundamentals of this structure have been observed at 1870 cm^{-1} and 1789 cm^{-1}.[8] These are characterized as the ν_1 symmetric stretch (vibrational symmmetry A_1) and the ν_4 (B_2) antisymmetric NO stretch, respectively. Although the high resolution spectrum of the ν_4 band has been observed, that of ν_1 has not since it lies near the NO fundamental at 1876 cm^{-1}; however, the $\nu_1+\nu_4$ combination band at 3630 cm^{-1} has been characterized.[8] The low frequency modes associated with the vdW bond are 90 (A_2), 170 (A_1), 198 (B_2), and 263 (A_1) cm^{-1}.

Brechignac et al.[9] reported photodissociation of nitric oxide clusters following excitation in the ν_4 band region. In our lab, photodissociation of the dimer and larger clusters has also been observed following excitation of the ν_1 band at 1869.79 cm^{-1}.[10] The photochemistry which is observed is:

$$(NO)_2 \xrightarrow{\;\;h\nu\;\;} (NO)_2^*(v=1) \xrightarrow{\;\;\tau_{vp}\;\;} 2\ NO(J,\Omega,\Lambda) + E_K$$

where final states are designated by their total angular momentum J, the electronic angular momentum Ω, and the lambda doublet component Λ. E_K is the relative kinetic energy of the fragments and $\tau_{vp} = k_{uni}^{-1}$ is the lifetime of the energized complex. Conservation of energy arguments imply a $(NO)_2$ vdW bond energy (see below) $D_0\sim$800 cm^{-1}. Photodissociation in the region of the stretching fundamentals therefore produces $\sim$1000 cm^{-1} of excess energy which must be disposed into product rotational, vibrational, and translational degrees of freedom. Due to the small NO rotational constant ($B=1.7cm^{-1}$), energy gap arguments would lead to the expectation of long lived vibrationally excited dimers.[1]

II. EXPERIMENTAL

II.A. Energy Distributions

Van der Waals complexes were formed by expansions of gas mixtures through the 0.75 mm diameter nozzle of a pulsed free jet. The valve was of the Gentry-Giese design[11] and produced molecular pulses of 70 μs full-width at half-maximum (FWHM) duration. In the product state distribution experiments, the complexes were excited by the pulsed output of a hybrid, line tunable CO_2 laser[12] operated on the 10.6 μm P(30) transition. The infrared pump pulses were of TEM_{00} single longitudinal mode character. The pulse length was controlled electro-optically to be a temporal square-wave of either 50 or 2 ns duration. Since for the nitric oxide dimer excitation wavelengths in the 5 μm spectral region are required, the temporally modified ir output was frequency doubled in $AgGaS_2$. Conversion efficiencies ~1% were obtained, providing pulses of up to several hundred μJ energy at 1869.79 cm^{-1}. The pump laser traversed the molecular beam at a distance of 15 nozzle diameters downstream. At a backing pressure of 5 atms, the on-axis molecular beam density in this region was ~10^{17} cm^{-3}.

At some short time after the excitation pulse (i.e., 40±10 ns), the nitric oxide fragments formed from the vibrationally excited vdW complexes were probed by laser excited fluorescence techniques (LEF).[5,10,12] The probe laser was a frequency doubled, Nd^{+3}:YAG-pumped dye laser with a 9 ns FWHM duration and either a 0.3 or 0.017cm^{-1} FWHM bandwidth at 44 200 cm^{-1}. When tuned to resonance with an A $NO(V'=0;J')$ <-- X $NO(V=0;J,\Omega,\Lambda)$ transition, the resulting LEF intensity is proportional to the density of NO fragments in the level (J,Ω,Λ) at the time t_d during which the probe laser passed through the beam chamber. The fluorescence was collected by f/0.8 optics, passed through spectral filters to discriminate against laser scatter and a spatial filter to define the viewing region and detected by a solar blind (CsTe) photomultiplier tube. The LEF signals were recorded using a gated integrator and were normalized shot-by-shot to the probe laser energy (see Fig.1.). The state distributions of the fragments were determined by scanning the wavelength of the probe laser. By sufficient narrowing of the bandwidth of the excitation laser (i.e., to 0.017 cm^{-1}) the kinetic energy of the fragments was determined by Doppler profile measurements.

The ir pump and uv probe lasers were aligned either colinear or at right angles. The Gaussian ir beam radius was $R_{ir}=1x10^{-2}$ cm. The radius of the probe laser was smaller than that of the ir beam, $R_{uv}=5x10^{-3}$ cm, such that in the co-linear experiments the results corresponded to photolysis products formed by pulses of uniform, known

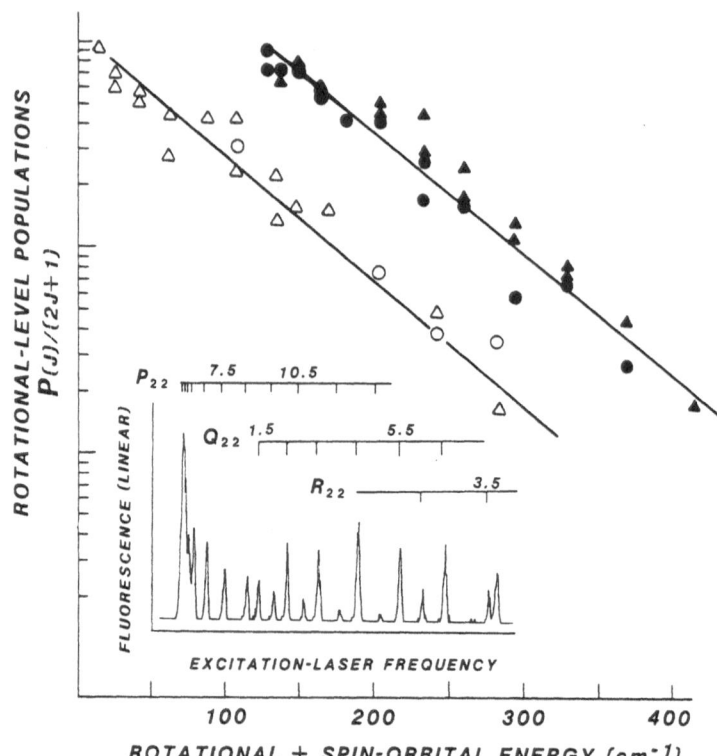

Fig.1. Portion of photofragment excitation spectrum probing NO fragments formed in the F_2 spin-orbit state and the resulting fragment state distribution obtained from the entire spectrum (see text).

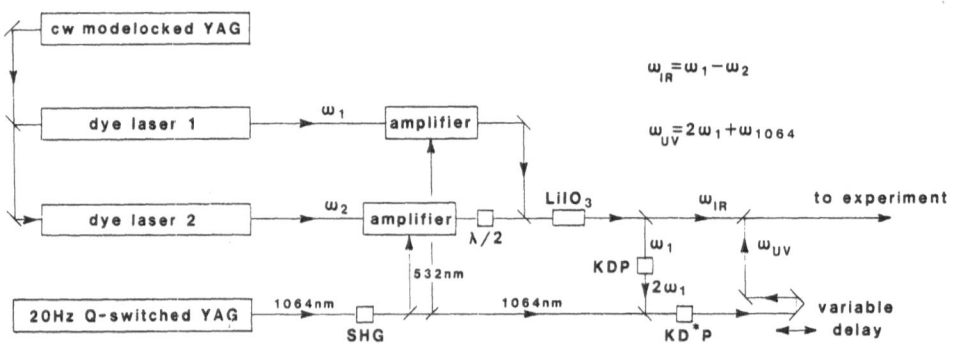

Fig.2. Apparatus for generating synchronous, tunable picosecond infrared pump and ultraviolet probe pulses.

intensity. In the perpendicular configuration independent control over both the orientation of laser electric vectors and propagation directions allowed for the assessment of fragment flux distributions and alignment.

II.B. Rates

The picosecond laser system, figure 2, used in these experiments employed a mode locked cw YAG laser to synchronously pump two tunable dye lasers, the outputs of which were amplified by pulsed dye amplifiers pumped by a frequency doubled YAG laser (2 ns pump pulses). The amplified outputs of these red-yellow lasers were each approximately 0.5 mJ per pulse. These pulses occured at 20 Hz and were transform limited in duration and spectral bandwidth. Presently, with 3-plate birefringent tuning elements in the dye laser cavities, the amplified pulses had 3 cm^{-1} bandwidths and 7.5 ps FWHM autocorrelations and 10 ps crosscorrelations. The ps pulses can be shortened in time, if necessary, at the expense of a broader spectral output.

To generate tunable ir pulses, difference frequency mixing between the two amplified dye laser outputs was done in $LiIO_3$ (for ω_{ir} = 1780 to 3500 cm^{-1}). The ir pulse energy obtained was about 10 μJ. Vibrational predissociation was observed as fragment appearance using the NO(fragment) uv LEF. Picosecond uv probe pulses near 226 nm were generated by mixing the frequency doubled output of one amplified dye laser with the 2 ns fundamental of the amplifier YAG in KD^*P. This leg of the ps system produced 5 to 10 μJ of uv energy. The time-resolved ir pump - LEF probe measurements of $(NO)_2$ fragmentation reported herein were performed using a dye laser wavelength of $\lambda_{dye\ 1}$ = 576.7 nm (ω_1=17 341 cm^{-1}) to produce uv probe pulses tuned to the P_{22} bandhead · probing fragments formed in low J-levels of the $NO(F_2)$ spin-orbit state. For ν_1 photodissociation, the wavelength of the second dye laser was tuned to $\lambda_{dye\ 2}$ = 646.4 nm as required to generate the difference frequency $\omega_1 - \omega_2$ = ω_{ir} = 1870 cm^{-1}, for ν_4 excitation, $\lambda_{dye\ 2}$ = 643.0 nm, gave ω_{ir} = 1789 cm^{-1}. The time evolution of the fragment concentration was measured by optically varing the delay time between ir pump and uv probe pulses. As in the ns experiments, the ir pump and optically-delayed probe pulses were focussed into a region of the pulsed supersonic jet sufficiently far from the nozzle that there were no collisions during the time scale of the experiment.

III. RESULTS AND DISCUSSION

In these experiments, the excitation sources were of low resolution. In the absence of fully resolved, assignable spectra, a recurring question is the identity of the parent species. A quadrupole mass spectrometer (QMS) was used in a differentially pumped molecular beam chamber to characterize the relative monomer/dimer/polymer concentrations produced by our valve for mixtures of NO in He and H_2 (see Figs. 3 and 4). Through such experiments, expansion gas mixtures were adjusted to obtain compositions which maximized dimer-to-monomer

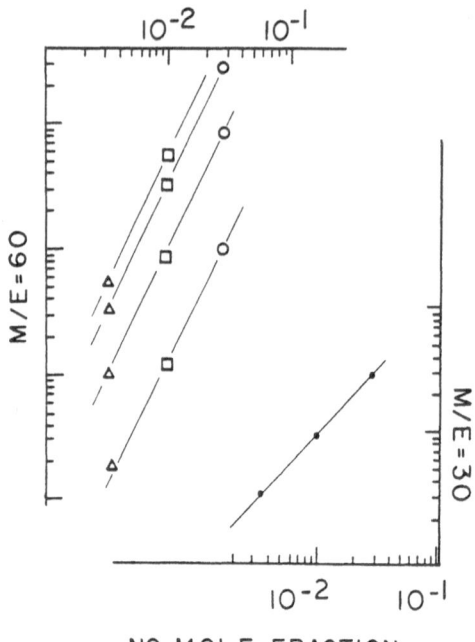

Fig.3. NO—mole fraction dependence of the m/e=60 mass spectrometer
signals for NO/He expansions at 8.3, 6.3, 3.7, and 1.8 atms. and
m/e=30 at 3.7 atms.

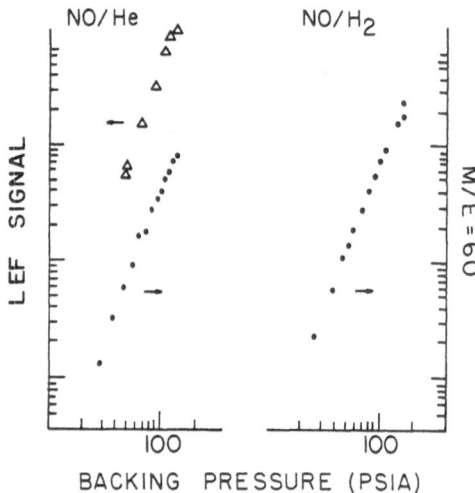

Fig.4. Backing pressure dependence of (dots) m/e=60 mass spectrometer
signals and (triangles) NO fragment LEF signals for 0.34% NO/He
expansions.

and dimer-to-polymer ratios. The QMS utilized an electron impact ionizer, and was operated with electron energies of 100 eV for the He expansions and 62 eV for H_2. The pressure dependence of the m/e=30 monomer signal was linear with backing pressure for $P_0 < 5$ atms, becoming sub-linear above this pressure. In addition, the m/e=30 signal exhibited a linear dependence on the NO mole fraction for a variety of mixtures within the range $10^{-3} < X_{NO} < 3 \times 10^{-2}$. For He expansions with backing pressures of $1.8 < P_0(atms) < 8.3$ the m/e=60 feature demonstrated a quadratic dependence on X_{NO}, at constant backing pressure, as required from a kinetics standpoint for the mole fraction scaling of the dimer. This quadratic dependence on X_{NO} was also observed in the LEF photolysis yield measurements, as shown in Fig.3.

Pressure dependence data, Fig.4., for the m/e=60 signal in expansions of 0.34% and 1% NO/He and 1% NO/H_2 showed $P_0^{2.7 \pm .3}$ scaling. To the extent that dimer formation is controlled by three body collisional events, the formation of dimeric species should ideally show a P_0^3 behavior, further implying that the m/e=60 ms peak represented the dimer species. We also measured the pressure dependence of the LEF-detected photodissociation yield experiments hoping to confirm that dimer photolysis was the predominant source of NO product. Results of such LEF yield vs P_0 experiments for both He and H_2 expansions gave $P_0^{2.7 \pm .3}$ scaling. The observed pressure dependence of the m/e=90 "polymer" feature was $\sim P_0^{3.8}$.

Our LEF diagnostics provide a straightforward technique for the determination of the photolysis yields. The $Q_{11}(0.5)$ transitions for both rare isotopic species [^{15}NO (0.37% natural abundance) and $N^{18}O$ (0.20 %)] are resolved. From these $Q_{11}(0.5)$ rare-isotope monomer intensities, the photo-fragment LEF intensities, and the fragment energy distributions (i.e., fragment partition function) we directly obtain the concentration of fragment species formed. Scaling the results for mole fraction, backing pressure, and ir energy to common values implies nearly equal dimer formation probabilities for the H_2 and He carriers.

III. A. Final State Distributions

Final state population distributions[10] were measured for different expansions excited by the frequency doubled output of the CO_2 laser at 1870 cm^{-1}. In particular, expansions of 1% NO in He at a backing pressure of 5.7 atms and 0.67% NO in H_2 at 8.3 atms gave rise to photolysis fragments that could be attributed predominantly to the fragmentation of the dimer. The data for fragment species with energies less that 450 cm^{-1} (>97% of all population lies in these levels) were well fit by the rotational temperatures $T_r(F_1)=101 \pm 11$ K and $T_r(F_2)=112 \pm 12$ K; the lower (F_1) and higher (F_2) energy spin-orbit states were fit as separate ensembles. The ratio of total population in the two spin-orbit states was $F_2/F_1 \doteq 0.9 \pm 0.2$. The population distributions for expansions in He and H_2 were essentially indistinguishable. Although the effects of state changing collisions became apparent at long time delays, the results were invariant for time delays of 10 to 150 ns.

The two spin-orbit states differ in energy, for a given value of
J, by an amount approximately equal to the 123 cm^{-1} spin-orbit
separation. The ratio of population in these two spin-orbit states
was substantially different from the value expected if the spin-orbit
and nuclear rotational degrees of freedom were equilibrated, as
observed for NO fragments formed from a number of thermal processes.[5]
If this condition were achieved, the populations in the F_2 state would
have been a factor of 5 lower than observed. That the ratio of spin-
orbit populations was near unity implies a strong spin correlation in
the dissociation mechanism which is consistent with the reaction
mechanism $(NO)_2(\nu_1)$ -->$NO(F_1)$ + $NO(F_2)$.

Many direct electronic photodissociation reactions result in an
alignment of the fragments.[13] For the NO A-X band system, high-J, Q
branch transitions are polarized parallel to the axis of nuclear
rotation. Thus a measure of the polarization dependence of the LEF
signal can in principle map out the alignment of the rotating
fragments, i.e., their m_J distribution. Using counter-propagating
lasers and monitoring LEF signal intensities from fragments in levels
J<9.5 the electric vector of the probe laser was rotated, with respect
to the electric vector of the pump laser, through 180° degrees using a
half-wave retardation plate. The results indicated a polarization
anisotropy[13] $|R|$<0.1; i.e., no significant alignment of the
fragments.

Kinetic energy distributions and fragment flux distributions
were measured by Doppler spectroscopy with a uv probe laser bandwidth
of 0.017 cm^{-1}. For an expansion of 0.67%NO in H_2 at a backing pressure
of 8.3 atms. in a geometry with the probe laser [tuned to the $Q_{22}(3.5)$
transition (E_{int}=145 cm^{-1})] aligned with its direction of propagation
perpendicular to that of the pump laser and to the electric vector of
the pump laser (i.e., the quantization axis), a Doppler profile of
nearly rectangular shape and with a FWHM of 0.165 cm^{-1} was observed
(see Fig.5.). Essentially identical Doppler profiles were recorded for
an expansion of 1% NO in He at 5.7 atms in this geometry, and for both
expansions in a geometry where the probe laser propagated in a
direction parallel to the electric vector of the pump laser.
Anisotropy in the fragment flux (as might be expected if the
vibrational predissociation process were prompt) should appear as a
pump-probe geometry dependence in the observed Doppler profiles. The
similarity of Doppler profiles obtained with the probe laser
propagating both parallel and perpendicular to the electric vector of
the pump laser indicated that the fragment flux was isotropic in space.
This is consistent with the predissociation occurring on a time scale
longer than the ~35 ps rotational period of the rotationally cold,
vibrationally excited dimer. More importantly, this spherical symmetry
in the flux allows for the derivation of a fragment velocity
distribution from these two sets of Doppler profiles.

Computer simulations of Doppler profiles were performed taking
into account the actual laser resolution, pump laser photon energy, and
final state distributions. Assuming the dimers in the expansion to be
essentially equilibrated with the beam temperature (T~1-5 K), and

allowing the internal energy of the co-fragment to be randomly chosen from the values available gave good fits to the observed FWHM=0.165±0.01 cm^{-1} near top-hat Doppler profiles. This gives an average per-fragment kinetic energy of 400±50 cm^{-1}, corresponding to a vdW bond energy of 800±150 cm^{-1}. Although the observed dependences of NO-fragment LEF and m/e=60 signals on NO mole fraction and backing pressure suggest these results are due to the vibrational predissociation of the nitric oxide dimer, it is the observed Doppler profiles that make the strongest argument for this. Photodissociation of either NO-He or NO-H$_2$ species can be excluded based on conservation of energy and momentum arguments.

Upon increasing the backing pressure with the 1% NO in He expansion above 5.7 atms, the observed Doppler profiles began to show evidence for the presence of fragments formed from the vibrational predissociation of polymer species. For an expansion of 3% NO in He at 9 atms the observed profiles were Gaussian in shape with a 0.090 cm^{-1} FWHM, independent of pump-probe geometry. Additionally, in contrast to the dimer results, under these expansion conditions the spin-orbit and nuclear rotational degrees of freedom appeared to be equilibrated, with only 15% of the total population being in F$_2$ states. Although most of the dimer had clearly been converted to trimer/polymer in such an expansion (as evidenced by the changes in fragment Doppler profiles and spin-orbit ratio), the mass spectrometer m/e=60 feature and NO fragment LEF signals continued to show $P_0^{2.7} X^2$ scaling.

Menoux et al.[8] have reported the results of ir intensity measurements of the dimer ν_1 fundamental and ($\nu_1 + \nu_4$) combination band in a static cell at temperatures of 118 to 138 K. From Van't Hoff plots, a value of the dimer heat of formation -2.25±0.25 kcal/mol (at 128 K) was derived. This corresponds to a vdW bond energy D_0=590±90 cm^{-1}. From the results of our energy distribution measurements, conservation of energy directly gives a value of D_0=800±150 cm^{-1}, in fair agreement with the ir intensity measurements.

III.B. Rates

Picosecond ir and uv pulses were used to measure the actual dissociation behavior of NO-dimers excited to the v=1 levels of the NO-stretches. The ir pulses were focussed to a 0.3 mm diameter beam waist to vibrationally excite the beam cooled dimers. The uv probe pulses traversed a variable optical delay line and were focussed to a 0.15 mm diameter waist colinear with the ir pump beam. Probe pulses, polarized at the magic angle and traversing the beam at time t_d, relative to the pump pulses, excited fragments formed in the J = 1.5-7.5 levels of the X$^2\Pi_{3/2}$ state (P$_{22}$ bandhead of the A-X system), resulting in LEF signals directly proportional to the number of NO fragments formed during the time t_D. Results were obtained for separate excitations of the (NO)$_2$ ν_1(v=1) and ν_4(v=1) bands at 1870 and 1789 cm^{-1} respectively. The dramatic difference in dissociation rates observed for these two modes of nearly equal energy is shown in figure 6.

If the ir laser excited an ensemble which dissociated with a single unimolecular rate constant, $k_{uni} = 1/\tau_{vp}$, the expected fragment

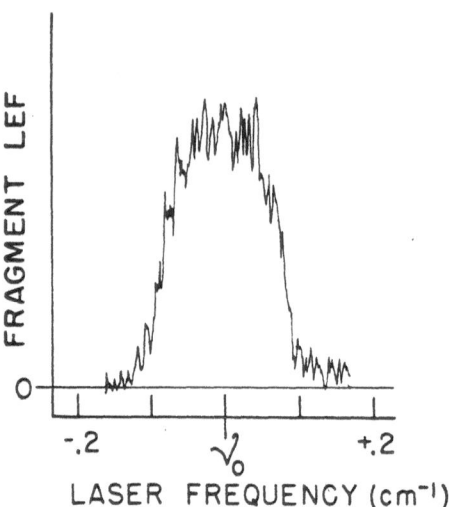

Fig.5. Doppler profile of NO(F$_2$,J=3.5) fragments formed from vibrational predissociation of the nitric oxide dimer excited at 1870 cm^{-1}.

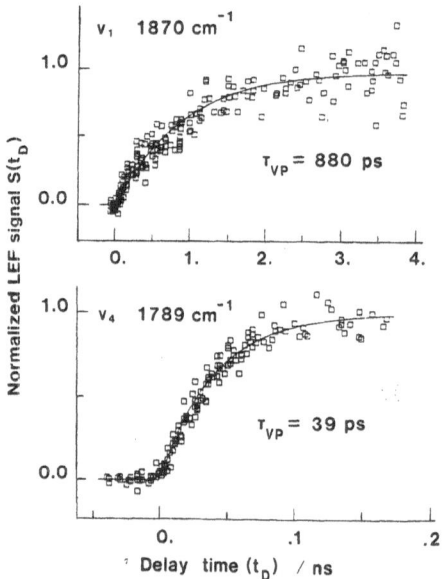

Fig.6. Picosecond infrared pump – time delayed picosecond ultraviolet probe laser measurements of the vibrational predissociation lifetimes of the nitric oxide dimer excited to the v=1 levels of its symmetric and antisymmetric stretches.

appearance behavior would be: $S(t_d) = S(\infty)[1-\exp(t_d/\tau_{vp})]$.
Of course non-exponential or multi-exponential decay could occur. Nonlinear least-squares fits of single exponential decays to the data,
yielded lifetimes of 880 ± 260 picoseconds for ν_1 and 39 ± 8 picoseconds
for ν_4 (a 39 ps lifetime would produce a 0.14 cm^{-1} FWHM homogeneous
linewidth). In the experiments these lifetimes represent average
values for a distribution of rotational levels. Assuming an infrared
bandwidth of 4-5 cm^{-1} (the convolution of two 3 cm^{-1} Gaussians) and the
thermalization of the dimers with the beam (T~1K), the pump laser
interacts with essentially all the NO-dimers in the molecular beam.
Thus, species in levels with $J''\leq3$, $K_a''\leq2$ account for 95% of the
photodissociation signal.

   That ν_4(1789 cm^{-1}) decays much faster than ν_1 (1870 cm^{-1}) is
inconsistent with standard statistical theories of unimolecular
reaction which predict unimolecular rates to increase with reactant
internal energy. On the other hand, models of vibrational
predissociation involving pure V-T energy transfer[14] correlate
increased lifetime with increased energy release. However,
predissociation from ν_1 only releases 81 cm^{-1} more energy than ν_4 and
theory quantitatively predicts a much smaller difference in rate
(factor of ≤2) between ν_1 and ν_4 than was observed. In the case of
$(NO)_2$, the predissociation might be nonadiabatic since the $^2\Pi$
configurations of the NO fragments combine to form several low-lying
electronic states in the dimer. Such nonadiabatic effects have been
suggested to explain the anomalously large collisional cross-section
observed for vibrational deactivation of NO(v=1) by NO(v=0).[15] The
different overall symmetries of the ν_1 and ν_4 levels of the dimer $^1A_1$
ground state may constrain coupling to low lying repulsive state(s),
causing the observed difference in lifetimes. Ongoing measurements of
product energy distributions following ν_4 dissociation and lifetimes
for states of different vibrational symmetries (e.g., combination bands
involving the vdW modes) might elucidate the source of the difference
in ν_1 and ν_4 predissociation rates.

IV. CONCLUSIONS

Vibrational predissociation of the stretching fundamentals of the
nitric oxide dimer has been observed. Excitation of ν_1 results in
fragments with little internal excitation. There is equal probability
of fragment formation in either NO-spin-orbit state implying a strong
spin correlation in the dissociation channel. Although ~75% of the
available energy goes into translational excitation of the fragments,
these vibrationally excited parents dissociated with a 880 ps lifetime.
Excitation of the lower energy ν_4 level resulted in a 39 ps lifetime.

V. REFERENCES

1. For a review, see K.C. Janda, Adv. Chem. Phys. **60**, 201 (1985).
2. M.A. Hoffbauer, K. Liu, C.F. Giese, and W.R. Gentry, J. Chem. Phys. **78**, 5567 (1983).
3. D.S. Bomse, J.B. Cross, and J.J. Valentini, J. Chem. Phys. **78**, 7175 (1983).
4. M.F. Vernon, J.M. Lisy, H.S. Kwok, D.J. Krajnovich, A. Tramer, Y.R. Shen, and Y.T. Lee, J. Phys. Chem. **85**, 3327 (1981).
5. D.S. King and J.C. Stephenson, J. Chem. Phys. **82**, 5826 (1985).
6. C.M. Western, P.R.R. Langridge-Smith, B.J. Howard, and S.E. Novick, Mol. Phys. **145**, 44 (1985).
7. S.G. Kukolich, J. Mol. Spectrosc. **98**, 80 (1983).
8. V. Menoux, R. LeDoucen, C. Haeusler, and J.C. Deroche, Can. J. Phys. **62**, 322 (1984).
9. Ph. Brechignac, S. DeBenedictis, N. Halberstadt, B.J. Whitaker, and S. Avrillier, J. Chem. Phys. **83**, 2064 (1985).
10. M.P. Casassa, J.C. Stephenson, and D.S. King, J. Chem. Phys. **85**, 2333 (1986).
11. W.R. Gentry and C.F. Giese, Rev. Sci. Insts. **49**, 595 (1978).
12. J.C. Stephenson and D.S. King, J. Chem. Phys. **78**, 1867 (1983).
13. P. Andresen, G.S. Ondrey, B. Titze, and E.W. Rothe, J. Chem. Phys. **80**, 2548 (1984).
14. J.A. Beswick and J. Jortner, Adv. Chem. Phys. **47**, 363 (1981).
15. J.T. Yardley, Introduction to Molecular Energy Transfer, Academic Press (New York, 1980).

COMPLEX FORMING REACTIONS IN NOBLE GAS CLUSTERS

D. Levandier, R. Pursel and G. Scoles,
Centre for Molecular Beams and Laser Chemistry,
University of Waterloo, Waterloo, Ontario, Canada, N2L 3G1.

ABSTRACT

We have succeeded in monitoring the complex-forming reaction between CH_3F and HCl in a noble gas cluster environment. A diluted mixture of CH_3F in Ar is expanded from a supersonic nozzle forming a beam of Ar clusters which contain one or more CH_3F molecules and which can be characterized by bolometric photoevaporation IR spectroscopy using line tunable CO_2 lasers. In the distance between the nozzle and the skimmer the clusters can also be exposed to a cross flux of HCl molecules. A fraction of the HCl molecules is picked-up by the clusters and the ensuing complex- forming reactions may be detected by monitoring the decrease (or lack thereof)'of the IR spectral peaks of the CH_3F containing species (CH_3F, $(CH_3F)_2$ and $(CH_3F)_3$) and the appearance of red shifted absorptions attributable to $(CH_3F)_p (HCl)_m$ complexes. The implications of these experiments for the study of chemical dynamics in a noble gas cluster environment are pointed out and the possibilities for future work discussed.

INTRODUCTION

The complete microscopic description of the relaxation behaviour and chemical activity of molecules in solution represents one of the major unresolved problems of chemical physics today.

While computer simulation, statistical theories and short-pulse laser spectroscopy are making inroads into the subject, a relatively recent line of work, made possible by the advent of molecular beam laser spectroscopy, studies the evolution of molecular behaviour in a cluster environment. Doing this as a function of the size of the cluster tries to bridge the gap between free-molecular and solvated behaviour in a quasi-continuous way. The bulk of the activity so far has been spectroscopic in nature, dealing mostly with small clusters[1-5]; large clusters are instead the focus of a relatively small number of studies (see ref. 6 to 10 and references therein). With respect to chemical activity, while several experiments have employed clusters as reagents (11), the majority of the papers dealing with what

525

A. Weber (ed.), Structure and Dynamics of Weakly Bound Molecular Complexes, 525–532.

happens <u>within</u> the clusters have studied reactions of the ion-molecule
type, a fact well documented by Garvey and Bernstein in their recent
paper on the subject[12]. There is, therefore, the need to study
reactions between neutral atoms and molecules imbedded in simple
clusters since the interactions prevailing in these systems are
different from those involving ions and if the molecules or atoms are
in their ground electronic states are also known or can be calculated
to a reasonable degree of accuracy. The prior knowledge of the forces
involved is naturally a prerequisite for a satisfactory study of
systems where the description of the dynamic behaviour requires complex
simulations and/or approximate statistical theories. As an example
of such calculations we will mention that recently Amar and Berne have
published a paper[13] in which they carry out the dynamical simulation of
the photodissociation/ recombination reaction of Br_2 in Argon clusters
while similar studies are being performed for Cl_2 by Gerber and
coworkers[14].

The present paper represents an experimental effort in the same
direction where, however, the reacting species are molecules instead
of atoms.

Recently we have demonstrated that[15] the method of bolometrically
detected photodissociation spectroscopy, introduced by us about ten
years ago[16], could also be used for large clusters thus making possible
the study of the evolution of the infrared spectrum of a molecule from
the isolated molecule state to the "matrix" type environment prevailing
in highly condensed beams. The latter conditions are obtained by
expanding a diluted mixture of the molecule of interest with a noble
gas (typically argon) from a supersonic nozzle at relatively high
pressures, while the former situation prevails of course at low
pressures. More recently we have also shown[17] that when the cluster is
large enough and the spectroscopically active molecule is deposited on
its surface the "surface spectrum" of the molecule can be obtained
showing features which are intermediate between the "matrix" spectrum
and the "isolated molecule" lines. The method by which the chromophore
is deposited on the cluster surface, called the pick-up technique, is
very simple. It consists of exposing the freshly formed clusters,
shortly after the expansion is completed and before they enter the
collimating skimmer, to a side flux of the chromophore molecules.
These are efficiently picked-up, probably by exchange reaction, by the
clusters causing perturbations which, when the clusters are large
enough, are not very important. Quite clearly by seeding a molecule
inside the cluster, and depositing another on its surface, the pick-up
technique can be used to study chemical reactions that eventually take
place between the two molecules, provided the reaction products also
show an infrared absorption in the region covered by the CO_2 lasers
used to characterize the beam. In this paper we wish to present the
results of a first successful study of this type in which $(CH_3F)_p$
species, seeded in argon clusters, are characterized by means of
infrared spectroscopy and these react with HCl molecules introduced in
the clusters by the pick-up technique. The products of the hydrogen
bonded complex forming reaction have been detected as a decrease in the
infrared peaks corresponding to CH_3F, $(CH_3F)_2$ and $(CH_3F)_3$ and the

appearance of red shifted features of which only that corresponding to
the $CH_3F \cdot HCl$ complex has been assigned with confidence. These
experiments quite clearly show the feasibility of using the pick-up
technique in order to extend our knowledge of chemical reactions in a
cluster environment. The detection method does not need to be limited
to the present one (infrared laser spectroscopy) since for different
seeded molecule-picked-up species combinations mass spectrometers,
laser induced fluorescence or merely photoluminescence detectors may
be a more productive choice. Detecting the reaction products at
different positions along the beam will quite clearly provide us with
the possibility of time resolving the physico-chemical process occurring in the cluster provided the characteristic times are not much
shorter than about 100 nsec, which corresponds to a spatial resolution
of about 0.01 cm along the beam. The implications of the present
results and the possibilities opened up by them will be discussed at
the end of the paper.

EXPERIMENTAL

    A schematic view of the apparatus is given in fig. 1. It consists
of a radiation source, a liquid He cooled Ge bolometer detector and a
differentially pumped supersonic molecular beam source.
    The radiation source is a tunable cw CO_2 laser (either $^{12}C^{16}O_2$ or
$^{13}C^{16}O_2$) operated on a single mode. The laser beam is attenuated such
that it has $\simeq$ 0.7W of power when entering the experimental chamber of
the apparatus.
    The molecular beam source is comprised of two differentially pumped
vacuum chambers, separated by a commercially available skimmer (Beam
Dynamics Model 1) with a 0.5 mm orifice. Typically the pressure in the
primary chamber ranges from 7×10^{-4} Pa to 3×10^{-1} Pa and the secondary
chamber operates in the 10^{-4} - 10^{-5} Pa range.
    The supersonic nozzle, which has an orifice of $\simeq 40 \mu m$, is mounted on
a 3-D translation stage. The distance from the nozzle to the skimmer
was kept constant at 10mm. When working on "pick-up" experiments an
effusive source (a multichannel source of ca.8,500 $10 \mu m$ holes $500 \mu m$
long) positioned, typically, 7mm downstream of the supersonic nozzle and
2-3mm below the molecular beam axis. The HCl gas used was a Matheson
Semiconductor gas of 99.995% purity; for the CH_3F/Ar mixtures Linde
99.999% purity argon and U.S. Services 97^+% purity CH_3F
(B.P. = $-78.5°C$) were used.
    In the secondary chamber the laser beam intersects the cluster beam
causing dissociation of the clusters. The main cluster beam is then
blocked while the fragment flux passes through a series of slits before
falling on the detector.
    The chopped laser beam is focused, using an anti-reflection coated
ZnSe lens, to a 0.5mm waist when passing through the molecular beam and
an external system of steering mirrors allows precise positioning such
that the signal from the bolometer is optimized.

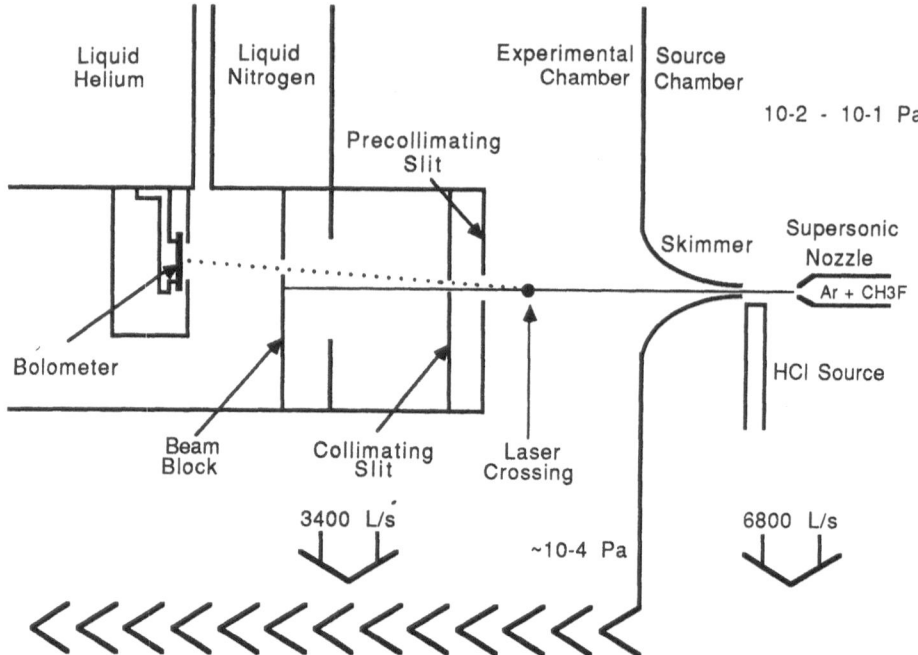

Figure 1 - Experimental arrangement.

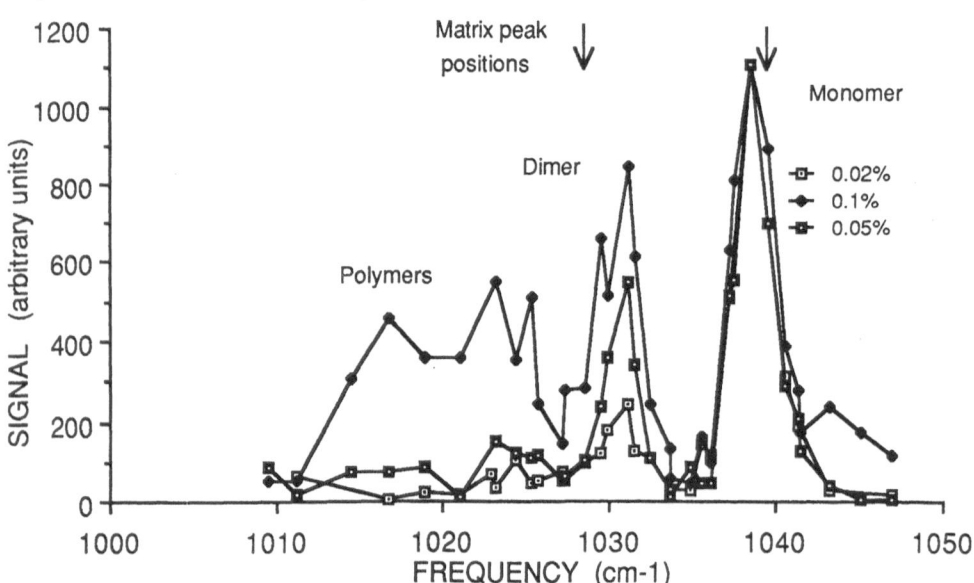

Figure 2 - The three spectra were generated by expanding CH_3F/Ar mixtures of the concentrations indicated, at 1800 kPa. The spectra are normalized to the signal of the 9R(30) $^{13}C^{16}O_2$ line. The arrows show the peak positions for the monomer and dimer in solid Ar matrices (ref. 18).

The detector used is a commercial liquid He cooled, doped Ge bolometer thermally connected to a gold-plated sapphire crystal of the dimensions 1 cm x 1cm x 1mm (Infrared Laboratories).

The cryostat maintains the operating temperature at 1.7K and in order to optimize the beam flux the bolometer can be rotated around the centre of the cryostat.

The modulated signal from the detector is passed through a lock-in-amplifier (Princeton Applied Research Model 126 or Stanford Research Systems Model SR510) and then stored on a microcomputer (Commodore PET). The spectrum is constructed by measuring the fragment flux produced upon dissociation at each laser line in the spectral region of interest. This flux is normalized for laser power and corrected for any evolution in the responsivity of the bolometer with time by normalization to the signal produced by dissociation using either the 9R(32) $^{13}C^{16}O_2$ or 9P(28) $^{12}C^{16}O_2$ line. Because of the near coincidence of these same lines (1039.3810 cm^{-1} and 1039.3693 cm^{-1}, respectively), the data obtained with one laser was normalized to those from the other by scaling the spectra at these points.

RESULTS AND DISCUSSION.

Three different measurements were carried out in order to monitor the reaction between CH_3F and HCl in the environment provided by the Ar cluster.

The relative intensities of monomer to higher order polymer peaks were measured over the concentration range 0.1% to 0.02% of CH_3F in Ar at a constant nozzle pressure of 1800 kPa. These data, as presented in fig. 2, are normalized to the monomer peak (at the 9(R)30 $^{13}C^{16}O_2$ line). Assignment of the monomer ($CH_3F.Ar_n$) and dimer (($Ch_3F)_2.Ar_n$) peaks to the two clearly resolved features occurring at 1038 cm^{-1} and 1031 cm^{-1} respectively is made by comparison with the IR spectra of CH_3F in solid argon matrices[18]. The unresolved structure, which appears to the red of the dimer, is attributed to contributions from higher polymers. The number of argon atoms depends on the nozzle backing pressure and, in the range 700 - 3450 kPa, is thought to have an average between 100 and 1000. The temperature of the clusters is estimated to be between 30K and 40K[19]. The "gaps" in the spectrum occur in the band gap of the 9μm branch of $^{13}C^{16}O_2$ (1011.2011 cm^{-1} - 1024.3677 cm^{-1}) and to the red of the 9P(52) $^{12}C^{16}O_2$ line (1014.5179 cm^{-1}) where CO_2 lines are inaccessible to us.

Comparing the relative signals for different CH_3F/Ar concentrations it can be seen that for the dimer the signal decrease is proportional to the decrease in concentration while the "polymer" band signal shows a far more dramatic concentration dependence. Matrix isolation studies performed by Barnes et al.[20] on the CH_3F trimer, tetramer and pentamer (which show absorption at 1022 cm^{-1}, 1016 cm^{-1} and 1008 cm^{-1}) lend support to this assessment.

In figs. 3 and 4 the results of HCl pick-up at constant CH_3F/Ar concentration (0.05%) but at two different expansion pressures (1800 kPa and 3100 kPa) are presented. In both instances the spectra shown were

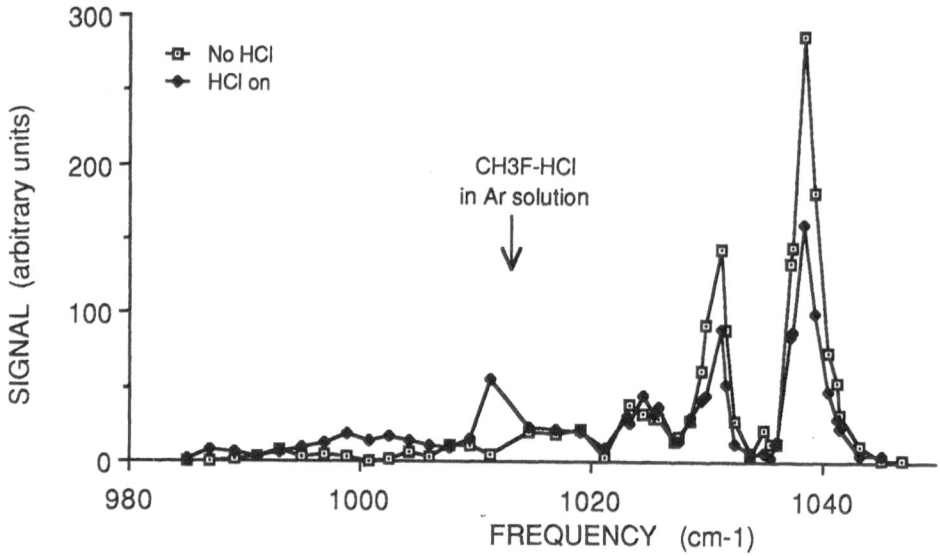

Figure 3 - These spectra were produced with 0.05% CH$_3$F/Ar expanded at
1800 kPa, either with or without an HCl pick-up beam, as indicated. The
arrow marks the peak position of the HCl-CH$_3$F complex in liquid Ar at
95K (ref. 21).

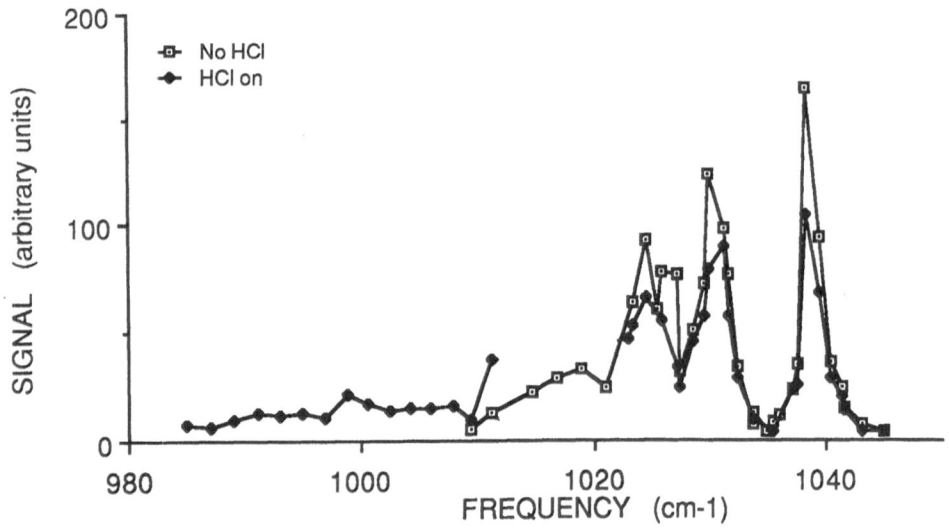

Figure 4 - These spectra were produced by expanding 0.05% CH$_3$F/Ar at
3100 kPa, with or without the HCl pick-up, as indicated.

recorded under identical conditions, first with the pick-up turned off, and then with a flux of HCl molecules present. At both pressures the monomer and dimer peaks are clearly reduced when the flux of HCl is present while a similar cross flux of argon produces only small changes in the spectra. Moreover, with the HCl flux present, fig. 3 shows clearly the appearance of new features at 1012 cm^{-1} and 1002 cm^{-1}. The IR spectrum of CH_3F/HCl complex in liquid Ar solutions taken by Barri and Tokhadze[21] suggests that the 1012 cm^{-1} peak can be assigned to a hydrogen bonded CH_3F-HCl species embedded in the Ar_n cluster, the complex formation causing a shift in the frequency of the monomer peak of 26 cm^{-1}. While the same general trend is seen, the magnitude the drop in the monomer signal is markedly less in the second pair of spectra taken at the higher nozzle pressure of 3100 kPa (and thus larger cluster size).

If it is assumed, as is reasonable, that there is no decrease in the efficiency of the pick-up technique with increasing cluster size, this comparison can be taken to imply that, while for the smaller clusters the CH_3F/HCl reaction is completed before interaction with the radiation, for the larger clusters this is no longer the case. This decrease in reactivity with cluster size is expected since the diffusion time of HCl into the larger clusters should be longer. We intend to test the validity of the assumption made above by carrying out spectral measurements at different distances downstream from the point of pick-up for pressures which indicate a decreased reactivity. However, since the bolometer should be kept at a fixed distance from the laser crossing (to avoid changing the fragment collection efficiency) the test described above requires substantial modifications to the apparatus and will have to be carried out at a later time. On the other hand, in case the reduced reactivity points will be shown to be stable at different points downstream from the point of pick-up the interesting possibility will arise that the reduced reactivity be generated by some kind of phase transition in the cluster. Since from other work done in our laboratory[22] the clusters are believed to be already solid at pressures as low as 700 kPa, the second possibility is believed to be less likely than the first.

These preliminary results on the dependence of reactivity on cluster size suggest that extending these measurements to a range of nozzle pressures could provide a useful method of studying the diffusion coefficient for HCl in Ar clusters. Furthermore the size dependence of the diffusion process can be studied by investigation of the reactivity with CH_3F of a series of compounds such as the hydrogen halides or the alkylhalides.

ACKNOWLEDGEMENTS

This work was carried out with the financial support of the Natural Science and Engineering Research Council and a grant of the Petroleum Research Fund of the American Chemical Society. It is a pleasure to thank K. Manning for technical assistance and Dr. J. McCombie for her help on completing the manuscript before the (second) deadline. One of us (DL) would like to acknowledge the support of an NSERC postgraduate scholarship during this work.

REFERENCES

1. W. Klemperer, J. Mol. Struct., $\underline{59}$, 161 (1980).
2. D.H. Levy, Adv. Chem. Phys., $\underline{47}$, 323 (1981).
3. W.R. Gentry in "Resonances", D.G. Truhler, Ed., ACS Symp. Series, No. 263, ACS, Washington, DC., 1984, page 289.
4. K.C. Janda, Adv. Chem. Phys., $\underline{60}$, 291 (1985) ; F.G. Celii and K.C. Janda, Chem. Rev., $\underline{86}$, 507 (1986).
5. R.E. Miller, J. Phys. Chem., $\underline{90}$, 3301 (1986).
6. A. Amirav, U. Even and J. Jortner, J. Chem. Phys., $\underline{75}$, 2489 (1981).
7. T.E. Gough, R.E. Miller and G. Scoles, J. Phys. Chem., $\underline{85}$, 4041 (1981).
8. R.E. Miller, R.O. Watts and A. Ding, Chem. Phys., $\underline{83}$, 155 (1984).
9. D.F. Coker, R.E. Miller and R.O. Watts, J. Chem. Phys., $\underline{82}$, 3554 (1985).
10. D.H. Levy, in "Quantum Dynamics of Molecules", Wooley, R.G., Ed., Plenum Press, N.Y. 1980.
11. See ref. 9 to 16 of J.F. Garevy and R.B. Bernstein, Chem. Phys. Lett., $\underline{126}$, 394 (1986).
12. J.F. Garvey and R.B. Bernstein, J. Phys. Chem., $\underline{90}$, 3577 (1986).
13. F.G. Amar and B.J. Berne, J. Phys. Chem., $\underline{88}$, 6720 (1984).
14. R.B. Gerber, private communication (1986).
15. T.E. Gough, D.G. Knight and G. Scoles, Chem. Phys. Lett., $\underline{97}$, 155 (1983).
16. T.E. Gough, R.E. Miller and G. Scoles, J. Chem. Phys., $\underline{69}$, 1588 (1978).
17. T.E. Gough, M. Mengel, P.A. Rowntree and G. Scoles, J. Chem. Phys., $\underline{83}$, 4958 (1985).
18. L.H. Jones and B.I. Swanson, J. Chem. Phys., $\underline{76}$, 1634 (1982).
19. J. Farges, M.F. deFeraudy, B. Rault and G. Torchet, J. Chem. Phys., $\underline{78}$, 5067 (1983) and J. Chem. Phys., $\underline{84}$, 3491 (1986).
20. A.J. Barnes, H.E. Hallam, J.D.R. Howells and G.F. Scrimshaw, Farad. Soc. Trans., $\underline{269}$, 738 (1973).
21. M.F. Barri and K.G. Tokhadze, Opt. Spectr. (USSR), $\underline{51}$, 70 (1981).
22. T.E. Gough, M. Mengel, P. Rowntree and G. Scoles (in preparation).

STRUCTURE AND DYNAMICS OF THE RARE GAS–HALOGEN VAN DER WAALS MOLECULES: PRODUCT STATE DISTRIBUTIONS FOR VIBRATIONAL PREDISSOCIATION OF $NeBr_2$

Joseph I. Cline,[a] Dwight D. Evard, Brian P. Reid, N. Sivakumar, Fritz Thommen[b] and Kenneth C. Janda
Department of Chemistry
University of Pittsburgh
Pittsburgh, Pennsylvania 15260
U. S. A.

ABSTRACT: A brief review of the structure and dynamics of the rare gas halogen van der Waals molecules is given. New results are presented for the lifetimes of several vibrational levels of the B state of $NeBr_2$. Also presented for the first time are the branching ratios into the possible product channels upon vibrational predissociation of $NeBr_2$. Most of the molecules have structures not too different from what one would obtain by bringing the rare gas atom into van der Waals contact with the two halogen atoms. ArClF and KrClF are exceptions to this trend. They are linear molecules with a van der Waals bond length substantially shorter than the sum of atomic van der Waals radii. The dynamics observed for $NeBr_2$ are not too different from what one would have predicted on the basis of earlier HeI_2 results. For lower vibrational levels the $\Delta v = -1$ channel predominates and only a modest fraction of the product kinetic energy is transferred to the rotational degree of freedom. For higher vibrational levels (near $v = 28$) the $\Delta v = -1$ channel closes and increasing values for Δv are observed. Analysis of this data yields a value for the X state D_0 of 70.5 ± 2.0 cm^{-1}.

1. Introduction

   The set of van der Waals molecules formed by joining a rare gas atom to a halogen or an interhalogen molecule has been intensively studied during the last thirteen years. These molecules provide interesting prototypical examples for how the structure and dynamics of a van der Waals molecule vary with the mass and atomic number of the

a. National Science Foundation Predoctoral Fellow
b. Present address, CIBA–Geigy, Basel, Switzerland

A. Weber (ed.), Structure and Dynamics of Weakly Bound Molecular Complexes, 533–551.

constituents, the interaction well depth and anisotropy and the vibrational frequencies of the molecule. Even for such a similar set of
molecules as the rare gas–halogens it is not easy to predict the molecular properties in advance of a measurement. Particularly in the case
of vibrational predissociation dynamics, the range of the results was
completely unexpected when these experiments were started. This paper
will briefly review the literature on this set of molecules and present
new results on the vibrational predissociation dynamics of NeBr$_2$.

2. Review of structure and dynamics of rare gas halogens

2.1 Molecular Structure

The first rare gas–halogen molecules to be studied were ArClF and
KrClF.[1,2] Molecular Beam Electric Resonance Spectroscopy (MBER) was
used to obtain precise rotational constants, hyperfine splittings and
Stark shifts for the ground electronic and vibrational levels of the
molecules. These constants were interpreted in terms of a linear
equilibrium structure in which the rare gas atom bonds to the Cl atom
of ClF. Since the atom–atom distances, 3.330 Å for Ar–Cl and 3.388 Å
for Kr–Cl, are somewhat shorter than the sum of the van der Waals radii
of the atoms, the linear structures were interpreted in terms of a
HOMO–LUMO model. This conclusion was later reinforced by analogous
bonding in HFClF[3] and HFCl$_2$[4] in which the heavy atoms are bonded in a
linear configuration and the bond angle of the hydrogen atom is
suggestive of a sp^3 hybridization. A series of LEF studies by Levy and
his students[5] determined both the structure and the vibrational predissociation lifetimes for HeI$_2$. The molecule was found to have a
triangular, or "T" shaped, structure as would result if atom–atom van
der Waals forces were the major determinant of its structure.

If ArClF is really a chemically bonded molecule while HeI$_2$ is
strictly a van der Waals molecule, then it is of great interest to
determine which molecular properties distinguish the two types of
bonding and to classify the rest of the series of rare gas halogens
into one group or the other. Helium is the least polarizable rare gas
atom and it is not surprising that it would form purely van der Waals
bonds. In fact the He–I$_2$ bond energy is only 18 cm^{-1}, 0.05 kcal/
mole.[6] The potential energy surface would probably best be described
as that of a hindered internal rotor. The ArI$_2$ bond energy is 235
cm^{-1}, 0.71 kcal/mole. Is this interaction strong enough so that a
linear, "chemical" structure results? This question remains
unanswered. Alternatively, the ClF molecule has a rather large dipole
moment. Could it be that the linear structure of ArClF is more a
result of electrostatic forces than the HOMO–LUMO interaction? How
can the structure be expected to vary as the rare gas atom or the
electronic state of the halogen molecule is changed? These are the
sort of simple structural questions that remain to be answered.

Table I contains a summary of the geometrical structures, van der
Waals bond lengths, and bond dissociation energies for the ground
electronic states of those members of this class of molecules for which

such information is presently available. For cases where the measurements have been made, it appears that the electronic ground state and the A and B electronic excited states for a given molecule have very similar bond lengths and geometries for the van der Waals bond. The similarity of structures is determined by rotational resolution of the spectra and the absence of vibrational progressions involving van der Waals modes. Apparently, however, the ion pair (E) state of the NeICl molecule has a structure which is quite different from the lower excited states. Lester et al.[7] have attributed structure in the E-A spectrum to vibrational progressions in both the van der Waals stretching and bending modes.

Table I. Structures and Bond Energies for Rare Gas-Halogen Molecules

Molecule	Structure Type	Bond Energy (cm^{-1})	Bond Length $(Å)^a$	Refs.
$HeCl_2$	T shape	9 ± 3	3.8 ± 0.4	8,9
$HeBr_2$	"	–	3.7 ± 0.2	10
HeI_2	"	18 ± 5	4.5 ± 0.1	5
$NeCl_2$	"	55 ± 5	3.57 ± 0.04	11,12
$NeBr_2$	"	70 ± 2	3.67 ± 0.01	13,14
NeI_2	"	75 ± 5	–	5
$ArCl_2$	"	175 ± 5	4.0 ± 1.0	15
ArI_2	–	235 ± 5	–	5
ArClF	linear	230	3.3301	1,16
KrClF	linear	290	3.3884	2,16
HeICl	bent	–	–	17
NeICl	bent	48.2 ± 0.6	–	17

a. For the "T" shaped molecule the bond length represents the rare-gas atom to halogen center of mass distance. For the other molecules, the nearest neighbor atom-atom distance is given. All lengths are the R_0 values.

As a point of reference for comparing relative bond lengths and energies, Table II gives the analogous values for some of the rare gas dimers. One very interesting comparison is that the well depth is about constant for Ne bonded to Ar, Kr or Xe, but increases by 50 percent going from $NeCl_2$ to NeI_2. On the other hand the difference in the bond lengths of $NeCl_2$ and $NeBr_2$ is about the same as the difference between the NeAr and NeKr bond lengths. The van der Waals radius of Cl is estimated to be about 0.1 Å less than that of Ar.[19] Thus the $NeCl_2$ bond length is not only considerably longer than that of ArClF, it is also longer than what one might have expected from simply letting the van der Waals radii of the constituent species come into close contact. It is apparent that ArClF has a substantially shorter bond than other members of this class of molecules. Although the $ArCl_2$ bond length has only been estimated from low resolution spectroscopy, it is unlikely that it is comparable to that of ArClF. Finally, it is clear that the ClF dipole moment is not the primary reason for the short, linear ArClF bond. The dipole moment of ICl is fifty percent larger than that of ClF, yet NeICl has a bent configuration of the atoms with a relatively long van der Waals bond length. Perhaps ArClF and KrClF do indeed exhibit incipient chemical bonding. It would be very exciting to obtain the structures of Ar and Kr bonded to the other interhalogen molecules to see if other examples of such bonding can be identified.

It is particularly interesting to note that the NeICl bond energy is not only much less than that of ArClF, it is even less than the $NeCl_2$ bond energy. How can this be understood? For rare gas pairs it is generally true that the well depth correlates with the least polarizable bonding partner. In the comparison of NeICl and $NeCl_2$, however, one might have expected the ICl dipole moment to add considerably to the bond strength. It is not clear how any of the attractive forces in the NeICl would cancel each other out. van der Waals forces, dipole induced dipole forces and quadrupole induced dipole forces are attractive for all geometries.

Table II.   Bond Lengths (top, Å) and Bond Energies (bottom, cm^{-1}) for Rare Gas Dimers[a]

	He	Ne	Ar	Kr
He	2.97			
	7.3			
Ne	3.21	3.15(3.1)		
	9.9	28.6(16.3)		
Ar	3.51	3.43	3.758(3.821)	
	19.	50.0(36.0)	97.82(84.75)	
Kr	3.75	3.58	3.9	4.0
	17.	52.(39.)	110	140.(126.)
Xe	4.15	3.75	4.2	4.2
	18.(8.)	52.(39.)	125	159.(148.)

a. Values given are R_e and D_e. When known, the R_0 and D_0 values are given in parentheses. All numbers are from Herzberg.[18]

Another interesting comparison is the series $HeCl_2$, $HeBr_2$ and HeI_2. It appears that the HeI_2 bond length is considerably longer than the other two. Although these three molecules each have their potential energy minimum at the "T" shaped structure, they are perhaps better thought of as hindered internal rotors. The numbers in Table I, however, are based on the rigid rotor approximation. Clearly, higher resolution spectroscopy is necessary to evaluate the effect of wide amplitude bending which may be partially responsible for the large difference between the $HeBr_2$ and HeI_2 bond lengths.

In summary, the data in Table I raise many interesting questions about the relative structures of the rare gas-halogen molecules. In almost every case, however, one wishes that the available data were less ambiguous. The only limits on our ability to attain higher precision are the limitations of time and money. In no case is the homogeneous linebroadening the effect that limits experimental resolution. This effect is overcome by studying transitions to lower vibrational levels of the excited electronic states.

2.2 Molecular Dynamics

Levy et al. initiated the study of molecular dynamics of the rare gas halogen molecules with their work on the vibrational predissociation of HeI_2, NeI_2 and ArI_2 in the B electronic excited state.[5] The HeI_2 results have received most of the attention because theory for that molecule is the easiest to perform.[20] It was found that vibrational predissociation of HeI_2, for vibrational levels up to v=27, is dominated by a $\Delta v = -1$ dissociation channel, and that the lifetimes of the complex ranged from 221 ps for v=12 to 38 ps for v=27. This has been interpreted with a V-T "half collision" mechanism which leads to the energy gap,[20] or momentum gap[21] law for the vibrational predissociation rate. An early indication that the energy gap law has at least qualitative value over a very large range of predissociation rates was the observation of metastable vibrational levels ($\tau > 10^{-5}$ s) for the ground electronic state of $NeCl_2$.[22]

The V-T mechanism for vibrational predissociation of HeI_2 and NeI_2 was supported by dispersed fluorescence experiments which revealed the product I_2 molecules to be rotationally cold.[5] For ArI_2 there is considerable rotational energy in the products.[23] This may be due to the fact that ArI_2 dissociates via a $\Delta v = -3$ sequential mechanism allowing the first two quanta of energy released from the I_2 stretch to be randomized over the van der Waals modes before the third quantum dissociates the molecule.

In the discussion below we show that the dynamics of $NeBr_2$ is not too different from what one might have expected based on the HeI_2 and NeI_2 results. Lester et al.[23] have shown that $HeICl$ and $NeICl$ have product rotational distributions that are consistent with an impulsive V-T,R predissociation mechanism from a bent nuclear configuration. Thus for these molecules it appears that the results are consistent with the same mechanism that applies to HeI_2, NeI_2, and $NeBr_2$ except that the bent structures of $HeICl$ and $NeICl$ give a nonzero average impact parameter leading to a peak in the rotational distribution at $J \neq$ zero. Recent results from our laboratory indicate that the mechanism for $NeCl_2$ vibrational predissociation is quite different from

that of the other molecules being considered.[11] Even though the primary predissociation channel for $NeCl_2$ (B state, v=11) is $\Delta v = -1$,
there is significantly greater rotational excitation of the products
than what one would have predicted on the basis of the half collision
mechanism. Interpretation via the half collision model would require
classical impact parameters of up to 8 Å! These results are still
preliminary and will be discussed in future publications.

3.1 Spectroscopy of $NeBr_2$

Low resolution spectra for many bands of $NeBr_2$ have previously been
reported and used to extract lifetimes for B state vibrational levels
with v > 16.[13] A high resolution spectrum for the B-X, 10-0 band was
reported and analyzed to obtain the molecular structure and B state
v = 10 level lifetime.[14] The molecule was found to have a "T" shaped
structure in both electronic states, with Ne atom to Br_2 center of mass
bond lengths of 3.67 (± .01) Å for the X state and 3.65(± 0.01) Å for
the B state. The similarity of the two bond lengths is surprising when
it is considered that the Br_2 length increases by 0.6 Å upon electronic
excitation. That the intermolecular potentials for the two electronic
states are very similar was shown by the fact that careful searching
revealed no excitation of van der Waals vibrational modes in the
electronic spectrum.

In this paper we report new results for the $NeBr_2$ B-X band for B
state vibrational levels v = 14, 16, 17 and 20. Although a frequency
stabilized c.w. dye laser was used to record these bands, lifetime
broadening vitiated any attempt to assign the rotational structure to
obtain accurate bond lengths for bands other than that of v = 10.
Levels below v = 10 had Franck-Condon factors so small that obtaining
useful signal to noise ratios was difficult with the Coherent 599-21
dye laser. Levels v = 11, 12, and 13 could not be studied because the
spectrum of Br_2 molecules severely overlapped those of $NeBr_2$. The
experimental technique was similar to that of references 12 and 14.

The spectra were analyzed using an asymmetric top, rigid rotor
spectrum simulation program which had six adjustable parameters: the
rare-gas atom to halogen center of mass distance in each of the two
electronic states, the band origin, the rotational temperature, a
Lorentzian linewidth to represent lifetime broadening and a Gaussian
linewidth to represent the laser width and the Doppler width. The
simulated spectrum was least squares fitted to the data to obtain the
values for the parameters as discussed below. The structure of the X
state was fixed at that given by the 10-0 band.[14] The laser linewidth
and the Doppler width were obtained by fitting nearby Br_2 transitions
for which spectral overlap and lifetime broadening do not exist. The
fitting parameters were the B state bond length, the rotational temperature, and the homogeneous linewidth. In each case the rotational
temperature was 2.7 (± 0.2) K. No significant variation of the upper
state bond length with vibrational quantum number was obtained. The
new linewidths are significant and are reported in Table III along with
values for other vibrational levels.

Table III. Predissociation lifetimes for NeBr$_2$, B state

v	$\Gamma(v)/cm^{-1}$ [a]	$\tau(v)/ps$	$\omega Br_2(v)/cm^{-1}$ [b]
10	0.015±0.005	355±100	132.0
14	0.051±0.005	105±10	116.4
16	0.057±0.006	95±9	108.2
17	0.081±0.008	65±6	104.1
20	0.151±0.015	35±4	91.7
21	0.40±0.1	13±10	87.6
22	0.75±0.2	7±3	83.5
24	0.95±0.2	6±1	75.3
25	1.09±0.2	5±1	71.4
26	2.38±0.4	2.2±0.5	67.4
27	3.12±0.4	1.7±0.3	63.5
28	2.50±0.4	2.1±0.3	59.7
29	2.88±0.4	1.8±0.3	56.0
30	1.88±0.4	2.8±0.3	52.3

a. Values for v=21-30 are from previous low resolution studies.[13]
b. This is the Br$_2$ vibrational quantum, i.e. $\omega(10)$ is equal to the energy of the v=10 level minus the energy of the v=9 level.

As has been observed for previously studied rare gas halogen molecules, the excited state lifetimes are in accord with an energy gap rule for vibrational levels between 10 and 27. Above v=27 the lifetime is roughly independent of v. We have previously assumed that its behavior is due to the closing of the $\Delta v = -1$ channel at v=27. This yields an X state, v=0, NeBr$_2$ bond energy of 73 cm^{-1}. It is shown below that this simple analysis is remarkably accurate.

Between v=10 and v=27 the linewidth grows very smoothly as a function of v. For v=10 clear rotational structure can be resolved. Above v=20, the homogeneous linewidth is so broad that no structure at all is seen in the vibrational band. At v=28, where the $\Delta v = -1$ dissociation channel closes, the linewidth no longer increases monotonically with increasing v. Since these changes in the linewidth as a function of the vibrational level are those predicted by the Beswick-Jortner or the Ewing theories, we feel certain that the major source of homogeneous broadening is the limitation of the complex lifetime by vibrational predissociation. Although the vibrational and rotational temperatures are very low, it is difficult to argue conclusively against dephasing as a major contributor to spectral broadening. It might seem that the recent determination of the NeICl A state, v=14, lifetime to be 3 ns[7] would indicate that the much shorter lifetimes we infer for NeBr$_2$ may be due to dephasing. The vibrational frequency of the NeICl A state v=14 level is 148 cm^{-1}. This is a higher frequency than any mode studied for NeBr$_2$. Also, the NeICl bond is somewhat weaker than that

of NeBr$_2$; 44 cm^{-1} versus 65 cm^{-1}. We believe that these two facts, in
conjunction with the energy gap law, will account for the longer lifetime of NeICl. A quantitative comparison of the dynamics of the two
molecules will require a better determination of the two potential
energy surfaces. Also, it will be desirable to make real time lifetime
determinations on the same molecule as has been studied with the line
broadening method.

Although the analysis of lifetime data in terms of an energy gap
law argues strongly for a V-T,R mechanism for vibrational predissociation, it is still not proved that the process is completely electronically adiabatic. In fact, Levy et al.[5] found that vibrational
predissociation is in direct competition with electronic predissociation for ArI$_2$. For ArBr$_2$ we have yet to observe fluorescence from
the B state and believe that electronic predissociation occurs on the
picosecond time scale. To be absolutely certain that the short lifetimes we observe are not due to mixing with other vibronic states, it
would be necessary to study the predissociation lifetimes for the
ground electronic state. In the case of NeCl$_2$ the v=1 level of the
ground electronic state was observed. In fact the lifetime is so long
that the molecule travels out of the detection region of the experiment before dissociating so that the lifetime cannot be measured. No
other vibrationally excited ground state levels of the rare gas halogens have been reported. If we use the momentum gap law as formulated
by Ewing to estimate the lifetime for NeBr$_2$, X state, v=1 (assuming
that the coupling is the same as for the B state) we obtain a value of
8 μs. This predicts that the vibrationally excited ground electronic
state NeBr$_2$ molecules should live long enough to be observed within
100 to 400 nozzle diameters of our supersonic expansion. Until
recently, the signal to noise ratio we were able to attain was not
sufficient to observe such states, and we had thought that the estimate for the lifetime was too long. The pump probe technique substantially improves the signal to noise ratio and the X state v=1 level
has now been observed.[11] Preliminary studies indicate that the lifetime is on the order of 10 μs. The close agreement between this value
and that predicted on the basis of excited state data gives us confidence that excited state dynamics are dominated by vibrational
effects.

4. Dispersed Fluorescence Studies of NeBr$_2$ Predissociation Dynamics

Measurements of linebroadening in van der Waals molecule absorptions give information about the rate of decay of the prepared state
but do not describe how the energy is distributed among the available
product channels. Studies of the predissociation of electronically
excited rare gas halogen complexes have exploited the well characterized B-X transitions of the halogens.[27] By dispersing the B→X
fluorescence of the halogen fragment of the the dissociation, its rotational and vibrational quantum state distribution is obtained. Here we
present data on the product state distribution of the Br$_2$ fragment from
the photodissociation of NeBr$_2$.

4.1 Experimental

A mixture of 9:1 Ne:He (Spectra Gases) was passed over a trap maintained at -40°C containing Br_2 (MCB, reagent grade, naturally occuring
isotopic abundances). $NeBr_2$ was prepared in a pulsed, free jet expansion through a 500 μm nozzle at stagnation pressures of 100-140 psig.
Care was taken to minimize the concentration of clusters containing
more than one Ne atom. Higher order clusters can be detetected in LEF
experiments as features blue shifted from the $NeBr_2$ bandhead.

The pulsed beam from an excimer (Lambda Physik EMG 201 MSC) pumped
dye laser (Lambda Physik FL 2002) crossed the expansion at more than 80
nozzle diameters downstream of the nozzle. The optical system for
collecting total fluorescence from emitting species in the jet, the
vacuum chamber, and pumping system used have been previously
described.[24] A cooled S-20 photomultiplier tube was used to detect the
fluorescence. Since Franck-Condon factors for the B-X transition of
Br_2 strongly favor emission to the red of the excitation frequency,[25]
scattered laser light was rejected by appropriate red pass filters. A
Spex 1269 monochromator was used to disperse product emission collected
by an optical system designed by B.A. Swartz and described elsewhere.[26]
A red-sensitive RCA 31034 photomultiplier tube was used to detect the
dispersed fluorescence. The preamplified photomultiplier signals were
recorded by a boxcar integrator (Stanford Research Systems). A
computer-based data acquisition system controlled the timing of the
valve, laser, and boxcar, scanned the dye laser and monochromator, and
performed signal averaging.

4.2 Data and Results

Product state distributions for the photodissociation of $NeBr_2$ were
obtained by tuning the excitation laser to the maximum of the $NeBr_2$ van
der Waals band and recording the emission spectrum of the $NeBr_2$ dissociation fragment. The $Ne^{79}Br_2$ species was studied since it is well
separated from the other isotopic species in the excitation spectrum.
The experiment can best be described using the following steps.

$NeBr_2(X,v''=0,J'') + hv_{laser} \rightarrow NeBr_2(B,v',J')$ (1) excitation

$NeBr_2(B,v',J') \rightarrow Ne + Br_2(B,v'-z,j)$ (2) vibrational
                                                          predissociation

$Br_2(B,v'-z,j) \rightarrow Br_2(X) + hv_{fluor}$ (3) product
                                                          fluorescence

The vibrational predissociation lifetime in step (2) was shown in
Sec. 3 above to range from 355 ps for $v'=10$ to 1.7 ps for $v'=27$. The
radiative lifetime for fluorescence of $Br_2(B)$ in step (3) is ~8 μs,[27]
so that a negligible amount of fluorescence occurs from the van der
Waals complex itself.

Figure 1 shows the Br_2 product emission spectra obtained upon excitation of several bands of $Ne^{79}Br_2$. Due to the weak fluorescence
signal of the Br_2 product, wide monochromator slits were required ($\sim$5
cm^{-1} resolution) and band contours, rather than resolved transitions of
the B-X emission, were obtained. For $NeBr_2(B,v' < 26)$ the dominant product vibrational channel populated is that in which one quantum of the
Br_2 stretch is used to dissociate the molecule. As higher vibrational
states of $NeBr_2$ are excited the product channels in which multiple Br_2
quanta are lost become increasingly more probable. Another qualitative
observation concerning the product distributions is the dramatic cutoff
of intensity in the one quantum loss channel for $v' > 28$.

There are three phenomena which complicate the extraction of quantitative product state distributions. One is the relatively long
fluorescence lifetime of the Br_2 B state. Inelastic collisions of the
Br_2 fragment with other species in the expansion prior to fluorescence
will alter the observed Br_2 state distribution from that initially produced by the dissociation. In such collisions there is the possibility
for rotational, vibrational, and electronic relaxation of the fragment
Br_2.[27] Far from the nozzle the number density of molecules in the
expansion varies as the inverse square of the distance from the nozzle
source. By increasing the distance of the excitation region from the
nozzle the number of collisions is reduced at the expense of
fluorescence intensity. The relative amount of vibrational relaxation
of Br_2 in this experiment was determined by exciting uncomplexed Br_2
in the expansion. At a position 80 nozzle diameters downstream of the
nozzle, where the number density is $\sim 10^{16}/cm^3$, the measured population
of vibrationally relaxed uncomplexed Br_2 never exceeded 8% of the
unrelaxed population. Collisions also affect the rotational state
distribution to some degree. The rotational temperature of
uncomplexed Br_2 emitting from the same vibrational state in which it
was prepared is identical to the 0.7 K rotational temperature of the
expansion within experimental error. However, vibrationally inelastic
collisions significantly alter the rotational state. For example, the
rotational temperature of $Br_2(B,v'=21)$ which has undergone an
inelastic collision in which $\Delta v' = -1$ is $\sim$8 K. Br_2 produced by the
photodissociation of $NeBr_2$ is expected to be translationally more
excited than uncomplexed Br_2. Russell, et al.[28] have performed a
careful study of the energy redistribution that occurs in low energy
collisions of $I_2(B)$ and He in a free jet expansion. They found that
there is a very high cross section for vibrational relaxation in low
energy collisions and that the cross section actually decreases with
increasing collision energy. This result implies that vibrational
relaxation of $Br_2(B)$ produced by van der Waals molecule predissociation
should be less likely than that measured for uncomplexed Br_2. In order
to confirm that the translationally excited $Br_2(B)$ rotational distribution was not significantly altered in vibrationally elastic collisions,
the 10$\leftarrow$0 B$\leftarrow$X band was studied by the laser pump-probe technique in
which the probe laser is delayed by only 10 ns so that the product
state distributions are not affected by collisions. Within experimental
uncertainty, the rotational distribution of the $v'=9$ Br_2 fragment was
the same in both measurements.

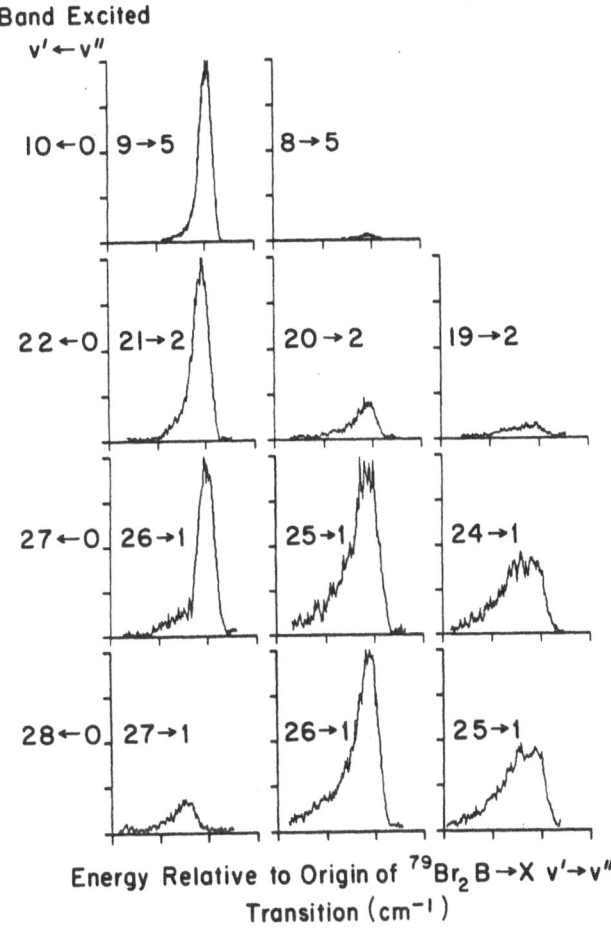

Figure 1: Br$_2$ emission spectra observed upon excitation of several NeBr$_2$ bandheads. The transition labels on the far left are the Br$_2$ stretching quantum numbers for the NeBr$_2$ B-X excitation transition used to obtain the emission spectra for each row. Each column of spectra correspond to a specific Δv relaxation channel: left column, $\Delta v = -1$; center column, $\Delta v = -2$; right column, $\Delta v = -3$. Each emission band is marked with the Br$_2$ vibrational quantum number of the B-X transition. For the $\Delta v = -1$ channel, the observed rotational band gets broader as the initial stretching quantum number increased from 10 to 26. For NeBr$_2$ $v'=27$, the observed emission band is narrower than the others, showing that the $\Delta v = -1$ channel is starting to close. For NeBr$_2$ $v'=28$ the $\Delta v = -1$ channel is completely closed. (The observed emission slightly to the red is due to isotopic contamination). See the text for further discussion.

Another complication in obtaining product state distributions is
due to the fact that the rate of electronic predissociation of the Br_2
B state depends on the j quantum number.[27],[29] This effect significantly reduces the lifetime of the excited state Br_2 fragments for the
higher j states populated in the dissociation. In order to minimize
the distortion of our spectra, fluorescence was averaged only for the
first 400 ns following the laser pulse. Since the j-dependent electronic predissociation rate constants are not known for each band studied,
we did not correct our distributions for this effect. The agreement of
the values for the rotational temperature of $Br_2(B,v'=9)$ from predissociation of $NeBr_2(B,v'=10)$ obtained by dispersed fluorescence and by the
pump-probe technique (which should be practically unaffected by
electronic predissociation) indicate that electronic predissociation
does not qualitatively change the results we obtain.

Finally, there are interferences arising from the presence of the
other Br_2 isotopic species present in the expansion. The broad
Lorentzian profiles of high v' bands prevent selective excitation of
van der Waals molecules associated with only one Br_2 species. Also,
due to the limited resolution of our monochromator, it is difficult to
isolate fluorescence from the $^{79}Br_2$ and the $^{79}Br^{81}Br$ species. Among
the bands studied, only the 10-0 and 22-0 bands are free from contamination of the dispersed $^{79}Br_2$ emission by the $^{79}Br^{81}Br$ species.
With these caveats in mind, the results presented here should be
accepted as a good qualitative description of the photodissociation
dynamics. A quantitative description of the final rotation quantum
number distribution remains to be determined in future laser pump-probe
experiments.

Table IV presents the product state distributions for the predissociation of $Ne^{79}Br_2$. Due to the uncertainties described above, only one
significant figure is reported. Populations are extracted from the
observed intensities, corrected for collisional vibrational relaxation
and fluorescence Franck-Condon factors as described in the Appendix.
Rotational temperatures are reported for the $\Delta v = -1$ channels of $NeBr_2$
v'=10 and v'=22.

Table IV. Vibrational product state disribution for the
photodissociation of $Ne^{79}Br_2$.

Band Excited	Relative Population				
	$\Delta v = -1$	$\Delta v = -2$	$\Delta v = -3$	$\Delta v = -4$	$\Delta v = -5$
10 ← 0	1[a]	0	–	–	–
22 ← 0	1[b]	0.2	0.1	0.1	0
27 ← 0	0.7	1	0.6	0.3	0.1
28 ← 0	0[c]	1	0.6	0.4	0.2

a) T_{rot} = 6K
b) T_{rot} = 5K
c) The weak fluorescence observed in this channel is
entirely due to the $^{79}Br^{81}Br$ species.

4.3 Discussion

Comparison of the product emission spectra for $Ne^{79}Br_2$ (B,v'=27) and $Ne^{79}Br_2$ (B,v'=28) shows a dramatic decrease in the apparent population of the $\Delta v'=-1$ channel. That the weak emission near where emission from $^{79}Br_2$ (B,v'=27) should occur is actually due to the $^{79}Br^{81}Br$ species is shown by the absence of any emission at the $^{79}Br_2$ band origin. Due to the very broad van der Waals features in this region, it is no longer possible to excite only one isotopic species. The $\Delta v'=-1$ channel has in fact closed for $Ne^{79}Br_2$ (B,v'=28). The amount of energy available in the $\Delta v'=-1$ channel for $Ne^{79}Br_2$ (B,v'=27) is the energy in a quantum of the Br_2 vibration quantum, $\omega_{27}=63.5$ cm^{-1}. The $\Delta v'=-1$ channel closes at v'=28 where $\omega_{28}=59.7$ cm^{-1}. The van der Waals bond energy for $Ne^{79}Br_2$ is bracketed to be $59.7 < D_0(B) < 63.5$ cm^{-1}. The bond energy in the ground electronic state is calculated by adding the blueshift[13] of the $Ne^{79}Br_2$ bandhead from the $^{79}Br_2$ bandhead to yield $68.5 < D_0(X) < 72.6$ cm^{-1}. This bond energy determination is close to an estimate of $D_0(B)=64\pm4$ cm^{-1} made by Swartz, et al. on the basis of an increasing lifetime for $NeBr_2$ (B,v' > 27).

There remains the question of why weak fluorescence is observed in the $\Delta v'=-1$ channel for $Ne^{79}Br^{81}Br$ (B,v'=28). It may appear contradictory that fluorescence from v'=27 of $^{79}Br^{81}Br$ should be detected upon excitation of a band which cannot populate v'=27 for $Ne^{79}Br_2$. For $NeBr_2$ v'=28 the full width at half maximum of the Lorentzian lineshape is ~3 cm^{-1}. When the excitation laser is positioned on the maximum of the $Ne^{79}Br_2$ band, it also excites the blue tail of the Lorentzian profile of the $Ne^{79}Br^{81}Br$ absorption at an energy which lies above the threshold for population of the $^{79}Br^{81}Br$ (B,v' =27) product. In a separate experiment where the monochromator is positioned at the origin of the $^{79}Br_2$ fluorescence (so that fluorescence of the other isotopic species is not detectable) and the excitation laser is tuned through the Lorentzian profile of $Ne^{79}Br_2$ (B,v'=28), a weak threshold occurs when the energy of the prepared state is sufficient to predissociate via the one quantum mechanism. This phenomenon is illustrated in Figure 2.

The rotational distributions obtained for the Br_2 fragment can be closely simulated by a Boltzman distribution, although our limited resolution does not preclude somewhat non-Boltzmann behavior. In fact, the distribution must have a definite termination at some maximum j quantum number, j_{max}, that is the largest j that can be populated according to conservation of energy. For the $Ne^{79}Br_2$ (B,v'=10) $\Delta v'=-1$ channel, 70 cm^{-1} is available for product rotational and relative translational energy and j_{max} ~35. The observed rotational temperature of 6 K has an average of 4 cm^{-1} of energy in rotation and gives practically no population of rotational levels approaching j=35. This contrasts strongly with the $NeCl_2$ (B,v'=11) rotational distribution in which every state kinematically and energetically allowed has significant population (available energy =87 cm^{-1}) and an average of 15 cm^{-1} of energy in rotation.[11] The very cold (< 1K) rotational distribution of the Br_2 (B,v'=26) product from $NeBr_2$ (B,v'=27) is due to the

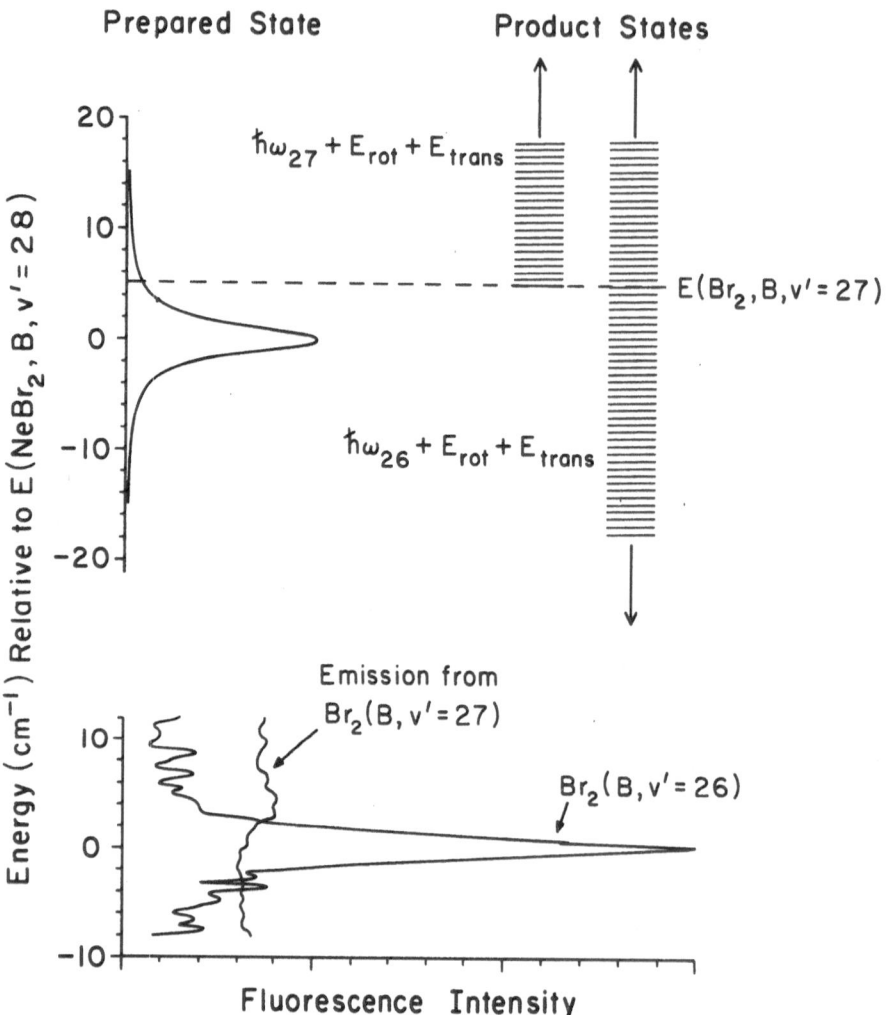

Figure 2: Direct measurement of the NeBr$_2$ bond energy. The top
figure schematically illustrates how the NeBr$_2$ B state, v=28 level
overlaps with the free Br$_2$ product channels. The bottom figure shows
the observed Br$_2$ emission intensity from Br$_2$, v=27 and Br$_2$, v=26 upon
tuning the excitation laser across the NeBr$_2$, v'=28 band. As
expected, detection of fluorescence from the $\Delta v = -2$ channel, Br$_2$,
v=26, yields the entire excitation band. Detection of fluorescence
from the $\Delta v = -1$ channel shows a distinct threshold in the blue tail
of the excitation band. The location of this threshold gives a direct
measurement of the Ne-Br$_2$, X state bond energy.

small amount of energy available for rotation near the closure of the one quantum channel.

For high v' states of NeBr$_2$ the vibrational product state distributions are qualitatively quite broad with appreciable intensity still present at $\Delta v = -5$ for excitation of NeBr$_2$ v'=27 and v'=28. It is apparent that the number of vibrational product channels with significant population increases with v'. A similar result is observed in NeI$_2$ where for NeI$_2$ (B,v'=21) more than 90% of the product population is present in the I$_2$ v'=20 channel[30] and for NeI$_2$ (B,v' ~ 33) where significant population is observed in several product channels.[31] In conclusion, the distribution of product energy upon dissociation of NeBr$_2$ is much as would be expected from consideration of the "half collision" theories of Beswick and Jortner and Ewing in that little rotational energy is imparted to the Br$_2$ fragment and the highest open vibrational product channel is favored in the dissociation.

Summary

The literature on the structure and dynamics of the rare gas-halogen molecules has been briefly reviewed. While there are still considerable gaps in our understanding of these molecules, they represent the best prototypical example of a case where structure and dynamics can be compared for a whole class of molecules. Of particular help at this time would be precision microwave results to help determine the shape of the van der Waals well near its minimum. Unfortunately there is no molecule for which both microwave and optical studies have been performed for comparison. Also, it would be desirable to obtain a direct measurement of the van der Waals vibrational frequencies. This is difficult for the lowest electronic states, because the structure changes very little upon electronic excitation which leads to small Franck-Condon factors for all but the 0-0 van der Waals bands.

New results have been presented for the spectroscopy and dynamics of NeBr$_2$. The v=10 level of the B electronic excited state decays via a $\Delta v = -1$ channel with a lifetime of 330 ps. Only 6 percent of the available product energy appears in the rotational degrees of freedom. For B state rotational levels between 26 and 28 the lifetime is down to 2 ps, and the closing of the $\Delta v = -1$ channel can be observed in the product state distribution. The data leads to a NeBr$_2$ X state, v=0, bond energy of 70.5 cm^{-2}.

Appendix

Vibrational state population ratios were extracted from the integrated rotational contours of each vibrational band of Br$_2$ using the equation

$$\frac{[Br_2(v'-z)]}{[Br_2(v')]} = \frac{I_{v'-z,v''}}{I_{v',v''}} \cdot \frac{q_{v',v''}}{q_{v'-z,v''}} \cdot \frac{v_{v',v''}^3}{v_{v'-z,v''}^3} \tag{1}$$

where $I_{v',v''}$ is the observed integrated intensity of the $v' \rightarrow v''$ band, $q_{v',v''}$ is the Franck-Condon factor,[25] and $\nu_{v',v''}$ is the frequency of the transition.

The derivation of the kinetic equations describing the relaxation of the initial state distribution following the laser pulse closely follows that of Sharfin, et al. for a cw laser experiment.[32] The relaxation processes for uncomplexed Br_2 are

$$Br_2(v_i') \xrightarrow{A(v_i',v'')} Br_2(v'') + h\nu_{fluor} \qquad \text{(2) fluorescence}$$

$$Br_2(v_i') + X \xrightarrow{Q(v_i')} 2Br + X \qquad \text{(3) quenching}$$

$$Br_2(v_i') + X \xrightarrow{R(v_i',v_f')} Br_2(v_f') + X \qquad \begin{array}{l}\text{(4) vibrational} \\ \text{relaxation}\end{array}$$

The duration of the excitation pulse is negligible with respect to the rate of these processes. The $Br_2(v_i')$ population, $[Br_2(v_i')]$, decays as

$$[Br_2(v_i')] = [Br_2(v_i')]_o\, e^{-\Gamma t}. \qquad (5)$$

$[Br_2(v_i')]_o$ is the concentration of $Br_2(v_i')$ immediately following laser excitation, t is the elapsed time since the laser pulse, and

$$\Gamma = \sum_{v''} A(v_i',v'') - \sum_{v_f' < v_i'} R(v_i',v_f')[X] - A(v_i')[X]. \qquad (6)$$

The rate Γ is assumed to be independent of the initial vibrational state, v_i', over a narrow range of initial states.

The population of $Br_2(v_i'-1)$ varies in time as

$$\frac{d[Br_2(v_i'-1)]}{dt} = -\Gamma[Br_2(v_i'-1)] + R(v_i',v_i'-1)[X][Br_2(v_i')] \qquad (7)$$

Applying the initial condition of $[Br_2(v_i'-1)]_{t=0} = 0$ gives

$$[Br_2(v_i'-1)] = [Br_2(v_i')]_o R(v_i',v_i'-1)[X]t\, e^{-\Gamma t} \qquad (8)$$

$$[Br_2(v_i'-2)] = R(v_i',v_i'-2)[X][Br_2(v_i')]_o t\, e^{-\Gamma t}$$

$$+ 1/2 R(v_i'-1,v_i'-2)R(v_i',v_i'-1)[X]^2[Br_2(v_i')]_o t^2\, e^{-\Gamma t} \qquad (9)$$

At a given observation time, τ, the vibrational state population ratios are

$$\left| \frac{[Br_2(v_i'-1)]}{[Br_2(v_i')]} \right|_{t=\tau} = R(v_i',v_i'-1)[X]\tau \equiv f_1 \tag{10}$$

$$\left| \frac{[Br_2(v_i'-2)]}{[Br_2(v_i'-1)]} \right|_{t=\tau} = R(v_i',v_i'-2)[X]\tau$$

$$+ \ 1/2R(v_i'-1,v_i'-2)R(v_i',v_i'-1)[X]^2\tau^2 \equiv f_2 \tag{11}$$

Now consider the relaxation of the population distribution which arises from the photodissociation of NeBr$_2$. At time t=0 the initial population of each vibrational state is $[Br_2(v_i'-z)]_o$ where z is the number of Br$_2$ vibrational quanta lost in the dissociation. Applying this set of initial conditions and assuming the relaxation rate constants are equal to those for uncomplexed Br$_2$ as discussed in section 4, the time dependent vibrational state populations are

$$[Br_2(v_i'-1)] = [Br_2(v_i'-1)]_o e^{-\Gamma t} \tag{12}$$

$$[Br_2(v_i'-2)] = [Br_2(v_i'-2)]_o e^{-\Gamma t} + [Br_2(v_i'-1)]_o R(v_i'-1,v_i'-2)[X]t \ e^{-\Gamma t} \tag{13}$$

$$[Br_2(v_i'-3)] = [Br_2(v_i'-3)]_o \ e^{-\Gamma t}$$

$$+ \ R(v_i'-1,v_i'-3)[X][Br_2(v_i'-1)]_o t \ e^{-\Gamma t}$$

$$+ \ 1/2R(v_i'-2,v_i'-3)R(v_i'-1,v_i'-2)[X]^2[Br_2(v_i'-1)]_o t^2 \ e^{-\Gamma t}$$

$$+ \ [Br_2(v_i'-2)]_o R(v_i'-2,v_i'-3)t \ e^{-\Gamma t}$$

At a given observation time, τ, the product population ratios are

$$\left| \frac{[Br_2(v_i'-2)]}{[Br_2(v_i'-1)]} \right|_{t=\tau} = \frac{[Br_2(v_i'-2)]_o}{[Br_2(v_i'-1)]_o} + f_1 \tag{14}$$

$$\left| \frac{[Br_2(v_i'-3)]}{[Br_2(v_i'-1)]} \right|_{t=\tau} = \frac{[Br_2(v_i'-3)]_o}{[Br_2(v_i'-1)]_o} + f_2 + f_1 \frac{[Br_2(v_i'-2)]_o}{[Br_2(v_i'-1)]_o} \tag{15}$$

where f_1 and f_2 are the values obtained in equations (10) and (11) for an experiment on uncomplexed Br$_2$.

The branching ratios $[Br_2(v_i'-z)]_o/[Br_2(v_i'-1)]_o$ are obtained from the experiment using equations (14) and (15). In our experiment the emission is observed for a short, but finite length of time. This requires equations (5), (8), (9) and (12)-(14) to be integrated over the observation time. The equations are then somewhat more algebraically complex, but the final ratios are identical.

Acknowledgements: This work was supported by the United States National Science Foundation. Acknowledgment is made to the donors of the Petroleum Research Fund administered by the American Chemical Society for partial support of this work. We would like to thank Prof. Marsha Lester for providing results on NeICl prior to publication.

References

1. S. J. Harris, S. E. Novick and W. Klemperer, **J. Chem. Phys.** 61, 193, (1974).
2. S. E. Novick, S. J. Harris, K. C. Janda and W. Klemperer, **Canadian J. Phys.** 53, 2007, (1975).
3. S. E. Novick, K. C. Janda and W. Klemperer, **J. Chem. Phys.** 65, 5115, (1977).
4. F. A. Baiocchi, T. A. Dixon and W. Klemperer, **J. Chem. Phys.** 77, 1632, (1982).
5. D. H. Levy, **Adv. Chem. Phys.** 47, 323, (1981).
6. W. Sharfin, P. Kroger, and S. C. Wallace, **Chem. Phys. Letts.**, 85, 81, (1982).
7. J. C. Drobits, J. M. Skene and M. I. Lester, J. Chem. Phys., 84, 2896, (1986).
8. The $HeCl_2$ well depth was obtained by solving for the D_0 level of the $HeCl_2$ potential obtained by differential scattering studies. B. P. Reid, unpublished results.
9. J. I. Cline, D. D. Evard, F. Thommen and K. C. Janda, **J. Chem. Phys.**, 84, 1165 (1986).
10. L. J. van de Burgt, J. P. Nicolai, and M. C. Heaven, **J. Chem. Phys.**, 81, 5514 (1984).
11. J. I. Cline, N. Sivakumar, D. E. Evard, and K. C. Janda, to be published.
12. D. D. Evard, F. Thommen, and K. C. Janda, **J. Chem. Phys.** 84, 3630 (1986).
13. B. A. Swartz, D. E. Brinza, C. M. Western, and K. C. Janda, **J. Phys. Chem.** 84, 6272 (1984).
14. F. Thommen, D. D. Evard and K. C. Janda, **J. Chem. Phys.** 82, 5295 (1985).
15. N. Sivakumar, J. I. Cline, D. D. Evard, and K. C. Janda, to be published.
16. The bond energies quoted for ArClF and KrClF are D_e values taken from ref. 2. These were obtained by fitting 6,12 type potentials to centrifugal distortion constants. The errors involved in these calculations are difficult to estimate.
17. J. C. Drobits and M. I. Lester, to be published.
18. K. P. Huber and G. Herzberg, Molecular Spectra and Molecular Structure IV. **Constants of Diatomic Molecules**, Van Nostrand, New York, 1979.
19. G. C. Pimentel and R. D. Spratley, **Understanding Chemistry**, Holden-Day Inc., San Francisco, 1971.
20. J. A. Beswick and J. Jortner, **Adv. Chem. Phys.** 47, 323 (1981).
21. G. E. Ewing, **Faraday Discuss. Chem. Soc.** 73, 325 (1982).

22. D. E. Brinza, B. A. Swartz, C. M. Western and K. C. Janda, **J. Chem. Phys.** 79, 1541 (1983).
23. J. M. Skene, J. C. Drobits and M. I. Lester, **J. Chem. Phys.** 85, 2329 (1986).
24. D. E. Brinza, C. M. Western, D. D. Evard, F. Thommen, B. A. Swartz, and K. C. Janda, **J. Phys. Chem.** 88, 2004 (1984).
25. J. A. Coxon, **J. Quant. Spectrosc. Radiat. Transfer,** 12, 639 (1972).
26. B. A. Swartz, Ph.D. Thesis, California Institute of Technology, 1984.
27. M. C. Heaven, Molecular Spectroscopy, **Chem. Soc. Spec. Per. Rep.,** The Chemical Society, London, in press.
28. T. D. Russell, B. M. DeKoven, J. A. Blazy, and D. H. Levy, **J. Chem. Phys.** 72, 3001 (1980).
29. M. Siese, E. Tiemann, and U. Wulf, **Chem. Phys. Lett.** 117, 208 (1985).
30. J. E. Kenny, K. E. Johnson, W. Sharfin, and D. H. Levy, **J. Chem. Phys.** 72, 1109, (1980).
31. J. A. Blazy, B. M. DeKoven, T. D. Russell, and D. H. Levy, **J. Chem. Phys.** 72, 2439 (1980).
32. W. Sharfin, K. E. Johnson, L. Wharton, and D. H. Levy, **J. Chem. Phys.** 71, 1292 (1979).

DYNAMICS OF ENERGY TRANSFER IN VAN DER WAALS MOLECULES OF s-TETRAZINE AND ARGON

Marc Heppener and Rudolf P.H. Rettschnick
Laboratory for Physical Chemistry, University of Amsterdam
Nieuwe Achtergracht 127, 1018 WS Amsterdam,
The Netherlands

ABSTRACT

   Energy conversion in the photoexcited van der Waals complex T.Ar of s-tetrazine and argon has been investigated in a supersonic jet by using the technique of laser-induced fluorescence. Time-resolved experiments on a picosecond timescale were performed in combination with steady-state measurements of the dispersed fluorescence. The temporal intensity profiles of selectively detected emission bands provide a detailed picture of the pathways (i.e. the proper succession of states) and the dynamics of intramolecular vibrational redistribution (IVR) and vibrational predissociation (VP) in the lowest excited singlet state of the complex. These processes have been studied by selectively pumping the levels $\overline{16a}^2$, $\overline{4}^1$, $\overline{6a}^1$ and $\overline{6b}^2$ of the complex (a bar is used to assign vibronic states of the complex). The rates of IVR are in the range $5 \times 10^7 - 2 \times 10^9 \mathrm{s}^{-1}$ and the rates of VP are between 5×10^8 and $3 \times 10^9 \mathrm{s}^{-1}$.

1. INTRODUCTION

   Weakly bound complexes of a medium-size polyatomic molecule and a rare gas atom are suitable prototype systems for studies of intramolecular vibrational relaxation (IVR). The low frequency van der Waals modes of the complex provide a stairway that may enable the vibrational energy to flow between different vibrational levels of the polyatomic molecule. IVR processes can be studied in detail by using the technique of laser-induced fluorescence if the absorption and emission spectra of the polyatomic molecule are well-resolved and if the fluorescence of the host molecule is not quenched due to the presence of the rare gas atom.
   The complex T.Ar of s-tetrazine and argon satisfies these conditions very well. Both IVR and VP (vibrational predissociation) of the photoexcited complex T.Ar have been studied by Levy and coworkers (1,2) and also in our laboratory (3-5). In this paper we present some recent results concerning the dynamics of IVR and VP processes that

A. Weber (ed.), Structure and Dynamics of Weakly Bound Molecular Complexes, 553–562.
© *1987 by D. Reidel Publishing Company.*

occur after pumping well-defined vibronic levels of the host molecule.

Much is already known about the spectroscopy of s-tetrazine and its van der Waals complexes containing rare gas atoms (2,4,6-10, and references cited in these papers). Selective excitation of vibronic levels of T.Ar can be achieved because the absorption bands of the complex are shifted with respect to the corresponding transitions in the bare molecule. These spectral shifts depend on the vibrational level of the complex (2,4). The structure of T.Ar has been determined from a rotationally resolved fluorescence excitation spectrum of the complex (10). The argon atom is situated on the out-of-plane C_2 axis of tetrazine at a distance of 0.344 nm from the molecular plane. This distance is somewhat less (0.340 nm) in the excited state S_1.

Generally, the fluorescence spectra of the complex show three different kinds of emission bands which originate from (i) the initially excited vibronic level of the complex, (ii) vibronic levels formed by IVR, and (iii) emission from vibronic levels of the bare tetrazine molecule which is released by rupture of the van der Waals band. Time-resolved measurements of selectively detected bands in the fluorescence spectra provide information about the pathways and the dynamics of IVR and VP processes.

2. EXPERIMENTAL PROCEDURE

The van der Waals complexes were produced in a free jet of argon, seeded with 0.03% of tetrazine. The beam of an argon laser-pumped dye laser was focussed at the central axis of the jet at distances between 0.5 and 2 mm downstream of the nozzle (nozzle diameter 0.10 mm). The fluorescence light was collected by a condensor   and focussed on the entrance slit of a 1.5 m monochromator (Jobin-Yvon THR 1500). The spectral bandwidth used in these experiments was between 2 and 5 cm^{-1} FWHM.

Time-resolved measurements were performed with excitation pulses of 7.5 ps duration from a dye laser which was synchronously pumped by a mode-locked argon ion laser. Temporal intensity profiles of selectively detected fluorescence bands were measured by means of time-correlated single photon counting. For the time-resolved measurements a microchannelplate photomultiplier with S 20 photocathode (Hamamatsu R 1564 U-01) was used.

Time constants were obtained from the intensity vs. time profiles with a home-written deconvolution program based on a non-linear least squares iterative reconvolution method (11). The time-resolution of this technique was about 25 ps.

3. RESULTS AND DISCUSSION

Fig. 1 shows the fluorescence excitation spectrum in the vicinity of the 0_0^0 band of tetrazine. The 0_0^0 transition of the bare molecule is located at 551.49 nm. The shape of this band with its pronounced P and R branches could be simulated by using a home-written rotational

contour analysis program, assuming a rotational temperature of 10 K.
The $\bar{0}_0^0$ transition of T.Ar is found 22 cm^{-1} to the red of the 0_0^0 band

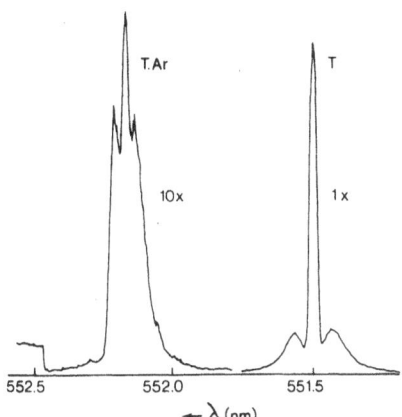

Figure 1. Fluorescence excitation spectrum of s-tetrazine and T.Ar.
Excitation bandwidth 0.7 cm^{-1}, argon backing pressure 1.4 atm.,
excitation position 0.5 mm from the nozzle.

of the free molecule. Transitions in T.Ar will be denoted always with
a bar on top of the assignment in order to distinguish them from
transitions in the free molecule. The relative intensity of the $\bar{0}_0^0$
band with respect to the 0_0^0 transition is 0.15. The contour of this
band could be simulated assuming a temperature of 10 K. This means
that the $\bar{0}_0^0$ excitation band contains no observable contributions from
sequence transitions of the van der Waals vibrational modes.
    The fluorescence spectrum obtained when the laser is tuned to the
Q-branch of the 0_0^0 transition consists of bands that originate all
from the excited level. This observation does not only hold for the
vibrational ground state but is found to be valid for all excited
vibronic levels in the isolated tetrazine molecule. Also all emission
bands obtained after excitation of the $\bar{0}^0$ level can be assigned to
resonant transitions. The lifetimes of the 0^0 level and the $\bar{0}^0$ level,
as obtained from time-resolved measurements are both equal to 740 ps.
It is clear from these observations that the influence of collision-
induced relaxation effects is well below the detection limit under the
conditions of our experiments.

16a^2 excitation

    The $16a_0^2$ transition in s-tetrazine is located at 536.37 nm. The
lifetime of the 16a^2 level is 1400 ps. The corresponding transition in
T.Ar is located at 536.59 nm. In this case the spectral shift is only
7.5 cm^{-1}, which indicates that the binding energy of the complex is
not independent of the vibrational state of the tetrazine molecule.

Similar effects are found for other vibrational modes in T.Ar (2,4).
These effects can be considered as due to perturbations imposed by the
argon atom on the vibrations of the tetrazine moiety. Table I lists
the frequencies of the vibrational modes in T and T.Ar which are involved in the present investigations.

TABLE I

Frequencies (in cm^{-1}) of four vibrational modes of
s-tetrazine (isolated and complexed molecule).

mode	S_0		S_1	
	T	T.Ar	T	T.Ar
16a	336	336	256	263
6a	736	737	704	703
4	802	801	580	578
2 x 6b	1282	1280	786	783

Data concerning the electronic ground state S_0 are obtained from
emission spectra, measured after 0_0^0 or $\overline{0}_0^0$ excitation respectively.
The S_1 data are obtained from fluorescence excitation spectra. The experimental error is less than 1 cm^{-1}. The fundamental frequency of the
6b mode cannot be obtained directly from the spectra because the
transitions $6b_0^1$ and $6b_1^0$ are forbidden (like in the case of the 16a
mode). In Table I the values for the twofold excitation of the 6b vibration are given since the $6b^2$ level is in strong Fermi resonance
with the nearby $6a^1$ level.

When the $\overline{16a}^2$ level of the complex is excited, the fluorescence
spectrum shows several bands which are not originating from the initially excited level. In the spectral region between 550 and 560 nm we
see in addition to the resonant transition $\overline{16a}_2^2$ two other bands that
can be assigned as $\overline{16a}_1^1$ and 0_0^0. The first one shows the result of IVR:
the energy of one 16a quantum is converted to vibrational energy of
the van der Waals vibrations. The assignment of this band should therefore include sequence transitions of these modes. They can be designated as σ (stretching vibration), β_x and β_y (bending vibrations in the
x and y directions parallel to the plane of the tetrazine molecule).
Instead of writing $16a_1^1\sigma_m^m\beta_{xn}^n\beta_{yp}^p$ we will use the abbreviation $\overline{16a}_1^1*$.
The sequence transitions give rise to a broadening of this band and to
a displacement of + 8 cm^{-1} with respect to the calculated position of
the (pure) $\overline{16a}_1^1$ band.

The second non-resonant band obtained after $\overline{16a}^2$ excitation is the
0_0^0 band of the bare tetrazine molecule. Here the effect of vibrational
predissociation is observed. The energy of two 16a quanta (512 cm^{-1})
is enough to break the van der Waals bond with a binding energy D_0 in
the order of 300 cm^{-1} (2), leaving behind a bare tetrazine molecule in
the 0^0 state and an argon atom. Part of the energy released in this
process is converted to translational energy of the fragments, the
remaining part is converted to rotational energy of the tetrazine molecule.

There are two possibilities for the deactivation pathways subsequent to $\overline{16a^2}$ excitation. Either the IVR and VP steps proceed independently or they occur sequentially. Time-resolved measurements of the decay curves of each of the three bands can answer this question since the two schemes predict a different behaviour of the temporal dependence of the emissions.

For the parallel process we expect for the emission intensities:

$$I(\overline{16a^2})(t) \sim \exp\{-(k_{16a^2}+k_{IVR}+k_{VP}).t\} \otimes P(t) \qquad (1.a)$$

$$I(\overline{16a^1}*)(t) \sim \exp\{-k_{16a^1}.t\} \otimes I(\overline{16a^2})(t) \qquad (1.b)$$

$$I(0^0)(t) \sim \exp\{-k_0^0.t\} \otimes I(\overline{16a^2})(t) \qquad (1.c)$$

with k_x^{-1} the lifetime found for level x in isolated tetrazine, k_{IVR} and k_{VP} the rate constants for the IVR and VP processes, respectively, and the symbol $\otimes$ standing for convolution. P(t) represents the laser pulse (as monitored). For the sequential process we expect:

$$I(\overline{16a^2})(t) \sim \exp\{-(k_{16a^2}+k_{IVR}).t\} \otimes P(t) \qquad (2.a)$$

$$I(\overline{16a^1}*)(t) \sim \exp\{-(k_{16a^1}+k_{VP}).t\} \otimes I(\overline{16a^2})(t) \qquad (2.b)$$

$$I(0^0)(t) \sim \exp\{-k_0^0.t\} \otimes I(\overline{16a^1}*)(t) \qquad (2.c)$$

With a homewritten program the decay curves have been analyzed in terms of the above expressions. It follows that the sequential scheme is the right one; the results are shown in Fig. 2. Thus the deactivation of the $\overline{16a^2}$ level proceeds as follows:

$$T.Ar(\overline{16a^2}) \xrightarrow{k_{IVR}} T.Ar(\overline{16a^1}*) \xrightarrow{k_{VP}} T(0^0)+Ar \qquad (3)$$

The rate constants can be calculated from the "effective" lifetimes of $\overline{16a^2}$ and $\overline{16a^1}*$ which are provided by the convolution procedure. These rate constants are tabulated in Table II. If we had known beforehand that the deactivation of this level proceed sequentially the rate constants for IVR and VP could also have been obtained from the relative intensities as measured in steady-state experiments. Values obtained in this way equal those provided by time-resolved measurements within experimental error.

4^1 excitation

Mode 4 in s-tetrazine has symmetry b_{2g} which implies that the 4_0^1 transition is symmetry-forbidden. The in-plane polarized band at 534.38 nm has been ascribed to the vibronically induced transition 4_0^1 by Brumbaugh and Innes (7). We have measured a lifetime of 660 ps for the 4^1 level. Dispersed emission monitored after excitation of the 4_0^1 transition in T.Ar (located at 535.05 nm) shows no traces of non-resonant bands. However, the lifetime of the 4^1 level appears to be 580 ps. The difference between these lifetimes exceeds the experimental error. Although the deactivation pathways cannot be established, an upper limit of $2 \times 10^8 s^{-1}$ can be set to the total deactivation rate constant.

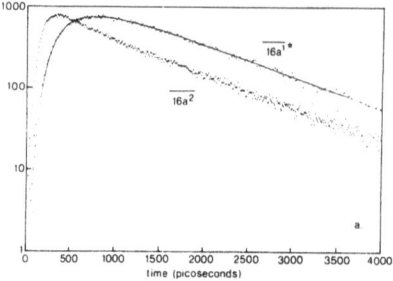 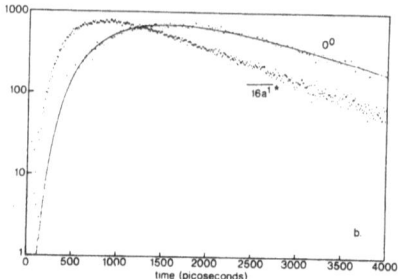

Figure 2. Fluorescence decay curves for emission originating from the levels $\overline{16a}^2$ and $\overline{16a}^1*$ of T.Ar and 0^0 of the free molecule tetrazine, obtained after excitation of the $\overline{16a}^2$ level (diagnostic emission bands $16a_2^2$, $16a_1^1*$ and 0_0^0). The solid lines represent the best fits of the experimental data, in accordance with the sequential deactivation scheme.

$6a^1$ excitation

Fig. 3 shows a segment of the fluorescence spectrum, obtained after excitation of the $\overline{6a}^1$ level of T.Ar. In addition to resonance emission bands, the spectrum exhibits transitions originating from the levels $\overline{16a}^1\overline{16b}^1*$ and $\overline{16a}^2*$ which are formed by IVR in the complex T.Ar and emission from the tetrazine fragment in its $16a^1$ state which is released by VP of the complex.

Information about the deactivation pathways (the proper succession of vibronic states) is obtained in the same way as described for the case of $\overline{16a}^2$ excitation. Although the situation is somewhat more complex in the present case, the same procedure can be applied. It could be shown that the decay curves are described by solutions of a set of coupled differential equations following from the scheme:

$$T.Ar(\overline{6a}^1) \xrightarrow{k_1} T.Ar(\overline{16a}^1\ \overline{16b}^1*) \xrightarrow{k_3} T.Ar(\overline{16a}^2*) \xrightarrow{k_4} T(16a^1)+Ar \quad (4)$$

with k_2 pathway shown above.

Since the IVR steps 1 and 2 are parallel, it is not possible to extract the rate constants k_1 and k_2 from the effective lifetime of the $\overline{6a}^1$ level only. However, with the additional information provided by the relative intensities of the $\overline{16a}_1^1\overline{16b}_1^1*$ and $\overline{16a}_2^2\overline{6a}_1^0*$ bands all constants in (4) can be obtained (12). These rate constants are given in Table II.

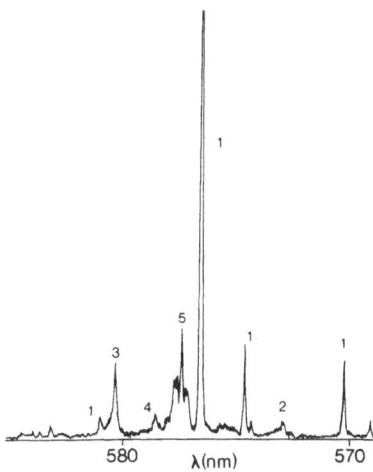

Figure 3. Dispersed fluorescence obtained while pumping the $\overline{6a}_0^1$ transition of T.Ar. The experimental conditions are the same as for Fig. 1. Assignments: 1, several resonant emission bands from the $\overline{6a}^1$ level, the main peak is the $\overline{6a}_2^1$ band; 2: $\overline{16a}_1^1 \overline{16b}_1^1*$; 3: $\overline{16a}_2^2 \overline{6a}_1^0*$; 4: $\overline{16a}_4^2*$; 5: $16a_1^1 6a_1^0$.

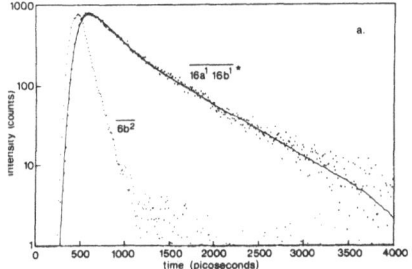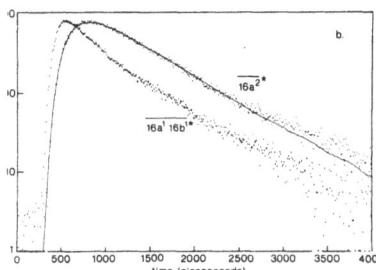

Figure 4. Fluorescence decay curves of emission originating from the levels $\overline{6b}^2$, $\overline{16a}^1 \overline{16b}^1*$ and $\overline{16a}^2*$ obtained by pumping the $\overline{6b}_0^2$ transition (diagnostic emission bands $\overline{6b}_0^2 \overline{6a}_1^0$, $\overline{16a}_1^1 \overline{16b}_1^1*$ and $\overline{16a}_2^2*$). The solid lines are the best fits of the experimental data, in accordance with the deactivation scheme given in the text.

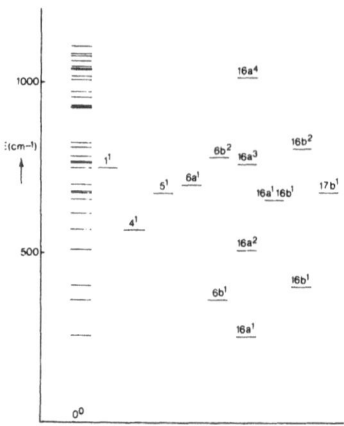

Figure 5. Energy diagram of the vibronic levels in the S_1 state of s-tetrazine up to 1100 cm^{-1}.

TABLE II

Rate constants of the deactivation processes described in this work

excited levels of T.Ar	levels of T.Ar formed by IVR	levels of T formed by VP	rate constants (in sec^{-1})
$\overline{16a^2}$ ────k_{IVR}────► $\overline{16a^1{*}}$ ────k_{VP}────► 0^0			$k_{IVR} = 4.6 \times 10^8$ $k_{VP} = 9.5 \times 10^8$
$\overline{4^1}$ ——— ———			$k_{tot} \lesssim 2 \times 10^8$
$\overline{6a^1}$ k_1→ $\overline{16a^1 16b^1{*}}$ $\Big) k_3$ k_2↘ $\overline{16a^2{*}}$ k_4→ $16a^1$			$k_1 = 5 \times 10^7$ $k_2 = 3.5 \times 10^8$ $k_3 = 4 \times 10^8$ $k_4 = 3.2 \times 10^9$
$\overline{6b^2}$ k_1→ $\overline{16a^1 16b^1{*}}$ k_3→ $16b^1$ k_{-2}↶ $\Big) k_2$ $\overline{16a^2{*}}$ k_4→ $16a^1$			$k_1 = 1.6 \times 10^9$ $k_2 = 1.9 \times 10^9$ $k_{-2} = 5 \times 10^8$ $k_3 = 5 \times 10^8$ $k_4 = 2.2 \times 10^9$

$6b^2$ excitation

The $6b^2$ level of s-tetrazine has a very short lifetime of only 45 ps. Although this reduces the fluorescence intensity considerably it was still possible to obtain good emission spectra while pumping the $\overline{6b}^2_0$ transition in T.Ar. In addition to resonant emission also fluorescence from the levels $\overline{16a}^1\,\overline{16b}^1*$ and $\overline{16a}^2*$ of the complex, as well as emission bands originating from the vibronic states $16a^1$ and $16b^1$ of the released tetrazine fragment have been observed.

The time dependent behaviour of the selectively detected emission bands of the dispersed fluorescence was again analyzed by using the iterative reconvolution method. It turns out that the decay of the level $\overline{16a}^1\,\overline{16b}^1*$ is biexponential. The short component is found to correspond with the risetime of the $\overline{16a}^2*$ emission whereas the long component is equal to the decay time of $\overline{16a}^2*$. This indicates a reversible IVR step. Analysis of the decay curves for all bands in the emission spectrum obtained after $\overline{6b}^2$ excitation leads to the following deactivation scheme (12):

$$\text{T.Ar}(\overline{6b}^2) \xrightarrow{k_1} \text{T.Ar}(\overline{16a}^1\,\overline{16b}^1*) \underset{k_{-2}}{\overset{k_2}{\rightleftharpoons}} \text{T.Ar}(\overline{16a}^2*)$$

$$\Big\downarrow k_3 \qquad\qquad\qquad\qquad \Big\downarrow k_4 \qquad\qquad (5)$$

$$\text{T}(16b^1)+\text{Ar} \qquad\qquad\qquad \text{T}(16a^1)+\text{Ar}$$

The values for the various rate constants, obtained primarily from time-resolved measurements, but with complementary use of relative intensities of bands in spectrally resolved emission, are given in Table II.

The experimental results collected in Table II demonstrate the mode selectivity of the IVR processes. Only a few of the energetically accessible vibronic states are involved in these processes (cf. Fig.5). In the IVR as well as in the VP processes the 16a mode plays an important role.
Actually, reversibility of IVR processes should be a common phenomenon. However, in most cases the rate constants will differ considerably so that either the long or the short component of the decay will be difficult to observe.

Acknowledgements

We wish to thank Mr. A.G.M. Kunst, Mr. W.G. Bouwman and Ing. D. Bebelaar for their valuable assistance in obtaining part of the data presented in this paper and Mr. G. Jansen for preparing the tetrazine samples. These investigations were supported by the Netherlands Founda-

tion for Chemical Research (SON) and were made possible by financial
support from the Netherlands Organization for Pure Research (ZWO).

References

1. J.E. Kenny, D.V. Brumbaugh and D.H. Levy, J.Chem.Phys.,$\underline{71}$ (1979)
   4757
2. D.V. Brumbaugh, J.E. Kenny and D.H. Levy, J.Chem.Phys., $\underline{78}$ (1983)
   3415
3. J.J.F. Ramaekers, J. Langelaar and R.P.H. Rettschnick, in: Picosecond Phenomena III, ed. K.B. Eisenthal, R.M. Hochstrasser,
   W. Kaiser and A. Laubereau (Springer-Verlag, Berlin, 1982), pp.
   264-268
4. J.J.F. Ramaekers, H.K. van Dijk, J. Langelaar and R.P.H.
   Rettschnick, Faraday Discuss. Chem. Soc. $\underline{75}$ (1983) 183
5. M. Heppener, A.G.M. Kunst, D. Bebelaar and R.P.H. Rettschnick,
   J.Chem.Phys., $\underline{83}$ (1985) 5341
6. K.K. Innes, L.A. Franks, A.J. Merer, G.K. Vemulapalli, T. Cassen
   and J. Lowry, J.Mol.Spectrosc., $\underline{66}$ (1977) 465
7. D.V. Brumbaugh and K.K. Innes, Chem.Phys., $\underline{59}$ (1981) 413
8. K.K. Innes, J.Chem.Phys., $\underline{76}$ (1982) 2100
9. R.E. Smalley, L. Wharton, D.H. Levy and D.W. Chandler, J.Chem.
   Phys., $\underline{68}$ (1978) 2487
10. C.A. Haynam, D.V. Brumbaugh and D.H. Levy, J.Chem.Phys., $\underline{80}$ (1984)
    2256
11. A.È. McKinnon, A.G. Szabo and D.R. Miller, J.Phys.Chem., $\underline{81}$ (1977)
    1564
12. M. Heppener, Thesis, University of Amsterdam (in preparation).

THEORY OF VAN DER WAALS PHOTOFRAGMENTATION DYNAMICS

J. Alberto Beswick
LURE, Laboratoire CNRS, CEA, MEN.
Université de Paris Sud
91405 Orsay
France

ABSTRACT. The quantum mechanical treatment of van der Waals photofragmentation dynamics is presented. This includes vibrational-rotational predissociation as well as electronic predissociation in which the optically excited van der Waals molecule undergoes a non-radiative transition to a lower electronic state.

1. INTRODUCTION

In addition to their importance in the determination of intermolecular interactions, van der Waals molecules provide a particularly useful means to study dynamical processes in chemistry. With the use of isentropic nozzle beam expansions and high resolution laser spectroscopy it is now possible to study in great detail the intramolecular dynamics of van der Waals molecules in vibrationally excited states of the ground electronic configuration and in electronically excited configurations.

An interesting feature of excited state intramolecular relaxation processes in such systems involves the breaking of the weak intermolecular bond. This new class of photochemical fragmentation via vibrational or electronic excitation of van der Waals molecules provides the half-collision analog of electronic and vibrational relaxation in collisions. Assume that an excited species A^* collides with another species B. Relaxation induced by the intermolecular interaction can be represented by:

$$A^* + B \longrightarrow A + B \tag{1}$$

In the experiments on van der Waals molecules on the other hand, the complex A...B is first formed in the ground state and then excited by a photon to give:

$$A...B + h\nu \longrightarrow A^*...B \longrightarrow A + B \tag{2}$$

It is clear that (2) can be viewed as the half-collision analog of (1), the systems and the interactions being the same. However, the

A. Weber (ed.), Structure and Dynamics of Weakly Bound Molecular Complexes, 563–571.
© *1987 by D. Reidel Publishing Company.*

relaxation rate in (1) is the result of an average over the impact
parameter and usually over a broad distribution of initially relative
kinetic energies. This situation makes very difficult the comparison
between theory and experiments. Consider on the other hand the photofragmentation experiment represented by Eq. (2). Initially the molecule
is cold and therefore the intermolecular distance and mutual orientation
of A and B are restricted to a much smaller domain. Since the total
angular momentum of the complex is also low, the number of partial waves
is considerably reduced as compared to the collisional case. Finally,
by the use of high resolution spectroscopy, the structure and energetics
of the complex can be studied in detail providing crucial information
on the relevant potential energy surfaces.

An extensive and exciting body of information is emerging from infrared and visible excitation experiments on single electronic potential
energy surface dynamics (1). This provides an unambiguous example for
vibrational and rotational predissociation of a polyatomic molecule.
In the investigation and interpretation of excited-state nonradiative
decay processes of medium-sized molecules it is extremely difficult to
make the distinction between intrastate vibrational predissociation and
interstate electronic relaxation, so that experimental evidence for
vibrational predissociation is sparse. The intramolecular dynamics in
excited states of van der Waals molecules has provided the first demonstration of the experimental manifestations of vibrational predissociation.
This work was pioneered by D. Levy and his collaborators in the studies
of photodissociation dynamics of $I_2...X$ (X = He, Ne, Ar) van der Waals
molecules (2). Many other experimental results on the dynamics of
vibrational predissociation of polyatomic van der Waals molecules were
obtained recently (see other papers in this volume).

It is also possible to study electronic predissociation in van der
Waals molecules. Jouvet and Soep, and Duval and Soep (3), have studied
fine structure predissociation in Hg...M (M = N_2, CO, CH_4, NH_3) molecules.
The complex was excited in the region of the $(6\ ^1S_0 \rightarrow 63P_1)$ transition
in Hg, and the population of the $63P_0$ level of the Hg fragment was
monitored by the use of a second laser. The process can then be
represented as:

$$\text{Hg}\ (6^1S_0)...M\ +\ h\nu\ \longrightarrow\ \text{Hg}\ (6^3P_1)...M\ \longrightarrow\ \text{Hg}(6^3P_0)\ +\ M \qquad (3)$$

and corresponds to a half-collision fine structure relaxation process.

It is the purpose of this chapter to present the general quantum
mechanical theoretical framework in which the aforementioned dynamical
processes can be described. In order to be able to deal with both
vibrational, Eq. (2), and electronic, Eq. (3), predissociation in a
unified way, we shall consider the case of an atom-diatom van der Waals
molecule in which the atom has non-zero electronic angular momentum.

It should be noted at this point that it is also possible to
study reactive processes in van der Waals molecules. Jouvet et al. (4)
have observed the fragments HgH and HgCl when exciting the van der Waals
complexes HgH_2 and $HgCl_2$, respectively.

2. GENERAL FORMALISM

Let us consider an atom A with non-zero electronic angular momentum interacting with a diatomic molecule BC with zero total electronic angular
momentum. The total Hamiltonian, after separation of the center of mass
motion, may be written:

$$H = -\frac{\hbar^2}{2m}\frac{d^2}{dR^2} - \frac{\hbar^2}{2\mu}\frac{d^2}{dr^2} + \frac{\ell^2}{2mR^2} + \frac{j^2}{2\mu r^2} + H_{el} + H_{so} \qquad (4)$$

where R is the intermolecular distance between the atom A and the center
of mass of the BC molecule, r is the internuclear distance of the BC
molecule, while ℓ and j are the two angular momentum operators associated
to R and r, respectively. Thus ℓ describes the "orbital" angular momentum of the fragments about the center of mass of the total system and j
is the rotational angular momentum of the BC molecule. In Eq. (4),
$m = m_A(m_B + m_C)/(m_A + m_B + m_C)$ is the reduced mass for the relative
motion of A and BC, while $\mu = m_B m_C/(m_B + m_C)$ is the reduced mass for
BC. Finally, H_{el} is the non-relativistic electronic Hamiltonian and
H_{so} is the spin-orbit interaction. We shall consider $H_{so} = g\, \underline{L}.\underline{S}$
with g being independent of R and r. We assume the atom A to be described
by a total electronic orbital angular momentum $\underline{L}$ and a total spin angular
momentum $\underline{S}$, with projections on the z-axis equal to Λ and Σ, respectively.
We define the total electronic angular momentum $\underline{J}_A = \underline{L} + \underline{S}$ and its
projection on the z-axis $\Omega_A = \Lambda + \Sigma$. Thus the electronic basis set
$|L,S,J_A,\Omega_A>$ (denoted $|J_A,\Omega_A>$ for simplicity) diagonalises H_{so}.
We now consider H_{el}. It can be written as:

$$H_{el} = H_{el}^{(0)} + H'_{el} \qquad (5)$$

where $H_{el}^{(0)}$ is the asymptotic non-relativistic electronic Hamiltonian,
while H'_{el} is the intermolecular interaction Hamiltonian which goes to
zero when R goes to infinity. We shall make the following simplifying
assumptions:
i) We consider only one electronic state of BC with total orbital and
spin angular momenta equal to zero ($^1\Sigma$ state).
ii) For the atom A we consider two manifolds (ground and excited states)
characterized by well defined quantum numbers L and S.
      Thus asymptotically there will be $(2L+1)(2S+1)$ degenerate eigenstates
of H_{el} in each one of the manifolds. This degeneracy is lifted by the
spin-orbit interaction and we get $2\min(L,S) + 1$ states $|J_A,\Omega_A>$ with
energies: $E_{J_A} + E_{BC}(r)$, where E_{J_A} are the fine structure atomic energies
of A, while E_{BC} is the potential energy for the BC molecule.
      For finite values of R these states will be coupled by the inter-
molecular electronic interaction H'_{el}. For the atom-atom case this term
is often referred to as the "radial coupling" (5).
      An additional interaction between the fine structure states (the
"rotational" coupling) is induced by the orbital centrifugal term in
Eq. (4), i.e., by ℓ^2. This coriolis interaction is important in inducing
electronic relaxation in many cases.
      It is convenient to use a coordinate system of reference in which
the z-axis lies along the $\underline{R}$ vector and the x-axis lies on the plane of

the molecule. With this choice the matrix elements of H'_{el} will depend only on R, r and θ (the angle between $\underline{R}$ and $\underline{r}$).

Let us denote by $|L,\Lambda,S,\Sigma>$ the eigenfunctions of the asymptotic Hamiltonian $H_{el}^{(0)}$which will be given by the direct product of the electronic wavefunction of the molecule M and the eigenfunctions of the atom A. The matrix elements of H'_{el} between these wavefunctions will be given by:

$$<L,\Lambda,S,\Sigma|H'_{el}|L,\Lambda',S,\Sigma'> \ = \ \delta_{\Sigma\Sigma'} \sum_K \ V_{\Lambda\Lambda'}^{(K)} (R,r) \ Y_{K,(\Lambda'-\Lambda)}(\theta,0) \qquad (6)$$

As discussed above, we define our basis set as those which diagonalise $H_{el}^{(0)}$ + H_{so}, where H_{so} is approximate by H_{so} = g $\underline{L}.\underline{S}$. We then have the Hund's case (c) wavefunctions:

$$|J_A,\Omega_A> \ = \ \sum_{\Lambda\Sigma} \ C(L,\Lambda,S,\Sigma;J_A,\Omega_A) \ |L,\Lambda,S,\Sigma> \qquad (7)$$

where C is a Clebsh-Gordan coefficient. Using (7), we get from (6):

$$<J_A,\Omega_A| \ H'_{el} \ |J'_A,\Omega'_A> \ = \ \sum_{\Lambda\Lambda'} \ \sum_\Sigma \sum_K C(L,\Lambda,S,\Sigma;J_A,\Omega_A) \ C(L,\Lambda',S,\Sigma;J'_A,\Omega'_A)$$

$$V_{\Lambda\Lambda'}^{(K)}(R,r) \ Y_{K,(\Lambda'-\Lambda)}(\theta,0) \qquad (8)$$

We now introduce the rotational angular momentum $\underline{j}$ of the diatomic molecule M and the total angular momentum $\underline{J}$ = $\underline{J}_A$ + $\underline{\ell}$ + $\underline{j}$. A body fixed basis set can be defined as:

$$|J,M,\Omega,j,J_A,\Omega_A> \ = \ \left(\frac{2J+1}{4\pi}\right)^{1/2} \ D_{M\Omega}^J(\phi_R,\theta_R,\phi) \ Y_{j,\lambda=\Omega-\Omega_A}(\theta,0) \ |J_A,\Omega_A> \qquad (9)$$

with Ω, the projection of J on the body fixed z axis, being equal to the sum of the projections of $\overline{j}$ and J_A. The Euler angles ϕ_R, θ_R, ϕ, specify the orientation of the body fixed with respect to the space fixed system of reference. Using (9) and (8) we finally obtain:

$$<J,M,\Omega,j,J_A,\Omega_A| \ H'_{el} \ |J',M',\Omega',j',J'_A,\Omega'_A> \ = \ \delta_{JJ'} \ \delta_{MM'} \ \delta_{\Omega\Omega'}$$

$$\sum_{\Lambda\Lambda'} \ \sum_\Sigma \sum_K \ (-)^K \ \left(\frac{2K+1}{4\pi}\right)^{1/2} \ V_{\Lambda\Lambda'}^{(K)}(R,r) \ C(L,\Lambda,S,\Sigma;J_A,\Omega_A) \qquad (10)$$

$$C(L,\Lambda',S,\Sigma;J'_A,\Omega'_A) \ C(j',0,K,0;j',0) \ C(j',\Omega-\Omega'_A,K,\Omega'_A-\Omega_A;j,\Omega-\Omega_A)$$

Moreover, it is convenient to work with wavefunctions of definite parity with respect to total inversion through the space fixed origin. They can be written:

$$|J,M,\Omega,j,J_A,\Omega_A,p> \ = \ C_{\Omega,\Omega_A} \left(|J,M,\Omega,j,J_A,\Omega_A> + p \ (-)^{J+J_A+L} \right.$$

$$\left. |J,M,-\Omega,j,J_A,-\Omega_A> \right) \qquad (11a)$$

with

$$C_{\Omega,\Omega_A} = \left[(2 - \delta_{0,\Omega}\,\delta_{0,\Omega_A}) \right]^{1/2} / 2 \tag{11b}$$

and $\Omega_A \geq 0$, $\Omega \geq 0$, $p = \pm 1$.

The next step in the calculation will be to expand the total wave-function including vibration as:

$$|\psi_{\alpha E}^{JMp}\rangle = \sum_{\Omega > 0} \sum_{vjJ_A\Omega_A \geq 0} \sum \phi_{\alpha E vjJ_A\Omega_A}^{J\Omega p}(R)\ \chi_{vj}(r)\ |J,M,\Omega,j,J_A,\Omega_A,p\rangle \tag{12}$$

where $\chi_{vj}(r)$ represents a vibrational wavefunction for the diatomic fragment M, and $\Phi(R)$ describes the relative radial motion of A and M. The functions Φ are determined by solving the coupled Schrödinger equations:

$$\left[-\frac{\hbar^2}{2m}\frac{d^2}{dR^2} + E_{vj} + E_{J_A} - E \right] \phi_{\alpha E vjJ_A\Omega_A}^{J\Omega p}(R) =$$

$$= - \sum_{v'j'J_A'\Omega_A'} \int dr\ \langle J,M,\Omega,j,J_A,\Omega_A,p| H_{el}' |J,M,\Omega,j',J_A',\Omega_A'\rangle\ \chi_{vj}(r)\ \chi_{v'j'}(r)$$

$$\times \phi_{\alpha E v'j'J_A'\Omega_A'}^{J\Omega p}(R) - \frac{1}{2\,m\,R2} \sum_{\Omega'\Omega_A'} \langle J,M,j,J_A,\Omega_A,p| \ell^2 |J,M,\Omega',j,J_A,\Omega_A',p\rangle$$

$$\times \phi_{\alpha E vjJ_A\Omega_A'}^{J\Omega' p}(R) \tag{13}$$

where E_{vj} are the ro-vibrational energies of the diatomic M fragment. The basic matrix elements of ℓ^2 which are needed in Eq. (13) can be obtained using:

$$\langle J,M,\Omega,j,J_A,\Omega_A| \ell^2 |J,M,\Omega',j,J_A,\Omega_A'\rangle = \hbar^2 \Big[J(J+1) + j(j+1) + J_A(J_A+1)$$

$$- 2\,\Omega^2 - 2\,\Omega_A^2 + 2\,\Omega\,\Omega_A \Big]\ \delta_{\Omega\Omega'}\ \delta_{\Omega_A\Omega_A'} -$$

$$- \Big[\langle J,M,\Omega,j,J_A,\Omega_A| (J_+j_- + J_-j_+ + J_+J_{A-} + J_-J_{A+}$$

$$- j_+J_{A-} - j_-J_{A+}) |J,M,\Omega',j,J_A,\Omega_A'\rangle \Big] \delta_{\Omega',\Omega\pm 1}\ \delta_{\Omega_A',\Omega_A\pm 1} \tag{14}$$

with the raising and lowering operators being defined as usual:

$$J_\pm |J,\Omega\rangle = \hbar \left[J(J+1) - \Omega(\Omega \mp 1) \right]^{1/2} |J,\Omega \mp 1\rangle$$

$$j_\pm |j,\lambda\rangle = \hbar \left[j(j+1) - \lambda(\lambda \pm 1) \right]^{1/2} |j,\lambda \pm 1\rangle \tag{15}$$

and J_A having the same behavior as j.

The close coupling equations (13) have to be solved for both the ground and the excited states of the system. Let denote by $|\psi J''M''p''>_{gEg}$ the total wavefunction for the ground state of the complex. Since this is a bound state of the molecule the wavefunction goes to zero as R goes to infinity. On the other hand the electronically excited wavefunction in which the fine structure transitions occur are continuum wavefunctions which obey to the following boundary conditions:

$$|\psi_{\alpha E}^{JMp}> \underset{R\to\infty}{\sim} e^{ik_\alpha R}|\alpha> + \sum_{\alpha'} S_{\alpha\alpha'}^* e^{-ik_{\alpha'}R}|\alpha'> \tag{16}$$

where α denote a particular state of the fragments, i.e.,

$$|\alpha> = \chi_{vj}(r) \ |J,M,\Omega,j,J_A,\Omega_A,p> \tag{17}$$

while $e^{\pm ik_\alpha R}$ are outgoing (ingoing) spherical waves with:

$$k_\alpha = \sqrt{2m(E - E_\alpha)} \ / \ \hbar \tag{18}$$

Finally, S in Eq. (16) is the usual scattering matrix. The asymptotic behavior (16) corresponds to the appropriate boundary conditions for the photodissociation case, namely, incoming waves in all channels and unit flux outgoing wave in channel α. With all these conditions fulfilled, we are now in position to write the photodissociation cross section. For a photon with energy $\hbar\omega$, the photodissociation cross section from an initial ground state g to a final continuum channel α with energy $E = E_g + \hbar\omega$, is given by:

$$\sigma_{g\to\alpha E} = \frac{4\pi^2\omega}{c} \ \left|<\psi_{gE_g}^{J''M''p''}| \ \underline{\mu}\cdot\underline{e} \ |\psi_{\alpha E}^{JMp}>\right|^2 \tag{19}$$

where it was assumed that the continuum wavefunctions are energy normalized. In Eq. (19) $\underline{\mu}$ is the dipole moment operator while $\underline{e}$ is the polarization vector of the photon.

The equations above constitute our final expressions for describing vibrational/rotational as well as fine structure predissociation in a triatomic van der Waals molecule of the type $A^*...BC$. The generalization to the case $A^*...M$ where M is a general polyatomic molecule can be obtained by replacing the rotational wavefunctions $|j,\lambda>$ in Eq. (9) by the appropriate asymmetric rotor wavefunctions.

The approximations involved in our treatment are the following:
i) The coupling to other electronic states are neglected. In particular, the interaction with the ionic surface should be considered when the ionization potential of one of the partners is low (intermediate ionic complex model (6)).
ii) The spin-orbit coupling was assumed to be constant and equal to the value for the isolated atom A.

Both are likely to be good approximations in many cases.

3. THE SLOW PREDISSOCIATION CASE

When predissociation is slow (i.e., when individual rovibrational lines can be resolved), perturbation theory can be applied. If the Hamiltonian for the electronic excited state is partitioned as:

$$H = H_0 + V \tag{20}$$

the photodissociation partial cross section (19) will assume the form:

$$\sigma_{g \to \alpha E} = \frac{4\pi^2 \omega}{c} \sum_s \left| \langle \Psi_{gE_g} | \underline{\mu} \cdot \underline{e} | \Psi_{sE_s}^{(0)} \rangle \right|^2 \frac{\left| \langle \Psi_{sE_s}^{(0)} | V | \Psi_{\alpha E}^{(0)} \rangle \right|^2}{(E - E_s)^2 + \Gamma_s^2} \tag{21}$$

where the E_s are the positions of the resonances and the Γ_s their linewidths given by the wellknown Golden-rule expression:

$$\Gamma_s = \pi \sum_\alpha \left| \langle \Psi_{sE_s}^{(0)} | V | \Psi_{\alpha E}^{(0)} \rangle \right|^2 ; \quad \text{for } E = E_s \tag{22}$$

The wave vectors $|\Psi_{sE_s}^{(0)}\rangle$ and $|\Psi_{\alpha E}^{(0)}\rangle$ in Eqs. (21) and (22) represent discrete and continuum functions of the zero-order Hamiltonian H_0. If the resonances are well resolved and very narrow, the sum over s in Eq. (21) reduces in good approximation to a single term when the total energy is in the vecinity of a particualr resonance. In that case one obtains the familiar Lorentzian lineshape:

$$\sigma_{g \to \alpha E} = \frac{4\pi^2 \omega}{c} \left| \langle \Psi_{gE_g} | \underline{\mu} \cdot \underline{e} | \Psi_{sE_s}^{(0)} \rangle \right|^2 \frac{\left| \langle \Psi_{sE_s}^{(0)} | V | \Psi_{\alpha E}^{(0)} \rangle \right|^2}{(E - E_s)^2 + \Gamma_s^2} \tag{23}$$

The total photodissociation cross section is then given by:

$$\sigma_{tot} = \sum_\alpha \sigma_{g \to \alpha E} = \frac{4\pi \omega}{c} \left| \langle \Psi_{gE_g} | \underline{\mu} \cdot \underline{e} | \Psi_{sE_s}^{(0)} \rangle \right|^2 \frac{\Gamma_s}{(E - E_s)^2 + \Gamma_s^2} \tag{24}$$

and the probability for dissociating into a particular channel α is:

$$P_\alpha = \frac{\sigma_{g \to \alpha E}}{\sigma_{tot}} = \frac{\pi \left| \langle \Psi_{sE_s}^{(0)} | V | \Psi_{\alpha E}^{(0)} \rangle \right|^2}{\Gamma_s} \tag{25}$$

What remains to be done is to specify H_0 and V in Eq. (20). Here we are interested in the situation where low lying vibrational levels in the van der Waals well corresponding to a particular fine structure level, predissociate to another fine structure lower limit.

This situation can only be met if the van der Waals binding energy is
smaller than the fine structure splitting. Incidentally, this consideration raises the question of the basis set used in the calculations.
As long as the coupled equations are solved exactly, it is irrelevant
to discuss this matter (it can only be more or less convenient from
the numerical point of view). On the other hand if approximate treatments based on perturbation schemes are implemented, the choice of the
appropriate basis set is crucial.

Since we are interested in the study of van der Waals complexes at
low temperature, case (b) wavefunctions are not appropriate. On the
other hand, the choice between case (a) and case (c) will depend on the
relative values of the van der Waals interaction and the spin-orbit
coupling. If the spin-orbit coupling is large compared with the
van der Waals interaction, then case (c) wavefunctions should be used.
This is exactly the case we are studying here. We therefore define H_0
as:

$$H_0 = \sum_{J_A \Omega_A} |J_A,\Omega_A> <J_A,\Omega_A| H |J_A,\Omega_A> <J_A,\Omega_A| \qquad (26)$$

i.e., the projection of the total Hamiltonian onto the atomic spin-
orbit fine structure states J_A,Ω_A. Thus J_A and Ω_A are good quantum
numbers in the zero-order H_0 space and V couples different $|J_A,\Omega_A>$ states.

3.1 APPLICATION

In order to provide some explicit example let us consider the case where
the excited states of the atom A is described by the quantum numbers
L = 1 and S = 1/2. We then have J_A = 1/2 with Ω_A = 1/2 and J_A = 3/2
with Ω_A = 1/2 and 3/2.

3.1a ATOM-ATOM VAN DER WAALS COMPLEX. When BC is an atom with electronic
angular momentum zero, the variables r and θ in our general expressions of section 2 disappear and in addition we have $\Omega = \Omega_A$. Thus
K = 0 in Eq. (6). From our general expressions we obtain the following
coupling scheme:

	$J_A'=3/2, \Omega_A'=3/2$	$J_A'=3/2, \Omega_A'=1/2$
$J_A=1/2, \Omega_A=1/2$	0	$<1/2,1/2|H_{el}'|3/2,1/2>$
$J_A=3/2, \Omega_A=1/2$	$<3/2,1/2|H_{rot}|3/2,3/2>$	

where we have denoted by H_{rot} the coriolis term in ℓ^2.

Thus, assuming the spin-orbit coupling positive, the state 3/2,1/2
can predissociate into the 1/2,1/2 channel by coupling through the H_{el}'
van der Waals interaction. On the other hand the 3/2,3/2 state is
not coupled directly to the 1/2,1/2 channel. It can predissociate in-
directly via the rotational coupling to the 3/2,1/2 state. We conclude
that the state 3/2,1/2 will predissociates much faster than the 3/2,3/2
in all cases where the coriolis coupling is weak.

3.1b ATOM-DIATOM VAN DER WAALS COMPLEX. From our general expressions
of section 2 we have now the following coupling scheme:

	$J_A'=3/2, \Omega_A'=3/2$	$J_A'=3/2, \Omega_A'=1/2$
$J_A=1/2, \Omega_A=1/2$	$<1/2,1/2|H_{el}'|3/2,3/2>$	$<1/2,1/2|H_{el}'|3/2,1/2>$
$J_A=3/2, \Omega_A=1/2$	$<3/2,1/2|H_{rot}|3/2,3/2>$	

and we note that now there is a direct coupling via H_{el}' between the
state 3/2,3/2 and the 1/2,1/2 channel. This is a direct consequence
of the removal of the $C_{\infty v}$ symmetry. The effect of this new coupling
between fine structure levels in the atom diatom case already been
considered in the study of collisional relaxation (7).

4. CONCLUSIONS

We have presented the theory of rovibrational and fine structure
predissociation of van der Waals complexes. The general treatment is
based on the solution of coupled equations within the fine structure
manifold. We have also considered the limiting case of slow predissociation and provide the expressions using perturbation theory. A particular
case, namely L=1 and S=1/2, has been analyzed. Recently (8), the same
treatment has been applied to the case of L=1 and S=1, which corresponds
to the experiments performed on Hg(^{3}P) (3). This provided a rationali-
zation of the different behavior observed for Hg...Ar and Hg...N2 molecules with respect to electronic predissociation.
    The theory presented here can be applied to many other systems
characterized by a spin-orbit coupling larger than the van der Waals
binding energy. We hope this will trigger the development of new
detailed experiments in this interesting area of intramolecular dynamics.

5. REFERENCES

1. a) K.C.Janda, Adv. Chem. Phys. 60, 201 (1985);
   b) R.E.Miller, J. Phys. Chem. 90, 3301 (1986).
2. D.H.Levy, Adv. Chem. Phys. 47, 323 (1981).
3. a) C.Jouvet and B.Soep, J. Chem. Phys. 80, 2229 (1984);
   b) C.Jouvet, Thèse, Université de Paris Sud (1985).
4. a) C.Jouvet and B.Soep, Chem. Phys. Lett. 96, 426 (1983);
   b) C.Jouvet and B.Soep, Laser Chem. 5, 157 (1985).
5. E.E.Nikitin, Theory of Elementary Atom and Molecular Processes in
   Gases, Clarendom Press, Oxford (1974).
6. E.A.Andreev and A.I.Voronin, Chem. Phys. Lett. 3, 488 (1969).
7. a) A.Hickman, J. Phys. B15, 3005 (1982);
   b) J.M.Mestdagh, P. de Pujo, J.Cuvellier, A.Binet and J.Berlande,
      J. Phys. B15, 663 (1982).
8. C.Jouvet and J.A.Beswick, to be published.

DISSOCIATION DYNAMICS OF THE He-I$_2$-Ne VAN DER WAALS COMPLEX

G. Delgado-Barrio, P. Villarreal, A. Varadé[(*)], N. Martín,
and A. García.
Instituto de Estructura de la Materia, C.S.I.C.
Serrano 123, 28006 - Madrid
SPAIN
(*) Present address: ETSI Minas, Rios Rosas 21, 28006-MADRID

ABSTRACT. The vibrational predissociation (VP) process of the van der Waals complex He-I$_2$-Ne, where the diatomic subunit is initially in an excited vibrational level of the electronic B state, is studied using an approximate quantum mechanical model and also a quasiclassical trajectory (QCT) method.
    Only coplanar configurations of the complex have been considered. The potential surface is assumed to be the addition of atom-atom interactions that are described by Morse functions. A good agreement between QCT and quantal calculations is found for the total rate of predissociation.

I. INTRODUCTION

    In recent years the predissociation of van der Waals (vdW) molecules is receiving an increasing attention as it becomes widely realized that van der Waals molecules provide tractable model systems for state-to-state studies of intramolecular energy redistribution. The greatest part of the experimental and theoretical studies on this field have been devoted to systems with one vdW bond and one or more chemical bonds[1], providing a detailed knowledge about their structure and dynamics and contributing to the understanding of intramolecular energy transfer and photodissociation processes. Also some complexes of I$_2$ with two or more rare gas atoms have been studied[2,3] in order to elucidate such questions as the importance of correlation between different dissociative steps and the competition in the flowing of energy from the diatomic molecule to each weak bond.
    In this paper we study the VP of the He-I$_2$-Ne van der Waals complex for collinear and perpendicular (coplanar) configurations. The hamiltonian may be expressed in relative coordinates as the sum of a hamiltonian for the "free" I$_2$ subunit and two triatomic hamiltonians corresponding to He-I$_2$ and Ne-I$_2$, plus a crossed kinetic term and the He-Ne interaction.
    As it was pointed out[3] a rigorous quantal treatment of four-body systems is prohibitively difficult. Hence, we use some reasonable

A. Weber (ed.), Structure and Dynamics of Weakly Bound Molecular Complexes, 573–582.
© *1987 by D. Reidel Publishing Company.*

approximations in Quantal and Classical Mechanics.

In the quantum mechanical model, and for low and intermediate vibrational diatomic excitations, $v \lesssim 40$, we may apply a diabatic distorted wave approach to describe the iodine vibrations within the complex. Representing the v-averaged Hamiltonian in a basis formed by the products of discrete solutions corresponding to He-I_2 and Ne-I_2, we can obtain after a diagonalization the tetraatomic discrete levels. As regards the continuum wavefunctions we shall consider two different continua, each one of them representing the breaking up of one vdW bond. Thus, we assume in the model that the interaction between both continua is negligible. Finally, the predissociation rates are obtained in the "Golden Rule" framework.

Concerning with the quasiclassical model, we use the energy of the quasibound state obtained in the quantal model in order to get the initial conditions for the trajectories. The relevant result of these calculation is again the rate of predissociation of the vdW complex. For obtaining this magnitude, an exponential law of decay is assumed.[4,5]

Concerning the potential energy, we express it as a sum of pairwise atom-atom interactions, where each one of them is a Morse function. This is a reasonable assumption because the VP process is mainly sensitive to the region of the well.

In next Section we describe in detail the quantal model, while Section III is devoted to the quasiclassical model and in Section IV the potential energy used in both models is presented. Finally, the calculations and results obtained through both models are discussed.

II. Quantal model

The hamiltonian for nuclear motion, after separation of the center of mass of the whole system, when He-I_2-Ne is in collinear or in perpendicular (coplanar) configurations, may be written as

$$H = -\frac{\hbar^2}{2\mu_{I_2}} \frac{\partial^2}{\partial r^2} + V_{I_2}(r) - \frac{\hbar^2}{2\mu_{He,I_2}} \frac{\partial^2}{\partial R_1^2} + U_{He,I_2}(r,R,)$$

$$-\frac{\hbar^2}{2\mu_{Ne,I_2}} \frac{\partial^2}{\partial R_2^2} + U_{Ne,I_2}(r,R_2) + \frac{\hbar^2}{2m_I} \frac{\partial^2}{\partial R_1 \partial R_2} +$$

$$+ W_{He,Ne}(R_1,R_2)$$

where r is the internuclear distance for I_2, R_1 and R_2 are the distances between He or Ne and the center of mass of I_2, respectively. $V_{I_2}(r)$ is the intramolecular potential interaction for the "free" diatomic molecule I_2 while U_{He,I_2}, U_{Ne,I_2}, $W_{He,Ne}$ are the van der Waals interactions between He and I_2, Ne and I_2, and He and Ne respectively.

The μ's are the reduced masses for I_2, He with I_2 and Ne with I_2 and m_I is the mass of the iodine atom. By using the diabatic approach for describing the iodine vibration, we can write the wavefunction for discrete states as

$$\Psi_{v,k}(r,R_1,R_2) = \chi_v(r)\,\psi_k(R_1,R_2) \qquad |2|$$

where $\chi_v(r)$ is the solution of the isolated I_2 Schrödinger equation for the

$$\left[-\frac{\hbar^2}{2\mu_{I_2}}\frac{\partial^2}{\partial r^2} + V_{I_2}(v)\right]\chi_v = E_{I_2}(v)\chi_v \qquad |3|$$

corresponding to a vibrational quantum number v, while $\psi_k(R_1,R_2)$ is the solution of

$$\left[-\frac{\hbar^2}{2\mu_{He,I_2}}\frac{\partial^2}{\partial R_1^2} + U_{He,I_2}(\bar{r},R_1) - \frac{\hbar^2}{2\mu_{Ne,I_2}}\frac{\partial^2}{\partial R_2^2} + \right.$$

$$\qquad |4|$$

$$\left. + U_{Ne,I_2}(\bar{r},R_2) + \frac{\hbar^2}{2m_I}\frac{\partial^2}{\partial R_1 \partial R_2} + W_{He,Ne}(R_1,R_2)\right]\psi_k(R_1,R_2)$$

$$= E_k\,\psi_k(R_1,R_2)$$

where $\bar{r}$ is the equilibrium bond length for I_2. The solution of equation $|4|$ is now straightforward by expansion of $\psi_k(R_1,R_2)$ in a basis set and diagonalization. We have chosen here as a basis the direct product of the discrete solutions for the triatomic problems so that

$$\psi_k(R_1,R_2) = \sum_{nm} a_{nm}^k\,\phi_n(R_1)\,\zeta_m(R_2)$$

where ϕ_n and ζ_m are the solutions for

$$\left[-\frac{\hbar^2}{2\mu_{He,I_2}}\frac{\partial^2}{\partial R_1^2} + U_{He,I_2}(\bar{r},R_1)\right]\phi_n(R_1) = E_n^{He,I_2}\,\phi_n(R_1) \quad |5|$$

and

$$\left[-\frac{\hbar^2}{2\mu_{Ne,I_2}} \frac{\partial^2}{\partial R_2^2} + U_{Ne,I_2}(\bar{r},R_2) \right] \zeta_m(R_2) = E_m^{Ne,I_2} \zeta_m(R_2)$$

Finally the energies for the discrete states are

$$E_{vk} = E_{I_2}(v) + E_k \qquad\qquad |6|$$

where E_k is not very different to the simple addition of E_k^{He,I_2} and E_m^{He,I_2} being a_{nm}^k the biggest coefficient in the expansion of $\psi_k(R_1,R_2)$.

We turn now to the determination of the continuum wave functions. After the diabatic approach for iodine vibration, we can write the wavefunction as

$$\Psi_{v,n\varepsilon}(r,R_1,R_2) = \chi_v(r)\, \psi_{n,\varepsilon}(R_1,R_2) \qquad\qquad |7|$$

where $\psi_{n,\varepsilon}(R_1,R_2)$ has two indices, one of them is discrete and describes the final state for the triatomic fragment, while ε is the kinetic energy between the fragments. The expression is

$$\psi_{n\varepsilon}(R_1,R_2) = \phi_n(R_1)\, \zeta_\varepsilon^n(R_2) \qquad\qquad |8|$$

where ϕ_n is a discrete solution of

$$\left[-\frac{\hbar^2}{2\mu_{He,I_2}} \frac{\partial^2}{\partial R_1^2} + U_{He,I_2}(\bar{r},R_1) \right] \phi_n(R_1) = E_n^{He,I_2} \phi_n(R_1) \qquad\qquad |9|$$

and ζ_ε^n is the energy normalized solution of

$$\left[-\frac{\hbar^2}{2\mu_{Ne,I_2}} \frac{\partial^2}{\partial R_2^2} + U_{Ne,I_2}(\bar{r},R_2) + \langle\phi_n| W_{He,Ne}(R_1,R_2)|\phi_n\rangle \right] \zeta_\varepsilon^n(R_2)$$

$$= \varepsilon\, \zeta_\varepsilon^n(R_2) \qquad\qquad |10|$$

We are now in position to calculate the linewidth of the level specified by vibrational quantum numbers v,k using the "golden-rule" approximation

$$\Gamma_{v,k} = \pi \sum_{v',n'} \left| V^{d-c}_{v,k;v',n',\varepsilon} \right|^2 \qquad |11|$$

where

$$V^{d-c}_{v,k;v',n',\varepsilon} = \langle \Psi_{v,k} | H | \Psi_{v',n',\varepsilon} \rangle$$

$$= \sum_{n,m} a^k_{nm} \langle \phi_n \zeta_m | \langle \chi_v | U_{He,I_2} + U_{Ne,I_2} | \chi_{v'} \rangle | \zeta_\varepsilon \phi_{n'} \rangle \qquad |12|$$

$$= \sum_{n,m} a^k_{nm} \{ \langle \zeta_m | \langle \chi_v | U_{He,I_2} | \chi_{v'} \rangle | \zeta_\varepsilon \rangle \, \delta_{nn'} +$$

$$+ \langle \zeta_m | \zeta_\varepsilon \rangle \langle \phi_n | \langle \chi_v | U_{Ne,I_2} | \chi_{v'} \rangle | \phi_{n'} \rangle \}$$

with the continuum function being calculated on the energy shell, i.e.,

$$\varepsilon = E_{I_2}(v) + E_k - E_{I_2}(v') - E^{He,I_2}_{n'} \qquad |13|$$

If we want to describe the breaking up of R_1 the equations $|7|$ to $|13|$ are the same interchanging R_1 by R_2 and U_{He,I_2} by U_{Ne,I_2}.

III. QUASICLASSICAL TREATMENT.

The QCT method has already been applied successfully to treat VP of He...I$_2$(B) in the collinear configuration[4] and also in a 3D model[5,6]. Recently, the VP of several tetraatomic complexes formed by I$_2$ and He and/or Ne have been studied by this procedure[3].

The classical Hamilton function associated to a planar configuration may be written, in the center of mass, as follows

$$H = \frac{P_r^2}{2\mu_{I_2}} + K \left[\frac{P_{R_1}^2}{2\mu_{He,I_2}} + \frac{P_{R_2}^2}{2\mu_{Ne,I_2}} + \frac{\cos(\gamma_1 + \gamma_2)}{2m_I} P_{R_1} P_{R_2} \right] +$$

$$+ V(r, R_1, R_2, \gamma_1, \gamma_2) \qquad |14|$$

where r, R_1, R_2 and μ are the same that those of the equation $|1|$; γ_1 and

γ_2 are the angles formed by the vectors R_1 and r, and R_2 and r, respectively, and the P's are the conjugate momenta to the respective coordinates. K is a quantity that depends on the masses and the orientations in the following way:

$$K = 2Mm_I \Big/ \Big[2m_I M + M_{He}m_{Ne}\sin^2(\gamma_1+\gamma_2)\Big] \qquad\qquad |15|$$

where $M = m_{He} + m_{Ne} + 2m_I$

Finally, $\dot{V}$ in eq. $|14|$ represents the full potential energy surface.

In the "Infinite order sudden" approximation (IOSA), the angular variables γ_1 and γ_2 are taken fixed. This approach is supported by the great difference existing between vibrational (fast) and rotational (slow) motions within each corresponding triatomic complex, together with the weakness of the He-Ne interaction. Hence, the usual Hamilton equations constitute a set of six first order coupled differential equations, that must be numerically solved once the relevant initial conditions have been established.

In order to follow closely the quantal treatment, we consider initially that the vibrational energy stored in the I_2 molecule is that corresponding to a quantum v state, $E_{I_2}(v)$. Hence, we choose the r distance according to the distribution of a Morse oscillator[7], that provides us directly the necessary momentum. As regards the choosing of the R_1 and R_2 distances, and the associated momenta, we take adventage of the fact that the tetraatomic quantal energy is close to the simple addition of triatomic energies. Thus, we start fixing each triatom at its corresponding quantal energy and obtain the relevant distances at random using an uniform distribution between the classical turning points[5]. Obviously, the total energy is not exactly conserved in the first step of integration due to the presence of the He-Ne interaction and the crossed kinetic terms. If this energy variation is higher than an established value, the calculation stops and a new set of initial conditions is generated. A trajectory is considered as dissociated when R_1 or R_2 becomes larger than a value $R_{máx}$, for which the corresponding vdW interaction is negligible. Anyway, the trajectories finish if the time exceeds a given $T_{máx}$. For a trajectory leading to fragmentation, the duration of the complex is estimated as the elapsed time from the begining till the last inner turning point in the corresponding coordinate was reached. This is a reasonable way of avoiding spurious long times caused by a very low exiting relative velocity between the fragments. Finally, the distribution of trajectories is fitted to a typical exponential law of decay[4,5] allowing us the calculation of the VP rate.

IV. POTENTIAL ENERGY SURFACE (PES).

The lack of available PES for the system under study forces us to use a functional form to describe it. Since a dumbbell model has yielded reasonable results for the VP of triatomic systems[8] and the He-Ne interaction is rather weak[9], we choose this model potential and thus represent the PES as an addition of atom-atom interactions, where each one of them is taken to be a Morse function. The interactions I-I[10] and He-Ne[3] are well known from the literature. This is not the case for the He-I and Ne-I potentials and a proper determination of the corresponding Morse parameters must be done. In fact, these parameters can dramatically affect the VP dynamics of the complex. For instance, when the parameters of Ref. (3) are used within our quantal approach, the collinear state specified by $v=28$, $k=0$ can not predissociate by the He-I$_2$($v=27$) + Ne channel, that is the most important exiting channel for breaking up the stronger vdW bond, as will be seen below. This results in a much higher probability to get the Ne-I$_2$ + He fragments than the He-I$_2$ + Ne ones. This is just the opposite behaviour to that found in the classical calculations[3]. The parameters used in this work are listed in Table I. They were fitted within the framework of a diabatic vibrational-adiabatic rotational model[11] in order to reproduce the available spectroscopic data[12] for the He-I$_2$(B)[13] and Ne-I$_2$(B)[14] triatomic systems.

Table I. Morse potential parameters.

	$D(cm^{-1})$	$\alpha(\text{Å}^{-1})$	$R_{eq}(\text{Å})$
I-I	5168.72	1.834	3.0247
I-Ne	44.	1.9	4.36
I-He	16.5	1.5	4.
He-Ne	9.94	1.45	3.21

V. CALCULATIONS AND RESULTS.

In order to apply the quantal model, we numerically solve equation $|3|$ for the isolated I$_2$ by means of a mixed Truhlar-Numerov procedure[15] using a grid of $N=8000$ points starting in $R_{in}=4.5$ a.u. and a step size of integration $h=0.001$ a.u. After that, we treat eqs. $|5|$, in the same way with the values $N=2000$, $R_{in}=4$. a.u. and $h=0.007$. We obtain five discrete levels for Ne-I$_2$ and two levels for He-I$_2$ in the perpendicular configuration, while these numbers reduce to four and one, respectively, in the collinear case. By using the products of these discrete functions as a basis, we solve variationally eq. $|4|$. Likewise, the continuum wavefunctions, eq. $|10|$, are obtained by a procedure described elsewhere[16]. All the necessary integrals are evaluated by standard numerical quadratures.

Concerning with the QCT model, we employ an Adams-Moulton integrator initiated by a fourth-order Runge-Kutta-Gill one[17]. The values of the necessary parameters are: time step of the integration $4.\times10^{-16}$s, $T_{máx}=6.\times10^{-11}$s, and $R_{máx}=12$ Å. Seven hundred trajectories were calculated for the perpendicular configuration and also varying γ_1 and γ_2, eq. $|14|$, at random[5].

We have applied the quantal model to study the collinear configuration of this system, in $v_{I_2} = 28$, and used the parameters of Ref. (3) to describe the PES.

This set of parameters yields the closing of the $\Delta v_{I_2} = -1$ channel to obtain the Ne atomic fragment and the associated rate becomes very low, $\Gamma_{Ne} = 1.4\times10^{-5}$ cm^{-1}, while the corresponding to He is $\Gamma_{He}=0.065$ cm^{-1}. Therefore, this quantal result is just the opposite to the classical one[3].

Henceforth, we shall use the set of parameters listed in Table I.

Table II. Quantal energies and rates (cm^{-1}) in collinear (C) and perpendicular (P) configurations.

	$C(v_{I_2}=25)$	$P(v_{I_2}=25)$	$C(v_{I_2}=28)$	$P(v_{I_2}=28)$
$E_k=0$	-39.617	-91.384	-39.617	-91.384
$E_m=0$	-33.093	-72.830	-33.093	-72.830
$E_n=0$	- 6.523	-18.534	- 6.523	-18.534
Γ_{He} ($\Delta v=-1$)	0.314	0.1619	0.393	0.211
Γ_{He} ($\Delta v=-2$)	0.002	0.0017	0.004	0.002
Γ_{Ne} ($\Delta v=-1$)	0.571	0.2685	0.799	0.367
Γ_{Ne} ($\Delta v=-2$)	0.001	0.0005	0.002	0.001
Γ_{He}	0.316	0.164	0.397	0.213
Γ_{Ne}	0.572	0.269	0.801	0.368
Γ	0.888	0.433	1.198	0.581

In Table II we present the quantal results for $v_{I_2} = 25,28$ in the collinear (C) and perpendicular (P) configurations. In all the cases, the tetraatomic k=0 level is very close to the simple addition of the triatomic m=n=0 energies, as can be expected due to the weakness of the coupling terms. As regards the rates of predissociation, several features can be stressed from the inspection of this table. Firstly, we find like in triatomic systems the propensity rule $\Delta v=-1$ for iodine after dissociation by the two possible channels studied here, while the contribution of $\Delta v=-2$ to the VP process is almost negligible. Of course, this situation strongly depends on the initial vibrational excitation of I_2 once the parameters describing the vdW interaction were choosen. On the other hand we observe that the collinear rates are always greater than the perpendicular ones. This result is similar to that found for He-I_2[11]. Also we note that the predissociation rates for He and Ne grow as a function of the vibrational quantum number v_{I_2}. The more remarkable feature of this system, as can be seen from this table, is that the lower stability corresponds to the stronger vdW bond.

Table III. Number of dissociated trajectories for each channel N_{He} and N_{Ne}, and classical total rate for perpendicular (P) and γ_1,γ_2 variables (2D) together with the Quantal P result in the last column. These calculations correspond to $v_{I_2}=28$.

	N_{He}	N_{Ne}	$\Gamma^{QCT}(cm^{-1})$	$\Gamma\ (cm^{-1})$
P	475	225	0.61	0.58
2D	351	283	0.18	

In order to obtain the QCT rates for predissociation we have fitted the non-dissociated trajectories as a function of the time to an exponential law. This fit was inadequate when the He-I₂-Ne complex dissociates He first or Ne first but it worked fairly good in the total dissociation case. This is the reason to present the number of dissociated trajectories for each channel in table III, instead the partial rates, together with the total QCT and quantal rates in the perpendicular configurations as well as in a 2D calculation. The agreement between total QCT and quantal rates is fairly good in the P configuration. If we estimate QCT partial rates from the number of dissociated trajectories for each channel, the result obtained is in complete disagreement with the quantal one (see Table II). Concerning the 2D calculation, we observe a strong increase of stability of the complex as compared with the P calculation. This effect has already been pointed out for triatomic systems[5,11]. Another effect is the compensation in the number of dissociated trajectories for each channel when we relax the angular motions.

To conclude, we have presented a quantal approximate model to treat the VP of the He-I₂-Ne vdW complex at fixed coplanar configuration. All predissociation rates obtained are in the same order of magnitude that those found with the quasiclassical model as well as with the few experimental data[2].

REFERENCES.

(1) See, e.g., K.C. Janda, Advan. Chem. Phys. LX, 201(1985) and references therein.
(2) J.E. Kenny, K.E. Johnson, W. Sharfin, and D.H. Levy, J. Chem. Phys. 72, 1109 (1980)
(3) G.C. Schatz, V. Buch, M.A. Ratner, and R.B. Gerber, J. Chem. Phys. 79, 1808 (1983)
(4) S.B. Woodruff and D.L. Thompson, J. Chem. Phys. 71, 376 (1979)
(5) G. Delgado-Barrio, P. Villarreal, P. Mareca and G. Albelda, J. Chem. Phys. 78, 280 (1983)
(6) J.A. Beswick, G. Delgado-Barrio, P. Villarreal and P. Mareca, Faraday Discuss. Chem. Soc. 73, 406 (1982); P. Villarreal, P. Mareca, G. Delgado-Barrio and J.A. Beswick, J. Mol. Struct. 120, 303 (1985)

(7) R.N. Porter, L.M. Raff and V.H. Miller, J. Chem. Phys. 63, 2214 (1975)

(8) J.A. Beswick and J. Jortner , Adv. Chem. Phys. 47, 363 (1981)

(9) C.H. Chen, P.E. Siska and Y.T. Lee, J. Chem. Phys. 59, 601 (1973)

(10) K.P. Huber and G. Herzberg, "Constants of Diatomic Molecules", Van Nostrand, N.Y., 1979

(11) J.A. Beswick and G. Delgado-Barrio, J. Chem. Phys. 73, 3653 (1980); M. Aguado, P. Villarreal, G. Delgado-Barrio, P. Mareca and J.A. Beswick, Chem. Phys. Lett. 102, 227 (1983)

(12) J.A. Blazy, B.M. DeKoven, T.D. Russell and D.H. Levy, J. Chem. Phys. 72, 2439 (1980)

(13) E. de Pablo, M.S. Guijarro, P. Villarreal, P. Mareca and G. Delgado -Barrio, An. Fis. A80, 210 (1984); P. Villarreal, E. de Pablo, P. Mareca and G. Delgado-Barrio, Abstracts of 2'ECAMP, 387 (1985)

(14) E. de Pablo, S. Miret-Artés, P. Mareca, P. Villarreal and G. Delgado-Barrio, J. Mol. Struct. 142, 505 (1986)

(15) G. Delgado-Barrio, A.M. Cortina, A. Varadé, P. Mareca, P. Villarreal and S. Miret-Artés, J. Compt. Chem. 7, 208 (1986)

(16) O. Roncero, S. Miret-Artés, G. Delgado-Barrio and P. Villarreal, J. Chem. Phys. 85, 2084 (1986)

(17) N. Martín, G. Delgado-Barrio, P. Villarreal, P. Mareca and S. Miret-Artés, J. Mol. Struct. 142, 501 (1986)

DYNAMICS OF THE AR...H_2 VAN DER WAALS COMPLEX: A DIABATIC DISTORTED
WAVE APPROACH INCLUDING DISCRETE-DISCRETE COUPLINGS

P. Villarreal, G. Delgado-Barrio, O. Roncero, E. de Pablo and
S. Miret-Artés.
Instituto de Estructura de la Materia, C.S.I.C.
Serrano 123, 28006 - Madrid
SPAIN

ABSTRACT. We calculate energy positions and widths of predissociating
levels of the Ar...H_2 complex in a diabatic vib-rotational distorted
wave decoupling scheme. For non-zero values of the total angular
momentum, discrete-discrete couplings are important for obtaining a
good estimation of these magnitudes. Also, we study the effect of the
diatomic vibrational state on the predissociation widths showing, as
expected, a progressive increment when the vibrational excitation
increases. Furthermore, the reliability of our approach is checked
by means of close-coupling calculations for some triatomic levels.

I. INTRODUCTION

Weakly bound X-BC systems, where X is a rare gas atom and BC a
conventional diatomic molecule, have received a great experimental
and theoretical attention in the last years. In this context, the
Ar...H_2 van der Waals complex is one of the most carefully studied
systems. This complex provides an excellent example of rotational
predissociation (RP) consisting in the flow of rotational energy stored
in the H_2 molecule to the weak bond leading to its fragmentation. Due
to the large vibrational and rotational spacings of the diatomic
partner, Ar...H_2 is a good candidate to be studied by using several
approximate methods[1,2]. These methods furnish physical pictures of
the predissociation process, involving less computational effort than
"exact" calculations, and also may be used in order to get reliable
potential surfaces.
   For a total angular momentum J=0, a diabatic decoupling scheme,
as proposed by Beswick et al[2] and after used in an entirely numerical
way[3], yielded successful results concerning with energy levels and
associated rates of predissociation for this complex. However, for
J ≠ 0, strong couplings appear between different sublevels of the
diatomic rotation, and thus the above approach is no longer adequate.
One way to improve this diabatic scheme is to use an adapted rotational
basis set[2] or, as recently shown[4], to include the discrete-discrete
couplings through a configuration interaction (CI) to properly describe

583

A. Weber (ed.), Structure and Dynamics of Weakly Bound Molecular Complexes, 583–591.
© 1987 by D. Reidel Publishing Company.

the discrete levels of the triatomic system.

In general, three orders of CI can be considered by including progressively the couplings among sublevels of each diatomic rotational state, different rotational states and, finally, different diatomic vibrational states. Due to the features of the Ar...H_2 system already mentioned, the first order of ÇI is expected to be the most important.

As it was pointed out[2,5] the vibrational predissociation (VP) is much slower than the RP for this system, and the H_2 subunit was frequently taken like a rigid rotor[3]. However, it is important to investigate the role played by the H_2 vibrational excitation on the intramolecular dynamics. For this purpose, we have used the very accurate anisotropic potential of Le Roy and Carley[6] that also depends on the diatomic bond length. In addition, this choice of the potential allows us to compare our results with those recently reported[7,8] and available experimental data[9].

The work is organized as follows. In Section II, we describe the diabatic and diabatic plus CI vib-rotational approximations. The results for Ar...H_2 are reported and discussed in detail in Section III. Some predissociating triatomic levels are also investigated by means of very precise scattering simulations[10,4a] in order to check the goodness of our treatment.

II. THEORY

The Hamiltonian for nuclear motion of a X...BC triatomic molecule, in the Born Oppenheimer approximation and after separation of the center-of-mass motion of the whole system, may be written as

$$H = - \frac{\hbar^2}{2\mu} \frac{\partial^2}{\partial R^2} + \frac{\hbar^2}{2\mu} \frac{\ell^2}{R^2} - \frac{\hbar^2}{2m} \frac{\partial^2}{\partial r^2} + \frac{\hbar^2}{2m} \frac{j^2}{r^2} + W(R,r,\theta) \tag{1}$$

where R is the distance between the atom and the diatomic center of mass, r is the internuclear distance of the diatom, θ is the angle between the vectors R and r, μ and m are the atom-diatom and the diatom reduced masses, respectively. Here, j and ℓ are the angular momentum operators associated with the diatomic rotation and the centrifugal rotation, respectively.

The relative weakness of the van der Waals bond suggests that the potential energy $W(R,r,\theta)$ of the whole system may be subdivided as

$$W(R,r,\theta) = V_d(r) + V(R,r,\theta) \tag{2}$$

where $V_d(r)$ is the intramolecular diatomic potential and $V(R,r,\theta)$ is the interpartner potential.

Then the Hamiltonian can be rewritten as

$$H = - \frac{\hbar^2}{2\mu} \frac{\partial^2}{\partial R^2} + \frac{\hbar^2}{2\mu} \frac{\ell^2}{R^2} + V(R,r,\theta) + H_d(r) \tag{3}$$

where $H_d(r)$ is the Hamiltonian for the BC diatomic molecule.

In the space-fixed frame[11], the total wavefunction may be expanded as

$$\psi^{JMp}(R,r) = \sum_{vj\ell} \zeta^{J}_{vj\ell}(R)\ \phi^{Jp}_{vj}(r)\ \chi^{JMp}_{j\ell}(\hat{R},\hat{r}) \qquad (4)$$

where J and M are the quantum numbers associated with the total angular momentum $\mathcal{J} = j + \ell$ and its third component, respectively; $p = (-1)^{**}(J+j+\ell)$ is the parity index, and $\chi^{JMp}_{j\ell}(\hat{R},\hat{r})$ are the angular basis functions.

Taking into account that the operator j^2 is diagonal in the space-fixed basis with eigenvalues equal to $\hbar^2 j(j+1)$ the functions $\phi^{Jp}_{vj}(r)$ are obtained by solving:

$$\left[-\frac{\hbar^2}{2m}\frac{d^2}{dr^2} + \frac{\hbar^2}{2mr^2}j(j+1) + V_d(r) - E_d^{vj} \right]\phi^{Jp}_{vj}(r) = 0 \qquad (5)$$

where E_d^{vj} are the vibration-rotation eigenvalues of the isolated diatom. Henceforth, the superscripts J,M and p drop out for simplicity.

Representing the Hamiltonian (3) in the basis formed by $\{\phi_{vj}(r)\ \chi^{JMp}_{j\ell}(\hat{R},\hat{r})\}$, the usual Schrödinger equation leads to the following system of coupled equations for the ζ functions:

$$\left[-\frac{\hbar^2}{2\mu}\frac{\partial^2}{\partial R^2} + V_{vj\ell,vj\ell}(R) - E \right]\zeta_{vj\ell}(R)$$

$$= \sum_{v'j'\ell'\neq vj\ell} V_{vj\ell;v'j'\ell'}(R)\ \zeta_{v'j'\ell'}(R) \qquad (6)$$

where:

$$V_{vj\ell;v'j'\ell'}(R) = \frac{\hbar^2}{2\mu R^2}\ \ell(\ell+1)\delta_{vv'}\delta_{\ell\ell'}\delta_{jj'} + E_d^{vj}\delta_{vv'}\delta_{\ell\ell'}\delta_{jj'} +$$

$$+ \langle\phi_{vj}(r)\chi_{j\ell}(\hat{R},\hat{r})|V(R,r,\theta)|\chi_{j'\ell'}(\hat{R},\hat{r})\phi_{v'j'}(r)\rangle \qquad (7)$$

The partition of the total wave function used by Beswick and Requena[5] is not adopted here because in a close-coupling calculation including different vibrational channels such approach yields non-vanishing couplings at $R \rightarrow \infty$

DIABATIC VIB-ROTATIONAL METHOD   (D)

     This approach is just a generalization of the diabatic rotational
method first proposed by Beswick and Requena[2] and after developed in
a numerical way[3]. This method neglects all nondiagonal $V_{vj\ell;v'j'\ell'}$
elements in eq.(6). This is a good approximation only if
these elements are small as compared with the diagonal ones, as
happens for J=0 in the Ar...H_2 system[2].
     The order zero energies and wavefunctions are then obtained by
solving for each channel in the discrete subspace the following
equation:

$$\left[-\frac{\hbar^2}{2\mu} \frac{d^2}{dR^2} + V_{vj\ell;vj\ell}(R) \right] \zeta^o_{vj\ell k}(R) = E^o_{vj\ell k} \zeta^o_{vj\ell k}(R) \tag{8}$$

where k is the quantum number associated to the stretching in the van
der Waals bond. Then, the wavefunction associated to a discrete level
becomes

$$\psi^{dis}_{vj\ell k}(\underset{\sim}{R},\underset{\sim}{r}) = \zeta^o_{vj\ell k}(R)\ \phi_{vj}(r)\ \chi_{j\ell}(\hat{R},\hat{r}) \tag{9}$$

     As regards the continuum wavefunctions describing the atomic plus
the diatomic fragments, we write

$$\psi^{cont}_{vj\ell\varepsilon}(\underset{\sim}{R},\underset{\sim}{r}) = \zeta^{cont}_{vj\ell\varepsilon}(R)\ \phi_{vj}(r)\ \chi_{j\ell}(\hat{R},\hat{r}) \tag{10}$$

where ζ are the energy normalized solutions of

$$\left[-\frac{\hbar^2}{2\mu} \frac{d^2}{dR^2} + V_{vj\ell;vj\ell}(R) \right] \zeta^o_{vj\ell\varepsilon}(R) = \varepsilon\ \zeta^o_{vj\ell\varepsilon}(R) \tag{11}$$

     The overall predissociation widths of a triatomic level (vjℓk)
can be estimated in the "Golden Rule" scheme as

$$\Gamma_{vj\ell k} = 2\pi \sum_{v'j'\ell'} |\langle \psi^{dis}_{vj\ell k}|H|\psi^{cont}_{v'j'\ell'\varepsilon}\rangle|^2 \tag{12}$$

DIABATIC VIB-ROTATIONAL PLUS CONFIGURATION INTERACTION METHOD (DCI)

     As it was pointed out[2,5], the D approach must be improved for J $\neq$ 0.
In order to relax the above assumption, we take into account the
interaction among the discrete levels as a generalization of the also
called DCI method in Ref.|4a|.In this way, we rewrite the wavefunction

as

$$\psi_n^{dis}(R,r) = \sum_{vj\ell k} a_{vj\ell k}^{(n)} \psi_{vj\ell k}^{dis}(R,r) \qquad (13)$$

and the problem reduces now to diagonalize the matrix representing the Hamiltonian (3) in the basis of the $\psi_{vj\ell k}^{dis}$ functions, i.e.,

$$H_{vj\ell k;v'j'\ell'k'} = \langle \psi_{vj\ell k}^{dis} | H | \psi_{v'j'\ell'k'}^{dis} \rangle \qquad (14)$$

$$= E_{vj\ell k}^{o} \delta_{vv'} \delta_{jj'} \delta_{\ell\ell'} \delta_{kk'} + \langle \zeta_{vj\ell k}^{o} | V_{vj\ell;v'j'\ell'}(R) | \zeta_{v'j'\ell'k'}^{o} \rangle \times$$

$$\times (1-\delta_{vv'} \delta_{jj'} \delta_{\ell\ell'})$$

Neglecting the continuum-continuum couplings, the predissociation widths are again obtained by means of eq. (12), where the discrete function is now that of eq.(13) and the continuum functions must be recalculated because the energy has changed.

III. RESULTS AND DISCUSSION

The three dimensional potential, $V(R,r,\theta)$, used is that of Ref. |6|. This potential has been expanded as

$$V(R,r,\theta) = \sum_{\lambda} \sum_{k} \xi^k P_\lambda(\cos\theta) v_{\lambda k}(R) \qquad (15)$$

where $\xi = (r-r_o)/r_o$ and r_o is the expectation value of r for the H_2 with v=0 and j=0. The potential surface has been expanded in four terms corresponding to $\lambda=0,2$ and k=0,1 then, the last addend in eq. (7) can be expressed as

$$\langle \phi_{vj}(r) \chi_{j\ell}(\hat{R},\hat{r}) | V(R,r,\theta) | \chi_{j'\ell'}(\hat{R},\hat{r}) \phi_{v'j'}(r) \rangle \qquad (16)$$

$$= \sum_{\lambda} \sum_{k} \langle \phi_{vj}(r) | \xi^k | \phi_{v'j'}(r) \rangle \langle \chi_{j\ell}(\hat{R},\hat{r}) | P_\lambda(\cos\theta) | \chi_{j'\ell'}(\hat{R},\hat{r}) \rangle v_{\lambda k}(R)$$

where [11]

$$\langle \chi_{j\ell}^{J}(\hat{R},\hat{r}) | P_{\lambda}(\cos\theta) | \chi_{j'\ell'}^{J}(\hat{R},\hat{r}) \rangle$$

$$= (-1)^{j+j'-J} \left| (2j'+1)(2j+1)(2\ell'+1)(2\ell+1) \right|^{\frac{1}{2}} \; \times$$

$$\times \left\{ \begin{matrix} J & \ell & j \\ \lambda & j' & \ell' \end{matrix} \right\} \left(\begin{matrix} j' & \lambda & j \\ o & o & o \end{matrix} \right) \left(\begin{matrix} \ell' & \lambda & \ell \\ o & o & o \end{matrix} \right) \tag{17}$$

where $(\begin{smallmatrix}:::\end{smallmatrix})$ and $\{\begin{smallmatrix}:::\end{smallmatrix}\}$ are the 3-j and 6-j coefficients, respectively.
To calculate $\langle \phi_{vj}(r) | \xi^{k} | \phi_{v'j'}(r) \rangle$ and the diatomic energies, E_d^{vj}, we have followed the numerical procedure proposed in Ref. $|12|$. The diatomic potential used, V_d, is that of Ref. $|13|$.

In Table I we present the results obtained for Ar...H_2 by means of the D and DCI approaches.

The DCI results include the configuration interaction among levels with the same v,j but different ℓ values. Therefore, the magnitudes associated to states with J=0 or j=0 remain unchanged. For the other states, the variation in energy position and width increase in the same direction as J or/and v. We want to stress two remarkable points from the inspection of this table: 1) The influence of the CI is higher on the widths than on the energies, reading a factor of ~ 2 between Γ^D and Γ^{DCI} for the level (v J j ℓ k) = (22200) and 2) For J=j=2, whatever be the v value, we find a change in the ordering of stability for the rotational sublevels when the CI is included.

We have also carried out calculations for some (v,J,j,ℓ,k) states including CI among levels with different j,ℓ(a) and v,j,ℓ(b) values, to see the importance of the succesive orders of interaction. As an illustration, we show the following examples:

$$(1,2,2,0,0) \quad E_a = -23.2971 \text{ cm}^{-1} \quad \Gamma_a = .0495 \text{ cm}^{-1}$$

$$E_b' = -23.2991 \text{ cm}^{-1} \quad \Gamma_b = .0496 \text{ cm}^{-1}$$

$$(1,2,2,2,0) \quad F_a = -19.3490 \text{ cm}^{-1} \quad \Gamma_a = .0306 \text{ cm}^{-1}$$

$$F_b = -19.3510 \text{ cm}^{-1} \quad \Gamma_b = .0307 \text{ cm}^{-1}$$

We may conclude that the main interactions occur among levels of the same v,j values, as was expected because the large vibrational and rotational spacings of the H_2 molecule.

As can be seen in Table I, the widths for levels differing only in the v quantum number increase as v increases, as an extension of the energy gap law[14] in RP processes. As regards the behaviour displayed by those levels differing in the excitation of the vdW stretching, the

Table I. Energies and widths (in cm^{-1}) for bound and quasibound levels of Ar...H$_2$ relative to its own threshold of H$_2$ calculated by D and DCI^2 methods.

J	j	ℓ	k	E^D	Γ^D	E^{DCI}	Γ^{DCI}
		v=0					
0	0	0	0	-21.9056		-21.9056	
		0	1	- .4575		- .4575	
	2	2	0	-19.6037	.0412	-19.6037	.0412
1	0	1	0	-20.7834		-20.7834	
		1	1	- .1067		- .1067	
	2	1	0	-21.5415	0.0186	-21.5518	.02044
		1	1	- .1694	.0019	- .1694	.0019
		3	0	-16.0983	.0229	-16.0880	.0212
2	0	2	0	-18.5521		-18.5521	
	2	0	0	-21.9939	.0108	-22.1623	.0163
		0	1	- .4676	.0014	- .4675	.0014
		2	0	-18.4377	.0142	-18.2893	.0105
		4	0	-11.6662	0.0174	-11.6462	0.0158
		v=1					
0	0	0	0	-22.9997	.4671 (-8)	-22.9997	.4671 (-8)
		0	1	- .5949	.7936 (-9)	- .5949	.7936 (-9)
	2	2	0	-20.8659	.1097	-20.8659	.1097
1	0	1	0	-21.8846	.3212 (-8)	-21.8846	.3212 (-8)
		1	1	- .2134	.4168 (-9)	- .2134	.4168 (-9)
	2	1	0	-22.7360	.0508	-22.7500	.0567
		1	1	- .2991	.0063	- .2988	.0063
		3	0	-17.3556	.0625	-17.3413	.0568
2	0	2	0	-19.6664	.9262 (-9)	-19.6664	.9262 (-9)
	2	0	0	-23.0901	.0315	-23.2978	.0496
		0	1	- .6065	.0045	- .6065	.0045
		2	0	-19.5290	.0427	-19.3491	.03064
		4	0	-12.9386	.0482	-12.9111	.0428
		v=2					
0	0	0	0	-24.1466	.4178 (-7)	-24.1466	.4178 (-7)
		0	1	- .7561	.8032 (-8)	- .7561	.8032 (-8)
	2	2	0	-22.2641	.2152	-22.2641	.2152
1	0	1	0	-23.0384	.2607 (-7)	-23.0384	.2607 (-7)
		1	1	- .3475	.4065 (-8)	- .3475	.4065 (-8)
	2	1	0	-24.0417	.1014	-24.0630	.1159
		1	1	- .4649	.0150	- .4649	.0150
		3	0	-18.7284	.1249	-18.7080	.1109
2	0	2	0	-20.8332	.1355 (-7)	-20.8332	.1355 (-7)
	2	0	0	-24.2399	.0660	-24.5190	.1100
		0	1	- .7696	.0107	- .7700	.0108
		2	0	-20.6586	.0975	-20.421	.0627
		4	0	-14.3166	.0975	-14.275	.0836

associated widths decrease an order of magnitude when going from
k=0 to k=1, in agreement with a previous work[8].

Also in this table, the widths for pure vibrational predissociation
may be observed for those levels with j=0, showing that the VP process
is in fact several orders of magnitude slower than the RP one.

In Table II we show some frequencies corresponding to transitions
(v=0, J,j,ℓ,k=0) → (1, J',j',ℓ',0) together with the experimental
data[9]. The discrepancies obtained are close to those found between
our numerical diatomic energies using the potential of Ref. |13| and
the experimental values for H_2[15].

Table II. Frequencies (in cm^{-1}) for the transitions (v=0,J,j,ℓ)→
          (v'=1,J',j',ℓ').

J	j	ℓ	J'	j'	ℓ'	(Ref. |9|)	(present work)
2	0	2	1	0	1	4157.880	4158.03161
1	0	1	0	0	0	4159.001	4159.14797
0	0	0	1	0	1	4161.256	4161.38516
1	0	1	2	0	2	4162.337	4162.48197
2	0	2	1	2	3	4499.125	4499.09061
1	0	1	2	2	2	4499.306	4499.31297
0	0	0	1	2	3	4502.486	4502.44416
1	0	1	2	2	4	4505.815	4505.75197

In order to check the reliability of the DCI method, we compare
in Table III some energies and widths with those obtained through
close-coupling calculations, in which we have used a very precise
method proposed in Ref. |10| to analyze isolated resonances and
successfully employed to study RP of van der Waals complexes formed
by a rare gas atom and a slow rotating diatomic molecule[4a].

Table III. Energies and widths (in cm^{-1}) for v=1, j=2 and k=0 relative
           to v=1, j=2 threshold of H_2.

ℓ	J	DCI	CC	Ref. |7|	Ref. |8|
0	2	-23.2990		-23.238	
		(.0496)		(.0394)	
1	1	-22.7510	-22.7229	-22.653	-22.6736
		(.0567)	(.0632)	(.0530)	(.0507)
2	0	-20.862	-20.8072	-20.727	
		(.1098)	(.1093)	(.1103)	
3	1	-17.341	-17.2890	-17.243	-17.2627
		(.05701)	(.0601)	(.0551)	(.0526)
4	2	-12.9110		-12.828	
		(.0429)		(.0408)	

In both types of calculations, we have considered a grid of 1500 points in the R coordinate starting from 3.5 a.u. and using a step-size of 0.05 a.u. For the J values examined we have included the channels $v=0,1,2$ and $j=0,2,4$ as well as all the possible ℓ values compatible with J,j and an even parity. We see that the accuracy of the DCI energies and widths are of the order of 0.1 and 0.01 cm^{-1}, respectively. We also present in this table recent results obtained by means of CC calculations[7] and the artificial channel method[8]. All the results are very close and we conclude that the DCI approach constitutes a very efficient way to study rotational predissociating levels of triatomic systems similar to Ar...H$_2$, even for total angular momentum values $J \neq 0$.

IV. REFERENCES

1.- R.J. Le Roy, G.C. Corey, and J.M. Hutson, Faraday Discuss. Chem. Soc. 73, (1982) 339

2.- J.A. Beswick and A. Requena, J. Chem. Phys. 72 (1980) 3018

3.- O. Roncero, S. Miret-Artés, G. Delgado-Barrio and P. Villarreal, J. Chem. Phys. 85 (1986) 2084

4.- a) P. Villarreal, G. Delgado-Barrio, O. Roncero, F.A. Gianturco and A. Palma, Phys. Rev. A (submitted)

   b) O. Roncero, P. Villarreal, S. Miret-Artés and G. Delgado-Barrio, Sixth European Study Conference on Dynamics of Molecular Collisions and half-collisions, 1986, p. 13

5.- J.A. Beswick and A. Requena, J. Chem. Phys. 73 (1980) 4347

6.- R.J. Le Roy and J.S. Carley, Adv. Chem. Phys. 42 (1980) 353

7.- J.M. Hutson and R.J. Le Roy, J. Chem. Phys. 83 (1985) 1197

8.- I.F. Kidd and G.G. Balint-Kurti, J. Chem. Phys. 82 (1985) 93

9.- R.W. McKellar, Faraday Discuss. Chem. Soc. 73 (1982) 89

10.- G. Delgado-Barrio, P. Villarreal, P. Mareca and J.A. Beswick, Int. J. Quantum Chem. 27 (1985) 173

11.- A.M. Arthurs and A. Dalgarno, Proc. Roy. Soc. A256 (1960) 540

12.- G. Delgado-Barrio, A.M. Cortina, A. Varadé, P. Mareca, P. Villarreal and S. Miret-Artés, J. Compt. Chem. 7 (1986) 208

13.- D.M. Bishop and S. Shih, J. Chem. Phys. 64 (1976) 162

14.- J.A. Beswick and J. Jortner, J. Chem. Phys. 68 (1978) 2277

15.- H.L. Buijs and H.P. Gush, Can. J. Phys. 49 (1971) 2366

David S. King
National Bureau of Standards
Molecular Spectroscopy Division
Gaithersburg, Md   20899 USA

During the past year we have seen a dramatic increase in the breadth and depth of experimental studies into the vibrational predissociation dynamics of van der Waals (vdW) molecules, and especially vdW dimers. The interest in this area has been fostered by intriguing, early results of frequency domain measurements, and has been facilitated by advances jointly in molecular beam and laser technologies (after all, most of these experiments are beam photodissociation experiments). Activity in the area of van der Waals vibrational predissociation dynamics encompasses 1) theory, 2) frequency domain measurements, and 3) time domain and energy distribution measurements. In the anticipation of relating very detailed dynamical information to the inter-molecular forces responsible for the van der Waals bonding it is unfortunate that, as yet, there is no single system upon which all three approaches have been successfully used.

Theoretically, the vibrational predissociation of a jet-cooled, weakly bonded vdW molecule presents a close, half collisional analogy to the reactions $A^* + B \longrightarrow A + B + \Delta E$. Quantum close coupled channel calculations on realistic potential surfaces have been applied to systems as diverse as $Rg-H_2$ to $Rg-C_2H_4$ (Rg = Rare gas) and Hg-M complexes [1]. These calculations predict microscopic state-to-state rates giving both the temporal decay behavior of the vibrationally excited parent and the most probable fragment energy distributions. The propensity rules emerging from such calculations imply that channels minimizing momentum transfer (both linear and angular momenta) will be most favorable, i.e., the reaction following these channels will proceed faster. One mechanism for the minimization of momentum transfer would be the vibrational excitation (where energetically possible) of the fragments. This accounts, in part for the much faster rates calculated for $Ne-C_2H_4(v_7) \longrightarrow Ne + C_2H_4(v_{10}) + \Delta E \sim 100$ cm^{-1} than for the channels leading to v=0 fragments. One practical impediment to relating theory to experiment is that the theoretical dimers are "cold," whereas experimental beam temperatures are often 10 - 50 K. Few dynamical experiments have been reported for individual

593

A. Weber (ed.), Structure and Dynamics of Weakly Bound Molecular Complexes, 593–597.

(v, low J) levels and exact calculations are really only feasible for initial states of very low angular momenta. Allowing the dimer ensemble a finite temperature results in a wide range of inter-molecular separations and orientations, necessitating the inclusion of a substantially larger number of partial waves in the calculation.

Frequency domain experiments of simple dimers for which individual ro-vibrational transitions are resolvable present a picture of the bonding geometries of these systems (cf. Structure section) and provide a venue for non-statistical behavior. In $(HF)_2$ and HCN-HF, for example, the vibrational densities of states will be sparse and observed spectral line broadening must be due to predissociation effects. In $(HF)_2$ a twenty five-fold difference in linewidth [2,3] is observed for excitations of the internal vs free HF stretching modes. Here, at least, the difference in lifetimes can be qualitatively understood in terms of the coupling of the excited vibrational modes to the hydrogen bonding coordinate. Similarly, HCN-HF exhibits a range of vibrational linewidths; linewidths dependent on vibrational mode rather than vibrational energy, and which are largely independent of rotational level [4].

The photodissociation spectra of more complex systems quite often give broad vibrational features, with no resolvable rotational structure. In such cases, the gross band structure, splittings and solvent shifts may provide information on the important bonding interactions. For instance, using a spectroscopic model for homo-dimers dominated by resonant dipole-dipole interactions, a contour analysis of $(SiF_4)_2$ spectra allows the monomer units free rotation; a similar analysis of $(SiH_4)_2$ spectra provides a picture of rigidly locked units [5].

The notion of order, so prevalent in conventional spectroscopies, breaks down at molecular levels sufficiently energetic to allow substantial tunneling or dissociation. With a few exceptions (e.g., inversion tunneling in NH_3), such levels for stable molecules are generally high, of order of their dissociation enthalpies (50-100 kcal/mol). Direct excitation to these states therefore requires high-order overtone excitations, being of low cross-section and making these states largely unaccessible experimentally. Most vdW molecules, on the other hand, are readily promoted to regions of large amplitude motion with the absorption of a single infrared photon [6,7]. Although these large amplitude motions must be important to the dimer dynamics, there generally appears to be poor coupling of the molecular pump modes to overtones of the low frequency vdW modes. This creates the possibility for bottlenecks to intramolecular vibrational redistribution (IVR), reminiscent of parallel observations in C-H stretch overtone work [8]. Details of this coupling will play an important role in determining subsequent mode-specific dynamics vs energy stochastization.

The ethylene dimer has long presented a battleground for proponents of IVR vs direct dissociation as the primary relaxation process [9].

Ethylene vdW complexes have been characterized with low resolution techniques to exhibit broad, apparently homogeneous, Lorentzian band contours of width 7-17 cm^{-1}. If these widths arise from lifetime broadening, they imply vibrational predissociation lifetimes of order 0.5 ps, just a few periods of the vdW bond stretching vibration. However, recent high resolution work has shown spectral structure at 0.0003 cm^{-1} resolution embedded within a broader continuum absorption [5,10]. This latter result implies a lifetime of ca. 15 ns. In a coupled channel calculation including both vibrations and rotations of the ethylene monomers, a vibrational predissociation lifetime of 1,000 ns has been calculated [11], in qualitative agreement with the longer "lifetimes" implied by the high resolution measurements.

An answer [9] to the conundrum posed by the frequency domain measurements on the ethylene dimer might lie in the radiationless transition theory set forth by Bixon and Jortner in 1968 [12]. That is, allowing both IVR and dissociation to proceed simultaneously and competitively. Consider a single vibrational state |s> that carries all oscillator strength from the ground state and that is coupled to a periodic array of bath states |l>; where these |l> levels have uniform energy spacings, vibronic couplings, and constant decay widths due to couplings to the dissociative continuum. When the coupling between levels |l> and the dissociative continuum (i.e., decay width) is less than that between |s> and |l> (i.e., IVR-width) and the width of levels |l> is less than their spacing, the spectrum will exhibit manifestations of both IVR and dissociation, i.e., be composed of a set of "narrow" lines contained within a broader Lorentzian envelope. Such a description makes certain predictions about the temporal behavior of these excited levels, predictions awaiting confirmation [9].

Real-time measurements of the rates and pathways of intra-molecular vibrational energy redistribution and dissociation have been determined for tetrazine-Ar complexes [13] in the tetrazine lowest excited singlet state, S_1. Although tetrazine S_1 has a radiative lifetime of ca. 100 ns, the vibrationless level and all low lying vibrational levels of S_1 exhibit fluorescence lifetimes of 40 to 2,000 ps. This emission is resonant, i.e., it originates from the initially excited vibronic level; there is no evidence of IVR in the bare monomer. The operative relaxation mechanism is, presumably, internal conversion from the S_1 vibrational manifold into isoenergetic levels of the ground electronic state, S_0. The wealth of molecular dynamics exhibited by the tetrazine-Ar system is displayed by the following set of state-specific relaxation mechanisms. Excitation of the vibronic level 4^1 in tetrazine-Ar results only in resonant emission with a 580 ps fluorescence lifetime, very close to the 660 ps lifetime of tetrazine 4^1. The relaxation mechanism following excitation of the tetrazine-Ar $16a^2$ level is more complex. Emission is observed from the initially excited $16a^2$ level of the complex, from isoenergetic levels of the complex containing one quantum of excitation of $16a^1$ plus 263 cm^{-1} of excitation in the vdW modes, and from the vibrationless level of the bare tetrazine monomer. Relaxation of the $16a^2$ level of the complex appears to involve first IVR and then predissociation, both processes proceeding at rates of ca. 10^9 s^{-1}. In contrast to the monomer photophysics where excitation of the monomer to level $16a^2$ produces no unrelaxed fluorescence, there appears to be no direct predissociation from level $16a^2$ of the complex. Whereas excitation of tetrazine-Ar $16a^2$ can energetically only produce vibrationally relaxed tetrazine S_1, excitation of the higher energy $6a^1$ mode of the complex can produce vibrationally excited monomer S_1 fragments. Intriguingly, the main relaxation channel following $6a^1$ excitation of the complex is first through IVR and then subsequent predissociation to yield S_1 monomer fragments in the level $16a^1$.

Overall, the relaxation rates and mechanisms for tetrazine-Ar S_1 species are dramatically mode specific, with the low frequency 16a and 16b vibrational modes playing a dominant role. However, the photophysics and photochemistry of the tetrazine monomer itself are complicated and mode specific, involving strong couplings of the vibronic levels of S_1 to either/both T_1 or/and S_0. For simplicity one would like to explore such processes proceeding along a ground state surface where non-adiabatic processes might be ignored.

A complete study has been performed for the ground electronic state vibrational predissociation of $v_1=1$ and $v_4=1$ levels of the nitric oxide dimer [14]. The NO-dimer has C_{2v} symmetry, with its symmetric stretching mode slightly higher in energy than the anti-symmetric stretch. Using independently tunable, synchronized ps infrared pump and ultraviolet probe lasers the lifetimes of each of these two levels was directly measured. Surprising, the lower energy anti-symmetric stretching mode predissociates with a unimolecular rate constant $(2.6 \times 10^{10}$ s$^{-1})$ 22 times faster than that observed for v_1. Simple arguments based on statistical theory of unimolecular decay or energy gap considerations do not come close to predicting such a variance. This mode-specific variance in predissociation rate is similar to that observed for $(HF)_2$. Unlike $(HF)_2$ however, where a simple physical picture of nuclear motions provides a qualitative understanding of this difference, ball-and-spring models of $(NO)_2$ do not provide a basis for expecting such mode-specificity. Ad hoc energy gap arguments might lead one to speculate that the excess energy (i.e., photon energy minus bond dissociation energy) were dissipated by rotational excitation of the fragments of the fast v_4 predissociation, thus minimizing momentum transfer, and that the fragments of the slower v_1 predissociation were forced (vis-a-vis dynamical constraints) to carry away excessive amounts of kinetic energy. That this postulate is not operative was demonstrated through fragment energy distribution measurements which indicate that the fragments from both initial vibrational levels of the dimer exhibit similar energy distributions, i.e., dissipating ca. 13% of the available excess energy into fragment rotation. Despite the large difference in rates, the predissociation mechanism out of these nearly isoenergetic vibrational levels is similar.

In summary, the study of molecular dynamics in weakly bound systems is a healthy, exciting field involving theorist, spectroscopist and photochemist. One intriguing aspect of this area is that it allows experimentalist and theorist to readily address questions with state-to-state detail about molecular dynamics at excitation energies near to- or exceeding dissociation or tunneling threshold -- in that region of wide amplitude molecular motion of particular interest to chemists.

REFERENCES

1) J.A. Beswick, These Proceedings, pp. 563-572, J.M. Hutson, D.C. Clary and J.A. Beswick, J. Chem. Phys. 81, 4474 (1984).
2) A.S. Pine, These Proceedings, pp. 93-105.
3) R.E. Miller, These Proceedings, pp. 131-140.
4) J.W. Bevan, These Proceedings, 149-169.
5) M. Snels, R. Fantoni and J. Reuss, These Proceedings, pp. 489-494.
6) J.T. Hougen, These Proceedings, pp. 191-200.
7) J. Reuss, These Proceedings, pp. 501-512.
8) K.V. Reddy, D.F. Heller and M.J. Berry, J. Chem. Phys. 76, 2814 (1982).
9) W.R. Gentry, These Proceedings, pp. 467-475.
10) K.G.H. Baldwin and R.O. Watts, Chem. Phys. Letts. 129, 237 (1986).
11) A.C. Peet, D.C. Clary and J.M. Hutson, Poster Paper No. 5, Thursday, September 25, 1987; Faraday Disc. Chem. Soc. 82/19 (1986).
12) M. Bixon and J. Jortner, J. Chem. Phys. 48, 715 (1968).
13) M. Heppener and R.P.H. Rettschnick, These Proceedings, pp. 553-562.
14) M.P. Casassa, J.C. Stephenson and D.S. King, These Proceedings, pp. 513-524.

Dr. V. Aquilanti
Department of Chemistry
Universita degli Studi di
 Perugia
Via Elce di Sotto 10
I-06100 Perugia
Italy

Prof. Dr. H. Baumgärtel
Institut für physikalische
 Chemie
Freie Universität Berlin
Takustrasse 3
1000 Berlin, 33,
West Germany

Dr. A. Beswick
Universite de Paris-Sud et
CNRS
LURE
Bat. 209C
91405 Orsay, Cedex
France

Professor J. W. Bevan
Department of Chemistry
Texas A&M University
College Station, TX 77843
USA

Dr. P. Botschwina
Fachbereich Chemie
Universität Kaiserslautern
Erwin Schrödinger Straße
D-6750 Kaiserslautern
West Germany

Dr. G. Brocks
Institut voor Theoretische
 Chemie
Katholieke Universiteit
Nijmegen
Toernooiveld
6525 ED Nijmegen
The Netherlands

Prof. Dr. U. Buck
Max-Planck-Institut für
Strömungsforschung
3400 Göttingen
West Germany

Dr. P. Casavecchia
Department of Chemistry
Universita degli Studi di
Perugia
Via Elce di Sotto 10
I-06100 Perugia
Italy

Mrs. Simonetta Cavalli
Department of Chemistry.
University of Perugia
Italy

Dr. Joe Cline
Department of Chemistry
University of Pittsburgh
Pittsburgh, PA 15260
USA

Prof. Dr. R. G. Colin
Faculty of Sciences:
Chemistry
Universite Libre de
Bruxelles
Ave. F. D. Roosevelt, 50
1050 Bruxelles
Belgium

Dr. G. Delgado-Barrio
Instituto de Estructura de
la Materia, CSIC
Serrano 119
28006 Madrid
Spain

Professor T. Dyke
Chemistry Department
University of Oregon
Eugene, OR 94703
USA

Professor C. E. Dykstra
Chemistry Department
University of Illinois
505 S. Mathews Avenue
Urbana, IL 61801
USA

Dr. R. Fantoni
ENEA, Dipartimento TIB
Divisione Fisica Applicata
C.R.E. Frascati, C.P. 65
00044 Frascati,
Roma
Italy

Dr. M. Faubel
Max-Planck-Institut für
 Strömungsforschung
3400 Göttingen
West Germany

Dr. G. T. Fraser
Molecular Spectroscopy
 Division
National Bureau of
Standards
Gaithersburg, MD 20899
USA

Mrs. Fortunata Gallese
General and Inorganic
 Chemistry Department
University of Rome
Italy

Professor W. R. Gentry
Chemistry Department
University of Minnesota
Minneapolis, MN 55455
USA

Prof. Dr. F. A. Gianturco
Chemistry Department
Faculty of Mathematical,
Physical and Natural
Sciences
Universita degli Studi di
 Roma
Piazza delle Scienze 5
00185 Roma
Italy

Dr. Elizabeth M. Gibson
The Royal Institution of
 Great Britain
21 Albemarle Street
London, W1X 4BS
United Kindgom

Dr. M. Herman
Faculty of Sciences:
Chemistry
Universite Libre de
Bruxelles
Ave. F. D. Roosevelt, 50
1050 Bruxelles
Belgium

Dr. J. T. Hougen
Molecular Spectroscopy
Division
National Bureau of
Standards
Gaithersburg, MD 20899
USA

Dr. B. J. Howard
Physical Chemistry
Laboratory
Oxford University
South Parks Road
Oxford, QX1 3QZ
United Kingdom

Dr. J. M. Hutson
Chemistry Department
Cambridge University
Lensfield Road
Cambridge, CB2 1EW
United Kingdom

Professor K. C. Janda
Department of Chemistry
University of Pittsburgh
Pittsburgh, PA 15260
USA

Dr. Anita C. Jones
The Royal Institution of
 Great Britain
21 Albemarle Street
London, W1X 4BS
United Kindgom

Dr. D. S. King
Molecular Spectroscopy
Division
National Bureau of
 Standards
Gaithersburg, MD 20899
USA

Professor W. A. Klemperer
Department of Chemistry
Harvard University
Cambridge, MA 02138
USA

Professor A. Legon
Department of Chemistry
Exeter University
Exeter, EX4 4QD
United Kingdom

Professor R. J. Le Roy
Department of Chemistry
University of Waterloo
Waterloo, Ont., N2L 3G1
Canada

Professor D. H. Levy
Department of Chemistry
The University of Chicago
Chicago, IL 60637
USA

Professor J. L. Lisy
Department of Chemistry
University of Illinois
Urbana, IL 61801
USA

Dr. Shi-Yi Liu
Chemistry Department
University of Illinois
Urbana, IL 61801
USA

Dr. Arlan W. Mantz
Spectra Physics
Laser Analytics Division
25 Wiggins Avenue
Bedford, MA 01730-2392
USA

Dr. Mark Marshall
Herzberg Institute of
 Astrophysics
National Research Council
 of Canada
Ottawa, Ont., K1A 0R6
Canada

Dr. A. R. W. McKellar
Herzberg Institute of
 Astrophysics
National Research Council
 of Canada
Ottawa, Ont., K1A 0R6
Canada

Dr. W. L. Meerts
Fysisch Laboratorium
Katholieke Universiteit
 Nijmegen
Toernooiveld
6525 ED Nijmegen
The Netherlands

Dr. A. Metropoulos
Theoretical and Physical
 Chemistry Institute
National Hellenic
 Research Foundation
Athens, 11653
Greece

Professor R. E. Miller
Department of Chemistry
University of North
 Carolina
Chapel Hill, NC 27514
USA

Prof. Dr. H. Møllendal
Department of Chemistry
University of Oslo
Apalvein
Blindern, Oslo 3
Norway

Professor J. S. Muenter
Department of Chemistry
University of Rochester
Rochester, NY 14627
USA

Dr. David Nelson
Department of Chemistry
Harvard University
Cambridge, MA 02138
USA

Dr. David Nesbitt
Joint Institute Laboratory
  for Astrophysics
University of Colorado
Boulder, CO 80309
USA

Professor S. E. Novick
Department of Chemistry
Wesleyan University
Middletown, CT 06457
USA

Dr. F. Palmer
Department of Chemistry
University of Nottingham
University Park
Nottingham, NG7 2RD
United Kingdom

Dr. A. S. Pine
Molecular Spectroscopy
Division
National Bureau of
Standards
Gaithersburg, MD 20899
USA

Dr. F. Pirani
Department of Chemistry
Universita degli Studi di
  Perugia
Via Elce di Sotto 10
I-06100 Perugia
Italy

Prof. Dr. Rudolph P. H.
Rettschnick
Laboratory for Physical
Chemistry
University of Amsterdam
Nieuwe Achtergracht 127
1018 WS Amsterdam
The Netherlands

Professor J. Reuss
Fysisch Laboratorium
Katholieke Universiteit
  Nijmegen
Toernooiveld
6525 ED Nijmegen
The Netherlands

Dr. Ruth Robinson
Department of Chemistry
University of California
Berkeley, CA 94720
USA

Dr. E. Rühl
Institut für physikalische
  Chemie
Takustrasse 3
1000 Berlin
West Germany

Prof. Dr. R. J. Saykally
Department of Chemistry
University of California
  Berkeley
Berkeley, CA 94720
USA

Prof. Dr. E. W. Schlag
Chemie Department
Technische Universität
München
Lichtenbergstrasse 4
D-8046 München
Bayern
West Germany

Professor G. Scoles
Department of Chemistry
University of Waterloo
Waterloo, Ont., N2L 3G1
Canada

Professor J. P. Simons
Department of Chemistry
University of Nottingham
University Park
Nottingham, NG7 2RD
United Kingdom

Professor E. Brian Smith
Physical Chemistry
Laboratory
South Parks Road
Oxford, OX1 3QZ
United Kingdom

Dr. M. Snels
ENEA, Dipartimento TIB
Divisione Fisica Applicata
C.R.E. Frascati, C.P. 65
00044 Frascati,
Roma
Italy

Professor B. Soep
Laboratoire de
Photophysique
Moleculaire
Bat. 213, Universite de
 Paris-Sud
91405 Orsay Cedex
France

Dr. R. D. Suenram
Molecular Spectroscopy
Division
National Bureau of
 Standards
Gaithersburg, MD   20899
USA

Dr. A. G. Taylor
The Royal Institution
 of Great Britain
21 Albemarle Street
London, W1X 4BS
United Kingdom

Prof. Dr. A. van der Avoird
Theoretical Chemistry
Institute
Faculty of Mathematics
 and Natural Sciences
Katholieke Universiteit
 Nijmegen
Toernooiveld
6525 ED Nijmegen
The Netherlands

Prof. A. J. C. Varandas
Departemento de Quimica
Universidade de Coimbra
3000 Coimbra
Portugal

Prof. F. Vecchiocattivi
Department of Chemistry
Universita degli Studi di
 Perugia
Via Elce di Sotto 10
I-06100 Perugia
Italy

Dr. M. Venanzi
Department of Chemistry
The University of Rome
Citta Universitaria
00185 Rome
Italy

Dr. P. Villarreal
Instituto de Estructura de
la Materia, CSIC
Serrano 119
28006 Madrid
Spain

Dr. J. Waite
National Hellenic
 Research Foundation
48 Vas. Constantinou Ave.
GR-116 35 Athens
Greece

Dr. Alfons Weber
Molecular Spectroscopy
Division
National Bureau of
Standards
Gaithersburg, MD   20899
USA

Dr. B. A. Wofford
Chemistry Department
Texas A&M University
College Station,TX 77843
USA

LIST OF CONTRIBUTORS

A

Amos,A.T.	263	Gianturco,F.A.	389
Aquilanti,V.	423	Gibson,E.M.	291
		Gwo,Dz.-Hung	85

B

H

Baumgartel,H.	303		
Beneventi,L.	441	Heppener,M.	553
Beswick,J.A.	563	Hougen,J.T.	191
Bevan,J.W.	149	Howard,B.J.	69
Bisling,P.	303	Huisken,F.	477
Boernsen,K.U.	251		
Botschwina,P.	181	J	
Breckenridge,W.H.	213		
Brocks,G.	337	Janda,K.C.	533
Brutschy,B.	303	Jones,A.C.	291
Buck,U.	477	Jouvet,C.	213

C

K

Casassa,M.P.	513	Kettley,J.C.	263
Casavecchia,P.	441	King,D.S.	513,593
Cline,J.I.	533	Klemperer,W.	455
Cohen,S.M.	263	Kolenbrander,K.D.	171

D

L

Delgado-Barrio,G.	573,583		
dePablo,E.	583	Lauenstein,Ch.	477
Duval,M.C.	213	Legon,A.C.	23
Dyke,T.R.	43	Levandier,D.	525
Dykstra,C.E.	319	Levy,D.H.	231
		Lisy,J.M.	171
E		Liu,Shi-yi	319

Evard,D.D.	533	M	

F

		Majewski,W.A.	279
		Marstokk,K.-M.	57
Fantoni,R.	489,495	Martin,N.	573
Faubel,M.	405	McKellar,A.R.W.	141
		Meerts,W.L.	279,495
G		Michael,D.W.	171
		Miller,R.E.	131
Garcia,A.	573	Miret-Artes,S.	583
Gentry,W.R.	467	Mollendal,H.	57
		Muenter,J.S.	3

605

LIST OF POSTER PAPERS

Tuesday, September 23, 1986

1.   ESTIMATES OF THE MOLECULAR QUADRUPOLE AND PARALLEL
      POLARIZABILITY OF BORON TRIFLUORIDE.
      Steward E. Novick, Wesleyan University, Middletown, CT
      06457, USA.

2.   DETERMINATION OF THE STRUCTURE OF HCl-BF$_3$.
      James M. LoBue, Jane K. Rice, Thomas A. Blake, and
      Stewart E. Novick, Department of Chemistry, Wesleyan
      University, Middletown, CT 06457, USA.

3.   NEARLY FREE INTERNAL ROTATION IN Ar-CH$_3$Cl.
      G. T. Fraser, R. D. Suenram, and F. J. Lovas, Molecular
      Spectroscopy Division, National Bureau of Standards,
      Gaithersburg, MD USA.

4.   ROTATIONAL SPECTRUM OF THE HYDROGEN BONDED FORMAMIDE-WATER
      AND FORMAMIDE-METHANOL COMPLEXES.
      R. D. Suenram, F. J. Lovas, G. T. Fraser, J. Zozom[*],
      and C. W. Gillies[*], Molecular Spectroscopy Division,
      National Bureau of Standards, and ([*]) Department of
      Chemistry, Rensselaer Polytechnic Institute, USA.

5.   MICROWAVE SPECTRA AND WEAK INTRAMOLECULAR HYDROGEN
      BONDING IN 3-BUTENE-1-THIOL AND N-METHYLALLYLAMINE.
      K.-M. Marstokk and Harald Møllendal, Department of
      Chemistry, The University of Oslo, P.O. Box 1033
      Blindern, N-0315 Oslo 3, Norway.

6.   BOUND AND ROTATIONAL RESONANCE STATES AND THE IR
      SPECTRUM OF N$_2$Ar.
      G. Brocks, Institute of Theoretical Chemistry,
      University of Nijmegen, Nijmegen, The Netherlands.

7.   CURRENT PROGRESS IN THE ROVIBRATIONAL ANALYSIS OF DCN··DF.
      M. W. Jackson, B. A. Wofford, and J. W. Bevan, Chemistry
      Department, Texas A&M University, College Station, TX
      77843, USA.

8.   FOURIER TRANSFORM INFRARED SPECTROSCOPY OF GAS PHASE
     HYDROGEN-BONDED INTERACTIONS.
     B. A. Wofford, M. W. Jackson, <u>J. W. Bevan</u>, W. B. Olson[*],
     and W. J. Lafferty[*] , Chemistry Department, Texas A&M
     University, College Station, TX 77843, and [*]Molecular
     Spectroscopy Division, National Bureau of Standards,
     Gaithersburg, MD 20899, USA.

9.   FTIR SPECTRA OF $(CO_2)_n$ CLUSTERS.
     J. Barnes and T. E. Gough, Centre for Molecular Beams and
     Laser Chemistry, Department of Chemistry, University of
     Waterloo, Waterloo, Ont. Canada.

10. DETERMINATION AND IMPLICATIONS OF NEW POTENTIAL ENERGY
     SURFACES FOR THE H_2-INERT GAS SYSTEMS.
     Robert J. Le Roy and Jeremy M. Hutson[*], Guelph-Waterloo
     Centre for Graduate Work in Chemistry, University of
     Waterloo, Waterloo, Ont., Canada, and (*)Department of
     Theoretical Chemistry, University of Cambridge, Cambridge,
     UK.

11. HIGH RESOLUTION CROSSED MOLECULAR BEAM STUDIES OF VAN DER WAALS
     FORCES.
     Laura Beneventi, <u>Piergiorgio Casavecchia</u>, and Gian
     Gualberto Volpi, Dipartimento di Chimica, Universita di
     Perugia, 06100 Perugia, Italy.

12. TRANSPORT PROPERTIES OF VdW MOLECULES COMPUTED FROM
     ACCURATE INTERACTIONS.
     F. A. Gianturco and M. Venanzi, Department of Chemistry,
     The University of Rome, Citta Universitaria, 00185 Rome,
     Italy.

13. SPECTROSCOPY IN THE VISIBLE AND NEAR ULTRAVIOLET REGION OF
     SOME ORGANIC MOLECULES AND THEIR VAN DER WAALS COMPLEXES.
     W. M. van Herpen, W. A. Majewski, and <u>W. L. Meerts</u>,
     Fysisch Laboratorium, Toernooiveld, 6525 ED Nijmegen, The
     Netherlands, and D. W. Pratt, Department of Chemistry,
     University of Pittsburg, PA, USA.

14. SLIT JET IR ABSORPTION SPECTROSCOPY OF MOLECULAR
     COMPLEXES.
     Christopher M. Lovejoy, Michael D. Schuder, and <u>David J.
     Nesbitt</u>, Joint Institute Laboratory for Astrophysics,
     National Bureau of Standards, and Department of Chemistry
     and Biochemistry, University of Colorado, Boulder, CO,
     USA.

15. SYMMETRY CLASSIFICATION OF THE ROTATION-VIBRATION ENERGY
LEVELS IN AMMONIA DIMER.
D. D. Nelson, Jr., G. T. Fraser(*) and W. A. Klemperer,
Department of Chemistry, Harvard University, and (*)
Molecular Spectroscopy Division, National Bureau of
Standards, Gaithersburg, MD, USA.

16. CO_2 LASER STARK SPECTROMETER FOR VAN DER WAALS MOLECULES.
Mark D. Marshall and A. R. W. McKellar, Herzberg Institute
of Astrophysics, National Research Council of Canada,
Ottawa, Ont., K1A OR6, Canada.

17. VIBRATIONAL SPECTROSCOPY OF ArHCl BY INTRACAVITY FAR
INFRARED LASER AND MICROWAVE-FAR INFRARED DOUBLE RESONANCE
SPECTROSCOPIES.
R. L. Robinson, D. Ray, D.-Gwo, and R. J. Saykally,
Department of Chemistry, University of California,
Berkeley, CA 94720, USA.

18. RADIO FREQUENCY ELECTRIC RESONANCE OF THE LOWEST-LYING II
BENDING STATE OF AR-HCl.
Mark D. Marshall[*], Arthur Charo, Helen O. Leung, and
William Klemperer, Department of Chemistry, Harvard
University, 12 Oxford Street, Cambridge, MA 02138, USA,
([*])Present address: Herzberg Institute of Astrophysics,
National Research Council Canada, Ottawa, Ont., K1A OR6,
Canada.

19. THE STATIC ELECTRIC HYPERPOLARIZABILITY OF WATER. EFFECT
OF INTERMOLECULAR INTERACTIONS ON THIS PROPERTY IN $(H_2O)_n$.
J. Waite, National Hellenic Research Foundation, 48 Vas.
Constantinou Avenue, GR-116 35 Athens, Greece.

Thursday, September 25, 1986

1. DYNAMICS AND PHASE TRANSITIONS OF SF_6-Ar_n CLUSTERS.
Robert J. Le Roy, Dieter Eichenauer, and John Shelley,
Guelph-Waterloo Centre for Graduate Work in Chemistry,
University of Waterloo, Waterloo, Ont., Canada.

2. INFRARED PHOTODISSOCIATION OF VAN DER WAALS COMPLEXES
PREPARED BY MOLECULAR BEAM SCATTERING.
F. Huisken, C. Lauenstein, R. Sroka, T. Pertsch, and U.
Buck, Max-Planck-Institut fur Strömungsforschung,
Göttingen, West Germany.

3.   IR DISSOCIATION OF WEAKLY BOUND VAN DER WAALS COMPLEXES OF
      $(SiF_4)_n$, $(SiH_4)_n$, $(C_2H_4)_n$, (n=2,3) IN THE 9-11 μm RANGE.
      M. Snels, R. Fantoni, and J. Reuss, University of
      Nijmegen, The Netherlands.

4.   INFRARED (9-11 μm) DISSOCIATION OF THE HYDROGEN BONDED
      CLUSTERS $(NH_3)_n$ (n $\geq$ 2) DETECTED BY BOLOMETRIC TECHNIQUE.
      M. Snels, R. Fantoni, and W. L. Meerts, University of
      Nijmegen, The Netherlands.

5.   VIBRATIONAL PREDISSOCIATION OF THE ETHYLENE DIMER.
      Andrew C. Peet, David C. Clary, and Jeremy M. Hutson,
      University Chemical Laboratory, Cambridge University,
      Cambridge, UK.

6.   PICOSECOND LASER STUDIES OF VIBRATIONAL PREDISSOCIATION OF
      VAN DER WAALS DIMERS.
      M. P. Casassa, J. C. Stephenson, and D. S. King, National
      Bureau of Standards, Molecular Spectroscopy Division,
      Gaithersburg, MD 20899, USA.

7.   DEPENDENCE OF THE ROTATIONAL PREDISSOCIATION OF AR$\cdots$H$_2$ ON
      THE DIATOMIC VIBRATION: A MODIFIED DIABATIC DISTORTED
      WAVE APPROACH.
      P. Villarreal, G. Delgado-Barrio, O. Roncero, E. de Pablo,
      and S. Miret-Artes, Instituto de Estructura de la Materia,
      Madrid, Spain.

8.   NON-BONDING ATOM-DIATOM POTENTIALS VIA A DOUBLE MANY-BODY
      EXPANSION METHOD.
      A. J. C. Varandas, University of Coimbra, Portugal.

9.   MOLECULAR BEAM STUDIES OF THE VAN DER WAALS INTERACTIONS
      OF OPEN-SHELL SYSTEMS: THE STRUCTURE OF RARE GAS OXIDES
      AND FLUORIDES.
      V. Aquilanti and F. Pirani, University of Perugia, Italy.

10.   PROPENSITY RULES FOR THE DISSOCIATION OF THE $H_4(C_{3v})$
      EXCITED STATE COMPLEX.
      A. Metropoulos, National Hellenic Research Foundation,
      Athens, Greece.

11.   PHOTOIONIZATION AND PHOTODISSOCIATION OF ORGANIC VAN DER
      WAALS COMPLEXES.
      E. Rühl, P. Bisling, B. Brutschy, and H. Baumgärtel, Freie
      Universität Berlin, West Germany.

12.   PRODUCT STATE DISTRIBUTION OF THE PHOTODISSOCIATION OF THE
      NeBr VAN DER WAALS MOLECULE.
      J. I. Cline, D. D. Evard, and K. C. Janda, University of
      Pittsburgh, Pittsburgh, PA, USA.

13.   SUPERSONIC JET SPECTROSCOPY OF COMPLEXES OF CARBAZOLE AND
       N-ETHYL CARBAZOLE WITH ALKYL CYANIDES.
       Anita C. Jones, A. G. Taylor, Elizabeth Gibson, and D.
       Phillips, The Royal Institution of Great Britain, London,
       UK.

14.   SOLVENT SHIFTS, SPECTROSCOPY AND STRUCTURE IN VAN DER
       WAALS COMPLEXES OF PERYLENE.
       A. T. Amos, S. M. Cohen, J. C. Kettley, T. F. Palmer, and
       J. P. Simons, University of Nottingham, UK.

15.   VIBRATIONAL FREQUENCY SHIFTS AND TRANSITION MOMENT
       ENHANCEMENTS IN HYDROGEN BONDED COMPLEXES.
       Shi-yi Liu and Clifford E. Dykstra, University of
       Illinois, USA.

16.   COLOR CENTER LASER SPECTROSCOPY OF SIMPLE HYDROGEN-BONDED
       COMPLEXES.
       E. K. Kyro, D. Bender, M. Eliades, D. Danzeiser, A. M.
       Gallegos, P. Shoja-Chagervand, and J. W. Bevan, Texas A&M
       Univesity, USA.

17.   KINETIC AND SPECTROSCOPIC INVESTIGATION OF THE
       $N_2O_4 \rightleftarrows 2NO_2$ CHEMICAL SYSTEM.
       M. Herman, J. Vander Auwera, and M. Roozendael,
       Laboratoire de Chimie Physique Moleculaire, Universite
       libre de Bruxelles, Bruxelles, Belgium.

18.   STRUCTURES OF THE Ar•••FORMAMIDE VAN DER WAALS COMPLEXES.
       R. D. Suenram, F. J. Lovas, G. T. Fraser, J. Zozom[*], and
       C. W. Gillies[*], National Bureau of Standards, and [*]
       Department of Chemistry, Rensselaer Polytechnic Institute,
       NY, USA.

19.   ROTATIONAL SPECTRUM AND STRUCTURE OF CF_3H-NH_3.
       G. T. Fraser, F. J. Lovas, R. D. Suenram, D. D. Nelson,
       Jr.,[*], and W. Klemperer[*], National Bureau of
       Standards, and [*] Department of Chemistry, Harvard
       University, Cambridge, MA, USA.

20.   REAL TIME MEASUREMENTS OF THE IR PHOTODISSOCIATION
       LIFETIME IN ETHYLENE DIMERS.
       A. Mitchell, W. R. Gentry, and C. F. Giese, Department of
       Chemistry, University of Minnesota, USA.

A

Allnatt, A. R.	443,445
Amar, F. G.	526
Amirav, A.	285
Andrews, L.	179
Applequist, J.	322
Ashton, L. J.	346
Aziz, R. A.	425,443,444,445

B

Barker, J. A.	444,446
Barnes, A. J.	529
Barri, M. F.	531
Beauchamp, J. L.	313
Benzel, M. A.	182,186,187
Berne, B. J.	526
Bernstein, R. B.	526
Beswick, J. A.	228,583,585
Bevan, J. W.	181,458,460
Bixon, M.	470,595
Bomse, D. S.	514
Boom, E. W.	86
Botschwina, P.	458
Brechignac, Ph.	514
Brickmann, J.	305
Brocks, G.	344
Brown, F. B.	359
Brunetti, B.	400
Buck, U.	469,496
Buckingham, A. D.	17,38,39, 320,461
Bunker, P. R.	459

C

Campbell, E. J.	29
Candori,	346,347,425 444,445,449
Carley, J. S.	584
Charo, A. C.	459
Claverie, P.	266
Connor, J. N.	437
Cool, T. A.	185
Coulson, C. A.	33,319
Curtiss, L. A.	186

D

Dallinga, C.	271
De Leon, R. L.	185
Dickinson, A. S.	401,403
Dunitz, J. D.	456
Duval,	564
Dyke, T. R.	83,198,457,459
Dykstra, C. E.	179,182,186 187,323,461

E

Eckart, C.	368
Eliason, M. A.	377
Ewing, G. E.	227,344,347

F

Fantoni, R.	511
Faubel, M.	451
Flygare, W.	4,6,24,332
Forster, M. S.	313
Fowler, P. W.	38,39,40,461
Fraser, G. T.	117,120,459,496
Frisch, M. J.	328
Fuchs, R. R.	451
Fung, K. H.	252,254

G

Garvey, J. F.	526
Gelbart, W. M.	470
Gentry, W. R.	515
Gerber, R. B.	526
Giese, C. F.	515
Grabenstetter, J. E.	346
Gutowsky, H.	18

H

Habitz, P.	451
Haenen, J. P.	504
Hager, J.	298
Halberstadt, N.	104
Harrington, M. M.	380

I